RELATIVISTIC DYNAMICS AND QUARK-NUCLEAR PHYSICS

RELATIVISTIC DYNAMICS AND QUARK-NUCLEAR PHYSICS

Edited by

Mikkel B. Johnson

Alan Picklesimer

Los Alamos National Laboratory

A Wiley-Interscience Publication

JOHN WILEY & SONS

New York / Chichester / Brisbane / Toronto / Singapore

***Library of Congress Cataloging-in-Publication Data*:**

Relativistic dynamics and quark-nuclear physics.

Papers based on lectures given at the Los Alamos Workshop on Relativistic Dynamics and Quark Nuclear Physics, held at the Los Alamos National Laboratory during June 1985.

"A Wiley-Interscience publication."

1. Particles (Nuclear physics)—Congresses.
2. Nuclear structure—Congresses. 3. Scattering (Physics)—Congresses. 4. Quarks—Congresses.
I. Johnson, Mikkel B. II. Picklesimer, Alan.
III. Los Alamos Workshop on Relativistic Dynamics and Quark-Nuclear Physics (1985 : Los Alamos National Laboratory)

QC793.R44 1986 539.7'21 86-15814
ISBN 0-471-85327-5

Printed in the United States of America

10 9 8 7 6 5 4 3 2 1

PREFACE

This book covers two rapidly growing areas in theoretical physics. The first part of the book describes recent developments in the relativistic formulations of nuclear many-body scattering and bound state problems within the meson/nucleon context. The contribution by Tandy develops the formalism for scattering, leading up to and including relativistic scattering theory. Next, Machleidt introduces the reader to application of meson theory in nuclear physics. Dubach, Horowitz, and Walker then show the successes of this theoretical approach in electron scattering, nuclear structure and scattering, and antiproton physics. More specialized discussions by other distinguished members of the nuclear physics community focus on the ongoing development of specific issues touched upon by the main contributors. This is valuable for nuclear physicists working in intermediate energies, but, more important, the amount of detail covered in each lecture makes it an invaluable asset for any beginning graduate student aiming for a career in nuclear physics.

The second part of the book discusses the more rapidly growing area of the overlap between nuclear and particle physics. The focus here is on the underlying dynamics of nuclear physics as currently embodied in quantum chromodynamics (QCD), the theory of quarks and gluons. Fundamental theory and applications to specific problems are given primary attention. Jaffe develops an application of QCD to the scattering of electrons from nuclei, the EMC effect. He introduces the reader to the kinematics in the target rest frame, the parton model and scaling, as well as to the subject of scaling violations. Isgur begins with the basic equations of QCD, introduces the concept of lattice gauge theory in the strong coupling limit, and then moves to applications to hadron spectra and nuclear forces. Polonyi develops in more detail the subject of lattice gauge theory in a series of applications beginning with ordinary quantum mechanics. Finally, Zahed and Brown discuss an important connection between QCD and meson theory, the theory of chiral skyrmions. As in the first part of the book, these discussions are followed by more specialized issues in ongoing research in the same areas. The material emphasizes basics and would therefore be of interest to experimentalists in the developing areas of nuclear physics: electron scattering, heavy ion scattering in search of the quark-gluon plasma, and scattering of light hadrons from nuclei.

Special topics courses such as this used to occupy a highly visible place in nuclear physics. We have in mind the Boulder lecture series, which ceased to exist some ten years ago. Since that time nuclear physics has witnessed rapid and continuous development of its fundamental precepts stimulated by new experimental facilities, together with improved experimental techniques and extended capabilities. On the theoretical side, new analytical sophistications have led to a partial synthesis of the traditional concepts of nuclear physics and more fundamental microscopic descriptions based on relativistic field theory. This synthesis represents an ongoing process that will be particularly pronounced in the coming decade as the field looks forward to experimental facilities in the multi-GeV energy range and to a clearer understanding of the relationship of nuclear phenomena to QCD. With the rapid development of nuclear physics, the time is right to renew the summer school tradition.

This book came into existence in response to these considerations. As the field enters this new era, we must attract a new generation of imaginative and highly motivated students to meet the challenges. This book was written to confront students with some of the subjects that are at the cutting edge of recent developments. The vitality of any field depends, in fact, upon the continuing infusion of vigor and enthusiasm provided by students. Summer schools provide a natural forum for promoting interactions between students and researchers, provide students with a uniquely accessible contact with current research in the field, and offer an effective means for communicating to students something of the excitement, the challenge, and especially the opportunities for further meaningful research. The pedagogical nature of the forum, coupled with the focus on detailed coverage of current research topics, both complements and strengthens the student's university graduate education.

The papers in this volume are based on lectures given at the Los Alamos Workshop on Relativisitic Dynamics and Quark Nuclear Physics held at the Los Alamos National Laboratory during the first two weeks of June 1985 and attended by 100 graduate students and senior researchers from the United States and abroad. The choice of Los Alamos as the location of the school was based on our belief that university graduate programs in nuclear physics could be enhanced by the confluence of students and researchers from national laboratories, enabling students to meet future colleagues and experience the research environment of the main experimental facilities. We hope also that Los Alamos sponsorship of this work represents the first of what will be continuing activity in this direction.

The contributors were asked to provide pedagogical, exciting, and stimulating lectures appropriate for graduate students at all levels. The seminars convey more of the élan of research; these are by nature less adapted to pedagogy, but are still tied to the fabric of the school. Both the lectures and seminars provide an accurate and critical assessment of the development and current status of the subject matter. The quality of the response of the authors, including the painstaking effort, devotion, and care with which they prepared their presentations, is evident in these proceedings. This book, therefore, adds to the literature an up-to-date and thorough description of two rapidly developing areas of nuclear physics. It also serves as a continuing resource for those who attended the summer school and conveys the essential content of the school to others.

We wish to express our gratitude, on behalf of the students and the organizers, to the contributors for an excellent job on their lectures and notes. We also wish to thank Louis Rosen of the Los Alamos Meson Physics Facility (LAMPF), who provided the financial support for this school, to the organizing and advisory committee for advice on the choice of lecturers, and to the LAMPF User's Group, which offered several scholarships for students. We are especially grateful to the many speakers who waived their expenses so that additional student scholarships could be offered. The central players were, of course, the students, who exceeded our expectations. We deliberately set a very grueling and challenging pace, and they responded with enthusiasm. Their interest and dedication were evident in their frequent and penetrating questions and in their pursuit of the topics during breaks, meals, and scheduled student-speaker sessions. Their avid and vigorous interest made the school worthwhile.

Our heartfelt appreciation goes to Roberta Marinuzzi and Maxine Joppa of the Visitor Liaison Office at LAMPF and to Theresa Tandy for their efficiency and dedicated behind-the-scenes efforts in the actual logistics and running of the school. Finally, we thank Beverly Talley for her careful assembly of the manuscript.

Mikkel B. Johnson
Alan Picklesimer
May 1986

CONTENTS

PART I — RELATIVISTIC DYNAMICS

PART II — QUARK-NUCLEAR PHYSICS

Morning Lectures

Afternoon Seminars

RELATIVISTIC DYNAMICS AND QUARK-NUCLEAR PHYSICS

PART I
RELATIVISTIC DYNAMICS

NONRELATIVISTIC AND RELATIVISTIC MULTIPLE SCATTERING

Peter C. Tandy
Kent State University
Kent, OH 44242

ABSTRACT

In this series of lectures the theoretical elements for treating the scattering of an intermediate energy projectile from a nucleus are reviewed. An audience of graduate students is assumed. The first lecture is introductory and sketches the formal development of one-body potential scattering theory and introduces the S-matrix, the T-matrix and the Lippmann-Schwinger equation. The application to nucleon-nucleus scattering viewed as a many- nucleon problem is considered, the microscopic optical potential is developed and the first order multiple scattering approximation is described. The second lecture continues with multiple scattering expansions, second order terms, and antisymmetry constraints. The third lecture treats relativistic scattering of a spin-1/2 particle from a static potential and discusses the role of negative energy states and the Z-graph mechanism. The Dirac-based approach to proton-nucleus scattering is discussed from the perspective of coupling to pair-effects and lecture four concludes with a development of the microscopic formalism for inelastic scattering.

LECTURE 1 - INTRODUCTION TO NON-RELATIVISTIC SCATTERING

Many aspects of our concept of nuclear physics are dominated by the "classical" approach which is to find the interaction between nucleons largely through the study of nucleon-nucleon scattering data and to use this information to calculate nuclear structure and nuclear reactions and compare these results with the body of experiment. The theoretical tool required to make the relation between the two-body system and many-body reactions is loosely referred to as "multiple scattering theory". In this lecture I will endeavor to acquaint you with the central ingredients in this theoretical formulation. Needless to say in the time at our disposal, we can at best only pick up the most elementary concepts.

We shall begin by assuming a standard graduate quantum mechanics course as the shared background of this audience. We shall be completely non- relativistic, and we shall assume that constituent particles interact through short-range two-body forces, i.e. to say the presence of a third particle does not affect the mutual interaction of the first two particles.

It may be worthwhile to emphasize that this starting point was by no means historically inevitable, nor is the logic well established today. Atomic physics may be said to have yielded its major results through the electronic Schroedinger equation

$$\frac{-h^2}{2m} \nabla^2 \Psi(\vec{r},t) - \frac{e^2}{r} \Psi(\vec{r},t) = ih\frac{\partial}{\partial t} \Psi(\vec{r},t)$$

for the electron-proton (Hydrogen atom) problem, and

$$\frac{-h^2}{2m} \sum_{i=1}^{z} \nabla_i^2 \Psi + \left(- \sum_{i}^{z} \frac{Ze^2}{r_i} + \frac{1}{2} \sum_{i\neq j}^{Z} \frac{e^2}{r_{ij}}\right)\Psi = ih\frac{\partial}{\partial t} \Psi$$

for the many-electron problem. The interpretation of this as,

$$H_{op}\Psi = E_{op}\Psi,$$

where H_{op} is the classical hamiltonian for the system with the kinetic energy operator replaced by $\frac{-h^2}{2m}\nabla^2$, leads to the usual extension to nuclear physics. Thus in the early days of nuclear physics the idea was simply to find the rest of the hamiltonian, the potential V, and then to proceed to calculate all the nuclear phenomena. This interpretation is by no means the only possibility, nor is the extension to nuclear physics a necessary one. For example, the above Schroedinger equation omits the important contributions to atomic physics that are provided by relativistic quantum electrodynamics. The nuclear counterpart of such a field description and the non-elementary nature of nucleons have been of major concern in nuclear physics in recent decades. Nevertheless, for the time being at least, we shall travel the path described by

$$H = \sum_{i} \frac{p_i^2}{2m} + \sum_{i<j} V_{ij}$$

as the nuclear hamiltonian for a collection of nucleons of mass m interacting pairwise. We wish to consider the problem of scattering of a nucleon from a nucleus assuming that the nuclear bound states are known. The bound state problem is discussed at this Workshop in a series of lectures by C. Horowitz.

TIME-DEPENDENT SCATTERING

In most applications of potential scattering theory, an explicitly time- independent formulation in terms of the stationary-state eigenfunction of H is employed. However, the various operators that enter, especially in approaches that employ integral equations and associated diagramatic expansions, are defined in the passage from the time-dependent formulation to the time-independent formulation. I feel it is worthwhile to summarize here the elements of that procedure for two reasons. Firstly, the time- dependent origins of various quantities are kept in mind, and secondly as a reminder that in problems where the interaction is not simply a (instantaneous) potential (such as field theory and relativistic problems where the interaction is time-dependent) one should persist with the time- dependent formulation.

We consider one-body potential scattering governed by $H=H_o + V$, where H_o is the kinetic energy operator. The time development equation $i\partial_t\Psi(t) = H\Psi(t)$ is satisfied by ($\hbar=1$)

$$\Psi(t) = e^{-iH(t-t_o)}\,\Psi(t_o) = e^{-iHt}\,\Psi(o) \quad . \tag{1.1}$$

This state is to describe scattering under the initial boundary condition

$$\Psi(t\to-\infty) = \Phi_i(t\to-\infty) \quad , \tag{1.2}$$

where $\Phi_i(t)$ is a non-interacting state governed by H_o and specified by the quantum numbers of the beam. Let $\Phi_f(t)$ be the non-interacting state specified by the quantum numbers selected by the detector. The states Φ_i, Φ_f develop in time according to $i\partial_t\Phi(t) = H_o\Phi(t)$ so that

$$\Phi(t) = e^{-iH_ot}\,\Phi(o) \quad . \tag{1.3}$$

The asymptotic stationary states $\Phi_i(o)$, $\Phi_f(o)$ are to be built from the eigenfunctions of H_o (e.g. an integral distribution of plane wave states) so that $\Phi(t)$ is a normalizable wave packet of finite spatial extent. (Modifications of this procedure are necessary in the case of a field theory in which V allows creation and destruction of particles. In that case the "bare" particle states of H_o must be supplemented by the effects of V that serve to add all orders of virtual emission and absorption of field quanta to obtain the physical mass of particles.) The fundamental quantity for describing scattering observables is the S-matrix element S_{fi} which is the probability amplitude that $\Psi(t)$ will finally have a component $\Phi_f(t)$, that is

$$S_{fi} = \lim_{t\to+\infty} \langle\Phi_f(t)|\Psi(t)\rangle \quad , \tag{1.4}$$

$$=\lim_{t\to+\infty} \langle\Phi_f(o)|e^{iH_ot}|\Psi(t)\rangle \quad , \tag{1.5}$$

$$=\langle\Phi_f(o)|S|\Phi_i(o)\rangle \quad . \tag{1.6}$$

To identify the operator S, we use Eq.(1.1) for $\Psi(t)$ and use the initial boundary condition to obtain

$$\Psi(t) = \lim_{t_o\to-\infty} e^{-iH(t-t_o)}\, e^{iH_ot_o}\, \Phi_i(o) \quad , \tag{1.7}$$

whence S is identified to be

$$S = \lim_{\substack{t\to+\infty \\ t_o\to-\infty}} U(t,t_o) = U(+\infty,-\infty) \quad , \tag{1.8}$$

where

$$U(t,t_o) = e^{iH_ot}\, e^{-iH(t-t_o)}\, e^{-iH_ot_o} \quad , \tag{1.9}$$

$$= U(t,o)\; U(o,t_o) \quad . \tag{1.10}$$

From Eqs.(1.6,8,10) the S-matrix element can be expressed as

$$S_{fi} = \langle\Psi_f^{(-)}|\Psi_i^{(+)}\rangle \quad , \tag{1.11}$$

where

$$|\Psi_i^{(+)}\rangle = U(o,-\infty)|\Phi_i(o)\rangle = \Omega_+|\Phi_i(o)\rangle \quad , \tag{1.12}$$

and

$$\langle\Psi_f^{(-)}| = \langle\Phi_f(o)|U(\infty,o) = \langle\Phi_f(o)|\Omega_-^+ \quad , \tag{1.13}$$

where the notation is meant to imply the proper time limits. Since time-independent states will always be calculated at t=o, we will write $\Phi_i = \Phi_i(o)$, etc.

The results of this development are simply related to the positive energy eigenfunction of H. Let the eigenstates of H_o be given by $H_o|\vec{k}\rangle = E_k|\vec{k}\rangle$ and let $|\Phi_i\rangle$ be the wave packet

$$|\Phi_i\rangle = \int d^3k|\vec{k}\rangle\langle\vec{k}|\Phi_i\rangle \quad . \tag{1.14}$$

Then it can be shown that, with a careful treatment of the limit[1,2],

$$|\Psi_i^{(+)}\rangle = \lim_{t_o\to-\infty} U(o,t_o)|\Phi_i\rangle = \int d^3k|\psi_k^{(+)}\rangle\langle\vec{k}|\Phi_i\rangle, \tag{1.15}$$

where $\psi_k^{(+)}$ is the positive energy eigenfunction of H that is given by the Lippmann-Schwinger[3] equation

$$|\psi_k^{(+)}\rangle = |\vec{k}\rangle + G_o(E_k^+)V|\psi_k^{(+)}\rangle \quad , \tag{1.16}$$

where the non-interacting Green's function is

$$G_o(E_k^+) = (E_k + i\eta - H_o)^{-1} \quad . \tag{1.17}$$

In Eq.(1.17), the limit $\eta \to o+$ must be taken after the equation is solved. This limit, which ensures that $\langle\vec{r}|\psi_k^{(+)}\rangle$ has outgoing spherical waves e^{ikr}/r outside of the interaction region, has its origin in the time limit of Eq.(1.15). The use of a wave packet state, Eq. (1.14), in which $\langle\vec{k}|\Phi_i\rangle$ becomes negligible for $\vec{k}$ significantly different from the mean momentum of the beam, is necessary for the time limit to converge.

In a similar manner Eq.(1.13) leads to

$$\langle\Psi_f^{(-)}| \equiv \lim_{t\to+\infty} \langle\Phi_f|U(t,o) = \int d^3p \langle\Phi_f|\vec{p}\rangle\langle\psi_p^{(-)}| , \tag{1.18}$$

where

$$|\psi_p^{(-)}\rangle = |\vec{p}\rangle + G_o(E_p^-)\, V|\psi_p^{(-)}\rangle \tag{1.19}$$

and $G_o(E_p^-)$ is defined in the same manner as Eq.(1.17), except that the energy parameter is $E_p - i\eta$, and the limit is again $\eta \to o+$. Outside of the interaction region $\langle\vec{r}|\psi_p^{(-)}\rangle$ has a scattered part that consists of incoming spherical waves e^{-ipr}/r. Use of Eqs. (1.15,18) in Eq.(1.11) for the S-matrix element leads to

$$S_{fi} = \int d^3p\, d^3k\, \langle\Phi_f|\vec{p}\rangle\, S_{pk}\, \langle\vec{k}|\Phi_i\rangle, \tag{1.20}$$

where

$$S_{pk} = \langle\vec{p}|S|\vec{k}\rangle = \langle\psi_p^{(-)}|\psi_k^{(+)}\rangle \quad . \tag{1.21}$$

Usually the experimental conditions are such that the spatial extent of the incident beam (and the "exit beam") is large compared to the range of V so that Φ_i are Φ_f are sharply peaked in momentum space[4]. Then the plane wave matrix element S_{pk} is a satisfactory description of S_{fi}, and we will now deal with the former, Eq. (1.21). Note that $\langle\psi_p^{(-)}|$ is the fully interacting state at t=o that would develop into $\langle\vec{p}|$ in the distant future, and $|\psi_k^{(+)}\rangle$ is the fully interacting state at t=o that has developed from $|\vec{k}\rangle$ in the distant past.

In the absence of interactions, we have $S_{pk} = \delta(\vec{p}-\vec{k})$. It is convenient to separate from S_{pk} the non-interacting piece so as to deal with a quantity, the transition amplitude, that describes true scattering. A procedure for this in terms of time limits is as follows:

$$S_{pk} = \langle\vec{p}|U(\infty,o)\ U(o,-\infty)|\vec{k}\rangle = \langle\vec{p}|U(\infty,o)|\psi_k^{(+)}\rangle \quad ,$$

$$= \lim_{t\to\infty} \langle\vec{p}|e^{iH_o t}e^{-iHt}|\psi_k^{(+)}\rangle = \lim_{t\to\infty} e^{i(E_p-E_k)t}\langle\vec{p}|\psi_k^{(+)}\rangle. \qquad (1.22)$$

Use of Eq.(1.16) then leads to

$$S_{pk} = \delta(\vec{p}-\vec{k}) + \lim_{\substack{t\to\infty \\ \eta\to o+}} \frac{e^{i(E_p-E_k)t}}{E_k+i\eta-E_p}\langle\vec{p}|V|\psi_k^{(+)}\rangle. \qquad (1.23)$$

With the help of the integral representations (for $\eta\to o+$)

$$(E+i\eta)^{-1} = -i\int_o^\infty e^{i(E+i\eta)\tau}d\tau \quad , \qquad (1.24)$$

$$2\pi i\delta(E) = (E-i\eta)^{-1} - (E+i\eta)^{-1} = i\int_{-\infty}^{+\infty} e^{iE\tau}e^{-\eta|\tau|}d\tau, \qquad (1.25)$$

we obtain

$$S_{pk} = \delta(\vec{p}-\vec{k}) - 2\pi i\delta(E_p-E_k)\ T_{pk}, \qquad (1.26)$$

where the transition amplitude is

$$T_{pk} = \langle\vec{p}|V|\psi_k^{(+)}\rangle = \langle\vec{p}|T(E_k^+)|\vec{k}\rangle. \qquad (1.27)$$

The second equality in Eq.(1.27) serves to define a transition operator T such that $V|\psi_k^{(+)}\rangle = T|\vec{k}\rangle$. The combination of this definition with the Lippmann-Schwinger equation [Eq.(1.16)] for $|\psi_k^{(+)}\rangle$ leads to the formal integral equation for the operator T

$$T(E^+) = V + VG_o(E^+)T(E^+), \qquad (1.28)$$

often called the Lippmann-Schwinger equation for T. Eqs. (1.26) and (1.27) show that all the observable quantities for a scattering process can be calculated from the plane wave matrix element $\langle\vec{p}|T(E^+)|\vec{k}\rangle$ under the (on-energy-shell) constraint $E_p = E = E_k$. Matrix elements of T in which one of these equalities is not satisfied (half-off-energy- shell), or both are not satisfied (fully-off-energy-shell), are still defined by Eq.(1.28) but they are not measureable quantities. Note that knowledge of the "half-shell" T-matrix element $\langle\vec{p}|T(E_k^+)|\vec{k}\rangle$ for $E_p \neq E_k$ is equivalent to knowledge of the scattering wave function within the range of the potential[5].

The derivation of the total scattering cross section and the differential scattering cross section in terms of T_{pk} can be found in standard texts[1,6] and we simply sketch the results here. The time rate of increase of the probability amplitude for the process $\vec{p}\leftarrow\vec{k}$, that is, the transition rate W_{pk}, is found from the time derivative of the magnitude squared of the right hand side of Eq. (1.22) before the time limit is taken[1]. The result is $W_{pk} = 2\pi\, \delta(E_p - E_k)|T_{pk}|^2$, $\vec{p} \neq \vec{k}$. The total scattering cross section σ_T is

$$\sigma_T = \frac{1}{J_i} \int d^3p\, W_{pk} = \frac{2\pi}{J_i} \int d\Omega_p\, p^2 dp\, \delta(E_p - E_k)|T_{pk}|^2, \tag{1.29}$$

where J_i is the incident flux. The differential cross section is

$$\frac{d\sigma}{d\Omega_p} = \frac{2\pi}{J_i} \int p^2 dp\, \delta(E_p - E_k)|T_{pk}|^2 = |F_{pk}|^2 , \tag{1.30}$$

where F_{pk} is the scattering amplitude and is given by $F_{pk} = K\, T_{pk}$ in which the kinematic factor K depends on the choice of normalization of plane wave states and upon the relation between energy and momentum of a free particle. For the delta function normalization of plane wave states that we have used, $\langle\vec{r}|\vec{k}\rangle = e^{i\vec{k}\cdot\vec{r}}/(2\pi)^{3/2}$, and for non-relativistic kinematics, one obtains for the kinematic factor $K = -(2\pi)^2 m$. In terms of the scattering amplitude, the large distance behavior of the stationary scattering wave is

$$\langle\vec{r}|\psi_k^{(+)}\rangle \xrightarrow[r\to\infty]{} e^{i\vec{k}\cdot\vec{r}} + F_{pk}(\theta,\phi)\frac{e^{ikr}}{r} . \tag{1.31}$$

DIAGRAMS

For the problem of non-relativistic nucleon-nucleus scattering in which the interaction is taken to be potentials acting (instantaneously) between nucleons, the vehicle we shall use is a suitable generalization of the Lippmann-Schwinger equation (1.28) for the operator T. The procedures and approximations involved are often described by diagrams. Such diagrams are related to the Feynman diagrams which have meaning in a more general context when the interaction is not instantaneous and particles can be created and destroyed (e.g. quantum electrodynamics). In order to provide some perspective to discussions throughout this Workshop that use diagrams to describe terms from iteration of an equation like Eq.(1.28), viz

$$T(E^+) = V + VG_o(E^+)V + VG_o(E^+)VG_o(E^+)V + \cdots , \qquad (1.32)$$

we sketch the relationship to Feynman-like diagrams for the simple case of one-body potential scattering. Diagrams have time as one of the dimensions and hence come from the time-dependent formulation of the S- matrix. A time derivative of Eq. (1.9) produces the result $i\partial_t U(t,t_o) = V_I(t)\, U(t,t_o)$ where $V_I(t) = e^{iH_ot}\, V\, e^{-iH_ot}$, that is, the interaction representation of V. An integral equation for $U(t,t_o)$ which embodies the boundary condition $U(t_o,t_o) = 1$ is

$$U(t,t_o) = 1 - i\int_{t_o}^{t} d\tau e^{-\eta|\tau|}\, V_I(\tau)\, U(\tau,t_o). \qquad (1.33)$$

where the limit $\eta \to o+$ is to be taken. The damping factor forces the convergence of the integral for large time intervals, a property that is automatic once wave packet matrix elements are taken[1,2]. Since we will not impose the wave packet condition explicitly, the damping factor is to be understood to be implicit in V. The S-matrix element, given by $S_{pk} = \langle\vec{p}|U(\infty,-\infty)|\vec{k}\rangle$, can then be expressed as

$$S_{pk} = \delta(\vec{p}-\vec{k}) - i\int_{-\infty}^{+\infty} d\tau\langle\vec{p}|V_I(\tau)U(\tau,-\infty)|\vec{k}\rangle. \qquad (1.34)$$

A series for S_{pk} in powers of the interaction is obtained by iterations of Eq.(1.33) so that we can write

$$S_{pk} = S^{(0)} + S^{(1)} + S^{(2)} + \cdots , \qquad (1.35)$$

where $S^{(i)}$ is the contribution containing V acting i times, and $S^{(0)} = \delta(\vec{p}-\vec{k})$. The first-order term is simply

$$S^{(1)}_{pk} = (-i)\int_{-\infty}^{+\infty} d\tau\langle\vec{p}|e^{iE_p\tau}V\, e^{-iE_k\tau}|\vec{k}\rangle \qquad (1.36)$$

$$= (-i)\int_{-\infty}^{+\infty} d\tau e^{i(E_p-E_k)\tau}e^{-\eta|\tau|}\, V_{pk}$$

$$= 2\pi i\ \delta(E_p-E_k)\ V_{pk}\ . \qquad (1.37)$$

Comparison with Eq.(1.26), which relates S to T, shows that the first-order term for T is produced by $S^{(1)}$. This is described by the first-order diagram in Fig. (1) which depicts the time-dependence occurring in Eq.(1.36). The interaction happens instantaneously at time τ, the initial free particle state at that time is $e^{-iE_k\tau}|\vec{k}\rangle$ and the final free particle state is $e^{-iE_p\tau}|\vec{p}\rangle$. The time point τ of the interaction is not measureable and must be integrated over, and the energy-conserving delta function is produced.

The second-order S-matrix element is

$$S_{pk}^{(2)} = (-i)^2 \int_{-\infty}^{+\infty} d\tau \int_{-\infty}^{\tau'} d\tau \; \langle\vec{p}|e^{iE_p\tau'} V \, e^{-iH_o(\tau'-\tau)} \, V \, e^{-iE_k\tau}|\vec{k}\rangle, \tag{1.38}$$

which is depicted by the second-order diagram in Fig. (1). Notice that the particle line connection between the time τ of the first interaction and the time τ' of the second interaction represents the free particle progagator $e^{-iH_o(\tau'-\tau)}$. The integration over all times $\tau'>\tau$ produces, after some manipulations and use of Eq.(1.24) with E replaced by E_k-H_o, the result

$$S_{pk}^{(2)} = -2\pi i \; \delta(E_p-E_k) \; \langle\vec{p}|VG_o(E_k^+)V|\vec{k}\rangle. \tag{1.39}$$

Thus after the time integrals in such diagrams are performed there emerges a direct correspondance with terms in the iteration of the time independent equation $T = V + VG_oT$. We can see why the diagrams and this equation are often used interchangeably. Generalizations of the above procedures to relativistic quantum mechanics leads to Feynman diagrams in which some particle lines may point backwards in time[1,7]. This situation is employed to interpret the phenomena in which the relativistic operator that plays the role of H_o in the above treatment can have negative energy eigenvalues.

NUCLEON-NUCLEUS SCATTERING

We now turn to elastic scattering of a nucleon from nuclei (composed of constituent nucleons regarded as elementary). We write the hamiltonian for the entire (A+1)-nucleon system as

$$H = h_o + H_A + \sum_{i=1}^{A} v_i \equiv H_o + V, \tag{1.40}$$

where now $H_o = h_o + H_A$, and

$$V = \sum_{i=1}^{A} v_i \quad , \tag{1.41}$$

with h_o defined to be the kinetic energy operator of particle 0, the projectile nucleon, H_A defined to be the hamiltonian of the A-nucleon target system and v_i defined to be the two-body interaction potential between the particle 0 and the ith nucleon of the target nucleus. With these definitions the Lippmann-Schwinger equation for nucleon-nucleus scattering becomes

$$T = V + V\,G_o\,T. \tag{1.42}$$

The transition amplitude between two different eigenstates of H_o having the same energy is

$$T_{fi} = \langle\Phi_f|T|\Phi_i\rangle \quad , \tag{1.43}$$

where $H_o|\Phi\rangle = E|\Phi\rangle$ for Φ_f and Φ_i. In terms of nuclear bound states ϕ we can write

$$|\Phi_i\rangle = |\phi_i\rangle|\vec{k}_i\rangle\ , \quad |\Phi_f\rangle = |\phi_f\rangle|\vec{k}_f\rangle, \tag{1.44}$$

where $h_o|\vec{k}\rangle = E_k|\vec{k}\rangle$ and $H_A|\phi_\alpha\rangle = \varepsilon_\alpha|\phi_\alpha\rangle$ for $\alpha = i$ or f, and $E = E_{ki} + \varepsilon_i = E_{kf} + \varepsilon_f$. The transition amplitude in Eq.(1.43) can be used for elastic or inelastic scattering from nuclei. The quantity G_o in Eq.(1.42) is

$$G_o(E^+) = (E+i\eta - H_o - H_A)^{-1}, \tag{1.45}$$

which is a complicated many-body propagator. Many-body integral equations such as Eq.(1.42) must be restructured before the limit $\eta\to o$ can be handled in a manageable way in order that the solution T is defined by the equation. Such topics are treated by techniques in many-body theory[8] and we will not go into them here. An example of the restructuring of Eq.(1.42) provided by the limit $\eta\to o$ in the case of the three-body problem is provided by the Faddeev equations[9]. Here we simply view Eq.(1.42) as a relation satisfied by T and we will restructure the equation to bring out the relationship between the many-body quantity T and the scattering operators for the subsystems. This is called a multiple scattering expansion.

FIRST-ORDER MULTIPLE SCATTERING

Use of Eq.(1.41) allows Eq.(1.42) to be written

$$T = \sum_i v_i + \sum_i v_i G_o T, \tag{1.46}$$

from which we may define

$$T_i = v_i + v_i G_o\,T, \tag{1.47}$$

with the result that $T = \sum_{i=1}^{A} T_i$. If we interate Eqs. (1.47), we see that

$$T_i = v_i + v_i G_o \sum_j v_j + \cdots \quad . \tag{1.48}$$

The first term is a single scattering (interaction) terms, the second term is a double scattering (interaction) term, and so on. We observe that each of the higher order terms on the right hand side of Eq. (1.48) has a component in which the multiple scatterings (interactions) take place on the same target nucleon. Separating these off we may write Eq.(1.48) as

$$\begin{aligned} T_i &= v_i + v_i G_o v_i + v_i G_o v_i G_o v_i + \cdots \\ &\quad + v_i G_o \sum_{j \neq i} v_i + \cdots \\ &= t_i + v_i G_o \sum_{j \neq i} v_j + \cdots \ , \end{aligned} \tag{1.49}$$

where the first series in Eq.(1.49) is the solution of

$$t_i = v_i + v_i G_o t_i = v_i + v_i \ G_o v_i + \cdots \ . \tag{1.50}$$

One should view the introduction of t_i as collecting all the terms that correlate the projectile with one nucleon of the target. In this sense the organization so produced is related to the Brueckner approach[10] for the bound state. A quantity for which this relationship is even closer will be discussed later. If the target were a dilute gas of nucleons, the approximation

$$T = \sum_i T_i \simeq \sum_i t_i , \tag{1.51}$$

makes a great deal of sense, since the projectile, once it comes close to a given target particle may multiply interact with that particle, but once it is ejected will, with a high degree of probability, "miss" all the other target particles. Further we note that as far as the scattering of the projectile is concerned, it makes very little difference how the interactions in Eq.(1.50) organize themselves, the only important question is how big is t_i. The approximation given in Eq. (1.50) is called the <u>single scattering approximation</u> to the transition amplitude.

In this approximation, the required amplitude described by t_i does not correspond to the solution of a nucleon-nucleon scattering problem. Because of the presence of H_A in the Green's function operator G_o of Eq. (1.50), the motion of nucleon i is governed not only by its interaction v_i with the projectile, but also by its interaction with the other constituents of the target. A further approximation can be envisaged in which H_A is assumed to simply set an energy scale so that the solution of Eq.(1.50) might be replaced by the solution of a free nucleon-nucleon scattering problem. With this interpretation of t_i, Eq.(1.51) is referred to as impulse approximation.

Formally, we may obtain the impulse approximation as the first term in a systematic development by using Eq. (1.50) to eliminate v_i in favor of t_i in Eq. (1.47). The result of this trivial algebraic substitution is

$$T_j = t_i + t_i G_o \sum_{j \neq i} T_j . \tag{1.52}$$

The lowest order approximation to the interated solution of Eq. (1.52) is the single scattering approximation, Eq.(1.51).

The pure single scattering approximation to the transition amplitude as discussed above is of limited validity in nuclear physics. The nucleus is not really a dilute gas since the density of nuclear matter, 0.17 nucleons $/f^3$, indicates that the average spacing between point nucleons is of the same size as the range of the nucleon-nucleon force (1.4f). Rather, it is helpful to consider the nucleus as a material medium with an index of refraction. In classical language, the orbit of the projectile through the medium is continuously altered as the projectile passes through the nuclear medium. For a finite, not necessarily uniform chunk of matter, the so called optical potential plays the familiar role of the index of refraction (or rather n-1). The optical potential is defined as the potential that describes the passage of a projectile of a given energy through the nucleus, with the nucleus treated as a passive medium. The term passive medium is taken to mean that the nucleus is treated as though it cannot be excited.

OPTICAL POTENTIAL

To carry through our discussion of the optical potential, as discussed above, we define a projection operator P_o which projects onto the target nuclear ground state. Likewise the operator Q_o projects onto the complementary space of target excited states (including break-up states), so that

$$P_o + Q_o = 1, \tag{1.53}$$

where

$$P_o = |\phi_o\rangle\langle\phi_o| \quad , \tag{1.54}$$

and $|\phi_o\rangle$ is the target ground state. For elastic scattering both the initial and final target states are the ground state so that

$$\langle\Phi_f|T|\Phi_i\rangle_{\text{elastic}} = \langle\Phi_f|P_o T\, P_o|\Phi_i\rangle_{\text{elastic}} \quad , \tag{1.55}$$

and hence for this purpose the operator $P_o T\, P_o$ is equivalent to the operator T. The optical potential may then be defined through the equation

$$P_o T P_o = P_o U P_o + P_o U P_o G_o P_o T P_o \quad , \tag{1.56}$$

or equivalently

$$T = U + U\, G_o P_o\, T. \tag{1.57}$$

We use the fact the G_o and P_o commute and that $P_o^2 = P_o$. Because of the projectors, which can be throught of as integrating out the internal degrees of freedom of the target, it is immediately obvious that Eq.(1.33) is a one-body relation, involving only the coordinates of the projectile.

To be more specific, we can write

$$\langle\Phi_f|U|\Phi_i\rangle = \langle\vec{k}'|\langle\phi|U|\phi_o\rangle|\vec{k}\rangle = \langle\vec{k}'|\hat{V}|\vec{k}\rangle \quad , \tag{1.58}$$

and, together with a similar equation defining $\hat{T} = \langle\phi_o|T|\phi_o\rangle$, we can see that Eq.(1.56) is just

$$\hat{T} = \hat{V} + \hat{V}\, \frac{1}{E^+ - \varepsilon_o - h_o}\, \hat{T} \quad , \tag{1.59}$$

the equation for one-body scattering due to the optical potential $\hat{V}$. Note that ε_o is the energy of the nuclear ground state and $E-\varepsilon_o$ is the energy of the incident projectile. The wave function associated with $\hat{T}$ is defined by $\hat{T}|\vec{k}\rangle = \hat{V}|\psi_k^{(+)}\rangle$ and satisfies $((E-\varepsilon_o)-h_o)|\psi_k^{(+)}\rangle = \hat{V}|\psi_k^{(+)}\rangle$. The relation between $|\psi_k^{(+)}\rangle$, the elastic scattering wave function and $|\Psi_k^{(+)}\rangle$ the exact wave function for the problem is easily found to be

$$|\phi_o\rangle|\psi_k^{(+)}\rangle = P_o|\Psi_k^{(+)}\rangle \quad . \tag{1.60}$$

Note that $|\Psi_k^{(+)}\rangle$ will contain outgoing waves in all possible final channels of the system (including rearrangement channels e.g. (p,d) reactions, and breakup channels e.g. (p,2p) reactions as well as channels where the nucleus is in an excited but bound state). In contrast, the elastic state in Eq. (1.60) contains outgoing waves in only the channel having the nucleus in its ground state.

To identify the microscopic content of the optical potential we now use Eqs. (1.57) and (1.42) to find a relation between U and V. The result is easily seen to be

$$U = V + V\, G_o Q_o U \quad , \tag{1.61}$$

where the presence of Q_o indicates that the internal degrees of freedom of the target are necessarily active, that is excited, in the reaction processes describing U. These two equations, Eq. (1.57) and (1.61) taken together are, of course, completely equivalent to Eq. (1.42), so that the separation of this problem into two parts accomplishes nothing as a formal device. Physically, however, the separation is of great value. In fact we shall see that the appropriate truncation of Eq.(1.61) can lead to a good approximations to T that are exceedingly difficult to realize without this separation. We now consider a single scattering approximation to this optical potential, using the method we previously applied to the transition operator of Eq. (1.47). From Eq.(1.61), we may express U as $\sum_i U_i$, and write

$$U_i = v_i + v_i G_o Q_o \sum_j U_j \quad . \tag{1.62}$$

Following Watson[11] we may then define an operator τ_i through the relation

$$\tau_i = v_i + v_i G_o Q_o \tau_i \quad . \tag{1.63}$$

In terms of this object, Eq.(1.62) is

$$U_i = \tau_i + \tau_i G_o Q_o \sum_{j \neq i} U_j \quad , \tag{1.64}$$

and thus is obtained the Watson[11] multiple scattering series for the optical potential operator

$$U = \sum_i \tau_i + \sum_{i \neq j} \tau_i Q_o G_o \tau_j + \sum_{i \neq j \neq k} + \cdots \tag{1.65}$$

FIRST-ORDER KMT OPTICAL POTENTIAL

The first-order (single scattering) approximation to the optical potential is the first term in Eq. (1.65). Let us see what is involved in calculating the nuclear elastic scattering due to this first term. We will take for granted that the operator t_i, which satisfies the same Eq.(1.63) except that Q_o is replaced by 1, is a more manageable and accessible object than τ_i. With $Q_o = 1 - P_o$, the relation between the two is easily seen to be

$$\tau_i = t_i - t_i P_o G_o \tau_i \quad . \tag{1.66}$$

We will also employ states which are antisymmetric in the target nucleon coordinates. Thus we can write $\langle\phi_o|\Sigma_i \tau_i|\phi_o\rangle = \langle\phi_o|A\tau|\phi_o\rangle$ where A is the mass number of the target. The nuclear scattering due to the first-order optical potential is then given by

$$T = A\tau(1+P_o G_o T) \quad . \tag{1.67}$$

With t_i as an input quantity, it is inefficient to first solve Eq. (1.66) and then solve Eq.(1.67). Since these two equations have essentially the same structure, they can be combined into one by substituting τ in terms of t. The result is that Eq. (1.67) becomes

$$T = At(1-P_o G_o \tau)(1+P_o G_o T) \quad ,$$

that is

$$T = At + (A-1)tP_o G_o T \quad , \tag{1.68}$$

or, equivalently in the optical potential form

$$\left[\frac{A-1}{A}T\right] = (A-1)t + (A-1)tP_o G_o \left[\frac{A-1}{A}T\right] \quad . \tag{1.69}$$

This is the first-order result of the multiple scattering development in the work of Kerman, McManus and Thaler (KMT)[12]. The procedure is to use a "scaled" optical potential

$$U' = (A-1)t, \tag{1.70}$$

solve the equation

$$T' = U' + U'P_o G_o T' \quad , \tag{1.71}$$

and then recover the physical transition operator $T = (A/A-1)T'$. The result will then be identical to the scattering produced by the true first-order (Watson) optical potential.

The advantage of the KMT approach becomes obvious if the many-body quantity t can be shown to be well approximated by the solution of the two-body projectile-nucleon scattering problem. Then U′ has a close connection with two-body phase shift data. On the other hand we note that the Watson quantity τ as given by Eq. (1.63) has a close connection with the Brueckner g-matrix operator[10] of bound-state perturbation theory. The operator Q_o in Eq. (1.63) allows only intermediate states in which the target is excited. In an independent particle model, intermediate states of the active target nucleon must be a state that is not occupied in the ground state configuration. In the limit of a nuclear matter approximation for the target, Q_o requires the intermediate state momentum of the active nucleon to be greater than k_F. A feature of the Brueckner g-matrix that is not present in the operator τ is a similar restriction on the states of the other colliding particle (here the projectile). This raises the question of the handling of antisymmetry in multiple scattering developments a topic we will return to later.

From the first order result one can employ a crude argument to suggest that the shape of the optical potential in position space should be dominated by the shape of the nuclear density. The one-body optical potential from Eq.(1.70) is $\hat{V}' = \langle\phi_o|U'|\phi_o\rangle = (A-1)\langle\phi_o|t|\phi_o\rangle$. If we assume that t behaves like the simplest two-body potential and depends only on the separation distance $r=|\vec{r}_1-\vec{r}_2|$ between the two nucleons, i.e. t(r), then on obtains

$$\hat{V}'(r) \simeq (A-1)\int d^3r' t(|\vec{r}-\vec{r}'|)\rho(r') \quad , \qquad (1.74)$$

where ρ is the one-nucleon density of the nucleus. This is like the Hartree mean field[13] as used in bound state problems except t replaces v. In the limit of a zero range for the NN interaction (and therefore t) we have $t(|\vec{r}-\vec{r}'|) \to \bar{t}\,\delta(\vec{r}-\vec{r}')$ and thus

$$\hat{V}'(r) \sim (A-1)\bar{t}\,\rho(r) \quad , \qquad (1.75)$$

where $\bar{t}$ is a complex number. Differences in shape between the optical potential $\hat{V}'(r)$ and $\rho(r)$ will arise among other things, from the finite range of t compared to that of ρ. In a momentum-space representation, Eq.(1.74) is

$$\hat{V}'(q) \simeq (A-1)\, t(E,q)\rho(q) \quad , \qquad (1.76)$$

where $\vec{q} = \vec{k}'-\vec{k}$, the momentum transfer suffered by the projectile. Now, in nucleon-nucleon scattering at energy E one can in principle infer t(E,q) from the data and construct an approximate optical potential from Eq. (1.76). This is the impulse approximation. The work of KMT[12] in 1959 showed that much of the nucleon-nucleus elastic (and inelastic as discussed later) scattering data could be understood from NN scattering phase shifts in terms of Eq.(1.76). There are many questions concerning the nature of corrections to Eq.(1.76) that have been investigated and remain to be investigated, but Eq. (1.76) is still the basic workhorse for studies of scattering of a variety of projectiles from nuclei.

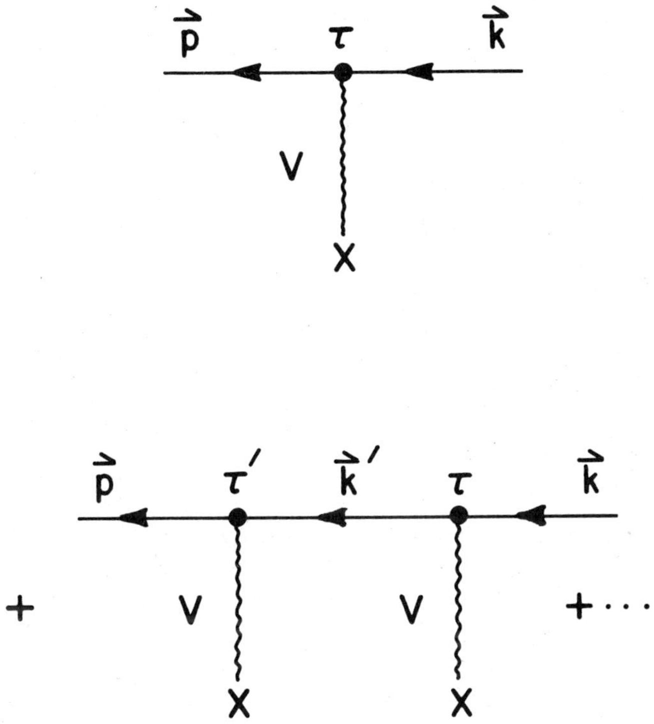

FIGURE 1. Feynman diagrams for the first-order and second-order contributions to the non-relativistic S-matrix for one-body potential scattering.

LECTURE 2 - NON-RELATIVISTIC MULTIPLE SCATTERING

REACTIVE CONTENT OF THE OPTICAL POTENTIAL

Before considering further developments of multiple scattering expansions, let us explore the meaning of the imaginary part of the optical potential. For one-body scattering from an hermitian potential ($V^{+}=V$) the conservation of probability flux results in the S-matrix being unitary $S^{+}S = 1$. The use of a complete set of plane wave states together with the substitution of Eq.(1.26) to express S in terms of T yields

$$(T^{+}-T)_{pk} = 2\pi i \int d^3k' (T^{+})_{pk'} \, \delta(E-E_{k'}) \, T_{k'k} \ . \tag{2.1}$$

For $\vec{p}=\vec{k}$, this becomes

$$-2 \, \text{Imag} \, (T_{kk}) = 2\pi \int d^3k' \delta(E_k-E_{k'}) |T_{k'k}|^2 \ . \tag{2.2}$$

The meaning of this equation becomes clear after division by the incident flux J_i. The right hand side becomes the total cross section σ_T, and the left hand side becomes $4\pi k^{-1}$ Imag (F_k (θ=o)), that is, proportioned to the forward scattering amplitude. This equality is often called the optical theorem. In operator form, Eq.(2.1) is

$$T^{+} - T = T^{+} \Delta G_o \, T \ , \tag{2.3}$$

where $\Delta G_o = G_o(E^{-}) - G_o(E^{+}) \equiv 2\pi i \ \delta(E-H_o)$, a relation which follows directly from the Lippmann-Schwinger equation for T when V is hermitian. Now in the many-body case, the elastic scattering Lippmann-Schwinger equation for T due to the optical potential U leads directly to the relation

$$T^{+}-T = T^{+}\Delta G_o P_o T + (T^{+}G_o^{+}P_o+1)(U^{+}-U)(1+P_oG_oT) \ , \tag{2.4}$$

that is, there is an extra term due to $U^{+}\neq U$, which can in principle be calculated from knowledge of U in terms of V. Let us do this from the equation $U=V+VG_oQ_oU$ in a very schematic way for the simple model in which the target can be excited but cannot dissociate. We use a label (o) for states of the whole system that have the target in the ground state, and the label n ($n\neq o$) for states in which the target is excited. Then we obtain, in the same way that Eq.(2.3) is obtained, the relation

$$(U^{+}-U)_{oo} = 2\pi i \sum_{n\neq o} (U^{+})_{on} \delta(E-E_n) U_{no} \ . \tag{2.5}$$

Substitution into Eq.(2.4) then yields

$$(T^{+}-T)_{oo} = 2\pi i\int\delta(E-E_o)|T_{oo}|^2 + 2\pi i\sum_{n\neq o}\int\delta(E-E_n)|T_{no}|^2 , \quad (2.6)$$

where $T_{no} = U_{no}(1+G_oP_oT)$ and the integral signs indicate integration over the direction of the final momentum of the projectile. After proper normalization this result becomes $4\pi k^{-1}\mathrm{Imag}(F(\theta=o)) = \sigma_{el} + \sigma_R$ where σ_{el} is the total elastic cross section and σ_R is the total reaction (here inelastic) cross section. The total reaction cross section arises because of the non-hermitian part (essentially the imaginary part) of the optical potential.

Of course in a complete analysis, the second term of Eq.(2.6) becomes all possible non-elastic channels including rearrangement and break-up channels. If we make a limited microscopic model for U, such as the first order multiple scattering approximation we have discussed, then the implicit non-elastic processes thereby included in σ_R are limited to just a few. A study of this "reactive content"[14] can provide an important check on the physical sense of any microscopic model for U that is proposed. In the case of the first-order KMT optical potential $(A-1)\langle\phi_o|t|\phi_o\rangle$ the implicit reactive content consists of nucleon knockout processes, and also nucleon pick-up processes if the effective two-body energy of t is allowed to properly couple to the integration variable of the above matrix element[14].

FIRST-ORDER APPLICATION

We sketch briefly what is involved in the calculation of the first-order KMT optical potential. Since t is a one-body operator in the target space, one obtains (for $\vec{q}=\vec{k}'-\vec{k}$)

$$\hat{V}'(\vec{k}';\vec{k}) = \frac{A-1}{A}\sum_{\alpha}\int d^3p \, \langle k';\vec{p}-\frac{\vec{q}}{2}|t_\alpha(\varepsilon)|\vec{k},\vec{p}+\frac{\vec{q}}{2}\rangle \, \rho_\alpha(\vec{p}-\frac{\vec{q}}{2};\vec{p}+\frac{\vec{q}}{2}), \quad (2.7)$$

where we have ignored recoil effects of order 1/A in the kinematics. Here ε is the energy of the two-body collision which we take to be a fixed number appropriate to the scattering of the beam from a free nucleon. The sum over α identifies the N neutrons and the (A-N) protons of the target and associates the t-matrix with the relevant one-nucleon density matrix ρ_α. Equation (2.7) is quite complicated because the integral is at least two- dimensional, the t-matrix is fully off the energy shell, the density matrix must be calculated from a nuclear structure model and the spin and isospin degrees of freedom must be explicitly treated. It is not surprizing that Eq.(2.7) has never been evaluated without further approximation. The usual approximations involve a factorization of the t-matrix outside of the integral at some fixed value of $\vec{p}$, for then one can employ

$$\int d^3p\ \rho_\alpha(\vec{p}-\frac{\vec{q}}{2};\ \vec{p}+\frac{\vec{q}}{2}) = \rho_\alpha(q) \quad , \tag{2.8}$$

which is the Fourier transform of the one nucleon density profile $\rho_\alpha(r)$. For protons this can be deduced (with some model dependance) from electron- nucleus scattering data. One way to define a sensible factorization is to expand t in a Taylor series in $\vec{p}$ about some $\vec{p}_o$ and then to choose $\vec{p}_o$ so that

$$\int d^3p\ (\vec{p}-\vec{p}_o)\,\rho_\alpha(\vec{p}-\frac{\vec{q}}{2};\vec{p}+\frac{\vec{q}}{2}) = o.$$

For a spin-zero nucleus, parity conservation and time reversal invariance leads to the choice $\vec{p}_o$=o. This is called optimum factorization and leads to[15]

$$\hat{V}'(\vec{k}',\vec{k}) = \frac{A-1}{A}\sum_\alpha \langle\vec{k}',-\frac{\vec{q}}{2}|t_\alpha(\varepsilon)|\vec{k},\frac{\vec{q}}{2}\rangle\ \rho_\alpha(q) \quad . \tag{2.9}$$

This is more feasible, but still the t-matrix is off the energy shell, i.e. not a measurable two-body quantity. The kinematics corresponding to Eq. (2.9) are shown in Fig.(2) where recoil effects are also retained. The two-body t-matrix depends on only two momenta which can be taken to be the initial and final relative momenta or alternatively the difference (which is $\vec{q}$) and the sum (which here is $(\vec{k}'+\vec{k})/2 = \vec{K}$). The alternative form of Eq. (2.9), which is

$$\hat{V}'(\vec{k}',\vec{k}) \equiv \hat{V}'(\vec{q};\vec{K}) = \frac{A-1}{A}\sum_\alpha t_\alpha(\varepsilon;\vec{q},\vec{K})\ \rho_\alpha(q) \quad , \tag{2.10}$$

displays the nonlocality of the first-order optical potential (in position- space a local form $\delta(\vec{r}'-\vec{r})\hat{V}'(\vec{r})$ is obtained only if there is no dependance on $\vec{K}$ in Eq. (2.10)). Almost all calculations previous to the last few years have employed the further assumption that t in Eq.(2.10) be replaced by the on-shell value t′ appropriate to physical NN scattering. That is, the conditions $\vec{q}\cdot\vec{K}$=o and $q^2 = K^2 = 4k_o^2$ are imposed, where k_o is the NN relative momentum for physical scattering at energy ε. Thus this prescription (often called the impulse approximation) is

$$\hat{V}'(\vec{k}',\vec{k}) \rightarrow \hat{V}'(q) = \frac{A-1}{A}\sum_\alpha t'_\alpha(\varepsilon,q)\ \rho_\alpha(q) \quad , \tag{2.11}$$

where t′ is taken from NN scattering data. However, only a limited range of q is covered by such NN data at a given energy and assumptions must be employed to extrapolate to higher q.

Simple considerations show that scattering beyond 60° depends upon the type of extrapolation employed. At high energies (>500 MeV) most of the presently accessible data is forward of such angles, but at lower energies the use of extrapolations is clearly unsatisfactory.

At this point we digress to discuss spin and recall that the optical potential has a spin-dependence that arises from the spin-dependence of t which we have not yet made explicit. From symmetry principles (invariance under rotations, time reversal, parity and Galilean frame transformations) the non-relativistic spin 1/2 ⊗ 1/2 t-matrix can be written in terms of the Pauli spin operators in the schematic form[16]

$$t = A + C\,(\vec{\sigma}_1+\vec{\sigma}_2)\cdot\hat{n} + M\,\vec{\sigma}_1\cdot\hat{m}\;\vec{\sigma}_2\cdot\hat{m} + G\,\vec{\sigma}_1\cdot\hat{n}\;\vec{\sigma}_2\cdot\hat{n}$$

$$+ H\,\vec{\sigma}_1\cdot\hat{l}\;\vec{\sigma}_2\cdot\hat{l} + D\,(\vec{\sigma}_1\cdot\hat{m}\;\vec{\sigma}_2\cdot\hat{l} + \vec{\sigma}_1\cdot\hat{l}\;\vec{\sigma}_2\cdot\hat{m}) \qquad (2.12)$$

where the unit vectors $\hat{n} = \hat{l} \times \hat{m}$ are determined by the momenta of the problem and A,C,M,G,H and D are invariant functions of the momenta (and isospin) only. In particular, for $t(\vec{q},\vec{K})$, we have $\hat{l} = \vec{q}$ and $\hat{m} = \vec{K}$ and hence $\hat{n} = \vec{k}' \times \vec{k}$. For any non-relativistic NN potential such as the Reid[17] or Paris[18] potentials, the off shell t-matrix can be calculated, and the functions A,C etc identified. For an on-shell t-matrix, symmetry principles show that D=o leaving five independent amplitudes[16]. When the form in Eq.(2.12) is substituted into Eq. (2.7) for the optical potential for a spin-zero nucleus, and the spin projections of the target nucleons are integrated over, one expects all terms of t that are linear in the target nucleon spin to be eliminated, leaving only the terms A and C. This does happen for a diagonal ground state matrix element provided the spin-zero nucleus is also spin-saturated. By this we mean that the spatial dependence of the density matrix is the same for spin-up and spin-down. This condition is most nearly realized for closed major shell nuclei. Within an optimum factorization approximation, the spin-saturation condition is not necessary, for time-reversal and parity invariance eliminate all terms except A and C. The spin-dependence of the optical potential is

$$\hat{V}'(\vec{k}',\vec{k}) = \hat{V}_c(\vec{q},\vec{K}) + i\vec{\sigma}\cdot\vec{k}' \times \vec{k}\;\hat{V}_{LS}(\vec{q},\vec{K}) \quad , \qquad (2.13)$$

and this form is not special to any approximation scheme but is just that form dictated by the symmetry principles. In the approximation we are discussing, $\hat{V}_c$ comes only from A and $\hat{V}_{LS}$ comes only from C. The above form, in the absence of dependence upon $\vec{K}$, corresponds in position space to the local form

$$\hat{V}'(\vec{r}) = \hat{V}_c(\vec{r}) + W_{LS}(\vec{r})\ \vec{L}\cdot\vec{S}, \qquad (W_{LS} = \frac{1}{r}\frac{d}{dr}\hat{V}_{LS}) \ , \qquad (2.14)$$

that is employed in standard phenomenological optical model analysis. Since $P_oH_oP_o$ is spin-independent, the nuclear scattering amplitude due to $\hat{V}'(\vec{k}',\vec{k})$ has the same form as Eq. (2.13), viz.,

$$F(q) = F_1(q) + \vec{\sigma}\cdot\hat{n}\ F_2\ (q) \ . \qquad (2.15)$$

The differential cross section is

$$\frac{d\sigma}{d\Omega} = \Sigma(q) = |F_1|^2 + |F_2|^2 \ , \qquad (2.16)$$

while the analyzing power (or polarization induced in the direction $\hat{n}$ for an incident unpolarized beam) is

$$A_y = 2\ \mathrm{Real}\ (F_1F_2^*)/\Sigma \ . \qquad (2.17)$$

An independent quantity that can be measured is the spin rotation function

$$Q = 2\ \mathrm{Imag}\ (F_1F_2^*)/\Sigma \ . \qquad (2.18)$$

There are numerous applications of the $t(q)\rho(q)$ form of the optical potential to be found in the literature and a representative list is provided in Ref.(19). The neutron densities which are much less well known than the proton densities are often adjusted to fit the elastic nuclear scattering data as best as possible. This can provide useful information on the neutron density differences in neighboring isotopes[20]. The descriptions of nuclear scattering data provided by the impulse approximation $t(q)\rho(q)$ above about 500 MeV is generally very good (to within about 10%) except in deep diffractive minima and at large q especially for light nuclei. The analyzing power (or polarization) data are described less well and generally only qualitative features are reproduced. There are many questions that need to be investigated in order to understand and remove the inadequacies of the simple $t(q)\rho(q)$ form of the optical potential. A representative list is:

1) off-shell and non-locality effects in t?

2) accuracy of factorization of the matrix element $\langle\phi_o|t|\phi_o\rangle$?

3) nuclear structure effects arising in spin-unsaturated nuclei?

4) second-order multiple scattering terms of the optical potential?

5) antisymmetrization and related medium effects?

6) extra degrees of freedom related to virtual meson production and the $\Delta \to N\pi$ resonance

7) extra degrees of freedom related to relativistic descriptions of the reaction mechanism and the structure of nuclei.

We shall address a selection of these questions, some only very briefly.

OFF-SHELL EFFECTS AND NONLOCALITY

Recently an investigation of these effects has been carried out [15] by employing the optimum factorized form given in Eq. (2.10). The off- shell t-matrix has been taken from the model of Love and Franey[21] or from calculations[22] from the Reid potential. The results from both models are quite similar for the purposes under discussion here. In Fig.(3) (taken from Ref.(15)) the solid line is the result from the optical potential that includes off-shell effects and their resulting nonlocality while the dashed line is the result from the impulse (on-shell) approximation $t(q)\rho(q)$. There are significant differences especially for A_y, but neither calculation is an accurate estimate of the theoretical uncertainty introduced by employing a factorization rather than computing the "full-folding" integral in Eq. (2.7). Since different treatments of the matrix element couple to different off-shell treatments, the shaded band can also be thought of as indicating the typical influence of off-shell and nonlocal effects. A more detailed description of these calculations can be found in Ref.(15). The corresponding situation at a lower energy and for a lighter target is shown[15] in Fig.(4). The influence of off-shell effects and the consequences of approximations to the full-folding integral are much larger and increase with momentum transfer. The summary of a number of such studies for scattering between 100 and 500 MeV is that off-shell and nonlocal effects need to be included for accurate work, especially at lower eneries and for the larger scattering angles, but the clear deficiencies in representations of spin-dependent observables are unlikely to be removed with the first- order theory under discussion at present. The only way to settle these questions is to perform the nuclear matrix element of t without approximation and this has been done by E.F. Redish and K. Stricker-Bauer[22] and calculations with such an optical potential are in progress[23]. As will be discussed at this workshop, there is compelling evidence that a first-order optical potential extended to a relativistic treatment of nucleons as Dirac particles can greatly improve the correspondance with data especially the spin-dependent observables. Before discussing that approach, we will describe a more traditional topic.

SECOND ORDER TERMS

The Watson multiple scattering expansion of the optical potential operator U is, from Eq.(1.65),

$$U = \sum_i \tau_i + \sum_{i \neq j} \tau_i Q_o G_o \tau_j + \cdots \quad . \tag{2.19}$$

We have discussed the first term which corresponds to the projectile interacting to all orders with one nucleon. The remaining terms can be classified according to how many nucleons the projectile interacts (to all orders) with. The result is called a spectator expansion[24] and is a finite series of A terms

$$U = \sum_i U_i^{(1)} + \sum_{i \neq j} U_{ij}^{(2)} + \sum_{i \neq j \neq k \neq i} U_{ijk}^{(3)} + \cdots + U^{(A)} , \tag{2.20}$$

where $U_i^{(1)} = \tau_i$. The second-order term involves scattering to all orders from a pair of target nucleons and is given by $U_{ij}^{(2)} = \tau_{ij} + \tau_{ji}$ where

$$\tau_{ij} = \tau_i Q_o G_o (\tau_j + \tau_{ji}) \quad . \tag{2.21}$$

In the absence of complications due to the many-body nature of the propagator G_o and the projector Q_o (which prevents the target ground state from appearing in intermediate reaction steps in τ_{ij} and τ_i), the required quantity τ_{ij} is obtainable from the solution of a three- body problem. For the matrix element $\langle \phi_o | \tau_{ij} | \phi_o \rangle$, one requires knowledge of the two-nucleon density matrix of the target. An important question to ask is whether the expansion is sufficiently convergent to warrant belief in first-order calculations. The presence of the operator Q_o makes it seem plausible that the answer is yes. To sketch this argument, we take just the double scattering part of τ_{ij} and consider

$$\hat{V}^{(2)} = \langle \phi_o | U^{(2)} | \phi_o \rangle \simeq A(A-1) \langle \phi_o | \tau_1 Q_o G_o \tau_2 | \phi_o \rangle \quad . \tag{2.22}$$

Evidently for this process, the projectile must excite the target by scattering from one nucleon and bring the target back to the ground state by scattering from a different nucleon. This process can only contribute through nucleon-pair correlations in the target state. One way to see that the effects on scattering should be small is to use the fixed scatterer approximation[25] and treat τ as depending only on momentum transfer. Then the nuclear structure quantity that enters is (for $\vec{q}=\vec{q}_1+\vec{q}_2$) the pair-correlation function

$$C_2(\vec{q}_1,\vec{q}_2) = \rho_2(\vec{q}_1,\vec{q}_2) - \rho(\vec{q}_1)\rho(\vec{q}_2) \quad , \tag{2.23}$$

where ρ_2 is the two-body density for momentum transfers $\vec{q}_1$ and $\vec{q}_2$ to be absorbed by a pair of nucleons

$$\rho_2(\vec{q}_2,\vec{q}_2) = \langle\phi_o| e^{-i\vec{q}_1\cdot\vec{r}_1}\, e^{-i\vec{q}_2\cdot\vec{r}_2}\, |\phi_o\rangle \quad , \tag{2.24}$$

and ρ is the one-body density for momentum transfer to be absorbed by one nucleon

$$\rho(\vec{q}_1) = \langle\phi_o| e^{-i\vec{q}_1\cdot\vec{r}_1}|\phi_o\rangle \quad . \tag{2.25}$$

In the limit of zero momentum transfer $C_2\to o$. As momentum transfer is increased one can expect the effects of the second order optical potential to eventually compete with the scattering mechanisms retained by use of a first- order optical potential. Calculations of second order optical potentials based upon Eqs. (2.22,23) have been made[19], mostly at energies above 500 MeV, with τ_1 and τ_2 replaced by the free NN t-matrix and eikonal approximations employed to compute the matrix element and to make local approximations. With reasonable estimates of the correlation function, the effects are generally small and decrease with increasing mass number of the target, and increase with momentum transfer. These effects do not tend to change the qualitative angular shape of spin-dependent observables. Effects on the differential cross section are typically[20] an increase in the height of diffraction maxima by about 15%.

It is worthwhile to contrast the mechanisms in first- and second- order optical potentials from the perspective of coherent contributions to scattering as this will illustrate the rationale for developing multiple scattering expansions in the first place. With a first- order plus second-order optical potential, the elastic transition operator is

$$T = A\tau + (A\tau)\, P_oG_o\, (A\tau) + \cdots$$

$$+ A(A-1)\,\tau\, Q_oG_o\, \tau + \cdots \quad , \tag{2.26}$$

where the first series comes from the first-order optical potential, and the second series comes from the second-order optical potential. To judge the relevance of a first-order approach, we should compare the relative magnitude of the quadratic terms displayed above. They look to be of the same order but that formula is deceptive. Consider the nuclear ground state $|\phi_o\rangle$ and a particular one-particle one-hole excited state $|\phi_n\rangle$ where both are Slater determinants properly normalized to unity. We need to compare $\langle\phi_o|\tau|\phi_o\rangle$ and $\langle\phi_n|\tau|\phi_o\rangle$. The inelastic matrix element is smaller than the first by a statistical factor of 1/A if we ignore for the moment differences in the spatial distributions of the states. This is because in the elastic matrix element, orthogonality prevents the single particle quantum numbers from changing, and all A of the single particle states contribute. In the inelastic matrix element, orthogonality permits a connection only between the "hole" and "excited" single particle states, and the relative weight is 1/A. If we keep only the first- order optical potential, the ratio of the omitted double scattering term (for T) to that retained is $R_T \sim 1/A^2$, which is $1/A^4$ for the cross section. Taking some account of the differences in spatial distributions of the states leads to the cross section ratio

$$R_\sigma \sim \left[\frac{1}{A}\,\frac{\rho_t(q)}{\rho(q)}\right]^4 \quad , \tag{2.27}$$

where ρ_t is the inelastic transition density and ρ is the ground state density. This is a large suppression factor for all but the lightest nuclei (e.g. taking $\rho_t \leqslant 10\rho$, and A=208, the second order effects would become equal to the first order effects only after the cross section has decreased by 10^{-5}).

This estimate is at best qualitative, since the full spectrum of intermediate inelastic excitations must be included. Then the pair correlation function becomes the centerpiece of the analysis. However, the above picture serves to illustrate that the reaction processes described by the second-order optical potential have significantly less coherence than the processes generated by the first-order optical potential. It is this large degree of coherence and the absence of orthogonality -based suppression effects introduced by Q_o, that suggests a large degree of confidence in the results from first-order optical potentials for forward angle scattering. This can be termed Watson coherence[11] and is the rationale for employing multiple scattering expansions.

ANTISYMMETRIZATION

In the matrix element $\langle\phi_o|t|\phi_o\rangle$ of the first-order KMT optical potential, the operator $t=v+vG_ot$ becomes the free NN t-matrix if the many- body quantity G_o is replaced by a two-body propagator. When the projectile is indistinguishable from the target particles it makes sense to use the properly antisymmetrized two-body t (which is the only one we know about anyway). However, the formulation of multiple scattering expansions that we have described so far does not explicitly build in the effects of the Pauli principle upon the projectile. There are essentially two distinct approaches to this problem.

The first approach, which seems more appropriate for the high energy regime (projectile momentum much larger than the fermi momentum k_F) is to consider the Pauli exchange processes as exchange forces which are treated on the same footing as the direct nuclear forces. One can in this case develop spectator type expansions for scattering operators in which the various terms have the projectile interating with one, two, three .... nucleons of the target[26]. The Pauli exchanges for the whole process can be distributed over the terms of this expansion so that each "few-body" problem that needs to be solved is antisymmetric with respect to the active particles[26]. As an illustration of this approach, an antisymmetric transition operator for nucleon-nucleus scattering can be defined by

$$\tilde{T}|\Phi_i\rangle = V\,\hat{A}|\Psi_i^{(+)}\rangle \quad , \tag{2.28}$$

where $\hat{A} = 1 - \sum_{j=1}^{A} E_{oj}$, and E_{oj} is an operator that exchanges all coordinates of the projectile and particle j of the target. Other quantities on the right hand side are the same as we have previously discussed. In operator form, we have

$$\tilde{T} = \sum_i v_{oi}\,(1-\sum_j E_{oj})(1+G_oT) \quad , \tag{2.29}$$

if Φ_i and G_o are understood to maintain proper antisymmetry amongst the target particles. The first two terms of a spectator expansion for T are[26]

$$\tilde{T} = \sum_i \tilde{t}_i + \sum_{i\neq j}(\tilde{t}_{ij} - \tilde{t}_i - \tilde{t}_j) + \cdots \,, \tag{2.30}$$

where

$$\tilde{t}_i = v_{oi}(1-E_{oi})(1+G_ot_i) \quad , \tag{2.31}$$

$$\tilde{t}_{ij} = (v_{oi} + v_{oj})(1-E_{oi}-E_{oj})(1+G_ot_{ij}) \quad . \tag{2.32}$$

and t_i and t_{ij} satisfy Eqs. (2.31) and (2.32) but without the exchange operators. The continuation to higher-order is obvious and the expansion is formally exact. When applied to the optical potential this approach yields for the first- order term

$$\tilde{\tau}_i = \tilde{t}_i - \tilde{t}_iP_oG_o\tilde{\tau}_i \quad , \tag{2.33}$$

which is the Watson operator for this case. In any application, G_o must be approximated. With a two-body approximation for G_o in Eq.(2.31), E_{oi} then commutes with G_o and $\tilde{t}_i$ becomes the properly antisymmetized t-matrix for NN scattering. Similarly $\tilde{t}_{ij}$ describes the antisymmetric three-body problem.

For low energies the physical relevance of this formulation is questionable. For $\tilde{t}_i$, the Pauli exchanges between the projectile and "struck" nucleon are included, as are the exchanges between all target particles but Pauli exchanges between the projectile and nucleons other than the ith one are omitted! When two nucleons of comparable energy are interacting in the neighborhood of a nucleus, both should be influenced in a similar way by the Pauli exclusion principle. In this respect one is reminded of the treatment of the pair-correlation contribution to the binding energy and the mean field for the nuclear bound state problem, that is Brueckner theory and the NN g-matrix[10]. This is the second approach we referred to, and has been employed for first-order optical potentials and also used as an effective NN interaction for inelastic nuclear scattering in recent years.[27] The differences between the NN operators t and g are often described as medium effects. However the relevant comparison is to be made between g and the first-order Watson optical potential $\tilde{\tau}$.

We briefly examine the differences and similarities between the Watson first-order optical potential (from perturbation theory of scattering states) and the Brueckner g-matrix (from perturbation theory of bound states). We will be very schematic since the bound state problem is treated carefully in other lectures at this workshop. Consider the many-fermion bound state problem $(E-H)\Phi=o$, with $H=K+V$, where K is the sum of the free particle energies and V is the sum of two body potentials. With addition and subtraction of W, which is a sum of identical one-body potentials, we can write $H=\bar{H}_o+\bar{V}$, where $\bar{H}_o=K+W$ and $\bar{V}=V-W$. For a perturbation treatment, the eigenstates of $\bar{H}_o$ are employed as the unperturbed basis states and $\bar{V}$ is the residual interaction to be treated in perturbation theory. Both $\bar{V}$ and $\bar{H}_o=\Sigma_i h_i$,(where $h_i=K_i+W_i$) are symmetric under the exchange of any particles. Given the antisymmetric solutions of

$$\bar{H}|\Phi_n\rangle = E_n|\Phi_n\rangle \quad , \tag{2.34}$$

we introduce the antisymmetric projection operators $\bar{P}=|\Phi_o\rangle\langle\Phi_o|$ and $\bar{Q}_o=\pi-P_o$ where Φ_o is the ground state of H_o, and π projects onto all antisymmetric states. Then the exact state may be expanded as

$$|\Phi\rangle = \bar{P}_o|\Phi\rangle + \bar{Q}_o|\Phi\rangle \quad . \tag{2.35}$$

Formal elimination of the component $\bar{Q}_o|\Phi\rangle$ from the equation $(E-H)|\Phi\rangle=o$ gives

$$(E-\bar{H}_o-\bar{P}_o\bar{U}\bar{P}_o)\bar{P}_o|\Phi\rangle = o \quad , \tag{2.36}$$

where

$$\bar{U} = \bar{V} + \bar{V}\,\bar{Q}_o\bar{G}_o(E)\bar{U} \quad , \tag{2.37}$$

and we have used $\bar{G}_o(E)=(E-\bar{H}_o)^{-1}$. Essentially $\bar{P}_o\bar{U}\bar{P}_o$ is the correction to the potential energy $\bar{P}_oW\bar{P}_o$ due to the residual interaction $\bar{V}$. The similarity of the formal structure of the equation for $\bar{U}$ to that discussed previously for an optical U is obvious.

For this bound state problem, $\bar{U}$ depends upon the single particle potential W, and the sought-after eigenvalue E, so there is a self-consistency to be implemented. Consider the lowest order approximation $\bar{U}\simeq\bar{V}=V-W$. Then the potential energy is $\bar{P}_o(\bar{U}+W)\bar{P}_o = \bar{P}_oV\bar{P}_o$, the Hartree-Fock mean field. Equation (2.37) then determines the single particle states to be used for Φ_o. The operator $\bar{Q}_o$ in Eq.(2.37) describes particle-hole excitations built upon Φ_o. With W now taken to be the Hartree-Fock mean field, it turns out that $\bar{P}_o\bar{V}\bar{Q}_o=o$ for one-particle one-hole excitations, and $\bar{P}_o\bar{V}\bar{Q}_o= \bar{P}_oV\bar{Q}_o$ thereafter, since $\bar{P}_oW\bar{Q}_o=o$ for higher-order excitations. At this stage the analysis becomes cumbersome without exploiting the power of second-quantization. We simply use a result from such an analysis[13]. To all orders in the interaction of a given pair of particles (ladder approximation), the mean field is to be calculated from $\bar{P}_og(E)\bar{P}_o$ where g is the Brueckner operator

$$g = \tilde{v} + \tilde{v}\bar{Q}_o\bar{G}_o(E)g \quad , \tag{2.38}$$

where $\tilde{v}_{ij}=v_{ij}(1-E_{ij})$, and $\bar{Q}_o$ projects upon two-particle two-hole antisymmetric states. The basis states to be used with this g-matrix operator are to be built from the single-particle eigenstates of the mean field in a totally antisymmetric way (e.g. as Slater determinants). The requirements of the Pauli principle need only be explicitly imposed by the operator $\bar{Q}_o$ in intermediate states. Note that $\bar{Q}_o$ forces both particles above the fermi level. The differences between g and $\tilde{\tau}$ reside in the differences between the projectors $\bar{Q}_o$ and Q_o and in the differences between the propagators $\bar{G}_o$ and G_o. Both $\bar{Q}_o$ and Q_o require the nucleus to be excited, however, the former affects both active nucleons and the latter only one. For the g-matrix, the propagator $\bar{G}_o$ has both active particles experiencing the mean field, while for $\tilde{\tau}$, the propagator G_o allows free propagation for the projectile. For a high energy projectile g and $\tilde{\tau}$ should be comparable.

It would not be correct to use such a g-operator in the context of scattering of a nucleon from a bound system of nucleons if the initial and final unperturbed basis states used for g have one particle in a plane -wave and the rest interacting. The corresponding unperturbed hamiltonian is not symmetric in all the particles, a property that is used to develop g in the first place. One can envisage applications to scattering in which the symmetric mean field hamiltonian $\bar{H}_o=\Sigma_i(K_i+W_i)$ is used to generate a continuum channel state in which one particle is in a scattering eigenstate of K+W and the rest are bound. A formalism essentially identical to the bound state perturbation theory is obtained and g appears as an effective potential in the ladder approximation as sketched above. This solves the problem of comparing the exact state which propagates according to exp(-iHt) with the less interacting state which propagates according to $\exp(-i\bar{H}_o t)$. However, the asymptotic hamiltonian for the scattering problem is $H_o=K_o+H_A$ (one particle is a free particle) and we must complete the problem by comparing the state which propagates as $\exp(-i\bar{H}_o t)$ with the true asymptotic state which propagates as $\exp(-iH_o t)$. This is a basic problem in adapting second-quantized field-theoretic methods of bound state perturbation theory to accommodate the boundary conditions required in scattering[28]. In the limit where the nucleus is treated as infinite nuclear matter, this problem disolves since the single particle eigenstates of both $\bar{H}_o$ and H_o are plane-waves. This is usually the method adopted in practice. In the g-matrix of Eq.(2.38), $\bar{Q}_o$ becomes in this limit a restriction that the intermediate state momenta of both nucleons be greater than k_F. After the equation is solved, the local density approximation is usually employed to approximate the case of finite nuclei by allowing k_F to depend upon distance according to the density profile $\rho(r)$. The resulting operator g is referred to as the density-dependent or medium-dependent effective two- body interaction for scattering[27]. The successes achieved within this approach need to be followed up by research into the finite nucleus corrections to the infinite nuclear matter assumption, since the intermediate states of the g-matrix can have low momenta as long as the states are orthogonal to occupied finite nuclear states.

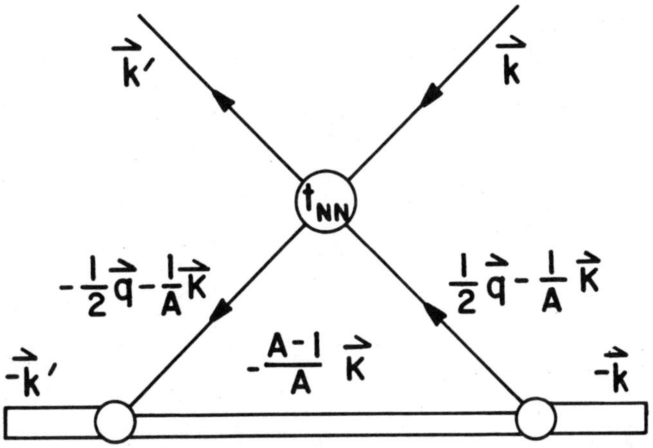

FIGURE 2. Kinematics for the first-order optical potential evaluated within the optimum factorization approximation. Target recoil effects are retained. [Taken from Ref.(15)].

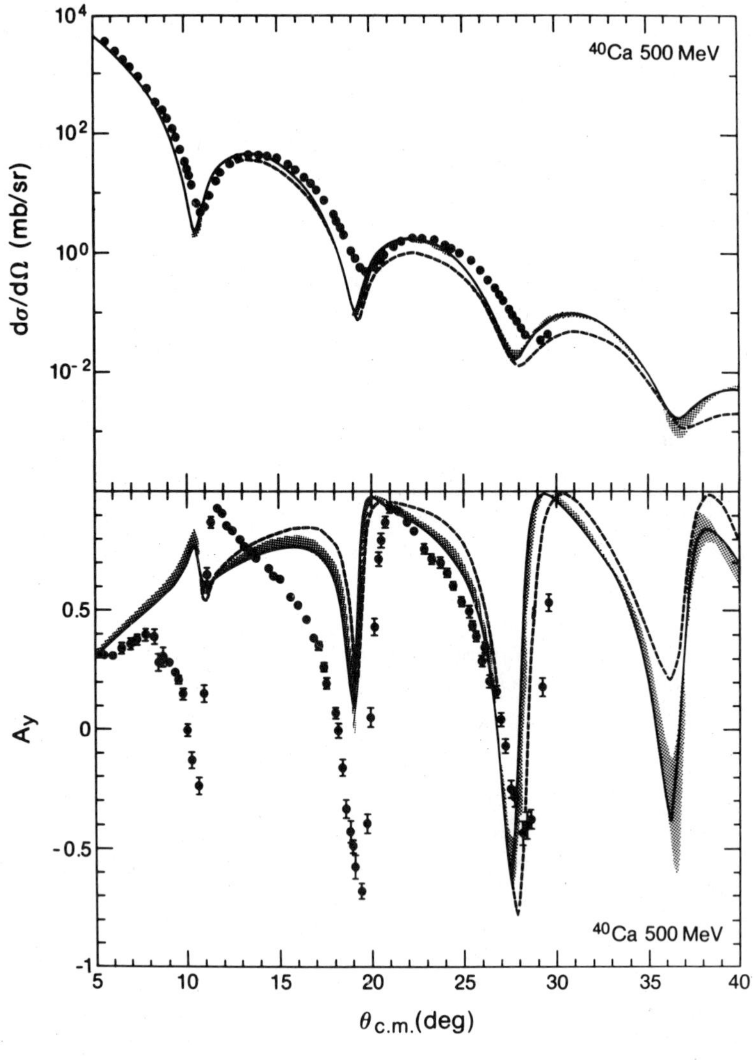

FIGURE 3a. The effect of off-shell and nonlocal properties of the non-relativistic first-order optical potential upon the cross section and analyzing power for proton scattering at 500 MeV from ^{40}Ca. The solid curve contains these effects while the dashed curve does not. The shading estimates typical uncertainties due to factorization. [Taken from Ref.(15)].

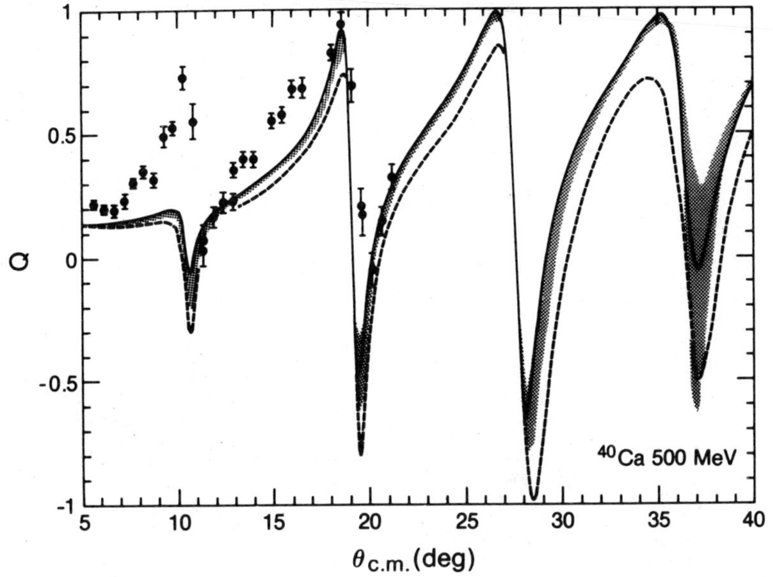

FIGURE 3b. The effect of off-shell and nonlocal properties of the non-relativistic first-order optical potential upon the spin rotation function for the case shown in Fig.(3a). [Taken from Ref.(15)].

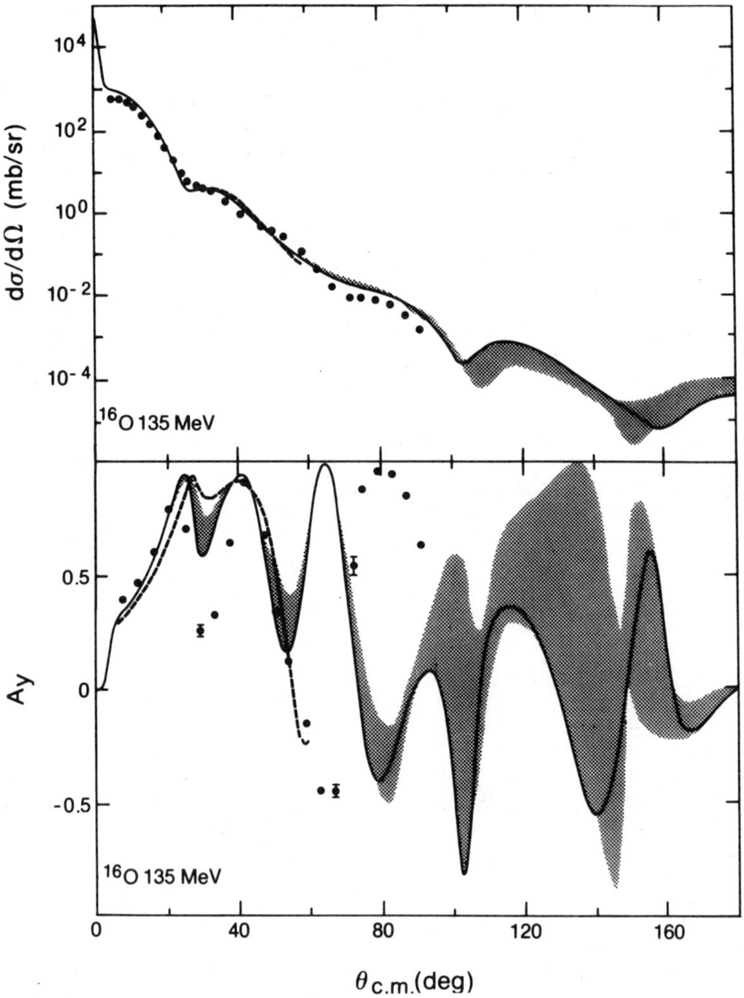

FIGURE 4. The effect of off-shell and nonlocal properties of the non-relativistic first-order optical potential for proton scattering from ^{16}O at 135 MeV. See also Fig.(3a). [Taken from Ref.(15)].

LECTURE 3 - RELATIVISTIC SCATTERING

ANTI-PARTICLES AND NEGATIVE ENERGY STATES

By way of introduction, we recall from Lecture 1 the time-dependent description of scattering of a structureless particle from a static potential. The second-order S-matrix element is

$$S^{(2)}_{pk}=(-i)^2\int_{-\infty}^{+\infty} d\tau'\int_{-\infty}^{\tau'} d\tau\langle\vec{p}|e^{iE_p\tau'}Ve^{-iH_o(\tau'-\tau)}Ve^{-iE_k\tau}|\vec{k}\rangle, \tag{3.1}$$

and we note that in the non-relativistic case, the free particle Hamiltonian has only positive energy eigenstates i.e., $H_o|\vec{k}\rangle = E_k|\vec{k}\rangle$, where $E_k > o$. A central feature of relativistic quantum mechanics is that the object that plays the role of H_o has both positive and negative energy eigenstates. For example, the Kemmer formulation[29] of the the Klein- Gordon (spin zero) wave equation and the Dirac (spin 1/2) wave equation have this feature. Let us now take the H_o in Eq. (3.1) to have such a property, and we use the notation

$$H_o|\vec{k}+\rangle = E_k|\vec{k}+\rangle \quad ,$$

$$H_o|\vec{k}-\rangle = -E_k|\vec{k}-\rangle \quad , \tag{3.2}$$

where $E_k > o$. We take the initial and final free particle states to have positive energy. We will be interested in the Dirac equation in which case the above free- particle states are four component objects to accommodate the Lorentz transformation structure of the problem. Likewise, the interaction V that we have been using must now be considered as a 4 x 4 matrix which describes its Lorentz structure. It is not necessary to make this structure of the problem explicit yet. To represent the intermediate state propagator $e^{-iH_o(\tau'-\tau)}$ in Eq.(3.1), we must expand in terms of the complete set of eigenstates of H_o, including both the positive energy states $|\vec{k}+\rangle$ and the negative energy states $|\vec{k}-\rangle$. Consider the expansion

$$e^{-iH_o(\tau'-\tau)} = \int d^3k \ |\vec{k}+\rangle e^{-iE_k(\tau'-\tau)}\langle\vec{k}+| \quad , \quad \tau'>\tau$$

$$= -\int d^3k|\vec{k}-\rangle e^{+iE_k(\tau'-\tau)} \langle\vec{k}-|, \quad \tau'<\tau \quad . \tag{3.3}$$

This satisfies the required equation $(i\partial_t-H_o)G_o(\tau'-\tau)\propto\delta(\tau'-\tau)$ for a Green's function or propagator. The choice of boundary conditions here is to let the negative energy states propagate backwards in time. This is the choice introduced by Feynman[7] in his quantum electrodynamics. Other choices are possible including letting both types of states propagate forward in time, but the physical interpretation is not so convenient. Whatever the choice, there is a new ingredient over and above the non- relativistic situation of positive energy states going forward in time. With the choice made above, Eq. (3.1) must be modified by adding a backward contribution $\int_{\tau}^{\infty} d\tau$. We can then write

$$S_{pk}^{(2)} \to S_{pk}^{(2)}(NR) + S_{pk}^{(2)}(R) \tag{3.4}$$

where the first term, which we refer to as the "non-relativistic" type of contribution is given as before by Eq. (3.1), and the new (relativistic) term is

$$S_{pk}^{(2)}(R)=(-i)^2\int_{-\infty}^{+\infty} d\tau'\int_{\tau'}^{+\infty} d\tau\langle\vec{p}|e^{iE_p\tau'}Ve^{-iH_o(\tau'-\tau)}Ve^{-iE_k\tau}|\vec{k}\rangle. \tag{3.5}$$

Here the incident particle state propagates until receiving an instantaneous interaction at time τ, meanwhile there has been a second interaction at an earlier time τ' from which the final particle state propagates. This process is described by the Feynman diagram in Fig. (5) and is to be contrasted to Fig. (1) which describes the "non-relativistic" piece. The propagation of a state with energy $E_{k'}$ and momentum $-\vec{k}'$ backward in time is interpreted as the antiparticle with energy $E_{k'}$ and momentum $\vec{k}'$ propagating forward in time. Thus, for the Z-graph in Fig. (5), the earlier interaction with the potential stimulates pair-production, and the produced antiparticle annihilates with the incoming particle via the later interaction with the potential. If instead of a static instantaneous potential, we had a time-dependent interaction expanded in terms of Fourier components weighted by $e^{i\omega t}$, then from the terms of different sign for ω, Fig. (5) would correspond to absorption of a field quanta to make a pair at the earlier time and emission of a field quanta from annihilation of a pair at the later time. In this case the possibility of transferring energy through the interaction allows physical pair production. For a static potential, the virtual production and annihilation processes must be treated in order to account fully for the effects of a H_o that has a positive and negative energy spectrum. Because we no longer have a true one-body problem, field- theoretic methods are required for a proper interpretation.

We continue with the example of a static potential. The "non-relativistic" piece of $S_{pk}^{(2)}$ leads, as discussed in Lecture 1 to

$$S_{pk}^{(2)}(NR) = -2\pi i\delta(E_p-E_k)\langle\vec{p}|V \frac{P_+}{E_k+i\eta-H_o} V|\vec{k}\rangle \quad , \tag{3.6}$$

where

$$P_+ = \int d^3k' \; |\vec{k}'+\rangle\langle\vec{k}'+| \tag{3.7}$$

is the projector onto the positive energy states of H_o. In the non-relativistic case of Lecture 1, we did not need such a projector since these are the only possible states. The relativistic piece can be put in the form

$$S_{pk}^{(2)}(R) = (-i)^2\int_{-\infty}^{+\infty} d\tau' e^{i(E_p-E_k)\tau'} e^{-\eta|\tau'|}\langle\vec{p}|V \; g(E_k,\tau')V|\vec{k}\rangle, \tag{3.8}$$

where

$$g(E_k,\tau') = \int_{\tau'}^{\infty} d\tau \; e^{i(E_k-H_o)(\tau'-\tau)} e^{-\eta|\tau|} \quad , \tag{3.9}$$

which, according to Eq.(3.3), will introduce P_- the projector onto the negative energy states of H_o given by

$$P_- = \int d^3k'|\vec{k}'-\rangle\langle\vec{k}'-| \quad . \tag{3.10}$$

The result of the integral for $g(E_k,\tau')$ is independent of τ' and, via Eq. (1.24) gives $-P_- \left[i(E_k-i\eta-H_o)\right]^{-1}$. The integral over τ' produces the energy conserving delta function and the result is

$$S_{pk}^{(2)}(R) = -2\pi i\delta(E_p-E_k)\langle\vec{p}|V \frac{P_-}{E_k-i\eta-H_o} V|\vec{k}\rangle \quad . \tag{3.11}$$

After combination of these results, the total second-order S-matrix element is

$$S_{pk}^{(2)} = -2\pi i\delta(E_p-E_k)\langle\vec{p}|V \; \hat{G}_o(E_k)V|\vec{k}\rangle \quad , \tag{3.12}$$

where the time-independent relativistic propagator for this problem of scattering from a static potential is

$$\hat{G}_o(E_k) = \frac{P_+}{E_k+i\eta-H_o} + \frac{P_-}{E_k-i\eta-H_o} \quad . \tag{3.13}$$

To all orders, the identified transition operator is

$$\hat{T}(E_k) = V + V\;\hat{G}_o(E_k)\hat{T}(E_k) \tag{3.14}$$

which is identical in structure to the non-relativistic Lippmann-Schwinger equation except for the modifications to the propagator. The corresponding scattering wave function defined by $V|\psi_k\rangle = \hat{T}|\vec{k}+\rangle$ is given by

$$|\psi_k\rangle = |\vec{k}+\rangle + \hat{G}_o(E_k)V|\psi_k\rangle \quad . \tag{3.15}$$

Note that for the first term of $\hat{G}_o$ in Eq. (3.13), H_o takes on all positive values and in the second term, H_o takes on all negative values. Thus there will be a pole singularity in the first term but not in the second term. The wave function $|\psi_k\rangle$ will have outgoing particle spherical waves but no outgoing antiparticle waves. If the incident state is an antiparticle, then E_k is negative in Eq. (3.13) and the above situation is reversed and the wave function $|\psi_k\rangle$ describes antiparticle scattering from the static potential V.

To make contact with more modern notation, it is useful to combine the forward and backward going pieces of $\hat{G}_o$ (identified by the different signs for $i\eta$) into one composite term by using the trick employed by Feynman[7]. In the momentum representation we have

$$\hat{G}_o(E) = \int d^3p \left\{ \frac{|\vec{p}+\rangle\langle\vec{p}+|}{E-(E_p-i\eta)} + \frac{|\vec{p}-\rangle\langle\vec{p}-|}{E+(E_p-i\eta)} \right\} \quad , \tag{3.16}$$

where, although $i\eta$ has opposite signs in the two terms, we see that we can treat the positive number E_p as though it has a vanishingly small negative imaginary part. Since we will use relativistic kinematics in which $E_p^2 = p^2+m^2$ so that $E_p = m + K.E.$, we can consider the rest mass m to have the small negative imaginary part. Thus we can write

$$\hat{G}_o(E) = (E-\hat{H}_o)^{-1} \tag{3.17}$$

where the eigenvalues of $\hat{H}_o$ are understood to be $\pm E_p$ where $E_p^2 = p^2+(m-i\eta)^2$.

THE DIRAC EQUATION

At this stage we need to become more specific and take $\hat{H}_o$ to be the Dirac hamiltonian for a free spin-1/2 particle. That is

$$\hat{H}_o = \vec{\alpha}\cdot\vec{\partial}/i + \beta m \tag{3.18}$$

where $\vec{\partial}$ is the gradient operator, $\vec{\alpha}=\gamma^o\vec{\gamma}$ and $\beta\equiv\gamma^o$. The representation employed for the Dirac gamma matrices can be found in the text by Bjorken and Drell[30]. With Feynman's expression of the boundary conditions, we have in the momentum representation

$$\hat{G}_o(E) = [E-\vec{\alpha}\cdot\vec{p}-\beta(m-i\eta)]^{-1} = [\not{p}-m+i\eta]^{-1}\gamma^o \ , \tag{3.19}$$

where $\not{p}=\gamma\cdot p=\gamma^{\mu}p_{\mu}=\gamma^o p_o-\vec{\gamma}\cdot\vec{p}$, and we have used $\gamma^o\gamma^o=1$. In the present (static) case we have the restriction $p^o=E$. It is convenient to absorb the γ^o into the interaction by defining $U=\gamma^o V$ and $T=\gamma^o\hat{T}$ so that Eqs. (3.14) and (3.15) have the covariant form

$$T = U + U\frac{1}{\not{p}-m+i\eta}T \ , \tag{3.20}$$

$$|\psi_k\rangle = |\vec{k}+\rangle + \frac{1}{\not{p}-m+i\eta}U|\psi_k\rangle \ , \tag{3.21}$$

where $T|\vec{k}+\rangle = U|\psi_k\rangle$. These equations are completely equivalent to the Dirac equation

$$(\not{p}-m-U)|\psi_k\rangle = o \tag{3.22}$$

when scattering boundary conditions are imposed. From Eqs. (3.19) and (3.16) we have for the Feynman propagator of this problem, the expansion

$$(\not{p}-m+i\eta)^{-1} = \int d^3p \left\{ \frac{|\vec{p}+\rangle\overline{\langle\vec{p}+|}}{E+i\eta-E_p} + \frac{|\vec{p}-\rangle\overline{\langle\vec{p}-|}}{E-i\eta+E_p} \right\} \tag{3.23}$$

where we have introduced the Dirac adjoint states

$$\overline{\langle\vec{p}\pm|} = \langle\vec{p}\pm|\gamma^o \tag{3.24}$$

which differ from the hermitian adjoint states of non-relativistic quantum mechanics by the factor γ^o.

FREE PARTICLE STATES AND NORMALIZATION

So far the free Dirac states we have been using are normalized according to

$$\langle\vec{k}'\pm|\vec{k}\pm\rangle = \overline{\langle\vec{k}'\pm|}\gamma^o|\vec{k}\pm\rangle = \delta(\vec{k}'-\vec{k}), \ \langle\vec{k}'-|\vec{k}+\rangle = o \ . \tag{3.25}$$

These states are

$$\langle\vec{r}|\vec{k}\pm\rangle = \frac{e^{i\vec{k}\cdot\vec{r}}}{(2\pi)^{3/2}}\, u(\vec{k}\pm) \quad , \tag{3.26}$$

where the Dirac spinors are

$$u(\vec{k}+) = N_k \begin{pmatrix} 1 \\ \dfrac{\vec{\sigma}\cdot\vec{k}}{\varepsilon_k} \end{pmatrix} \tag{3.27}$$

$$u(\vec{k}-) = N_k \begin{pmatrix} \dfrac{-\vec{\sigma}\cdot\vec{k}}{\varepsilon_k} \\ 1 \end{pmatrix} , \tag{3.28}$$

with $\varepsilon_k = E_k+m$ and $N_k^2 = \varepsilon_k/2E_k$. These states are not normalized in a way that makes evident the Lorentz transformation properties of probability densities. This is a price we pay for discussing the relativistic situation by emphasizing only the enlargement of the spectrum of H_o. The foregoing development may be recast into an equivalent form in which any probability density that may be extracted from intermediate stages of the calculation transforms as the fourth component of a Lorentz 4-vector. The required basis states are

$$\langle\vec{r}|\vec{k}\pm) \equiv \langle\vec{r}|\vec{k}\pm\rangle\left(\frac{E_k}{m}\right)^{1/2} = \frac{e^{i\vec{k}\cdot\vec{r}}}{(2\pi)^{3/2}}\, u_\pm(\vec{k}), \tag{3.29}$$

where the spinors $u_\pm$ are given by Eqs. (3.27) and (3.28) except that now $N_k^2=\varepsilon_k/2m$. In the rest frame, both sets of spinors are identical. A Lorentz boost to the frame where the particle has momentum $\vec{k}$ produces spinors that have the normalization of the $u_\pm$. This normalization is $u_\pm^\dagger(\vec{k})u_\pm(\vec{k})=\bar{u}_\pm(\vec{k})\gamma^o u_\pm(\vec{k})=E_k/m$ so that probability densities transform as E_k/m to compensate the Lorentz contraction of spatial volume elements by m/E_k. The normalization can also be expressed as $\bar{u}_\pm(\vec{k})u_\pm(k)=\pm 1$. In the notation of Bjorken and Drell[30] $u(\vec{k})=u_+(\vec{k})$ and $v(-\vec{k})=u_-(\vec{k})$. In this basis, the expansion of the propagator is

$$(\not{p}-m+i\eta)^{-1} = \int d^3p\, \frac{m}{E_p}\left\{ \frac{|\vec{p}+)\overline{(\vec{p}+|}}{E+i\eta-E_p} + \frac{|\vec{p}-)\overline{(\vec{p}-|}}{E-i\eta+E_p} \right\} , \tag{3.30}$$

and $d^3p(m/E_p)$ is Lorentz invariant. The final results of the scattering problem in a given frame are the same, for either of these choices of basis. We will continue with the previous basis as expressed by Eqs. (3.25) through (3.28) since one can easily convert to the invariantly normalized basis when translating findings from one Lorentz frame to another.

INTEGRAL EQUATIONS FOR RELATIVISTIC SCATTERING

With the use of Eq. (3.23) in Eq. (3.20), the relativistic scattering problem involving a static interaction can now be cast in the interesting form[31] (suppressing the 3-momentum variables and integrations)

$$T^{++} = U^{++} + U^{++}\Gamma_+T^{++} + U^{+-}\Gamma_-T^{-+}$$

$$T^{-+} = U^{-+} + U^{-+}\Gamma_+T^{++} + U^{--}\Gamma_-T^{-+}, \tag{3.31}$$

where the sought after scattering amplitude is

$$T^{++}(\vec{k}';\vec{k}) = \overline{\langle\vec{k}'+|T|\vec{k}+\rangle} \quad . \tag{3.32}$$

The required matrix elements of the interaction are

$$U^{ab}(\vec{k}';\vec{k}) = \overline{\langle\vec{k}'a|U|\vec{k}b\rangle} \tag{3.33}$$

where the labels a,b can be either + or −. Equations (3.31) are a set of coupled-channel integral equations and the integration variable is the 3-momentum $\vec{p}$ of the intermediate state of the particle. Since for convenience, we have not explicitly included the Pauli two-component spinors in our basis states, the quantities in Eqs. (3.31) to (3.33) are operators in Pauli spin-space, i.e. 2x2 matrices. This together with the ± labels makes evident the 4-component nature of the problem. The positive and negative energy components of the free-particle Feynman propagator are

$$\Gamma_\pm(p) = (E\pm i\eta\mp E_p)^{-1} \quad . \tag{3.34}$$

Although we are dealing with an unfamilar representation of the Dirac equation, there are decided advantages to be gained by following this path especially in considerations which contrast the physical content of relativistic and non-relativistic dynamics. Consider the Dirac wave function which, in the present basis, is given by

$$|\psi_k\rangle = \int d^3p\{\,|\vec{p}+\rangle\, \psi_+(\vec{p}) + |\vec{p}-\rangle\, \psi_-(\vec{p})\} \quad . \tag{3.35}$$

The component functions $\psi_\pm$ satisfy

$$\psi_+(\vec{p}) = \delta(\vec{p}-\vec{k}) + \sum_{\alpha=\pm}\int d^3p' \Gamma_+(p)\ U^{+\alpha}(\vec{p},\vec{p}')\ \psi_\alpha(\vec{p}')$$

$$\psi_-(\vec{p}) = \sum_{\alpha=\pm}\int d^3p' \Gamma_-(p)\ U^{-\alpha}(\vec{p},\vec{p}')\ \psi_\alpha(\vec{p}') \quad , \tag{3.36}$$

and they relate to the particle and antiparticle content of the state. For the simple example of a Dirac electron scattering from a static charge, these equations were derived and discussed in the review paper by Feshbach and Villars[32]. The relation to a field-theoretic description of this problem is also discussed in that reference. Essentially, the unit-charge electron-position field solution (relative to the true electron-position vacuum) has an electron component ψ_+ and a position component ψ_- as given by the above equations. The choice of free Dirac solutions as the basis states for expanding the full problem allows the scattering boundary conditions to be implemented. At large distances, the static interaction U is not operative, the electron-position field is free and described in terms of independent one-particle states, and only the particle component with energy $E^2=k^2+m^2$ matches the incoming and outgoing "beam" requirements. In the vicinity of the interaction, the particle and antiparticle components mix.

The contrast between the relativistic and non-relativistic dynamics of the scattering process is provided by the appearance of the free antiparticle components of the basis in the relativistic case. The Hilbert space of the problem is doubled compared to the non-relativistic situation. If we ignore the negative energy component Γ_- of the propagator, then the dynamical description for the T-matrix reduces to $T^{++}=U^{++}+U^{++}\Gamma_+T^{++}$ which corresponds to the non-relativistic form (with relativistic kinemetics). A more precise comparison cannot be made until the interaction U^{++} is specified. In the case of an electron scattering from a static charge Ze, the Lorentz structure of the interaction operator U to be used in the Dirac equation is $U=e\gamma^\mu q^{-2}J_\mu$ where the 4-current of the static charge is $J^\mu=(Ze,\vec{o})$, and q is the 3-momentum transfer. The matrix element of U connecting positive energy Dirac spinors i.e. $U^{++}(\vec{k}',\vec{k})$ from Eq. (3.33), is not simply the Coulomb electric interaction Ze^2/q^2 but includes also the magnetic effects through the (relativistic) Thomas spin-orbit term. We will discuss this more explicitly later. If we take the Schrodinger approach to include the Coulomb central interaction and the spin-orbit term (as is usual), then the extra dynamical feature incorporated through a relativistic (Dirac) treatment is the appearance of the negative energy component Γ_- of the propagator. In this sense the distinction between the mechanisms of relativistic and non-relativistic scattering of a particle from a static potential can be discussed in terms of Z-graph terms.

This point of view becomes even more cogent when nucleon-nucleus scattering in the framework of an impulse approximation is considered. When the interaction operator U is deduced from empirical information on projectile- nucleon free scattering, the positive energy free Dirac states for the projectile have to combine with the Lorentz structure to produce an object that is intimately related to nucleon-nucleon scattering data. Thus from this organization $U^{++}(\vec{k};\vec{k})$ is necessarily very closely related to the non-relativistic impulse approximation for the interaction in the Schrodinger based approach to the same problem. (Differences will arise if different descriptions of the internal structure of the target nucleus are employed). The integral equations (3.31) for the Dirac scattering problem can be cast into a single integral equation for T^{++} having the typical non-relativistic form by formal elimination of T^{-+}. The result is

$$T^{++} = W^{++}+W^{++}\Gamma_+T^{++}, \tag{3.37}$$

where the effective interaction is

$$W^{++} = U^{++}+U^{+-}\frac{1}{\Gamma_-^{-1}-U^{--}} U^{-+} \quad ,$$

$$= U^{++} + U^{+-}\Gamma_-U^{-+} + \cdots. \tag{3.38}$$

The meaning of this is that given an interaction operator U for the Dirac equation, then exactly the same scattering is produced by using W^{++} in the Lippmann-Schwinger equation. From knowledge of U, W^{++} can be calculated. Furthermore, the component state ψ_+, the "particle" sector of the Dirac state from Eq. (3.35), is the wave function produced by a use of the interaction W^{++}. That is, in momentum space,

$$(E-E_p-W^{++})\psi_+ = o \quad . \tag{3.39}$$

We see that use of U^{++} in a non-relativistic wave equation will give a scattering amplitude that is unable to reproduce the full Dirac solution to the extent that the terms beyond the first one in Eq. (3.38) are significant. These are the Z-graph terms that characterize the essential difference between relativistic and non-relativistic dynamics. Note that if the Dirac interaction operator U is specified as proportional to a certain spatial shape factor, then due to the Z-graph effects, the equivalent Lippmann-Schwinger interaction W^{++} will have a different spatial dependence obtained in a non-linear way from the original shape factor. Also if U is local, W^{++} (and U^{++}) is nonlocal.

A GENERAL STATIC INTERACTION

For a spin-1/2 particle, the most general operator U describing interaction with a static field is, in momentum space

$$U = S + \gamma\cdot K\ V_1 + \gamma\cdot q q\cdot K\ V_2 + \sigma^{\mu\nu} q_\mu K_\nu\ T \quad . \tag{3.40}$$

Here the employed 4-momenta are $q=k'-k$ and $K=k'+k$, where $k^\mu=(E,\vec{k})$ is the initial momentum of the particle and $k'^\mu=(E,\vec{k}')$ is the final momentum. The four Lorentz invariant amplitudes S, V_1, V_2 and T are, in general, functions of the three momentum-space Lorentz scalars of the problem namely q^2, K^2 and $(q\cdot K)^2$, and are invariant under parity and time- reversal. The structure given for U is the general form consistent with transformations under parity, time-reversal and Lorentz transformations. The structure of the V_2 term is necessary since both $\gamma\cdot q$ and $q\cdot K$ are time-reversal odd. Note that, in general, all terms except the scalar term involve nonlocalities (dependence upon $\vec{K}$), even if the amplitudes are local. When matrix elements of U are taken between Dirac free spinors, the results can be represented by just two types of Lorentz structure in each of the sectors U^{++}, U^{-+}, U^{--}. For example, in the U^{++} sector, use of the Gordon form of the current

$$\bar{u}(\vec{k}'+)[\gamma^\mu = \frac{K^\mu + i\sigma^{\mu\nu}q_\nu}{2m}]u(\vec{k}+) \quad , \tag{3.41}$$

shows that the tensor term can be converted to a scalar and vector. Further, the space parts of Lorentz vector terms can be converted to scalars and fourth-component (γ^0) vectors by repeated use of the free Dirac equation $\vec{\gamma}\cdot\vec{k}\ u(\vec{k}+) = (\gamma^0 E_k - m)u(\vec{k}+)$ and its conjugate. Thus one can always write

$$U^{++}(\vec{k}',\vec{k}) = \bar{u}(\vec{k}'+)[\hat{S}(\vec{k}',\vec{k}) + \gamma^0\hat{V}(\vec{k}',\vec{k})]u(\vec{k}+) \quad , \tag{3.42}$$

where the effective scalar and vector amplitudes $\hat{S}$ and $\hat{V}$ are linear combinations of S, V_1, V_2 and T. Note that $\hat{S}$ and $\hat{V}$ will be nonlocal even if the original four amplitudes were local. The above arguments can be repeated for the sectors U^{+-} and U^{--} with the result that these matrix elements of the general U can be represented by matrix elements of $\hat{S}_{+-}+\gamma^0\hat{V}_{+-}$ and $\hat{S}_{--}+\gamma^0\hat{V}_{--}$ respectively. Time reversal dictates that $\hat{S}$ and $\hat{V}$ in the +− and −+ sectors are identical. The three pairs of effective scalar and vector amplitudes are not independent but are linear combinations of S, V_1, V_2 and T from Eq.(3.40). The ability to use a Lorentz scalar and fourth- component Lorentz vector in each sector of the problem is a direct consequence of the symmetries applied to the spin-1/2 structure of the problem. A single effective scalar and a single effective vector amplitude can be employed throughout all sectors of the problem to the extent that any two of the amplitudes S, V_1, V_2 or T are very small. We note that in the nuclear matter limit any dependence upon orientation drops out and Eq. (3.40) reduces to two terms $U=S+\gamma^0 2EV_1$.

To make contact with a more familar language from nonrelativistic nuclear interactions, consider the evaluation of the matrix element for $U^{++}(\vec{k}',\vec{k})$ given in Eq. (3.42). One finds that

$$U^{++}(\vec{k}',\vec{k}) = U_c(\vec{k}',\vec{k}) + i\vec{\sigma}\cdot\vec{k}'\times\vec{k}\; U_{LS}(\vec{k}',\vec{k}), \tag{3.43}$$

where

$$U_c(\vec{k}',\vec{k}) = N_{k'}N_k\{\hat{S}+\hat{V} + \frac{\vec{k}'\cdot\vec{k}}{\varepsilon_{k'}\varepsilon_k}(\hat{V}-\hat{S})\} \quad , \tag{3.44}$$

and

$$U_{LS}(\vec{k}',\vec{k}) = N_{k'}N_k\,\frac{(\hat{V}-\hat{S})}{\varepsilon_{k'}\varepsilon_k} \quad . \tag{3.45}$$

This is exactly the form of a spin-1/2 nuclear interaction in the Pauli-Schrodinger representation, and corresponds to the central plus spin-orbit form given earlier in Eq. (2.13) of Lecture 2 where the nonrelativistic optical potential is discussed. Note that for an electromagnetic interaction (Lorentz vector only), the spin-orbit term generated this way corresponds to the Thomas form generated by retaining the lowest order coupling of upper and lower Dirac wave function components as is discussed in standard texts[30].

The matrix elements of U in the other sectors of the problem can be found in Ref(31). Consider now the consequences of taking for the quantities U_c and U_{LS} typical nuclear physics strengths (e.g. from a nonrelativistic shell model) for the interaction of a nucleon with a nucleus i.e. U_c~50 Mev and U_{LS}~ 5 MeV. Then Eqs. (3.44) and (3.45) produce, for the Lorentz components, the strengths $\hat{S}$~-400 MeV and $\hat{V}$~+350 MeV. These are the characteristic strengths found in relativistic mean field treatments of nuclear matter and finite nuclei within the Walecka model[33] which employs vector (ω) and scalar (σ) meson fields. It would be surprizing if this relationship did not occur since the two different representations of interaction strengths that we are discussing are both fitted to empirical properties relating to nuclear matter. We refer to the lectures by C.J. Horowitz at this Workshop for an exposition of relativistic mean field theory models for nuclear matter and finite nuclei. One should not gain the impression that the outcome of relativistic mean field theory is summarized by the reproduction of the empirical Schrodinger strengths U_c~50 MeV and U_{LS}~5 MeV. Rather such a formulation provides an excellent and unique laboratory for investigations in terms of interacting nucleon and meson fields and couplings to the quantum vacuum in a controlled (renormalizable) way. Such opportunities, of course, are outside the scope of treatments that start from a one-body wave equation. The point we do want to emphasize is that the Lorentz strengths resulting from such approaches combine, within the Dirac equation, so that the dominant interaction sector U^{++} is extremely close to the familar Schrodinger-Pauli interaction of nuclear physics. However, the solutions of the Dirac equation will differ from nonrelativistic approaches to the extent that the coupling to negative energy plane wave states is significant. Since the propagator Γ_- that controls this coupling (in the case of a static interaction) is never larger than 1/2m, the U^{++} sector is guaranteed to be the dominant sector of the problem.

SCATTERING FROM A NUCLEUS

Does the foregoing one-body scattering discussion have any relevance to the nucleon-nucleus elastic scattering problem? For a spin-zero nucleus, the symmetries of the problem limit the form of an optical potential interaction U to the four terms given in Eq. (3.40) if the nucleus is treated as a fixed static source. More realistically, if the motion of the nucleus is to be described as well, the general form of U contains, in principle, ten invariant amplitude functions[34]. The reason for this increase in complexity over the previous simple case is that now there are three (instead of two) independent 4-momenta with which to build invariants. Let $k(k')$ and $p(p')$ be the initial (final) 4-momenta of the projectile and nucleus respectively. Then the three independent momenta can be chosen as $q=k'-k$, $K=k'+k$ and $P=p'+p$ (under the total momentum constraint $k'+p'=k+p$). Examples of extra terms now allowed for U are $\gamma \cdot P$ and $\sigma^{\mu\nu} K_\mu P_\nu q \cdot K$ among others. If the nucleus is treated as being always on mass shell, i.e. $p^2=p'^2=M_A^2$, then the ten invariant amplitudes reduce to the eight described in Ref.(35). In this situation, the relative energies $(k-p)^0$ and $(k'-p')^0$ are no longer independent variables but are determined by the 3-momenta and total energy of the problem. In the projectile-nucleus zero-momentum frame, the interaction can then be described by a function $U(E,\vec{k}';\vec{k})$. The kinematics for this situation is shown in Fig.(6). The corresponding Dirac wave equation in momentum space is

$$\left(\gamma^0(E-E_A(k))-\vec{\gamma}\cdot\vec{k}-m\right)\psi(\vec{k}) = \int d^3\vec{k}'\, U(E,\vec{k},\vec{k}')\,\psi(\vec{k}'). \qquad (3.46)$$

This has the same form as for the problem of scattering from a static interaction that we have previously discussed, except that nuclear recoil effects are retained. As before, an effective scalar and fourth-component vector can be used in each of the three independent sectors ++, --, +-, to represent the interaction. These effective scalars and vectors can be found in Ref. (35). In the limit where $M_A \to \infty$, the eight invariant components[35] of this U reduce to the four given in Eq.(3.40).

The question which now arises is can one infer the important features of a microscopic relativistic optical potential $U(E,\vec{k}',\vec{k})$ for Eq. (3.46) with guidance from the content and form of non-relativistic multiple scattering expansions? Such a heuristic approach is necessary at present since the analog of the manipulations discussed in Lectures 1 and 2 for defining a multiple scattering expansion is not presently available within the context of a relativistic quantum field theory. The closest candidate at present is the relativistic mean field model[33] of the bound state problem (nuclear matter and finite nuclei). However, the model lagrangians employed so far in work of that type do not contain sufficient meson degrees of freedom to identify an object capable of realistically describing nucleon-nucleon scattering. The advantage of renormalizability of these simple models of nuclear matter has so far proved insurmountalby difficult to extend to the scattering situation with a meson content comparable to modern one boson exchange models of nucleon-nucleon scattering. We refer to the lectures at this workshop by R. Machleidt for an exposition of one boson exchange models of the nucleon-nucleon interaction.

Consider the following assumption for a relativistic extension of the first-order KMT optical potential

$$U' = \frac{A-1}{A} \langle \bar{\phi}_o | \Sigma_{i=1}^{A} t_{oi}^{D} | \phi_o \rangle \ , \qquad (3.47)$$

where now ϕ_o is the target ground state in a Dirac representation and t_{oi}^{D} is an operator for the free NN t-matrix also in a Dirac representation. The elastic scattering operator is to be calculated from $T=T'(A/A-1)$, where

$$T' = U' + U' \frac{1}{\not{k}-m+i\eta} T' \ . \qquad (3.48)$$

In accordance with the kinematics displayed in Fig. (6) or Eq. (3.46), the time component of the intermediate state momentum k is $k^o=E-E_A(\vec{k})$.

As discussed earlier, in the ++ sector the matrix element of Eq.(3.47) can be put in the form

$$U'^{++}(\vec{k}',\vec{k}) = \bar{u}(\vec{k}'+)\{\hat{S}(\vec{k}',\vec{k}) + \gamma^o\hat{V}(\vec{k}',\vec{k})\}u(\vec{k}+) \ , \qquad (3.49)$$

as is always allowed by the underlying symmetries. We wish to describe here the consequences for scattering of adopting the simple attitude that the same Lorentz vector and scalar components of the interaction are operative throughout all sectors of the problem. In this respect the assumption corresponds to that of vector-scalar dominance found so successful in the initial version of the relativistic mean field description of ground state nuclei. The effective $\hat{S}$ and $\hat{V}$ could, in principle, be calculated from a one-boson exchange model of t^D for NN scattering. However, for the ++ sector, some amount of the required information is already contained in the nonrelativistic optical potential due to the form of Eq.(3.47).

It is instructive to analyze the implications of Eq.(3.47) from the following perspective. Within a single particle model for the nuclear ground state, each (four component) Dirac state can be expanded as

$$|\phi_o\rangle = \int d^3p \, \{ |\vec{p}+\rangle\phi_+(\vec{p}) + |\vec{p}-\rangle\phi_-(\vec{p})\}, \tag{3.50}$$

where, in this representation, ϕ_+ is dominant and ϕ_- is necessarily small. If, for the moment, we ignore ϕ_-, then the matrix element of U′ in the ++ sector is

$$U'^{++}(\vec{k}';\vec{k})=(A-1)\int d^3p' d^3p \; \phi_+^\dagger(\vec{p}')\overline{\langle\vec{k}'|}\overline{\langle\vec{p}'+|}t^D|\vec{p}+\rangle|\vec{k}+\rangle\phi_+(\vec{p}). \tag{3.51}$$

The two-body matrix element of t^D that now appears must necessarily describe free NN scattering from the initial state $(\vec{p},\vec{k})$ to the final state $(\vec{p}'\vec{k}')$. Thus Eq.(3.51) is identical in form to the corresponding non-relativistic first-order KMT optical potential given in Eq.(2.7) of Lecture 2. Furthermore, U'^{++} will be very close to the nonrelativistic optical potential, if the ++ density matrix in Eq. (3.51) is very close to the density matrix employed in Eq.(2.7). This correspondance cannot be perfect since the diagonal point nucleon density that is employed in the nonrelativistic approach should be compared with the vector density

$$\rho_V(q) = \sum_{\alpha=\pm} \int d^3p \{u^\dagger(\vec{p}-\tfrac{\vec{q}}{2},\alpha)u(\vec{p}+\tfrac{\vec{q}}{2},\alpha)\phi_\alpha^\dagger(\vec{p}-\tfrac{\vec{q}}{2})\phi_\alpha(\vec{p}+\tfrac{\vec{q}}{2})\} , \tag{3.52}$$

of a relativistic treatment. Even if ϕ_- were small enough to be neglected, the product of the two spinors is not unity except at q=o.

In order to focus only on differences between the relativistic and nonrelativistic treatments of the projectile motion in the scattering situation, we will intentionally ignore the differences between the $U'^{++}(\vec{k}';\vec{k})$ of Eq. (3.51) and the nonrelativistic optical potential that relate to dynamics of the target states. Then Eqs. (3.44) and (3.45) provide an estimate of $\hat{S}$ and $\hat{V}$ from knowledge of the Schrodinger central and spin-orbit parts of the first-order KMT nonrelativistic optical potential. Solutions of the Dirac equation with interaction $U'=\hat{S}+\gamma^o\hat{V}$ then provide a useful illustration of the influence of the effects arising solely from the relativistic propagation of the projectile. That is, the complete Dirac optical potential is taken to be given by

$$U'^{ab}(\vec{k}';\vec{k}) = \bar{u}(\vec{k}'a)\{\hat{S}+\gamma^o\hat{V}\}u(\vec{k}\ b) \quad , \tag{3.53}$$

where a and b can each be + or −. Any difference between the results so calculated and the nonrelativistic results originate from the negative energy plane-wave intermediate states of the projectile. Such Z-graph effects do account for the major improvements in the spin-dependent observables from proton nucleus elastic scattering when the Dirac equation is employed. An illustration is provided by Fig. (7) in which the relativistic calculation is carried out exactly as described above. The momentum-space integral equations (3.31) are solved to obtain these results[31]. The nonrelativistic calculation, shown for comparison purposes, is obtained by setting U'^{+-} to zero. No parameters are adjusted in these calculations. The NN t-matrix is taken from NN data through the Love-Franey representation[21] and the point proton and neutron densities are taken to be equal and deduced from electron scattering analysis.

To test the significance of the Z-graph mechanism in relativistic scattering, the Schrodinger equivalent interaction W^{++}, given in Eq. (3.38), was calculated to first-order in the relativistic Z-graph addition to U'^{++}. That is the two terms given in the second expression of Eq. (3.38) are added and the Lippmann-Schwinger Eq. (3.37) is solved. For the case of Fig. (7) the results are indistinguishable from the Dirac equation results which are the solid curves. The inescapable conclusion is that for elastic scattering at intermediate energies the extra dynamical feature provided by a Dirac description of the projectile motion can be accurately summarized by the single Z-graph addition to the interaction for a Schrodinger-based description of the projectile motion.

This observation is not particularly dependent upon the nature of the simple procedure described above for obtaining the effective Lorentz scalar and vector components of the interaction. For example, use of a complete Dirac description of single particle states of the target would produce somewhat different scalar and vector components of the nucleon-nucleus interaction. There are indications that relativistic features of the target states correlate strongly with the behavior of the spin-dependent observables in the region of the first diffractive minimum of Fig. (7). Empirically it is found that a 1% increase in the strength of the scalar component of the optical potential would convert the solid curve of Fig. (7) into an almost perfect reproduction of the forward- angle spin-dependent data[31]. Even for this case, the employment of W^{++} to first-order in the Z-graph produces results which are indistinguishable from the Dirac equation solutions as is evident in Fig. (8). At lower energies, the first-order contribution of the Z-graph to W^{++} is not sufficient to give such a high quality reproduction of the Dirac solutions. This is illustrated in Fig. (9), and for comparison purposes the Dirac and nonrelativistic results are compared in Fig. (10). Although the relativistic calculation for the analyzing power is in better qualitative agreement with the data, the use of a free NN t-matrix to describe a first- order optical potential at this low energy is subject to question. Medium effects as implemented through the use of a Brueckner g-matrix are found to be strong and lead to a much improved correspondence with data. This is discussed in the lectures by C.J. Horowitz at this Workshop.

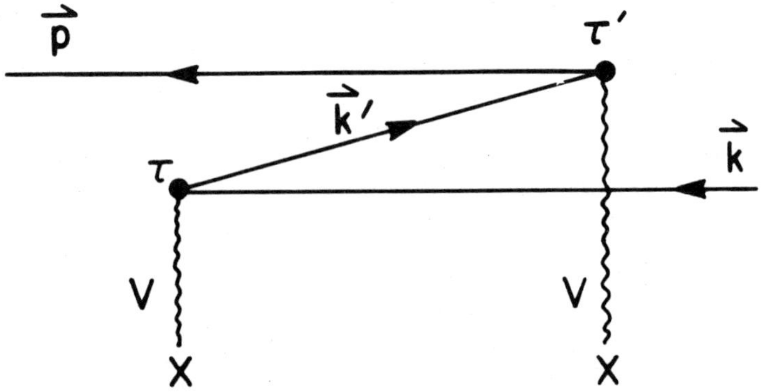

FIGURE 5. The Feynman diagram (Z-graph) for the relativistic contribution to the second-order S-matrix for scattering from a static potential.

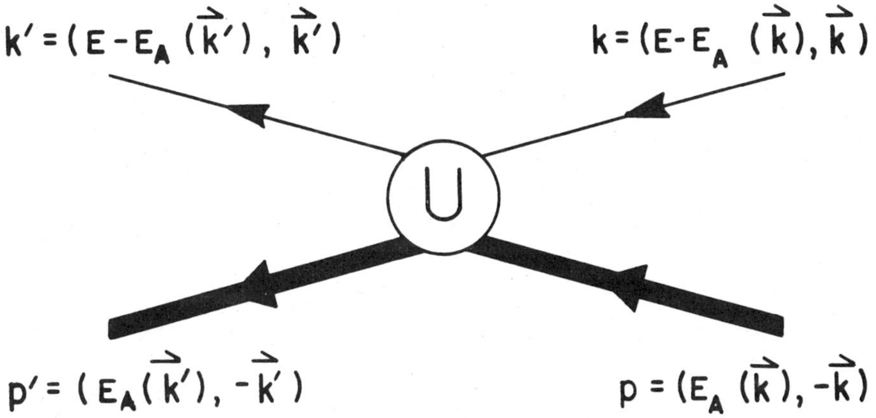

FIGURE 6. Kinematics in the zero-momentum frame for a projectile-nucleus interaction $U(E,\vec{k}',\vec{k})$ with the nucleus constrained to be on-mass-shell.

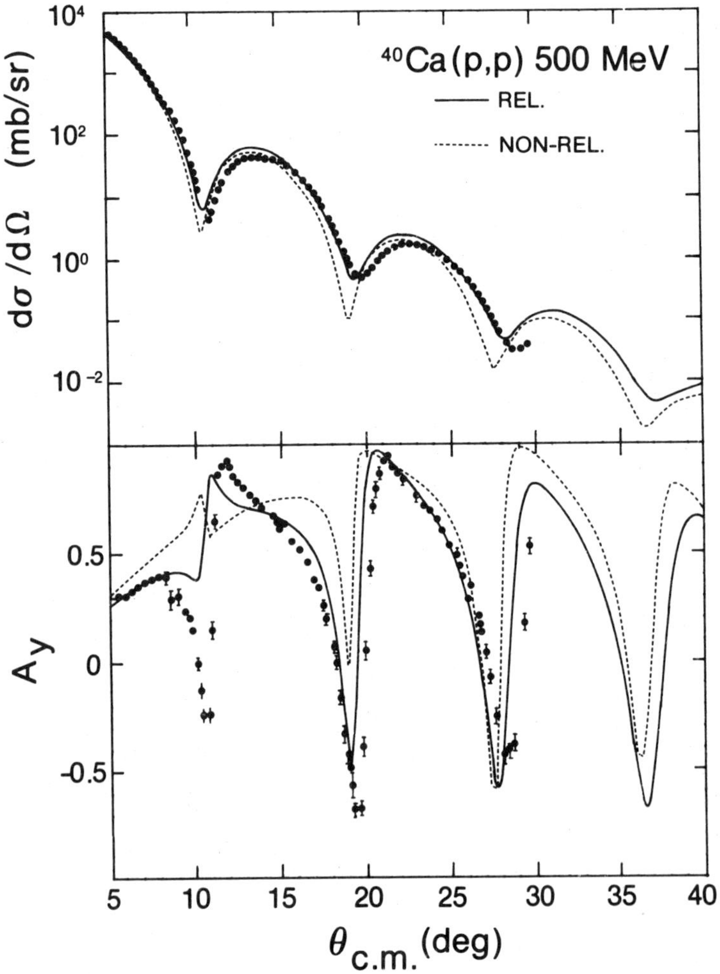

FIGURE 7a. Cross section and analyzing power for proton scattering from ^{40}Ca at 500 MeV. Compared are relativistic and non-relativistic calculations employing first-order optical potentials as described in the text. [Taken from Ref.(31)].

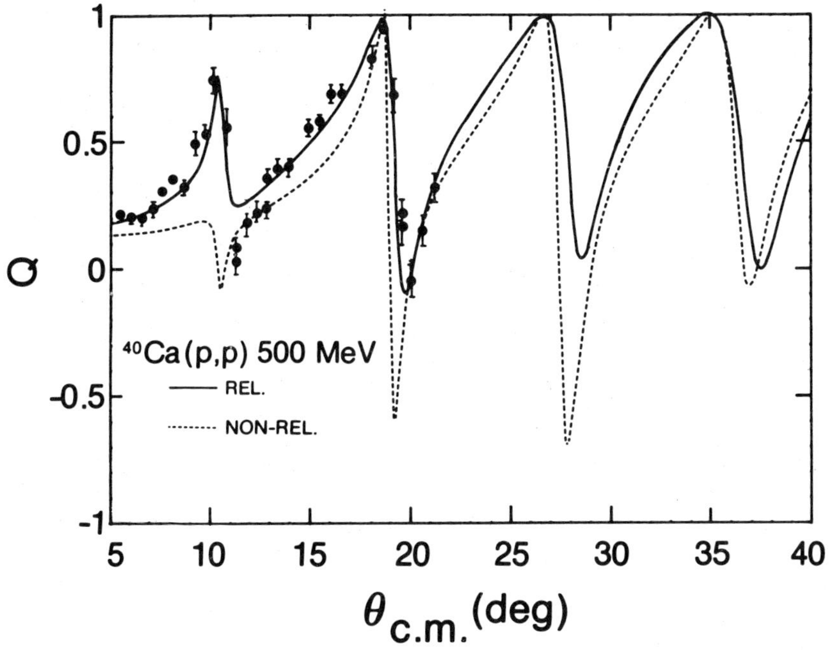

FIGURE 7b. Spin rotation function for proton scattering from ^{40}Ca at 500 MeV. Compared are relativistic and non-relativistic calculations employing first-order optical potentials as described in the text. [Taken from Ref.(31)].

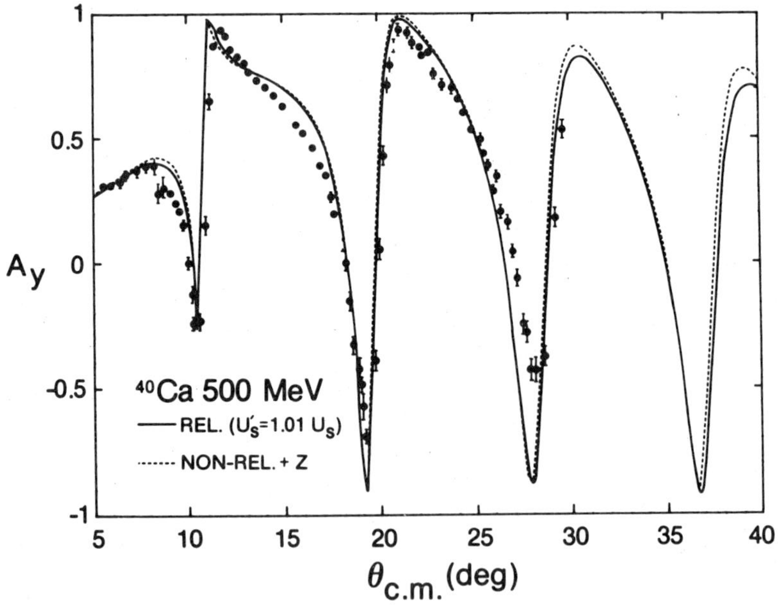

FIGURE 8. Demonstration that retention of the lowest-order Z-graph mechanism in the equivalent Lippmann-Schwinger interaction W^{++} (dashed curve) represents the dominant relativistic effect contained in the solution of the Dirac equation at 500 MeV (solid curve). See text of Lecture 3.

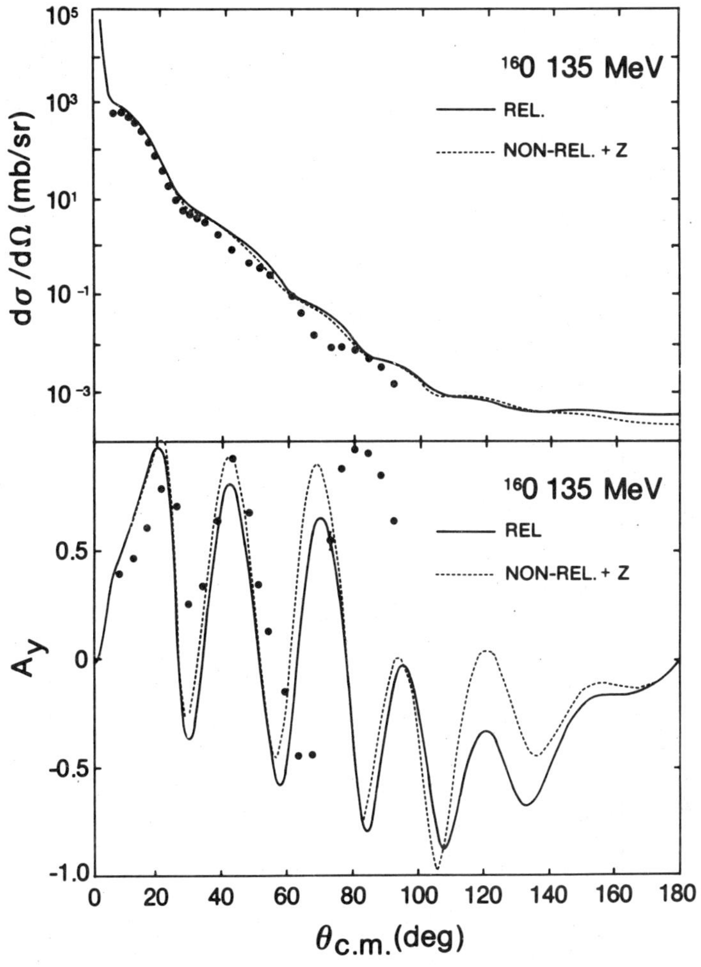

FIGURE 9. Cross section and analyzing power for proton scattering from ^{16}O at 135 MeV. The solid curve arises from use of a first-order optical potential in the Dirac equation. The dashed curve retains only the single Z-graph relativistic contribution to the equivalent Lippmann-Schwinger interaction W^{++}.

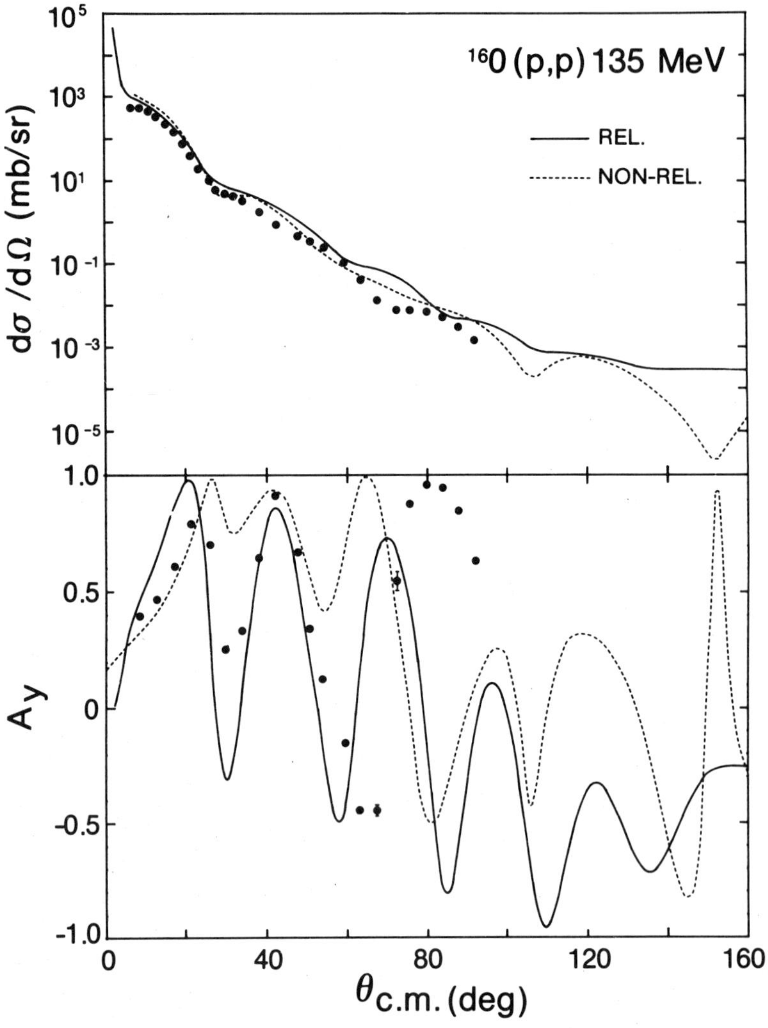

FIGURE 10. Comparison of the relativistic and non-relativistic first-order optical potential results for proton scattering from ^{40}Ca at 135 MeV.

LECTURE 4 - OVERVIEW OF ELASTIC AND INELASTIC SCATTERING

PAIR ENHANCEMENT IN ELASTIC NUCLEAR SCATTERING

We have discussed the Z-graph mechanism as the feature distinguishing relativistic and nonrelativistic descriptions of the projectile motion in elastic proton-nucleus scattering. One may well ask why such a mechansim has not been found to be an indispensible part of descriptions of NN scattering that proceed through relativistic one boson exchange interaction models. It is usual in such approaches to retain only the positive energy Dirac spinors for descriptions of the intermediate states. We recall that pseudoscalar (γ^5) coupling for the πN vertex produces strong pair effects that give an incorrect behavior for the πN scattering length[36]. Likewise, descriptions of NN scattering data are made difficult without adjustments of meson-nucleon coupling constants beyond acceptable limits[36]. The alternative pseudovector ($\gamma^5 \not{q}/2m$) coupling, which is identical only for onshell matrix elements and produces much weaker pair effects, leads to a more acceptable behavior on both counts. Calculations of NN scattering via the Bethe-Salpeter equation with and without the inclusion of the negative energy nucleon spinors have shown that the NN data does not clearly call for their inclusion. They can be omitted without deterioration in the description of data if small adjustments are made to meson-nucleon coupling strengths[37].

To see why pair-effects are apparently playing a more important role in elastic N-Nucleus scattering than in NN scattering, consider the multiple scattering contribution to the physical N-Nucleus transition operator T=T′(A/A-1) that results from Eqs. (3.47) and (3.48). To second order in U′ the result is

$$T = \frac{A}{A-1}U' + \frac{A}{A-1}U'\frac{1}{\not{K}-m+i\eta}\,U' \quad , \tag{4.1}$$

$$= \langle\overline{\phi_o}|\,T^{(2)}\,|\phi_o\rangle \quad ,$$

where after use of Eq. (3.47), the operator $T^{(2)}$ can be written in the form

$$T^{(2)} = \sum_{i=1}^{A} t^D_{oi} + \sum_{i\neq j}^{A} t^D_{oi}\frac{|\phi_o\rangle\langle\overline{\phi_o}|}{\not{K}-m+i\eta}\,t^D_{oj} \quad . \tag{4.2}$$

For the second term, the contribution of the negative energy projectile states adds to the familar nonrelativistic quadratic contribution from a first-order optical potential the extra term

$$T^{(2)}_Z = \sum_{i\neq j}^{A} t^D_{oi}\frac{|\phi_o\rangle P_-\langle\overline{\phi_o}|}{\not{K}-m+i\eta}\,t^D_{oj} \quad . \tag{4.3}$$

For target ground state to ground state matrix elements this term, illustrated in Fig. (11), is coherent with respect to participation by the target nucleons since the intermediate state of the target is also the ground state. The statistical weight of this term is A(A-1) and the contribution will be largest at low momentum transfer. This is to be contrasted to the behavior of a second order multiple scattering term of the optical potential where orthogonality effects due to intermediate state excitation of the target cause a suppression at low momentum transfer as discussed in Lecture 2. In the limit of a zero-range NN t-matrix, the nuclear shape dependence of the Z-graph contribution will be approximately the Fourier transform of $\rho^2(r)$. Because of absorption due to other open channels in N-Nucleus scattering, geometrical properties of the nuclear form factor are much more important there than are geometric properties of the form factors in NN scattering. It is not surprizing that pair effects show up strongly in spin-dependent observables in the region of diffractive minima where interferences between very small amplitudes can magnify small effects.

To have reasonable confidence that couplings to pair effects in the Dirac equation description of N-Nucleus scattering are represented in an adequate way, one must avoid the adoption of an operator having a Lorentz structure which is designed only to reproduce known information about positive energy matrix elements. There is no guarantee that negative energy matrix elements of such an operator are realistic. The discussion we have presented makes such an assumption at the level of the projectile- nucleus interaction operator. The original work[38] on the Dirac impulse approximation made such an assumption at the level of the NN t-matrix operator t^D. There the form adopted for t^D is[38]

$$P_{++}t^DP_{++} = P_{++}\{F_S 1 + F_V\gamma_1\cdot\gamma_2 + F_P\gamma_1^5\gamma_2^5 + F_A\gamma_1^5\gamma_1\cdot\gamma_2^5\gamma_2$$

$$+ F_T\sigma_{2\mu\nu}\}P_{++} \quad , \tag{4.4}$$

where P_{++} is the projector onto that part of the Hilbert space in which both particles are in positive energy plane wave Dirac states. The five invariant amplitude functions F are all that is necessary to reproduce the five Wolfenstein amplitudes of the onshell NN t-matrix of the Pauli representation. Since this was the only constraint, the functions F could be taken to be local. When ground state nuclear matrix elements are taken in the optimum factorization approximation, the symmetries of parity, rotational invariance and time reversal eliminate all terms except 1, γ_1^o and σ_1^{oi} where i=1,2,3. There are fifteen other sectors of the two-particle Hilbert space besides P_{++}, and although not all of them are independent, the number of independent invariant amplitudes is not less than 44 for on mass shell particles. This is because on-shell NN amplitudes are a necessary but not a sufficient constraint on the relativistic operator t^D. There are many invariants that can be added to the above five that vanish for matrix elements in the ++ sector but not in the other sectors. The occurrence of such invariants is intimately related to nonlocality. In principle all this information is obtainable from a one boson exchange model of the NN interaction. The extent to which a simple form such as that in Eq. (4.4) will suffice for all matrix elements of t^D is presently being investigated by Tjon and Wallace[39], and we refer to the presentation by S.J. Wallace at this Workshop for further elaboration.

INELASTIC SCATTERING

The Distorted Wave Impulse Approximation (DWIA) is the main theoretical tool for microscopic treatments of the inelastic scattering of an intermediate energy projectile from nuclei. The employed matrix element is taken to be

$$T_{fi} \simeq \langle\chi_1^{(-)}|\langle\phi_1|\Sigma_i t_i|\phi_o\rangle|\chi_o^{(+)}\rangle \quad , \tag{4.5}$$

where ϕ_o and ϕ_1 are the initial and final nuclear states, and the states $\chi_o^{(+)}$ and $\chi_1^{(-)}$ are one-body elastic scattering states at the appropriate energies. The operator t_i describes projectile-nucleon scattering and is usually taken to be the free t-matrix or the Brueckner g-matrix. One can envisage relativistic extensions of this matrix element in the same manner as for the elastic optical potential matrix element. The presentation at this Workshop by J.R. Shepard describes recent work in that line. We will comment briefly on relativistic aspects at the end of this Lecture.

In order to appreciate the many successes in applications of the DWIA and to be equipped to address the short comings from a wider perspective, I feel it is worthwhile to discuss in what sense Eq. (4.5) is an approximation, and indeed what it is an approximation to. The origins of Eq. (4.5) as the first term in an explicitly defined multiple scattering expansion are not made evident in standard texts. The development given here is to be found in Ref. (40). As for elastic scattering in Lecture 1, we begin from the Hamiltonian $H=H_o+\Sigma_i v_i$ where $H_o=h_o+H_A$. The transition operator T for elastic and inelastic scattering is given by

$$T = U + U\, P_o G_o T \tag{4.6}$$

where the (elastic optical potential) operator U is

$$U = V + VQ_oG_oU \quad , \tag{4.7}$$

with the definition $V=\Sigma_i v_i$. The projectors are defined as before, that is $P_o=|\phi_o\rangle\langle\phi_o|$ and $Q_o=1-P_o$ where ϕ_o is the target ground state. The factorization of the inelastic matrix element into a form containing the elastic scattering state as the initial distorted wave is easily accomplished. We have

$$T_{fi} = \langle\Phi_1|T|\Phi_o\rangle \quad , \tag{4.8}$$

$$= \langle\Phi_1|U(1+P_oG_oT)|\Phi_o\rangle \quad , \tag{4.9}$$

$$= \langle\Phi_1|U|\phi_o\rangle|\chi_o^{(+)}\rangle \quad , \tag{4.10}$$

where $\chi_o^{(+)}$ is the one-body elastic scattering state, which in Lecture 1 was denoted by $\psi_k^{(+)}$. Here the final state Φ_1 is a product of the final nuclear state ϕ_1 and the final plane wave for the projectile and Φ_o is similarly defined. Comparison of Eq. (4.10) with Eq. (4.5) shows that the DWIA can be viewed in terms of approximations to $\langle\Phi_i|U$. In order to discuss perturbative treatments of this quantity that result in one factor being a distorted wave, it is useful to characterize the various quantities in Eq. (4.10) in terms of sums of scattering processes or diagrams. The definition

$$|\phi_o\rangle|\chi_o^{(+)}\rangle = (1+P_oG_oT)|\Phi_o\rangle \quad , \tag{4.11}$$

together with the perturbation series for T produced by iterations of Eq. (4.6), allows this initial state to be characterized as the sum of all possible scattring processes or diagrams that begin and end with the nucleus in the ground state. Since we are dealing with a scattering state, we must include the non-interacting piece, that is the operator 1 in Eq. (4.11).

Considering the definition of U given in Eq. (4.7), we may characterize the remaining factor $\langle\Phi_1|U|\phi_o\rangle$ of Eq. (4.10) as the sum of all possible diagrams that start with the nucleus in the ground state and end in the excited state without the ground state ever appearing in intermediate steps. Now, in each of these contributions, there will be factors that connect the excited state ϕ_i with itself. It is helpful to collect together the sum of all possible matrix elements that are diagonal in the excited nuclear state. In doing so, we will have factorized $\langle\Phi_1|U$ so that there appears a new final state which has a characterization similar to that of Eq. (4.11) for the initial distorted wave. This new final state, being defined in terms of diagonal nuclear matrix elements will be a one-body distorted wave in which the nucleus is "frozen" in the excited state ϕ_1. Since a large infinite class of the elementary interactions that make up $\langle\Phi_1|U$ (in the sense of a perturbation series) will have been removed, by this procedure, to form the new state, a new operator will appear to describe the inelastic excitation. This excitation operator for the full distorted wave basis should be much more amenable to a first-order perturbation treatment since all strong contributions which are diagonal in nuclear states have been removed.

To implement the above mentioned factorization, we introduce the excited nuclear state projector

$$P_1 = |\phi_1\rangle\langle\phi_1| \quad , \tag{4.12}$$

and define $\hat{U}$ as that part of U containing no intermediate P_1 states, that is,

$$\hat{U} = U - UP_1G_o\hat{U} \; . \tag{4.13}$$

Then we see that

$$\langle\Phi_1|U = \langle\Phi_1|(UP_1G_o+1)\hat{U} \; , \tag{4.14}$$

$$= \langle\chi_1^{(-)}|\langle\phi_1|\hat{U} \; , \tag{4.15}$$

and the inelastic matrix element now has the exact distorted wave form

$$T_{fi} = \langle\chi_1^{(-)}|\langle\phi_1|\hat{U}|\phi_o\rangle|\chi_o^{(+)}\rangle \; . \tag{4.16}$$

The final distorted wave is given by

$$\langle\chi_1^{(-)}|\langle\phi_1| = \langle\Phi_1|(UP_1G_o+1) \; , \tag{4.17}$$

which, after use of Eq. (4.15), can be put in the form of the Lippmann-Schwinger equation

$$\langle\chi_1^{(-)}|\langle\phi_1| = \langle\Phi_1| + \langle\chi_1^{(-)}|\langle\phi_1|\hat{U}P_1G_o \; . \tag{4.18}$$

The one-body potential that generates $\langle\chi_1^{(-)}|$ is $\langle\phi_1|\hat{U}|\phi_1\rangle=\hat{U}_{11}$ say. Then Eq. (4.18) in a more conventional form is

$$|\chi_1^{(-)}\rangle = |\vec{k}_f\rangle+G_o(E^-)\hat{U}_{11}{}^{\dagger}|\chi_1^{(-)}\rangle \; . \tag{4.19}$$

The one-body potential that generates $|\chi_o^{(+)}\rangle$ is $\langle\phi_o|U|\phi_o\rangle = U_{oo}$ the elastic scattering optical potential. The one-body effective interaction that generates the transition is $\langle\phi_1|\hat{U}|\phi_o\rangle = \hat{U}_{10}$, that is, the off-diagonal matrix element of the same operator whose diagonal matrix elements generate the final distorted wave. Because of the close connection between U and $\hat{U}$, given by Eq. (4.13), a multiple scattering expansion of U can be used to define a multiple scattering expansion of $\hat{U}$. The physical pictures associated with microscopic treatments of elastic and inelastic scattering may thereby be coordinated. It should be emphasized that at this stage, no approximations have been made.

In terms of the residual interaction V, the combination of Eqs. (4.13) and (4.7) yields

$$\hat{U} = V + VQ_oQ_1G_o\hat{U} \quad . \tag{4.20}$$

where $Q_1 = 1-P_1$. With the decomposition $V = \Sigma_i v_i$, a multiple scattering expansion for $\hat{U}$ can be developed in the same way that such expansions for U are developed in Lectures 1 and 2. A Watson multiple scattering expansion for $\hat{U}$ is easily seen to be

$$\hat{U} = \sum_i \hat{\tau}_i + \sum_{i\neq j} \hat{\tau}_i Q_1 Q_o G_o \hat{\tau}_j + \cdots, \tag{4.21}$$

where the ladder operator $\hat{\tau}_i$ that correlates the projectile to all orders with one target nucleon is

$$\hat{\tau}_i = v_i + v_i Q_1 Q_o G_o \hat{\tau}_i \quad , \tag{4.22}$$

or, in terms of the corresponding elastic channel Watson operator τ_i discussed earlier,

$$\hat{\tau}_i = \tau_i - \tau_i P_1 G_o \hat{\tau}_i \quad . \tag{4.23}$$

In terms of the t-matrix t_i which satisfies Eq. (4.22) with the projectors Q_1 Q_o removed, we have

$$\hat{\tau}_i = t_i - t_i P_o G_o \hat{\tau}_i - t_i P_1 G_o \hat{\tau}_i \quad . \tag{4.24}$$

With the two-body approximation in which t_i is taken to be the free NN t-matrix, the required quantity can easily be calculated. The first-order or single-scattering distorted wave approximation is

$$T_{fi}^{(1)} = \langle\chi_1^{(-)}|\langle\phi_1|\Sigma_i\hat{\tau}_i|\phi_o\rangle|\chi_o^{(+)}\rangle \quad . \tag{4.25}$$

The DWIA requires two further approximations: 1) $\hat{\tau}_i \simeq t_i \simeq t(\text{free})$, and 2) the final distorted wave $\chi_1^{(-)}$ is approximated by an elastic scattering wave usually from the ground state of the nucleus.

There are several points to be noted at this stage. The DWIA does not include all first-order scattering contributions to the excitation in the multiple scattering or ladder summation sense. There are significant orthogonality blocking mechanisms in the $\hat{\tau}_i$ of Eq. (4.24) that are not included in t_i. The use of a Brueckner g-matrix for $\hat{\tau}_i$ will account for the $-P_o$ projector but not for the $-P_1$ projector of Eq. (4.24). This latter term is a direct consequence of avoiding double counting of mechanisms (propagation in P_1 states) that are already included through use of a final channel distorted wave. The unsuitability of the bound state perturbation origins of the g-matrix for applications to inelastic scattering are likely to be more serious here than in the case of elastic scattering. We note that most DWIA calculations of proton inelastic scattering overestimate the cross section by a factor of almost 2. One effect of the $-P_1$ term in Eq. (4.24) is to decrease the strength of the effective interaction compared to that used in the DWIA. The other point to be noted is that the final distorted wave that appears in the single-scattering distorted wave matrix element of Eq. (4.25) is not the wave function that describes "elastic scattering" from the excited nucleus. The presence of the projector Q_o in Eq. (4.20) prevents the final channel distorting potential $\langle\phi_1|\hat{U}|\phi_1\rangle$ from being the optical potential for projectile scattering from the excited nucleus. If instead $\langle\phi_1|U|\phi_1\rangle$ were to be used as a final channel distorting potential, then Eq. (4.17) indicates that the required final distorted wave is obtained exactly by working to first-order in this interaction[40]. Thus the absorptive effect of the final channel distortion would be less than is currently employed in the standard implementation of the DWIA. This matter should be investigated closely with regard to the general overestimate of inelastic cross sections that is common with present uses of DWIA. When the final nuclear state has $J \neq o$, the final channel distorting potential has a more general spin dependence than the optical potential for spin zero ground states and the consequences of this effect for spin-dependent observables have never been systematically investigated.

We have discussed a multiple scattering expansion of the off-diagonal matrix element $\langle\phi_1|\hat{U}|\phi_o\rangle$, which describes the transition in the distorted wave basis. Similar expansions can also be applied to the distorting potentials $\langle\phi_o|U|\phi_o\rangle$ and $\langle\phi_1|\hat{U}|\phi_1\rangle$ so that the complete distorted wave matrix element can be analyzed in multiple scattering form. With first-order truncations in each of the three interactions a so-called "consistent" DWIA is produced. The details of this development are to be found in Ref. (40). The result is that Eq. (4.25) becomes

$$T_{fi}^{(1)} \simeq \langle\hat{\chi}_1^{(-)}|\langle\phi_1|\Sigma_i t_i|\phi_o\rangle|\hat{\chi}_o^{(+)}\rangle \ , \qquad (4.26)$$

where the distorted waves $\hat{\chi}_1$ and $\hat{\chi}_o$ are now generated by the first-order KMT pseudopotentials $(A-1)\langle\phi_1|\tau|\phi_1\rangle$ and $(A-1)\langle\phi_o|t|\phi_o\rangle$ respectively. The occurrence of the two-body operator τ instead of t in the first-order distorting potential for $\hat{\chi}_1$ is a consequence of the fact that all intermediate P_o states are included through use of the elastic scattering state as the initial channel distorted wave.

Studies of the DWIA in a Dirac representation, namely

$$T_{fi} \simeq \langle\overline{\chi_1^{(-)}}|\langle\overline{\phi_1}|\Sigma_i t_i^D|\phi_o\rangle|\chi_o^{(+)}\rangle \ , \qquad (4.27)$$

where the states are four-component objects are presently underway[41]. Because of the non-zero angular momentum transfer to the nucleus in most inelastic excitations, essentially the complete tensor character of the two-body t-matrix in the target space is required. A utilization of the dominance of Lorentz scalar and vector components is not adequate to describe the spin structure of inelastic matrix elements. A basic question that will need to be faced is whether a Lorentz invariant structure for t^D based on the five Fermi invariants of Eq. (4.4) will suffice, or whether the larger number of more general Lorentz invariants having explicit nonlocalities from contractions of Dirac gamma matrices with the two-body momenta are important. Sizable values for P-A, the difference between polarization and analyzing power, have recently been measured for inelastic spin-coupled transitions[42]. The realization that nonlocalities of the two-body t-matrix are required for analysis of this sensitive quantity[43], points to the need for careful consideration of the many small invariant amplitudes that are possible in a 16x16 matrix representation of a two-body operator in Dirac basis states. We note that the topic of nonlocalities is somewhat ambiguous, since even if the invariant amplitudes F_i in Eq. (4.4) are local (functions of q only), the Dirac plane wave matrix elements of t^D are nonlocal. It is not clear at present whether the important nonlocalities reside in the Dirac basis states or the Lorentz invariant amplitudes. A calculation of t^D from a relativistic interaction can, in principle, provide an answer to this question that is model-dependent but less so than fitting a postulated form to onshell two-body scattering data. One important constraint to be observed in these considerations is that if the states in Eq. (4.27) are expanded in terms of positive and negative energy free particle Dirac spinors, then the positive energy state matrix elements of t^D that appear are the same t-matrix elements that occur in the Schrodinger-based DWIA before any assumption of locality (for ease of computation) is imposed. The extra degree of freedom that enters in a Dirac-based DWIA is then seen to be the amplitudes for coupling to the negative energy components of the states.

ACKNOWLEDGEMENT

These Lectures have drawn heavily upon discussions and collaborations over the last few years with M. V. Hynes, A. Picklesimer and R. M. Thaler. The support of Los Alamos National Laboratory where these Lectures were prepared is gratefully acknowledged. Partial support was provided by the National Science Foundation under Grants No. PHY83-05745 and No. PHY85-05736.

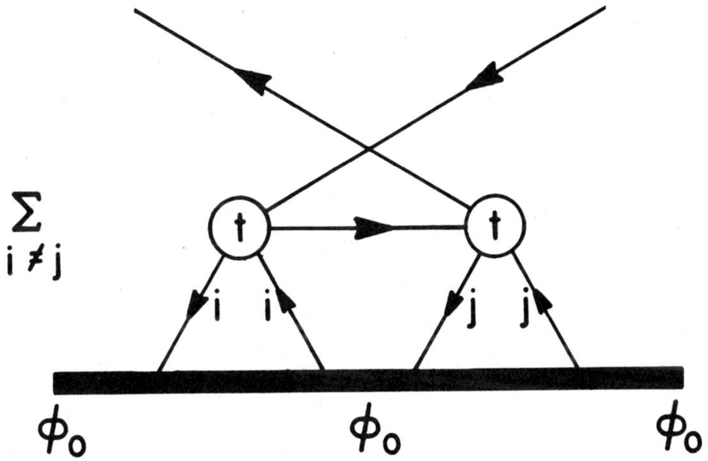

FIGURE 11. Three-body nature of the lowest-order relativistic mechanism (Z-graph) that arises from use of a first-order relativistic optical potential. See text of Lecture 4.

REFERENCES

1. S. S. Schweber, An Introduction to Relativistic Quantum Field Theory, (Row, Peterson and Co., New York, 1961).

2. J. R. Taylor, Scattering Theory, (John Wiley, New York, 1972).

3. B. A. Lippmann and J. Schwinger, Phys. Rev. 79, 469 (1950).

4. L. Rodberg and R. M. Thaler, Introduction to the Quantum Theory of Scattering, (Academic Press, N.Y., 1967).

5. E. F. Redish, G. J. Stephenson Jr., and H. S. Picker, Phys. Rev. C5, 707 (1972).

6. M. Goldberger and K. M. Watson, Collison Theory, (J. Wiley, New York, 1964).

7. R. P. Feynman, Quantum Electrodynamics, (W. A. Benjamin, 1962); Phys. Rev. 76, 749 (1949); 76,769 (1949).

8. A. Picklesimer, P. C. Tandy and R. M. Thaler, Ann. Phys. (N.Y.) 145, 207 (1983).

9. L. D. Faddeev, Sov. Phys. JETP 12, 1014 (1961); Mathematical Aspects of the Three-Body Problem in Quantum Scattering Theory (English Transl. by Israel Program for Scientific Translations, Jerusalem, Sivan Press 1965).

10. K. A. Brueckner, Phys. Rev. 100, 36 (1955); H. A. Bethe and J. Goldstone, Proc. Roy. Soc. Ser. A238, 551 (1956); K. A. Brueckner and J. L. Gammel, Phys. Rev. 109, 1023 (1958).

11. K. M. Watson, Phys. Rev. 89, 575 (1953); N. C. Francis and K. M. Watson, Phys. Rev. 92, 291 (1953).

12. A. K. Kerman, H. McManus, and R. M. Thaler, Ann. Phys. (N.Y.) 8, 551 (1959).

13. A. L. Fetter and J. D. Walecka, Quantum Theory of Many-Particle Systems, (McGraw-Hill, New York, 1971).

14. P. C. Tandy, E. F. Redish, and D. Bollé, Phys. Rev. Lett. 35, 921 (1975); Phys. Rev. C16, 1924 (1977).

15. A. Picklesimer, P. C. Tandy, R. M. Thaler and D. H. Wolfe, Phys. Rev. C29, 1582 (1984); Phys. Rev. C30, 1861 (1984).

16. L. Wolfenstein and J. Ashkin, Phys. Rev. 85, 947 (1952).

17. R. V. Reid, Ann. Phys. (N.Y.) 50, 411 (1968).

18. M. Lacombe et al., Phys. Rev. D12, 1495 (1975); Phys. Rev. C21, 861 (1980).

19. E. Lambert and H. Feshbach, Ann. Phys. (N.Y.) 76, 80 (1973); E. Boridy and H. Feshbach, Ann. Phys. (N.Y.) 109, 468 (1977); L. Ray, Phys. Rev. C20, 1857 (1979); S. J. Wallace, in Advances in Nuclear Physics, vol 12, J. W. Negele and E. Vogt, eds. (Plenum, N.Y. 1984).

20. L. Ray, Phys. Rev. C19, 1855 (1979) L. Ray et al; Phys. Rev. C23, 828 (1981).

21. W. G. Love and M. A. Franey, Phys. Rev. C24, 1073 (1981).

22. K. Stricker-Bauer and E. F. Redish, in preparation.

23. A. Picklesimer, E. F. Redish, K. Stricker-Bauer and P. C. Tandy, in preparation.

24. E. R. Siciliano and R. M. Thaler, Phys. Rev. C16, 1322 (1977).

25. D. J. Ernst, J. T. Londergan, G. A. Miller and R. M. Thaler, Phys. Rev. C16, 537 (1977).

26. A. Picklesimer and R. M. Thaler, Phys. Rev. C23, 42 (1981).

27. H. V. von Geramb, The Interaction Between Medium Energy Nucleons in Nuclei - 1982, AIP Conference Proceedings No. 97, ed. H. O. Meyer (AIP, N.Y. 1983), P. 44.

28. J. S. Bell and E. J. Squires, Phys. Rev. Lett 3, 96 (1959).

29. N. Kemmer, Proc. Roy. Soc. (London) A173, 91 (1939).

30. J. D. Bjorken and S. D. Drell, Relativistic Quantum Mechanics, (McGraw-Hill, New York, 1966).

31. M. V. Hynes, A. Picklesimer, P. C. Tandy and R. M. Thaler, Phys. Rev. Lett. 52, 978 (1984); Phys. Rev. C31, 1438 (1985).

32. H. Feshbach and F. Villars, Rev. Mod. Phys. 30, 24 (1958).

33. J. D. Walecka, Ann. Phys. (N.Y.) 83, 491 (1974); C. J. Horowitz and B. D. Serot, Nucl. Phys. A368, 503 (1981).

34. A. Picklesimer and P. C. Tandy, in preparation.

35. L. S. Celenza and C. M. Shakin, Phys. Rev. C28, 1256 (1983).

36. S. Weinberg, Phys. Rev. Lett. 16, 879 (1966); 18, 188 (1967); S. L. Adler, and Y. Dothan, Phys. Rev. 151, 1267 (1966).

37. M. Zuilhof and J. A. Tjon, Phys. Rev. C24, 736 (1981).

38. J. R. Shepard, J. A. McNeil, and S. J. Wallace, Phys. Rev. Lett. 50, 1443 (1983); B. C. Clark, S. Hama, R. L. Mercer, L. Ray and B. D. Serot, Phys. Rev. Lett. 50, 1644 (1983).

39. J. A. Tjon and S. J. Wallace, Phys. Rev. Lett. 54, 1357 (1985); and submitted to Phys. Rev. c.

40. A. Picklesimer, P. C. Tandy and R. M. Thaler, Phys. Rev. C25, 1215 (1982); Phys. Rev. C25, 1233 (1982).

41. J. R. Shepard, E. Rost, and J. Piekarewicz, Phys. Rev. C30, 1604 (1984).

42. T. A. Carey et al, Phys. Rev. Lett. 49, 266 (1982).

43. W. G. Love and J. R. Comfort, Phys. Rev. C29, 2135 (1984).

THE MESON THEORY OF NUCLEAR FORCES AND NUCLEAR MATTER

R. Machleidt
TRIUMF, Vancouver, B.C., Canada, and
Institute für Theoretische Kernphysik der Universität Bonn, West Germany

ABSTRACT

The meson theory of nuclear forces is reviewed both historically and pedagogically. A comprehensive and consistent meson-exchange model for the nucleon-nucleon (NN) interaction is developed. The various meson-exchange contributions to the nuclear force are discussed and their power in quantitatively explaining NN scattering data demonstrated. One-boson-exchange (OBE) approximations to this full meson theory are derived for practical as well as historical reasons. The nature and quality of this useful and traditional approximation is examined. The full meson-exchange model as well as the simplified OBE potential are then applied to nuclear matter with particular stress laid on Brueckner theory. Past (failed) attempts to explain the empirical nuclear matter properties are surveyed. The relativistic Dirac-Brueckner approach to nuclear matter is motivated and explained. The predictive power of this new approach is demonstrated by the results for the nuclear matter saturation. The exceptional quantitative nature of the relativistic meson theory for both the free NN-interaction and nuclear matter is considered a strong motivation for further applications to other fields of nuclear physics, as e.g. nucleon-nucleus scattering and the structure of finite nuclei.

*Supported in part by Deutsche Forschungsgemeinschaft

CONTENTS

PREFACE I

The general purpose of these lectures is to serve as an introduction. The specific audience addressed are graduate students in theoretical as well as experimental nuclear physics — in accordance with the goal of this workshop. However, every teacher and researcher in the field of nuclear and intermediate energy physics not specialized to the particular topics of these lectures may find useful information in these notes.

The special purpose of these lectures is manifold. In line with the introductory character I will first undertake a brief excursion into history and then give a pedagogical introduction which is supposed to provide the reader the qualitative and physical understanding of the meson theory of the nuclear force. These chapters can be understood with little familiarity in the field. In addition, my intention is to lead the reader up to the latest status in the two topics addressed in this article, namely the nucleon-nucleon (NN) interaction and nuclear matter from a consistent meson theoretic point of view. In principle, a detailed knowledge of the subtleties of field theoretic perturbation theory is required for our advanced theory of the nuclear force. However, by banning most of the formalism into appendices and through the suggestive character of the Feynmann diagram language the essential points of this serious and detailed description of the NN interaction will also be understandable to the non-expert. References are provided for those who want to go beyond the discussion presented here. Concerning nuclear matter, we will also start with a short introduction and a brief review of the theoretical attempts of the past for which we will in addition quote some appropriate literature. More thoroughly we will describe the modern relativistic approach to nuclear matter and its excellent results.

Finally, these lectures want to point into the future and at the same time also serve practical purposes. Therefore, a representation of the nuclear force in the simple and (for numerical calculations) handy relativistic one-boson-exchange (OBE) approximation in momentum space with all necessary formulae and parameters will be given. This furnishes the future researcher with the necessary equipment needed to start his/her own work in relativistic nuclear structure physics. Future goals in research in this field are suggested and the appropriate background literature is given.

In conjunction with practical aspects, also a non-relativistic configuration space OBE potential will be provided. In spite of the approximations necessary to derive this simplest version of a meson theoretic NN potential, it may be useful for simple qualitative and quantitative comparisons with the recent and increasing attempts, in which the nuclear force is derived from the quark model. The advantage of the non-relativistic approximation is that the contributions of each meson-exchange can be represented in the conventional terms of a central, spin-spin, spin-orbit and tensor force. This facilitates certain physical discussions and the comparison with results from alternative approaches to a microscopic understanding of the nuclear force. This latter aspect is particularly relevant with regard to the second part of this workshop.

PREFACE II

These lectures are embedded in a workshop on relativistic dynamics and quark-nuclear physics. Therefore, it appears appropriate to indicate the relationship of these lectures to the other material being presented in the course of these two weeks.

The mean field theory (MFT), which is covered in the lectures by Chuck Horowitz, is historically one of the first (comprehensive) attempts of a relativistic theory for the nuclear many body problem. One of its great merits is that it first demonstrated the importance of relativistic saturation effects. However, as a first attempt one cannot expect it to be perfect. Therefore, there are naturally some significant drawbacks. One is that MFT has no connection to the free NN interaction. The parameters of the model are fitted directly to the empirical nuclear matter properties; consequently their values are such that free NN-scattering is far beyond a quantitative description. Further, MFT takes only the nucleon, a ω-meson and a fictitious σ-boson as relevant degrees of freedom into account. From a serious and comprehensive theory for the NN-interaction one knows, however, that the π- and ρ-meson are of fundamental importance. Moreover, a realistic description of the intermediate range attraction by the 2π-exchange (which makes the fictitious σ-boson obsolete) requires definitely virtual Δ-isobar excitation in intermediate states. This adds another important degree of freedom to nuclear physics. In these lectures we will gradually develop a quantitative theory for the NN interaction which takes all these essential degrees of freedom into account. After that, we will turn to nuclear matter in a relativistic framework which starts from the free NN interaction. The "Dirac-Brueckner" approach, as this framework came to be known, was created in the spirit of MFT (adding another historical credit to that theory), but at the same time overcomes the fundamental problems and deficiencies in which MFT was trapped. In fact, we will finally explain the empirical saturation properties of nuclear matter quantitatively in terms of the free NN interaction, i.e. in a parameter-free way.

From the relativistic analysis of nucleon-nucleus scattering (lectures by Bunny Clark and Jim Shepard) one has strong empirical evidence for large common scalar and vector fields in the nuclear medium, which are, so far, assumed just phenomenologically. For an understanding of the origin of these fields by microscopic derivation from the free nuclear force, we will present the framework.

John Dubach indicated in his lectures about "Electronuclear Reactions and Meson Exchange Currents (MEC)" that all MEC calculations up to date are rather incomplete and contain serious inconsistencies with regard to the nuclear force. The comprehensive meson-exchange model to be presented in these lectures will provide the basis for consistent and "complete" MEC calculations of the future.

Finally, there is of course a most intimate relationship between these lectures and the general topic of the second part of this workshop, namely quark-nuclear physics. In fact, the latter topic provides in a certain sense the theoretical foundation for a(n) (effective) meson theory being understood as an appropriate approximation to low energy QCD. As this aspect is of fundamental importance, we will handle it specially in the following introduction.

1. INTRODUCTION

Nowadays it is widely accepted that quantum chromodynamics (QCD)[1] is the fundamental theory of strong interactions. On that basis, the nucleon-nucleon (NN) interaction has to be considered as completely determined by the underlying quark-gluon dynamics. However, due to the mathematical problems raised by the non-perturbative character of QCD in the low energy regime, we are far away from a quantitative understanding of the NN force from this point of view.

Closely related and of even broader relevance is the problem of the confinement of hadrons. Here, the intractability of low energy QCD* is usually circumvented by the ad hoc introduction of (picking up one of the most popular euphemisms in modern physics) "QCD inspired" models, e.g. bag or potential models. There are naturally large uncertainties in the details of these models. For example, in the context of bag models, a crucial question is the size of the confinement radius R. Should R turn out to be small ($R \lesssim 0.5$ fm) as e.g. in the little bag,[3] there would be enough space for conventional hadrons like nucleons, mesons and isobars to represent the essential degrees of freedom for a wide range of nuclear physics phenomena, and meson exchange would be a valid picture. In that case, the appropriate procedure is to construct the nuclear force from meson-nucleon and meson-isobar vertices, these being understood as effective descriptions of complicated n-quark reactions. Hadron masses, coupling constants and vertex-form factors, which are the physical parameters of such a meson theory, are then left to be ultimately explained by QCD.[4,5]

Genuinely new quark-gluon-processes, which cannot even effectively be taken into account by meson-exchange, may occur only for overlapping hadrons. Therefore, their role also depends decisively on the hadron size. As discussed, for small radii ($R \lesssim 0.5$ fm) they should have a negligible influence provided the consideration is restricted to phenomena involving comparatively low energies and momentum transfer, as e.g. NN scattering up to a laboratory energy of about 300 MeV and nuclear binding energies. In this case, the introduction of meson-nucleon (-isobar) vertex form factors would be a sufficient accounting of the inner structure of the hadrons and its consequences.

The situation may be substantially different if the bag radius is large ($R \gtrsim 1$ fm) as e.g. is the case in the MIT-bag[6] or the cloudy bag model.[7] In a very naive interpretation these radii

---

*Presently the only promising treatment of low energy QCD is lattice gauge theory.[2] However, because of computational restrictions on present day computers, lattice QCD is not yet a practical tool for every day nuclear physics.

may tolerate just the pion, leave no space for the heavier scalar and vector bosons, and instead require the inclusion of genuine quark-glon exchanges as dominant contributions. It is worth mentioning that there has been very little success in understanding low energy nuclear forces in such a picture.

For instance, in the work of Maltman and Isgur[8] an attempt is made to describe the deuteron from six quarks with chromodynamics. However, in the end it turns out that the empirical deuteron properties (i.e. the binding energy, the quadrupole and magnetic moment and the asymptotic D/S state ratio) can only be explained when an artificial pion tail is attached, which has nothing to do with the original ansatz of that model. In fact, this pion tail provides almost the total contribution to all deuteron properties. Therefore, this work essentially fails to confirm that the deuteron properties can be deduced from a naive quark-gluon picture.

An appropriate case for the test of large "quarkish" contributions to the nuclear force would be NN scattering, especially in P-waves (L=1), since in addition to the general short-range repulsion the spin-orbit (LS) force comes into play here in a decisive way. It is known, however, that one-gluon-exchange creates too little LS force to account for the empirical NN scattering data.[9-11] The tensor force also turns out to be far too small.[10,11] The pion tail could be invoked for help again, yet, if the large bag radii are applied consistently they cut down the pionic tensor force too much. Moreover, all known models which attempt to derive the nuclear force from the quark model create little or no intermediate range attraction, so that they obtain either only repulsion[12] or, to repair for this defect, artificially add the attraction of scalar boson exchange,[13,14] which, however, contradicts the basic assumptions of those models.

The common source for these problems and contradictions may be that the idea of a bag is interpreted too naively in these models, namely merely by classical geometrical considerations. The fact that it costs ≈1 GeV (the string tension) to separate colored sources by 1 fm may be more relevant to considerations in nuclear physics than the bag radius.[15] In other words, in low energy nuclear physics, even with overlapping bags, the colour singlet exchange may very well continue to be the predominant process.

So far we have discussed only quark models (complemented by perturbative QCD). An alternative description of low energy phenomena is offered by the Skyrmion model. Following a proposal by t'Hooft,[16] QCD is generalized from SU(3) to an $SU(N_c)$ gauge group. In this generalization, $1/N_c$ is the coupling constant. If one assumes confinement, then, in the large N_c limit, QCD is supposedly equivalent to a local field theory of mesons. Further, it can be shown that in this theory the baryons arise as soliton solutions of the meson field equations,[17] an idea which was already suggested by Skyrme[18] about twenty years ago. In this

"Skyrme model" the exact size of the bag radius plays a subordinate role and can in fact be rather small.[19,20] This feature is most beautifully described by the magical picture of the "Cheshire Cat"[21] (here, the bag wall) which tends to fade away when examined closely, leaving behind only its grin (here, the confinement in principle).[22] Vector bosons can be introduced into the Skyrmion model in a natural way.[23,24] More details about this alternative description of low energy phenomena are given in a special lecture.[25]

Thus, from the point of view of the models discussed, which in some sense claim to approximate QCD, it appears that an effective meson theory may very well be the appropriate representation for the NN interaction in the domain of nuclear physics.

In addition, there is traditionally strong phenomenological evidence for a meson theory* of nuclear forces. In fact, meson-exchange is presently the only quantitative model for the NN interaction. Such an accurate representation of the nuclear force is for example needed as starting point for the large field of nuclear structure physics. This adds a further argument in favour of pursuing a meson theoretic description of nuclear physics — an argument which is quite different from those given before, as it is purely "practical." Yet, it is as compelling as the need for a quantitative nuclear structure theory.

We conclude these introductory remarks with the hope that the reader has been left some idea of the various ways in which meson theory is relevant to present day nuclear physics.

Nevertheless, we will briefly turn to the past now.

2. HISTORICAL EXCURSION

We are now going to undertake a short excursion into the history of nuclear forces, with special emphasis on meson theory. It will mean that we shall pass, essentially, through four phases of theoretical developments. The first leads from the very early ansatzes for the nuclear force until the discovery of the pion in 1947/48, and, due to the lack of empirical evidence in those days, is bound to be of highly speculative character. The following period is naturally, prevailingly, devoted to the pion. The third period starts with the discovery of the vector bosons in the early

---

*We should note that nowadays the term "meson theory" is strictly speaking incorrect, since a "meson theory" in the fundamental sense of the word does not exist. (QCD may turn out to be a theory.) However, for historical reasons, this term is well established — due to the fact that originally it was really believed to be a theory — and we will continue using it. More precisely, one should speak of something like an "effective meson model."

1960's and, therefore, deals with models applying several different bosons. Another important topic of that period is dispersion theory, developed alternatively to field theory, since the latter had fallen somewhat in disgrace for some time. The finale starts in the early 1970's and is marked by the development of absolutely accurate meson theoretic potentials, essentially, along two lines: dispersion relations and field theory. Both turn out to be quite successful, bringing history — at least in the framework of meson theory — to a good ending.

2.1 The Speculative Start

Quite naturally, the theory of nuclear forces begins in 1932 right after the discovery of the neutron by Chadwick[26] suggesting a special force between protons and neutrons to explain nuclear binding. This same year, Heisenberg[27] publishes his first ansatz for the nuclear force, in which he already introduces the isospin formalism. Majorana[28] follows shortly after. Proton-proton scattering experiments develop rapidly during the 1930's up to 1 MeV laboratory energy.[29] They soon indicate that an additional force, besides the electrostatic Coulomb force, must exist also between two protons, leading finally to the hypothesis of the "charge independence" of nuclear forces.[30]

The first fundamental idea for a deeper origin of the nuclear force appears as soon as 1935. Yukawa[31] suggests that a particle with an "intermediate" mass (compared to electron and proton), therefore called "meson", could be responsible for the interaction energy between proton and neutron. The short (finite) range nature of the nuclear force is already well known at this time from the saturation properties of finite nuclei; the massive character of the exchanged particle predicted by Yukawa provides the force with such a finite range. In the first consideration[31] a charged scalar boson in classical field theory is assumed to act between proton and neutron only, in accordance with Heisenberg's[27] first assumptions. Soon Yukawa himself reconsiders his proposal in the framework of quantized field theory,[32] and further extends his idea with a group of collaborators.[33]

What nobody can possibly foresee at this time: the stage for a half-century struggle of hope and desperation is set. The known forces in these days are just the Coulomb and the gravitational force both having a very simple form. Naturally, one expects first something equally simple for the nuclear potential, e.g. just one central-force Yukawa: $Ce^{-\mu r}/r$. However, even just phenomenologically, the nuclear force shall finally turn out to be much more complicated due to its spin dependence. In addition, field theory — the framework within which the nuclear force is to be derived — shortly runs into fundamental mathematical problems with itself and with its application. Both will act soon as a seemingly never-ending source of oppressing problems and startling surprises, which together with some great discoveries and successes gives the history of nuclear forces the touch of a detective novel.

It first starts encouragingly: In 1937 a "meson" is found in cosmic ray, the muon.[34] It is interpreted (as we know now, incorrectly) as the particle predicted by Yukawa, particularly since its mass (≈106 MeV) appears to be roughly right, according to what is known about the approximate range of the nuclear force. This discovery arouses considerable interest in Yukawa's idea; and because of that, the misinterpretation has a lucky side. So, Kemmer[35] feels inspired to suggest a rich variety of possible meson fields including pseudoscalar, axial-vector and tensor (for an explanation of these terms see Sec. 3), after Proca,[36] in 1936, has already considered vector fields. Also a "symmetric theory" (the ancient term for a theory with iso-vector bosons, i.e. bosons of three charge states: +, −, neutral) is proposed by Kemmer[37] and Bhabha[38] to account for the known hypothesis of charge independence.[30] This suggestion is made in spite of the experimental fact that at this time only charged "mesons" (μ^+, μ^-) can be found. These (in lowest order) cannot be exchanged between like nucleons, and therefore seriously violate charge independence. (Some authors express the hope that the two meson exchange may balance that.) Moreover, in 1939, Kemmer[39] discusses an entirely new, higher dimensional meson field equation, which has recently been revived in the context of meson-nucleus scattering.[40]

Wick[41] gives a concrete picture of how to relate the mass of a particle, m, to the range, R, of the force caused by its exchange, based on Heisenberg's uncertainty relation:

$$\Delta E \cdot \Delta t \approx \hbar \qquad (2.1)$$

With $\Delta E = mc^2$, the energy required to create the mass of the particle* (since the spontaneous process of "virtual" particle creation violates energy conservation), and the particle moving with the speed of light, c, allowing it the time $\Delta t = R/c$ to "stay in the air", one obtains for the range:

$$R \approx \hbar/mc \qquad (2.2)$$

which is identical to the Compton wave length of that particle. (We shall see later, Sec. 5.2, that this range estimate is not to be taken too literally, as, in practice, it underestimates the range substantially.)

---

*We neglect the kinetic energy of the exchanged particle assuming that its momentum $kc \ll mc^2$.

The experimental discovery of the quadrupole moment of the deuteron by Kellog and coworkers[42] in 1939 gives rise to theoretical considerations of increasing sophistication. It is derived that vector fields create a tensor force leading to a quadrupole-moment in the deuteron, but with the wrong sign compared to experiment. This problem is soon overcome by also including pseudoscalar fields. In these "mixed meson theories", in which vector and pseudoscalar fields are assumed simultaneously (Møller and Rosenfeld[43]: with meson of equal mass; Schwinger[44]: with the vector meson heavier than the pseudoscalar), the problematic r^{-3} singularity in the tensor force can be removed. With the Schwinger force, about 1/3 of the empirical quadrupolemoment can be reproduced.[45] Concerning the singularity problem, Bethe,[46] in 1940, suggests the use of a "cut-off" for small r, which can be interpreted as assuming extended meson sources (i.e. extended nucleons). This idea and the problems it raises, especially with regard to the relativistic invariance of the theory, are examined in length, apart from others, by Pauli. An account of this question, as well as many other topical issues in meson theory of this time (e.g. the "strong coupling theory"), can be found in his lectures given at the MIT in fall 1944.[47] Most interestingly, Pauli concludes, from the fact that the pseudoscalar "symmetric" theory predicts the right sign for the deuteron quadrupole moment, that this is most likely the correct theory — long before the pion is found and its spin and parity determined. Also quite early it is recognized by some physicists that vector and scalar fields create a spin-orbit force (Breit[48] (1937/38), Rosenfeld[49]). Empirical evidence for this is seen in the spectra of light nuclei. E.g. Rosenfeld* in 1948: "The occurrence of a rather large spin-orbit coupling in ^{5}He may be regarded as an indication of the existence of mesons of spin one."

In 1948/49 A.E.S. Green[51] takes up again the Kemmer idea[37] of a rich variety of meson fields. He shows that the problem of short-range singularities can also be overcome when generalized meson fields with higher derivatives in the Lagrangian are used. Part of his work is rejected from publication for being too "speculative."[52]

For further details about this first, "speculative" period of meson theory (i.e. before the discovery of the pion), we refer the interested reader to the above-mentioned lectures by Pauli,[47] and the book by Wentzel[53] which contains an informative chapter on this topic. An extremely thorough account of the considerations in these days, including many contemporary details, is given in the book by Rosenfeld about nuclear forces,[50] published in 1948.

Experiment finishes this period: In 1947 Conversi, Pancini and Piccioni[54] show that the μ does not interact strongly with

* Ref. 50, p.368.

nuclei, and therefore is, in fact, not a meson — it is a lepton. The same year, a meson with a mass of about 140 MeV, which does interact strongly, the "pion", is found in cosmic ray by Occhialini and collaborators,[55] and shortly after in the Berkeley cyclotron laboratory.[56] The final and conclusive confirmation of the reality of mesons is given in 1949 by the Swedish Academy of Science by awarding the Nobel prize to Yukawa.

2.2 The pion battle

Quite understandably, the new reality of a strongly interacting meson motivates vigorous theoretical efforts to describe the nuclear force, now, by the pion only. This will become the program of the following decade — the 1950's. Naturally, it starts with high expectations and great enthusiasm (but will end, however, in deep disappointment). The success of renormalization has just put field theory on firm grounds. The pion appears to be the particle for the strong interaction in analogy to the role of the photon in quantum electrodynamics (QED). Considering the great quantitative successes of QED, it is hard to set a limit on the expectations for strong interaction theory.

The work of Japanese physicists deserves our special attention for this time, as — in the tradition of Yukawa — they probably did the most in this field and at the same time may have also been ignored the most outside their country. In 1951, Taketani and associates[57] (TNS) present their historic suggestion to subdivide the nuclear force into three regions. The far-sighted character of this suggestion is best indicated by the fact that this subdivision is still in use today. TNS speak of a "classical" (long range, $r \gtrsim 2$ fm),* a "dynamical" (intermediate range, 1 fm $\lesssim r \lesssim$ 2 fm) and a "phenomenological" or "core" (short range, $r \lesssim 1$ fm) region. In the classical region the longest range part of the potential, namely, the one-pion-exchange (OPE) is dominant (the pion having the smallest mass of all mesons or multi-meson configurations). In the dynamical region the two-pion-exchange (TPE) becomes important. Finally, in the core region everything can contribute: in particular, multi-pion exchange, heavy mesons, and (from today's point of view) quark-gluon exchange. This classification has been of utmost practical importance because it allows a step-by-step exploration of the two-nucleon interaction. Thus it is possible, to first calculate the longer range parts of the potential, and correlate the results with experiments sensitive to just that region. (We will do this in Sec. 4.) In this way, the whole problem — with all its oppressive complexity — does not have to be faced at once. By

---

*r denotes the distance between (the centres of) two nucleons.

the way, in the light of TNS e.g. the use of extended sources ("cut-offs") does not appear so forbidden anymore!

In the decade under consideration, the one-pion exchange becomes experimentally well established as the long range part of the nuclear force.* The evidence comes from the analysis of NN scattering data and the deuteron. Speaking in dispersion theoretic terms (see Sec. 2.3) — the scattering amplitude in the non-physical region of the complex cos θ plane (θ being the scattering angle in the CM system) has two symmetrically situated poles associated with contribution of one-pion intermediate states. The experimental data is extrapolated to these poles to yield the pion nucleon coupling constant. This procedure can be applied both near 0° and 180° for the scattering angle θ. The πN coupling constant obtained in this way agrees with that known from πN scattering.[59] An equivalent procedure can be applied in the framework of the (partial wave) phase-shift analysis, where for sufficiently high orbital angular momentum L (corresponding to large distances) OPE alone is assumed. It turns out that the χ^2 for the phase-shift solutions is a minimum for the correct pion mass and coupling constant (as known from πN scattering).[60,61] In the case of the deuteron the evidence for the one-pion-exchange comes from the quadrupole moment which is almost entirely explained by OPE.[62,63,64] It is also realized that the asymptotic D/S state, sometimes called η, is determined by far from OPE.[64,62] However, reliable measurements of η will not appear before 1980. In fact, nowadays, η may be considered as the most precise and compelling evidence for the reality of the pion in the nuclear force.[65]

As the OPE contribution to the nuclear force combines all the pleasant features a physicist may wish from a theory — such as easy to evaluate and most satisfactory to explain data — so does the TPE evolve in an opposite way. It is painful to evaluate and for a long time it has not even been doing well in correlating data. The calculations are not only extremely complicated and tedious, but, in addition, they are beset, for a long time, by a number of serious ambiguities, which lead to quantitatively rather different results causing serious controversies. The many efforts of pion theoretical potentials of the 1950's are usually divided into two groups: The Taketani-Machida-Onuma (TMO)[66] and the Brueckner-Watson (BW)[67] type. In the former case an S-matrix is evaluated directly from meson field theory, from which in turn a potential is derived. In contrast, the BW method is based on an expansion in the particle number and derives a potential directly. The main differences between the two approaches are that the box-diagram (Fig. 2.1(a)) and the pair terms (Fig. 2.1(c) and (d)) are

*For a survey on the related Japanese work see Ref. 58, especially p.32 therein.

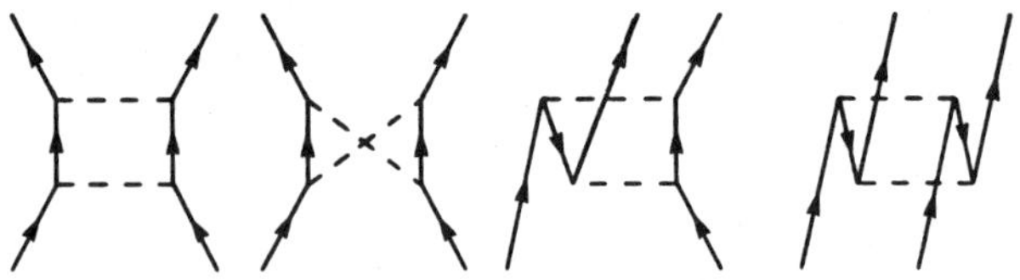

FIG. 2.1. Some two-pion-exchange contributions to the NN interaction. (a) is called a box diagram, (b) a crossed box. (c) and (d) contain "pair terms." There are also pair term diagrams with crossed pion exchange, which we have not shown here. Full lines denote nucleons, dashed pions. The underlying time axis is vertical, pointing upwards into the future.

always included in TMO whereas BW excludes the box from the beginning and can also at will leave out the pair terms. For this "pair suppression" there is evidence from πN scattering in S-waves where the pair term leads to (by about one order of magnitude) too large a scattering length. Therefore it has been suggested that the suppression of virtual pair terms might be a general feature of meson theory.[68] In addition, BW find an almost exact cancellation of the one-pair (Fig. 2.1(c)) and the two-pair (Fig. 2.1(d)) contribution to the NN interaction. A further source of discrepancies in the pion theories of the 1950's are ambiguities in the subtraction of the iterated OPE necessary to extract a potential.[69] Apart from the general uncertainty in the results, the spin-orbit force derived from TPE turns out to be too weak (by one order of magnitude) than experimentally needed.[70]

The balance of this decade is given in a quote by Goldberger[71] from 1960: "There are few problems in nuclear theoretical physics which have attracted more attention than that of trying to determine the fundamental interaction between two nucleons. It is also true that scarcely ever has the world of physics owed so little to so many. In general, in surveying the field, one is oppressed by the unbelievable confusion and conflict that exists. It is hard to believe that many of the authors are talking about the same problem or, in fact, that they know what the problem is."

Already in 1953, Bethe — in an article for the Scientific American[72] — estimates that in the preceding quarter century more man hours of work have been devoted to the NN problem than to any other scientific question in the history of mankind. In fact, a second quarter century will follow in which about one order of magnitude more man hours are invested in this problem.

The interested reader will find an excellent review of this period in the book by Moravcsik,[73] the review article by

Phillips[73] and a more comprehensive account in Bethe and de Hoffmann.[74] The enormous work by the Japanese physicists is summarized in two Supplements.[58,75]

Fortunately, there is also another line of research on the NN interaction during the 1950's; and this line, developing in almost complete independence from the theoretical efforts discussed so far, is more successful. The goals are much more modest — it is the attempt to give a simple phenomenological description of the NN potential.

Data from NN scattering up to about 400 MeV accumulate rapidly during this period and the phase-shift analyses improve both through better theoretical constraints and an increasing data base.* Gradually, sufficient empirical information on the NN interaction up to 300 MeV laboratory energy (and especially around that latter energy[77]) is available for a phenomenological potential description.

The basic aim of a potential description of the two-nucleon interaction is twofold. One is to provide an economical summary of the data for comparison with potential-like results from theory (e.g. meson theory). The other aim of a phenomenological potential is to serve as an input for nuclear structure calculations.

From general invariance considerations, already carried out in the 1940's,[78] the most general form of a non-relativistic potential can be deduced. It consists of a central, spin-spin, tensor, spin-orbit and quadratic spin-orbit term.** The first attempt to fit the NN scattering data with just the first three terms fails.[79] Another attempt, which includes in addition the spin-orbit force, is quite successful.[80] It is the first (semi-) quantitative NN-potential ever constructed — known by the names of Gammel and Thaler[80] and published in 1957. This potential uses a hard core at small distances suggested by the trend of the $^1S_0$ phase-shift to turn negative for laboratory energies $\gtrapprox$ 250 MeV. (A hard core was first considered by Jastrow[81] in 1951 to account for the seemingly isotropic differential cross section in pp scattering at 340 MeV.) The spin-orbit force is demanded by the triplet P-phase-shifts. Its relatively short-range character is evident from the fact that F-phase shifts do not show much spin-orbit splitting anymore. The important role of the spin-orbit force in nuclei was already forseen in the late 1930's (Breit[48]) and assumed in the late 1940's (Rosenfeld,† Mayer and Jensen[82]).

---

*A review with particular stress on the experimental aspect is found in the book on the NN interaction by Wilson.[76]

**For more details see Ref. 78 and Moravcsik, Ref. 73, p.46.

†Ref. 50, p.368.

Another early semi-quantitative potential is that of Signell, Zinn and Marshak.[83]

A few years later improved phenomenological hard-core potentials follow, notably the Hamada-Johnston[84] and the Yale[85] potentials. Both use all five terms for a potential mentioned above. In contrast to Gammel-Thaler, these latter potentials also fit the deuteron accurately since they include the OPE potential. The "Reid" hard- and soft-core potentials[86] appear in the late 1960's. The soft-core version becomes one of the most applied potentials in nuclear physics of the 1970's. Phenomenological potentials, as we have them discussed here, use in general on the order of 50 parameters to fit the data.

2.3 The OBEP and "Dispersive" Phase

Now, we take up again, the description of the meson theoretic development of the nuclear force. It is now 1960. This time is essentially characterized by two points: the failure of the pion field-theoretic program and the accumulation of rich phenomenological knowledge about the NN force — particularly concerning a short-range repulsion and a spin-orbit force. It is obvious that this has to revive the old idea of vector meson exchange (which has both the features just mentioned) discussed already in the late 30's (Breit[48]) and in the 40's (Rosenfeld[49,50]). As vector boson exchange is not renormalizable, it had been discarded for a number of years. Nambu,[87] Breit,[88] Sakurai[89] and Frazer and Fulco[90] reconsider the possibility of vector bosons with regard to the NN interaction as well as in conjunction with the electromagnetic form factor of the nucleon. Their presentiment is soon confirmed: in 1961 the ω-meson is discovered[91]; the ρ follows shortly after.[92] Both are spin one bosons, the ρ being a 2π- and the ω a 3π-resonance with masses of approximately 770-780 MeV.

The discovery of heavy mesons leads to the one-boson-exchange (OBE) model for the NN interaction, which breaks the dead-lock situation in meson theory. The OBE model is based on the old idea that the nuclear force is meson mediated. However, it tries to take advantage of the observations in high energy physics that two or more particles (e.g. pions) as a group like to behave most of the time as if they were forming a single particle with a definite mass and definite intrinsic quantum numbers (i.e. they are "correlated" or even form a resonance). It is hoped and, in fact, assumed in the OBE model, that the uncorrelated many-pion exchange (i.e. with no interaction in between the pions "in the air") may be negligible (apart from iterative contributions which are generated by the unitarizing equation). (In contrast, the calculations of the 1950's were concerned only with the uncorrelated plural-pion exchange.) This basic assumption is certainly too extreme and we will see how and why the OBE-model can make up for that.

There is also a very pragmatic reason for this model. As we see from the work of the 1950's, the explicit evaluation of multi-pion exchanges is very difficult and ambiguous. In contrast, the calculation of one-particle exchanges is relatively straightforward.

All known OBE models include a large contribution from one or two isoscalar scalar bosons with their mass in the range of 400 to 800 MeV, to provide the indispensable intermediate range attraction for the nuclear force. These particles are supposed to represent 2π-S-wave resonances. Indeed, during the 1960's and early 70's, the Particle Data Group listed in their meson table an alleged isoscalar scalar 2π-S-wave resonance (called σ or ε) in the above mass range.[93] This low mass ε has disappeared from the tables since 1976[94] in favour of an ε(1200) (which cannot be used by OBE potentials to provide intermediate range attraction and, therefore, is useless within that model). Thus, on this point of the OBE model, confirmation does not happen. However, as the model is very successful, it seems to indicate that in the case of the NN-interaction even uncorrelated 2π-exchange may be well simulated by a boson of definite mass and zero width. Still, from a more fundamental point of view this is not satisfactory and therefore we will come back to this point later.

Some of the first OBE potentials (OBEP) are those of Bryan and coworkers,[95] Wong and collaborators,[96] Japanese theorists,[97,98] McKean[99] and A.E.S. Green and associates,[100] who could pick up the threads of earlier work.[51,52] One of the great successes of all these models is that they need in the order of only 10 parameters to fit the NN data (in contrast to phenomenological potentials which require about 50). These parameters are meson-nucleon coupling constants and cutoff parameters.

The improvement of the OBE model continues into the 1970's. In fact, OBEPs which accurately describe all NN data up to about 300 MeV and the deuteron are not provided until the middle 1970's. Such potentials, represented in configuration space (r-space), are constructed by the Nijmegen group[101] and Sprung and coworkers[102] — the latter being, however, semiphenomenological. The OBE concept further improves by considering three-dimensional relativistic reductions of the Bethe-Salpeter[103] equation and by working in momentum space to avoid the drastic approximations necessary to obtain analytic r-space expressions. The most accurate work along this line is done by the Bonn group.[104-106] Tjon and collaborators[107] finally solve the four-dimensional Bethe-Salpeter equation in the ladder approximation constructing an OBEP in that framework.

The work up to 1971 is well summarized by Moravcsik[108] with a complete list of the bibliography. Good reviews on this subject are the contribution by S. Ogawa et al. in the Supplement Ref. 75 and the Proceedings of the First International Conference on the Nucleon-Nucleon Interaction held in Gainsville, Florida, in 1967

(Ref. 109), which also reflects some of the enthusiasm of the early OBEP-years.

I mentioned before that quite apart from the quantitative success of the OBE model in fitting the empirical NN data, such models cannot be conceptually accepted to form a complete theory, as it is hard to believe that the uncorrelated plural-particle exchange should be totally negligible. The longest range component of such exchanges, and therefore the most important of that kind, is the two-pion-exchange (TPE). How to take into account the TPE more accurately, or even "completely," is the other main topic — besides the OBE model — of the 1960's.

Of course, this topic is not new. It was one main goal of the 1950's, and it failed at that time. In retrospect, and with the new knowledge which accumulated during the 1960's, we now understand the reason for this failure: the old program did not consider correlations between the exchanged pions or pion-resonances, which, as the OBE model has demonstrated, play a crucial role in the two-nucleon force.

Naturally, the new goal must be to include "all" correlated and uncorrelated multi-particle exchanges; first, for two pions.

There are two conceptually very different ways to actually calculate these contributions: by field theory and by dispersion relations. It is psychologically quite understandable, that after the failure of the field theoretic program in the previous decade, there is not much motivation to try this again. Therefore, during this period, there is only little work along this line. Fortunately, field theory is not dead forever; in fact, it will revive intensely later on.

A completely alternative approach to multi-particle exchange is now pursued, which has become known as "dispersion relations." Apart from the disappointment and the doubt about field theoretic techniques there is also a second positive motivation for this new approach: it can take correlated and uncorrelated multi-particle exchange into account within the same framework.

Since in the actual work on the NN interaction presented later in these lectures, the dispersion theoretic approach is never used (we will present a field theoretic model), however, results gained from dispersion relations are sometimes quoted, so let us here, very briefly, outline what this approach is about. For an excellent introduction see Moravcsik,* and for more comprehensive explanations see Mandelstam[110] or Chew.[111]

As this new approach is born out of a frustration with field theory — with its formidable problems of renormalization, convergence, selection of diagrams (e.g. pair terms or not) and how to develop a potential concept — dispersion theory attempts to avoid these shortcomings from the beginning trying to deal with

*Ref. 73, p.104.

physically observable quantities only. Lagrangians, Hamiltonians and potentials do not occur anymore. Instead, the new theory deals directly with reaction amplitudes. In fact, one of its great advantages (especially with regard to the NN interaction) is, that it can relate different measurable reactions to each other. In this way it provides the framework for a consistency check between such different processes as NN, πN and nucleon-electron scattering (nucleon electromagnetic form factors).*

The principal framework of dispersion relations is based on three fundamental assumptions: causality, unitarity and crossing symmetry. From the first the analyticity of the reaction amplitude is concluded. The third allows us to relate processes which differ from each other only by the interchange of some incoming and outgoing particles of the reaction. One-particle exchange appears as a pole in the reaction amplitude. In this way, dispersion theory gives empirical evidence for the reality of meson-exchange. (We discussed this for the pion before.)

To calculate the 2π-exchange contribution to the NN interaction with the help of dispersion theory, one proceeds as described schematically in Fig. 2.2. The total "diagram," (a), is divided into two halves, (b), each of which, when considered in the t-channel [i.e. looking horizontally from left to right into the diagram (c)], represents an amplitude $N\bar{N}\rightarrow 2\pi$. This amplitude is constructed from empirical input from πN scattering, (d),** and $\pi\pi$ scattering [shaded circle in (e)].

Work along this line starts as early as 1960 (e.g. Amati et al.[112]), and soon after many groups are actively engaged in this field. (For a comprehensive review of the past literature see Moravcsik.[108]) Nevertheless it takes until the end of the decade for quantitative

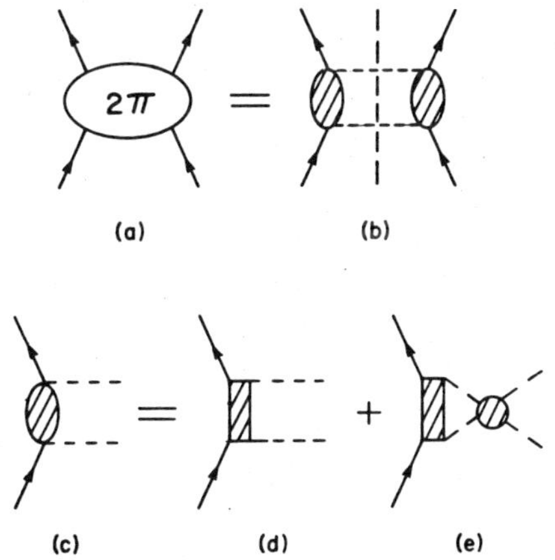

FIG. 2.2. 2π-exchange contribution to the NN interaction as viewed by dispersion theory and explained in the text. (Full lines represent nucleons, dashed lines are pions.)

*See e.g. Signell and Durso, Ref. 109, p.635 and earlier references therein.

**The shaded rectangle in (d) stands for the nucleon and all possible πN resonances [e.g. $\Delta(1232)$].

results (see e.g. Signell and Durso,† Furuichi,†† Binstock,[113] Chemtob et al.,[114] Vinh Mau et al.[115]). All these results indicate that the 2π-exchange is able to provide the intermediate range attraction of the nuclear force. Still, these efforts are far from constructing a full quantitative NN potential.

Thus finishes the OBEP and "dispersive" period of the 1960's.

2.4 The finale

The finale takes place essentially in the 1970's and early 1980's. It is concerned with the ultimate goal of finally providing an absolutely quantitative nuclear force, with a derivation based on something which can be (almost) called a theory. (Phenomenological potentials deliberately and OBEPs for reasons given in Sec. 2.3 do not fulfil this requirement.)

Again, the work proceeds along two lines: dispersion theory and field theory. Both lines turn out to be finally very successfull. Therefore, in the end there will be two accurate NN potentials with a sound theoretical basis: The Paris and the Bonn potential.

However, let us proceed in chronological order. First, the final disperion theoretic efforts.

In continuation of the work of Chemtob et al.,[114] which was mentioned at the end of the last subsection, the Stony Brook Group[116] constructs a potential, in which the dispersion theoretic result for the 2π-exchange is complemented by one-π and one-ω exchange. For short distances the eikonal form factor is used.[117] The fit to the phase-shifts of NN scattering is semi-quantitative. The Paris group publishes a potential that same year.[118] In this case the short range part of the potential ($r \lesssim 1$ fm) is treated purely phenomenologically. For the 2π-exchange contribution derived by dispersion relations both groups agree quantitatively. Further refinements and a convenient representation of the potential are left to the Paris group, which publishes their final version in 1980.[119] The potential is expanded in Yukawa functions of multiples of the pion mass.

A brief review about the dispersion theoretic approach to the NN interaction is given by Vinh Mau.[120] Detailed explanations can be found in the book by Brown and Jackson.[121]

Let us now turn to the field theoretic line. After a decade of prevailing abstinance, for reasons explained earlier, the field theoretic approach is revived by the work of Lomon and Partovi.[122] They evaluate the 2π-exchange Feynmann diagrams with nucleons and

---

†Ref. 109, p.635

††Ref. 75, p.190

represent the result in the framework of the relativistic three-dimensional reduction of the Bethe-Salpeter equation[103] proposed by Blankenbecler and Sugar.[123] It is a non-static approach to the 2π-exchange. By using this well-defined framework the ambiguity is absent of how to construct and subtract the iterated one-pion-exchange, when defining a potential. So, in 1970 Lomon and Partovi finally showed how to do the 1950's program correctly. Their quantitative result is neither close to BW[67] nor to TMO.[66] In the following years Lomon and coworker extend their calculations by also introducing a σ-boson.[124]

For a "complete" field theoretic model for the 2π-exchange more diagrams than those in Ref. 122 and 124 must to be included. From dispersion theory one knows that isobars in intermediate states lead to substantial contributions (see Signell and Durso,* Sugawara and von Hippel,[125] Tamagaki**). A field theoretic model has to take that into account. Further, 3π and 4π exchange should be considered. The program aiming for such a comprehensive and "complete" field theoretic model for the NN interaction is pursued by the Bonn group. Over a number of years this group evaluates step by step the 2π-exchange diagrams[126-129] and finally also the relevant 3π and 4π exchanges.[127,130,131] The complete set of diagrams gives in a quantitative description of all known NN data (up to about 300 MeV) and the deuteron with very high precision.[132,133] (More about this in Sec. 4.)

There are several advantages to a field theoretic model. First, it determines the off-shell behaviour of the potential in a well-defined way. As dispersion theory deals with reaction amplitudes, which are always on-shell, the off-shell behaviour remains undetermined in such an approach and is left to guess-work or using the simplest ansatz (e.g. Yukawa's). The "complete" set of diagrams, which a field theoretical model provides, is a necessary basis for the consideration of related processes, e.g. meson-exchange current corrections to electromagnetic properties. Moreover, when applying the nuclear force in the many-body problem, medium effects alter the nuclear force. They can be evaluated on the basis of a field theoretic model. The same is true for many-body forces. (More about this in Part II of these lectures.)

Herewith finishes our excursion into history. It started with Heisenberg[27] and Yukawa[32] and took us through half a century. It is a story of excitement, error, depression and especially of hard, hard work which, however, finally has led to a successful end.

---

*Ref. 109, p.635

**Ref. 109, p.629

After history has given us a general background, let us now get to know the details of the meson theory of the NN interaction. This will be done in the next two sections.

3. A PEDAGOGICAL INTRODUCTION

In this section we want to look more closely into meson-exchange. Our goal is to understand the principle features meson-exchange can predict for the NN problem. With "understand" we mean here really understand from the point of view of theory, and that is, we want to directly and explicitly derive things starting from first principles. This sounds very ambitious, but we will see that this can be done quite easily (for the lowest order contributions, to which we will restrict ourselves in this section).

As theory is no end unto itself, we have to compare the results of our theoretical derivations with experiment, to see if anything is predicted which agrees with reality. For this purpose, we want to recall first, what is known empirically about the NN interaction. In this section the discussion can be restricted to essentially qualitative considerations. The logic behind this is, that we had better make sure, first, that our theory offers the right qualitative features, before going to a quantitative approach. (The latter will be done in Sec. 4.) Moreover, the stress in this section is on physical understanding; a qualitative consideration can serve that purpose much better than the chaos of precise numbers and curves (of which we will get enough, anyhow, in Sec. 4).

3.1 Facts About the Nuclear Force

Of course, we cannot mention all the facts. For a comprehensive description of the NN interaction the reader is referred to the literature. The first two books devoted exclusively to the two-nucleon problem are still today excellent introductions into this field (Moravcsik,[73] Wilson[76]). An excellent introductory review has been written by Signell.[134] We further recommend the book by Brown and Jackson.[121] Conventional nuclear physics text-books also have in general a useful introductory chapter on the two-nucleon problem, see e.g. Segrè.[135]

Here, we will pick only the most remarkable and outstanding empirical properties of the nuclear force. These are essentially five, and we will go through them one by one indicating also some evidence for each of them.

1. Nuclear forces are of short range (finite range). "Short" is meant here in comparison with the Coulomb or gravitational force. Early evidence for this was seen from the saturation

properties of nuclei. When going from the A=4 nucleus (A=number of nucleons in a nucleus), namely helium, upwards to higher A nuclei, one realizes that the binding energy per nucleon in a nucleus is about constant. The density remains also roughly the same as the radius of heavy nuclei is proportional to $A^{1/3}$. If the nuclear force were of large range, like e.g. the Coulomb force, the potential energy per particle would increase with A and so would the density. On the other hand, for light nuclei ($A \lesssim 4$) the binding energy per particle does grow. The deuteron is bound by 2.2 MeV, ^3H by 8.5 MeV. It is best, to analyze this in terms of energy per "bond." Thus, the binding energy per bond is about 2 MeV in the two-nucleon system and 3 MeV for the triton. In ^4He we have ≈4.5 MeV per bond (28 MeV total binding). The conclusion is, when nucleons are pulled closer to each other by more bonds (due to more particles) also the energy <u>per bond</u> increases (up to saturation). From this Wigner,[136] in 1933, concluded that the nuclear force has to be of short range (namely shorter than the deuteron diameter of about 4 fm and about equal to the radius of the alpha particle of about 1.7 fm).

<u>2. The Nuclear Force is Attractive in its Intermediate Range.</u> "Intermediate" is meant here with respect to the total range of the nuclear force, which we now subdivide according to TNS[57] in short, intermediate and long range. The proof for the attractive character of the nuclear force (at least, in a certain range) is given by the fact of nuclear binding. The range of this attraction can be obtained — now, more precisely than in point 1 — by considering the central density of heavy nuclei, which is known from electron scattering experiments from nuclei. This density (≈ nuclear matter density) is about 0.17 fm^{-3} providing for each nucleon a volume with a radius of about 1 fm. Thus, the average distance between the centers of two nucleons in the interior of a nucleus is about 2 fm. This should be about the effective range of the attraction. Further evidence for the partially attractive character of the nuclear force comes from scattering. The $^1S_0$-phase-shift is positive (equivalent to attraction) for a laboratory energy (E_{LAB}) $\lesssim$ 250 MeV, see Fig. 3.1. (For an explanation of the term of phase-shifts and how to obtain them from scattering data and, further, for the "spectroscopic" notation used for defining partial wave NN states, as e.g. "$^1S_0$," see Messiah,* Moravcsik** or Wilson***.)

*Ref. 138, p.385

**Ref. 73, p.68

***Ref. 76, p.110

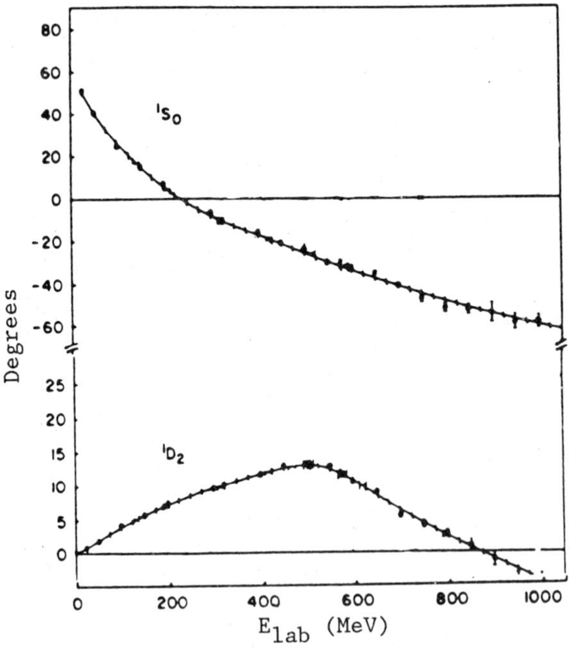

FIG. 3.1. Empirical $^1S_0$ and $^1D_2$ phase-shifts from NN scattering. The curves are an energy-dependent phase-shift analysis. The error bars represent results from an energy-independent analysis. (From Arndt et al.[137])

3. The Nuclear Force has a Repulsive Core. Such an assumption would help explaining the saturation of nuclear forces and the constant nuclear density. But this argument is not compelling proof for a repulsive core as saturation can also be generated in other ways (e.g. by "exchange" forces[27] or by Pauli and relativistic effects (see Part II)). Historically, as we pointed out in Sec. 2.2, a repulsive core was first suggested by Jastrow[81] in explaining the isotropy of proton-proton (pp) differential cross sections at E_{LAB} = 340 MeV. However, the best argument is the behaviour of the $^1S_0$ and $^1D_2$ phase-shifts as a function of increasing energy. The latter stays positive (equivalent to attraction) up to about 800 MeV (see Fig. 3.1) whereas the $^1S_0$-phase-shift turns negative (equivalent to repulsion) around 250 MeV. As an S-state (orbital angular momentum L=0, no centrifugal barrier) feels the innermost region of the force, whereas in a D-state (L=2) the nucleons are kept apart through the centrifugal barrier, one may conclude that a repulsion of short range is indicated. The rule of thumb, that relates the range of a potential, R, to the highest orbital angular momentum state, L_{MAX}, which is still affected by that range, is given by the semi-classical formula*:

$$L_{MAX} \approx R \cdot p \tag{3.1}$$

*From now on we use natural units:
$\hbar$=c=1; conversion factor: $\hbar \cdot c$=197.3 MeV•fm

where p, the momentum of a nucleon in the centre of mass (CM) frame of the NN system, is related to E_{LAB} by:

$$E_{LAB} = \frac{2p^2}{M} \tag{3.2}$$

with M the mass of the nucleon. For E_{LAB} = 250 MeV, the turning point in the $^1S_0$ phase-shift, we have $p \approx 1.7\ fm^{-1}$. From this we obtain with $L_{MAX} \lesssim 1$:

$$R \lesssim 0.6\ fm \tag{3.3}$$

for the radius of the repulsive core.

4. <u>There is a Tensor Force.</u> The most striking evidence for this is seen in the deuteron: the quadrupole moment,[42] the magnetic moment (which requires a D-state contribution) and the asymptotic D/S ratio.[65] The fact that for $^3S_1$ a bound state exists and for $^1S_0$ does not, is also an indication (the tensor force operator is zero in singlet (S=0) states; in triplet states it provides attraction, especially, through its second order contribution). Further evidence is given by the non-vanishing ϵ_J "mixing parameters" obtained from a phase-shift analysis (Fig. 3.2). This parameter is proportional to the matrix element for a transition from L = J − 1 to L′ = J + 1. Of all operators, which can occur in the most general form of a non-relativistic

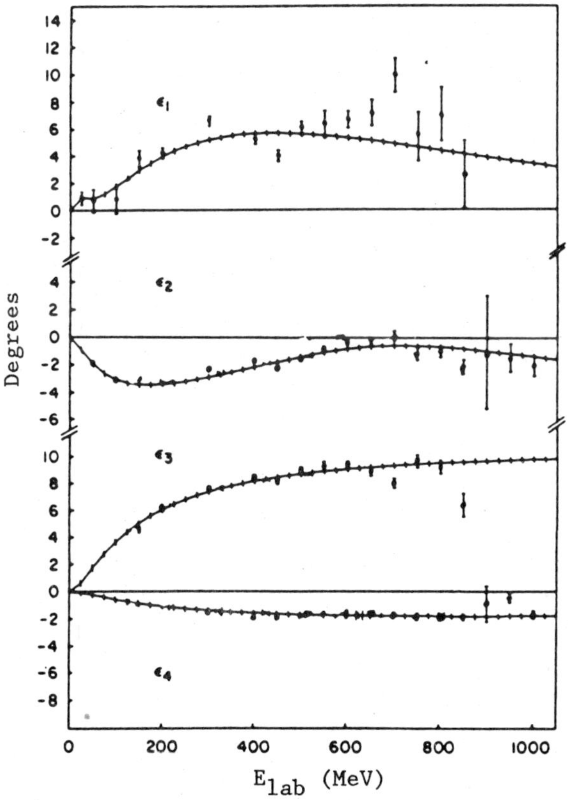

FIG. 3.2. Empirical ϵ_J mixing parameters for J=0 to 4. Notation as in Fig. 3.1. (From Arndt et al.[137])

potential,[78,*] only the tensor operator has non-vanishing matrix elements for this kind of transition.

5. There is a Spin-Orbit Force. This was first observed in the single particle model for nuclei,[48,50,82] though this refers to an "effective" nuclear force in the nucleus, which is not identical with the "free" NN interaction we are discussing here. However, these two forces should be related; in fact in principal the effective force could be derived from the "free" within a many-body theory (see e.g. Part II of these lectures). Clear evidence came from the first reliable phase-shift analysis of NN scattering;[77] the triplet P-waves can only be explained by assuming a strong spin-orbit force,[80,61] see Fig. 3.3. (In singlet states, $S=0$, and S-states, L=0, there is no spin-orbit force, as the $\vec{L}\cdot\vec{S}$ operator has vanishing matrix elements.)

Speaking in more general terms, point 4 and 5 are dealing with two special cases of spin-dependences in the nuclear force. In fact, there is more of it, namely, in addition, there exists a spin-spin force. However, the empirical evidence for this in the free NN interaction is not so outstanding as for the tensor and

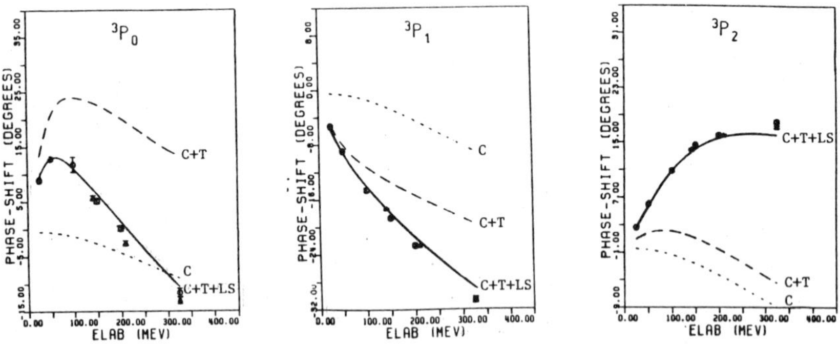

FIG. 3.3 Theoretical phase-shifts for triplet P-states calculated with (full line) and without (dashed line) a spin-orbit force. In addition, the phase-shifts as predicted by the central force (= central + central spin-spin) only are shown (dotted). The difference between the dotted and dashed line is due to the tensor force. It is clearly seen that a variation of the tensor force could not make up for the discrepancies, which occur between theory and experiment, when the LS force is omitted. The error bars represent the results from energy-independent phase-shift analyses.

---

*See also Ref. 73, p.46.

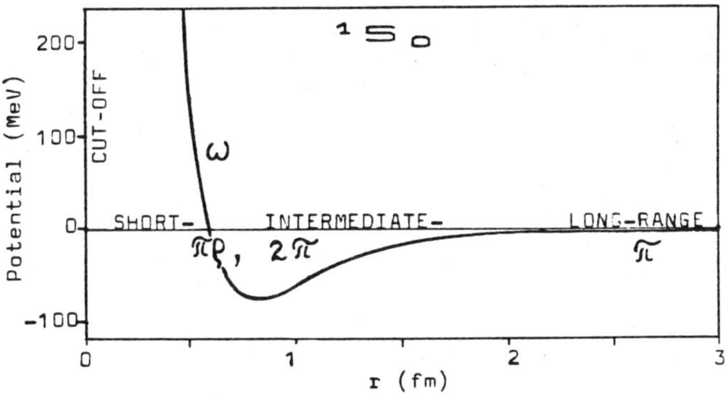

FIG. 3.4. r-dependence of the central force in the $^1S_0$-state. The greek letters refer to meson-exchange contributions to the nuclear force being explained in Sec. 4.

spin-orbit force. Therefore, we have not listed it here separately.

From our discussion of the empirical features of the nuclear force, a r-dependence of the central force like that plotted in Fig. 3.4 is to be expected.

This finished our summary of the empirical facts about the nuclear force. With these facts in the back of our mind, we can now turn to first attempts of a theoretical explanation.

3.2 The Historic Idea

In the 1930's the best established and most striking (and for a long time also the only reliably known) feature of the nuclear force was its short-range nature[136] by which it is distinguished from all other forces well-known at that time, namely Coulomb and gravitation. That is why, the first theoretical attempts concentrated on how to derive a force of finite range from some more fundamental idea. Yukawa[31] did this in 1935 by constructing a strict analogy to quantum-electro-dynamics (QED). His first consideration[31] was carried out in the framework of classical field theory, and we will repeat it here.

In QED a field of particles with zero mass, the photons, is assumed fulfilling a field equation, which is — in static approximation (and considering the fourth component of the field only) — the Laplace equation of classical electrodynamics:

$$-\Delta V(\vec{x}) = e\delta(\vec{x}) \tag{3.4}$$

($\Delta \equiv \nabla^2$, Laplace-Operator)*
The solution is:

$$V(r) = \frac{e}{4\pi}\frac{1}{r}\,; \qquad r \equiv |\vec{x}| \tag{3.5}$$

the familiar Coulomb potential.

In analogy, in meson theory a field of particles with non-zero mass, μ, the mesons, is assumed, fulfilling a field equation, which is the Klein-Gordon equation:

$$(\Box + \mu^2)\phi(x) = g\bar{\psi}(x)\psi(x) \tag{3.6}$$

Assuming the nucleons, represented in Eq. (3.6) by their fields $\psi(x)$, as infinitely heavy and fixed at the origin, we obtain the static approximation of this equation:

$$(-\Delta + \mu^2)\phi(\vec{x}) = g\delta(\vec{x}) \tag{3.7}$$

The solution is:

$$\phi(r) = \frac{g}{4\pi}\frac{e^{-\mu r}}{r} \tag{3.8}$$

the "Yukawa" potential. Through the exponential, which is a direct consequence of the massive character of the particles, this potential has the desired finite range. For zero-mass, μ=0, we are back to the Coulomb potential.

This consideration was the birth of particle physics.

Simple and traditional estimates of the range, R, identify it with

$$R = \frac{1}{\mu} \tag{3.9}$$

From this a range of 1.4 fm is predicted for the pion, which is certainly too small; in fact, at that range the pion just starts to become dominant. This argument and further results displayed in Sec. 5.2 indicate that the conventional range estimate, Eq. (3.9), is generally too small. The essential reasons for this are that the coupling constants are large and that the final nuclear potential is a result of strong interferences of large contributions. In practice an "empirical" factor of 3 to 4 should be applied to $1/\mu$ in Eq. (3.9).

---

*For notation see Ref. 138, 139.

3.3 Quantized Field Theory, Perturbation Theory and Feynman Diagrams

As we saw, the very first, "historic" consideration was done in the framework of classical field theory. Of course, in a correct treatment one has to use quantized field theory. This has been developed first for QED. Dealing with interacting fields, perturbation theory is applied, which is reasonable for a coupling constant of 1/137. The contributions from perturbation theory are most conveniently represented by Feynman diagrams.[139,140]

In the case of nucleons and mesons ("meson theory") the coupling constants are in the order of 10. Therefore, it must appear questionable if a perturbation expansion is appropriate. Nevertheless, it is customary to use perturbation theory and, consequently, to consider the various possible contributions in the language of Feynman diagrams. (An exception is the old "strong coupling theory,"[47] which, however, on the other hand has the disadvantage of being based on classical field theory.) A justification for the use of perturbation theory can be given in terms of the Taketani program[57] (compare Sec. 2.2). Contributions of increasing order, which may finally be divergent, are of shorter and shorter range. Therefore, for the long and intermediate range, there is definitely only a finite number of contributions. Thus, one may have confidence in the predictions for these ranges. In the very short range part of the force, due to the quark-structure of hadrons, meson-exchange cannot be taken seriously, anyhow. For that reason, in most meson theories, one allows for a partly phenomenological treatment of the short distances by the introduction of vertex form factors, which, in a certain sense, take the extended structure of hadrons effectively into account. Fortunately, since the nuclear force is repulsive at short distances, the phenomenology of the very short range is "masked" behind a repulsive wall. Thus, one expects that, at least, for energies typical for nuclear physics, the uncertain part of the nuclear force at very short distances and the special way, in which it may be treated in a particular model, is insignificant.

For the reasons given, we follow here the conventional treatment and consider meson-exchange in perturbation theory; that is, more practically speaking, we will be dealing with Feynman diagrams. Obviously, we shall start with the lowest order: the one-boson-exchange contribution to NN scattering. (In fact, this is the only order we will deal with in this section.) The respective Feynman diagram is depicted in Fig. 3.5. Since we are working in the center of mass (CM) system of the two interacting nucleons the momenta of the two incoming particles are $\vec{q}$ and $-\vec{q}$, the outgoing $\vec{q}\,'$ and $-\vec{q}\,'$. We consider the diagram as Born contribution to a real scattering process. Thus, the nucleons are "on their mass shell" (real physical nucleons), i.e.

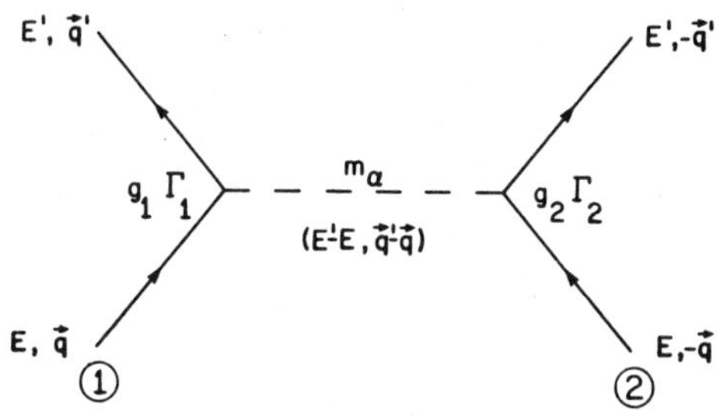

FIG. 3.5. Feynman diagram for the one-boson-exchange contribution to NN scattering considered in the CM frame. Full lines denote nucleons, the dashed line a boson with mass m_α. The underlying time axis is vertical, pointing upwards into the future.

$$E = \sqrt{M^2 + \vec{q}^{\,2}}$$

$$E' = \sqrt{M^2 + \vec{q}'^{\,2}} \; ; \qquad M \equiv \text{nucleon mass} \tag{3.10}$$

Further, the process takes place "on the energy shell," i.e. energy is conserved; consequently the energy of the nucleons before, E, and after, E', the scattering process must be the same:

$$E' = E \tag{3.11}$$

According to the "Feynman rules"[139] the depiction, Fig. 3.5, is converted into the formula:

$$\frac{g_1 \bar{u}_1(\vec{q}\,')\Gamma_1 u_1(\vec{q}) P_\alpha g_2 \bar{u}_2(-\vec{q}\,')\Gamma_2 u_2(-\vec{q})}{q^2 - m_\alpha{}^2} \tag{3.12}$$

where the left half of the numerator represents the left part of the diagram; for the right respectively.

An outgoing nucleon is represented by a Dirac spinor,* e.g. for particle 1:

$$u_1(\vec{q}) = \sqrt{\frac{E+M}{2E}} \begin{pmatrix} 1 \\ \dfrac{\vec{\sigma}_1 \cdot \vec{q}}{E+M} \end{pmatrix} \tag{3.13}$$

(Here and in the following we suppress spin-indices and spin functions.) An outgoing nucleon is represented by an adjoint Dirac spinor; again for particle 1:

---

*For notation and underived material see Ref. 139.

$$\bar{u}_1(\vec{q}\,') = u_1^\dagger(\vec{q}\,')\gamma^0$$

$$= \sqrt{\frac{E'+M}{2E'}}\left(1, \frac{\vec{\sigma}_1\cdot\vec{q}\,'}{E'+M}\right)\begin{pmatrix}1 & 0\\ 0 & -1\end{pmatrix}$$

$$= \sqrt{\frac{E'+M}{2E'}}\left(1, -\frac{\vec{\sigma}_1\cdot\vec{q}\,'}{E'+M}\right) \tag{3.14}$$

The normalization of the Dirac spinors eq. (3.13), (3.14) is:

$$u^\dagger(\vec{q})u(\vec{q}) = 1 \tag{3.15}$$

For reasons of simplicity and consistency with later chapters of these lectures, we have chosen this normalization. As a consequence Eq. (3.12) represents already the T-matrix (of NN scattering in Born approximation), and not the invariant amplitude, $\mathfrak{M}$. (The relation between them is in our case: $T(\vec{q},\vec{q}) = M^2/E^2\,\mathfrak{M}(\vec{q},\vec{q})$). The T-matrix is an adequate starting point for deriving a non-relativistic potential, which is our objective in this section (for qualitative and pedagogical considerations). For a more accurate and relativistic treatment, which is required and used in the case of reliable and quantitative work, see Sec. 4 and 5. The dashed (meson) line in Fig. 3.5 represents the propagator which in Eq. (3.12) appears as

$$\frac{P_\alpha}{q^2-m_\alpha^2} \tag{3.16}$$

where $q^2 = (E'-E)^2 - (\vec{q}\,'-\vec{q})^2 = -(\vec{q}\,'-\vec{q})^2$, (using Eq. (3.11)) is the square of the four-momentum transferred by the meson. Thus, we have for the propagator:

$$\frac{P_\alpha}{-(\vec{q}\,'-q)^2 - m_\alpha^2} \tag{3.17}$$

In simple cases the numerator of the propagator, denoted here by P_α, is just 1, so that we can forget about it. For vector boson exchange, however, it is:

$$P_\alpha = -g_{\mu\nu} + \frac{q_\mu q_\nu}{m_\alpha{}^2} \tag{3.18}$$

Since the vector bosons couple to a conserved nucleon current the second term will become zero in the actual calculations. Thus, we can use for vector bosons exchange:

$$P_V = -g_{\mu\nu} \tag{3.19}$$

The last pieces which are still unexplained in Fig. 3.5 and Eq. (3.12) are the "vertices" $g_1\Gamma_1$ and $g_2\Gamma_2$. These are obtained from the interaction Lagrangian (densities).*

In fact, logically we should have begun with the interaction Lagrangians, as they are the starting point for the development of the field theoretic perturbation theory, the lowest order result of which (for NN and excluding renormalization) is our Feynman diagram Fig. 3.5. In any case, the respective interaction Lagrangians for Fig. 3.5 are:

$$\mathcal{L}_i = g_i\bar{\psi}\Gamma_i\psi\phi_\alpha; \qquad i=1,2 \tag{3.20}$$

where $(\bar{\psi})\psi$ is the (adjoint) nucleon Dirac field and ϕ_α the meson field operator. Comparison of Eq. (3.20) with (3.12) shows in an obvious way, how to obtain the vertex from a Lagrangian in a simple case. So far, the Γ_i have been just unexplained general symbols.

Now, where do we get the explicit Lagrangians from? Here, we are allowed to use our fantasy. In fact, everything is allowed for $\mathcal{L}$, as long as we make sure, that it is hermitean and obeys certain symmetries; e.g. it should be a Lorentz scalar. In general, however, the fields themselves and/or at most, their first derivatives are used to "build" an interaction Lagrangian.** We have to know the transformation properties of the boson field, ϕ_α, since this has to be "counterbalanced" by a suitable choice for $\bar{\psi}\Gamma_i\psi$ such that the whole $\mathcal{L}$ fulfills the required symmetries. How this works in practice, we will see right now, through the example of some simple, but important cases.

3.4 Various Boson Fields and their Role in NN

In this section we want to go systematically through some of the simplest boson fields and their simplest couplings. In each case we shall consider the one-boson-exchange diagram and derive from it explicitly what it predicts for the NN interaction.

---

*We will leave out the term "densities" in what follows, though strictly speaking we will always be dealing with Lagrangian densities here.

**We do not discuss here questions of renormalizability; we are dealing with "effective" Lagrangians, anyhow, see Introduction.

3.4.1 The pseudoscalar (ps) field

Pseudoscalar means that the field, ϕ_{ps}, switches sign in the case of either a space or a time reflection. Particles with negative intrinsic parity, e.g. the π and η, have this property.* To "counterbalance" this we have to find an expression $\bar{\psi}\Gamma\psi$ (compare Eq. (3.20)) which has the same property as ϕ_{ps}, to obtain a scalar for the whole expression for the interaction Lagrangian. The simplest case with this property is** $\bar{\psi}\gamma_5\psi$. Thus

$$\mathcal{L}_{ps} = g_{ps}\bar{\psi}i\gamma_5\psi\phi_{ps} \tag{3.21}$$

(The i is needed for the hermicity, as γ_0 and γ_5 anti-commute.) The one-boson-exchange contribution, Fig. 3.5, for this interaction is according to Eq. (3.12) and (3.16):

$$g_{ps}^2 \frac{\bar{u}_1(\vec{q}\,')i\gamma_5 u_1(\vec{q})\bar{u}_2(-\vec{q}\,')i\gamma_5 u_2(-\vec{q})}{-(\vec{q}\,'-\vec{q})^2 - m_{ps}^2} \tag{3.22}$$

We will evaluate the left half of the numerator explicitly here (using Eq. (3.13) and (3.14)):

$$\bar{u}_1(\vec{q}\,')i\gamma_5 u_1(\vec{q}) = i\sqrt{\frac{(E'+M)(E+M)}{4E'E}}\left(1,\ -\frac{\vec{\sigma}_1\cdot\vec{q}\,'}{E'+M}\right)\begin{pmatrix}0 & 1\\ 1 & 0\end{pmatrix}\begin{pmatrix}1\\ \frac{\vec{\sigma}_1\cdot\vec{q}}{E+M}\end{pmatrix}$$

$$= i\sqrt{\frac{(E'+M)(E+M)}{4E'E}}\left(\frac{\vec{\sigma}_1\cdot\vec{q}}{E+M} - \frac{\vec{\sigma}_1\cdot\vec{q}\,'}{E'+M}\right)$$

$$= \frac{i}{2E}\,\vec{\sigma}_1\cdot(\vec{q}-\vec{q}\,') \tag{3.23}$$

where in the last step we applied Eq. (3.11) ("on-shell"). Repeating the analogous calculation for the right half of the numerator of Eq. (3.22) and putting everything together, we obtain for the whole diagram the following "momentum space potential":

---

*For more about this aspect and the experimental evidence see e.g. Segrè, Ref. 135, p.757.

**See Ref. 139, p.26.

$$\hat{V}_{ps}(\vec{k}) = -\frac{g_{ps}^2}{4M^2}\frac{\vec{\sigma}_1\cdot\vec{k}\;\vec{\sigma}_2\cdot\vec{k}}{\vec{k}^2+m_{ps}^2} \tag{3.24}$$

where $\vec{k} \equiv \vec{q}\,' - \vec{q}$ and the approximation, $E \approx M$, is assumed. Rewriting this as

$$\hat{V}_{ps}(\vec{k}) = -\frac{1}{12M^2}\frac{g_{ps}^2}{\vec{k}^2+m_{ps}^2}\left\{\vec{\sigma}_1\cdot\vec{\sigma}_2 + S_{12}(\hat{k})\right\} \tag{3.24a}$$

with

$$S_{12}(\hat{k}) \equiv 3\vec{\sigma}_1\cdot\hat{k}\vec{\sigma}_2\cdot\hat{k} - \vec{\sigma}_1\cdot\vec{\sigma}_2; \qquad \hat{k} \equiv \frac{\vec{k}}{|\vec{k}|} \tag{3.24b}$$

it becomes obvious that we have created a spin-spin and a tensor force. We Fourier transform into coordinate space:

$$V_{ps}(\vec{k}) = \frac{1}{(2\pi)^3}\int d^3k\; e^{-i\vec{k}\cdot\vec{x}}\; V_{ps}(\vec{k})$$

$$= +\frac{1}{(2\pi)^3}\frac{g_{ps}^2}{4M^2}(\vec{\sigma}_1\cdot\vec{\nabla})(\vec{\sigma}_2\cdot\vec{\nabla})\int d^3k e^{i\vec{k}\cdot\vec{x}}\frac{1}{\vec{k}^2+m_{ps}^2}$$

$$= \frac{g_{ps}^2}{4\pi}\frac{1}{4M^2}(\vec{\sigma}_1\cdot\vec{\nabla})(\vec{\sigma}_2\cdot\vec{\nabla})\frac{e^{-m_{ps}r}}{r}; \qquad r \equiv |\vec{x}| \tag{3.25}$$

where the integral has been solved easily by applying Cauchy's integral formula, and $\vec{\nabla}$ is the differential operator with respect to $\vec{x}$. The differentiation yields

$$(\vec{\sigma}_1\cdot\vec{\nabla})(\vec{\sigma}_2\cdot\vec{\nabla})\frac{e^{-m_{ps}r}}{r} = \frac{1}{3}\vec{\sigma}_1\cdot\vec{\sigma}_2\;\nabla^2\frac{e^{-m_{ps}r}}{r} + S_{12}(\hat{r})\frac{m_{ps}^2}{3}$$

$$\times\left(1+\frac{3}{m_{ps}r}+\frac{3}{(m_{ps}r)^2}\right)\frac{e^{-m_{ps}r}}{r} \tag{3.26}$$

and $$\vec{\nabla}^2\frac{e^{-m_{ps}r}}{r} = m_{ps}^2\frac{e^{-m_{ps}r}}{r} - 4\pi\,\delta(\vec{x}) \tag{3.27}$$

(compare Eq. (3.7), static Klein-Gordon equation). In the following we leave out the $\delta(\vec{x})$ function term, as it drops out as soon as extended source (cut-offs) are used which is customary in meson theory and physically reasonable. Thus, our final result for the coordinate space potential is:

$$V_{ps}(\vec{x}) = \frac{g_{ps}^2}{4\pi}\frac{m_{ps}^2}{12M^2}\left\{\vec{\sigma}_1\cdot\vec{\sigma}_2 + S_{12}(\hat{x})\left(1 + \frac{3}{m_{ps}r} + \frac{3}{(m_{ps}r)^2}\right)\right\}\frac{e^{-m_{ps}r}}{r} \tag{3.28}$$

with

$$S_{12}(\hat{x}) \equiv 3\vec{\sigma}_1\cdot\hat{x}\;\vec{\sigma}_2\cdot\hat{x} - \vec{\sigma}_1\cdot\vec{\sigma}_2; \qquad \hat{x} = \frac{\vec{x}}{r} \tag{3.29}$$

which is called the tensor operator. Its matrix elements are zero for singlet (S=0) states and its expectation value vanishes for S-states (L=0, spherical states). The important property is that it has non-vanishing matrix elements for triplet states of non-diagonal L.* The final balance is: a weak spin dependent central force and a strong tensor force.

The best known pseudoscalar field is the pion. There exist three charge states of the pion: +, −, neutral or with other words, its isospin is one; it is an isovector particle. In such a case the Lagrangian Eq. (3.21) is slightly extended:

$$\mathcal{L}_{ps} = g_{ps}\bar{\psi}i\gamma_5\vec{\tau}\psi\cdot\vec{\phi}_{ps} \tag{3.30}$$

where the three components of $\vec{\phi}_{ps}$ are operators in isospin space, as there are now three charged states. $\vec{\tau}$ is the usual isospin operator for isospin 1/2 particles, here the nucleons. $\vec{\tau}\cdot\vec{\phi}_{ps}$ is an invariant in isospin space. By that, the charge-independence of the interaction is guaranteed.** As a consequence, for isovector particle exchange, the Feynman diagram Eq. (3.22) and the potentials derived, Eq. (3.24) and (3.28), obtain a factor $\vec{\tau}_1\cdot\vec{\tau}_2$.

In summary let us stress the important points. We started with a boson field for which we assumed that it was pseudoscalar (equivalent to a particle with negative intrinsic parity which is observed in nature, e.g. for π, η). Consequently, we had to use the γ_5-coupling (as the simplest possibility to comply with certain indispensible symmetries). A small calculation of a few

*For more details see e.g. Ref. 138, Vol. II, p.579.

**For more details see Ref. 139, p.222.

lines then leads directly to a tensor force. In this way it is easily understood that, starting from first principles, the pion creates a tensor force.

Note also that the γ_5-coupling projects small components of the Dirac spinors onto large components, Eq. (3.23). Therefore, it is, in its analytical structure, a "weak" coupling. The reason, why the pion, nevertheless, is non-negligible, is the small mass of the pion, which strengthens the potential (note, that the meson mass squared appears in the denominator of the Feynman diagram, Eq. (3.12)). In fact, the simple rule of thumb, to roughly compare the strength of two OBE contributions of the same kind, is to consider:

$$\frac{g_\alpha^2}{m_\alpha^2} \tag{3.31}$$

From this argument it is now obvious that a heavy ps-particle leads to very small contributions. Examples are the η (m_η=549 MeV) and the η' ($m_{\eta'}$=958 MeV).

We mentioned before that for a certain field, in general, several (in principal infinitely many) couplings are possible. So, for a ps-field a derivative coupling is also commonly considered, the pseudovector (pv) coupling:

$$\mathcal{L}_{pv} = \frac{f_{ps}}{m_{ps}} \bar{\psi}\gamma_5\gamma^\mu\psi\partial_\mu\phi_{ps} \tag{3.32}$$

The resulting left vertex is

$$\Gamma_{pv} = -\frac{f_{ps}}{m_{ps}} i\gamma_5\gamma^\mu(q'-q)_\mu \tag{3.33}$$

($i\partial_\mu$ is the momentum operator; $(q'-q)_\mu$ the four-momentum of the exchanged meson.)

Application in the Feynman diagram, Eq. (3.12), leads, in the numerator, to expressions like $\gamma^\mu q_\mu u_1(\vec{q})$ and $\bar{u}_1(\vec{q}')\gamma^\mu \vec{q}'_\mu$. The Dirac equation* allows to simplify these:

$$\gamma^\mu q_\mu u_1(\vec{q}) = M u_1(\vec{q}) \tag{3.34}$$

$$\bar{u}_1(\vec{q}')\gamma^\mu q'_\mu = M\bar{u}_1(\vec{q}') \tag{3.35}$$

*See Ref. 139, p.283

With these replacements the upper left part of the Feynman diagram yields:*

$$f_{ps} \frac{2M}{m_{ps}} u_1(\vec{q}')i\gamma_5 u_1(\vec{q}) \tag{3.36}$$

When compared to Eq. (3.22) it turns out that this is exactly the same result as for ps coupling, provided we relate the coupling constants as

$$g_{ps} = f_{ps} \frac{2M}{m_{ps}} \tag{3.37}$$

In this consideration, as in all of this section, the nucleons are on their mass shell. In such a case the Dirac equations, Eq. (3.34), (3.35), apply, and we saw that, then, the ps and pv couplings are equivalent. Off-shell this is in general not true, compare Sec. 5.1

As ps and pv couplings are equivalent on-shell, we can derive our non-relativistic form of OPE also by starting from the pv coupling Eq. (3.32) and proceed as follows; let us consider only the important part of the vertex Eq. (3.33) (abbreviating $k \equiv q' - q$):

$$\begin{aligned} \tilde{\Gamma}_{pv} &= \gamma_5 \gamma^\mu k_\mu \\ &= \gamma_5 \gamma^o k_o + \gamma_5 \gamma^i k_i; \qquad i=1 \to 3 \\ &= \begin{pmatrix} 0 & 1 \\ 1 & 0 \end{pmatrix} \begin{pmatrix} 0 & \sigma^i \\ -\sigma^i & 0 \end{pmatrix} k_i \\ &= \begin{pmatrix} \vec{\sigma}\cdot k & 0 \\ 0 & -\vec{\sigma}\cdot\vec{k} \end{pmatrix} \end{aligned} \tag{3.38}$$

where we use $k_o = E' - E = 0$ (Compare Eq. (3.11)).

Non-relativistic approximation means assuming $|\vec{q}|$, $|\vec{q}'| \ll M$ and therefore neglecting the small (lower) components in the Dirac spinor Eq. (3.13); i.e.

$$u_1(\vec{q}) \approx \begin{pmatrix} 1 \\ 0 \end{pmatrix}; \qquad \bar{u}_1(\vec{q}') \approx (1,0) \tag{3.39}$$

Sandwiching the vertex Eq. (3.38) with these "Dirac" spinors and recollecting the constant factors yields:

*Note the γ_5 and γ^μ anticommute.

$$\bar{u}_1(\vec{q}\,')\Gamma_{pv}u(\vec{q}) \approx -i\,\frac{f_{ps}}{m_{ps}}\,\vec{\sigma}\cdot\vec{k} \tag{3.40}$$

a (non-relativistic) pseudoscalar in three-dimensional space. Repeating the same consideration for the right vertex (note that the momentum carries an opposite sign on the right) and taking Eq. (3.37) into account leads again to the momentum space OPE Eq. (3.24). In this way the non-relativistic character of the derivation is more obvious.

3.4.2 The Scalar (s) Field

It has the simplest interaction Lagrangian one can possibly think of (for meson nuclear coupling):

$$\mathcal{L}_s = g_s\bar{\psi}\psi\phi_s \tag{3.41}$$

The one-scalar-boson exchange contribution is:

$$g_s^2\,\frac{\bar{u}_1(\vec{q}\,')u_1(\vec{q})\bar{u}_2(-\vec{q}\,')u_2(-\vec{q})}{-(\vec{q}\,'-\vec{q})^2 - m_s{}^2} \tag{3.42}$$

Again, we evaluate explicitly the left half of the numerator:

$$\bar{u}_1(\vec{q}\,')u_1(\vec{q}) = \sqrt{\frac{(E'+M)(E+M)}{4E'E}}\left(1,\ \frac{-\vec{\sigma}_1\cdot\vec{q}\,'}{E'+M}\right)\begin{pmatrix}1\\ \dfrac{\vec{\sigma}_1\cdot\vec{q}}{E+M}\end{pmatrix}$$

$$= \sqrt{\frac{(E'+M)(E+M)}{4E'E}}\left\{1 - \frac{\vec{\sigma}_1\cdot\vec{q}\,'\vec{\sigma}_1\cdot\vec{q}}{(E'+M)(E+M)}\right\}$$

$$= \sqrt{\frac{(E'+M)(E+M)}{4E'E}}\left\{1 - \frac{\vec{q}\,'\cdot\vec{q} + i\vec{\sigma}_1\cdot(\vec{q}\,'\times\vec{q})}{(E'+M)(E+M)}\right\} \tag{3.43}$$

where we used the well-known vector identity: $\vec{\sigma}\cdot\vec{a}\vec{\sigma}\cdot\vec{b} = \vec{a}\cdot\vec{b} + i\vec{\sigma}\cdot(\vec{a}\times\vec{b})$. Defining: $\vec{k} \equiv \vec{q}\,' - \vec{q}$ and $\vec{p} \equiv 1/2(\vec{q}\,'+\vec{q})$ we obtain now for the expression Eq. (3.43):

$$\sqrt{\frac{(E'+M)(E+M)}{4E'E}}\left\{1 - \frac{\vec{p}^{\,2} - \frac{1}{4}\vec{k}^2 + i\vec{\sigma}_1\cdot(\vec{k}\times\vec{p})}{(E'+M)(E+M)}\right\} \tag{3.44}$$

Leaving out terms of second and higher order in momentum, setting $E' \approx E \approx M$ and repeating the analogous calculation on the right, the final result for the full diagram Eq. (3.42) is:

$$\hat{V}_s(\vec{k}) = -\frac{g_s^2}{\vec{k}^2+m^2}\left\{1 + \frac{\frac{1}{2}(\vec{\sigma}_1+\vec{\sigma}_2)\cdot(-i)(\vec{k}\times\vec{p})}{2M^2}\right\} \tag{3.45}$$

The first term on the right hand side is a strong attractive central force, the second a spin-orbit force. This becomes even more obvious after a Fourier transformation into $\vec{r}$-space.

For the total spin we introduce $\vec{S} \equiv 1/2(\vec{\sigma}_1+\vec{\sigma}_2)$ and in preparing the Fourier transform we consider:

$$\frac{1}{(2\pi)^3}\int d^3k\ \vec{S}\cdot(-i)(\vec{k}\times\vec{p})\ \frac{e^{i\vec{k}\cdot\vec{x}}}{\vec{k}^2+m_s^2}$$

$$= -\vec{S}\cdot(\vec{\nabla}\times\vec{p})\ \frac{1}{(2\pi)^3}\int d^3k\ \frac{e^{i\vec{k}\cdot\vec{x}}}{\vec{k}^2+m_s^2}$$

$$= -\ \vec{S}\cdot(\vec{\nabla}\times\vec{p})\ \frac{e^{-m_s r}}{4\pi r}$$

$$= \vec{S}\cdot\vec{L}\ m_s^2\left\{\frac{1}{m_s r} + \frac{1}{(m_s r)^2}\right\}\frac{e^{-m_s r}}{4\pi r} \tag{3.46}$$

with $\vec{L} \equiv \vec{x}\times\vec{p}$.

Thus, for the full r-space scalar exchange potential one obtains:

$$V_s(\vec{x}) = -\frac{g_s^2}{4\pi}\left\{1 + \vec{L}\cdot\vec{S}\ \frac{m_s^2}{2M^2}\ \frac{1}{m_s r} + \frac{1}{(m_s r)^2}\right\}\frac{e^{-m_s r}}{r} \tag{3.47}$$

Let us repeat, the scalar meson-exchange causes a strong attractive central force and a spin orbit force. From the explicit derivation we realize that the strong central force is due to the fact that the scalar coupling projects large components of the Dirac spinors on large components. The negative over-all sign is a consequence of having a second order in the coupling constant. The spin-orbit force can be traced back to the small components of the Dirac spinors. Therefore, it is a genuine relativistic effect.

Comparing with the ps coupling potential, Eq. (3.28), we see that the latter is a factor

$$\frac{m_{ps}^2}{4M^2} \tag{3.48}$$

smaller than the scalar case considered in this subsection. For the pion, the mass of which is about 1/7 M, this results into a factor 1/200. Note, however, that this result is only valid for a range in which the Yukawa is about one.

3.4.3 The Vector (v) Field

A vector boson has spin one, like a photon, and is represented by a four vector field. To form a Lorentz scalar one can couple it to another four vector, e.g. the nucleon Dirac current in analogy to the coupling of a photon to an electron:

$$\mathcal{L}_v = g_v \bar{\psi}\gamma_\mu \psi \phi_v^\mu \tag{3.49}$$

The evaluation of the one-(vector-)boson exchange is, by now, straightforward for us:

$$g_v^2 \frac{\bar{u}_1(\vec{q}\,')\gamma_\mu u_1(\vec{q})(-g^{\mu\nu})\bar{u}_2(-\vec{q}\,')\gamma_\nu u_2(-\vec{q})}{-(\vec{q}\,'-\vec{q})^2 - m_v^2} \tag{3.50}$$

We will consider the γ_0 term only:

$$\bar{u}(\vec{q}\,')\gamma_0 u(\vec{q}) = \sqrt{\frac{(E'+M)(E+M)}{4E'E}} \left(1, \frac{-\vec{\sigma}_1\cdot\vec{q}\,'}{E'+M}\right)\begin{pmatrix} 1 & 0 \\ 0 & -1 \end{pmatrix}\begin{pmatrix} 1 \\ \frac{\vec{\sigma}_1\cdot\vec{q}}{E+M} \end{pmatrix}$$

$$= \sqrt{\frac{(E'+M)(E+M)}{4E'E}} \left\{1 + \frac{\vec{\sigma}_1\cdot\vec{q}\,'\vec{\sigma}_2\cdot\vec{q}}{(E'+M)(E+M)}\right\} \tag{3.51}$$

The further calculation is pretty much the same as in the scalar case and therefore not repeated here.

$$\hat{V}_v(\vec{k}) = \frac{g_v^2}{\vec{k}^2+m_v^2}\left\{1 - 3\,\frac{\vec{S}\cdot(-i)(\vec{k}\times\vec{p})}{2M^2}\right\} \tag{3.52}$$

From the γ^0 term one obtains only a factor $-1/2M^2$ for the spin-orbit term as in the scalar case, however, the inclusion of $\vec{\gamma}$ leads to the factor $-3/2M^2$. Coordinate space potential:

$$V_v(\vec{x}) = \frac{g_v^2}{4\pi}\left\{1 - \vec{L}\cdot\vec{S}\,\frac{3m_v^2}{2M^2}\left(\frac{1}{m_v r} + \frac{1}{(m_v r)^2}\right)\right\}\frac{e^{-m_v r}}{r} \tag{3.53}$$

We find a strong repulsive central force and a spin-orbit force which has the same sign as in the scalar case, but is by a factor of 3 stronger (if $m_S = m_V$). For the coupling discussed the ω-meson is the most famous example. The repulsion is plausible by analogy with the one-photon-exchange between like charges which also causes repulsion. The "charge" in the case of nucleons is the baryon number, which 1 for nucleons and (-1) for anti-nucleons.

The detailed reasons for the strong central force and the spin-orbit force are principally the same as in the scalar case.

In analogy to the magnetic dipole coupling of the photon to the nucleon explaining the anomalous magnetic moments of the nucleon, one can in the case of the vector bosons also consider a so-called tensor coupling (not to be confused with the non-relativistic tensor force, S_{12}):

$$\mathcal{L}_t = \frac{f_V}{2M} \bar{\psi}\sigma_{\mu\nu}\psi\partial^\mu\phi_V^\nu \tag{3.54}$$

with

$$\sigma_{\mu\nu} \equiv \frac{i}{2} [\gamma_\mu,\gamma_\nu] \tag{3.55}$$

The vertex is:

$$\Gamma_{t,\nu} = -i \frac{f_V}{2M} \sigma_{\mu\nu}(q'-q)^\mu \tag{3.56}$$

The non-relativistic reduction of this coupling proceeds similiar as in the pv case considered at the end of subsection 3.4.1. Let us cast the vertex (without constant factors) into a more transparent form (note that it is a four vector):

$$\begin{aligned}
\tilde{\Gamma}_t^\nu &= \sigma^{\mu\nu}k_\mu \\
&= \left(\sigma^{io}k_i,\ \sigma^{ij}k_i\right); \qquad i,j,\ell = 1 \rightarrow 3 \\
&= \left(\sigma^{io}k_i,\ \epsilon^{ij\ell}\begin{pmatrix}\sigma^\ell & 0\\ 0 & \sigma^\ell\end{pmatrix} k_i\right) \\
&= \left(-i\begin{pmatrix}0 & \sigma^i\\ \sigma^i & 0\end{pmatrix} k_i,\ -\begin{pmatrix}(\vec{\sigma}\mathrm{x}\vec{k}) & 0\\ 0 & (\vec{\sigma}\mathrm{x}\vec{k})\end{pmatrix}\right)
\end{aligned} \tag{3.57}$$

(See the Appendix A of Ref. 139 for the details of the identities used in this derivation; also $k_o = E' - E = 0$ (on-shell) is used.) Sandwiching this vertex with the non-relativistic "Dirac spinors," Eq. (3.39), and recollecting the constant factors yields:

$$\bar{u}_1(\vec{q}\,')\Gamma_t^{\nu} u_1(\vec{q}) \approx + i \frac{f_v}{2M} \vec{\sigma} x \vec{k} \tag{3.58}$$

Note that this coupling is of the "spin-transverse" type as compared to the "spin-longitudinal" $\vec{\sigma}\cdot\vec{k}$ coupling found for ps. For the full OBE diagram one finally obtains in the approximation considered here:

$$\hat{V}_t(\vec{k}) = \frac{-f_v^2}{4M^2} \frac{(\vec{\sigma}_1 x \vec{k})\cdot(\vec{\sigma}_2 x \vec{k})}{\vec{k}^2+m_v^2} \tag{3.59}$$

Since

$$(\vec{\sigma}_1 x \hat{k})\cdot(\vec{\sigma}_2 x \hat{k}) = \vec{\sigma}_1\cdot\vec{\sigma}_2 - \vec{\sigma}_1\cdot\hat{k}\vec{\sigma}_2\hat{k} \tag{3.60}$$

the Fourier transformation into configuration space can be done with the help of the formulae we use in the case of ps-coupling, compare Eq. (3.25)-(3.27). The result is:

$$V_t(\vec{x}) = \frac{f_v^2}{4\pi}\frac{m_v^2}{12M^2}\left\{2\vec{\sigma}_1\cdot\vec{\sigma}_2 - S_{12}(\hat{x})\left(1 + \frac{3}{m_v r} + \frac{3}{(m_v r)^2}\right)\right\}\frac{e^{-m_v r}}{r} \tag{3.61}$$

It is important to stress that the tensor force obtained has the opposite sign compared to the ps case. (Therefore, the ρ-meson, which is a representative of the tensor coupling Eq. (3.54), damps the strong tensor force of the pion at short distances. Also, the ρ, like the π, is an isovector particle, so that a factor $\vec{\tau}_1\cdot\vec{\tau}_2$ should be attached to Eq. (3.61)).

It is worth mentioning that the relationship between vector mesons and the photon is, in fact, much more intimate as just by analogy.

In the "vector dominance" model for the electro-magnetic form factor for the nucleon it is assumed that the photon prevailingly couples to the nucleon by "going through" a vector boson as "mediator." As a consequence of this the strong interaction tensor couplings of the vector bosons and the anomalous moments of the nucleon are related. In fact, thinking strictly within that model the latter gives the evidence for a strong tensor coupling of vector bosons and the nucleon. For more details about this interesting topic see Ref. 141.

3.5 Summary of the Qualitative Features of Boson-Exchange in NN

The main features we deduced systematically in the previous subsections are summarized in Table 3.1.

We should finish this section by going back to its beginning. We listed and explained five important properties of the nuclear force which are strongly indicated (or even proven) by experimental finding. When we compare these five points now with Table 1, we see that in fact for each empirical property of the nuclear force one can think of at least one boson field which would offer an explanation. With other words, boson theory tenders all the qualitative features necessary to account for the empirical facts known about the nuclear force.

The fact that just the three fields discussed in this section seem to provide "all you need," is, or course, tempting. When you give in to the temptation, you have the one-boson-exchange (OBE) model. In this model one does not care too much about the question whether each boson applied would really exist. What one essentially tries to do is to exploit the favourable features of the simple one-particle-exchange as much as possible. The amazing quantitative success, which can be achieved within that simple approach, provides a posteriori the justification. This is about the philosophy of the old OBE model. However, from a more

TABLE 3.1

Various meson-nucleon couplings and their consequences for the NN interaction as deduced from the OBE contribution

Coupling	Type of Forces	
ps	spin-spin (weak)	tensor (strong)
s	attractive central (strong)	spin-orbit
v	repulsive central (strong)	spin-orbit (same sign as s)
t	spin-spin (weak)	tensor (sign opposite to ps)

Abbreviations: pseudoscalar (ps), scalar (s), vector (v), tensor (t).

Note: the signs assumed in this table for s and v refer to isoscalar bosons of that type.

fundamental point of view such a concept is not satisfactory. Therefore, we will pursue in the next section a more serious and systematic line. Still, when that line has been carried through to a successful end, we will raise the question, if a simple parametrization of the full result is possible. This will, in fact, lead us then to the OBE model, however, now, on a sound foundation. Thus, the approximations to be made and their consequences can be checked carefully and a physical interpretation of the simplified terms representing the OBE model can be given. Such OBE potentials (OBEP) will be discussed in Sec. 5. In Sec. 5.2, in which we present a coordinate-space OBEP, the contributions of the different bosons to central, tensor and spin-orbit force will be demonstrated quantitatively.

4. A SERIOUS MESON-EXCHANGE MODEL FOR THE NN-INTERACTION

After all these preliminaries of the various kinds (history, qualitative considerations etc.) we have a good physical background and (hopefully) also a strong motivation to finally dive into precise theoretical and quantitative work. That is, we want to derive, now, the NN interaction systematically step by step from meson theory and compare our results accurately with the experimental data. It is our intention to work on a basis, which is well founded by our knowledge and experience from meson and nuclear physics, and to try for an approach that is "complete" insofar as it includes all (mesonic) processes which by any reasonable consideration are relevant to the NN problem. In addition, the derivation is aimed to be reliable and unambiguous from the field theoretic point of view.

Let us state in more detail what we mean by some of these points. First and most obviously, we want to use existing mesons only (and not allow for fictitious bosons). Secondly, we will also include plural meson exchange. It will be important to take uncorrelated and correlated multi-particle exchange into account. Thirdly, according to the mass-range relation we will start at long range and then step by step include all relevant diagrams with an exchanged mass up to about the cut-off mass used in the meson-nucleon vertex functions. The use of such "cutoffs" is natural in the light of the discussion given in the Introduction. They suppress meson-exchange for small distances. In fact, originally such form factors were introduced in meson theory in a purely ad hoc way supplying a sufficient fall-off for high momenta, necessary to get a solution of the scattering equation. Nowadays, due to the quark structure of hadrons, the form factor is an in principle founded concept, being related to the hadron size. Anticipating part of our results, we obtain cut-off masses in the range of 1.2-1.5 GeV. Consequently we include diagrams up to a total exchanged mass of about 1 GeV. Finally, we will be concerned with the subtleties of field theory. Certainly, we will not use the static approximation and will take meson retardation (recoil effects) into account completely.

There are many reasons why we pursue this careful and comprehensive approach to the NN interaction. First there is a more fundamental or "intrinsic" reason: namely, one wants to know if and to which extent meson theory alone is able to provide a quantitative model for the NN interaction, the resulting vertices to be determined by QCD. Further, the field theoretic approach used, provides an unambiguously defined off-shell behaviour for the nuclear force, which is important e.g. for Bremsstrahlung and in the nuclear many body problem. The underlying formalism and the set of diagrams contributing to the NN interaction is a sound basis for the consistent determination of meson-exchange current contributions to the electromagnetic properties of nuclei (e.g. the deuteron, ^{3}He, etc.). Moreover, the explicit field theoretic description of the contributions to the NN interaction allow for a reliable evaluation of the medium effects on the nuclear force when applied in the nuclear many body problem (more about this in Part II of these lectures). Finally, the model is an excellent basis for the consideration of charge-independence and charge-symmetry breaking of the nuclear force due to the mass differences between the charge states of mesons, nucleons and isobars.[142] Also, the real part of the $N\bar{N}$ interaction (implied by the NN interaction due to G-parity) can be determined in an unambiguous way. Several of the aspects mentioned here will become more clear in the end of this section when the model is displayed explicitly.

4.1 "Formalities"

The model we are going to develop in this section is based on field theory. It is therefore appropriate to note a few details on how we will treat the field theoretic perturbation theory explicitly. For a discussion of the issue, if perturbation theory is at all adequate, see Sec. 3.3 and Sec. 4.5.

Our general scheme is to start from a field theoretic Hamiltonian H and to treat nucleons, isobars (essentially Δ) and mesons on an equal footing. Consequently the Hamiltonian does not contain a NN potential, but NN-meson and NΔ-meson vertices. These are gained from the interaction Lagrangians which we list here:

$$\mathcal{L}_{pv} = \frac{f_{ps}}{m_{ps}} \bar{\psi}\gamma_5\gamma^\mu\psi\partial_\mu\phi_{ps}$$

$$\mathcal{L}_s = g_s\bar{\psi}\psi\phi_s \qquad \text{(4.1 Cont'd)}$$

$$\mathcal{L}_V = g_V\bar{\psi}\gamma_\alpha\psi\phi_V^\alpha + \frac{f_V}{4M}\bar{\psi}\sigma_{\mu\nu}\psi(\partial^\mu\phi_V^\nu - \partial^\nu\phi_V^\mu)$$

$$\mathcal{L}_{N\Delta\pi} = \frac{f_{N\Delta\pi}}{m_\pi}\bar{\psi}\vec{T}\psi_\mu\partial^\mu\vec{\phi}_\pi + \text{h.c.}$$

$$\mathcal{L}_{N\Delta\rho} = i\,\frac{f_{N\Delta\rho}}{m_\rho}\bar{\psi}\gamma_5\gamma_\mu\vec{T}\psi_\nu(\partial^\mu\vec{\phi}_\rho^\nu - \partial^\nu\vec{\phi}_\rho^\mu) + \text{h.c.} \tag{4.1}$$

(In the first lines the well-known isospin dependence which has to be incorporated for isospin one mesons is suppressed; "h.c." stands for "hermitean conjugate.") ψ_μ denotes the field operator for the Δ-isobar (Rarita-Schwinger spinor*). More explanations can be found in Ref. 127 and 133.

We will treat H in time-ordered perturbation theory.[53,145] Since this "old fashioned" perturbation theory corresponds to standard many body theory, it will allow us to go from the two-body problem to the many body problem in a well-defined way. By that it provides the adequate framework to take medium effects consistently into account.[146] Also, in the time-ordered formalism, meson retardation (the recoil effect) is treated correctly. It turns out that fourth-order diagrams** can differ by about a factor of two, depending if meson retardation is taken into account or not. This clearly indicates the importance of recoil effects in the evaluation of e.g. the 2π-exchange.[127]

We shall leave out negative-energy intermediate states in fourth and higher order contributions. There are several arguments for this "pair suppression" (apart from the "historical" ones which we mentioned already in Sec. 2.2): It has been shown by Zuilhof and Tjon[147] in a covariant calculation using the Bethe-Salpeter equation in the ladder approximation that the contributions from anti-nucleon intermediate states are small provided pseudovector coupling is used for the NNπ vertex, which, however, is suggested as an effective coupling because of chiral invariance.[148] Furthermore, according to quark-model arguments the nucleon-antinucleon ($N\bar{N}$) vertex is considerably suppressed compared to the NN vertex.[149] The general covariance of the theory is destroyed by leaving out anti-particles. However, in NN, relativistic aspects like meson retardation are much more important than those more formal ones, as covariance.

In time-ordered perturbation theory all possible time-orderings of a diagram have to be taken into account separately.

---

*See e.g. Ref. 143, p.288; and Ref. 144.

**This refers to the order in the coupling constant.

For brevity, however, we will not always display them all. Therefore, the diagrams depicted subsequently have to be considered as an abbreviated notation, just standing for all possible time-ordering (except for Z-graphs).

The non-iterative (irreducible) diagrams contributing to the nuclear force form the kernel ("driving force") of an integral equation which determines the T-matrix of NN scattering from which the phase-shifts are obtained. With other words the irreducible contributions are iterated in the calculation to any degree. More details about the formalism can be found in Ref. 133.

4.2 The Mesons

The latest status of mesons with masses below 1350 MeV is given in Table 4.1. We list non-strange mesons only. The reason is that strange particle exchange would convert nucleons into strange baryons, since strangeness is conserved in strong interaction; therefore it is excluded from our considerations of NN.

In building now a meson theory of the NN interaction, the way not to proceed is to slam all 14 mesons listed in the table together and pray that everything works out fine. Even if it would work out, we would not know why. As we want to have a detailed physical understanding for what is going on, we will have to proceed in a more careful and reasonable way. First, we should scan the meson list, Table 4.1, critically.

Some mesons are most probably essentially $s\bar{s}$ states (s= strange quark). In the quark picture, they can couple to the nucleon by gluon exchange only (without accompanying quark exchange), which is a suppressed process ("Zweig forbidden"[151]). These arguments apply to the ϕ, which is $s\bar{s}$ to 99.9%,[152] and at least in part to the S^*, which, however, may also be a $K\bar{K}$ state. For these reasons the ϕ and S^* are omitted from further considerations of the NN problem. For other mesons there are arguments, why they contribute little to the nuclear force. E.g. the η and especially the η' are heavy pseudoscalar particles which couple by γ_5; for that reason their contribution is very small (compare Sec. 3.4.1). The δ-meson with a mass of 983 MeV has only a weak contribution because of its relatively heavy mass. Note, that the meson propagator (for zero momentum transfer) is proportional to $1/m_\alpha^2$, with m_α the meson mass. Thus, the size of a one-boson-exchange contribution goes generally down with the square of the meson mass. Therefore, the "nest" of mesons in the area of 1200 to 1300 MeV will contribute only little. In addition, the coupling constants of those mesons are rather moderate.* However, the main argument why the very heavy mesons are not very relevant in NN is related to the $\omega(783)$. The one-ω-exchange is

*See Ref. 143, p.335.

TABLE 4.1

NON-STRANGE MESONS
WITH MASSES LESS THAN 1350 MeV AND THEIR PROPERTIES*

Name	J^P	I^G	Mass (MeV)	Full Width (MeV)	Dominant Decay Mode
$\pi^{\pm}$	0^-	1^-	139.57	0	$\mu^{\pm}\nu$
π^{0}	0^-	1^-	134.96	0	$\gamma\gamma$
η	0^-	0^+	548.8	0.001	$\gamma\gamma$, $3\pi^0$
ρ	1^-	1^+	769	154	2π
ω	1^-	0^-	782.6	9.9	3π
η'	0^-	0^+	957.6	0.3	$\eta\pi\pi$
S^*	0^+	0^+	975	33	2π, $K\bar{K}$
δ	0^+	1^-	983	54	$\eta\pi$, $K\bar{K}$
ϕ	1^-	0^-	1020	4	K^+K^-
B	1^+	1^+	1234	150	$\omega\pi$
f	2^+	0^+	1274	178	2π
A_1	1^+	1^-	1275	315	$\rho\pi$
D	1^+	0^+	1283	26	$\eta\pi\pi$, 4π
ε	0^+	0^+	1300	200-600	2π
A_2	2^+	1^-	1318	110	$\rho\pi$

J≡Spin, P≡Parity, I≡Isospin, G≡G-Parity

repulsive (see Sec. 3.4.3) and has a large coupling constant ($g_\omega^2/4\pi \approx 10$-$20$). By that it masks the contributions of all heavier mesons, which are, according to the mass-range relation, of shorter range.

An additional argument for leaving out the very heavy mesons refers to the fact that the meson-nucleon form factor is in the range of 1.2 to 1.5 GeV; it does not make sense to take meson-exchange serious in a range, to which phenomenological ad hoc modifications are applied anyhow and in which the extended structure of the hadrons must come into play.

*from Ref. 150.

The final balance of the critical scan of the long list of mesons is that essentially only 3 mesons survive: π, ρ and ω. The η and δ are sometimes included in boson-exchange models, however, only for purposes of fine tuning when fitting the NN data.

From Sec. 3 we know that with these three survivor mesons we obtain the tensor force, the short-range repulsion and the LS force.* So, already on the one-boson-exchange level several essential features of the nuclear force are explained (by existing mesons). Still, one important property is missing: the intermediate range attraction. An isoscalar scalar meson with an intermediate mass does not exist. However, as we discussed before, we have to consider plural meson exchange anyhow, and we will see how that provides us with the missing parts.

In the calculation of multi-meson exchange we will proceed systematically, i.e. in the spirit of the TNS[57] program (compare Sec. 2.2). The one-pion-exchange and its properties we discussed already in length. Thus, the next step is to consider the exchange of two pions to which we will turn now.

4.3 The 2π-exchange

Our model for the 2π-exchange contribution to the NN interaction is shown in Fig. 4.1. The essential features of the model are that it takes the effects from nucleon resonances (isobars) and direct $\pi\pi$ interaction into account, which are both phenomena well-known from πN scattering and other related processes. The low-lying (J=3/2, I=3/2) Δ-resonance with a mass of 1232 MeV is of particular

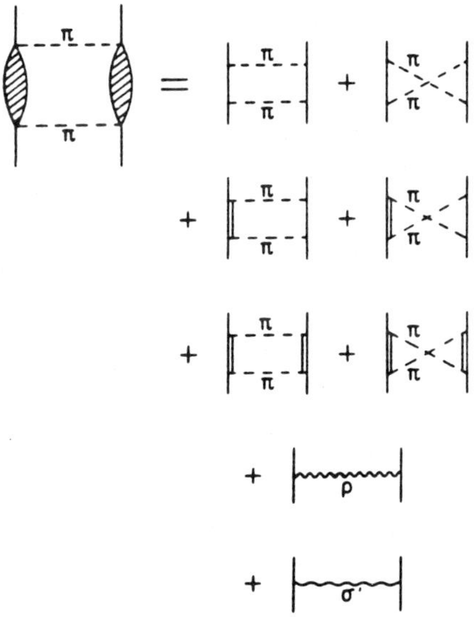

FIG. 4.1. Our model for the 2π-exchange contribution to the NN-interaction. A full line represents a nucleon and a double line a Δ(1232)-isobar. Further explanations are to be found in the text.

*The character of a particle and therefore also the character of its coupling to the nucleon is obtained from the second column in Table 4.1 (J^P): $0^+ \leftrightarrow$ scalar; $0^- \leftrightarrow$ pseudoscalar; $1^- \leftrightarrow$ vector; $1^+ \leftrightarrow$ axial vector.

importance. The effect of other resonances have been considered, too, in the literature and found to be negligible; e.g. the $P_{11}(1440)$ is considered in Ref. 153 and the $F_{15}(1688)$ in Ref. 154. Therefore, we confine ourselves to the $\Delta(1232)$ in our model.

The various crossed box diagrams, which have been left out in most field theoretic models, since they are hard to evaluate, have to be taken into account, as they are non-negligible and help to provide an isoscalar character for the 2π-exchange contribution (at least in high partial waves). An almost isoscalar character is suggested by results from dispersion theory.*

The six upper diagrams in Fig. 4.1 represent the uncorrelated 2π exchange. As mentioned before there may also be strong correlations between the two pions when "in the air". How to picture this is indicated in Fig. 4.2. If the two interacting pions are in relative P-wave, a resonance occurs: the ρ-meson. On the other hand, the strong $\pi\pi$ interaction in relative S-wave does not lead to a resonance. However, Durso et al.[153] have shown that the three diagrams on the left hand side of Fig. 4.2 may well be approximated by the exchange of a scalar isoscalar boson with a broad mass distribution, which we will denote by σ'. In the quoted work the $\pi\pi$-S-wave interaction is determined by a fit to the empirical $\pi\pi$-S-wave phase shifts. It is then applied in the NN process as indicated in Fig. 4.2. In this way a definite mass (namely, 660 MeV), full width (525 MeV) and a coupling constant of $g_{\sigma'}^2/4\pi = 13\pm3$ is obtained for σ'. We use the quoted mass distribution and a coupling constant of 10.

The alternative way to derive the 2π-exchange contribution to the NN-interaction is through dispersion theory, in which empirical information of πN- and $\pi\pi$-scattering is used in order to evaluate the amplitude $N\bar{N} \to \pi\pi$.[120,**] We perform a quantitative comparison with the results from this latter approach in higher partial waves of NN-scattering (equivalent to long and intermediate range) in which OPE and 2π are the only contributions

FIG. 4.2. Correlated 2π-exchange contributions considered in the work of Durso et al.[153] The circled S symbolizes the $\pi\pi$-S-wave interaction adjusted to empirical $\pi\pi$-S-wave scattering. Further notation as in Fig. 4.1.

---

*For a more thorough discussion of this aspect see Ref. 155.

**See also Sec. 2.3

apart from a small one-ω-exchange contribution. We further compare our model with the empirical phase shifts of NN-scattering. Both comparisons are given in Fig. 4.3.

Figure 4.3 demonstrates a close agreement of our 2π-exchange model with the empirical NN-phase-shifts as well as with the results from dispersion theory. This double agreement confirms that our model is physically most reasonable.

After this successful check of our 2π-exchange model in high partial waves, we now proceed to states of lower angular momentum (equivalent to shorter ranges), which will clearly exhibit the need for the inclusion of additional processes. Nevertheless, let us first consider the results obtained only from the contributions we have taken into account so far.* In Fig. 4.4 the dashed line (ππ 1.3) demonstrates that the 2π-exchange is too attractive in most cases. As these partial waves are sensitive to short range effects, we vary the vertex form factor (cutoff) of the pion and show the effect in that figure. (This variation has no effect in the higher partial waves considered in Fig. 4.3 before.) It is seen that the attraction can be reduced in this way (see e.g. $^1S_0$ curve "ππ 0.9"), however, a simultaneous fit of the $^1P_1$ and the $^3P_1$-phase-shift can never be achieved in this way, as a stronger cutoff lowers the $^3P_1$ and raises the $^1P_1$, although both phase-shifts had to be lowered for a closer agreement with the empirical phase-shift data. Also a consistent fit of $^1S_0$ and $^3P_1$ is obviously imposssible: When the $^1S_0$-phase-shift is correct (namely for Λ =0.9 GeV), the $^3P_1$ is still considerably too high.

Our conclusion states that a model consisting of one-meson- and 2π-exchange only is unable to describe the empirical NN-data if the concept of meson-nucleon vertex form factors is applied in a consistent way, and no other short-range contributions or phenomenological modifications are allowed.

---

*Note, that in low partial waves we iterate all contributions in the integral equation for the T-matrix; in this case the iterative 2π-exchange diagram (upper left in Fig. 4.1) is left out in the kernel.

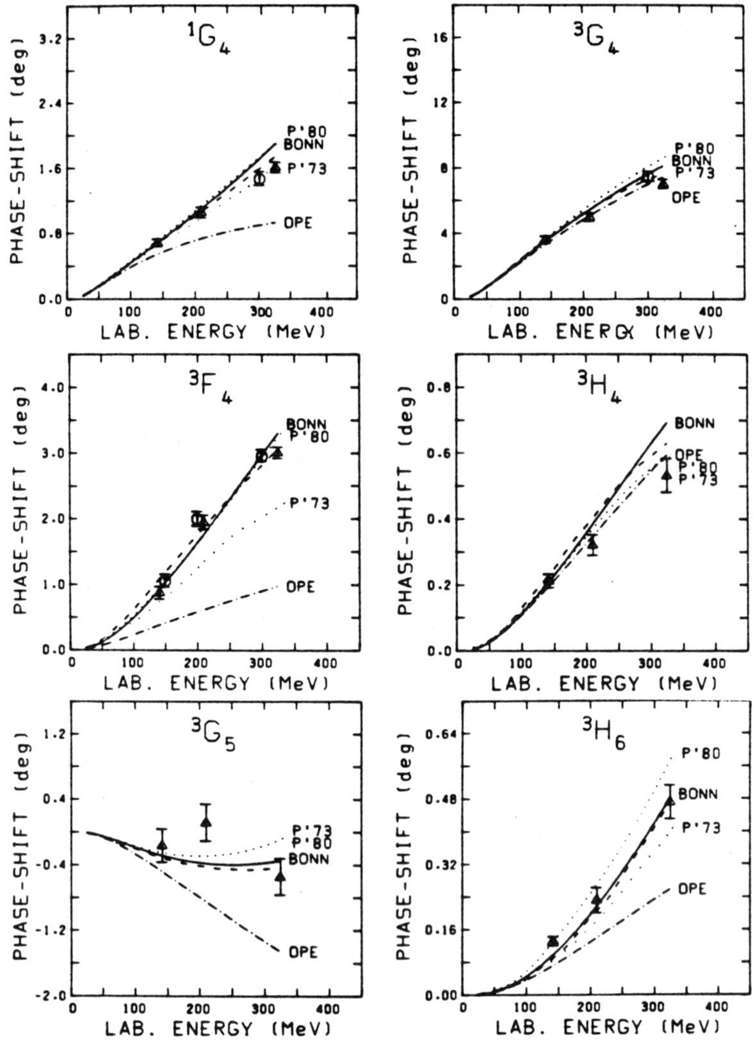

FIG. 4.3. Some higher angular momentum partial wave phase-shifts of NN-scattering. The full line labeled BONN contains all contributions from our model for the 2π exchange shown in Fig. 4.1 plus OPE and one-ω-exchange ($g_\omega^2/4\pi=5.7$). The dotted lines represent the results from dispersion theory, P'73 is taken from Ref. 156 and P'80 from Ref. 119. The dashed line is the energy-dependent analysis of Arndt et al.[137] The energy-independent analyses (error bars) are taken from Ref. 137 (octagon) and Ref. 157 (triangle). The OPE contribution is displayed in dash-dot. Note that all phase-shifts in this figure are evaluated in Born approximation as the effect of the iteration of the kernel is negligible in these high partial waves.

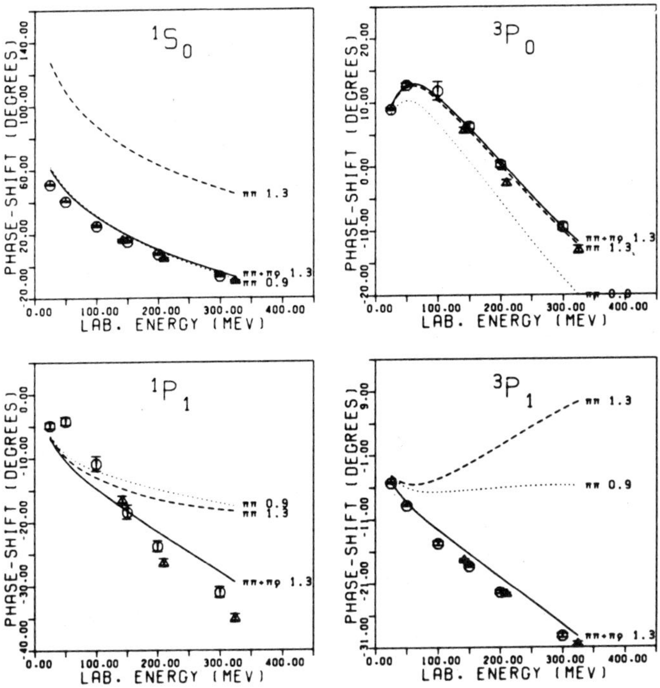

FIG. 4.4. 2π-exchange contributions to the phase-shifts of four different low angular momentum partial waves with two different choices of the cutoff-mass, Λ_π, regularizing the pion-vertex, namely Λ_π=1.3 GeV (denoted by ππ 1.3) and Λ_π=0.9 GeV (ππ 0.9). For the former case the effect of the additional πρ-contribution (compare Sec. 4.4) is also demonstrated (ππ+πρ 1.3). In addition all curves contain the one-meson-exchange of π, ω and δ.

4.4 The πρ-exchange

The failure of the model, developed so far, in low partial waves is not a disaster. In systematically and stepwise building up the nuclear force from meson-exchange, starting from long range and proceeding gradually to shorter ranges, we have up till now included only one- and two-pion exchanges "completely" (and the ω which is a 3π resonance). The next step is to take 3π-exchanges (apart form the ω) systematically into account.

One remarkable contribution of that kind is the exchange of π and ρ. We know that on the one-meson-exchange level π and ρ play the roles of opponents bearing opposite signs for the tensor force. For that same reason, it is to be expected that the πρ-contribution will in general have the opposite sign to 2π.

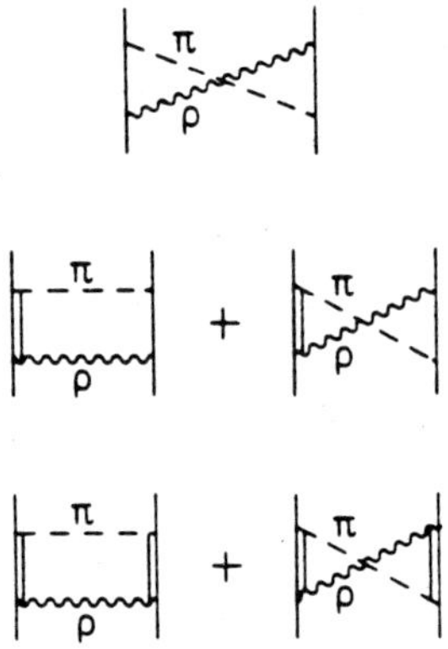

FIG. 4.5. πρ-contributions to the NN-interaction considered in this work.

Figure 4.5 displays the contributions in analogy to the diagrams of uncorrelated 2π-exchange. (Note, that only irreducible diagrams are included in the kernel as the iterative ones are generated automatically in the T-matrix equation; all stretched box diagrams are also taken into account, though they are not shown in Fig. 4.5.)

The effect of all πρ-diagrams is demonstrated in Fig. 4.4 curve "ππ+πρ 1.3" in comparison to curve "ππ 1.3." Clearly, this contribution is physically of utmost importance in quantitatively describing the low partial waves of NN scattering, or with other words for the short range part of the interaction.

The fact that the πρ-contributions are of relevance has been pointed out already in the work of Durso et al.[155] In that work, however, the πρ-contribution was suggested as an instrument to reduced the ω-coupling, which is always rather large in OBE-models compared to its SU(3) prediction:*

$$\frac{g_\omega^2}{4\pi} = 9\,\frac{g_\rho^2}{4\pi} \approx 5 \qquad (4.2)$$

Our findings are that this suggestion is not realistic for two reasons: First, the πρ-contribution sometimes varies tremendously from state to state (compare e.g. the contribution in $^3P_0$ and $^3P_1$ of Fig. 4.4) which is not the case with the ω; second, the over-attraction of the 2π-exchange in low partial waves is such that in addition to a rather large ω-coupling further repulsion is needed. Therefore, the effect of the πρ-contribution is to function as a counterpart of the 2π-exchange contributions and not to partially provide the general short range repulsion of the NN-interaction

*See e.g. Ref. 121, p.236.

which, namely, should be about equally strong in all partial waves.

At this stage of the development of our model it is instructive to ask the question what the fictitious scalar isoscalar σ-boson of the former one-boson-exchange (OBE) models stood for. The old belief that together with the ρ it replaced the sum of all correlated and uncorrelated irreducible 2π-exchange contributions is true only for high partial waves (where the πρ-contributions are negligible because of their short-range); in low partial waves the 2π-exchange contribution apart from the ρ appears by no means to be isoscalar scalar, whereas after its smoothening out by the πρ-exchanges an approximation by a one-σ-exchange appears possible, as demonstrated in Fig. 4.6.

With other words, the 2π-exchange provides also rather short-ranged contributions which are adequately counterbalanced only by the πρ-exchanges.

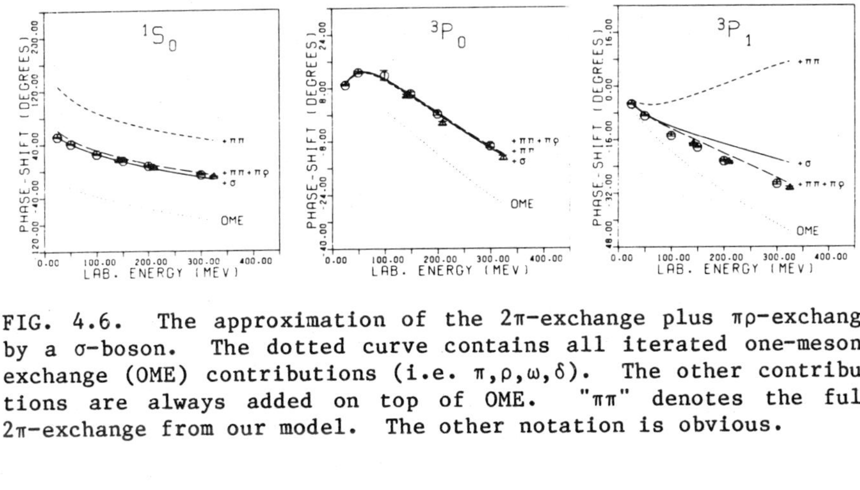

FIG. 4.6. The approximation of the 2π-exchange plus πρ-exchange by a σ-boson. The dotted curve contains all iterated one-meson-exchange (OME) contributions (i.e. π,ρ,ω,δ). The other contributions are always added on top of OME. "ππ" denotes the full 2π-exchange from our model. The other notation is obvious.

4.5 Final Results for the NN-System

In addition to the contributions discussed so far we also include further irreducible 3π- and 4π-exchanges in an approximate way. They turn out to be of little importance as they cancel each other to a large extent. This also indicates a kind of convergence with an increasing number of pion exchanges, if the diagrams are grouped in an appropriate way. For more details concerning this point see Ref. 133.

In Table 4.2 we give the meson parameters used in the final description of the NN data. We compare them to information from other sources and generally observe a close agreement. Note that not all parameters are used as fit parameters. The NNπ coupling

TABLE 4.2

MESON PARAMETERS APPLIED IN OUR MODEL AND FROM OTHER SOURCES

vertex	meson-mass m_α(MeV)	$\frac{g_\alpha^2(t=m_\alpha^2)}{4\pi}$	$g_\alpha^2(t=0)/4\pi$; (f_V/g_V)	coupling constants from other sources or comments	cutoff-mass Λ_α(GeV)	n_α
$NN\pi$	138.03	14.4[a]	14.08	14.28±0.18 πN-scattering, Ref. 158 14.52±0.40 pp forward dispersion relation, Ref. 159 NN-phase-shift analyses: 14.25 Ref. 157 (Bugg) 14.5 Ref. 137 (Arndt)	1.3	1
ρ	769.0	0.84	0.41;(6.1)	0.55±0.06 (6.1±0.6) fit to $N\bar{N}\to\pi\pi$ partial waves, Ref. 160	1.4	1
ω	782.6	20	10.6	12.0 Ref. 161 8.1±1.5 Ref. 162	1.5	1
δ	983	2.82	1.62	-	2.0	1
σ'	550	5.7	4.57	ππ-S-wave interaction, zero-width fit to the result of Ref. 153	1.7	1
$N\Delta\pi$	138.03	0.224	0.218	quark model value[b]	1.2	1
ρ	769	20.45	4.86	quark model value[c]	1.4	2

$$g_\alpha(t) \equiv g_\alpha^o \left\{\frac{\Lambda_\alpha^2-m_\alpha^2}{\Lambda_\alpha^2-t}\right\}^{n_\alpha};$$ nucleon mass: M=938.926 MeV; mass of Δ-isobar: 1232 MeV.

$$g_\alpha^o \equiv g_\alpha(t=m_\alpha^2)$$

[a] $g_\pi^2 = \left(\frac{2M}{m_\pi}\right)^2 f_\pi^2$; [b] $f_{N\Delta\pi}^2 = \frac{72}{25} f_\pi^2$; [c] $f_{N\Delta\rho}^2 = \frac{72}{25} g_\rho^2 \left(\frac{m_\rho}{2M}\right)^2 (1+f_\rho/g_\rho)^2$

(See Ref. 163 for details of the quark model derivation.)

was taken from empirical source: the $N\Delta\pi$- and $N\Delta\rho$-coupling from the quark model. The essential fit parameters are the pion- and rho-cutoff-masses and the omega- and NN-rho-vector-coupling constants as well as the δ-meson coupling. The excellent results for the NN-system are displayed in Table 4.3 for the deuteron and low energy scattering and in Fig. 4.7 for NN-scattering up to a laboratory energy of 325 MeV. It turns out that the np-phase-shifts and np-observables are described better by our model than by popular (semi-) phenomenological potentials in spite of the fact that our model has only about 1/10 of the number of free parameters of those other models. For a detailed discussion of our results we refer the interested reader to Ref. 133.

TABLE 4.3

DEUTERON AND LOW ENERGY PARAMETERS PREDICTED BY OUR MODEL (THEORY) AND FROM EXPERIMENT (EXPERIMENT).

		Theory	Experiment	
Deuteron:				
binding energy, E_B(MeV)		2.22465	2.224644±0.000046	Ref.143
D-state probability, P_D(%)		4.25	5±2	
quadrupole-moment, Q_D(fm^2)		0.281	0.2860±0.0015	Ref. 143
asymptotic S-state, A_S		0.9046	0.8846±0.0008	Ref. 65
asymptotic D/S-state, η		0.0267	0.0271±0.0004	Ref. 65
root-mean-square radius (fm)		2.0016	1.9635±0.0045	Ref. 143
$\Delta\Delta$-probability (%)		0.5		
np low energy scattering				
singlet:	a_s(fm)	-23.740	-23.748±0.010	Ref. 143
	r_s(fm)	2.766	2.75 ±0.05	
triplet:	a_t(fm)	5.427	5.424±0.004	
	r_t(fm)	1.755	1.759±0.005	

In the theoretical results quoted here the nucleonic wave function of the deuteron has been normalized to unity for simplicity. In a more refined consideration of the deuteron, Δ- and mesonic components should be separated out. They would, on the one hand, reduce the normalization of the nucleonic wave function and by that the theoretical result of most deuteron quantities cited, but, on the other hand, add meson-current contributions which will at least partly compensate the former effect.

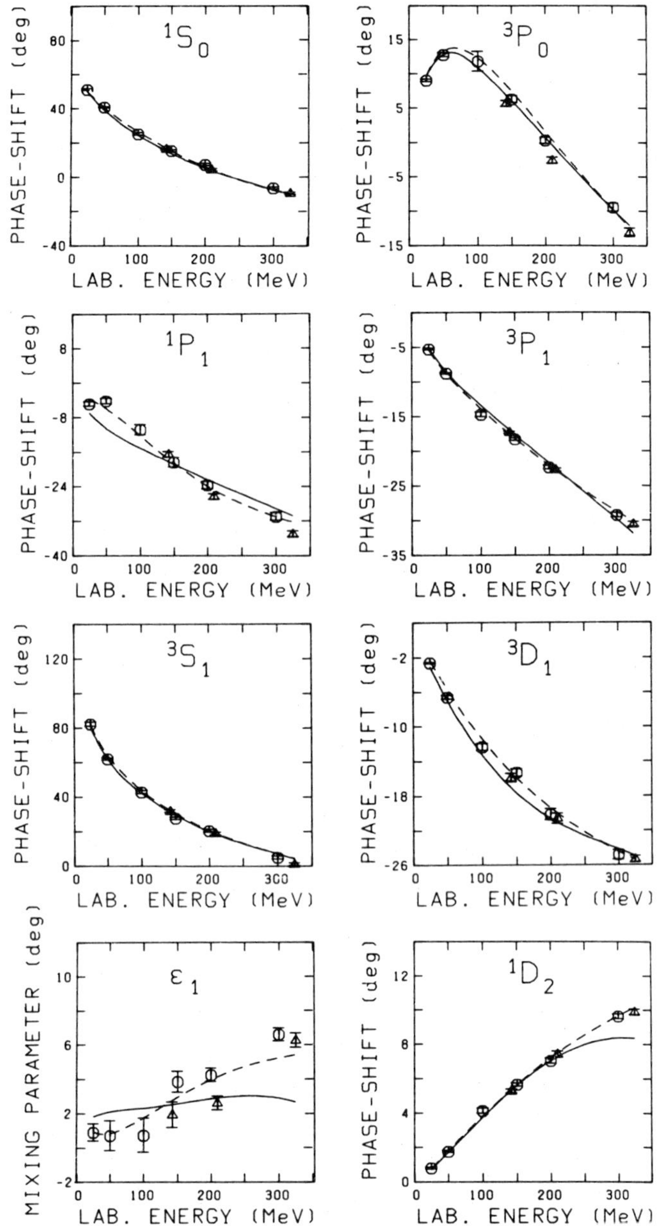

FIG. 4.7 (Continued on next two pages)

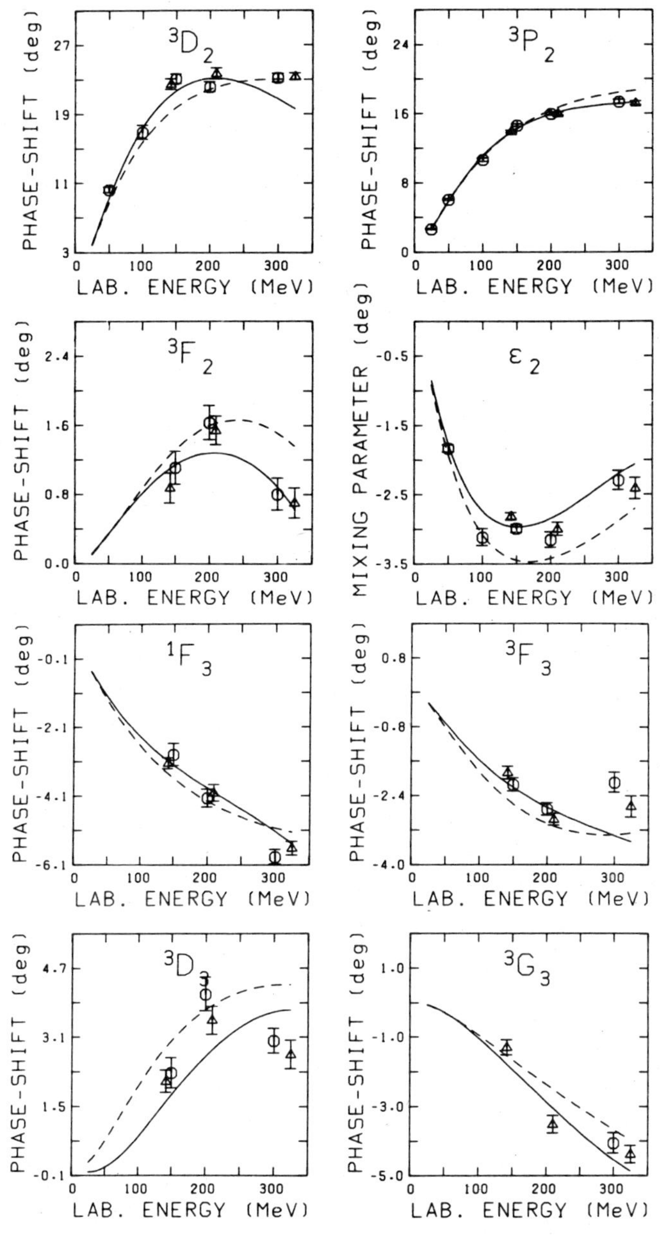

$^3D_2$
$^3P_2$
$^3F_2$
ε_2
$^1F_3$
$^3F_3$
$^3D_3$
$^3G_3$
PHASE-SHIFT (deg)
MIXING PARAMETER (deg)
LAB. ENERGY (MeV)

FIG. 4.7. (Cont'd)

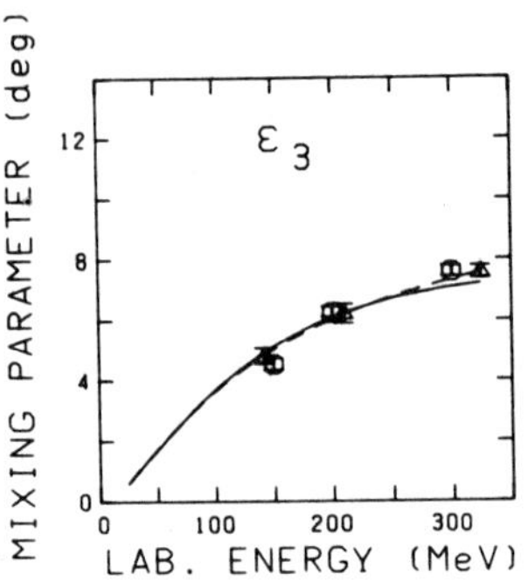

FIG. 4.7. np phase-shifts. The full line represents our predictions. The dashed line refers to the energy-dependent phase-shift analysis of Arndt et al.[137] Error bars denote the energy-dependent analyses of Bugg and coworkers[157] (triangle) and of Arndt et al.[137] (octagon).

5. THE PARAMETRIZATION OF THE NUCLEAR FORCE BY OBE-TERMS

Once one has shown that meson theory, performed consistently works quantitatively, one can think about a simple parametrization of that result, which would make applications in nuclear structure physics more easy. We will present two parametrizations by OBE terms, one relativistic presented in q-space, and one non-relativistic in r-space.

However, one should keep in mind that one has to make sacrifices to obtain a simpler version of the nuclear force. For example, the off-shell behaviour may be altered, the medium effects may be changed. But not in all applications in nuclear structure physics all subtleties of the nuclear force play a role. Therefore, there is not <u>the</u> nuclear force, there are various approximations to the nuclear force, which differ by their degree of sophistication. And depending on the problem under consideration, a more crude or more refined representation of the nuclear force is adequate.

5.1 A Relativistic OBEP in q-space

In this section we will present a relativistic one-boson-exchange potential (OBEP) in momentum space which approximates the full model discussed in Sec. 4. It is particularly suitable for relativistic nuclear structure calculations.

In most applications it is convenient if the potential is energy independent. (The time-ordered formalism in which the full model of Sec. 4 was developed, always results in an energy dependent potential.)

Traditionally relativistic energy independent potentials are obtained within the framework of so-called 3-dimensional reductions of the 4-dimensional Bethe-Salpeter equation,[103] which reads in operator notation:

$$T = \bar{V} + \bar{V}\hat{G}T \tag{5.1}$$

where T is the scattering amplitude, $\bar{V}$ the sum of all connected two-nucleon irreducible diagrams and $\hat{G}$ the relativistic two nucleon propagator.

The solution of Eq. (5.1) raises formidable mathematical and numerical problems.[107] Therefore so-called 3-dimensional relativistic reductions which can be handled more easily have been suggested. They are reviewed and discussed in the work of Woloshyn and Jackson[164] and in Ref. 121. The basic idea is to replace Eq. (5.1) by a set of coupled equations:

$$T = W + WgT \tag{5.2a}$$

$$W = \bar{V} + \bar{V}(\hat{G}-g)W \tag{5.2b}$$

where the propagator g is chosen such that Eq. (5.2a) reduces to a 3-dimensional integral equation. It is a common practice to leave out the second term on the right hand side of Eq. (5.2b) assuming that the old and new propagators are sufficiently close to keep that term small. This is an obvious desire as the inclusion of the full Eq. (5.2b) would spoil the significant simplification which is the whole purpose of the reduction.

Here, we will choose the Thompson equation[165]:

$$T(\vec{q}\,',\vec{q}) = \bar{V}(\vec{q}\,',\vec{q}) - \int d^3k\, \bar{V}(\vec{q}\,',\vec{k})\, \frac{M^2}{E_k{}^2}\, \frac{\Lambda_+^{(1)}(\vec{k})\, \Lambda_+^{(2)}(-\vec{k})}{2E_k-2E_q-i\varepsilon}\, T(\vec{k},\vec{q}) \tag{5.3}$$

with $\pm\vec{q}$ ($\pm\vec{q}\,'$) the initial (final) momenta of the interacting nucleons in the cm frame, M the nuclear mass, $E_k = \sqrt{M^2 + \vec{k}^2}$, $E_q = \sqrt{M^2 + \vec{q}^2}$, and $\Lambda_+^{(i)}(\vec{k})$ the positive energy projection operators of the i-th nucleon with momentum $\vec{k}$.

The following arguments are in favour of that equation:

(i) In the model calculations of Ref. 164 the Thompson results are the closest to those gained from the full Bethe-Salpeter equation compared to all other 3-dimensional reductions which can be cast into the form Eq. (5.2a) and are discussed in that paper. (Note that Thompson is in fact case "F" of Ref. 164 and not case "D" as stated in that work.)

(ii) Using the Thompson equation and the meson parameters of Tjon[107] we almost exactly reproduce the $^1S_0$ phase shifts which he obtains by solving the full Bethe-Salpeter equation. (This is, in fact, also true when the Blankenbecler-Sugar[123] equation is used.)

(iii) Thompson (like Blankenbecler-Sugar) does not include a retardation-like term in the meson propagator, i.e. for the exchange of a scalar boson the propagator is:

$$\frac{-1}{m_\alpha^2+(\vec{q}\,'-\vec{q})^2} \tag{5.4}$$

where the notation is explained in Fig. 3.5 and m_α denotes the mass of the exchanged boson. In the former work[104] of our group at Bonn we used to include a retardation-like term in the propagator:

$$\frac{-1}{m_\alpha^2+(\vec{q}\,'-\vec{q})^2 - (E_{q'}-E_q)^2} \tag{5.5}$$

However, the correct retardation is best considered in time-ordered perturbation theory, see Fig. 5.1, in which we have the following meson propagator:

$$\frac{-1}{\omega_\alpha(\omega_\alpha+E_{q'}+E_q-Z)} \tag{5.6}$$

with $\omega_\alpha \equiv \sqrt{m_\alpha^2 + (\vec{q}\,'-\vec{q})^2}$ the meson energy and $Z=2E_q$ the starting energy of the two interacting nucleons. (For more details see Ref. 166.) Eq. (5.6) more explicitly:

$$\frac{-1}{\sqrt{m_\alpha^2+(\vec{q}-\vec{q}\,')^2}\left(\sqrt{m_\alpha^2+(\vec{q}-\vec{q}\,')^2}+E_{q'}-E_q\right)} = \frac{-1}{m_\alpha^2+(\vec{q}-\vec{q}\,')^2+(E_{q'}-E_q)\sqrt{m_\alpha^2+(\vec{q}-\vec{q}\,')^2}} \tag{5.7}$$

As the intermediate momenta $\vec{q}\,'$ are prevailingly larger than $\vec{q}$ (e.g. in a 4th order diagram) the retardation term (last term on the right in the propagator Eq. (5.7)) is positive in most cases, whereas in Eq. (5.5) it is always negative and therefore wrong. It is just an artifact of that particular 3-dimensional reduction. For the reasons given, a meson propagator without the retardation like term of Eq. (5.5) is to be preferred.

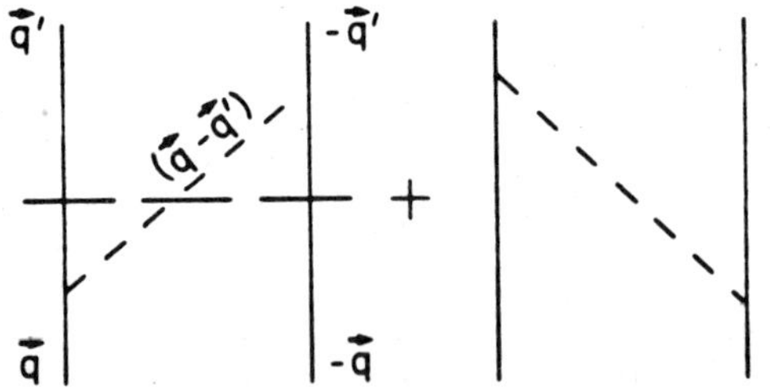

FIG. 5.1. A one-boson-exchange in time-ordered perturbation theory. The long dashed line indicates the states involved in the propagator.

Anticipating some of the nuclear matter formalism which we will introduce in Sec. 8, we want to indicate another fatal feature of the propagator Eq. (5.5). In the Dirac-Brueckner approach to nuclear matter one replaces

$$E_q \rightarrow \tilde{E}_q = \sqrt{\tilde{M}^2 + \vec{q}^{\,2}}$$

and

$$E_{q'} \rightarrow \tilde{E}_{q'} = \sqrt{\tilde{M}^2 + \vec{q}^{\,\prime 2}} \tag{5.8}$$

with $\tilde{M} < M$.

This replacement blows up the retardation-like term in Eq. (5.5) and because of its negative sign enhances the propagator. As a consequence the (attractive) second order in V is increased leading to substantially more attraction in nuclear matter. In fact, this increase of attraction is so disastrous that there is no saturation anymore. However, this is all an unphysical effect which only arises by taking all the incidental details of the propagator Eq. (5.5) too seriously.

There is a way to estimate the medium effects on the meson propagator correctly. It is again best considered in time ordered perturbation theory, Fig. 5.1 and Eq. (5.6). The replacement (5.8) applied to Eq. (5.6) weakens the propagator which leads to less attraction in nuclear matter (in contrast to the pseudo-effect discussed in the previous paragraph). This effect has been evaluated quantitatively by our group already some time ago (see Ref. 167).

In a recent paper[168] we also showed that the inclusion of the pion self-energy in the propagator more than compensates the medium effect discussed so far. Therefore we will not consider propagator effects here and use a meson propagator without retardation or retardation-like terms.

It should be noted that the arguments (ii) and (iii) given in this section also apply to the Blankenbecler-Sugar equation, which means that it could be used equally well. Therefore the final decision for Thompson was made by purely aesthetic aspects:

With our normalization of the Dirac spinors,

$$u(\vec{q})^\dagger\, u(\vec{q}) = 1, \tag{5.9}$$

which is the proper one for nuclear matter, the R-matrix version of the Thompson equation, Eq. (5.3), assumes the simple form:

$$R(\vec{q}\,',\vec{q}) = V(\vec{q}\,',\vec{q}) - \mathbb{P}\int d^3k\, \frac{V(\vec{q}\,',\vec{k})R(\vec{k},\vec{q})}{2E_k - 2E_q} \tag{5.10}$$

P denotes the principal value. This equation has exactly the form of the familiar Lippmann-Schwinger equation with the non-relativistic energies replaced by relativistic ones. For more details see Appendix A.

For the V in Eq. (5.10) we construct a relativistic OBEP with coupling constants and cutoff parameters as close as possible to the full model of Sec. 4. The diagrams of 2π and $\pi\rho$ exchange are replaced by a scalar isoscalar boson with the mass 550 MeV. The parameters of this OBEP are given in Table 5.1. The explicit momentum space formulae can be found in Appendix A. How the phase-shifts compare to the full model can be seen in Fig. 5.2.

TABLE 5.1

PARAMETERS OF A RELATIVISTIC OBEP IN THE FRAMEWORK OF THE THOMPSON EQUATION REPRESENTED IN q-SPACE ("OBEP A")

Meson	J^P	I	$\frac{g_\alpha^2}{4\pi}$	$\frac{f_V}{g_V}$	Meson Mass (MeV)	Λ_α(GeV)
π	0^-	1	14.6[a]		138.03	1.3
ρ	1^-	1	0.95	6.1	769	1.3
ω	1^-	0	20.0	0.0	782.6	1.5
δ	0^+	1	4.9973		983	1.5
η	0^-	0	3.0[a]		548.8	1.5
σ	0^+	0	7.8749		550	2.0

[a] $g_\alpha^2 = \left(\frac{2M}{m_\alpha}\right)^2 f_\alpha^2$; for π and η the pv coupling is applied.

Nucleon mass: M=938.926 MeV.

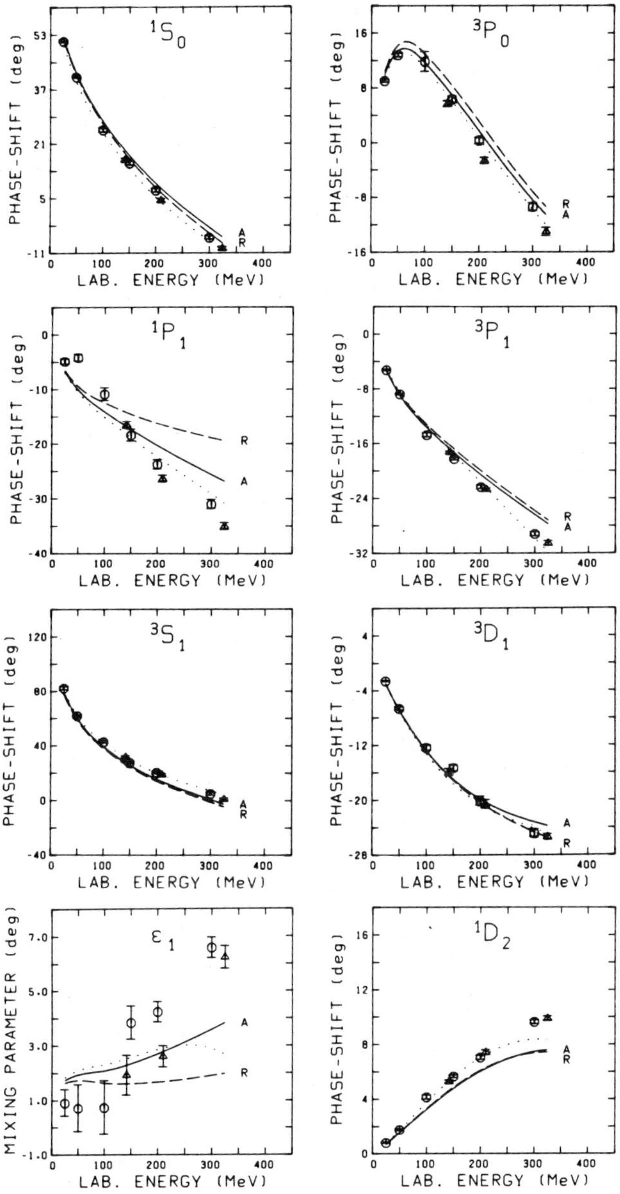

FIG. 5.2 (Continued on next page)

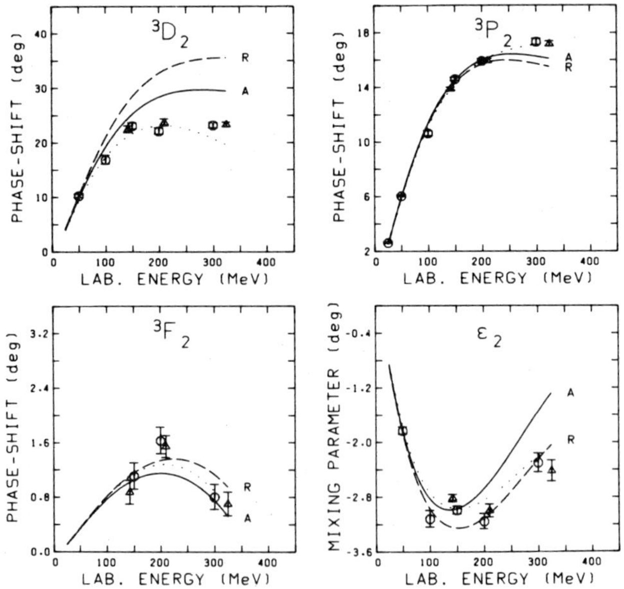

FIG. 5.2. Phase-shift predictions from two OBEPs in comparison to the results from the full model presented in Sec. 4 (dotted line). A (full line): relativistic momentum-space OBEP as discussed in Sec. 5.1 and defined in Table 5.1 and Appendix A. R (dashed line): non-relativistic r-space OBEP as discussed in Sec. 5.2 and defined in Appendix B and Table. 5.2.

5.2 A Non-Relativistic OBEP in r-Space

For purely practical reasons we present in this subsection a non-relativistic OBEP in r-space. Its derivation proceeds along the lines explained in Sec. 3. The advantages of this kind of representation is that one can consider explicitly the contributions to the central, tensor and spin-orbit force. Further, it is illustrative to explicitly see the r-dependence of the various mesonic contributions to the different terms of the potential.

The parameters are given in Table 5.2 and the explicit formulae in Appendix B. The phase-shift fit is compared to other models in Fig. 5.2. Obviously one has to make sacrifices in the quality of the fit when using the OBE approximation to the full meson theory.

The contributions of the various mesons to different parts of the force are plotted in Fig. 5.3-5.5.

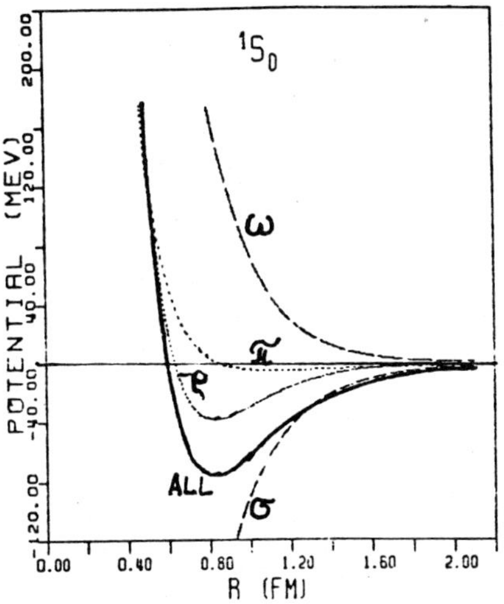

FIG. 5.3. r-space potential in the $^1S_0$ state (full line) and the contributions of the four most important mesons.

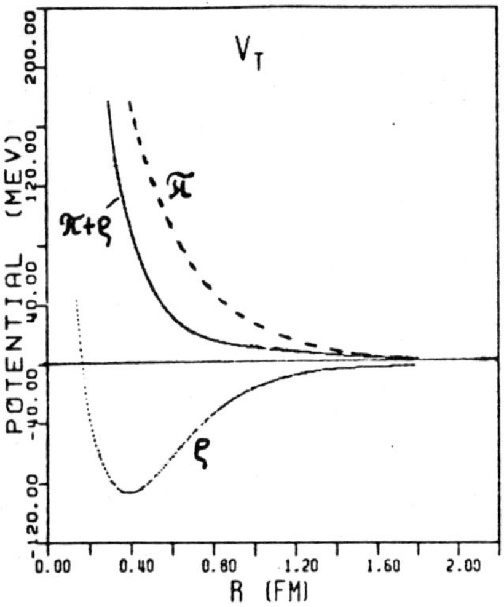

Tensor potential for π, ρ, and the sum of both.

TABLE 5.2

PARAMETERS OF A NON-RELATIVISTIC OBEP IN r-SPACE ("OBEP R")

Meson	J^P	I	$\frac{g_\alpha^2}{4\pi}$	$\frac{f_v}{g_v}$	Meson Mass (MeV)	Λ_α(GeV)
π	0^-	1	14.6		138.03	1.3
ρ	1^-	1	0.95	6.1	769.0	1.3
ω	1^-	0	20.0	0.0	782.6	1.5
δ	0^+	1	3.7064		983.0	1.5
η	0^-	0	3.0		548.8	1.5
σ	0^+	0	8.0568		550.0	1.75

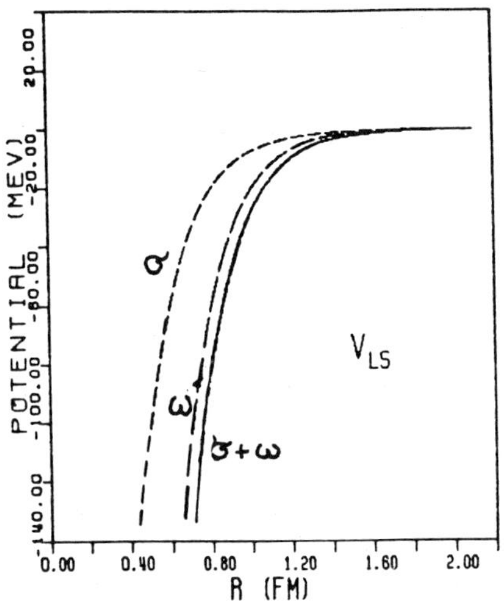

FIG. 5.5. Spin-orbit potential with the contributions from σ and ω.

6. GENERAL REMARKS ABOUT NUCLEAR MATTER

Nuclear matter is defined to be a hypothetical uniform system of nucleons of specified neutron and proton density interacting via the strong force without electromagnetic interactions. Here, we will consider only the case in which the neutron and proton densities are equal (symmetric nuclear matter). Our most precise quantitative information concerning nuclear matter comes from nuclear structure. The energy per nucleon in nuclear matter, E/N, can be deduced from the volumn term in the Bethe-Weizäcker mass formulae.[135] Semi-empirical information on the saturation energy and density of nucleons can be obtained from calculations of the charge distribution of closed shell nuclei.[169-171] Thus, nuclear matter is determined to have a binding energy per particle

$$E/N = -16.0 \pm 0.5 \text{ MeV} \tag{6.1}$$

and a saturation density

$$\rho_o = 0.16 \pm 0.015 \text{ fm}^{-3} \tag{6.2}$$

corresponding to a Fermi momentum*

$$k_F = 1.35 \pm 0.05 \text{ fm}^{-1} \tag{6.3}$$

---

* $\rho_o = \frac{2}{3\pi^2} k_F^3$

Another important quantity is the compression modulus (sometimes incorrectly called compressibility)

$$\mathcal{K} = k_F^2 \frac{d^2}{dk_F^2} E/N(k_F) \tag{6.4}$$

where the derivatives are to be taken at the equilibrium. It specifies the stiffness of nuclear matter with respect to bulk compression. Empirical information on $\mathcal{K}$ can be deduced from the systematics of the isoscalar monopole vibration, or breathing mode, in nuclei.[172] Thus, one obtains

$$\mathcal{K} \approx 210 \text{ MeV} \tag{6.5}$$

Introductions into nuclear matter theory are to be found in Ref. 173-176.

7. CONVENTIONAL BRUECKNER THEORY OF NUCLEAR MATTER

The basic quantity of Brueckener theory is the reaction matrix, G, satisfying the Brueckener-Bethe-Goldstone integral equation which reads in operator notation:

$$G = V - V \frac{Q}{e} G \tag{7.1}$$

and can be written diagrammatically as shown in Fig. 7.1. V, again, denotes the NN-potential (or more general: the kernel), Q the Pauli projection operator in nuclear matter which prevents nucleons from scattering into occupied intermediate states: 1/e is the two-nucleon propagator in the medium. Equation (7.1) is

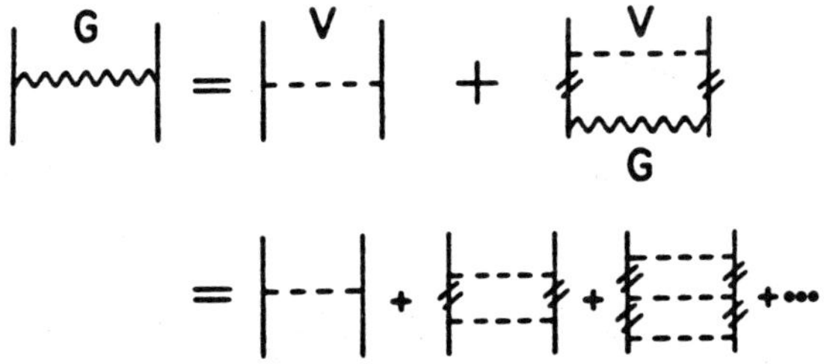

FIG. 7.1. Diagramatic representation of the Brueckner integral equation and the G-matrix. The double slash on intermediate lines indicates the change of the nucleon propagator and the Pauli-blocking in the medium.

defined in strict analogy to scattering the only difference being the Pauli-projector and the change of the nucleon propagator in the nuclear medium.

The goal of Brueckner theory is to evaluate the energy per nucleon in nuclear matter as a function of the density which is in lowest order in G:

$$E/N(k_F) = \frac{1}{N}\langle T\rangle + \frac{1}{2N}\langle G\rangle \tag{7.2}$$

where the first term on the right hand side stands for the average kinetic energy of the nucleons below the Fermi surface defined by the Fermi momentum, k_F, the second term indicates the average potential energy due to all effective two nucleon interactions. The explicit form of Eq. (7.1) is:

$$\langle \vec{k}_3\vec{k}_4|G(k_F,w)|\vec{k}_1\vec{k}_2\rangle = \langle \vec{k}_3\vec{k}_4|V|\vec{k}_1\vec{k}_2\rangle - \sum_{\vec{k}_m,\vec{k}_n} \langle \vec{k}_3\vec{k}_4|V|\vec{k}_m\vec{k}_n\rangle \cdot \frac{Q(k_F,\vec{k}_m,\vec{k}_n)}{\varepsilon(\vec{k}_m)+\varepsilon(\vec{k}_n)-w} \langle \vec{k}_m\vec{k}_n|G(k_F,w)|\vec{k}_1\vec{k}_2\rangle \tag{7.3}$$

where spin and isospin indices as well as the exchange terms, which we do include, have been suppressed. The single-particle energy of a nucleon in nuclear matter is defined by:

$$\varepsilon(\vec{k}_i) = T(\vec{k}_i) + U(\vec{k}_i) \tag{7.4}$$

with

$$U(\vec{k}_i) = \sum_{|k_j|<k_F} \langle \vec{k}_i\vec{k}_j|G[k_F,w=\epsilon(\vec{k}_i)+\epsilon(\vec{k}_j)]|\vec{k}_i\vec{k}_j-\vec{k}_j\vec{k}_i\rangle \tag{7.5}$$

Equation (7.2) is explicitly:

$$\frac{E}{N}(k_F) = \frac{1}{N}\sum_{|\vec{k}|<k_F} \langle \vec{k}|T|\vec{k}\rangle + \frac{1}{2N}\sum_{|\vec{k}_1|,|\vec{k}_2|<k_F} \langle \vec{k}_1\vec{k}_2|G[k_F,w=\epsilon(\vec{k}_1)+\epsilon(\vec{k}_2)]|\vec{k}_1\vec{k}_2-\vec{k}_2\vec{k}_1\rangle \tag{7.6}$$

where T denotes the kinetic-energy operator.

In conventional Brueckner theory non-relativistic energies are used, i.e. $T(\vec{k}_i) = \vec{k}_i^{\,2}/2M$ (in analogy to the non-relativistic

Lippmann-Schwinger equation of scattering) and any phenomenological NN-potential (i.e. Reid[86]) is suitable to be used for V.

In the first attempts for a relativistic extension of Brueckner theory (which we will subsequently call: conventional relativistic Brueckner theory) Eq. (7.1) was applied with the relativistic kinematical factors and energies which occured in the analogous scattering equation (e.g. Blankenbecler-Sugar or Thompson, compare Sec. 5.1.) Phenomenological potentials are in general unsuitable for this approach as at least the most popular ones like Reid[86] or Paris,[119] are defined for the non-relativistic Schrödinger equation. Therefore OBEPs as well as more comprehensive meson-exchange models have been used at this stage. As we know, an OBE consists of the sum of single meson-exchanges in which the nucleons are naturally represented by four-component free Dirac spinors, $u(\vec{q})$, i.e. the spinors satisfy the free Dirac equation:

$$(\not{q}-M)u(\vec{q}) = 0 \tag{7.7}$$

Within the scheme of conventional relativistic Brueckner theory unchanged free spinors are taken over into the nuclear matter calculation. Therefore the potential applied to nuclear matter is exactly the same as in NN-scattering and that is why the results from this first attempt of a relativistic approach were not substantially different from those of non-relativistic Brueckner theory.

Figure 7.2 contains a survey of results obtained in the two approaches discussed so far. Only the saturation minima of the saturation curves, $E/N(k_F)$, are given. The squares stand for calculations using a free spectrum for nucleons above the Fermi surface ("gap"). The difference in the results displayed in Fig. 7.2 is mainly due to a different strength of the tensor force contained in the NN-potentials applied. This strength is best measured by the %-D-state, P_D, in the deuteron which the potentials give rise to. For Hamada-Johnston[84] it is 6.97% (the highest point in the plot), for HM2[106] we have P_D=4.32% (the lowest point).

The circles indicate results in which a continuous choice is applied for the single particle spectrum of the nucleons.[177,178] These results are generally 4-5 MeV more attractive compared to gap-calculation and appear to simulate the 3- and 4-body correlation.[179,180] Finally, Δ(1232)-isobars and their special medium effects are included in results symbolized by a small triangle.

It is easily seen that each type of calculation has its own "Coester line" (so-called after Ref. 181). Though improvements with respect to the empirical area can be observed when extensions within the conventional scheme are applied (continuous choice,

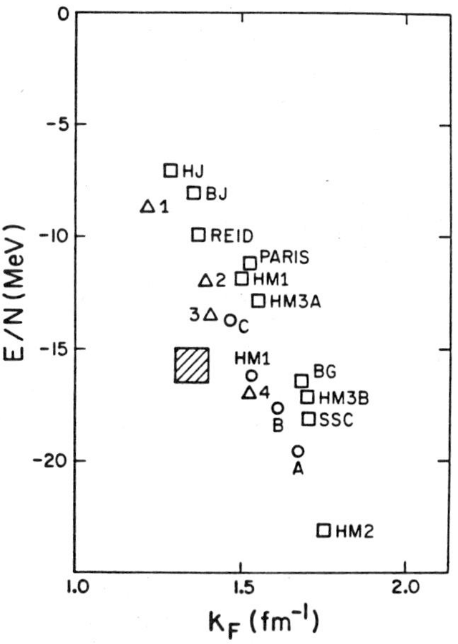

FIG. 7.2. Energy per nucleon in nuclear matter, E/N, as a function of the Fermi momentum k_F. The shaded square represents the empirical nuclear matter saturation. A brief history of results obtained in nuclear matter Brueckner theory is given by the small squares, circles and triangles which indicate the saturation minima of different conventional approaches explained in the text. The abbreviations for the potentials applied are given below: HJ: Hamada-Johston[84]; BJ: Bethe-Johnson[182]; REID: Reid-soft-core-potential[86]; PARIS: Ref. 119; HM1; Holinde-Machleidt(1)[105]; HM3A, HM3B: Ref. 183; BG: Bryan-Gersten[184]; SSC: Sprung-de Tourreil super-soft-core potential C[185]; HM2: Holinde-Machleidt(2)[106]; Δ1,2: Ref. 186; Δ3: Ref. 187; Δ4: Ref. 188.

Δ's), all bands of results clearly miss the empirical range. This summarized a long standing problem in nuclear matter theory.

8. THE DIRAC-BRUECKNER APPROACH TO NUCLEAR MATTER

8.1 The Idea and the Formalism

In this section we will explain a special extension of conventional relativistic Brueckner theory, subsequently called the Dirac-Brueckner approach.

The basic idea of this new approach (first introduced by Shakin and coworkers[189]) is that, as in the mean field theory,[190] one realizes that the nucleons in nuclear matter are exposed to a strong common scalar and vector field and therefore by no means free particles. Consequently free Dirac spinors satisfying the free Dirac Eq. (7.7) cannot be an adequate representation of a nucleon in nuclear matter. Instead, spinors obtained in a Dirac equation containing the strong common potentials should be used:

$$(\not{k}-M-\Sigma)u(\vec{k}) = 0 \tag{8.1}$$

with

$$\Sigma = A(\vec{k}) + \gamma^0 B(\vec{k}) \tag{8.2}$$

the self-energy operator in nuclear matter. $A(\vec{k})$ and $B(\vec{k})$ represent the scalar and vector potential in nuclear matter respectively.

As it turns out that the k-dependence of A and B is very weak (namely, when expanding $A(\vec{k}) = A_0 + A_1(\vec{k}^2/k_F^2)$ and $B(\vec{k}) = B_0 + B_1(\vec{k}^2/k_F^2)$ one gets: $A_1/A_0 \approx B_1/B_0 \approx 0.05$), they can be assumed to be constant to a good approximation.

With $\tilde{M} \equiv M + A$ and $\tilde{E} \equiv \sqrt{\tilde{M}^2+\vec{k}^2}$ the Dirac spinor satisfying Eq. (8.1) is simply:

$$\tilde{u}(\vec{k},s) = \sqrt{\frac{\tilde{E}+\tilde{M}}{2\tilde{E}}}\begin{pmatrix} 1 \\ \dfrac{\vec{\sigma}\cdot\vec{k}}{\tilde{E}+\tilde{M}} \end{pmatrix} \chi_s \tag{8.3}$$

where the normalization $\tilde{u}^\dagger\tilde{u}=1$, appropriate for nuclear structure calculations, is applied, as throughout this work.

The self-energy is given by Eq. (7.5) with

$$\Sigma(\vec{k}) = \bar{\tilde{u}}(\vec{k})\Sigma\,\tilde{u}(\vec{k}) = U(\vec{k}) \tag{8.4}$$

As Σ depends on the G-matrix (compare Eq. (7.5)) and the G-matrix depends on Σ via the single particle energies and the Dirac spinors defined in Eq. (8.1) a self-consistency is required.

Formally all formulae of conventional Brueckner theory Eq. (7.3)-(7.6), still apply; however, their explicit meaning is now more refined, namely:

$$|\vec{k}\rangle = \tilde{u}(\vec{k})$$

$$\langle\vec{k}| = \bar{\tilde{u}}(\vec{k}) \tag{8.5}$$

(This means that the potential V in Eq. (7.3) is obtained from that used in free scattering (which is given in Appendix A) by replacing: $M \to \tilde{M}$ and $E \to \tilde{E}$.)

$$T = \vec{\gamma}\cdot\vec{k} + M \tag{8.6}$$

and a term -M should be added to Eq. (7.6).

$$\varepsilon(\vec{k}) = \langle\vec{k}|T|\vec{k}\rangle + \langle\vec{k}|\textstyle\sum|\vec{k}\rangle$$

$$= \frac{M\tilde{M}+\vec{k}^2}{\tilde{E}} + \frac{\tilde{M}}{\tilde{E}}A + B$$

$$= \tilde{E} + B \tag{8.7}$$

Assuming A and B constant, they can be determined once $\varepsilon(\vec{k})$ has been evaluated for two different $\vec{k}$. $k=(1/2)k_F$ and $k=k_F$ is an appropriate choice ($k\equiv|\vec{k}|$). The explicit formulae for nuclear matter Brueckner theory are given in more detail in Appendix C.

Special care has to be taken with the ρ-exchange potential being derived from the Lagrangian:

$$\mathcal{L}_{NN\rho} = g_\rho\bar{\psi}\gamma^\mu\psi\phi_\mu + \frac{f_\rho}{4M}\bar{\psi}\sigma^{\mu\nu}\psi(\partial_\mu\phi_\nu-\partial_\nu\phi_\mu) \tag{8.8}$$

As $f_\rho/4M$ is a coupling constant which could as well be defined by $f'_\rho/4m_\rho$, where m_ρ denotes the mass of the ρ-meson, the M in Eq. (8.8) must not be replaced by $\tilde{M}$. If that was done incorrectly, there would be no saturation in nuclear matter, as we will see later.

8.2 Former Calculations

The first work within the approach outlined in the previous subsection was done by Shakin and coworkers.[189] However, they take the new effect only in first order perturbation theory into account. Single particle energies and wave functions (Dirac spinors) in the medium are not determined selfconsistently, though this is a substantial requirement in Brueckner theory. Further, outdated one-meson-exchange potentials were used and a 3-dimensional relativistic reduction of the Bethe-Salpeter equation which, if applied consistently, in the Dirac approach, results in a disaster in nuclear matter due to an unphysical retardation term in the meson propagators (compare Eq. (5.5) and subsequent discussion). Also, the pseudo-vector coupling of the pion to the nucleon is required for the Dirac approach; however, the potentials applied by the Brooklyn group were constructed and fitted to the NN data using the pseudoscalar coupling. For all these reasons the results are not conclusive.

Horowitz and Serot[191] have recently and independently from us[192] demonstrated how to perform the Brueckner selfconsistency in the Dirac approach both correctly and elegantly. As they intend to show relative effects only, they do not use a quantitative nuclear force in their calculations.

The work quoted so far leaves open the question as to what the predictions of the Dirac-Brueckner approach to nuclear matter are when performed consistently and correctly.

Further, all past work applied only OBEP, in which the fictitious σ-boson causes a large relativistic medium effect. Therefore the critical question to be asked is: Will the relativistic saturation effects survive when the fictitious term is omitted and realistic and explicit 2π- and πρ-exchanges introduced instead?

8.3 Results for the Full Meson-Exchange Model

All questions raised in the last subsection can be answered by performing a fully selfconsistent Dirac-Brueckner calculation as outlined in Sec. 8.1 with our full quantitative meson-exchange model introduced in Sec. 4.

This is what we will do in this subsection. To keep the computing time within responsible limits we have to do some minor simplifications. We use box-diagrams only and include the crossed 2π-exchange diagrams with intermediate Δ-isobars effectively by increasing the NΔπ-coupling constant from $f^2_{N\Delta\pi}/4\pi$ = 0.23 to 0.36. We make sure that the absolute size of the uncorrelated and correlated 2π-exchange is kept in agreement with the original model and that the fit of the NN-phase-shifts is retained.

The result is displayed in Fig. 8.1 by the full line labelled FULL MODEL. The empirical nuclear matter saturation properties are obviously described quantitatively by our result. The compressibility is 250 MeV. This is a very satisfactory result, since in the mean field theory[190] 550 MeV are obtained for $\mathcal{K}$.

For the particular examinations in this subsection we also construct an OBEP using time-ordered perturbation theory, which is applied in the full model, in order to compare the nuclear matter results gained with an OBEP with those from the full meson exchange model. The objective is to find out if the relativistic effect is the same in both cases. As the relativistic effect is also off-shell dependent the same type of perturbation theory has to be applied for both models. Therefore, the OBEP presented in Sec. 5.1 (OBEPA) is not suitable for this direct comparison.

The results with this OBEP in time ordered perturbation theory are also shown in Fig. 8.1 (dash-dot curves). It is seen that the OBEP does indeed have about the same "Dirac effect" as

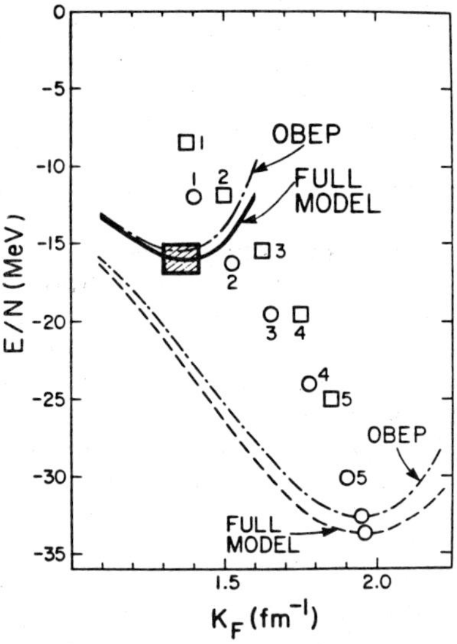

FIG. 8.1. Energy per nucleon, E/N, in nuclear matter versus the Fermi momentum, k_F. The full curve shows the result for the relativistic Dirac-Brueckner calculation using the model described in the text. The dashed curve is obtained with that same model in conventional Brueckner theory. The dash-dot curves refer to the respective calculations using the OBE-approximation to the full model in time-ordered perturbation theory. The squares and circles denote the saturation minima of conventional Brueckner calculations of the past using various different models for the NN-interaction. Squares are used when, for the particle spectrum above the fermi surface, free nucleon energies were used. Circles with the same number as a square refer to the corresponding calculations using a continuous single particle spectrum. The shaded rectangle denotes the empirical range of nuclear matter saturation.

the full model. More details about this comparison are to be found in Ref. 193.

After this result we feel justified to further examine the relativistic effect within the OBE model.

As mentioned before, it is more practical to work with energy independent potentials in nuclear structure. Therefore, we constructed in Sec. 5.1 a relativistic energy-independent OBEP. (The OBEP used in this subsection for reasons of performing an adequate comparison is energy-dependent since it is based on time-ordered perturbation theory.) The OBEP of Sec. 5.1 will be applied to nuclear matter in the next subsection.

8.4 Results with (Energy Independent) Relativistic OBEPs

The results for the Dirac-Brueckner approach to nuclear matter using the relativistic energy-independent OBEP presented in Sec. 5.1 ("OBEPA") are displayed in Fig. 8.2 by the full line labelled "A". The full curves labelled "B" and "C" refer to such calculations using variations of OBEPA with an increased tensor force (by using a larger cutoff mass for the πNN vertex).

The reason why we have performed the calculations with three different OBEPs is that the characteristic results for a certain type of many-body approach is not a point in the energy versus density plot; it is a band (a Coester band). There is not <u>the</u> nuclear force, there are several potentials, which describe the NN data equally well and never-the-less differ, namely, essentially in the strength of the tensor force. This difference leads to characteristic variations in the nuclear matter results, which always have a Coester-band structure.

Therefore, the reasonable question to be asked is if the band characteristic for one theory is oriented such that it would pass through the empirical area. Obviously our new approach provides additional strongly density-dependent repulsion such that the empiricial result can be met.

On the background of the failure of the many desperate attempts of the past to explain the empirical nuclear matter saturation, our result is not trivial.

The role of the various mesons in the relativistic saturation mechanism is demonstrated in Fig. 8.3. The effect of ω, σ and π is about equally strong, the π acting mainly through its second order contribution

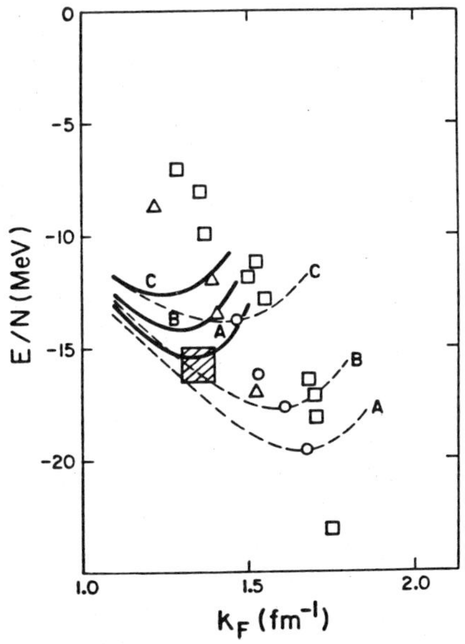

FIG. 8.2. Same as Fig. 7.2, but including the results of this subsection. For curves labeled "A" the OBEP presented in Sec. 5.1 and Table 5.1 is applied. Label "B" and "C" refers to two variations of that OBEP with an increased tensor force. The full lines denote Dirac-Brueckner results, the dashed curves stand for calculations with free spinors ("conventional relativistic Brueckner theory").

to the $^3S_1$-state. The ρ and especially the η and δ have very little effect. The consequence of using the wrong treatment of ρ-exchange and of the ps-coupling for the π are also demonstrated in Fig. 8.3. Both lead to unphysical results showing no saturation in nuclear matter.

The constant part of the scalar potential, A_0, and of the vector potential, B_0, are displayed versus the Fermi-momentum, k_F, in Fig. 8.4 and 8.5 respectively. Generally we find a smoother density-dependence of these quantities compared to other authors.

More details about these calculations and more results are given in Ref. 194.

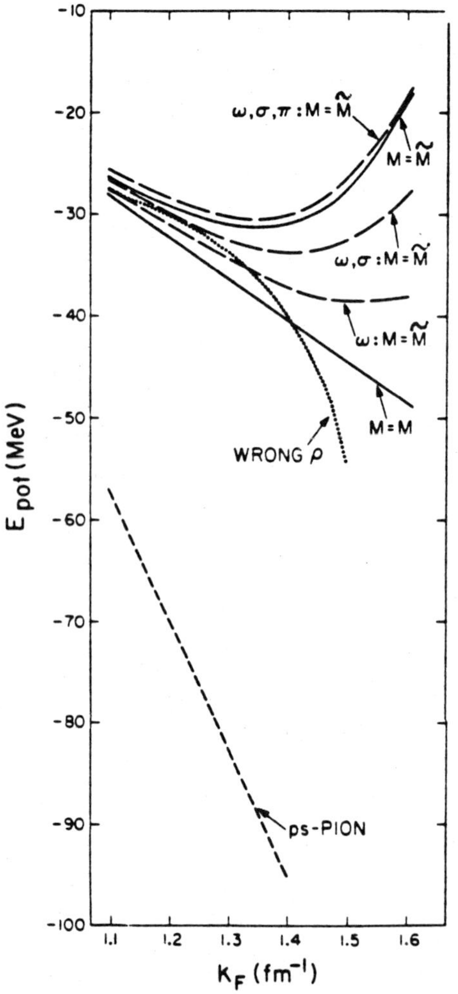

FIG. 8.3. "Potential energy," E_{pot}, (i.e. second term on r.h.s. of Eq. (7.6)) versus Fermi momentum, k_F, for the present OBEP. Starting with the conventional result labeled "M=M" the single mesons are switched successively on to the Dirac-Brueckner approach indicated by "$M=\tilde{M}$." The dotted curve shows the result when the M occurring in the Lagrangian for the ρNN-interaction, Eq. (8.8), is incorrectly replaced by $\tilde{M}$. The short dashed curve is obtained by using the ps-coupling for the pion.

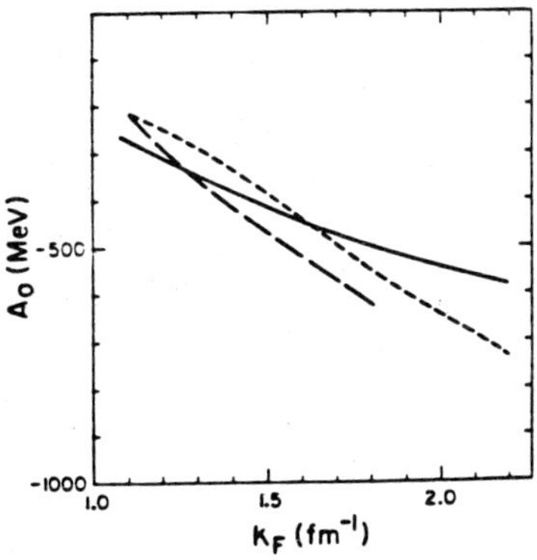

FIG. 8.4. Constant part of the scalar potential, A_0, versus Fermi-momentum, k_F. The full line displays the results using the present OBEP. The long and short dashed curve are from Ref. 189 and Ref. 190 respectively.

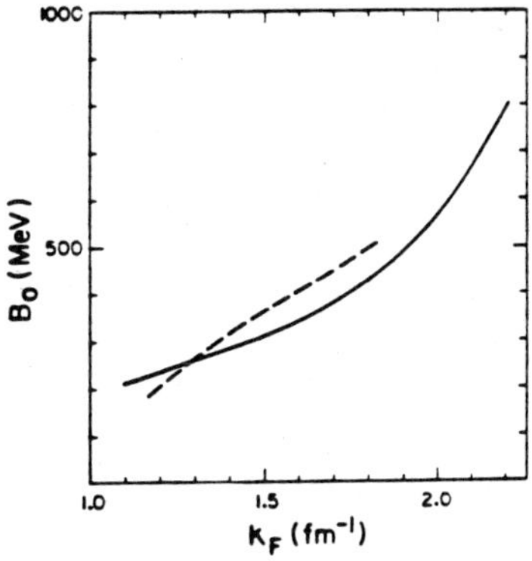

FIG. 8.5. Constant part of the vector potential, B_0, versus Fermi-momentum, k_F. The full line is obtained from the present OBEP; the dashed from the work of Ref. 189.

9. SUMMARY, CONCLUSIONS AND OUTLOOK

Briefly summarizing we want to state:

(i) Meson theory performed consistently and without fictitious terms is a quantitative theory for the NN-interaction. The necessity for the introduction of additional degrees of freedom is not indicated up to $E_{LAB} \approx 300$ MeV. The work presented is the first clean check of these questions.

(ii) The comprehensive field theoretic model for the NN interaction developed in this work presents a sound basis for studying several important issues in nuclear physics, like meson-current corrections, medium modifications of the nuclear force and others.

(iii) The empirical nuclear matter saturation can be explained quantitatively within a relativistically extended Brueckner theory applying a realistic and comprehensive meson-exchange model for the nuclear force. This is a parameter free calculation.

(iv) The contributions to the NN-interaction stemming from 2π- and $\pi\rho$-exchanges can well be approximated by a suitably chosen σ-boson. Even the relativistic medium effects of those explicit diagrams are simulated well by the σ. This justifies the use of OBEP in future relativistic nuclear structure calculations.

(v) Thus, the relationship between the free NN-interaction and nuclear structure in a relativistic framework is successfully established.

An important open question is the contributions from 3- and 4-body correlations in nuclear matter within the Dirac-Brueckner approach.

On the basis of the success of the Dirac-Brueckner approach to nuclear matter, future work should be devoted to the following topics: a) nucleon-nucleus scattering and b) structure of finite nuclei in a relativistic approach based on forces tested successfully in nuclear matter. Moreover, it is worthwhile to check the relativistic approach in other subfields of nuclear structure physics.

ACKNOWLEDGEMENT

The author gratefully acknowledges the collaboration with K. Holinde, R. Brockmann and C. Elster. Many thanks are due to Byron Jennings, Walter Wilcox and Leslie Williams for advice on several chapters. Finally, I thank Denise Mason for converting the hieroglyphs of the manuscript into an aesthetical typescript. It is a pleasure to thank TRIUMF for the warm hospitality and the support in the years 1984 and 1985, in which part of the work was done.

APPENDIX A: FORMULAE FOR THE RELATIVISTIC MOMENTUM SPACE OBEP AND THE RELATED TWO-BODY EQUATIONS[194]

Defining helicity states $|\lambda\rangle$ by

$$\frac{\vec{\sigma}\cdot\vec{q}}{2|\vec{q}|}\,|\lambda\rangle = \lambda|\lambda\rangle \tag{A1}$$

and abbreviating $\lambda = \pm 1/2$ by $\pm$, we consider six independent NN amplitudes:

$$\begin{aligned}
V_1^J &\equiv \langle ++|V^J(q',q)|++\rangle \\
V_2^J &\equiv \langle ++|V^J(q',q)|--\rangle \\
V_3^J &\equiv \langle +-|V^J(q',q)|+-\rangle \\
V_4^J &\equiv \langle +-|V^J(q',q)|-+\rangle \\
V_5^J &\equiv \langle ++|V^J(q',q)|+-\rangle \\
V_6^J &\equiv \langle +-|V^J(q',q)|++\rangle
\end{aligned} \tag{A2}$$

where the arguments (q',q) have been suppressed on the left hand side of all six equations. For convenience, we will later use the following combinations of helicity amplitudes:

$$\begin{aligned}
{}^{0}V^J &\equiv V_1^J - V_2^J \\
{}^{1}V^J &\equiv V_3^J - V_4^J \\
{}^{12}V^J &\equiv V_1^J + V_2^J \\
{}^{34}V^J &\equiv V_3^J + V_4^J \\
{}^{55}V^J &\equiv 2V_5^J \\
{}^{66}V^J &\equiv 2V_6^J
\end{aligned} \tag{A3}$$

We will represent the OBE contributions in these partial wave helicity amplitudes, the principal derivation of which is described in the review article by Erkelenz.* Therefore, we will not repeat it here. Note, however, that in the present work there are a few small, but decisive changes with regard to the quoted reference, namely in the present work

---

*Ref. 104, p.197-208.

(i) the meson propagator is (e.g. for a scalar meson):

$$\frac{1}{-(\vec{q}\,'-\vec{q})^2-m_\alpha^2} \tag{A4}$$

(ii) for the pseudoscalar mesons (π,η) the pseudovector (pv) coupling is used.

(iii) Due to the zeroth component of the meson momentum being zero the ρ-exchange is different from that of Ref. 104.

Because of these alterations the formula of Ref. 104 do not apply in all details. Therefore, to avoid any misunderstandings, we give below the final formulae explicitly for the different couplings involved in the OBE model presented in Sec. 5.1. We use the following definitions and abbreviations:

$$E \equiv \sqrt{\vec{q}^{\,2}+M^2}\ ,\ E' \equiv \sqrt{\vec{q}\,'^2+M^2} \tag{A5}$$

$$Z_\alpha \equiv \frac{\vec{q}\,'^2+\vec{q}^{\,2}+m_\alpha^2}{2q'q} \tag{A6}$$

$$Q_J^{(1)}(Z_\alpha) \equiv Z_\alpha Q_J(Z_\alpha) - \delta_{Jo}$$

$$Q_J^{(2)}(Z_\alpha) \equiv \frac{1}{J+1}\left(JZ_\alpha Q_J(Z_\alpha) + Q_{J-1}(Z_\alpha)\right)$$

$$Q_J^{(3)}(Z_\alpha) \equiv \sqrt{\frac{J}{J+1}}\left(Z_\alpha Q_J(Z_\alpha) - Q_{J-1}(Z_\alpha)\right)$$

$$Q_J^{(4)}(Z_\alpha) \equiv Z_\alpha Q_J^{(1)}(Z_\alpha) - \frac{1}{3}\,\delta_{J1}$$

$$Q_J^{(5)}(Z_\alpha) \equiv Z_\alpha Q_J^{(2)}(Z_\alpha) - \frac{2}{3}\,\delta_{J1}$$

$$Q_J^{(6)}(Z_\alpha) \equiv Z_\alpha Q_J^{(3)}(Z_\alpha) + \frac{\sqrt{2}}{3}\,\delta_{J1} \tag{A7}$$

with $Q_J(Z_\alpha)$ the Legendre functions of the second kind, which occur in the analytic solutions of the integral for the partial wave decomposition

$$I_\alpha^{(0)} \equiv \int_{-1}^{+1} dt\ \frac{P_J(t)}{(\vec{q}\,'-\vec{q})^2+m_\alpha^2} = \frac{1}{q'q}\,Q_J(Z_\alpha) \tag{A8}$$

with $P_J(t)$ the Legendre polynomials. The integral representations of Eq. (A7) are:

$$I_\alpha^{(1)} \equiv \frac{Q_J^{(1)}(Z_\alpha)}{q'q} = \int_{-1}^{+1} dt \, \frac{tP_J(t)}{(\vec{q}\,'-\vec{q})^2+m_\alpha^2}$$

$$I_\alpha^{(2)} \equiv \frac{Q_J^{(2)}(Z_\alpha)}{q'q} = \int_{-1}^{+1} dt \, \frac{\frac{J}{J+1}\, tP_J(t) + \frac{1}{J+1}\, P_{J-1}(t)}{(\vec{q}\,'-\vec{q})^2+m_\alpha^2}$$

$$I_\alpha^{(3)} \equiv \frac{Q_J^{(3)}(Z_\alpha)}{q'q} = \sqrt{\frac{J}{J+1}} \int_{-1}^{+1} dt \, \frac{tP_J(t)-P_{J-1}(t)}{(\vec{q}\,'-\vec{q})^2+m_\alpha^2}$$

$$I_\alpha^{(4)} \equiv \frac{Q_J^{(4)}(Z_\alpha)}{q'q} = \int_{-1}^{+1} dt \, \frac{t^2P_J(t)}{(\vec{q}\,'-\vec{q})^2+m_\alpha^2}$$

$$I_\alpha^{(5)} \equiv \frac{Q_J^{(5)}(Z_\alpha)}{q'q} = \int_{-1}^{+1} dt \, \frac{\frac{J}{J+1}\, t^2P_J(t) + \frac{1}{J+1}\, tP_{J-1}(t)}{(\vec{q}\,'-\vec{q})^2+m_\alpha^2}$$

$$I_\alpha^{(6)} \equiv \frac{Q_\alpha^{(6)}(Z_\alpha)}{q'q} = \sqrt{\frac{J}{J+1}} \int_{-1}^{+1} dt \, \frac{t^2P_J(t)-tP_{J-1}(t)}{(\vec{q}\,'-\vec{q})^2+m_\alpha^2} \qquad \text{(A9)}$$

These formulae can be verified with the help of the recurrence relations

$$tP_J(t) = \frac{J+1}{2J+1} P_{J+1}(t) + \frac{J}{2J+1} P_{J-1}(t) \qquad \text{(A10)}$$

and

$$Z_\alpha Q_J(Z_\alpha) = \frac{J+1}{2J+1} Q_{J+1}(Z_\alpha) + \frac{J}{2J+1} Q_{J-1}(Z_\alpha) + \delta_{Jo}$$

Instead of using the Legendre functions of the second kind and their combinations Eq. (A7), one can as well evaluate the integrals Eq. (A8) and (A9) numerically, since that can be done fast and reliably. This latter method has the advantage that the products of propagator and cutoffs need not to be decomposed (see below); the cutoffs are just included in the integrands of Eq. (A8) and (A9).

Pseudovector coupling (π,η):

$$\mathcal{L}_{pv} = \frac{f_{ps}}{m_{ps}} \bar{\psi}\gamma_5\gamma^\mu\psi\partial_\mu\phi_{ps} \tag{A12}$$

$$^0V^J_{pv} = C_{pv}\left(F^{(0)}_{pv}I^{(0)}_{ps} + F^{(1)}_{pv}I^{(1)}_{ps}\right)$$

$$^1V^J_{pv} = C_{pv}\left(-F^{(0)}_{pv}I^{(0)}_{ps} - F^{(1)}_{pv}\ I^{(2)}_{ps}\right)$$

$$^{12}V^J_{pv} = C_{pv}\left(F^{(1)}_{pv}I^{(0)}_{ps} + F^{(0)}_{pv}I^{(1)}_{ps}\right)$$

$$^{34}V^J_{pv} = C_{pv}\left(-F^{(1)}_{pv}I^{(0)}_{ps} - F^{(0)}_{pv}I^{(2)}_{ps}\right)$$

$$^{55}V^J_{pv} = C_{pv}F^{(2)}_{pv}I^{(3)}_{ps}$$

$$^{66}V^J_{pv} = -C_{pv}F^{(2)}_{pv}I^{(3)}_{ps} \tag{A13}$$

with

$$C_{pv} = \frac{f^2_{ps}}{2\pi^2 m^2_{ps}} \frac{M^2}{E'E}$$

and

$$F^{(0)}_{pv} = E'E - M^2 + \frac{1}{4M^2}(E'-E)^2(E'E+3M^2)$$

$$F^{(1)}_{pv} = -\left[M^2 - \frac{1}{4}[E'-E]^2\right]\frac{\underline{q}'\underline{q}}{M^2}$$

$$F^{(2)}_{pv} = -\frac{1}{M}(E'-E)\left(\frac{1}{4}(E'-E)^2+E'E\right)$$

Scalar coupling (σ,δ):

$$\mathcal{L}_s = g_s\bar{\psi}\psi\phi_s \tag{A14}$$

$$^0V^J_s = C_s\left(F^{(0)}_s I^{(0)}_s + F^{(1)}_s I^{(1)}_s\right)$$

$$^1V^J_s = C_s\left(F^{(0)}_s I^{(0)}_s + F^{(1)}_s I^{(2)}_s\right)$$

$$^{12}V_s^J = C_s\left(F_s^{(1)}I_s^{(0)} + F_s^{(0)}I_s^{(1)}\right)$$

$$^{34}V_s^J = C_s\left(F_s^{(1)}I_s^{(0)} + F_s^{(0)}I_s^{(2)}\right)$$

$$^{55}V_s^J = C_sF_s^{(2)}I_s^{(3)}$$

$$^{66}V_s^J = C_sF_s^{(2)}I_s^{(3)} \qquad \text{(A15)}$$

with

$$C_s = \frac{g_s^2}{8\pi^2 E'E}$$

and

$$F_s^{(0)} = -(E'E + M^2)$$

$$F_s^{(1)} = q'q$$

$$F_s^{(2)} = E'+E$$

Vector and tensor coupling (ρ,ω):

$$\mathcal{L}_v = g_v\bar{\psi}\gamma_\mu\psi\phi^\mu + \frac{f_v}{2M}\,\psi\sigma_{\mu\nu}\psi\partial^\mu\phi_v^\nu \qquad \text{(A16)}$$

One obtains three terms which are characterized by their coupling constants.

Vector-vector:

$$^{0}V_v^J = C_v(2E'E - M^2)I_v^{(0)}$$

$$^{1}V_v^J = C_v\left(E'E\cdot I_v^{(0)} + q'qI_v^{(2)}\right)$$

$$^{12}V_v^J = C_v\left(2q'q\cdot I_v^{(0)} + I_v^{(1)}\right)$$

$$^{34}V_v^J = C_v\left(q'q\cdot I_v^{(0)} + E'EI_v^{(2)}\right)$$

$$^{55}V_v^J = -C_vEI_v^{(3)} \qquad \text{(A17) Cont'd}$$

$$ {}^{66}V_v^J = -C_v E' I_v^{(3)} \tag{A17} $$

with

$$ C_v = \frac{g_v^2}{4\pi^2 E'E} $$

Vector-tensor:

$$ {}^{0}V_{vt}^J = C_{vt} M \left((q'^2+q^2) I_v^{(0)} - 2q'q I_v^{(1)} \right) $$

$$ {}^{1}V_{vt}^J = C_{vt} M \left(-(q'^2+q^2) I_v^{(0)} + 2q'q I_v^{(2)} \right) $$

$$ {}^{12}V_{vt}^J = C_{vt} M \left(6q'q I_v^{(0)} - 3(q'^2+q^2) I_v^{(1)} \right) $$

$$ {}^{34}V_{vt}^J = C_{vt} M \left(2q'q I_v^{(0)} - (q'^2+q^2) I_v^{(2)} \right) $$

$$ {}^{55}V_{vt}^J = C_{vt} (E'q^2 + 3Eq'^2) I_v^{(3)} $$

$$ {}^{66}V_{vt}^J = C_{vt} (Eq'^2 + 3E'q^2) I_v^{(3)} \tag{A18} $$

with

$$ C_{vt} = \frac{g_v \left(\frac{f_v}{M}\right)}{8\pi^2 E'E} $$

(Note, that in relativistic nuclear matter calculations when $M \to \tilde{M}$ the factor (f_v/M) has to be kept unchanged as it represents the tensor coupling constant.)

Tensor-tensor:

$$ {}^{0}V_t^J = C_t \Big\{ (q'^2+q^2)(3E'E+M^2) I_v^{(0)} + \left[q'^2+q^2-2(3E'E+M^2)\right] q'q I_v^{(1)} - 2q'^2q^2 I_v^{(4)} \Big\} $$

$$ {}^{1}V_t^J = C_t \Big\{ \left[4q'^2q^2+(q'^2+q^2)(E'E-M^2)\right] I_v^{(0)} + 2(E'E+M^2) q'q \cdot I_v^{(1)} - (q'^2+q^2+4E'E) q'q \cdot I_v^{(2)} - 2q'^2q^2 \cdot I_v^{(5)} \Big\} $$

(A19) Cont'd

$$ {}^{12}V_t^J = C_t \left\{ [4M^2-3(q'^2+q^2)]q'qI_V^{(0)} + [6q'^2q^2-(q'^2+q^2)(E'E+3M^2)]I_V^{(1)} + 2(E'E+M^2)q'q\cdot I_V^{(4)} \right\} $$

$$ {}^{34}V_t^J = C_t \left\{ -[q'^2+q^2+4E'E]I_V^{(0)} - 2q'^2q^2I_V^{(1)} + [4q'^2q^2+(q'^2+q^2)(E'E-M^2)]I_V^{(2)} + 2(E'E+M^2)q'q\cdot I_V^{(5)} \right\} $$

$$ {}^{55}V_t^J = C_tM \left\{ [E'(q'^2+q^2)+E(3q'^2-q^2)]I_V^{(3)} - 2(E'+E)q'q\cdot I_V^{(6)} \right\} $$

$$ {}^{66}V_t^J = C_tM \left\{ [E(q'^2+q^2)+E'(3q^2-q'^2)]I_V^{(3)} - 2(E'+E)q'q\cdot I_V^{(6)} \right\} \qquad \text{(A19)} $$

with

$$ C_t = \frac{(f_V/M)^2}{32\pi^2E'E} $$

(The note below Eq. (A18) applies again.)

Note, that for the isovector bosons π, δ and ρ the corresponding potentials have to be multiplied by a factor $\vec{\tau}_1\cdot\vec{\tau}_2$. The use of cutoffs means that the propagator

$$ \frac{1}{(\vec{q}\,'-\vec{q})^2+m_\alpha^2} $$

has to be replaced by

$$ \frac{1}{(\vec{q}\,'-\vec{q})^2+m_\alpha^2}\left(\frac{\Lambda_\alpha^2-m_\alpha^2}{(\vec{q}\,'-\vec{q})^2+\Lambda_\alpha^2}\right)^2 \qquad \text{(A20)} $$

This expression can be applied directly in the case of the numerical integration of Eq. (A8) and (A9). When using the Legendre functions of the second kind Eq. (A20) has to be replaced by

$$ \frac{1}{(\vec{q}\,'-\vec{q})^2+m_\alpha^2} - \frac{\Lambda_{\alpha,2}^2-m_\alpha^2}{\Lambda_{\alpha,2}^2-\Lambda_{\alpha,1}^2}\,\frac{1}{(\vec{q}\,'-\vec{q})^2+\Lambda_{\alpha,1}^2} + \frac{\Lambda_{\alpha,1}^2-m_\alpha^2}{\Lambda_{\alpha,2}^2-\Lambda_{\alpha,1}^2}\,\frac{1}{(\vec{q}\,'-\vec{q})^2+\Lambda_{\alpha,2}^2} \qquad \text{(A21)} $$

with

$$ \Lambda_{\alpha,1} = \Lambda_\alpha + \varepsilon $$

$$\Lambda_{\alpha,2} = \Lambda_\alpha - \varepsilon; \qquad \varepsilon << \Lambda_\alpha$$

(ε=10 MeV is a suitable choice.)

R-matrix equation (Thompson) for spin singlet states:

$$^{o}R^{J}(q',q) = {}^{o}V^{J}(q',q) - \mathbb{P}\int_0^\infty k^2 dk \frac{^{o}V^{J}(q',k)\,^{o}R^{J}(k,q)}{2E_k - 2E_q} \tag{A22}$$

where

$$^{o}V^{J} = \sum_{\alpha=\pi,\eta,\sigma,\delta,\omega,\rho} {}^{o}V^{J}_{\alpha} \tag{A23}$$

and

$$E_k \equiv \sqrt{M^2+\vec{k}^2}\ , \quad E_q \equiv \sqrt{M^2+\vec{q}^2}\ .$$

$\mathbb{P}$ denotes the principal value.

For the coupling constants, masses and cutoff masses of the various mesons summed up in Eq. (A23) see Table 5.1.

Phase-shift relation:

$$tg\,^{o}\delta^{J}(q) = -\frac{\pi}{2} qE_q\,^{o}R^{J}(q,q) \tag{A24}$$

with

$$E_{LAB} = \frac{2q^2}{M}$$

The formulae for the spin triplet states are given in Ref. 104 (pp.225-228); note, however, that those formulae have to be modified concerning the two-nucleon propagator (compare Eq. (A22)) and concerning the factor in the phase-shift relation (compare Eq. (A24)).

The transformation of the partial-wave helicity amplitudes, in which the potential is explicitly presented here, into a LSJ-state basis, which is commonly used in nuclear structure calculations, can be done with the help of the following formulae:

Notation:

$$V^{JST}_{L',L}$$

Spin singlet:

$$V^{JoT}_{J,J} = {}^{o}V^{J}$$

Spin triplet uncoupled:

$$V^{J1T}_{J,J} = {}^{1}V^{J}$$

Spin triplet coupled:

$$V^{J1T}_{J-1,J-1} = \frac{1}{2J+1}\left\{ J\,{}^{12}V + (J+1)\,{}^{34}V^{J} + 2\sqrt{J(J+1)}\;{}^{56}V^{J} \right\}$$

$$V^{J1T}_{J+1,J+1} = \frac{1}{2J+1}\left\{ (J+1)\,{}^{12}V^{J} + J\,{}^{34}V^{J} - 2\sqrt{J(J+1)}\;{}^{56}V^{J} \right\}$$

$$V^{J1T}_{J-1,J+1} = \frac{\sqrt{J(J+1)}}{2J+1}\left\{ {}^{12}V^{J} - {}^{34}V^{J} - \frac{J}{\sqrt{J(J+1)}}\;{}^{55}V^{J} + \frac{(J+1)}{\sqrt{J(J+1)}}\;{}^{66}V^{J} \right\}$$

$$V^{J1T}_{J+1,J-1} = \frac{\sqrt{J(J+1)}}{2J+1}\left\{ {}^{12}V^{J} - {}^{34}V^{J} + \frac{(J+1)}{\sqrt{J(J+1)}}\;{}^{55}V^{J} - \frac{J}{\sqrt{J(J+1)}}\;{}^{66}V^{J} \right\}$$

with

$${}^{56}V^{J} = V^{J}_{5} + V^{J}_{6}$$

APPENDIX B: EXPRESSIONS FOR A NON-RELATIVISTIC OBEP IN R-SPACE

Following the lines of Sec. 3, however, keeping terms up to momentum squared, the one-meson-exchange contributions are first simplified in q-space, such that they can be transformed analytically into r-space. Such q-space expressions are:[195,133]

pseudoscalar mesons (π,η)

$$V_{ps}(\vec{k},\vec{p}) = -\frac{g_{ps}^2}{4M^2}\frac{\vec{\sigma}_1\cdot\vec{k}\vec{\sigma}_2\cdot\vec{k}}{\vec{k}^2+m_{ps}^2} ;$$

$\vec{k} \equiv \vec{q}\,'-\vec{q}$ with $\vec{q}\,'$ and $\vec{q}$ as in Fig. 3.5.

scalar mesons (σ,δ)

$$V_s(\vec{k},\vec{p}) = -\frac{g_s^2}{\vec{k}^2+m_s^2}\left[1-\frac{\vec{p}^2}{2M^2}+\frac{\vec{k}^2}{8M^2}-\frac{i}{2M^2}\vec{S}\cdot(\vec{k}\times\vec{p})\right] ;$$

where $\vec{p} \equiv 1/2(\vec{q}\,'+\vec{q})$ and $\vec{S} = 1/2(\vec{\sigma}_1+\vec{\sigma}_2)$.

vector mesons (ω,ρ)

$$\begin{aligned} V_v(\vec{k},\vec{p}) = \frac{1}{\vec{k}^2+m_v^2}\Bigg\{ & g_v^2\left[1+\frac{3}{2}\frac{\vec{p}^2}{M^2}-\frac{\vec{k}^2}{8M^2}+\frac{3i}{2M^2}\vec{S}\cdot(\vec{k}\times\vec{p})\right. \\ & \left. -\vec{\sigma}_1\cdot\vec{\sigma}_2\frac{\vec{k}^2}{4M^2}+\frac{1}{4M^2}\vec{\sigma}_1\cdot\vec{k}\vec{\sigma}_2\cdot\vec{k}\right] \\ & +\frac{g_vf_v}{2M}\left[-\frac{\vec{k}^2}{M}+\frac{4i}{M}\vec{S}\cdot(\vec{k}\times\vec{p})-\vec{\sigma}_1\cdot\vec{\sigma}_2\frac{\vec{k}^2}{M}+\frac{1}{M}\vec{\sigma}_1\cdot\vec{k}\vec{\sigma}_2\cdot\vec{k}\right] \\ & +\frac{f_v^2}{4M^2}\left[-\vec{\sigma}_1\cdot\vec{\sigma}_2\vec{k}^2+\vec{\sigma}_1\cdot\vec{k}\vec{\sigma}_2\cdot\vec{k}\right]\Bigg\} \end{aligned}$$

After a Fourier transformation one obtains the following r-space potentials:

pseudoscalar mesons (π,η)

$$V_{ps}(r) = \frac{1}{12}\frac{g_{ps}^2}{4\pi}m_{ps}\left\{\left(\frac{m_{ps}}{M}\right)^2 Y(m_{ps}r)\vec{\sigma}_1\cdot\vec{\sigma}_2 + Z(m_{ps}r)S_{12}\right\}$$

scalar mesons (σ,δ)

$$V_s(r) = \frac{-g_s^2}{4\pi} m_s \left\{ \left[1 - \frac{1}{4}\left(\frac{m_s}{M}\right)^2 \right] Y(m_s r) + \frac{1}{4M^2} \left[\nabla^2 Y(m_s r) + Y(m_s r)\nabla^2 \right] + \frac{1}{2} Z_1(m_s r)\vec{L}\cdot\vec{S} \right\}$$

vector mesons (ω,ρ)

$$\begin{aligned} V_v(r) = \frac{g_v^2}{4\pi} m_v & \left\{ \left[1 + \frac{1}{2}\left(\frac{m_v}{M}\right)^2 \right] Y(m_v r) - \frac{3}{4M^2} \left[\nabla^2 Y(m_v r) + Y(m_v r)\nabla^2 \right] \right. \\ & \left. + \frac{1}{6}\left(\frac{m_v}{m}\right)^2 Y(m_v r)\vec{\sigma}_1\cdot\vec{\sigma}_2 - \frac{3}{2} Z_1(m_v r)\vec{L}\cdot\vec{S} - \frac{1}{12} Z(m_v r) S_{12} \right\} \\ & + \frac{1}{2}\frac{g_v f_v}{4\pi} m_v \left\{ \left(\frac{m_v}{M}\right)^2 Y(m_v r) + \frac{2}{3}\left(\frac{m_v}{M}\right)^2 Y(m_v r)\vec{\sigma}_1\cdot\vec{\sigma}_2 \right. \\ & \left. - 4Z_1(m_v r)\vec{L}\cdot\vec{S} - \frac{1}{3} Z(m_v r) S_{12} \right\} \\ & + \frac{f_v^2}{4\pi} m_v \left\{ \frac{1}{6}\left(\frac{m_v}{M}\right)^2 Y(m_v r)\vec{\sigma}_1\cdot\vec{\sigma}_2 - \frac{1}{12} Z(m_v r) S_{12} \right\} \end{aligned}$$

with

$$Y(x) = e^{-x}/x$$

$$Z(x) = \left(\frac{m_\alpha}{M}\right)^2 \left(1 + \frac{3}{x} + \frac{3}{x^2}\right) Y(x)$$

$$Z_1(x) = \left(\frac{m_\alpha}{M}\right)^2 \left(\frac{1}{x} + \frac{1}{x^2}\right) Y(x)$$

$$M = 938.926 \text{ MeV}/c^2$$

Apply a factor $\vec{\tau}_1\cdot\vec{\tau}_2$ for π, ρ, and δ. For the application of the cutoffs see Eq. (A21) from which the consequences for the r-space potential can be drawn easily. The parameters for the potential given in this appendix are in Table 5.2. The full potential is defined by the sum of the six one-meson-exchange contributions.

APPENDIX C: NUCLEAR MATTER FORMULAE FOR THE RELATIVISTIC DIRAC BRUECKNER APPROACH

The calculations are done in partial wave decomposition. The Brueckner equation, Eq. (7.3), for an uncoupled partial wave state is:

$$G_L(q',q,A,k_F) = \tilde{V}_L(q',q,A) - \int_0^\infty k^2 dk\, \tilde{V}_L(q',k,A)\, \frac{Q(k,k_F)}{e(k,q,A)}\, G_L(k,q,A,k_F) \tag{C1}$$

The extension to coupled cases is straightforward. The CM momentum, P, is set to zero for consistency with the potential and the way it was derived. q,q' and k are relative momenta; e.g. $q = 1/2|\vec{k}_1-\vec{k}_2|$ with $\vec{k}_1$ and $\vec{k}_2$ as in Eq. (7.3). The wiggle on top of the potential, $\tilde{V}_L$, indicates that in the expressions for the OBEP of Appendix A one has to do the following replacements:

$$M \rightarrow \tilde{M} = M + A$$

and

$$E \rightarrow \tilde{E} \equiv \sqrt{\tilde{M}^2+\vec{q}^{\,2}} ,$$

$$E' \rightarrow \tilde{E}' \equiv \sqrt{\tilde{M}^2+\vec{q}^{\,\prime 2}} \tag{C2}$$

due to the fact that the Dirac spinors Eq. (8.3) are used in nuclear matter. Q denotes the Pauli projector which for P=0 assumes the simple form:

$$Q(k,k_F) = \begin{cases} 0 & \text{for } k < k_F \\ 1 & \text{for } k > k_F \end{cases} \tag{C3}$$

The energy-denominator in the Brueckner equation is:

$$e(k,q,A) = 2\tilde{E}_k - 2\tilde{E}_q \tag{C4}$$

with

$$\tilde{E}_k \equiv \sqrt{\tilde{M}^2+\vec{k}^2} \quad \text{and} \quad \tilde{E}_q \equiv \sqrt{\tilde{M}^2+\vec{q}^{\,2}}$$

The scalar potential A remains to be determined. This can be done with the help of the following formualae. From Eq. (8.7) and (8.4) we know that

$$\varepsilon(k) = \frac{M\tilde{M}+k^2}{\tilde{E}_k} + U(k) \tag{C5}$$

with

$$k \equiv |\vec{k}|$$

and

$$U(k) = \sum_{J,S,T,L} (2J+1)(2T+1) \times \left\{ \left[4 \int_0^{1/2(k_F-k)} q_0^2 dq_0 + \int_{1/2(k_F-k)}^{1/2(k_F+k)} q_0^2 dq_0 \frac{k_F^2-(2q_0-k)^2}{2q_0 k} \right] G_{LL}^{JST}(q_0,q_0,A,k_F) \right\} \tag{C6}$$

(J ≡ total angular momentum, T ≡ total isospin.)

On the other hand, because of the ansatz Eq. (8.2), we also have:

$$\varepsilon(k) = \tilde{E}_k + B; \qquad 0 \leqslant k < \infty \tag{C7}$$

("continuous choice")

Computing ε for two different k ($k_1 = (1/2k_F)$ and $k_2 = k_F$ is an appropriate choice) according to Eq. (C5) and Eq. (C6) and then applying Eq. (C7) determines A and B.

The procedure to be done separately for each k_F is as follows: One starts with a guess for A (e.g. −600 MeV), solves the Brueckner equation Eq. (C1) to obtain G and evaluates U(k) (Eq. (C6)) with those G for two different k. From that one obtains with the help of Eq. (C5) and (C7) a new value for A (and B). With the new A one starts all over again till the old and the new A agree sufficiently well, i.e. selfconsistency is achieved. From the G-matrix elements, computed with the selfconsistent A, one evaluates the energy per nucleon according to the following formula (compare Eq. (7.6) and Eq. (8.5)-(8.7)):

$$\frac{E}{N}(k_F) = \frac{3}{k_F^3}\int_0^{k_F} q_o^2 dq_o \frac{q_o^2+M\tilde{M}}{\tilde{E}_{q_o}} + \sum_{JSTL}(2J+1)(2T+1)$$

$$\times \int_0^{k_F} q_o^2 dq_o \left(2-3\frac{q_o}{k_F}+\frac{q_o^3}{k_F{}^3}\right) G_{LL}^{JST}(q_o,q_o,A,k_F) - M \qquad (C8)$$

It is most practical to use in all calculations units of MeV for momenta, masses and energies (conversion factor: $\hbar c$ = 197.3286 MeV•fm). The potential V, $\tilde{V}$ and the G-matrix are then in units of MeV^{-2}.

Concerning the derivation of some of the formulae given in this Appendix and some computational hints see Ref. 196 and 197.

REFERENCES

1. For an introduction into QCD see e.g.: W. Marciano and H. Pagels, Phys. Reports 36, 137 (1978).
2. For a recent review of lattice gauge theories see: M. Creutz, L. Jacobs and C. Rebbi, Phys. Reports 95, 201 (1983); and Janos Polonyi, Lectures at this workshop.
3. G.E. Brown and M. Rho, Phys. Lett. 82B, 177 (1979); G.E. Brown, M. Rho and V. Vento, Phys. Lett. 84B, 383 (1979).
4. A. Mittal and A.N. Mitra, Phys. Rev. D29, 1408 (1984).
5. Yu You-Wen and Zhang Zong-Ye, Nucl. Phys. A426, 557 (1984).
6. T. DeGrand, R.L. Jaffe, K. Johnson and J. Kiskis, Phys. Rev. D12, 2060 (1975).
7. A.W. Thomas, S. Théberge and G.A. Miller, Phys. Rev. D24, 216 (1981); A.W. Thomas, Advances in Nucl. Phys. 13, 1 (1983).
8. K. Maltman and N. Isgur, Phys. Rev. D29, 952 (1984); N. Isgur, Lectures at this workshop.
9. K. Holinde, Phys. Lett. 157B, 123 (1985).
10. C.S. Warke and R. Shanker, Phys. Rev. C21, 2643 (1980).
11. J. Burger and M. Hofmann, Phys. Lett. 148B, 25 (1984); and preprint, Erlangen, 1985.
12. M. Harvey, J. Letourneux and B. Corazo, Nucl. Phys. A424, 428 (1984); N. Mankoc-Borstnik et al., Nucl. Phys. A395, 349 (1983), and contribution to this workshop.
13. A. Faessler, F. Fernanadez, G. Lübeck and K. Shimizu, Nucl. Phys. A402, 555 (1983).
14. O. Morimatsu, K. Yazaki and M. Oka, Nucl. Phys. A424, 412 (1984).
15. R.L. Jaffe, Lectures at this workshop.
16. G. t'Hooft, Nucl. Phys. B72, 461 (1974); B75, 461 (1974).
17. E. Witten, Nucl. Phys. B160, 57 (1979).
18. T.H.R. Skyrme, Proc. Roy. Soc. A260, 127 (1961); Nucl. Phys. 31, 556 (1962).
19. V. Vento, M. Rho, E. Nyman, J.H. Jun and G.E. Brown, Nucl. Phys. A345, 413 (1980).
20. G.E. Brown , A.D. Jackson, M. Rho and V. Vento, Phys. Lett. 140B, 285 (1984).
21. Lewis Carroll, Alice's Adventures in Wonderland (Macmillan, London, 1865).
22. S. Nadkarin, H.B. Nielsen and I. Zaked, "Bosonization Relations as Bag Boundary Conditions," preprint Niels Bohr Institut, NBI-HE-84-41, 1984.
23. G. Adkins and C.R. Nappi, Phys. Lett. 137B, 251 (1984).
24. M. Rho, Proc. Int. School of Physics "Enrico Fermi," 18-23 June 1984 (Varenna, Italy); K. Iketani, Kyushu University Preprint, 1984.
25. I. Zahed, Lectures at this workshop.
26. J. Chadwick, Proc. Roy. Soc. (London), A136, 692 (1932).
27. W. Heisenberg, Z. Physik 77, 1 (1932).

28. E. Majorana, Z. Physik 82, 137 (1933).
29. M. Tuve, N. Heydenburg and L. Hafstad, Phys. Rev. 50, 806 (1936).
30. G. Breit, E. Condon and R. Present, Phys. Rev. 50, 825 (1936).
31. H. Yukawa, Proc. Phys. Math. Soc. Japan 17, 48 (1935).
32. H. Yukawa, Proc. Phys. Math. Soc. Japan 19, 712 (1937); H. Yukawa and S. Sakata, Proc. Phys. Math. Soc. Japan 19, 1084 (1937).
33. H. Yukawa, S. Sakata and M. Taketani, Proc. Phys. Math. Soc. Japan 20, 319 (1938); H. Yukawa, S. Sakata, M. Kobayasi and M. Taketani 20, 720 (1938); the papers by Yukawa and other Japanese physicists from the first days in meson theory up to about 1950 are reprinted in Progr. of Theor. Phys. (Kyoto), Suppl. 1 and 2 (1955).
34. S.H. Neddermeyer and C.D. Anderson, Phys. Rev. 51, 884 (1937); J.C. Street and E.C. Stevenson, Phys. Rev. 51, 1005A (1937).
35. N. Kemmer, Proc. Roy. Soc. (London) A166, 127 (1938).
36. A. Proca, J. Phys. Radium 7, 347 (1936).
37. N. Kemmer, Proc. Cambridge Phil. Soc. 34, 354 (1938).
38. H.J. Bhabha, Proc. Roy. Soc. (London) A166, 501 (1938).
39. N. Kemmer, Proc. Roy. Soc. (London) A173, 91 (1939).
40. B. Clark, Contribution to this workshop.
41. G. Wick, Nature 142, 993 (1938).
42. J. Kellog, I. Rabi, N.F. Ramsey and J. Zacharias, Phys. Rev. 56, 728 (1939); 57, 677 (1940).
43. C. Møller and L. Rosenfeld, Kgl. Danske Vid. Selskab, Math-fys. Medd. 17, No. 8 (1940).
44. J. Schwinger, Phys. Rev. 61, 387 (1942).
45. J.M. Jauch and N. Hu, Phys. Rev. 65, 289 (1944).
46. H.A. Bethe, Phys. Rev. 57, 260, 390 (1940); for an application of Bethe's idea see also: W. Rarita and J. Schwinger, Phys. Rev. 59, 436, 556 (1941).
47. W. Pauli, Meson Theory of Nuclear Forces, (Interscience, New York, 1946).
48. G. Breit, Phys. Rev. 51, 248 (1937); G. Breit and J.R. Stehn, Phys. Rev. 53, 459 (1938).
49. L. Rosenfeld, Nature 156, 141 (1945).
50. L. Rosenfeld, Nuclear Forces (North-Holland Publ. Comp., Amsterdam, 1948).
51. A.E.S. Green, Phys. Rev. 73, 26 (1948); 75, 1926 (1949); 76, 460, 870 (1949).
52. A.E.S. Green, unpublished; private communication.
53. G. Wentzel, Quantum Theory of Fields (Interscience, New York, 1949).
54. M. Conversi, E. Pancini and O. Piccioni, Phys. Rev. 71, 209, 557 (1947).

55. G.P.S. Occhialini, C.F. Powell, C.M.G. Lattes and Muirhead, Nature 159, 186, 694 (1947); C.M.G. Lattes, G.P.S. Occhialini and C.F. Powell, Nature 160, 453, 486 (1947).
56. E. Gardner and C.M. Lattes, Science 107, 270 (1948).
57. M. Taketani, S. Nakamura and M. Sasaki, Prog. Theor. Phys. (Kyoto) 6, 581 (1951).
58. Progr. Theoret. Phys. (Kyoto), Suppl. 3 (1956).
59. P. Cziffra, M.H. MacGregor, M.J. Moravcsik and H.P. Stapp, Phys. Rev. 114, 880 (1959).
60. M.J. Moravcsik, Nucleon-Nucleon Scattering and the One-Pion-Exchange, UCRL 5886-T (1958).
61. G. Breit, M.H. Hull, K.E. Lassila and K.D. Pyatt, Phys. Rev. 120, 2227 (1960); G. Breit, Rev. Mod. Phys. 34, 776 (1962).
62. D.Y. Wong, Phys. Rev. Lett. 2, 406 (1959).
63. N.K. Glendenning and G. Kramer, Phys. Rev. 126, 2159 (1962).
64. J. Iwadare, S. Otsuki, R. Tamagaki and W. Watari, Progr. Theor. Phys. (Kyoto) 15, 86 (1956); 16, 455 (1956); for a summary see the contribution by these authors in Ref. 58, p.32.
65. T.E.O. Ericson and M. Rosa-Clot, Nucl. Phys. A405, 497 (1983).
66. M. Taketani, S. Machida and S. Onuma, Prog. Theor. Phys. (Kyoto) 7, 45 (1952).
67. K.A. Brueckner and K.M. Watson, Phys. Rev. 90, 699 (1953); 92, 1023 (1953).
68. K.A. Brueckner, M. Gell-Mann and M. Goldberger, Phys. Rev. 90, 476 (1953).
69. A. Klein, Prog. Theor. Phys. (Kyoto) 20, 357 (1958).
70. N. Hoshizaki and S. Machida, Prog. Theor. Phys. (Kyoto) 27, 288 (1962).
71. M.L. Goldberger, Proc. Midwestern Conf. on Theor. Phys., Purdue University, Lafayette, Ind. (USA), April 1960, pp.50-63.
72. H.A. Bethe, Scientific American, 189, 58 (1953).
73. M.J. Moravcsik, The Two-Nucleon Interaction (Clarendon Press, Oxford, 1963); for a more condensed review on the 1950's see also: R.J.N. Phillips, Repts. Progr. in Phys. 22, 562 (1959).
74. H.A. Bethe and F. de Hoffmann, Mesons and Fields, Vol. II, Mesons (Row, Peterson and Comp., Evanston, Ill., 1955).
75. Prog. Theor. Phys. (Kyoto), Suppl. 39 (1967).
76. R. Wilson, The Nucleon-Nucleon Interaction (Interscience, New York, 1963).
77. H.P. Stapp, T.J. Ypsilantis and N. Metropolis, Phys. Rev. 105, 302 (1957).
78. L. Eisenbud and W. Wigner, Proc. Nat. Acad. Wash. 27, 281 (1941); see also: S. Okubo and R.E. Marshak, Ann. of Phys. (N.Y.) 4, 166 (1958).

79. J.L. Gammel, R.S. Christian and R.M. Thaler, Phys. Rev. 105, 311 (1957).
80. J.L. Gammel and R.M. Thaler, Phys. Rev. 107, 291, 1337 (1957).
81. R. Jastrow, Phys. Rev. 81, 165 (1951).
82. M.G. Mayer, Phys. Rev. 75, 1969 (1949); 78, 16 (1950); O. Haxel, J.H.D. Jensen and H.E. Suess, Phys. Rev. 75, 1766 (1949).
83. P. Signell, R. Zinn and R. Marshak, Phys. Rev. Lett. 1 (1958) 416.
84. T. Hamada and I.D. Johnston, Nucl. Phys. 34, 382 (1962); for an improved version see: T. Hamada, Y. Nakamura and R. Tamagaki, Prog. Theor. Phys. 33, 769 (1965).
85. K.E. Lassila, M.H. Hull, H.M. Ruppel, F.A. McDonald and G.E. Breit, Phys. Rev. 126, 881 (1962).
86. R.V. Reid, Ann. of Phys. (N.Y.) 50, 411 (1968).
87. Y. Nambu, Phys. Rev. 106, 1366 (1957).
88. G. Breit, Proc. Natl. Acad. Sci. (U.S.), 46, 746 (1960); Phys. Rev. 120, 287 (1960).
89. J.J. Sakurai, Ann. Phys. (N.Y.) 11, 1 (1960); Phys. Rev. 119, 1784 (1960).
90. W.R. Frazer and J.R. Fulco, Phys. Rev. 117, 1609 (1960).
91. B. Maglich, L. Alvarez, A. Rosenfeld and C. Stevenson, Phys. Rev. Lett. 7, 178 (1961).
92. See e.g. C. Alff et al., Phys. Rev. Lett. 9, 322 (1962); Refs. therein.
93. Particle Data Group, Phys. Lett. 50B, 74 (1974); find a complete list of earlier references from the 1960's therein.
94. Particle Data Group, Rev. Mod. Phys. 48, 5114 (1976).
95. R.A. Bryan, C.R. Dismukes and W. Ramsey, Nucl. Phys. 45, 353 (1963); R.A. Bryan and B.L. Scott, Phys. Rev. 135, B434 (1964).
96. A. Scotti and D.Y. Wong, Phys. Rev. Lett. 10, 142 (1963); Phys. Rev. 138, 145 (1965).
97. N. Hoshizaki, I. Lin and S. Machida, Prog. Theor. Phys. 27, 288 (1961).
98. S. Sawada, T. Ueda, W. Watari and M. Yonezawa, Prog. Theor. Phys. 28, 991 (1962).
99. R.S. McKean, Jr., Phys. Rev. 125, 1399 (1962).
100. A.E.S. Green and R.D. Sharma, Phys. Rev. Lett. 14, 380 (1965); A.E.S. Green and T. Sawada, Nucl. Phys. B2, 267 (1967).
101. M.M. Nagels, T.A. Rijken and J.J. de Swart, Phys. Rev. D17, 768 (1978).
102. R. de Tourreil, B. Rouben and D.W.L. Sprung, Nucl. Phys. A242, 445 (1975).
103. E.E. Salpeter and H.A. Bethe, Phys. Rev. 84, 1232 (1951).
104. K. Erkelenz, Phys. Reports 13C, 191 (1974).
105. K. Holinde and R. Machleidt, Nucl. Phys. A247, 495 (1975).

106. K. Holinde and R. Machleidt, Nucl. Phys. A256, 479 (1976).
107. J. Fleischer and J.A. Tjon, Nucl. Phys. B84, 375 (1975); Phys. Rev. D15, 2537 (1977); D21, 87 (1980); M.J. Zuilhof and J.A. Tjon, Phys. Rev. C22, 2369 (1980).
108. M.J. Moravcsik, Rep. Prog. Phys. 35, 587 (1972).
109. Rev. Mod. Phys. 39, 495 (1967); for the status of meson theory in that year see especially p.594.
110. S. Mandelstam, Rep. on Prog. in Phys. 25 (1962).
111. G.F. Chew, S-Matrix Theory of Strong Interactions (Benjamin, New York, 1961).
112. D. Amati, E. Leader and B. Vitale, Nuovo Cim. 17, 68; 18, 409, 458 (1960).
113. J. Binstock, Phys. Rev. D3, 1139 (1971); J. Binstock and R.A. Bryan, Phys. Rev. D4, 1341 (1971).
114. M. Chemtob, J.W. Durso and D.O. Riska, Nucl. Phys. B38, 141 (1972).
115. R. Vinh Mau et al., Phys. Lett. 44B, 1 (1973).
116. A.D. Jackson, D.O. Riska and B. Verwest, Nucl. Phys. A249, 397 (1975).
117. R. Woloshyn and A.D. Jackson, Nucl. Phys. A185, 131 (1972).
118. M. Lacombe et al., Phys. Rev. D12, 1495 (1975).
119. M. Lacombe et al., Phys. Rev. C21, 861 (1980).
120. R. Vinh Mau, in Mesons in Nuclei, Vol. I., eds. M. Rho and D. Wilkinson (North-Holland, Amsterdam, 1979), p.151.
121. G.E. Brown and A.D. Jackson, The Nucleon-Nucleon Interaction (North-Holland, Amsterdam, 1976).
122. M.H. Partovi and E.L. Lomon, Phys. Rev. D2, 1999 (1970).
123. R. Blankenbecler and R. Sugar, Phys. Rev. 142, 1051 (1966).
124. F. Partovi and E.L. Lomon, Phys. Rev. D5, 1192 (1972); E.L. Lomon, Phys. Rev. D14, 2402 (1976); D22, 229 (1980).
125. H. Sugawara and F. von Hippel, Phys. Rev. 172, 1764 (1968).
126. K. Holinde and R. Machleidt, Nucl. Phys. A280, 429 (1977).
127. K. Holinde, R. Machleidt, M.R. Anastasio, A. Faessler and H. Müther, Phys. Rev. C18, 870 (1978).
128. K. Holinde, R. Machleidt, M.R. Anastasio, A. Faessler and H. Müther, Phys. Rev. C19, 948 (1979).
129. K. Holinde, R. Machleidt, A. Faesler, H. Müther, and M.R. Anastasio, Phys. Rev. C24, 1159 (1981); X. Bagnoud, K. Holinde and R. Machleidt, Phys. Rev. C24, 1143 (1981).
130. K. Holinde and R. Machleidt, Nucl. Phys. A372, 349 (1981).
131. X. Bagnoud, K. Holinde and R. Machleidt, Phys. Rev. C29, 1792 (1984).
132. R. Machleidt, in: Quarks and Nuclear Structure, ed. K. Bleuler, Lecture Notes in Physics (Springer Verlag, Heidelberg, 1984), Vol. 197, p.352.
133. R. Machleidt, K. Holinde and C. Elster, to be published.
134. P. Signell, Adv. Nucl. Phys. 2, 223 (1969).

135. E. Segrè, Nuclei and Particles (W.A. Benjamin, London, 1977).
136. E. Wigner, Phys. Rev. 43, 252 (1933).
137. R.A. Arndt et al., Phys. Rev. D28, 97 (1983).
138. A. Messiah, Quantum Mechanics (North-Holland, Amsterdam, 1966), Vol. I and II.
139. J.D. Bjorken and S.D. Drell, Relativistic Quantum Mechanics, (McGraw-Hill, New York, 1964).
140. J.D. Bjorken and S.D. Drell, Relativistic Quantum Fields (McGraw-Hill, New York, 1965).
141. F. Iachello, A.D. Jackson and A. Lande, Phys. Lett. 43B, 191 (1973); and earlier references therein.
142. E.M. Henley and G.A. Miller, in Mesons in Nuclei, eds. M. Rho and D.H. Wilkinson (North Holland, Amsterdam, 1979), p.406.
143. O. Dumbrajs et al., Nucl. Phys. B216, 277 (1983).
144. D. Lurié, Particles and Fields (Interscience, New York, 1968).
145. S.S. Schweber, An Introduction to Relativistic Quantum Field Theory (Row-Peterson, New York, 1961), pp.415-435.
146. D. Schütte, Nucl. Phys. A221, 450 (1974).
147. M.J. Zuilhof and J.A. Tjon, Phys. Rev. C24, 736 (1981).
148. S. Weinberg, Phys. Rev. Lett. 18, 188 (1967).
149. S.J. Brodsky, in Quarks and Nuclear Forces, eds. D. Fries and B. Zeitnitz, Springer Tracts in Modern Physics, Vol. 100, p.34.
150. Particle Data Group, Rev. Mod. Phys. 56, S1 (1984).
151. F.E. CLose, An Introduction to Quarks and Partons (Academic Press, London, 1979).
152. S. Godfrey and N. Isgur, Phys. Rev. D32, 189 (1985).
153. J.W. Durso, A.D. Jackson and B.J. Verwest, Nucl. Phys. A345, 471 (1980).
154. Q. Ho-Kim and D. Turcot, Phys. Rev. C22, 1352 (1980).
155. J.W. Durso, M. Saarela, G.E. Brown and A.D. Jackson, Nucl. Phys. A278, 445 (1977).
156. R. Vinh Mau et al., Phys. Lett. 44B, 1 (1973).
157. R. Dubois et al., Nucl. Phys. A377, 554 (1982).
158. R. Koch and E. Pietarinen, Nucl. Phys. A336, 331 (1980).
159. P. Kroll, Physics Data, 22, 1 (1981).
160. E. Pietarinen, Helsinki University, HU-/TFT-17-17.
161. W. Grein, Nucl. Phys. B131, 255 (1977).
162. W. Grein and P. Kroll, Nucl. Phys. A338, 332 (1980).
163. G.E. Brown and W. Weise, Phys. Reports 22C, 281 (1975).
164. R.M. Woloshyn and A.D. Jackson, Nucl. Phys. B64, 269 (1973).
165. R.H. Thompson, Phys. Rev. D1, 1738 (1970).
166. D. Schütte, Nucl. Phys. A221, 450 (1974).
167. K. Kotthoff, R. Machleidt and D. Schütte, Nucl. Phys. A264, 484 (1976).

168. R. Machleidt and K. Holinde, Phys. Lett. 152B, 295 (1985).
169. J.W. Negele, Phys. Rev. C1, 1260 (1970).
170. X. Campi and D.W.L. Sprung, Nucl. Phys. A194, 401 (1972).
171. C.J. Horowitz and B.D. Serot, Nucl. Phys. A368, 503 (1981).
172. D. Youngblood et al., Phys. Rev. Lett. 39, 1188 (1977).
173. B.D. Day, Rev. Mod. Phys. 39, 719 (1967).
174. H.A. Bethe, Ann. Rev. Nucl. Sci. 21, 93 (1971).
175. B.D. Day, Rev. Mod. Phys. 50, 495 (1978).
176. J.W. Negele, Rev. Mod. Phys. 54, 913 (1982).
177. J. Jeukenne, A. Lejeune and C. Mahaux, Nucl. Phys. A245, 411 (1975).
178. R. Machleidt and K. Holinde, Nucl. Phys. A350, 396 (1980).
179. B.D. Day, Phys. Rev. C24, 1203 (1981).
180. B.D. Day, Phys. Rev. Lett. 47, 226 (1981).
181. F. Coester, S. Cohen, B.D. Day and C.M. Vincent, Phys. Rev. C1, 769 (1970).
182. H.A. Bethe and M.B. Johnson, Nucl. Phys. A230 1 (1974).
183. K. Holinde, Phys. Reports 68, 121 (1981), Table 1 column 1 and 2.
184. R.A. Bryan and A. Gerstan, Phys. Rev. D6, 341 (1972).
185. R. de Tourreil and D.W.L. Sprung, Nucl. Phys. A201, 193 (1973).
186. Ref. 178: Δ1 is the result obtained in this work with the continuous choice for the single particle potential; for Δ2 the mesonic effects have been ignored.
187. K. Holinde and R. Machleidt, Nucl. Phys. A280, 429 (1977); for the result quoted in Fig. 7.2 a continuous choice is used.
188. M.R. Anastasio et al., Phys. Rev. C18, 2416 (1978); the result quoted in Fig. 7.2 refers to model MDFPΔ1 applied with a continuous choice and without mesonic effects.
189. M.R. Anastasio et al., Phys. Reports 100, 327 (1983).
190. B.D. Serot and J.D. Walecka, Advances in Nuclear Physics, eds. J.W. Negele and E. Vogt (Plenum Press, New York, 1985) Vol. 16, to be published.
191. C.J. Horowitz and B.D. Serot, Phys. Lett. 137B, 287 (1984).
192. R. Brockmann and R. Machleidt, Phys. Lett. 149B, 283 (1984).
193. R. Machleidt and R. Brockman, Phys. Lett. to be published.
194. R. Brockmann and R. Machleidt, to be published.
195. C. Elster, private communication.
196. M. Haftel and F. Tabakin, Nucl. Phys. A158, 1 (1970).
197. M.I. Haftel, Ph.D. Thesis, University of Pittsburgh (1969).

ELECTRONUCLEAR REACTIONS AND MESON EXCHANGE CURRENTS

John Dubach
University of Massachusetts
Amherst, MA 01003

ABSTRACT

The electromagnetic structure of the nucleus is described in terms of nucleon and meson degrees of freedom. The basic description of electron scattering in the formalism of Quantum Electrodynamics (QED) is reviewed, first in general terms and then with specific applications to nuclear physics. Construction of the nucleon current operators is briefly discussed. The calculational techniques for including the electromagnetic currents due to the exchange of mesons are then presented. The historical evidence for these "meson exchange currents" is reviewed and the present status of the confrontation between theory and experiment is summarized. Finally, speculation about the future directions in this field is offered.

INTRODUCTION

At this point in our history the Nuclear Physics community is attempting to make the transition from discussing the nucleus in terms of nucleons and mesons to describing it in terms of the more fundamental quarks and gluons. One key step in this process will be to obtain a working understanding of when quark degrees of freedom must be invoked and when (and why) the "old-fashioned" nucleons and mesons provide an adequate approximation (thereby making it difficult to isolate the quark effects). In these few lectures I will concentrate on the information obtained from electron scattering and describe the attempts to understand it in terms of nucleon and meson degrees of freedom. It is my intention to present this material as dispassionately as I can so that the reader may judge for him/herself the success or failure of modern theoretical calculations; only briefly in the conclusions will I present my own bias and perspective.

ELECTRON SCATTERING - A PRIMER

Introduction

Electromagnetic probes in general, and electron scattering in particular, have traditionally provided some of the most valuable insights into the structure of the nucleus and the dynamic processes within it. These probes can certainly be expected to play an important role in elucidating the questions raised in the discussion of the transition from nucleon/meson to quark degrees of freedom. (Indeed, this was one of the primary motivations for the Continuous Electron Beam Accelerator Facility (CEBAF), now in the detailed design state). Before I touch upon these questions, however, I should begin by reviewing the formalism of the field.

The analysis of electron scattering has evolved over the past thirty years. It has developed into a fairly standardized formalism (barring varying conventions for certain bothersome multiplicative factors) and can be found in detail in a number of places, e.g., Refs. 1-4. (Required reading for the serious student of the field is the series of lectures[4] given by Dirk Walecka at Argonne two years ago.) I will here only attempt to outline the basic derivations, trying to emphasize the physics input to and the physical interpretation of the formalism. The reader is referred to Refs. 1-4 for further details and references to the original work. Some recent reviews discussing the application of this formalism to nuclear physics problems may be found in Refs. 5-7.

The advantages of using electromagnetic probes to study the detailed structure of an object are fairly obvious. One understands electromagnetism far better than the other interactions. Indeed, Quantum Electrodynamics (QED), the theory we use to describe electromagnetism, is the most complete and most accurate theory of modern physics. In addition, the basic electromagnetic coupling strength is the fine structure constant, $\alpha \simeq 1/137$, a rather small number. As a result, not only do we have a well understood interaction, we also have a theory in which it is practicable to do a calculation since a perturbation series in α can be expected to rapidly converge.

Electron scattering has been the preferred electromagnetic probe for a number of practical reasons. First and foremost, the kinematic freedom to vary beam energy, scattering angle, etc. allows us to probe the target in a much more complete way than is obtained from, e.g., electromagnetic decay properties and static moments. As we shall see, this kinematic freedom allows us to directly measure and map out in detail the charge and current densities within the target. For nuclear physics studies, electron scattering in particular and electromagnetic probes in general have the advantage that they are penetrating probes; that is, unlike hadronic probes such as pions, they are not quickly absorbed or scattered in the nuclear medium and they therefore allow for direct probing of the entire nuclear volume.

One of the advantages of electron scattering is also one of the disadvantages. Because the probe is relatively weak, one calculates with confidence in perturbation theory and with some faith that the target is little modified because of the presence of the probe. However, for this same reason the cross sections to be measured are also substantially smaller than those obtained for nuclei with hadronic probes. This requires considerable attention to technique, background suppression, etc. on the part of my experimental colleagues (not to mention a great deal of patience). Using present "state-of-the-art" techniques, cross sections as small as 10^{-39}-10^{-40} cm^2/ster have been measured as part of studies of nuclear structure[8]. These cross sections are approaching the level one measures in neutrino scattering experiments[1]. Another disadvantage to the electromagnetic probes is that they are only sensitive to the electromagnetic structure of the target. In particular, for nuclear physics we are in fact interested in the complete hadronic structure (i.e., proton and neutron distributions) of the nucleus. This can, of course, be inferred from the electromagnetic structure in some model-dependent way, but it is not as clean a determination as one would like. There will always be room in the nuclear physics arsenal for the other probes[1].

General Description; Relativistic Treatment

I begin with a general description of electron scattering in a completely relativistic treatment using only the rules of QED, maintaining complete Lorentz covariance, and imposing only one condition, that of current conservation.

The lowest order Feynman diagram describing the electron scattering process is shown in Fig. 1. Using the usual Feynman rules (e.g., Refs. 9,10) one gets the three basic ingredients: i) an electron spinor, $u(k_i s_i)$ from the incoming lepton leg, a $\bar{u}(k_f s_f)$ from the outgoing lepton leg, and a factor $e\gamma_\mu$ at the lepton-photon vertex, thus giving the electron four-current, $j_\mu \sim e\bar{u}(k_f s_f)\gamma_\mu u(k_i s_i)$; ii) a photon propagator $\sim 1/q^2$; and iii) the nuclear current $\sim e\langle f|J_\mu|i\rangle$. This last term is written as the matrix element of the nuclear current operator between initial and final nuclear states to remind us that it is, in general, a very complex object. Indeed, it is this complexity which we wish to probe with electron scattering.

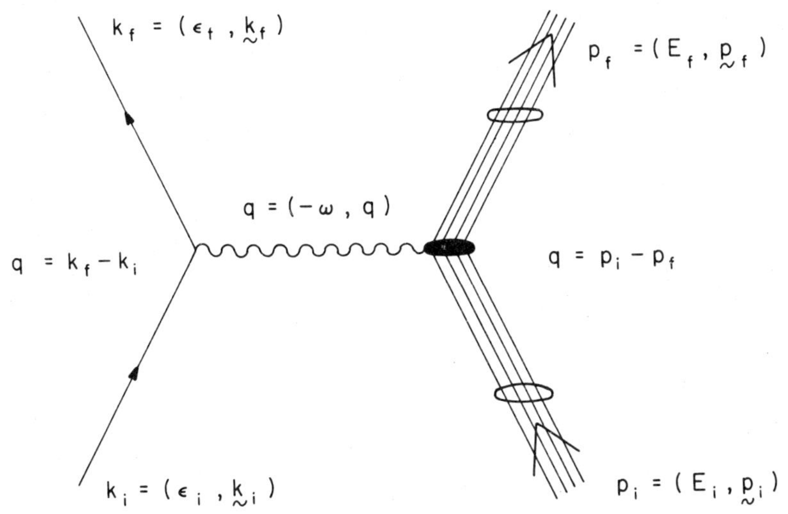

Fig. 1. One-photon exchange contribution to electron scattering.

Combining the above pieces, worrying about all of the proper factors of i, 2π, etc., one obtains the scattering matrix element corresponding to Fig. 1:

$$< k_f s_f \{f\} | S | k_i s_i \{i\}> =$$
$$(2\pi)^4 \delta^{(4)}(p_i - p_f - q)\frac{e^2}{\Omega}\, \bar{u}(k_f s_f)\gamma_\mu u(k_i s_i)$$
$$\frac{1}{q^2} < f | J_\mu(0) | i > \tag{1}$$

where $\{i\}$ and $\{f\}$ refer to a complete set of quantum numbers specifying the initial ($|i>$) and final ($|f>$) nuclear states. Box normalized electron spinors have been used with a box volume Ω. Finally, the four-dimensional delta function simply expresses energy/momentum conservation in the process.

Equation 1 is, in fact, all we need to begin our discussion of electron scattering. However, it may be instructive for the student to derive this expression from the usual insertion of electromagnetism into quantum mechanics (whenceforth, of course, derive the Feynman rules for QED). To begin, consider the interaction of a (relativistic) electron with external electric and magnetic fields which we shall eventually recognize as deriving from the nucleus. The "minimal" substitution, consistent with gauge invariance,

$$p^{\mu} \rightarrow p^{\mu} - eA^{\mu} \quad , \tag{2}$$

when applied to the Dirac equation of a free particle allows us to identify the "perturbing" interaction Hamiltonian as

$$H = - ej^{\mu}(x)A_{\mu}(x) \tag{3}$$

where $j^{\mu}(x) = i\bar{\Psi}_f(x)\gamma^{\mu}\Psi_i(x)$ is the lepton current and $A_{\mu}(x)$ is the four-vector electromagnetic potential of the external field.

Treating the process within first-order time-dependent perturbation theory gives us the matrix element of the scattering operator

$$\begin{aligned} \langle k_f s_f |S| k_i s_i\rangle &= \langle k_f s_f | -i \int H(x) d^4x | k_i s_i \rangle \\ &= - e \int d^4x \, \bar{\Psi}_f(x) \gamma^{\mu} A_{\mu}(x) \Psi_i(x) \quad . \end{aligned} \tag{4}$$

For plane-wave, box-normalized electrons

$$\Psi_i(x) = (\Omega)^{-1/2} u(k_i s_i) e^{ik_i \cdot x} \tag{5}$$

giving

$$\begin{aligned} &\langle k_f s_f |S| k_i s_i\rangle = \\ &\frac{-e}{\Omega} \bar{u}(k_f s_f)\gamma^{\mu}u(k_i s_i) \int e^{ik_i \cdot x - ik_f \cdot x} A_{\mu}(x) d^4x = \\ &\frac{-e}{\Omega} \bar{u}(k_f s_f)\gamma^{\mu}u(k_i s_i) \int e^{-iq \cdot x} A_{\mu}(x) d^4x \end{aligned} \tag{6}$$

where $q = k_f - k_i$ is the four-momentum transfer <u>to</u> the electron (be careful of the sign of q).

Now, $A_{\mu}(x)$ is the electromagnetic potential arising from the nuclear currents for a nucleus undergoing a transition from initial state $|i\rangle$ to final state $|f\rangle$. Thus, Maxwell's equations tell us

$$\Box A_{\mu}(x) = e\langle f| J_{\mu}(x) |i\rangle \tag{7}$$

where, again, the notation $\langle f| J_{\mu}(x) |i\rangle$ is used to emphasize that the nuclear current can, in general, be a rather complicated one owing to the structure of the nucleus.

The nuclear states can be taken as eigenstates (in a Heisenberg representation) of the four-momenta p_i and p_f. This gives

$$\Box A_\mu(x) = e\langle f|J_\mu(0)|i\rangle e^{i(p_i-p_f)\cdot x} \quad . \tag{8}$$

Fourier-transforming both sides with $e^{-iq\cdot x}$ gives

$$\int e^{-iq\cdot x}\Box A_\mu(x)d^4x = e\langle f|J_\mu(0)|i\rangle \int e^{i(p_i-p_f-q)\cdot x}d^4x \quad . \tag{9}$$

Partially integrating the left-hand side twice and recognizing the right-hand side as a representation of the delta function gives

$$-q^2\int e^{-iq\cdot x}A_\mu(x)d^4x =$$
$$e\langle f|J_\mu(0)|i\rangle(2\pi)^4\delta^4(p_i-p_f-q) \quad . \tag{10}$$

Substitution of Eq. 10 into the left-hand side of Eq. 6 then reproduces the desired result, Eq. 1.

We form the cross section simply by applying Fermi's Golden Rule using a density of final states for inclusive electron scattering (only the electron detected),

$$\rho_f = \frac{\Omega}{(2\pi)^3}\, d\underset{\sim}{k}_f \tag{11}$$

and an incident flux of

$$\frac{1}{\Omega}v_{rel} = \frac{1}{\Omega}\,\frac{|k_i\cdot p_i|}{\varepsilon_i E_i} \tag{12}$$

(where I have taken the mass of the electron to zero). In so doing (and remembering that with our box normalization we should replace $(2\pi)^3\ \delta(\underset{\sim}{p}_i - \underset{\sim}{p}_f - \underset{\sim}{q})$ with $\delta(\underset{\sim}{p}_i,\underset{\sim}{p}_f+\underset{\sim}{q})$, after some algebra, we can write

$$d\sigma = \frac{4\alpha^2}{q^4}\,\frac{d\underset{\sim}{k}_f}{2\varepsilon_f}\,\frac{1}{|k_i\cdot p_i|}\,\eta^{\mu\nu}W_{\mu\nu} \tag{13}$$

where α is the fine-structure constant and we define the "lepton response tensor"

$$\eta_{\mu\nu} = -2\varepsilon_i\varepsilon_f\bar{u}(k_is_i)\gamma_\nu u(k_fs_f)\bar{u}(k_fs_f)\gamma_\mu u(k_is_i) \tag{14}$$

and the "nuclear response tensor"

$$W_{\mu\nu} = (2\pi)^3 E_i \langle i|J_\nu(0)|f\rangle\langle f|J_\mu(0)|i\rangle \; \delta^4(p_i - p_f + q) \quad . \qquad (15)$$

For scattering of unpolarized electrons from unpolarized nuclei, with no final state polarizations detected (I will briefly mention the case of polarization later) we replace $\eta_{\mu\nu}$ and $W_{\mu\nu}$ by their values averaged over initial spins and summed over final:

$$\eta_{\mu\nu} \rightarrow \bar{\eta}_{\mu\nu} = \frac{1}{2} \sum_{s_i} \sum_{s_f} \eta_{\mu\nu} \qquad (16)$$

$$W_{\mu\nu} \rightarrow \bar{W}_{\mu\nu} = \overline{\sum_i} \sum_f W_{\mu\nu} \quad . \qquad (17)$$

Using the positive-energy projection operator

$$\sum_s u(p,s)\bar{u}(p,s) = (\not{p} + m)/2m \quad , \qquad (18)$$

and some algebra allows us to write (neglecting the electron mass, m_e)

$$\bar{\eta}_{\mu\nu} = k_{i\mu}k_{f\nu} + k_{i\nu}k_{f\mu} - k_i \cdot k_f \delta_{\mu\nu} \quad . \qquad (19)$$

The most general tensor we can make out of the available nuclear quantities (assuming the current is a polar vector, i.e., tacitly requiring parity conservation) must evidently take the form (with some foresight in notation, a la ref. 4):

$$W_{\mu\nu} = W_1 \delta_{\mu\nu} + W_2 \frac{p_{i\mu}p_{i\nu}}{M_T^2} + A \frac{q_\mu q_\nu}{M_T^2}$$
$$+ B \frac{(p_{i\mu}q_\nu + p_{i\nu}q_\mu)}{M_T^2} + C \frac{(p_{i\mu}q_\nu - p_{i\nu}q_\mu)}{M_T^2} \qquad (20)$$

where W_1, W_2, A, B, and C are 'structure functions' which depend only on the Lorentz scalars q^2 and $q \cdot p_i$.

We now impose the requirement that the nuclear current be conserved:

$$\frac{\partial}{\partial x_\mu} <f|J_\mu(x)|i> = \frac{\partial}{\partial x_\mu} e^{i(p_i-p_f)\cdot x} <f|J_\mu(0)|i>$$

$$= iq^\mu <f|J_\mu(0)|i> = 0 \quad , \tag{21}$$

i.e., that

$$q^\mu \overline{W}_{\mu\nu} = \overline{W}_{\mu\nu} q^\nu = 0 \quad . \tag{22}$$

(One observes trivially that $q^\mu \bar{\eta}_{\mu\nu} = \bar{\eta}_{\mu\nu} q^\nu = 0$, $m_e = 0$.) This requirement gives four conditions (since $p_{i\mu}$ and q_μ are independent) which require $C = 0$ and determines A and B in terms of W_1 and W_2. The result is that $\bar{W}_{\mu\nu}$ has as its most general form consistent with current conservation:

$$\overline{W}_{\mu\nu} = W_1(q^2, q\cdot p_i)[\delta_{\mu\nu} - \frac{q_\mu q_\nu}{q^2}]$$

$$+ W_2(q^2, q\cdot p_i)\frac{1}{M_T^2}(p_{i\mu} - \frac{p_i\cdot q}{q^2} q_\mu)(p_{i\nu} - \frac{p_i\cdot q}{q^2} q_\nu) \quad . \tag{23}$$

Neglecting the electron mass and recognizing that $q^2 = 4\varepsilon_i\varepsilon_f \sin^2(\theta/2)$ (where θ is the angle between k_i and k_f) one finds that

$$\bar{\eta}^{\mu\nu}\overline{W}_{\mu\nu} = 2\varepsilon_i\varepsilon_f\,[2\sin^2(\theta/2)\,W_1 + \cos^2(\theta/2)\,W_2] \quad . \tag{24}$$

Thus, after some further algebra, one achieves the desired result

$$\frac{d\sigma}{d\Omega_f d\varepsilon_f} = \frac{\sigma_M}{M_T}[W_2 + 2W_1 \tan^2(\theta/2)] \tag{25}$$

where

$$\sigma_M = \frac{\alpha^2 \cos^2\theta/2}{4\varepsilon_i^2 \sin^4\theta/2} \tag{26}$$

is the Mott cross section for scattering of a relativistic electron from a point charge.

Equation 25 is, therefore, the most general expression for the (e,e') scattering cross section that preserves Lorentz invariance and is consistent with current conservation (all only to lowest order in α).

I should remark in passing that one can also treat[4,11] weak interactions along the same lines (within, if you like, a single W, Z exchange approximation). In this case the full nuclear current contains both vector and axial-vector pieces and the analog to Eq. 20 will contain additional terms. There also will be no statement of current conservation similar to Eq. 21 (although one often uses CVC and assumes nuclear isospin to be a good quantum number). Thus the analysis of the weak interaction processes proceeds as above, but the final expression will generally involve more nuclear structure functions.

I should also observe at this point that there will be a neutral current contribution to the electron scattering process (with the photon exchange of Fig. 1 replaced by a Z^0 exchange). Since the weak interaction is so much weaker than the electromagnetic interaction, the Z^0 exchange contribution is almost always negligible. Indeed, one must work very hard to find effects of the weak interaction term. Usually this involves studying a parity-violating process, such as measuring asymmetries in polarized electron scattering[12,13]. This is, however, beyond the scope of the present discussion and I will simply neglect the neutral current contribution.

Higher Order Corrections

Equation 25 has been derived keeping only terms to first order in α. Quantum Electrodynamics supplies a prescription for calculating the higher order corrections as long as one's stamina holds out. In practice, these corrections are often small and can be neglected or at least approximated in some crude way. There are occasions, however, when, depending on one's application, it may be necessary to explicitly calculate these correction terms.

Representative diagrams of the higher-order processes are shown in Fig. 2. Because of its small mass, the electron tends to bremsstrahlung low energy real photons (Fig. 2a) rather profusely during the scattering process. These electrons thus lose more energy than one would expect from the simple kinematics of the electron-nucleus scattering. This effect appears in an experiment as a "radiation tail" spreading a given peak in the spectrum of scattered electrons out to higher energy loss. This is usually corrected for by the experimentalist by simply fitting over the radiative tail when determining a cross section. For broader peaks or when multiple peaks overlap considerably, the "radiative corrections" must be unfolded based upon detailed QED calculations.

The diagrams of Figs. 2b-d represent radiative corrections due to virtual photons and, in the terminology of QED, are

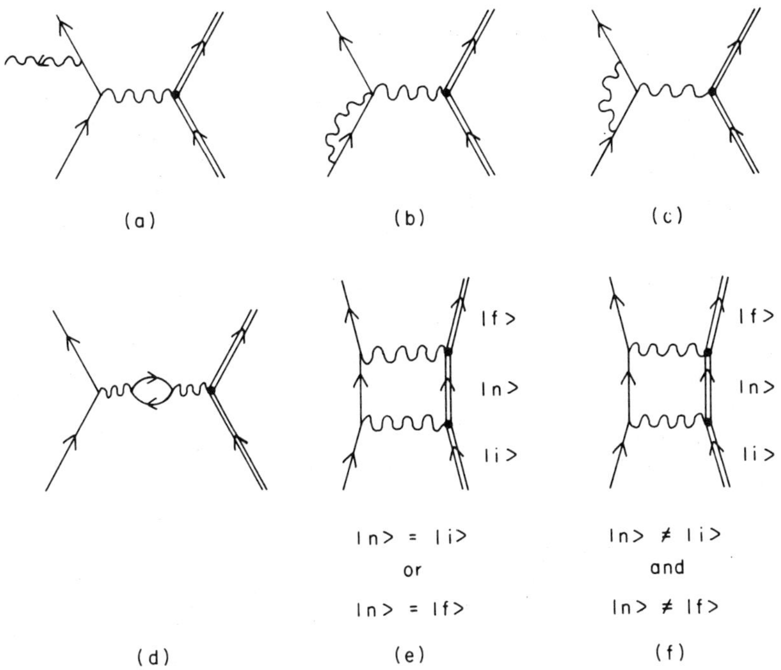

Fig. 2. Some higher-order (multiple-photon) processes relevant to electron scattering.

usually referred to as self-energy corrections, vertex corrections, and vacuum polarization, respectively. These terms represent corrections to Eq. 25 that modify the simple form of that equation. The calculation of these processes is again straightforward, though cumbersome; the details are outside the scope of the present discussion.

Figure 2e represents a process often called a "Coulomb" correction where, in addition to the scattering of interest, the electron also scatters elastically from the nucleus in either its initial or final state. Again, in reality a higher order QED effect, it is usually handled in practice by treating the distortion of the incoming and outgoing electron waves in the Coulomb field of the nucleus using standard Distorted Wave Born Approximation (DWBA) techniques. (Equation 25 is, in these terms, a Plane Wave Born Approximation (PWBA) result.) The key feature to note here is that since the distortion effects are dominated by scattering from the full charge of the nucleus, the Coulomb corrections represent corrections of order $Z\alpha$ and so become quite significant for heavier nuclei.

Finally, there is the graph of Fig. 2f which differs from that of Fig. 2e only in that the intermediate nuclear state is other than the initial or final state of the nucleus. That is, this diagram represents a scattering through two distinct (one virtual) excited states of the nucleus. This process is usually referred to as a "dispersion correction". In the spirit of QED, it would be hard to distinguish the processes of Figs. 2e and 2f. In practice, however, they are distinguished by the techniques used to calculate them. Of most importance is the fact that for the dispersion corrections, the "second scattering" involves a transition and hence is often governed not by Z protons but by the single nucleon (or, at most, a few nucleons) that takes part in the transition. The presence of the distortion corrections seems to be an undeniable experimental fact, but an experimental verification of the calculations of dispersion effects remains a difficult task. (There is recent hope that they may be deduced in a measurement now underway at NIKHEF (Amsterdam).)

Finally, in practice, one must also consider multiple scatterings of the electrons from different nuclei in the target. But, as a theorist, I am happy to live in a world where targets are infinitesimally thick and such effects do not exist! (They can, of course, be deduced in an experiment by the seemingly trivial process of varying the thickness of the target.)

Having thus briefly discussed some of the more important higher order corrections to the single-photon exchange approximation, I will assume that the form of Eq. 25 is sufficient for our present discussion. If it were really necessary, we could include the higher order QED effects.

Considerations of Nuclear Physics

Nuclear physicists are prone to construct models of nuclear structure wherein they predict charge and current densities within the nucleus. Electron scattering is the natural tool to use to confirm or deny such predictions. While Eq. 25 is indeed the general expression for the cross section and the structure functions W_1 and W_2 are indeed the quantities which one deduces from an experiment, I have yet to relate these in detail to the charge and current densities. This can, of course, be done in a general way, but I will take note of the most frequent aspects of the applications to nuclear structure and tailor the formalism accordingly. In particular, nuclear physicists frequently deal with discrete states characterized by a definite angular momentum and a definite parity. Most present models of nuclear structure also do not consider relativistic effects (although, of course, such effects are receiving considerable attention at the present time as we shall learn from other presentations to this workshop). I shall, therefore, endeavor to recast Eq. 25 in a form more amenable to common pursuits in nuclear structure physics. We have not the time to work through these derivations in detail. Instead, I will outline the key steps, define the key quantities, and try to emphasize the physics motivation for the derivations. The reader should be able to fill in the intermediate steps for himself; Refs. 1-4 should help considerably in this regard.

We begin by relaxing the restraints of relativity, working in only the three space dimensions. We can do so by backing up to Eq. 7, again performing the Fourier transform, but this time taking only the time-dependence out of the nuclear matrix element:

$$\begin{aligned}\int e^{-iq\cdot x}\Box A_\mu(x)\, d^4x &= \\ \int e \langle f|J_\mu(x)|i\rangle e^{-iq\cdot x} d^4x &= \\ 2\pi e\delta(E_i - E_f - \omega) J_\mu(\underset{\sim}{q})\end{aligned} \tag{27}$$

where ω is the energy transferred from the electron to the nucleus. I have defined

$$J_\mu(\underset{\sim}{q}) = \int e^{-i\underset{\sim}{q}\cdot\underset{\sim}{x}} \langle f|J_\mu(x)|i\rangle\, d^3x \quad , \tag{28}$$

that is, the usual three-dimensional Fourier transform of the matrix element of the nuclear charge and current operator.

It will be convenient to work in a coordinate system where the z-axis is chosen to lie along q, i.e., where $\hat{z} = \underset{\sim}{q}/|\underset{\sim}{q}|$. In such a coordinate system the exponential factor in Eq. 28 has the rather simple expansion

$$e^{-i\underset{\sim}{q}\cdot\underset{\sim}{x}} = \sum_{L=0}^{\infty} \sqrt{4\pi}\, [L](-i)^L j_L(qx) Y_{L0}(\Omega_x) \ . \tag{29}$$

where j_L is a spherical Bessel function, Y_{L0} is a spherical harmonic, and I use the shorthand notation $[L] = (2L+1)^{1/2}$. If I then define the quantity

$$M_{LM}(q) = \int j_L(qx) Y_{LM}(\Omega_x) \rho(\underset{\sim}{x})\, d^3x \quad , \tag{30}$$

the charge density (the μ=0 component of $J_\mu(q)$) takes on the simple form

$$\rho(\underset{\sim}{q}) = \sum_{L=0}^{\infty} 4\pi\, [L](-i)^L M_{L0}(q) \quad . \tag{31}$$

I will claim without proof that M_{LM} is an irreducible tensor operator in the nuclear coordinate space. I may thus apply the Wigner-Eckart theorem to the nuclear matrix elements of this operator. (The reader who is not fluent in angular momentum algebra may wish to consult the book by Edmonds (Ref. 14).) In so doing I am now in the position to take full advantage of the angular momentum quantum numbers of the initial and final nuclear states in limiting the sum on L, in establishing selection rules for a given transition, etc., and thereby simplifying any discussion of electron scattering to discrete states.

The analysis for the three-vector current proceeds along similar lines although it is somewhat complicated due to the vector character of the current. Again, one chooses a coordinate system with z-axis along q. In this case we express the three-current in a spherical basis of unit vectors $\underset{\sim}{e}_{\underset{\sim}{q},z}$ along the z-axis and $\underset{\sim}{e}_{\underset{\sim}{q},\pm 1}$ transverse to the z-axis. The current is then expanded

$$\underset{\sim}{J}(\underset{\sim}{q}) = \sum_{\lambda=z,\pm 1} J_\lambda(\underset{\sim}{q})\, \underset{\sim}{e}^{\dagger}_{\underset{\sim}{q},\lambda} \tag{32}$$

where

$$J_\lambda(\underset{\sim}{q}) = \underset{\sim}{e}_{\underset{\sim}{q},\lambda} \cdot \underset{\sim}{J}(\underset{\sim}{q}) \tag{33}$$

The three components of $\underset{\sim}{J}$ are not independent, however, because current conservation requires

$$\underset{\sim}{q} \cdot \underset{\sim}{J} = -\omega\rho \quad , \tag{34}$$

which allows us to eliminate one component, J_z in terms of ρ:

$$J_z = -\omega\rho / |\underset{\sim}{q}| \quad . \tag{35}$$

Thus, given the charge density, ρ, only two of the components of $\underset{\sim}{J}$ are independent. We have chosen to eliminate J_z leaving the two transverse components, $J_{\pm 1}$.

We again expand[14] the exponential function in Eq. 28 this time including the basis vector $e_{q\lambda}$:

$$\underset{\sim}{e}_{\underset{\sim}{q}\lambda} e^{-i\underset{\sim}{q}\cdot\underset{\sim}{x}} = -\sqrt{2\pi} \sum_{J>0} (-i)^J [J]$$

$$[\lambda j_J(qx) \underset{\sim}{Y}^{\lambda}_{JJ1}(\Omega_x) + \frac{1}{q} \underset{\sim}{\nabla} \times (j_J(qx) \underset{\sim}{Y}^{\lambda}_{JJ1}(\Omega_x))] \tag{36}$$

where $\underset{\sim}{Y}^{M}_{JJ1}$ is the vector spherical harmonic[14]

$$\underset{\sim}{Y}^{M}_{JJ1} = \sum_{m,m'} \langle Jm1m' | J1JM\rangle \, Y_{Jm}(\Omega_x) \, \underset{\sim}{e}_{m'} \; . \tag{37}$$

It is also related to the spherical harmonic by operation with the orbital angular momentum operator:

$$\underset{\sim}{L} Y_{JM} = \sqrt{J(J+1)} \, \underset{\sim}{Y}^{M}_{JJ1} \tag{38}$$

Defining the "transverse electric" and "transverse magnetic" operators (again they are irreducible tensor operators)

$$T^{el}_{JM}(q) = \frac{1}{q}\int d^3x \, \underset{\sim}{\nabla} \times [j_J(qx)\underset{\sim}{Y}^{M}_{JJ1}(\Omega_x)]\cdot \underset{\sim}{J}(\underset{\sim}{x}) \tag{39}$$

$$T^{mag}_{JM}(q) = \int d^3x \, j_J(qx)\underset{\sim}{Y}^{M}_{JJ1}(\Omega_x)\cdot \underset{\sim}{J}(\underset{\sim}{x}) \quad , \tag{40}$$

one can write the λ component of the three-current as

$$J_\lambda(q) = -\sqrt{2\pi} \sum_{J>0} (-i)^J [J] \, (T^{el}_{J\lambda} + \lambda T^{mag}_{J\lambda}) \quad . \tag{41}$$

Having thus expanded the charge and current operators in terms of irreducible tensor operators we can now rotate them to any arbitrary coordinate system using the Wigner rotation matrices ("D-matrices").

Forming the contraction of the lepton and nuclear response tensors, averaging over initial spins and summing over final, the quantity we need becomes

$$\overline{\eta}^{\mu\nu}\overline{W}_{\mu\nu} = \frac{1}{2J_i+1} \sum_{M_f \, M_i} [2(J_\mu Q^\mu)(J^*_\nu Q^\nu) + \frac{1}{2} q^2 J_\mu J^\mu] \tag{42}$$

where

$$Q_\mu = \frac{1}{2}(k_i + k_f)_\mu \quad . \tag{43}$$

The product of two ρ's is

$$\sum_{M_i M_f} |\rho(q)|^2 = 4\pi \sum_J |\langle J_f \| M_J(q) \| J_i \rangle|^2 \tag{44}$$

and we observe that

$$\sum_{M_i M_f} J_\lambda(q)\rho(q)^* = 0 \qquad \lambda = \pm 1 \qquad (45)$$

To form the product of two J's requires only to observe that T^{el} and T^{mag} have different parity. Thus we find

$$\sum_{M_i M_f} J_\lambda(q) J_{\lambda'}(q)^* =$$

$$2\pi \delta_{\lambda\lambda'} \sum_{J=1}^{\infty} [|<J_f\|T_J^{el}\|J_i>|^2 + |<J_f\|T_J^{mag}\|J_i>|^2] . \qquad (46)$$

When all of the algebraic dust settles, one arrives at the final operational expression

$$\frac{d\sigma}{d\Omega} = 4\pi \sigma_M [\frac{q^4}{\tilde{q}^4} F_L(q)^2 + (\frac{q^2}{2\tilde{q}^2} + \tan^2(\theta/2)) F_T(q)^2] \qquad (47)$$

where the "longitudinal" and "transverse" form factors are defined by

$$F_L(q)^2 = \sum_{J=0}^{\infty} |<J_f\|M_J(q)\|J_i>|^2 / (2J_i + 1) \qquad (48)$$

and

$$F_T(q)^2 = \sum_{J>0} [(|<J_f\|T_J^{el}(q)\|J_i>|^2$$

$$+ |<J_f\|iT_J^{mag}(q)\|J_i>|^2)] / (2J_i + 1) . \qquad (49)$$

The longitudinal form factor is also sometimes called the "charge" or "Coulomb" form factor and the transverse form factor is sometimes, especially for elastic scattering, called the "magnetic" form factor. The unwary reader should be aware that different authors sometimes use different conventions regarding factors of 4π, Z, and even overall factors of q^2.

Applications

We have, after some labor, now arrived at an expression for the electron scattering cross section which can be easily used to analyze and interpret measurements. The cross section is expressed in terms of the nuclear matrix elements of (the multipole projection of the Fourier transform of) the charge and current density operators. The multipolarity of the irreducible tensor operators and the definite quantum numbers of the (discrete) nuclear states can be used to establish the usual selection rules. Contact can now be directly made with the usual formalism for γ-decay (the decay $B(E\lambda)$'s are related to the matrix elements of M_λ and T_λ^{el} while the $B(M\lambda)$'s are related to the matrix elements of T_λ^{mag}). We have "lost sight" of the fully relativistic treatment we began with, but one could (although it is not simple) go back to the fully relativistic formalism if necessary.

Since the longitudinal and transverse form factors enter Eq. 47 multiplied by different kinematic factors, one can use electron scattering to separate the two. In particular, by taking data at various beam energies and scattering angles chosen to keep the momentum transfer, q, fixed, one can plot the measured cross sections as a function of $[q^2/(2\underset{\sim}{q}^2) + \tan^2(\theta/2)]$ and deduce F_L^2 and F_T^2. This process, frequently called a "Rosenbluth separation", allows one to determine the contributions due to the nuclear charge and the nuclear currents separately (F_L^2 depends on ρ alone while F_T depends only on $\underset{\sim}{J}$). A related method requires the measurement of the (e,e') cross section at a scattering angle, θ, of 180° (recall the form of the Mott cross section, Eq. 26) so that only the transverse form factor contributes. This information can then be combined with data taken at another scattering angle for the same q to determine the longitudinal form factor.

Equation 47 applies to the case of unpolarized electron scattering from unpolarized nuclear targets. If one were to specify an initial or final state polarization, the simplicity obtained in Eqs. 44-46 by summing over M_i and M_f would not be realized. Eq. 47 would contain additional pieces depending on interferences between the Coulomb, electric, and magnetic multipoles and between different multipolarities. However, the additional "kinematic" freedom obtained by varying the direction and manner of polarization can be used in a "super-Rosenbluth" separation to disentangle many of these interferences and, thereby, to gain additional information about the charge and current distributions. This method holds much promise for future studies once the necessary technology of polarized targets is established. Similar formalisms apply to the (e,e'X) reactions (where, in some cases at least, one may think of using the measurement of X to determine the final nuclear polarization). The reader is referred to the work of Donnelly and Raskin[15] for complete details of the application of the above formalism to the case where information about nuclear polarizations is available. I will confine further discussion to the simple form of Eq. 47.

It is not my purpose here to review the vast body of existing electron scattering data. Instead, I show only two examples (both elastic scattering) to demonstrate some of the points discussed above. Figure 3, from Refs. 5 and 16, shows the charge density of ^{208}Pb as deduced from electron scattering data (and some muonic atom data). ^{208}Pb is a spin-zero nucleus so that only one multipole, M_0, contributes in this case. In order to complete the Fourier transform of Eq. 31 to construct ρ one in principle needs data for all q's. However, in this case, data were available only for q < 500 MeV/c. This incompleteness is one of the major contributions to the error bars on the density in Fig. 3. Nonetheless, the accuracy obtained is still quite astounding.

Figure 4, taken from the review article of Ref. 7, shows an example of a somewhat more complicated case, that of the (elastic) transverse form factor of ^{93}Nb. The ground state spin of this nucleus is 9/2 so transverse multipoles with J= 1-9 contribute in this case (for elastic scattering only, time-reversal eliminates the even multipoles). Thus, the task of unfolding the data to deduce the nuclear current density is much more complicated. The comparison between theory and experiment is, therefore, usually made directly in terms of predicted and measured form factors.

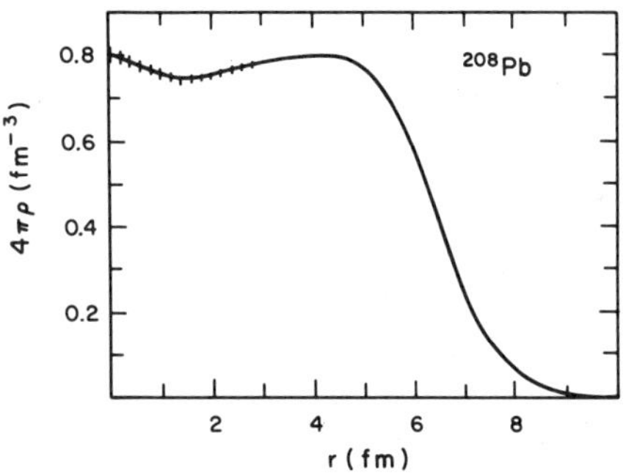

Fig. 3. The charge density of ^{208}Pb as deduced[16] from electron scattering and muonic atom data.

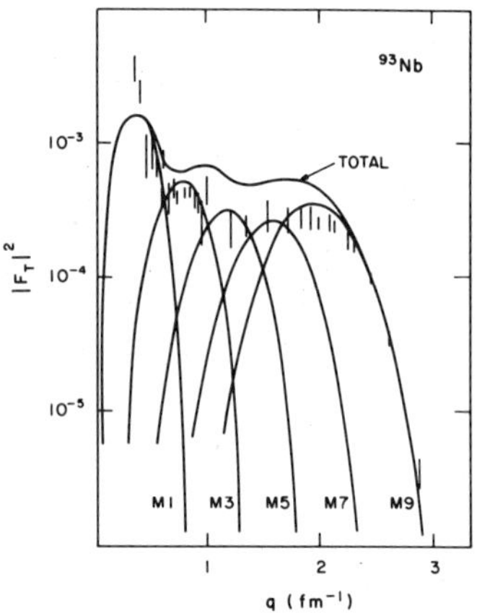

Fig. 4. Elastic magnetic (transverse) electron scattering form factor for ^{93}Nb. The measurements and the theoretical curve are discussed in some detail in Ref. 7.

Rather than continue to duplicate the discussion of the many excellent review articles available, let me instead turn to a brief discussion of an important theoretical point bearing upon the applications of electron scattering to nuclear physics. We have been careful above to point out that it is the <u>matrix elements</u> of the nuclear charge and current operators that enter the expressions. Whenceforth does one obtain these matrix elements in practice? In some models, one may parameterize the matrix elements themselves, but, in most applications (particularly for shell models), one assumes the charge and current operators to take on the form:

$$\rho(\mathbf{x}) = \sum_{i=1}^{A} e_i \delta(\mathbf{r}_i - \mathbf{x}) \tag{50}$$

$$\mathbf{j}(\mathbf{x}) = \sum_{i=1}^{A} e_i \frac{\mathbf{p}_i}{M} \delta(\mathbf{r}_i - \mathbf{x}) \tag{51}$$

$$\boldsymbol{\mu}(\mathbf{x}) = \sum_{i=1}^{A} \mu_i \frac{\boldsymbol{\sigma}_i}{2M} \delta(\mathbf{r}_i - \mathbf{x}) \tag{52}$$

$$\mathbf{J}(\mathbf{x}) = \mathbf{j}(\mathbf{x}) + \boldsymbol{\nabla} \times \boldsymbol{\mu}(\mathbf{x}) \quad . \tag{53}$$

The full nuclear operators are thus taken to be sums over the individual nucleon operators. These are, in turn, local operators for point nucleons with <u>free</u> nucleon properties. The full three-vector current operator (Eq. 53) contains contributions from both the "convection" current (Eq. 51) arising from the motion of the nucleon charges and the "magnetization" or "spin" current (Eq. 52) due to the intrinsic magnetic moments of the nucleons.

The form of Eqs. 50-53 derives from our understanding of the <u>free</u> Dirac current for the nucleon:

$$\langle p_f, \sigma_f, \tau_f | J_\mu | p_i, \sigma_i, \tau_i \rangle =$$

$$\frac{i}{\Omega} \bar{u}(p_f \sigma_f) \eta_f^\dagger [F_1 \gamma_\mu + F_2 \sigma_{\mu\nu} q^\nu] \eta_i u(p_i \sigma_i)$$

$$= \frac{1}{\Omega} \eta_f^\dagger \chi_f^\dagger F_1 \chi_i \eta_i \qquad (\mu = 0)$$

$$= \frac{1}{\Omega} \eta_f^\dagger \chi_f^\dagger \left[F_1 \frac{(\mathbf{p}_i + \mathbf{p}_f)}{2M} + (F_1 + 2mF_2) \frac{i \mathbf{q} \times \boldsymbol{\sigma}}{2M}\right] \chi_i \eta_i \qquad (\mu \neq 0) \tag{54}$$

Here η and χ are non-relativistic two-component isospin and spin state vectors. One of the burning questions in nuclear physics for the last few years concerns this assumption of free nucleon properties. Are the nucleons modified in some fundamental way when they are within the nuclear medium? Are binding effects and/or the relativistic corrections to Eq. 54 important in the case where the relatively weak binding results from the cancellation of two (or more) rather strong potentials? The meson exchange currents I will discuss in the second half of these lectures are, indeed, one correction to the form of Eqs. 50-53. But the question of the correct form of the nuclear current operator is not completely answered. I expect we shall hear more about this from some of the other lecturers.

If one does assume the form of a sum over single-nucleon operators, the demands upon a model of the nuclear structure are quite simple and the calculations remain tractable. In this case the full matrix elements are just sums over single particle matrix elements

$$\langle f|O^{(1)}|i\rangle = \sum_{\alpha\beta} \Psi_{\alpha\beta;fi} \langle\alpha|O^{(1)}|\beta\rangle \tag{55}$$

where the "one-body density matrix", Ψ, contains all the necessary structure information. Even the most complex shell model calculations will require only a few values for Ψ since the active orbitals, α and β, are usually quite limited. In some simple models it is even possible to invert the process and determine Ψ from experiment[17].

In addition to defining the one-body density matrix, the structure model should also be expected to supply the form of the single-particle wave functions to be used in evaluating the matrix elements of Eq. 55. In practice, however, these are often taken to be of some phenomenological form, e.g., harmonic oscillator, or Woods-Saxon single particle wave functions. The skeptical reader should indeed question this practice particularly for higher q where one is looking at these wave functions in finer detail.

One also has to recognize that in most models the coordinates r_i (i = 1 to A) are not independent but must at least sum to the center-of-mass coordinate of the nucleus. This gives rise to a correction that must be applied if Eqs. 50-53 are to be used. In addition, the nucleon is not in reality a point particle but has a size and charge/current distribution of its own (F_1 and F_2 in Eq. 54) which must also be folded in. Common practice[1] is to take both of these corrections into account by multiplying the nuclear form factors by "center-of-mass" and "single-nucleon" form factors which are themselves functions of q. It is my opinion, however, that this is a practice that needs to be re-examined, particularly as the measurements are pushed to higher and higher momentum transfers.

Thus, while the formalism of electron scattering is straightforward, there are a number of pitfalls that can be encountered in applying it. Over the years, various approximations have been made, usually simply for the sake of expediency. After a while these approximations have been accepted as valid. I believe it is time to take another look at some of these. Fortunately, a number of people, including some of the other lecturers, have been doing just that.

Response Surfaces

Electron scattering determines the cross section of Eqs. 25 and 47 as a function of both the energy transferred to the nucleus and the momentum transferred to the nucleus. Thus, it determines the entire "response function" of the nucleus in the (q,ω) plane. In concluding my primer on electron scattering, let me remind the reader of the rich diversity of nuclear phenomena that are accessible by varying both q and ω.

Figure 5 shows a (theorist's simplification of a) cut across the response surface at constant q. At $\omega = 0$ (neglecting recoil of the nucleus) one sees an elastic scattering peak including a "radiative tail". At low q and forward angles this peak will be quite large due to the charge scattering from all of the protons in the nucleus. At very backward angles, the magnetic elastic contribution will dominate the cross section, but it will be much smaller than the charge contribution since

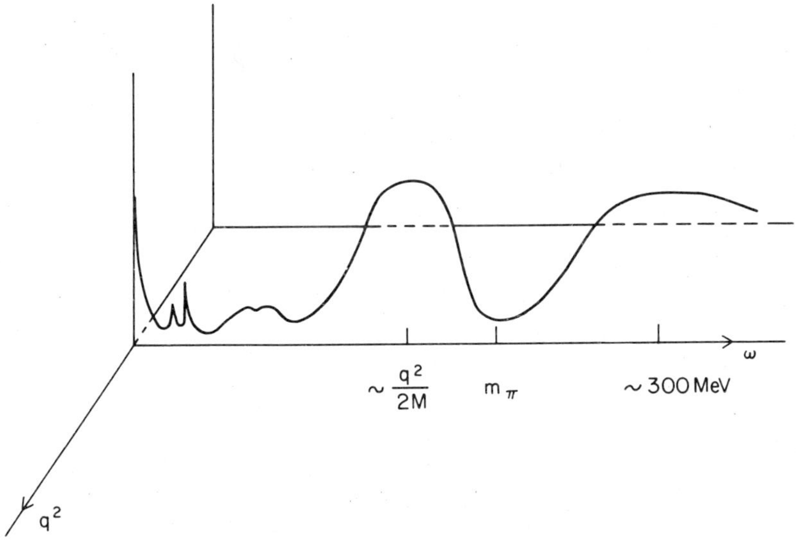

Fig. 5. A somewhat idealized constant-q cut across the response surface for electron scattering from nuclei.

magnetic scattering takes place only from nucleons with unpaired spins (and, hence, will be zero for spin-zero nuclei). Of course, the detailed q-dependence of the elastic form factors is determined by the details of the charge and current densities of the nuclear ground state.

For ω less than a few MeV most stable nuclei will display a number of sharp peaks corresponding to excitation of discrete excited states in the nucleus. Again, the details of the q-dependence of the cross section will specify the details of the charge and current densities, in this case, transition charge and current densities. Final states of differing spin will have considerably different form factor shapes reflecting in part the different q-dependences of the different orders of the spherical Bessel functions in Eqs. 30, 39, and 40, as selected by the Wigner-Eckart theorem. Thus, the response surface in the region of inelastic scattering to particle-stable states will show considerable and rapid variation as a function of both q and ω.

Once ω gets above particle threshold, the inelastic peaks widen out, the density of excited states increases, and the response surface smooths out somewhat. There may still be some distinct peaks because the selection rules for particle emission may occasionally suppress the particle-decay width. But, in general, the region above particle emission threshold is difficult to analyze. At slightly higher ω, however, one enters the region of the broad strongly-excited overlapping collective states, the "giant resonance" region. This is a region with a long history in nuclear physics. I will only comment that these resonances are most certainly excited in electron scattering. The polarized measurements and the (e,e'X) measurements discussed above can be expected to play important roles in future studies of this region as one attempts to unfold the overlapping resonances of different spin.

The largest feature seen in Fig. 5 is the broad peak centered around $\omega \simeq q^2/2M_n$. This is easily understood if one imagines the nucleus to be a structureless "holding area" for non-interacting nucleons. The electrons then simply scatter from the individual nucleons as if they were free and unbound. In that case the energy lost by the electron would simply be the kinetic energy of the recoiling nucleon, i.e., $q^2/2M_n$. This "as if" elastic scattering is termed "quasi-elastic" scattering or "quasi-free" scattering. The peak is, in fact, slightly shifted from $\omega = q^2/2M_n$ due to the binding of the nucleons. The width is easily understood by including the Fermi motion of the nucleons in considering the "quasi"-elastic scattering. This region of the response surface has been examined quite closely in the last few years and it has become quite clear that, while the description I have given above is qualitatively correct, there may be quite interesting physics to be learned in this region. Electron scattering, and in particular the separation of the longitudinal and transverse responses, is playing an important role in elucidating this physics. Unfortunately, I have no time to discuss this in further detail; I am sure some of the other lecturers will touch upon it.

As one increases ω further, the next "event" is pion production threshold. Above this threshold the cross section increases considerably and is dominated at $\omega \simeq 300$ MeV by pion production predominantly through excitation of the Δ-resonance in the nucleus. Again, there is very interesting physics here that I will simply not have the time to discuss.

Finally, I should mention that the region between the quasi-elastic region and the onset of pion production, the so-called "dip" region, has also been receiving considerable attention of late. The experimental cross section is higher (by perhaps a factor of two) than the standard theories predict. There have been a number of suggestions for "new" processes to explain this region but, once again, I must leave these for the other lecturers.

As I have said, a similar formalism applies to all electron scattering from composite objects. One of the main themes of this workshop is the role of quarks in nuclear physics. This becomes a viable field of study because we now recognize that the nucleon itself is a composite object made up of quarks. I therefore conclude my introduction to electron scattering by showing in Fig. 6 the (q,ω) response surface for electron scattering from the nucleon. In this case one also sees an elastic peak with a radiative tail. Then as one increases ω, the cross section vanishes until one reaches the first excited state of the nucleon, the $\Delta(3,3)$. Other excited states of the nucleon appear above the background at even higher energies until finally one reaches the equivalent of the "quasielastic" peak. Since the quark constituents of the nucleon apparently are not weakly bound, this region is not interpreted in the same way as in the nuclear case, and the term "quasielastic" is inappropriate. It is, instead, usually referred to as the "deep inelastic" region. Indeed, it is the study of the energy and momentum transfer dependences of the cross sections in this region that provides us with some of the most convincing evidence that the nucleon is indeed a composite object composed of multiple (three) quarks!

MESON EXCHANGE CURRENTS AND ELECTRONUCLEAR REACTIONS

Introduction - What and Why

About fifty years ago, Yukawa[18] postulated a description of the nucleon-nucleon interaction in terms of meson exchange.

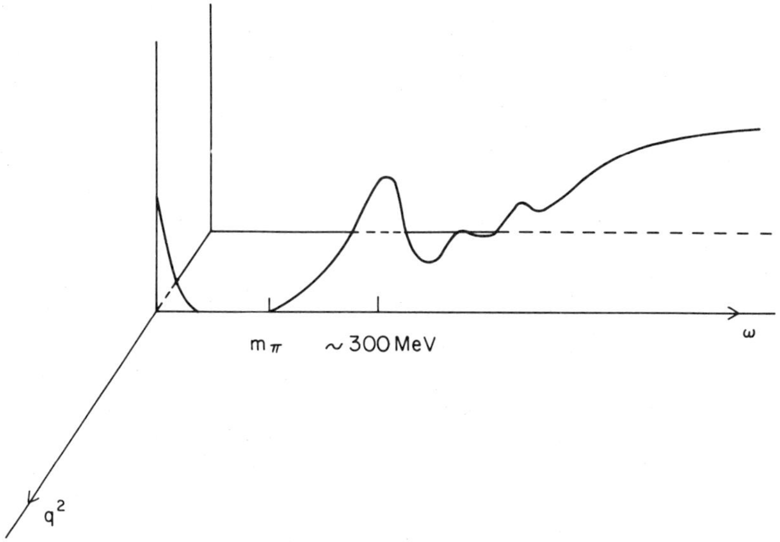

Fig. 6. A plot similar to that of Fig. 5, except now for electron scattering from the nucleon.

His one-pion exchange model has survived the years and remains as the popular mechanism for explaining the longest range portion of the nucleon-nucleon potential. The shorter range part of that interaction has, of course, since been formulated in terms of other exchanges: rho-meson exchange, omega-meson exchange, multiple pion exchanges, sigma-meson exchange, etc. with good qualitative but perhaps only modest quantitative success.

It was recognized (see e.g., Ref. 19) soon after Yukawa's proposed theory that, in addition to describing the nucleon-nucleon interaction, a pion exchange model would have implications for the electromagnetic properties of the nucleus. The full nuclear current densities could no longer be described solely in terms of the location, motion, and properties of the individual nucleons. The exchanged pions themselves carry charge and thus their presence would modify the nuclear current densities.

In principle, one would want to directly describe the mesonic contributions to the electromagnetic current in terms of the properties of the pion field within the nucleus. In practice, this is a very difficult problem. However, the mesons do not exist entirely independently of the nucleons. They are present in the nucleus because they are being exchanged between two (or more) nucleons. One can thus describe these "meson exchange currents" in two different ways. The first takes the nucleons as sources of the meson field. This meson field will pervade the nucleus in some equilibrium way and the electromagnetic properties can be derived from the field equations. (Recall how the "meson cloud" picture is used in describing the anomalous magnetic moments of the nucleon.) The second approach is to recognize that one nucleon is the source of the meson and another is its sink. Thus the contribution of the exchanged mesons to the electromagnetic properties of the nucleus can be written as an equivalent current operator in the nucleon space which depends not on the quantum numbers and variables of individual nucleons but on the quantum numbers and variables of pairs of nucleons. That is, the equivalent meson current operator is a two-nucleon current operator. In principle, of course, in going from the meson field to the equivalent operator in nucleon space, one may also give rise to three-nucleon,...,A-nucleon currents. These are, however, usually assumed to be small.

Since I have just spent some time describing how electron scattering is the ideal probe for studying the charge and current densities of the nucleus, I should now like to devote the second half of my lectures to discussing the interpretation of existing experimental data in terms of nucleon and meson degrees of freedom. In the back of my mind is the question of whether or not there is in this data undeniable evidence for the existence of quark effects in nuclei. Before I touch upon this question, however, let me first sketch how the meson exchange currents (MEC) are evaluated and then review the relevant experimental information. For more information and more detail, the reader is referred to the excellent three volume series, Mesons in Nuclei[20], for more than one could ever want to know about the subject.

How

There have been a number of approaches to the evaluation of the MEC. However, the most practical approach and the approach almost universally used for detailed calculations relies on evaluating what one guesses to be the most important Feynman diagrams describing the interactions of the nucleons and the exchanged mesons. Figure 7 shows a fairly complete set of the diagrams that have been considered in past calculations.

The diagram of Fig. 7a is the one which you would all think of first. Here we have, as I have just described, two nucleons interacting by exchanging a meson with that meson then interacting with an external electromagnetic field (indicated by

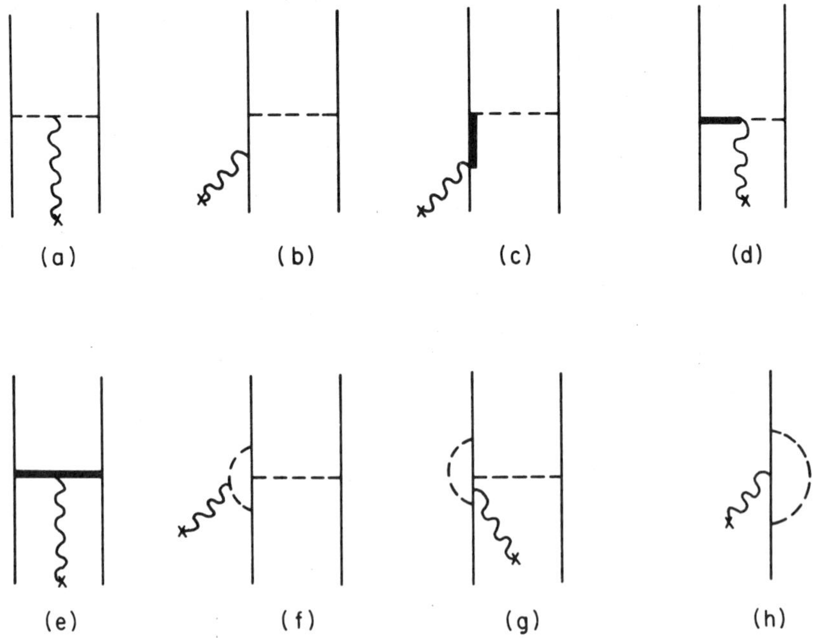

Fig. 7. Feynman diagrams frequently used to describe meson exchange currents.

an "x"). Figure 7e is similar except that instead of an exchanged pion (dashed line) a heavier meson such as a ρ or ω (heavy line) is exchanged. Figure 7a is sometimes called the "pionic" graph while Figs. 7a and 7e together are often called "true" exchange current graphs. Figure 7d is similar to Figs. 7a and 7e except that in this case the external field induces a meson "reaction" changing, e.g., a pion into a rho.

Figure 7b also represents a fairly obvious exchange current, although in this case one has to take care to identify the legitimate exchange current contributions. Figure 8 demonstrates this point by showing the first few terms in a time-ordered expansion of the Feynman diagram of Fig. 7b. I would argue that the first four terms (Figs. 8a-8d) are valid exchange current diagrams since, if one cuts them horizontally above the point of photon attachment, the meson exchange is required to return the system to a system of two nucleons. Figure 8e fails this test, however, and instead simply represents an interaction of a nucleon with the external field <u>followed</u> in time by an interaction with another nucleon. But this last process is exactly what is, in principle, described by the structure model based on interacting nucleons alone. Thus, in practice, Fig. 7b is usually broken down in the series shown in Fig. 8 with the diagrams of Figs. 8a-8c giving rise to the "pair" exchange current. The diagram of Fig. 8d is called the "recoil" exchange current. And, once again, the diagram of Fig. 8e should not be included. The diagram of Fig. 8 makes most sense in a model with pseudoscalar pion-nucleon coupling; in other models one evaluates the same diagram, then often referred to as the "seagull" diagram, in a different way. The result, at least to lowest order in relativity, is the same, however.

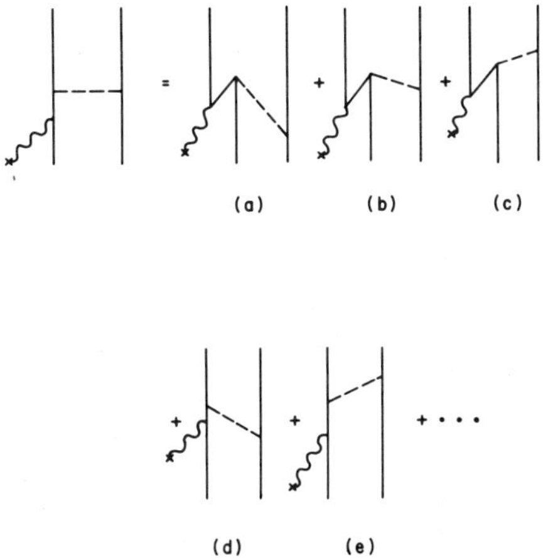

Fig. 8. A time-ordered expansion of the diagram of Fig. 7b.

Figure 7c represents the electroexcitation of a resonance (usually the $\Delta(3,3)$) within the nucleus. Again, cutting the graph above the point of photon attachment gives one nucleon and the resonance state so that a meson exchange is required to return the system to one with two nucleons. Thus this graph may also be considered an exchange current unless one explicitly includes the isobars as part of the nuclear wave function (in which case the isobar-nucleon system is part of the structure calculation). In this case, one must be very careful to avoid double counting.

Finally, the diagrams of Figs. 7f-7h represent "renormalization" diagrams describing how the nucleon's properties are themselves modified by the pion field. (Figure 7h is, in essence, describing part of the "pion cloud" often held responsible for the anomalous magnetic moments of the nucleons.)

I shall now set out to demonstrate how a two-nucleon current operator is obtained from consideration of such graphs. Before I begin, however, I should mention at least one other common approach[21,22] wherein one writes down the general expression for the exchange currents in terms of the invariants of the problem and all available variables (momenta, Pauli spin matrices, etc.) and then uses general principles such as current conservation, unitarity, time-reversal invariance, etc., to reduce the number of free parameters. While this is an appealing approach there are not, in general, enough restrictions one can place on the general form and one is left to rely on at least some phenomenological information. It is, nonetheless, a very instructive approach and anyone who intends to examine the problem of exchange currents in detail should study it.

I now proceed to explicitly evaluate the diagrams of Fig. 7a and Fig. 7b (at least the parts corresponding to the diagrams of Figs 8a-8c), that is, to obtain the "pair" and "pionic" exchange currents. Since these diagrams contain only a single pion, and since the pion is the longest ranged of the mesons we consider, one might expect these currents to be the most important, at least at relatively low momentum transfer. We shall see below that this is indeed usually the case.

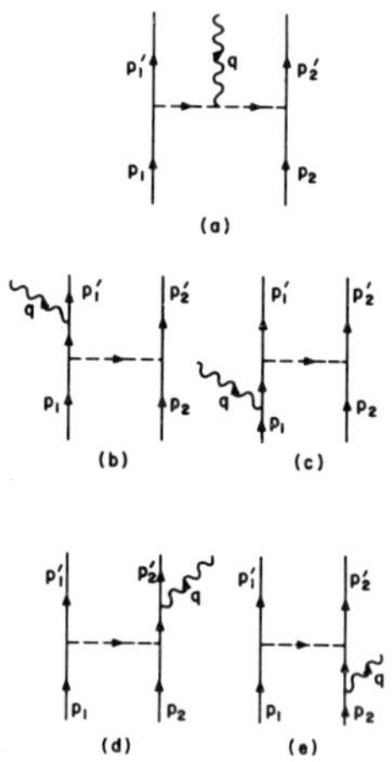

Fig. 9. Labelled graphs for the one-pion exchange currents we explicitly evaluate in the present discussion.

Figure 9 defines the notation I shall use with the incoming nucleons labelled by p_1 and p_2 and the outgoing nucleons labelled by p_1' and p_2'. As in the discussion of electron scattering above, q is the momentum transferred to the nucleus. To obtain the exchange currents, one simply uses the usual Feynman rules, evaluates the scattering matrix element for each of the diagrams, and then argues that this matrix element must have the general form:

$$\langle p_1' p_2' | S | p_1 p_2 \rangle = -2\pi i \delta(E) j^{\mu}(q) V_{\mu}(q) \tag{56}$$

in an external electromagnetic field V_μ (in this case deriving from the scattering electron).

The S-matrix element corresponding to Fig. 7a is thus

$$\langle p_1'p_2'|S|p_1p_2\rangle = \frac{1}{(2\pi)^6}\frac{M^2}{(E_1'E_2'E_1E_2)^{1/2}}$$

$$\bar{u}(p_1')\sqrt{2}\, g\gamma_5 \bar{u}(p_1)\bar{u}(p_2')\sqrt{2}\, g\gamma_5 u(p_2)$$

$$\frac{i}{(p_1'-p_1)^2-\mu^2+i\varepsilon}\; e_\pi\, (p_1-p_1'+p_2'-p_2)_\mu \; \frac{i}{(p_2'-p_2)^2-\mu^2+i\varepsilon}$$

$$(-i)^3 V^\mu (2\pi)^4 \delta^4(p_1+p_2-q-p_1'-p_2') \tag{57}$$

where I have assumed a pseudoscalar pion-nucleon coupling and where

$$f^2 = (g\mu/2M)^2/4\pi \simeq 0.08 \tag{58}$$

is the pion-nucleon coupling. The charge carried by the exchanged pion may be written in terms of the isospin operators of the two nucleons as

$$e_\pi = \frac{ie}{2}[\underset{\sim}{\tau}^{(1)} \times \underset{\sim}{\tau}^{(2)}]_3 \quad . \tag{59}$$

The current so deduced is (in momentum space)

$$j_\mu(p_1,p_1',p_2,p_2',q) = i\delta^3(\underset{\sim}{p}_1+\underset{\sim}{p}_2-\underset{\sim}{q}-\underset{\sim}{p}_1'-\underset{\sim}{p}_2'/(E_1E_2E_1'E_2')^{1/2}$$

$$eg^2M^2/(2\pi)^3\; [\underset{\sim}{\tau}^{(1)} \times \underset{\sim}{\tau}^{(2)}]_3$$

$$\bar{u}(p_1')\gamma_5 u(p_1)\bar{u}(p_2')\gamma_5 u(p_2)\; (p_1-p_1'+p_2'-p_2)_\mu$$

$$\frac{1}{[(p_1'-p_1)^2-\mu^2+i\varepsilon][(p_2-p_2')^2-\mu^2+i\varepsilon]} \quad . \tag{60}$$

Since the structure models in which we are likely to imbed these currents are nonrelativistic, there is no sense in keeping the full relativistic expression. We may thus approximate

$$\bar{u}(p_1')\gamma_5 u(p_1) \rightarrow \frac{i}{2M}\,\underset{\sim}{\sigma}_1 \cdot (\underset{\sim}{p}_1'-\underset{\sim}{p}_1) \quad . \tag{61}$$

The product of the two pion propagators can be rewritten using a Feynman parameterization[23]

$$\frac{1}{(p_1-p_1')^2-\mu^2+i\varepsilon}\;\frac{1}{(p_2-p_2')^2-\mu^2+i\varepsilon} =$$

$$\int_0^1 dz\; [((p_1-p_1')^2-\mu^2+i\varepsilon)z + ((p_2-p_2')^2-\mu^2+i\varepsilon)(1-z)]^{-2}$$

$$= \int_{-1/2}^{1/2} dv\; [(\underset{\sim}{p}_1'-\underset{\sim}{p}_1+\underset{\sim}{q}(1-2v)/2)^2 + L_\pi^2 - i\varepsilon]^{-2} \tag{62}$$

where

$$L_\pi = [\mu^2 + \underset{\sim}{q}^2(1-4v^2)/4]^{1/2} \quad . \tag{63}$$

To take our result over to coordinate space, we perform a Fourier transform for each of the nucleon momenta, and identify the "local" exchange current operator $j_\mu(\underset{\sim}{x}_1,\underset{\sim}{x}_2,q)$ accordingly:

$$j_\mu(\underset{\sim}{x}_1,\underset{\sim}{x}_2,\underset{\sim}{x}_1',\underset{\sim}{x}_2',\underset{\sim}{q}) = \frac{1}{(2\pi)^6}\int d\underset{\sim}{p}_1 d\underset{\sim}{p}_2 d\underset{\sim}{p}_1' d\underset{\sim}{p}_2'$$

$$e^{i(\underset{\sim}{p}_1'\cdot\underset{\sim}{x}_1'+\underset{\sim}{p}_2'\cdot\underset{\sim}{x}_2'-\underset{\sim}{p}_1\cdot\underset{\sim}{x}_1-\underset{\sim}{p}_2\cdot\underset{\sim}{x}_2)} j_\mu(\underset{\sim}{p}_1,\underset{\sim}{p}_1',\underset{\sim}{p}_2,\underset{\sim}{p}_2',\underset{\sim}{q})$$

$$= j_\mu(\underset{\sim}{x}_1,\underset{\sim}{x}_2,\underset{\sim}{q})\;\delta^3(\underset{\sim}{x}_1-\underset{\sim}{x}_1')\delta^3(\underset{\sim}{x}_2-\underset{\sim}{x}_2') \; . \tag{64}$$

Carrying out these steps for the current of Eq. 60, after some algebra, we find

$$\underset{\sim}{j}(\underset{\sim}{x}_1,\underset{\sim}{x}_2,\underset{\sim}{q}) = e\left(\frac{f}{\mu}\right)^2[\underset{\sim}{\tau}^{(1)}\times\underset{\sim}{\tau}^{(2)}]_3\;(\underset{\sim}{\sigma}_1\cdot\underset{\sim}{\nabla}_1)(\underset{\sim}{\sigma}_2\cdot\underset{\sim}{\nabla}_2)$$

$$\int_{-1/2}^{1/2} dv\;(-iqrv + \underset{\sim}{x})\;\frac{e^{-x}}{x}\;e^{-i\underset{\sim}{q}\cdot(\underset{\sim}{R}-v\underset{\sim}{r})}$$

$$j_0 \sim O(p^2/M^2) \simeq 0 \tag{65}$$

where $\underset{\sim}{x} = L_\pi \underset{\sim}{r}$, $\underset{\sim}{r} = \underset{\sim}{x}_1-\underset{\sim}{x}_2$, and $\underset{\sim}{R} = (\underset{\sim}{x}_1+\underset{\sim}{x}_2)/2$.

This, then, is the result for the pionic exchange current. It depends on the coordinates, spins, and isospins of the two nucleons involved in the exchange. The functional form certainly resembles that of the one-pion-exchange potential (OPEP) although the range depends in a non-trivial way on the momentum transfer. One can crudely picture the Feynman parameter integral as representing the attachment of the photon to the pion at all points along the pion line (with $v=-1/2$ at one nucleon and $v=+1/2$ at the other). The two key points to notice for later discussion are i) to lowest order in p/M, the charge operator vanishes and only the three-vector current remains, and ii) again to lowest order in p/M, the isospin structure of the current is purely isovector.

For the current of Fig. 9b, one finds in a similar fashion

$$j_\mu^{(b)}(p_1,p_1',p_2,p_2',q) = \delta^3(\underset{\sim}{q}+\underset{\sim}{p}_1'+\underset{\sim}{p}_2'-\underset{\sim}{p}_1-\underset{\sim}{p}_2)\, eg^2M^2/(16\pi^3)$$

$$[\tau_+^{(1)}\tau_-^{(2)} + \tau_3^{(2)} + \tau_3^{(1)}\tau_3^{(2)}]/(E_1E_1'E_2E_2')^{1/2}$$

$$\bar{u}(p_2')\gamma_5 u(p_2)\frac{1}{(p_2-p_2')^2-\mu^2+i\varepsilon}$$

$$\bar{u}(p_1')\gamma_\mu \frac{1}{\gamma\cdot(p_1'+q)-M+i\varepsilon}\gamma_5 u(p_1)\,. \tag{66}$$

To retain only the pair diagrams of Figs. 8a-8c, we take only the negative-frequency part of the intermediate nucleon propagator by making the replacement

$$\frac{1}{\gamma\cdot p-M+i\varepsilon} \to \frac{1}{2E_p}\,\frac{\beta(E_p+\underset{\sim}{\alpha}\cdot\underset{\sim}{p})-M}{p_0+E_p-i\varepsilon}\,. \tag{67}$$

Again we make the non-relativistic reduction of this current by making use of Eq. 61 and taking

$$\bar{u}(p_1')\gamma_\mu[\beta(E_{p_1'+q} + \underset{\sim}{\alpha}\cdot(\underset{\sim}{p}_1'+\underset{\sim}{q})-M]\gamma_5 u(p_1)$$

$$\begin{aligned} &\to i\underset{\sim}{\sigma}_1\cdot\underset{\sim}{q} && (\mu = 0)\\ &\to 2iM\underset{\sim}{\sigma}_1 && (\mu = i)\quad . \end{aligned} \tag{68}$$

The result obtained, after including the permutations corresponding to Figs. 9b-9e, is

$$\underset{\sim}{j}^{(b+c+d+e)}(\underset{\sim}{x}_1,\underset{\sim}{x}_2,\underset{\sim}{q}) = -ef^2[\underset{\sim}{\tau}^{(1)}\times\underset{\sim}{\tau}^{(2)}]_3$$

$$[\underset{\sim}{\sigma}_1\cdot\hat{\underset{\sim}{r}}\,\underset{\sim}{\sigma}_2\, e^{-i\underset{\sim}{q}\cdot\underset{\sim}{x}_2} + \underset{\sim}{\sigma}_1\,\underset{\sim}{\sigma}_2\cdot\hat{\underset{\sim}{r}}\, e^{-i\underset{\sim}{q}\cdot\underset{\sim}{x}_1}]\,(1+x_\pi)\frac{e^{-x_\pi}}{x_\pi^2} \tag{69}$$

where, again to lowest order in p/M, we find $j_0 \simeq 0$.

The structure of the pair current is somewhat simpler than that of the pionic current primarily because the electromagnetic interaction occurs with one nucleon (rather than the pion) so that the photon coupling depends primarily on the coordinates, etc. of that nucleon. Again, however, we see the structure of OPEP reflected in this current. As in the case of the pionic current, we discover that the charge operator vanishes in the non-relativistic reduction and that the current is a purely isovector object.

Many authors have restricted themselves to considering only the pair and pionic currents in their calculations. The main reason for this choice is the argument that the matrix elements of the shorter-range exchanges will be suppressed due to the nucleon-nucleon correlations which tend to keep the nucleons apart. The pair and pionic currents also have the very attractive feature that, together, they obey current conservation:

$$\nabla' \cdot j(x') + i[H, \rho(r')] = 0 \cdot \tag{70}$$

If the Hamiltonian consists of a one-body kinetic energy, T, and a two-body potential, V, and the charge density remains a one-body operator (as it does in the non-relativistic reduction of the pair and pionic currents), i.e., if

$$\rho(r') = \rho^{(1)}(r') = \sum_{i=1}^{A} e \frac{1}{2} (1+\tau_3^{(i)}) \delta(r' - x_i) , \tag{71}$$

then the statement of current conservation of Eq. 70 can be written as separate statements for the one- and two-body currents:

$$\nabla' \cdot j^{(1)}(r') + i[T, \rho(r')] = 0 \tag{72}$$

$$\nabla' \cdot j^{(2)}(r') + i[V, \rho(r')] = 0 \ . \tag{73}$$

The one-body current of Eqs. 51-53 obeys Eq. 72. The sum of the pair and pionic currents can be shown explicitly to obey Eq. 73 if the interaction potential is taken to be the OPEP potential

$$V = V^{OPEP} = -\left(\frac{f}{\mu}\right)^2 (\tau^{(1)} \cdot \tau^{(2)}) (\sigma_1 \cdot \nabla_1) (\sigma_2 \cdot \nabla_2) \frac{e^{-\mu r}}{r} \ . \tag{74}$$

Thus, not only does the model that takes the pionic and pair currents seem to give us the most important exchange current contribution, it also obeys current conservation if the interaction is chosen to be consistent with the one-pion exchange approximation.

The isobar current corresponding to Fig. 7c is divergenceless and, as such, does not contradict Eq. 73 if it is included with the pionic and pair currents. A number of authors therefore choose to include the isobar current in their calculations of the longest-range effects. Other authors choose to include isobar configurations directly in their nuclear wave functions and care must be taken to avoid double counting the isobar contributions. The values of the $\gamma N\Delta$ and $\Delta N\pi$ couplings to be used in evaluating Fig. 7c are a bit uncertain or at least model dependent. Thus, variations in the overall magnitude of the isobar current of at least 30% can be found in the literature.

Evaluation of the other graphs of Fig. 7, the shorter-range exchanges, can be more difficult. The form of the current for, e.g., vector-meson exchange, is simply much more complicated. The appropriate coupling constants are often not known well enough to evaluate the currents with confidence. Any explicit calculation of the matrix elements of these currents can be expected to be sensitive to the form of the above-mentioned nucleon-nucleon correlations and may, therefore, be somewhat ambiguous. I will not go any further in my explicit evaluation of the diagrams of Fig. 7. Other authors have looked at the other diagrams and these results may be found in the literature (see Ref. 23). The examples of the pair and the pionic current should be sufficient to give us some understanding of the approach that is usually taken. The reader should keep in mind as I review below some results of calculations that there is less unanimity concerning the details of some of the shorter-range exchange currents than for the pion exchange currents.

The reader should also keep in mind that, like the one-body currents of Eqs. 50-52, the exchange currents have so far been evaluated for point nucleons and point mesons. As in the case of the one-body currents, they must be corrected for the finite size and internal charge and current distributions of the nucleons and mesons. Again, as in the one-body case, this is usually accomplished by multiplying these contributions by a phenomenological fit to the measured data on the form factor of the nucleon and meson (most authors assume the nucleon form factor for all exchange currents).

However, unlike the one-body case, the exchange currents also directly involve strong interaction processes, i.e., meson-nucleon coupling, and one has to question whether meson-nucleon form factors should be applied at each of the hadronic vertices. This question is no more easily answered in the case of exchange currents than for any other strong interaction process. Almost all authors, if they include this effect, take vertex functions of the form

$$v(q) = \frac{\Lambda^2}{\Lambda^2+q^2} \quad . \tag{75}$$

But these authors vary considerably in their choice of Λ, with some taking values as low as 300 MeV/c and others values as high as 2500 MeV/c. We shall see below how different values of Λ affect some of the calculated results. Nonetheless, it is important to remember that exchange currents are, in fact, strong interaction effects and that they are plagued by many of the uncertainties that plague the analysis of other strong interaction processes.

Inclusion of the exchange currents in a structure calculation is, in principle a straightfoward matter. One can define, in analogy to the one-body density matrix of Eq. 55, a two-body density matrix $\Psi_{\alpha\beta\gamma\delta}$ such that

$$<f|O^{(2)}|i> = \sum_{\alpha\beta\gamma\delta} \Psi_{\alpha\beta\gamma\delta;fi} <\alpha\beta|O^{(2)}|\gamma\delta> \quad . \tag{76}$$

Since the operator now involves two nucleons, the number of combinations and hence the number of required two-body density matrix elements will be substantially larger than for the one-body density. The relevant density matrix elements are also not simply confined to the active (valence) orbitals in the calculation since the exchange currents can involve a meson exchange between a nucleon in a valence orbital and one in a core orbital.

Let me now turn to a comparison of calculations of these meson exchange currents with electron scattering measurements. As in the discussion above, I do not intend this to be a complete scholarly review of all the work in the field. I have instead selected some examples which I hope will provide a basic appreciation for the present status of the comparison of theory with experiment. I apologize to my many colleagues whose work I have omitted. I hope I have provided sufficient references to enable the reader to track down the rest of this work if he so desires.

Results, A=2

The most natural place to begin looking for the effects of meson exchange currents is in the two-nucleon system. One has considerable confidence in our ability to describe the structure of this system. One also recognizes that the deuteron is a fairly diffuse object so that one might expect only the longer range effects to be important. Unfortunately, the deuteron ground state is isospin-zero. Therefore elastic electron scattering will not involve the one-pion exchange currents which we discussed above since they are purely isovector operators. Thus, for elastic scattering at least, one must turn to the shorter-range exchanges and to the higher order relativistic corrections to the pion exchanges. For both of these we have less confidence in our abilities to correctly evaluate them.

Nonetheless, the deuteron is a sufficiently important system to warrant considerable attention.

Figure 10 shows the situation for elastic electron scattering from deuterium[24-27]. In the notation we have used above, $B(Q^2)$ is the transverse form factor and $A(Q^2)$ is a combination of the longitudinal and transverse form factors. The exchange currents are expected to be most important for the three-vector current, so our attention should be focussed on B. Figure 10 also shows the results of two calculations[28], one with one-body currents only while the other includes the meson exchange currents. In this case the important terms seem to be the $\gamma\pi\rho$ current of Fig. 7d and some of the higher-order, isoscalar terms in the pair current. Figures 11 and 12 show similar curves but only over the range of momentum transfers for which data exists. The calculations in Fig. 12 are those of Ref. 29. In general, the agreement between theory and experiment is quite good although the large theoretical uncertainties associated with the calculation of isoscalar exchange currents in general make such a conclusion a bit tenuous. There is currently an experiment under way at SLAC which expects to measure B down to perhaps 10^{-9}. Inspection of Fig. 10 suggests that this new experiment will certainly provide a much more stringent test of the MEC predictions.

Elastic electron scattering from the deuteron is therefore a bit disappointing if one is

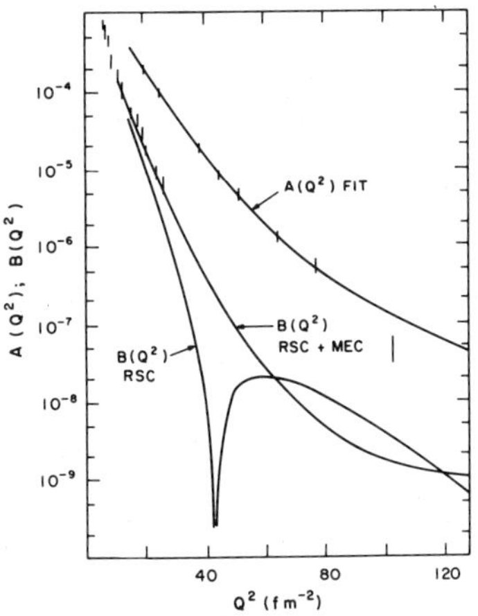

Fig. 10. Elastic electron scattering form factors for the deuteron.

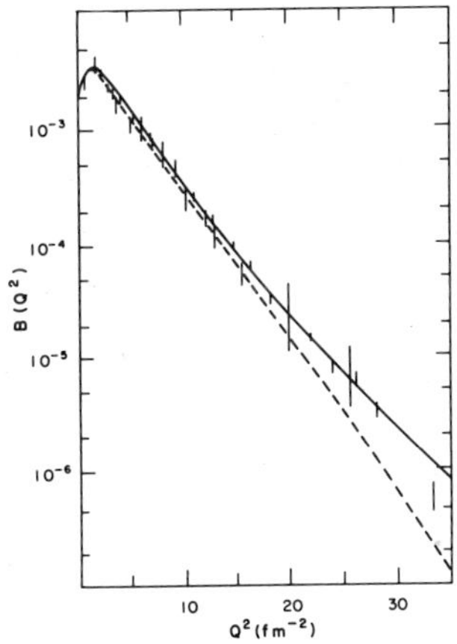

Fig. 11. Transverse form factors for electron scattering from the deuteron. Solid line includes MEC.

looking for convincing evidence for exchange currents. However, nature has not been completely unkind to us in the two-nucleon system. In particular, at energies just above breakup threshold one finds a region where the np pair is dominantly in an isovector state. Electroexcitation of this region may thus be expected to involve the non-relativistic one-pion exchange currents we have discussed above. This is indeed apparently the case. Figure 13 shows the electrodisintegration cross section (averaged over the first 3 MeV above threshold). The data[26,27] clearly lie above the one-body calculation. However, the calculation[30] including the one-pion exchange currents we have discussed certainly remedies this situation and provides a very successful description of this data. This calculation, now more than ten years old, provided much of the impetus to the study of exchange currents. It remains, as far as I know, the most convincing piece of evidence that the meson exchange currents do exist in nuclei and that they seem to behave in the manner which is predicted by the methods we have been discussing.

Figure 14 shows the situation for electrodisintegration of the deuteron out to somewhat higher momentum transfers including the new data of Ref. 31. The calculations[32] again show nearly perfect agreement with the data over the entire range of momentum transfer measured. In this case, one can see the dramatic effect of the MEC in filling in the diffraction minimum predicted with nucleons alone. One also sees that the one-pion exchange currents dominate out to perhaps 15 fm^{-2}

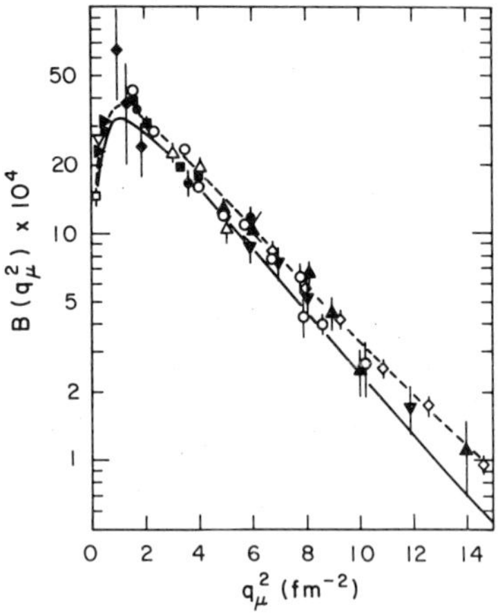

Fig. 12. Transverse form factor for elastic scattering from the deuteron. Dashed line includes MEC.

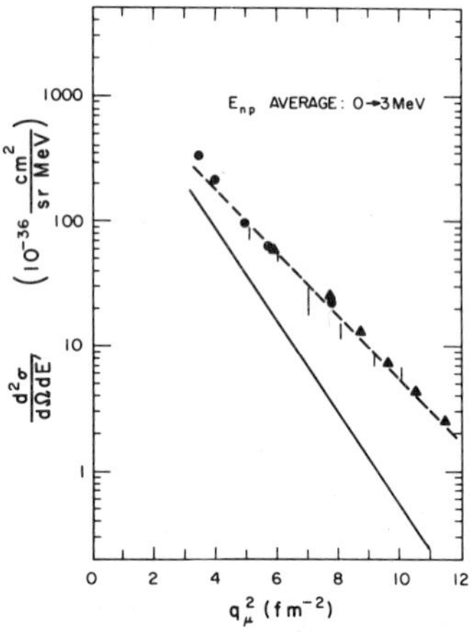

Fig. 13. Deuteron electrodisintegration cross section. Dashed line includes MEC.

and that it becomes necessary to include the shorter-range ρ exchange to describe the data out to 25 fm^{-2}. There is some hint that even more processes must be included above 25 fm^{-2}, but the evidence is as yet hardly compelling. At such large momentum transfers one is also fairly well embroiled in questions regarding the forms of electromagnetic and hadronic vertices, so that the theoretical uncertainties will increase the theoretical "error bars" that should be mentally added to the curves of Fig. 14.

Finally, for completeness I should add that some years ago Riska and Brown[33] used MEC to account for a 10% discrepancy between calculation and experiment for the total cross section for np radiative capture. This is far from the order of magntitude corrections seen in electrodisintegration, but it was also historically significant in establishing credibility for exchange currents since the discrepancy was long standing and had withstood many other attempts to resolve it.

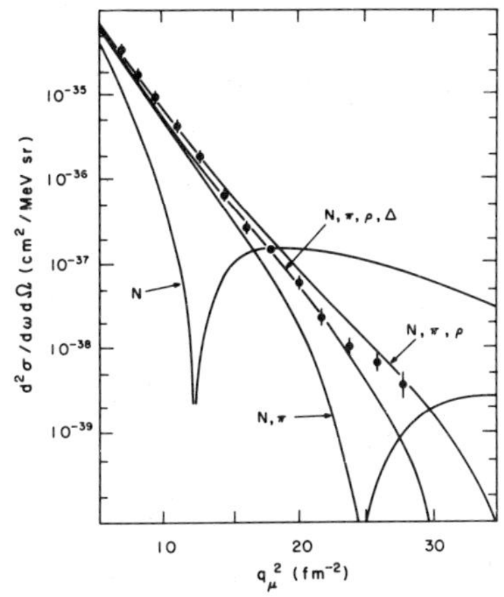

Fig. 14. Deuteron electro-disintegration cross section.

As I finish my discussion of the A=2 system, I must conclude that the evidence from electrodisintegration is quite compelling and that the (leading order at least) meson exchange currents are correctly describing the physics of the system. Even though the theoretical uncertainties are much greater for the elastic form factor, these calculations also seem to encompass the general trends in the data. The measurements to much higher q now underway at SLAC will certainly prompt the theorists to attempt to better understand this situation.

Results, A=3

Turning now to the next largest nuclear system, the A=3 system, we find that nature has been much kinder to us. Not only is the ground state of the three-body system T=1/2, so that the isovector exchange currents can contribute, we also find that both members of the isodoublet, ^3H and ^3He, exist as stable nuclei. Thus, by comparing measurements for ^3H and ^3He, one should be able to separate the isoscalar and isovector contributions. Calculations of the structure of the three-body nuclei are substantially more complicated than for the deuteron, but modern technology has begun to provide us with consistent, numerically stable solutions to the Faddeev equations. We may, therefore, have some faith in our ability to predict the one-body currents within the A=3 nuclei.

There are many calculations of the elastic electron scattering form factors for the three-body system. As we shall see below, exchange currents do play an important role in these form factors. Therefore, almost every structure calculation that has attempted to compare to the electron scattering form factors has had to include the exchange current effects.

Friar has pointed out a number of times (e.g., in Ref. 20) that extreme care must be taken if one is to include the effects of exchange currents beyond the non-relativistic limit we have discussed above. Essentially, one can perform a Foldy-Wouthuysen transformation that eliminates the exchange current contribution if an appropriate interaction is chosen. That is, there is a consistency requirement that must be met between the form of the currents and the form of the interaction used in the structure calculations. To the best of my knowledge no one has taken the care necessary to ensure this consistency condition. The MEC contributions to the charge operator are of this ilk. I will show the "standard" results for the charge form factors but the reader is duly warned that they may not be completely sensible for this reason.

As I have said, there are indeed a large number of calculations of exchange currents in the three-body system available. With one exception I shall show the results of Bornais et al.[34, 35] because they are, I believe, the most detailed and perhaps the most instructive.

Figure 15 shows the experimental data for the longitudinal form factor for elastic electron scattering from ^{3}He. Also shown are the calculations[34] with and without the MEC. There is little evidence for the necessity of including the MEC for q less than about 3 fm^{-1} (we are not plotting this against q^2 as we did for the two-nucleon systems), but in the region of the second maximum, for momentum transfers between 3 and 6 fm^{-1}, the MEC do seem to supply a much need boost in the calculated form factor. The MEC also seem to predict more structure at about 7 fm^{-1}, structure that is not obviously present in the data. In fact, in this region the one-body curve seems to do a better job in fitting the trend of the data. I will let the reader judge for him/herself whether this is strong evidence for the accuracy of the MEC calculations.

Figure 15 also shows the effect of including hadronic vertices in the calculation. $\Lambda = 4$ fm^{-1} is an intermediate value in the range that is usually adopted and, of course, the $\Lambda = \infty$ curve is the calculation without hadronic vertices. The effect is rather small until a q of about 7-8 fm^{-1} (50-60 fm^{-2}).

Figure 16 shows the contributions of the various diagrams to Bornais' calculations. Even for the charge operator, the pair term with pion exchange dominates over almost all of the other graphs. The interference between the many terms is obviously very complex and highly q-dependent.

The magnetic form factor for the A=3 nuclei is perhaps more interesting since here we have less theoretical uncertainty in the prediction of the exchange currents. The dominant pieces should be the non-relativistic pion exchanges for which our study of the two-nucleon systems has established some credibility. Figure 17 (taken from Ref. 25) shows the experimental data compared to calculations with and without the MEC. In this case there seems to be clear evidence for the exchange currents over almost the entire range of momentum transfers measured.

Figure 18 shows the decomposition of Bornais' calculations into the contributions from the different diagrams. Here we see that the pion exchange pair, pionic, and isobar contributions do dominate over most of the momentum transfer range shown. At higher momentum transfers, however, the interferences with the shorter-range exchanges can become quite significant.

Note in this case that the exchange current contributions do not go to zero as q goes to zero. This of course means that the MEC represent sizable modifications to the magnetic moments. Indeed, the study of magnetic moments was one of the first places where exchange

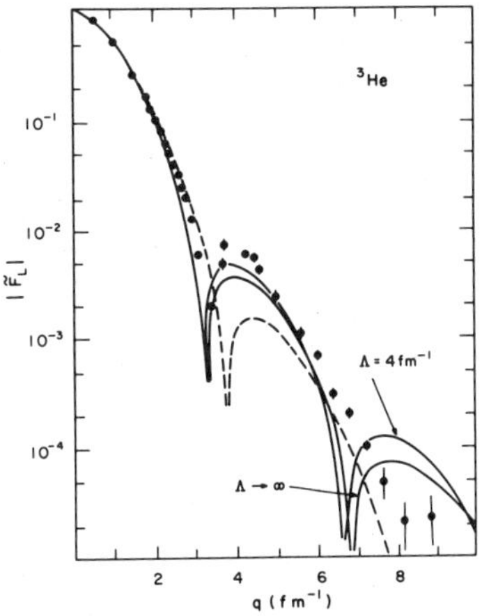

Fig. 15. Longitudinal form factor for elastic scattering from ^{3}He. $\tilde{F}$ indicates a form factor defined in a slightly different convention from the present work.

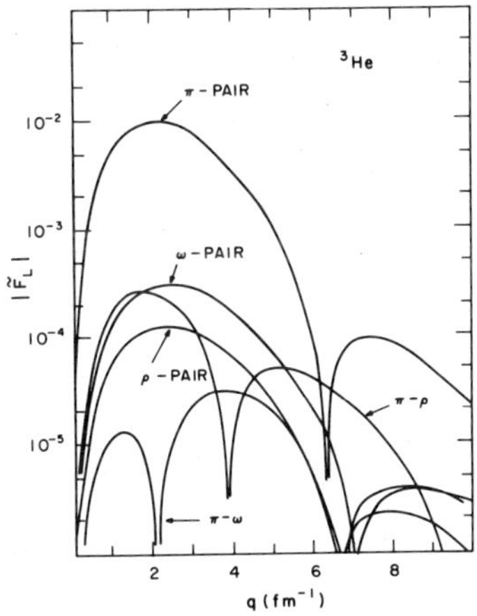

Fig. 16. Contributions[34] from different MEC processes. All are negative relative to the one-body contribution for $q \cong 0$.

currents were considered[36]. Today's "best" calculated values[37] for magnetic moments range from -2.08 nm to -2.18 nm for ^{3}He (to be compared to a one-body value of about -1.76 nm and an experimental value of -2.128 nm) and from 2.85 nm to 2.98 nm for ^{3}H (compared to 2.56 nm and 2.979 nm). Taking differences and sums of these moments suggests that a discrepancy of a few percent remains in both the isoscalar and isovector moments.

Finally, Fig. 19 shows the comparison between data and the calculations of Bornais for the magnetic form factor of tritium. Again there seems to be good evidence that the addition of the MEC to the calculations make a vast improvement in the agreement of theory with the data. Since tritium is such a difficult target to deal with in the laboratory, the data on both the ^{3}H charge and magnetic form factors do not extend to as high a momentum transfer as necessary to make a definitive statement about exchange currents. Recent measurements at Saclay will remedy this situation. While I have not yet seen this data, I am told through the grapevine that the new data is in fair agreement with the calculations. Hopefully, with the new data, it will be only a few months before we can see results of the long-awaited analysis of the three-body form factors in terms of isoscalar and isovector densities. Such a decomposition should offer considerable insight into the validity of the MEC calculations.

To summarize the A=3 system, then, it appears that the MEC play very important roles in the magnetic form factors for the ground states of both ^{3}He and

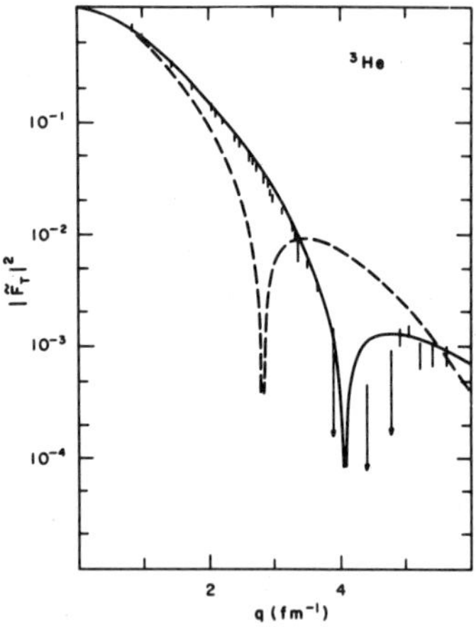

Fig. 17. Transverse form factor for elastic scattering from ^{3}He. Solid line includes MEC.

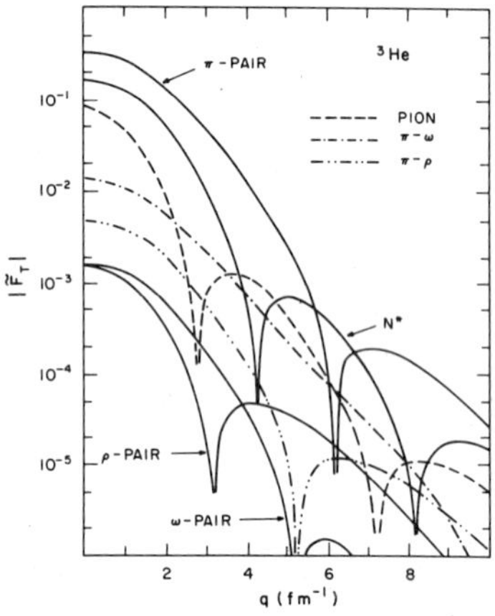

Fig. 18. Contributions[34] of various MEC processes to the transverse elastic form factor of ^{3}He. The pion, ρ-pair, and πργ, contributions are of opposite sign to the others for q≅0.

^3H. And in both cases it appears that the MEC calculations do supply the physics missing from the one-body only calculations. For the charge form factors the situation is less convincing. There are certainly important MEC effects in the curves presented here and they seem to generally be in the direction of improvement with data. My own theoretical bias is that these calculations must be viewed with some skepticism until some of the questions concerning the proper treatment of relativity have been answered. Nonetheless, the three-body system seems to offer similar compelling evidence for the exchange currents as we saw in some of the A=2 experiments.

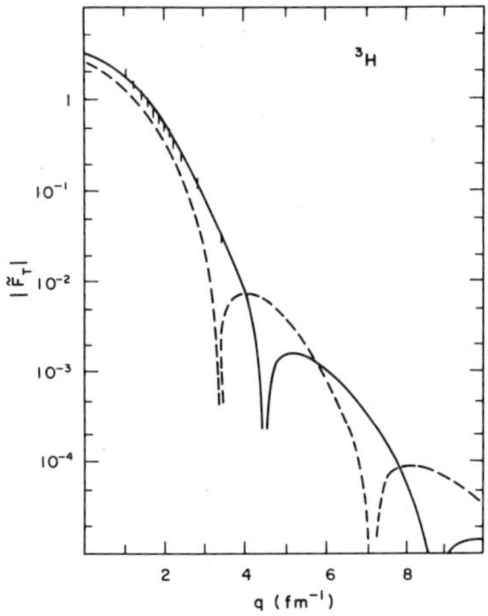

Fig. 19. Transverse form factor for elastic scattering from ^3H. Solid line includes MEC.

Results, A>3

As we turn to more complicated nuclear systems the role of the MEC becomes less clear. The problem here is that there is usually so much uncertainty in the treatment of the nuclear structure of these systems that it becomes much more difficult to disentangle MEC effects from structure effects. Thus it is hard to point to any compelling evidence for MEC in A>3 systems. At this point, I prefer to take a different tack. It does seem from our look at the A=2 and A=3 systems that the MEC play an important role and that the method we have described above seems adequate for evaluating the MEC effects at least for the isovector three-current MEC. Therefore, in the heavier systems, I would prefer to put some faith in our ability to calculate the MEC effects and instead ask to what extent consideration of the MEC causes us to reconsider our conclusions regarding the structure.

I will therefore restrict my discussion to two "typical" cases. The first of these, shown in Fig. 20, shows the magnetic form factors[38] for excitation of "stretched" transitions in ^{14}C. A stretched transition is usually a transition to a rather high spin state (in this case 4^- states) where one can use angular momentum selection rules to limit the number of orbitals involved in the transition. Within a 0 $\hbar\omega$ shell model space such transitions are usually limited to a single one-body density matrix element. In the case of ^{14}C this is the amplitude for the $1p_{3/2} \rightarrow 1d_{5/2}$ transition. For such transitions, the magnitude of the electron scattering form factor is, at least in this simple model, a direct measure of the one contributing one-body density matrix element.

The MEC will offer a different mechanism for excitation of such states and so will modify such an interpretation. Figure 20 shows the results of fitting the data by varying the relevant one-body density matrix element but with MEC included in the evaluation of the form factor. Also shown is the result using the fitted value of $\Psi_{\alpha\beta}$ but with the exchange currents turned off. The conclusion is that the MEC make a difference of 15-20% in the magnitude of the form factor but make very little difference to the overall shape. If one is attempting to confront the theoretical models for the structure of these transitions at a level of accuracy below, say 25%, it should be clear that MEC must be considered.

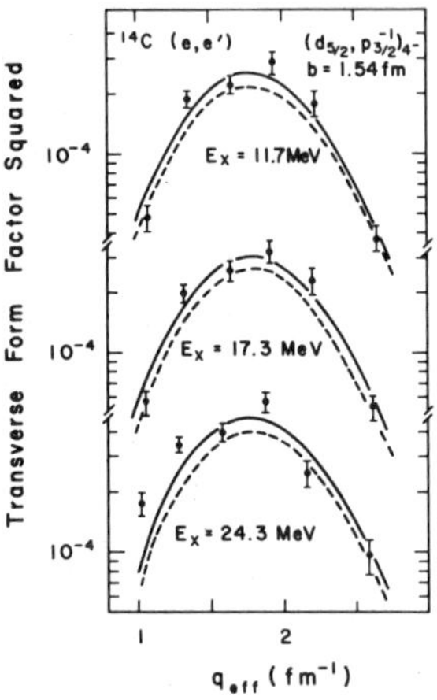

Fig. 20. Transverse form factors for excitation of 4^- states in ^{14}C.

Figure 21 also shows another "typical" case. This is for inelastic scattering[39] from the ground state of ^7Li to a J = 7/2 excited state. There are three points to notice. First, for the usual E2 and M3 multipoles in the vicinity of their maximum, the MEC contribution is quite modest, perhaps 10%. This is a typical magnitude of the MEC effects for many transitions. Secondly, the MEC do begin to play a more important role at higher q where they remain relatively strong while the one-body terms are rapidly decaying. Unfortunately, it will probably remain impossible to separate such MEC effects from uncertainties in the nuclear single-particle

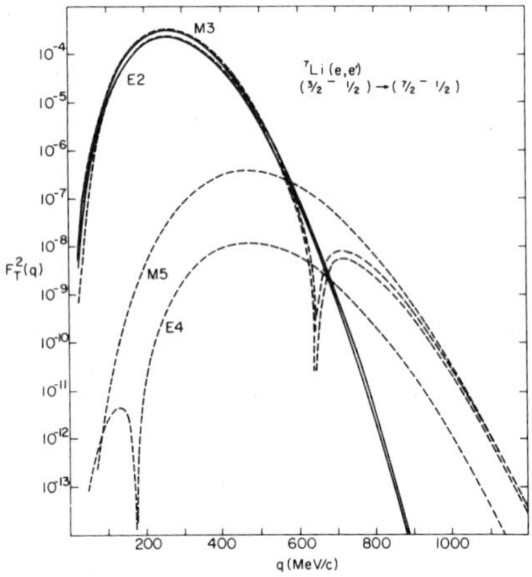

Fig. 21. Transverse form factor for excitation of $7/2^-$ state in ^7Li. Dashed lines include MEC.

wave functions. The third point to notice is that in this case the exchange currents are solely responsible for the introduction of higher multipoles into the form factor. The spins of the states involved in the transitions will allow for multipole operators of rank 2 through 5. In a 1p-shell description of ^{7}Li, the highest spin available in a single orbit is only 3/2. Therefore the one-body part of the transition can proceed only through multipolarity 2 or 3. Since the MEC are two-body processes, they can involve the spins of two nucleons whose angular momenta can couple to give the full range of multipolarities. These effects are small in the total transverse form factor and will probably never be seen. However, if one is able to use the methods of polarization or (e,e'X) to separate multipoles, there may be hope to observe this MEC-induced effect. (Even then, considerations of structure effects involving orbitals out of the 1p-shell may hopelessly cloud the issue.)

There has been much work on a number of A>3 systems. I will only mention two references[40,41] as a means of pointing the reader in this direction. The basic conclusions as to the size of the MEC and their implication for problems of nuclear structure for A>3 are generally similar to those I have discussed.

I would also be remiss if I did not at least mention the role of exchange currents in the "dip" region above the quasielastic peak. As I mentioned before, there has been considerable mystery as to the magntitude of the cross section in this region, particularly at large angles where the transverse response dominates. It was suggested[42] some years ago that MEC might be responsible for this enhanced cross section. Calculations to date generally find some enhancement but not nearly enough to explain the data. Since this is a topic of considerable controversy at the moment and since I expect we will hear a bit more about it from other speakers at this workshop, I will go no further.

Finally, to end this discussion of MEC effects in the heavier systems, I would like to briefly point out one case where exchange currents might be playing a more important role than the calculations would indicate. This is in low-multipolarity magnetic transitions at relatively high q (2.5-3.5 fm^{-1}). Figure 22 shows the measured[43] elastic magnetic form factor for ^{13}C. Since the ground state spin is 1/2, only a single multipole, the M1, will contribute. The data show very good agreement with shell model calculations over the first diffraction maximum, but as one goes to higher q, the data exceed any of the shell model predictions. Reasonable variations in the form of the single-particle wave functions and reasonable variations in the shell model assumptions do not seem to be able to explain the effect. Since, as we saw in Fig. 21, the MEC dominate at high q, one might expect them to supply the necessary strength in this case. But explicit evaluation of the pion-exchange currents we have discussed here shows that, while the effect of the MEC is in the right direction, it falls far short of the necessary enhancement.

The plot thickens a bit if one looks at other nuclei, particularly the case of ^{14}N shown in Fig. 23. In this case[44] the ground state is J=1, but T=0. The elastic magnetic form factor will again only depend on the M1 multipole, but it can only come from isoscalar operators. The effect discussed for ^{13}C is missing for the ^{14}N elastic form factor although it may be beginning to become noticeable at the highest q. On the other hand, the first excited state of ^{14}N is a J=0, T=1 state; excitation of this state again proceeds only through the M1 multipole, but in this case via a purely isovector operator. The effect seen in ^{13}C is also clearly visible in this data. This isovector/isoscalar selection rule is suggestive of the structure of the MEC operators derived above. However, explicit evaluation of the MEC simply do not explain the effect.

When this effect is finally understood I personally expect the explanation to come from considerations of nuclear structure. Nonetheless, it is intriguing to end my discussion of MEC in heavier nuclei with the slight suggestion that there may be important, as yet

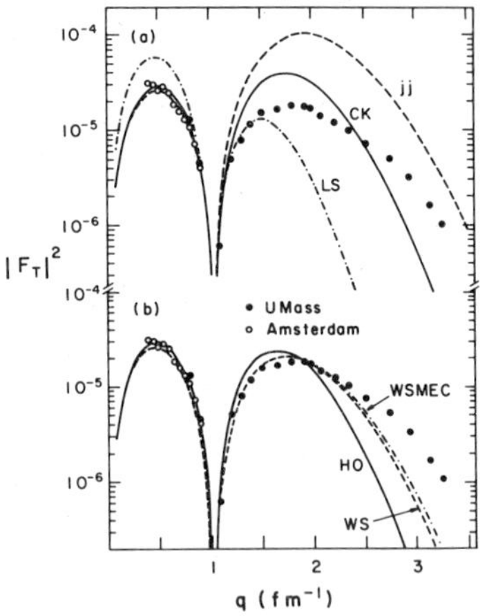

Fig. 22. Transverse elastic form factor for ^{13}C. Dashed-dotted line in lower curve includes MEC. See Ref. 43 for further details.

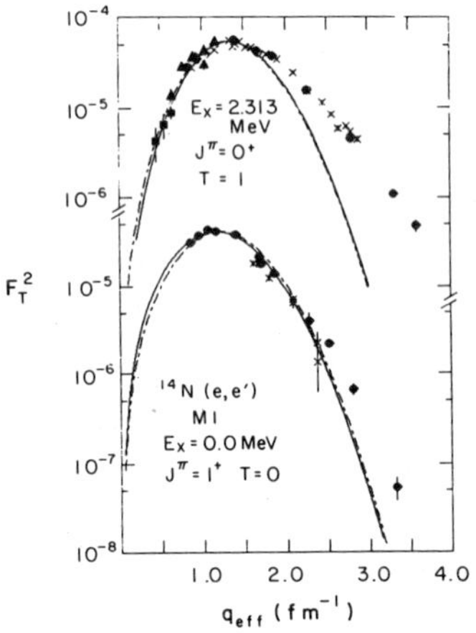

Fig. 23. Transverse form factors for ^{14}N.

undiscovered, signatures of the MEC to be considered.

Discussion

I would have to conclude from what I have shown you that all tests of the MEC to date have been consistent with experiment. In a few cases, the evidence seems undeniable and I believe even the strongest skeptic would have to agree that we seem to have a reasonable handle on the underlying physics. There are still many theoretical uncertainties involved in the evaluation of the MEC. The proper inclusion of electromagnetic and hadronic vertices remains to be defined. The details of some of the shorter-range exchanges are still quite model dependent. Since the tests involving isoscalar transitions and charge form factors can potentially provide considerable constraints on the MEC, perhaps the most important theoretical question to resolve is the connection to relativity and to other corrections of relativistic order. This will be a difficult question to resolve at least in practice if not in principle.

The new and long-awaited measurements of both the magnetic form factor of the deuteron and the elastic form factors for tritium may help refine our ideas of how to calculate the effects of the MEC. However, I tend to doubt that they will substantially change our conclusions. The new generation of electron scattering experiments involving polarized nuclei or coincidence measurements are almost assuredly also going to play an important role in our future understanding of MEC.

CONCLUSIONS

Let me briefly conclude by reminding you that agreement with data is far from proof positive that a theory is the correct one. I have shown you a number of measurements of the electromagnetic current distributions within nuclei. We have been able to explain all of them in terms of nucleon and meson degrees of freedom. I see no compelling evidence, no "smoking gun", indicating the presence of quark effects in these electromagnetic distributions. On the other hand, I am hardly prepared to disavow my belief in an underlying quark model for the structure of and interaction between nucleons. If I had a viable quark model that would allow me to calculate these distributions, I would be happy to use it. But, I suspect such a model will lie far in the future.

I think the most surprising thing about the results I have shown you is not that the model based on nucleons and mesons is successful at all, but that it is apparently successful to such high momentum transfers. If we measure electron scattering cross sections at a q of 10 fm^{-1}, we are, after all, probing the current distributions with a resolution of something like 0.1 fm. Even multiplied by π, this is still much smaller than what we think of as the nucleon's radius. So, where's the quarks?

As we attempt to construct an explanation of the nucleus in terms of the underlying quark structure, we must understand the transition from the 1 fm size scale of the nucleon to the 0.01 fm (?) size scale dominated by quark/gluon descriptions. It is this intermediate region that is the business of the nuclear physicists of the future. I am beginning to suspect that we may all be surprised by the smallness of the size at which this transition takes place.

ACKNOWLEDGMENTS

I would like to thank the organizers of this workshop for putting together an excellent program; I have learned a lot and I believe the students have as well. I would also like to acknowledge the students for their patient attention and interest and for their discussion both in and out of the lectures. Finally, I would like to thank the organizers for their patience with my unconscionable tardiness in submitting this manuscript.

REFERENCES

1. T. deForest and J.D. Walecka, Adv. in Phys. 15, 1 (1966).
2. H. Uberall, Electron Scattering From Complex Nuclei (Academic Press, New York, 1971), two volumes.
3. J.S. O'Connell, "Electromagnetic Nuclear Interactions", National Bureau of Standards report NBSIR 82-2547 (September, 1982).
4. J.D. Walecka, "Electron Scattering", Argonne National Laboratory report ANL-83-50 (January, 1984).
5. T.W. Donnelly and J.D. Walecka, Ann. Rev. Nucl. Sci. 25, 329 (1975).
6. J. Heisenberg and H.P. Blok, Ann. Rev. Nucl. Part. Sci. 33, 569 (1983).
7. T.W. Donnelly and I. Sick, Rev. Mod. Phys. 56, 461 (1984).
8. R.L. Huffman, Ph.D. Thesis, University of Massachusetts, (1985).

9. J.D. Bjorken and S.D. Drell, Relativistic Quantum Mechanics (McGraw-Hill, New York, 1964).

10. S.S. Schweber, An Introduction to Relativistic Quantum Field Theory (Row, Peterson, and Co., 1961).

11. J.D. Walecka in Muon Physics, Eds. V.W. Hughes and C.S. Wu (Academic Press, New York, 1975).

12. G. Feinberg, Phys. Rev. D12, 3575 (1975).

13. B.D. Serot, Nucl. Phys. A322, 408 (1979).

14. A.R. Edmonds, Angular Momentum in Quantum Mechanics (Princeton University Press, Princeton, 1960).

15. T.W. Donnelly and A.S. Raskin, Ann. Phys., to be published.

16. J.L. Friar and J.W. Negele, Nucl. Phys. A212, 93 (1973).

17. M.K. Singham, Phys. Rev. Lett. 54, 1642 (1985).

18. H. Yukawa, Proc. Phys.-Math. Soc. Japan, 17, 48 (1935).

19. A.J.F. Siegert, Phys. Rev. 52, 787 (1937).

20. M. Rho and D. Wilkinson, eds., Mesons in Nuclei (North-Holland Publishing, Amsterdam, 1979).

21. R.K. Osborn and L.L. Foldy, Phys. Rev. 79, 795 (1950).

22. P. Stichel and E. Werner, Nucl. Phys. A145, 257 (1974).

23. J.J. Sakurai, Advanced Quantum Mechanics (Addison-Wesley, Reading, 1967).

24. R. Arnold, Proc. of the Int. Conf. on Nuclear Physics with Electromagnetic Interactions, Mainz, 1979; Lecture Notes in Physics, vol 108 (Springer-Verlag, Berlin, 1979).

25. B. Frois, Nucl. Phys. A434, 57c (1985).

26. S. Auffret, J.M. Cavedon, J.C. Clemens, B. Frois, D. Goutte, M. Huet, Ph. Leconte, J. Martino, Y. Mizuno, X.H. Phan, S. Platchkov, and I. Sick, Phys. Rev. Lett. 54, 649 (1985).

27. B.L. Parker, Ph.D. thesis, University of Massachusetts, (1985).

28. M. Gari, quoted in Ref. 25; M. Gari and H. Hyuga, Nucl. Phys. A264, 409 (1976).

29. W. Leidemann and H. Arenhovel, Nucl. Phys. A393, 385 (1983).

30. J. Hockert, D.O. Riska, M. Gari, and A. Huffman, Nucl. Phys. A217, 14 (1973).

31. J.C. Clemens, as quoted in Ref. 25.

32. J.F. Mathiot, Nucl. Phys. A412, 201 (1984).

33. D.O. Riska and G.E. Brown, Phys. Lett. 38B, 193 (1972).

34. R. Bornais, Ph.D. thesis, Universite de Montreal, (1981).

35. E. Hadjimichael, B. Goulard, and R. Bornais, Phys. Rev. C27, 831 (1983).

36. F. Villars, Helv. Phys. Acta 20, 476 (1947).

37. E.L. Tomusiak, M. Kimura, J.L. Friar, B.F. Gibson, G.L. Payne, and J. Dubach, Phys. Rev., to be published.

38. M.A. Plum, R.A. Lindgren, R.S. Hicks, R.L. Huffman, B. Parker, G.A. Peterson, J. Alster, J. Lichtenstadt, M.A. Moinester, and H. Baer, Phys. Lett. 137B, 15 (1984).

39. J. Dubach, J.H. Koch, and T.W. Donnelly, Nucl. Phys. A271, 279 (1976).

40. B. Desplanques and J.-F. Mathiot, Phys. Lett. 116B, 82 (1982).

41. T. Suzuki and H. Hyuga, Nucl. Phys. A402, 491 (1983).

42. T.W. Donnelly, J.W. Van Orden, T. deForest, W.C. Hermans, Phys. Lett. 76B, 393 (1978).

43. R.S. Hicks, J. Dubach, R.A. Lindgren, B. Parker, and G.A. Peterson, Phys. Rev. C26, 339 (1982).

44. R.L. Huffman, R.S. Hicks, J. Dubach, B. Parker, M.A. Plum, G. Lahm, R. Neuhausen, and J.C. Bergstrom, Phys. Lett. 139B, 249 (1984).

*RELATIVISTIC MESON-NUCLEON MODELS: APPLICATIONS TO BOUND AND SCATTERING STATES**

C. J. Horowitz
Massachusetts Institute of Technology
Cambridge, MA 02139

ABSTRACT

In this series of lectures the motivation for applying meson nucleon field theories to nuclear physics is discussed. The Walecka model with nucleons interacting with neutral scalar and vector mesons is introduced. This model is solved in the mean field approximation for infinite nuclear matter and closed shell nuclei. Binding energies, single particle spectra, mean meson fields and baryon, scalar and charge densities are shown. Next, vacuum fluctuation corrections are examined. These arise from the change in mass of Dirac sea nucleons because of the strong scalar field.

The second half of the lectures deals with nucleon-nucleus scattering. First, a simple direct plus exchange model is developed to provide a Lorentz invariant representation of the NN amplitudes. Ambiguities in the use of these amplitudes for nucleon-nucleus scattering are discussed. Finally, relativistic Brueckner theory is introduced to examine correlation corrections to the mean field results and to include Pauli blocking corrections to the relativistic impulse approximation. Extensive results for 200 MeV elastic proton scattering from closed shell nuclei are presented. Calculations including Pauli blocking and using a pseudovector representation of the NN amplitudes provide an excellent quantitative description of all measured spin observables, both analyzing power and spin rotation function.

I. INTRODUCTION

One would like to have a consistent theoretical framework for extrapolating known nuclear information to nuclear matter under extreme conditions of temperature, density or flow, conditions present in the center of neutron stars or in energetic heavy ion collisions. Therefore, Walecka[1], in 1974, constructed a model relativistic quantum field theory of nucleons interacting with vector and scalar meson fields.

---

* This work was supported in part by funds provided by the United States Department of Energy (D.O.E.) under contract #DE-AC02-76ER03069.

Such a field theory explicitly incorporates meson degrees of freedom, thereby avoiding the approximation of a static NN potential. Furthermore, the correct causal propagation of the nucleons and the retarded propagation of the mesons generating the interactions are included naturally. Thus, the theory provides a minimal framework for describing extreme conditions where relativity is clearly important.

Interestingly, relativity may also be important for nuclei under normal conditions. For example, the spin-orbit force has long been understood as a relativistic correction to atomic physics. Presumably it is also a relativistic effect in nuclear physics where the spin-orbit potential is much stronger and forms an important part of the shell model.

Large relativistic effects in nuclei may arise from strong potentials comparable in strength to half the nucleon mass. Any nonrelativistic reduction must involve not only expanding in powers of the momentum divided by the nucleon mass but also powers of each potential over the mass. Here, second or higher powers may lead to interesting relativistic corrections.

There is no such thing as *just* relativistic kinematics for strongly interacting systems. In addition, one must specify how the interactions behave under a Lorentz transformation. Historically, it was assumed that the optical potential U either transformed completely as a Lorentz scalar or as a vector. This assumption of a homogeneous transformation is known as minimal relativity and forms the basis for nonrelativistic optical models with relativistic kinematics. [Note, relativistic kinematics are clearly needed, for even a 200 MeV proton is traveling at 2/3 c.] Minimal relativity should be strongly criticized because it is completely arbitrary.

Instead, there is growing evidence that U contains both important Lorentz scalar S and vector V terms. We can write,

$$U = S + \gamma_0 V \quad , \tag{1.1}$$

where γ_0 is a Dirac matrix. The scalar potential is believed to be almost -450 MeV and is attractive, while V is about 350 MeV repulsive. The nonrelativistic central potential (of about -50 MeV) involves cancellations between S and V, while the spin-orbit potential is a coherent sum of "Thompson presession" contributions from both S and V.

Relativistic mean field calculations[2] relate S and V to large scalar (sigma) and vector (omega) meson fields. The calculations then provide a good description of nuclear saturation and charge densities of closed shell nuclei. Dirac optical model fits to proton nucleus scattering [3] also obtain strong potentials in order to reproduce analyzing power data. Finally (as discussed in sections 5 and 6), relativistic impulse approximation (RIA) calculations[4] find strong potentials coming from large scalar and vector pieces of a Lorentz invariant representation of the NN amplitudes. These RIA calculations provide an excellent description of elastic scattering.

In section 2 we introduce a quantum field theory for the many-nucleon problem and discuss important philosophical questions. For example, given the nucleon's substantial internal structure, why describe it with a single Dirac equation? Also, given this substructure, should the meson-nucleon field theory be renormalizable?

In section 3 the field theory is solved in a mean field approximation (MFT) for infinite nuclear and neutron matter. Section 4 extends these MFT results to finite nuclei. Here binding energies, Dirac wavefunctions, spectra, mean meson fields and baryon, scalar and charge densities are presented for doubly magic nuclei.

Vacuum fluctuation corrections to the MFT are examined in section 5. These corrections arise from the shift in mass of all of the nucleons (even those in the Dirac sea) because of the strong scalar field. The scalar potential S shifts the nucleon's mass from M to M^*, where the effective mass M^* is

$$M^* = M + S \quad . \tag{1.2}$$

This may be as small as 1/2 the free nucleon's mass. This shift affects not only positive energy valence nucleons but also the negative energy Dirac sea. Thus, the energy density of the vacuum may change in the presence of a strong scalar field. Clearly, this effect is absent in the traditional nonrelativistic approach to nuclear physics.

The remainder of these lectures deals with nucleon-nucleus scattering. The relativistic impulse approximation (RIA) described in section 6 relates the NN amplitudes to nucleon-nucleus scattering observables. In order to compare NN amplitudes with meson-nucleon field theories, section 6 presents a simple direct plus exchange model of the NN amplitudes. This model also allows one to examine ambiguities in the form of the NN amplitudes.

Finally, section 6 introduces relativistic Brueckner theory to examine correlation corrections to the mean field theory and to examine Pauli blocking corrections to the RIA. In Brueckner theory, the relativistic two nucleon problem is solved in the strong potentials of the rest of the medium. The result is a reaction matrix (or G matrix) which is the product of a bare interaction times the correlated two nucleon wavefunction. The reaction matrix should reduce to the free NN amplutude in the zero density limit. However, the G matrix allows one to examine medium modifications to the free NN interaction due to the strong potentials and Pauli blocking (since some intermediate states are not available for the two nucleons in the medium).

Extensive elastic proton scattering calculations from closed shell nuclei are presented in section 6. These calculations illustrate the effects of Pauli blocking and ambiguities in the NN amplitudes. If Pauli blocking is included and a so-called pseudovector form of the NN amplitudes is used, the calculations provide an excellent description of all measured spin observables.

Section 7 goes on to present conclusions. No attempt has been made to provide a complete set of historical references. Instead, one is referred to the volume by Serot and Wakecka[5] which provides a reasonably complete introduction to relativistic field theories for nuclear physics and may be a useful starting point for graduate students. Students interested in starting to work in this field may find a simple relativistic Hartree code useful. Therefore, the code which was used for the calculations in section 4 has been documented and is available from the author.

There does not yet exist a complete review article on relativistic descriptions of proton-nucleus scattering, although many of the original RIA results are mentioned in the paper by Hoffman and Ray.[6] One can also consult the talks of this summer school by B. Clark, P. Tandy and S. Wallace.

II. QUANTUM HADRODYNAMICS

The simplest way to describe a strongly interacting many-body system, if one insists on Lorentz covariance, is with a relativistic quantum field theory (RQFT). We would like to develop a RQFT for nuclear physics in the *hadron* (meson and baryon) coordinates. The use of hadron rather than quark and gluon coordinates requires comment:

(a) Hadrons are the particles observed in experiments. So, a theory at the hadron level may be more easily related to physical observables. It is an important problem in nuclear theory to construct the appropriate collective coordinates and determine their dynamics. There are many indications in nuclear physics that the appropriate coordinates are hadrons. For example, the success of the shell model and interpretations of proton scattering with an NN interaction and nucleon densities suggest the role of nucleon coordinates. While meson exchange currents in the photo-disintegration of the deuteron indicate the role of meson coordinates.

(b) To provide evidence for the role of quark degrees of freedom one needs accurate predictions of what hadrons alone will give. To sharpen any "signatures" of quark dynamics, one needs to contrast quark and hardon predictions. Walecka has chosen the name quantum hadrodynamics (QHD) for any field theory of the nuclear many-body problem using hadron degrees of freedom.

There are a number of advantages of a RQFT over the conventional nonrelativistic potential description of nuclear physics:

(a) Proper relativistic propagation of nucleons and retarded interactions are included automatically.

(b) Static NN potentials need not be introduced. Traditionally, one chooses a potential fit to NN scattering. Then one hopes that any errors made by oversimplifying the complex (nonlocal, perhaps relativistic) two nucleon dynamics with this potential will somehow cancel when the potential is used in the many-body system. We know of no reason this need be the case just because the potential fits the NN phase shifts.

(c) Meson degrees of freedom are incorporated explicitly. In principle, a RQFT will automatically include meson exchange current corrections to the electromagnetic interactions in a consistent fashion. These extra degrees of freedom require comment. Nonrelativistically, one develops a scattering theory with an interaction representation which turns on and off potentials between the projectile and target nucleons. In a RQFT, the interaction representation is normally defined by turning on and off meson-nucleon vertices. Therefore, nonrelativistic multiple scattering theory does not simply generalize to the relativistic case (at least not for a RQFT). One can't even define a target Hamiltonian. One tries to say something like, "The target Hamiltonian has a nucleon interacting with an identical meson if and only if that meson 'came from' a target rather than projectile nucleon". Clearly, the meson degrees of freedom introduce large differences between relativistic and nonrelativistic approaches.

We are going to require the RQFT be *renormalizable*. This assures there is a well-defined framework for calculating corrections to any approximation. In contrast, in a nonrenormalizable theory, as one calculates to higher and higher order, in the couplings one must keep adding more unconstrained counterterms to assure finite results.

In addition, a renormalizable theory minimizes the sensitivity of physical results to very high momentum behavior. Given the complexity of the nucleon's substructure, one would like physical results to be insensitive to the very short distance behavior. [If results depend critically on short distances then a hadron rather than quark model may not be useful.]

A nonrenormalizable theory might have derivative couplings which bring extra factors of the momentum at each meson nucleon vertex. Therefore, loop diagrams will have very large contributions from high intermediate momenta. If these contributions are simply cutoff with *ad hoc* meson-nucleon form-factors (which somehow describe the nucleon substructure) results may be very sensitive to the cutoff scheme.

Finally, the constraint of renormalizability limits the choice of theories.

We emphasize renormalizability is our choice for some of the dynamical input which may be just as important (and could be wrong) as our choice of which meson fields to include.

We introduce the simple Walecka model[1] which has nucleons interacting with uncharged scalar and vector meson fields. The three fields are:

(1) The baryon field ψ which describes $(p, n, \ldots)$ and has mass M.

(2) The scalar field ϕ which describes sigma mesons of mass m_s.

(3) The vector field V^μ for the omega meson of mass m_v. The scalar field has a Yukawa coupling of strength g_s to the baryon scalar density,

$$\mathcal{L}_I^1 = g_s \overline{\psi}\psi\phi \tag{2.1}$$

while the vector meson couples with strength g_v to the baryon current,

$$\mathcal{L}_I^2 = -g_v \overline{\psi}\gamma^\mu \psi V_\mu \tag{2.2}$$

The total Lagrangian density is,

$$\begin{aligned}\mathcal{L} = \overline{\psi}[\gamma_\mu(i\partial^\mu - g_v V^\mu) - (M - g_s\phi)]\psi + \frac{1}{2}[\partial_\mu\phi\partial^\mu\phi - m_s^2\phi^2] \\ - \frac{1}{4}F^{\mu\nu}F_{\mu\nu} + \frac{1}{2}m_v^2 V_\mu V^\mu\end{aligned} \tag{2.3}$$

where $F^{\mu\nu}$ is the field strength tensor

$$F_{\mu\nu} = \partial_\mu V_\nu - \partial_\nu V_\mu \tag{2.4}$$

and our metric conventions are from Bjorken and Drell.[7] This theory is renormalizable for it is like massive QED with an additional scalar interaction.

As *motivation* for the model, one can consider the static potential limit of very heavy nucleons $M \longrightarrow \infty$ interacting with a one-boson-exchange potential (OBEP) from scalar and vector exchange. Important: this limit is not quantitative (for the nucleons may have effective masses near m_s) but does illustrate qualitative features of the model. In this limit the nucleons feel a potential,

$$V(r) = \frac{g_v^2}{4\pi}\frac{e^{-m_v r}}{r} - \frac{g_s^2}{4\pi}\frac{e^{-m_s r}}{r} \tag{2.5}$$

I.e., there is attraction from scalar and repulsion from vector exchange. If $g_v > g_s$ and $m_v > m_s$ then Eq. (2.5) will be attractive at medium distances and repulsive at short distances, qualitatively reproducing the main features of the NN interaction.

This simple model does not have a pion field. It can be added [see for example Ref. 5]. However, the strong spin and isospin dependence of the pi-N interaction causes the effects of the pi to average to zero (to lowest order) in nuclear matter.

Indeed in general, more meson and baryon fields can be added to Eq. (2.3) in an attempt to provide a more detailed description of all hadron degrees of freedom. Here we use Eq. (2.3) to illustrate the main features of QHD in as simple a fashion as possible.

The equations of motion for the fields can be derived from the Lagrangian using,

$$\frac{\partial}{\partial x^\mu} \frac{\partial \mathcal{L}}{\partial(\partial q_i/\partial x^\mu)} - \partial\mathcal{L}/\partial q_i = 0 \quad , \tag{2.6}$$

where q_i is one of the generalized coordinates (the fields ψ, ϕ, V_μ).

$$\left[\partial_\mu \partial^\mu + m_s^2\right] \phi = g_s \overline{\psi}\psi \tag{2.7}$$

$$\partial_\mu F^{\mu\nu} + m_v^2 V^\nu = g_v \overline{\psi}\gamma^\nu \psi \tag{2.8}$$

$$\left[\gamma^\mu(i\partial_\mu - g_v V_\mu) - (M - g_s\phi)\right]\psi = 0 \tag{2.9}$$

First, Eq. (2.7) is the Klein-Gordon equation with a scalar source. Next, Eq. (2.8) looks like massive QED with the baryon current,

$$B_\mu \equiv \overline{\psi}\gamma_\mu \psi \tag{2.10}$$

rather than the electromagnetic, as source. One can show the baryon current is conserved,

$$\partial_\mu \overline{\psi}\gamma^\mu \psi = 0 \tag{2.11}$$

Finally, Eq. (2.9) is the Dirac equation with the scalar and vector fields introduced in minimal fashion.

One can use the model to extrapolate from known nuclei to nuclear matter under extreme conditions of density, temperature or flow. This is needed to describe neutron stars, which may have central densities up to an order of magnitude greater than normal nuclei. Another application is in relativistic heavy ion collisions.

One is interested in the equation of state (the pressure as a function of the energy density $p(e)$) for nuclear and neutron matter. This can be obtained from the stress energy tensor $T_{\mu\nu}$,

$$T_{\mu\nu} = -g_{\mu\nu}\mathcal{L} + \frac{\partial q_i}{\partial x^\nu} \frac{\partial \mathcal{L}}{\partial(\partial q_i/\partial x^\mu)} \tag{2.12}$$

(which is summed over the $q_i = \psi, \phi, V$). Using Eqs. (2.3) and (2.7) – (2.9), one has

$$T_{\mu\nu} = \frac{1}{2}\left[-\partial_\lambda\phi\partial^\lambda\phi + m_s^2\phi^2 + \frac{1}{2}F_{\lambda\sigma}F^{\lambda\sigma} - m_v^2 V_\lambda V^\lambda\right] g_{\mu\nu} \\ + i\bar{\psi}\gamma_\mu\partial_\nu\psi + \partial_\mu\phi\partial_\nu\phi + \partial_\nu V^\lambda F_{\lambda\mu} \tag{2.13}$$

This can be compared to the form of the observed $T_{\mu\nu}$ for a uniform system,[5]

$$\langle T_{\mu\nu}\rangle = (e+p)\ u_\mu u_\nu - p\ g_{\mu\nu} \tag{2.14}$$

where p = pressure, e = energy density and u_μ = four velocity of fluid, $u_\mu^2 = 1$ and for a fluid at rest

$$u_\mu = (1, \underset{\sim}{0}) \tag{2.15}$$

This gives,

$$p = \frac{1}{3}\langle T_{ii}\rangle \\ e = \langle T_{00}\rangle \tag{2.16}$$

So the equation of state can be calculated from $T_{\mu\nu}$.

As written, the field Eqs. (2.7) – (2.9) are very complicated because the fields must satisfy the canonical commutation relations. For example,

$$\left[\pi(\underset{\sim}{x}, t),\ \phi(\underset{\sim}{x}', t)\right] = i\delta(\underset{\sim}{x} - \underset{\sim}{x}') \tag{2.17}$$

where π is the momentum conjugate to ϕ,

$$\pi \equiv \frac{\partial\mathcal{L}}{\partial(\partial\phi/\partial t)} \tag{2.18}$$

Furthermore, the couplings are large so perturbative approaches may not be useful.

To make progress, one needs an approximate solution about which one can calculate corrections. In the next section, we describe the mean field theory (MFT) where the field operators are replaced by their classical expectation values. The MFT can be solved exactly and should become increasingly valid as the baryon density increases.

Before proceeding, it is worth commenting on the use of a single field ψ and the Dirac Eq. (2.9) to describe a nucleon (given the nucleon's internal structure). First, much of the nucleon's substructure can be described (in the model) with meson clouds. Even though there are point couplings in the Lagrangian (Eq. (2.3)) the nucleons are still dressed with (virtual) meson clouds.

Second, the use of a Dirac equation to describe an extended object does not require the nucleon to be much smaller than one over its mass. It only requires that the nucleon be smaller than one over momentum transfers of nuclear interest, *i.e.*, smaller than the nucleus.

It is true that Eq. (2.9) is implicitly making assumptions about negative energy components and momenta of order M. However, it is hoped that results of nuclear interest will not be critically sensitive to these assumptions.

One is free to try and construct a phenomenology with nucleons described by some other equation (perhaps the Klein-Gordon equation with the spin added by hand). However, one should recognize that you are not avoiding making assumptions about negative energies and order M momemtum components. One is simply making different arbitrary assumptions than in the Dirac equation.

It is an important and general problem, how do you describe a composite relativistic fermion with a collective "center of mass" coordinate. We choose to use a Dirac equation, but one must check with experiment to see if this makes any sense. As discussed in later sections, the Dirac equation seems to provide an excellent description of proton-nucleus *spin* observables. In any case, we must be careful to consider only small momentum transfers.

III. MEAN FIELD THEORY

We now wish to apply the Walecka model in the mean field approximation to infinite nuclear matter. The results of this section are from [Refs. 1, 5]. At high density, the source terms on the right hand side of the meson field Eqs. (2.7) – (2.8) become large and one can replace the field operators by their expectation values (which are classical fields):

$$\phi \to \langle \phi \rangle \equiv \phi_0 \tag{3.1}$$

$$V_\mu \to \langle V_\mu \rangle \equiv \delta_{\mu 0} V_0 \tag{3.2}$$

For a static uniform system ϕ, V_μ are independent of x_μ. Rotational invariance implies $\langle \underset{\sim}{V} \rangle$ vanishes.

This mean field approximation is the analog of using the classical Maxwell's equations to describe an electromagnetic field with a large number of quanta. Once we make the mean field approximation, the theory can be solved exactly. We will discuss corrections to the mean field theory (MFT) in sections 5 and 6.

The meson field equations can be solved for ϕ_0, V_0.

$$\phi_0 = \frac{g_s}{m_s^2}\langle\overline{\psi}\psi\rangle \equiv \frac{g_s}{m_s^2}\rho_s \tag{3.3}$$

$$V_0 = \frac{g_v}{m_v^2}\langle\psi^+\psi\rangle \equiv \frac{g_v}{m_v^2}\rho_b \tag{3.4}$$

Here, the baryon density $\rho_b = \langle\psi^+\psi\rangle$ and the scalar density $\rho_s = \langle\overline{\psi}\psi\rangle$. The baryon density is a constant of the motion, so V_0 is known. In contrast, ϕ_0 will depend on the interactions, so Eqs. (3.3) must be solved self-consistently (since $\rho_s(\phi_0)$).

The Dirac equation (Eq. (2.9)) is now linear and can be solved exactly.

$$\left[i\partial_\mu\partial^\mu - g_v\gamma^0 V_0 - M^*\right]\psi = 0 \tag{3.5}$$

Here we have defined the effective mass M^*,

$$M^* \equiv M - g_s\phi_0 \tag{3.6}$$

because the Dirac equation in the field ϕ_0 looks like a free equation with $M \to M^*$. Important: the condensed scalar field shifts the mass of all of the nucleons. This is just like a Higgs field changing the mass of the W and Z bosons.

We look for plane wave solutions to Eq. (3.5),

$$\psi(\underset{\sim}{x}, t) = U(k, \lambda)e^{i[\underset{\sim}{k}\cdot\underset{\sim}{x} - \epsilon(k)t]} \tag{3.7}$$

$$\left[\underset{\sim}{\alpha}\cdot\underset{\sim}{k} + \beta M^*\right]U(k, \lambda) = (\epsilon - g_v V_0)U(k, \lambda) \tag{3.8}$$

Eq. (3.8) is just like a free Dirac equation with M replaced by M^* and the vector field shifting where the energy is measured from. Therefore, it has four solutions corresponding to spin up and spin down particles of both positive and negative energy. To find the eigenvalue $\epsilon(k)$ we can square the Dirac equation (Eq. (3.8)) and use the anticommutation relations of the Dirac matrices.

$$\epsilon^{\pm} = g_v V_0 \pm (\underset{\sim}{k}^2 + M^{*2})^{1/2} \tag{3.9}$$

The 4 component wavefunctions can be found from the standard results for mass M^*. For positive energy spinors one has,

$$U(k, \lambda) = \left[\frac{E_k^* + M^*}{2M^*}\right]^{1/2} \left[\begin{array}{c} 1 \\ \frac{\sigma\cdot k}{E_k^* + M^*} \end{array}\right] X_\lambda \tag{3.10a}$$

where X_λ is a two component Pauli spinor of spin projection, λ, $\underset{\sim}{\sigma}$ are the 2×2 Pauli matrices and,

$$E_k^* = \left(\underset{\sim}{k}^2 + M^{*2}\right)^{1/2} \tag{3.10b}$$

We note that the lower components in Eq. (3.10) are of order $k/(E^* + M^*)$ and this is substantially enhanced from the free spinor $k/(E + M)$ if M^* is less than M. Eq. (3.10) is a function of ϕ_0 only. The vector field V_0 only shifts the energy, it does not change the wavefunction.

Eq. (3.10) for both spins and similar solutions $V(k, \lambda)$ for antinucleons form a complete set of states so the nucleon field can be expanded,

$$\psi(\underset{\sim}{x}, t) = \frac{1}{\sqrt{V}} \sum_{k,\lambda} \left[A_{k\lambda} U(k, \lambda) e^{i\underset{\sim}{k}\cdot\underset{\sim}{x} - \epsilon^+ t} + B_{k\lambda}^+ V(k, \lambda) e^{-i\underset{\sim}{k}\cdot\underset{\sim}{x} - \epsilon^- t} \right] \tag{3.11}$$

where we have quantized in a box of volume V. The coefficients A_k, (B_k) can be interpreted as destruction operators for quasinucleons (antinucleons) of mass M^*. The field ψ will satisfy the proper anticommutation relations if,

$$\left\{A_{k\lambda}^+, A_{k'\lambda'}\right\} = \delta_{kk'}\delta_{\lambda\lambda'} \tag{3.12}$$

and likewise for B, B^+. Note, these are annihilation and creation operators for quasinucleons in the fields V_0, ϕ_0 not free nucleons.

In the MFT, the Lagrangian (Eq. (2.3)) is reduced to,

$$\mathcal{L}_{MFT} = \overline{\psi}\left[i\gamma_\mu\partial^\mu - \gamma^0 g_v V_0 - M^*\right]\psi - \frac{1}{2}m_s^2\phi_0^2 + \frac{1}{2}m_v^2 V_0^2 \tag{3.13}$$

where the only quantum variable left is ψ. The stress energy tensor (Eq. (2.13)) becomes

$$T_{MFT}^{\mu\nu} = i\overline{\psi}\gamma^\mu\partial^\nu\psi - g^{\mu\nu}\left(\frac{1}{2}m_v^2 V_0^2 - \frac{1}{2}m_s^2\phi_0^2\right) \tag{3.14}$$

From this the energy density e is

$$e = \psi^+ \left[-i\,\underset{\sim}{\alpha} \cdot \underset{\sim}{\nabla} + \beta M^* + g_v V_0\right]\psi - \frac{1}{2}m_v^2 V_0^2 + \frac{1}{2}m_s^2\phi_0^2 \tag{3.15}$$

and we can construct a Hamiltonian H.

$$H = \int d^3x\; e \tag{3.16}$$

By using the expansion for ψ (Eq. (3.11)) along with the Dirac equation (Eq. (3.8)) and the anticommutation relations (Eq. (3.12)) we can write H in terms of A and B,

$$H = \sum_{k,\lambda} E_k^* \left(A_k^+ A_k - B_k B_k^+\right) + g_v V_0 \hat{B} - V\left(\frac{1}{2} m_v^2 V_0^2 - \frac{1}{2} m_s^2 \phi_0^2\right) \qquad (3.17)$$

where we have suppressed the spin index λ and used $\hat{B}$ for the baryon density,

$$\hat{B} = \sum_{k,\lambda} \left(A_k^+ A_k + B_k B_k^+\right) = \sum_{k,\lambda} \left(A_k^+ A_k - B_k^+ B_k\right) + \sum_{k,\lambda} 1 \qquad (3.18)$$

Here, the last infinite term can be dropped since we only measure relative to the vacuum.

$$\hat{B} = \int d^3x \left[\psi^+\psi - \langle 0|\psi^+\psi|0\rangle\right] = \sum_{k,\lambda} \left[A_k^+ A_k - B^+ B_k\right] \qquad (3.19)$$

But in the expression for the energy one has a term

$$\sum_{k,\lambda} E_k^*(A^+A - BB^+) = \sum_{k,\lambda} E_k^*(A^+A + B^+B) + \sum_{k,\lambda} \sqrt{k^2 + M^{*2}} \qquad (3.20)$$

Here, subtracting the vacuum expectation value still leaves an infinite contribution since the free vacuum has mass M so $E \neq E^*$. We define the vacuum fluctuation contribution to H, δH as,

$$\delta H = \sum_{k,\lambda} \left(\sqrt{k^2 + M^2} - \sqrt{k^2 + M^{*2}}\right) \qquad (3.21)$$

Then, the Hamiltonian becomes

$$H = H_{MFT} + \delta H \qquad (3.22)$$

with the mean field H

$$H_{MFT} = \sum_{k,\lambda} E_k^*(A^+A + B^+B) + g_v V_0 \hat{B} - V\left(\frac{1}{2} m_v^2 V_0^2 - \frac{1}{2} m_s^2 \phi_0^2\right) \qquad (3.23)$$

We shall omit δH for now and return to consider these vacuum fluctuations in section 5. The problem for H_{MFT} is now solved since it is diagonal in the normal modes A, B. The grouand state of H_{MFT} consists of filling nucleon levels up to k_F.

$$e = \gamma \int^{k_F} \frac{d^3k}{(2\pi)^3} E_k^* + g_v V_0 \rho_b - \frac{1}{2} m_v^2 V_0^2 + \frac{1}{2} m_s^2 \phi_0^2 \tag{3.24}$$

Here γ = spin, isospin degeneracy = 4 for nuclear and 2 for neutron matter, and the baryon density is

$$\rho_b = \frac{\gamma k_F^3}{6\pi^2} \tag{3.25}$$

Now use the vector meson field equation to eliminate V_0,

$$V_0 = \frac{g_v}{m_v^2} \rho_b \tag{3.26}$$

But for ϕ_0, we need the scalar density. From Eq. (3.10) one has

$$\overline{U}(k,\lambda)U = \frac{M^*}{E_k^*} U^+ U \tag{3.27}$$

so

$$\rho_s = \langle \overline{\psi}\psi \rangle = \sum_{k,\lambda}^{k_F} \overline{U}U = \gamma \int^{k_F} \frac{d^3k}{(2\pi)^3} \frac{M^*}{E_k^*} \tag{3.28}$$

I.e., the scalar density is Lorentz contracted with respect to the baryon density by an amount $M^*/E^* = \sqrt{1 - v^2/c^2}$ where v is a nucleon's velocity.

$$\rho_s = \langle (1 - v^2/c^2)^{1/2} \rangle \rho_b \tag{3.29}$$

The average in Eq. (3.29) is over the Fermi sea.

Using $M^* = M - g_s \phi_0$, the scalar field equation (Eq. (3.3)) becomes a self-consistency condition for M^*.

$$M^* = M - \frac{g_s^2}{M_s^2} \gamma \int^{k_F} \frac{d^3k}{(2\pi)^3} \frac{M^*}{E_k^*} \tag{3.30}$$

This is a simple transcendental equation for M^* at each value of k_F. Once M^* has been determined, the energy is given by Eq. (3.24) or

$$e = \gamma \int^{k_F} \frac{d^3k}{(2\pi)^3} E_k^* + \frac{1}{2} \frac{g_v^2}{m_v^2} \rho_b^2 + \frac{1}{2} \frac{m_s^2}{g_s^2} (M - M^*)^2 \tag{3.31}$$

The pressure can be determined from $1/3\langle T_{ii}\rangle$ or equivalently by,

$$P = \rho_b^2 \frac{\partial}{\partial \rho_b}\left(\frac{e}{\rho_b}\right) \tag{3.32}$$

which is simple thermodynamics.

In the high density limit $k_F \to \infty$, Eq. (3.30) shows the nucleons become *massless*,

$$M^* \underset{k_F\to\infty}{\longrightarrow} M\left[1+\frac{g_s^2}{m_s^2}\frac{\gamma k_F^2}{4\pi^2}\right]^{-1} \tag{3.33}$$

and the energy becomes,

$$e/\rho_b \underset{k_F\to\infty}{\longrightarrow} \frac{1}{2}\frac{g_v^2}{m_v^2}\rho_b + \frac{3}{4}k_f + \ldots \tag{3.34}$$

In this limit the energy is dominated by the vector repulsion and the next term is the kinetic energy of a massless fermi gas. The scalar attraction has been shut off because of the Lorentz contraction of ρ_s, which is the source of the scalar field.

The pressure from Eqs. (3.32, 34) becomes,

$$p \underset{k_F\to\infty}{\longrightarrow} e \tag{3.35}$$

This is the stiffest possible equation of state. If p exceeds e then the speed of sound will become acausal (greater than c). Thus, the MFT predicts that neutron matter will become very stiff at high densities.

For infinite nuclear matter there are two ratios which are important,

$$C_s^2 = \frac{g_s^2}{m_s^2}M^2 = 266.9, \quad C_v^2 = \frac{g_v^2}{m_v^2}M^2 = 195.7 \tag{3.36}$$

These can be chosen so that nuclear matter saturates (the curve of e/ρ_b vrs. k_F has a minimum) at the empirical binding energy and density.

$$\frac{e}{\rho_b} = M - 15.75 \text{ MeV} \tag{3.37}$$

$$\frac{\partial}{\partial \rho_b}\left(\frac{e}{\rho_b}\right)\Big|_{k_F^0} = 0 \tag{3.38}$$

$$k_F^0 = 1.42\ Fm^{-1} \tag{3.39}$$

This gives the values in Eq. (3.36)

Note in the Hartree approximation, it is only possible to reproduce nuclear saturation because of the Lorentz contraction of ρ_s with respect to ρ_b. A nonrelativistic Hartree calculation will not saturate for any values of C_s, C_v. Because of this contraction, a relativistic theory, even if it includes correlations, will saturate at a lower density and binding energy than a corresponding nonrelativistic calculation (see for example [5]).

Figure 1 shows M^* determined from Eq. (3.30) vrs. k_F for nuclear and neutron matter. The effective mass is seen to drop rapidly with density. At k_F^0,

$$M^* = 0.56\ M \tag{3.40}$$

M^* is reduced almost in half.

One can identify the mean meson fields with scalar and vector optical potentials (Eq. (1.1)),

$$\begin{aligned} S &= -g_s\phi_0 \approx -400 \text{ MeV} \\ V &= g_v V_0 \approx 350 \text{ MeV} . \end{aligned} \tag{3.41}$$

The values listed are at normal density k_F^0. Thus, choosing the couplings to reproduce nuclear saturation leads to strong optical potentials.

Figure 2 shows the energy per nucleon of nuclear and neutron matter. Neutron matter is unbound in this approximation. Figure 3 shows the proton and Figure 4 the neutron matter equation of state. At low densities, the local minima in Figure 2 for neutron matter leads to a liquid vapor phase transition indicated by the dashed lines (with the horizontal line showing the phases in equilibrium). At high density the pressure approaches e.

This equation of state can be used in the TOV equations[8] of general relativity to determine the structure of neutron stars. The mass of a star of given central density is shown in Figure 5. The relatively stiff mean field equation of state can support a maximum stellar mass of almost 2.6 $M_\odot$ ($M_\odot$ is the mass of the sun).

IV. FINITE HARTREE

In this section we extend the MFT to finite nuclei. The results shown are from [2]. We start with the MFT Lagrangian but now include spatial dependence for the mean fields

$$\begin{aligned} \mathcal{L}_{MFT} = \overline{\psi}\left[i\gamma_\mu\partial^\mu - g_v V_0(r)\gamma^0 - (M - g_s\phi_0(r))\right]\psi \\ - \frac{1}{2}\left[(\nabla\psi_0)^2 + m_s^2\phi_0^2\right] + \frac{1}{2}\left[(\nabla V_0)^2 + m_v^2 V_0^2\right] \end{aligned} \tag{4.1}$$

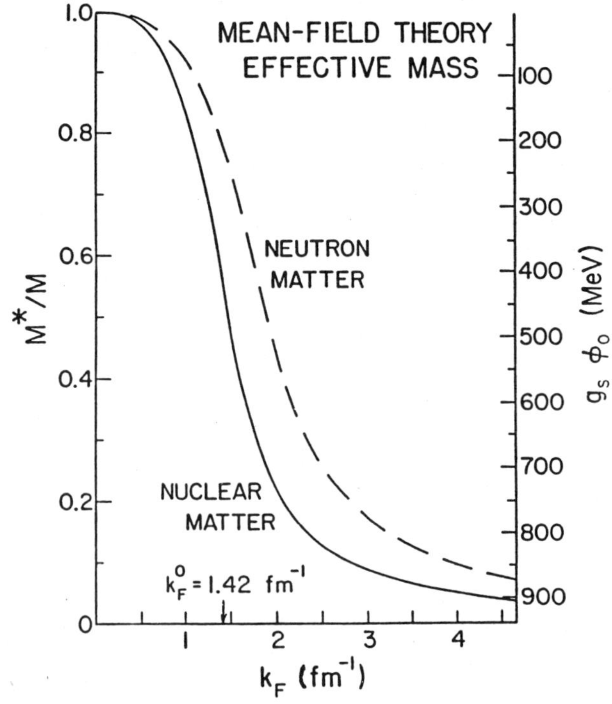

Fig. 1: Effective mass M^* vrs. density from Eq. (3.30).

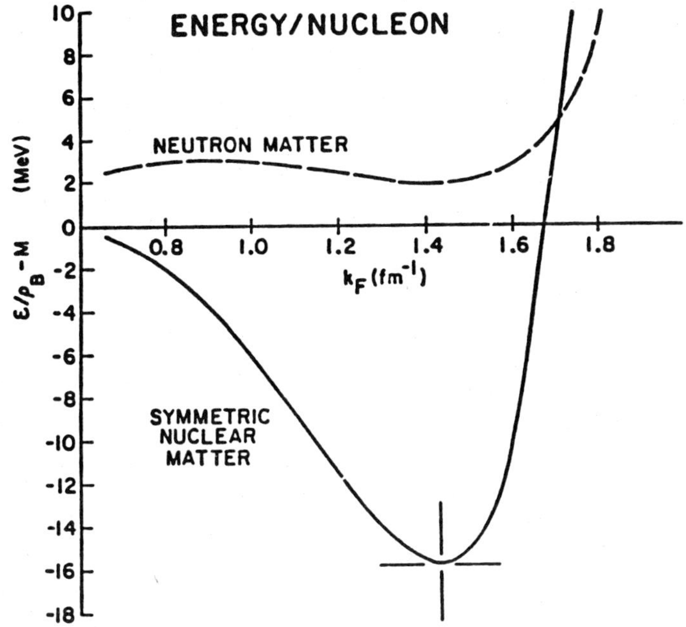

Fig. 2: Energy per nucleon vrs. density for nuclear and neutron matter, Eq. (3.31), using the parameters in Eq. (3.36).

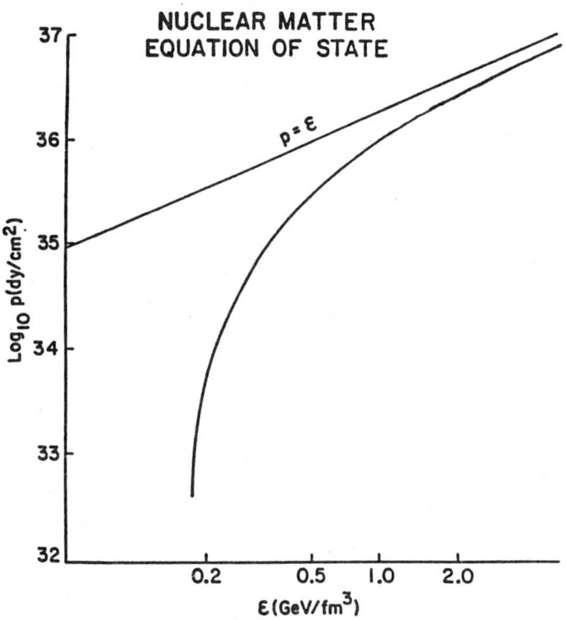

Fig. 3: Nuclear matter equation of state, pressure Eq. (3.32) vrs. energy density Eq. (3.21). The causality limit is the $p = \epsilon$ line.

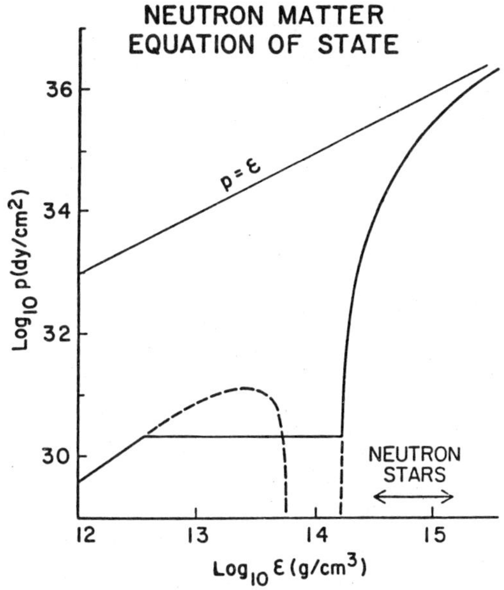

Fig. 4: Neutron matter equation of state.

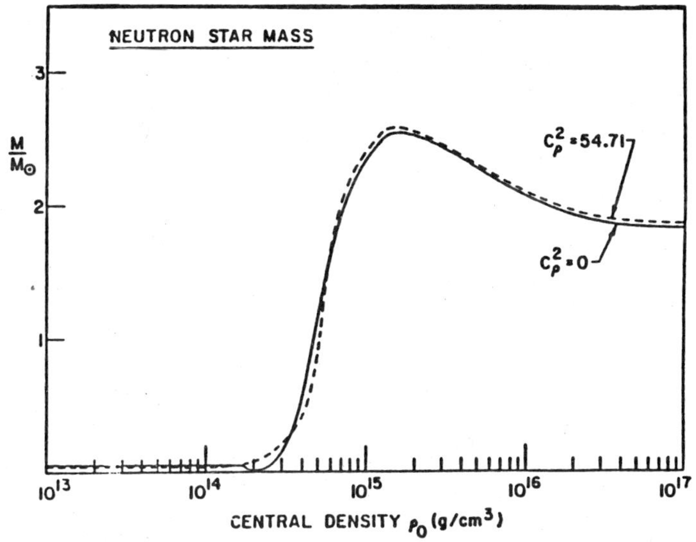

Fig. 5: Neutron star mass (in solar masses) vrs. central density (solid curve) for the equation of states in Fig. 4.

For spherically symmetric nuclei with angular momentum zero, the meson fields will only be a function of r. Furthermore, there is no three vector field $\underset{\sim}{V}(\underset{\sim}{r}) \propto \hat{r}V(r)$ because of current conservation. The source of $\underset{\sim}{V}$ is the baryon current which can neither flow away from or towards the origin in a static spherically symmetric nucleus.

The effective mass is now position dependent. We expect $M^*(r)$ to be about $1/2M$ inside the nucleus and rise back to M at the surface.

The Dirac equation reads

$$\left[i\gamma^\mu\partial_\mu - g_v\gamma^0 V_0(r) - M^*(r)\right]\psi = 0 \tag{4.2}$$

We look for eigenvalues

$$\begin{aligned} h\psi &= E\psi \\ h &\equiv -i\alpha\cdot\nabla + g_v V_0(r) + \beta M^*(r) \end{aligned} \tag{4.3}$$

where h is a single particle Dirac Hamiltonian. To solve Eq. (4.3) we need a set of operators which commute with h. The orbital angular momentum is not a good quantum number,

$$[h, \underset{\sim}{L}] \neq 0 \tag{4.4}$$

but the total angular momentum $\underset{\sim}{J}$ is.

$$\begin{aligned} &\underset{\sim}{J} = \underset{\sim}{L} + \underset{\sim}{S}, \qquad \underset{\sim}{S} = \frac{1}{2}\underset{\sim}{\Sigma}, \\ &[h, \underset{\sim}{J}] = 0, \qquad \underset{\sim}{\Sigma} = \begin{bmatrix} \underset{\sim}{\sigma} & 0 \\ 0 & \underset{\sim}{\sigma} \end{bmatrix} \end{aligned} \tag{4.5}$$

Also, the operator K commutes.

$$\begin{aligned} K &\equiv \gamma^0\left(\underset{\sim}{\Sigma}\cdot\underset{\sim}{J} - \frac{1}{2}\right) \\ [K, h] &= 0 \end{aligned} \tag{4.6}$$

One can show[9] that K has eigenvalues - $\mathcal{K}$ such that

$$K\psi = -\mathcal{K}\psi, \qquad \mathcal{K} = \pm\left(j + \frac{1}{2}\right) \tag{4.7}$$

We write the four component wavefunction,

$$\psi_{nKM} = \begin{bmatrix} iG_{nK}(r)/r & \Phi_{KM} \\ -F_{nK}(r)/r & \Phi_{-KM} \end{bmatrix} \tag{4.8}$$

where n is the radial quantum number and K determines both the j quantum number and the upper and lower orbital angular momentum quantum numbers. Here

$$\Phi_{KM} = \sum_{m_\ell, m_s} \langle \ell m_\ell \tfrac{1}{2} m_s | jm \rangle Y_{\ell m_\ell} X_{m_s} \tag{4.9}$$

with

$$j = |K| - \frac{1}{2} \tag{4.10}$$

and

$$\ell = \begin{cases} K & K > 0 \\ -(K+1) & K < 0 \end{cases} \tag{4.11}$$

Thus, the lower components with Φ_{KM} replaced by Φ_{-KM} will have an ℓ value which differs by one from the upper components. Therefore, Φ_{KM} and Φ_{-KM} have opposite parities.

The Dirac equation (4.3) becomes a coupled first order equation for the upper G and lower F wavefunctions.

$$\begin{aligned} \left(\frac{\partial}{\partial r} + \frac{K}{r}\right) G - [E - g_v V_0 + M^*] F = 0 \\ \left(\frac{\partial}{\partial r} - \frac{K}{r}\right) F + [E - g_v V_0 - M^*] G = 0 \end{aligned} \tag{4.12}$$

Given these wavefunctions one can sum over all the occupied states to obtain the baryon and scalar densities (we assume doubly closed shell nuclei).

$$\rho_b(r) = \sum_a^{\text{occ.}} \frac{2j_a + 1}{4\pi r^2} \left[G_a^2 + F_a^2\right] \tag{4.13}$$

$$\rho_s(r) = \sum_a \frac{2j_a + 1}{4\pi r^2} \left[G_a^2 - F_a^2\right] \tag{4.14}$$

$$\int_0^\infty dr \left[G^2 + F^2\right] = 1 \tag{4.15}$$

here we used

$$\sum_{m=-j}^{j} \Phi_{KM}^+ \Phi_{KM} = \frac{2j+1}{4\pi} \tag{4.16}$$

The scalar density is Lorentz contracted by $-2F(r)^2$. The meson field eqs. (2.7 - 2.8) become,

$$\left[\frac{\partial^2}{\partial r^2} + \frac{2}{r}\frac{\partial}{\partial r} - m_s^2\right]\phi_0 = -g_s\rho_s \tag{4.17}$$

$$\left[\frac{\partial^2}{\partial r^2} + \frac{2}{r}\frac{\partial}{\partial r} - m_v^2\right]V_0 = -g_v\rho_b \tag{4.18}$$

To solve this coupled set of equations for a given nucleus: (a) Guess some initial meson fields (in practice simple Wood-Saxon wells); (b) solve Eq. (4.12) in these fields for all occupied states; (c) sum up the squares of the F and G wavefunctions to calculate ρ_s, ρ_b using Eqs. (4.13, 14); (d) solve for new meson fields using Eqs. (4.17, 18); (e) iterate at step (b) until one has a self-consistent set of Dirac wavefunctions G, F and meson fields (in practice, about 10 iterations).

The parameters in the model are g_s, g_v, m_s and m_v. The vector meson we identify with the omega at

$$m_v = 783 \text{ MeV} \tag{4.19}$$

while g_v and the ratio g_s/m_s are adjusted to reproduce nuclear saturation. Finally, the scalar meson mass determines the range of the interactions and primarily affects the surface thickness in finite nuclei. We adjust m_s to reproduce the charge radius in ^{40}Ca. This gives

$$m_s = 520 \text{ MeV} \tag{4.20}$$

The calculation then provides:

(a) Single particle Dirac wavefunctions.

(b) Single particle spectra (the Dirac eigenvalues).

(c) Point proton and neutron baryon and scalar densities.

(d) Charge densities. We approximate the charge density by folding the single nucleon form factor $\rho_{sn}(r)$ with the point proton density. This takes into account the proton size.

$$\rho_{ch}(r) = \int d^3s\ \rho_b^{(\text{proton})}(\underset{\sim}{r} + \underset{\sim}{s})\rho_{sn}(s) \tag{4.21}$$

(e) Mean meson fields.

Figure 6 shows the baryon and scalar densities in Pb^{208}, also shown is $M^*(r)$. The distance from M down to $M^*(r)$ is minus the scalar meson field.

The scalar density is nearly equal to ρ_b in the surface where momenta are low and $M*$ is large. However, inside the nucleus momentum components are larger and M^* is smaller leading to much higher velocities. Therefore, ρ_s becomes Lorentz contracted with respect to ρ_b and the system saturates.

Figure 7 shows the unoccupied $1h\,9/2$ proton wavefunction in Pb^{208} (the valence wavefunction for Bi^{209}). The lower component $F(r)$is very small in the surface but becomes sizable inside where M^* is small. Next, Figure 8 shows the charge density in Pb^{208}. The agreement with experiment is basically good except for slightly large shell oscillatons. Charge densities for other closed shell nuclei 0^{16}, Ca^{40}, Zr^{90}, and the difference between $Ca^{48} - Ca^{40}$ are shown in Figures 9 – 12. Again, the agreement with experiment is good (comparable to more sophisticated nonrelativistic calculations).

The single particle spectrum for Ca^{40} (approximated by the Dirac eigenvalues in Eq. (4.3)) is shown in Figure 13. The spin-prbit splitting agrees well with experiment. This splitting is simply Thomas precession in the strong scalar and vector fields. Furthermore, the field strengths were determined to reproduce nuclear saturation. No parameters were fit to the spectrum.

Finally, neutron densities are shown in Figure 14 for Pb^{208} and Figure 15 for $Ca^{40} - Ca^{48}$. There is fair agreement with crude neutron densities extracted in a model dependent way from proton scattering cross sections.

V. VACUUM FLUCTUATIONS

In this section, one-baryon-loop fluctuation contributions to the vacuum energy are included in relativistic Hartree calculations of closed shell nuclei (see [10]). First, we calculate the δH contribution (omitted in section 3) for infinite nuclear matter. Then we apply the results to finite nuclei in a local density approximation. Vacuum fluctuations reduce the scalar density in the nuclear interior by as much as 15%.

The vacuum fluctuation correction to the Dirac sea because of the shift M to M^* is formally divergent (see Eq. (3.21))

$$\Delta e_{\text{vac}} = -\gamma \int \frac{d^3k}{(2\pi)^3} \left(\sqrt{k^2 + M^{*2}} - \sqrt{k^2 + M^2}\right) \tag{5.1}$$

It may be rendered finite by introducing appropriate counterterms in the Lagrangian of the form,

$$\delta\mathcal{L} = \sum_{i=1}^{4} \alpha_i \phi^i \tag{5.2}$$

where the α_i are (infinite) constants chosen to keep physical properties of the theory constant. For example, the ϕ^2 term α_2 is chosen to insure that

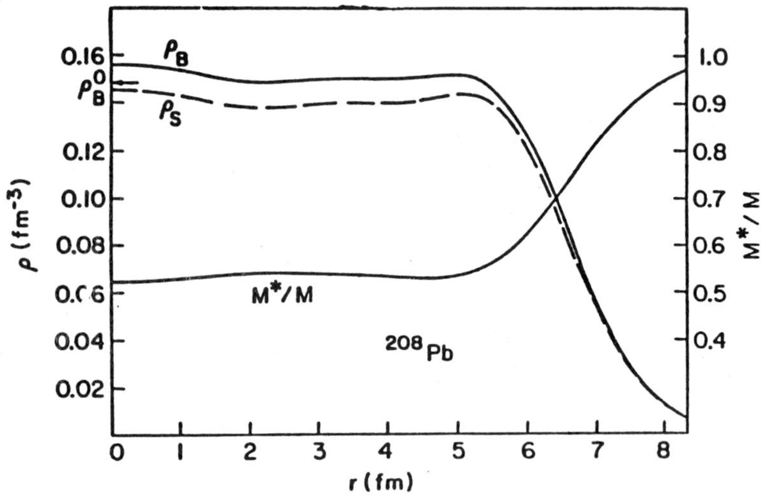

Fig. 6: Baryon ρ_B and scalar ρ_S densities in Pb^{208}. Also shown is the effective mass M^* read from the right hand scale.

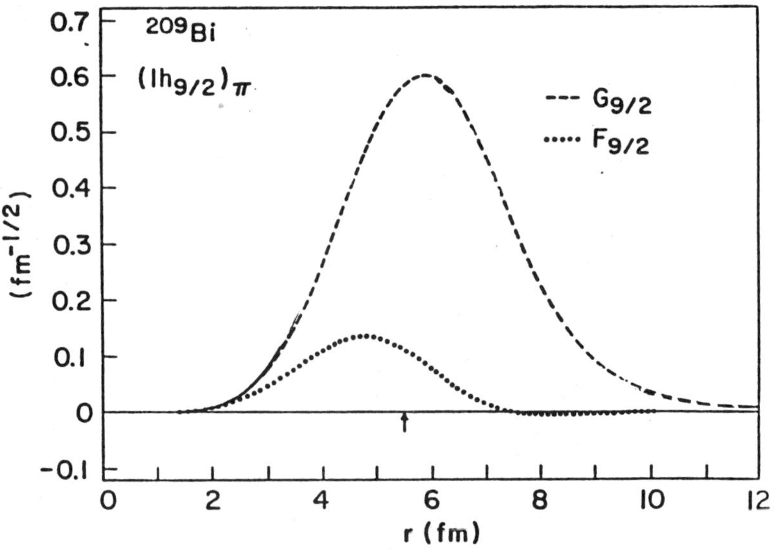

Fig. 7: Proton wavefunction for the $1h\,9/2$ level in Bi^{209}. The upper component G is dashed while the lower F is dotted, see Eq. (4.8).

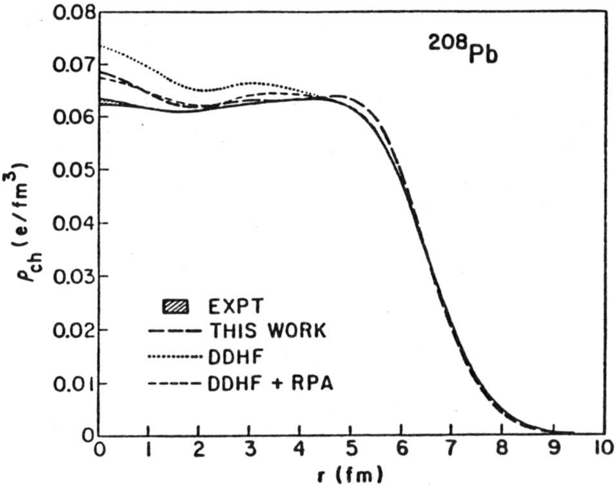

Fig. 8: Charge density (Eq. (4.21)) for Pb^{208}. Also shown are two typical nonrelativistic calculations.

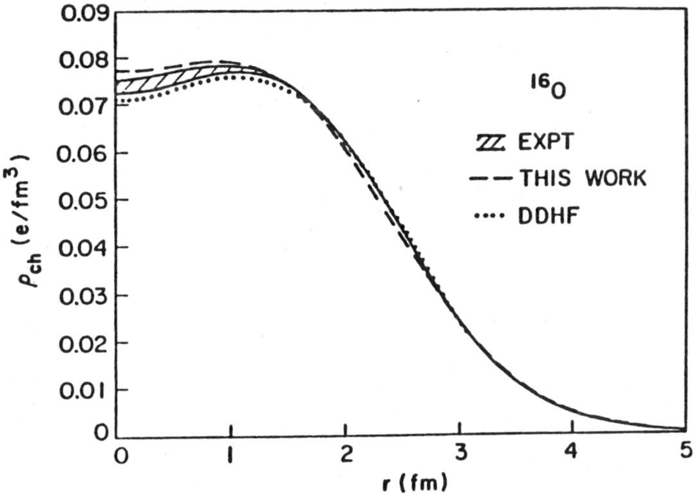

Fig. 9: Charge density in O^{16}.

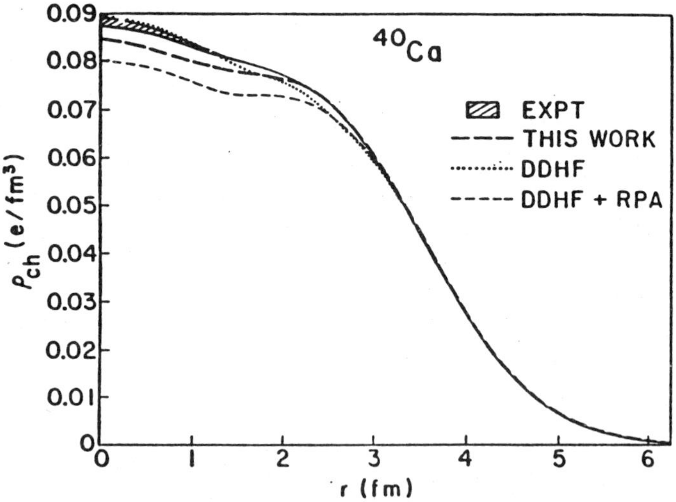

Fig. 10: Charge density in Ca^{40}.

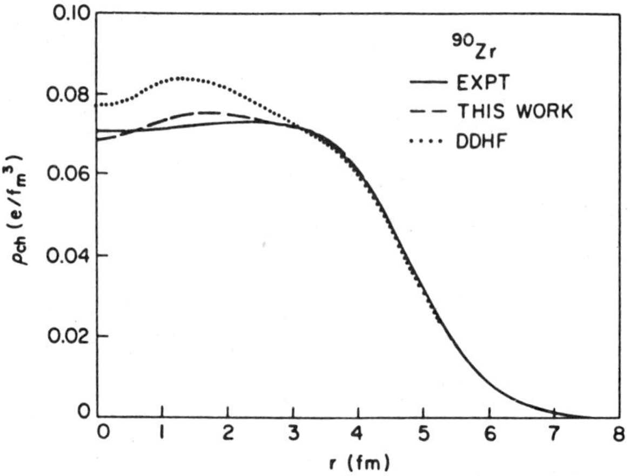

Fig. 11: Charge density in Zr^{90}.

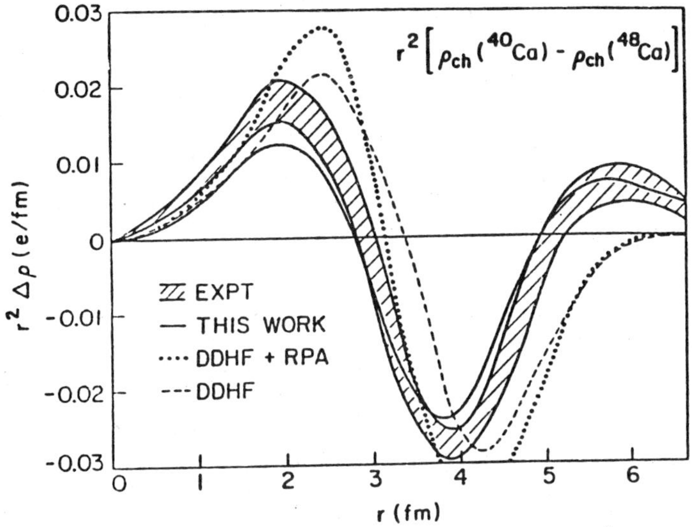

Fig. 12: Charge density difference, Ca^{40}–Ca^{48} times r^2.

the scalar meson has its physical mass m_s. The renormalization procedure of Chin and Walecka[11] leads to a finite correction to the energy density:

$$\Delta e_{\text{vac}} = \frac{-1}{4\pi^2}\left[M^{*4}\, ln\, \frac{M^*}{M} + M^3(M - M^*) - \frac{7}{2}M^2(M - M^*)^2 \right. \\ \left. + \frac{13}{3}M(M - M^*)^3 - \frac{25}{12}(M - M^*)^4\right] \tag{5.3}$$

The renormalization minimizes the vacuum fluctuation corrections by removing the first four terms (of Eq. (5.1)) in an expansion in powers of $M - M^*$. This is equivalent to setting the *renormalized* cubic and quartic scalar meson self-couplings to zero in nuclear matter. *I.e.*, we could have had ϕ^3 and ϕ^4 terms in the Lagrangian (Eq. (2.3)) to start with. Even given that we chose to leave them out we still needed ϕ^3 and ϕ^4 counterterms to render Δe_{vac} finite. We chose these counterterms to leave no ϕ^3 and ϕ^4 meson self-interactions when vacuum fluctuations are included. This minimizes explicit many-nucleon interactions. Note: Δe_{vac} is insensitive to the short-distance structure of the baryons, as it arises simply from the change in mass of the nucleons in the uniform scalar field $g_s\phi_0$.

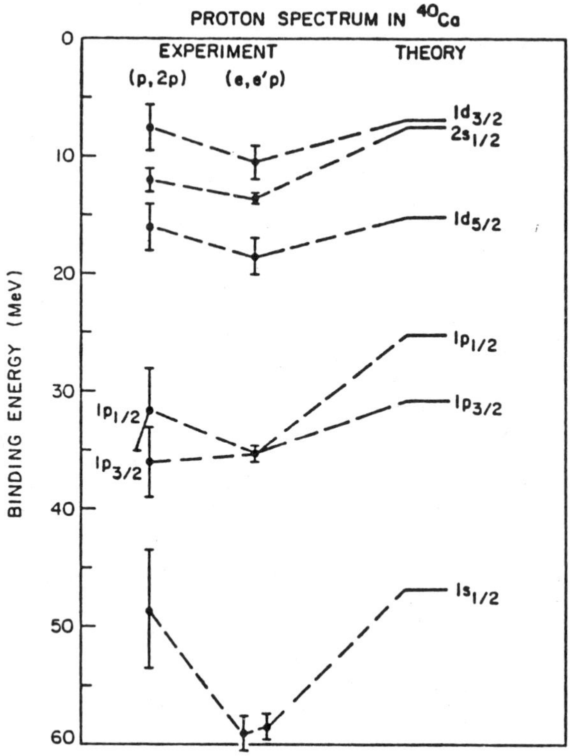

Fig. 13: Single particle spectrum, Dirac eigenvalues of Eq. (4.12), for Ca^{40}.

The correction (Eq. (5.3)) is incorporated into the calculation of finite nuclei using a local density or Thomas-Fermi approximation. We begin by adding a vacuum energy correction

$$\Delta E_{\text{vac}} = \int d^3r \; \Delta e_{\text{vac}}(r) \tag{5.4}$$

to the total energy of a relativistic Hartree calculation. Here Δe_{vac} is given by Eq. (5.3) with a spatially dependent effective mass

$$M^* = M - g_s \phi_0(r) \tag{5.5}$$

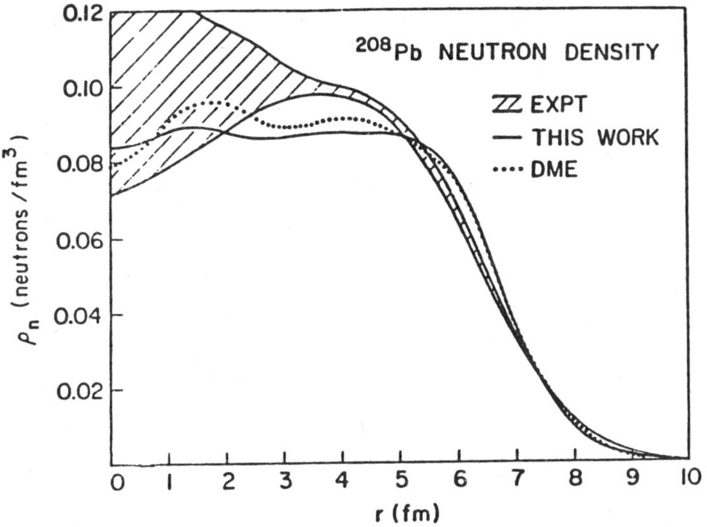

Fig. 14: Neutron density in Pb^{208}. The experimental error band has been deduced from 800 MeV proton scattering.[18]

The scalar meson field equation is obtained by functionally differentiating the total energy (including ΔE_{vac}) with respect to $\phi_0(r)$. The contribution $\delta \Delta E_{\text{vac}}/\delta\phi_0(r)$ can be identified as a vacuum fluctuation correction to the nuclear scalar density, leading to

$$\left[\frac{\partial^2}{\partial r^2} + \frac{2}{r}\frac{\partial}{\partial r} - m_s^2\right]\phi_0(r) = -g_s\rho_s(r) \tag{5.6}$$

where the scalar density in Eq. (4.17) has been modified

$$\rho_s(r) = \rho_s^{(+)}(r) + \Delta\rho_s^{\text{vec}}(r) \tag{5.7}$$

Here the vacuum contribution to the scalar density $\Delta\rho_s^{\text{vac}}$ is

$$\Delta\rho_s^{\text{vac}} = \frac{\delta\nabla E_{\text{vac}}}{\delta\phi_0} = -\frac{1}{\pi^2}\left[M^{*3}\, ln\,\frac{M^*}{M} + \frac{M^3}{3} - \frac{3}{2}M^2M^* + 3MM^{*2} - \frac{11}{6}M^{*3}\right] \tag{5.8}$$

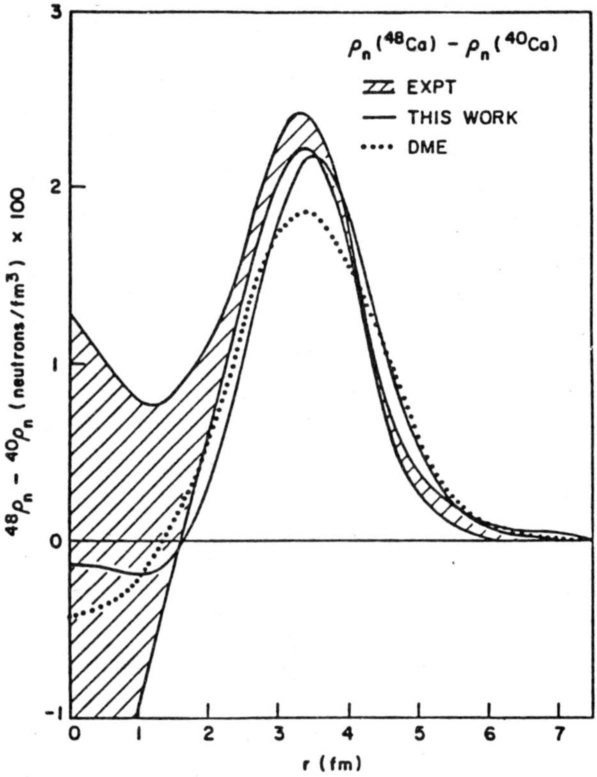

Fig. 15: Neutron density difference Ca^{48}–Ca^{40}, see Fig. 14.

and the positive energy (valence) contribution to ρ_s is given by Eq. (4.14)

$$\rho_s^{(+)} = \sum_a^{\text{occupied}} \frac{2j_a + 1}{4\pi r^2} \left[G_a^2 - F_a^2\right] \tag{5.9}$$

Note, we are making a hybrid approximation. The A valence nucleons are treated explicitly in a Hartree approximation to correctly include shell effects, while the infinite number of Dirac sea nucleons are only treated statistically.

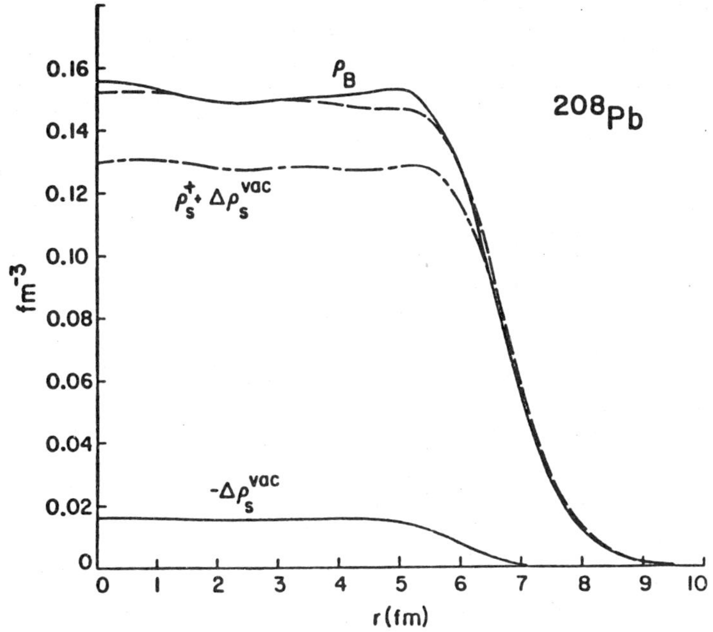

Fig. 16: Baryon and scalar densities in Pb^{208}. The solid line is the baryon density while the dashed line is for the calculation in Section 4 without vacuum fluctuations. The total scalar density Eq. (5.7) is dot-dashed while minus the vacuum contribution, Eq. (5.8) is also shown.

With the present approximations, vacuum fluctuations do not modify the nucleon Dirac or vector meson field equations of Section 4. Furthermore, the conservation of baryon number implies

$$\int d^3r \; \rho_b(r) = A \tag{5.10}$$

and there are no vacuum corrections to $\rho_b(r)$. In contrast, the normalization of $\rho_s(r)$ depends on the interactions. *I.e.*, virtual $N - N$ bar pairs can modify ρ_s but not ρ_b. Thus, the difference between ρ_s and ρ_b may be sensitive to vacuum fluctuations.

The parameters in the model can be adjusted as before to reproduce nuclear saturation including ΔE_{vac}. When this is done the charge and baryon densities are essentially the same as in Section 4. This is shown in Fig. 16 for Pb^{208}.

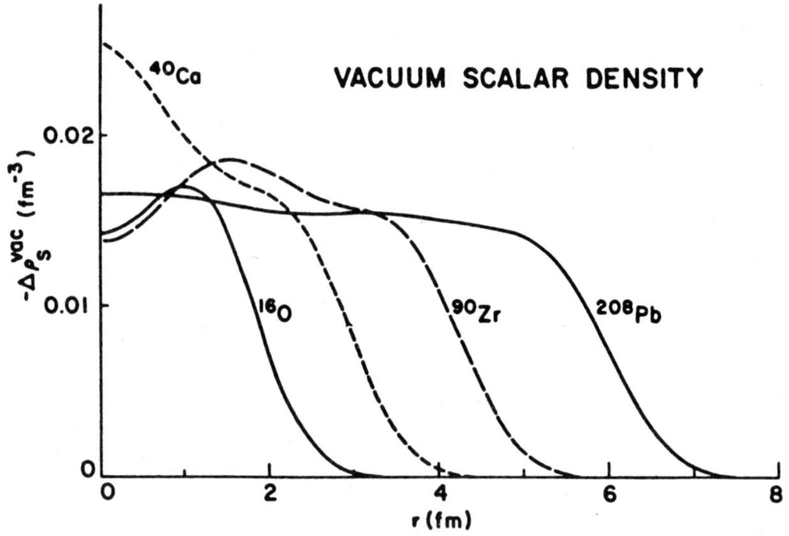

Fig. 17: Vacuum contribution to the scalar density Eq. (5.8) for closed shell nuclei. The different curves show the sensitivity to shell fluctuations in the scalar field strength.

The vacuum flucatuation correction to ρ_s, $\Delta\rho_s^{\rm vac}$ is seen in Figs. (16)-(17) to have the following properties: (1) It is negative, because the scalar density of the negative energy sea decreases (relative to the free vacuum) when M is replaced by M^* ($\langle M$); (2) It vanishes like $(M - M*)^4$ as M^* goes to M. This reflects the renormalization conditions: all results must reduce to the free vacuum as $\phi_0 \to 0$. Thus $\Delta\rho_s^{\rm vac}$ is important only in the nuclear interior and vanishes in the nuclear surface; (3) It is a sensitive function of M^*. For M^* about 0.7 M,

$$\Delta\rho_s^{\rm vac} \approx -0.15\rho_s \quad , \tag{5.11}$$

$\Delta\rho_s^{\rm vac}$ is a sizable fraction of ρ_s.

The difference betwen ρ_s and ρ_b is due to interesting relativistic effects. We now speculate on observing the difference in proton scattering. In the relativistic impulse approximation (see Section 6) the optical potential is given by,

$$U = -\frac{4\pi i p}{M}\left[F_s\rho_s + \gamma_0 F_v \rho_b\right] \tag{5.12}$$

Here, F_s and F_v are Lorentz invariant NN amplitudes. Eq. (5.12) has strong cancelling scalar $F_s\rho_s$ and vector $F_v\rho_b$ potentials. Therefore, one is very sensitive to the difference between ρ_s and ρ_b.

We conclude, in elastic proton scattering there are four target densities: neutron and proton, baryon and scalar. Only one of these is related to the charge density known from electron scattering. Because of sensitive cancellations between scalar and vector optical potentials one may be more sensitive to the difference between ρ_s and ρ_b than to a coherent change in both the neutron scalar and baryon densities. The difference between ρ_s and ρ_b is due to interesting relativistic affects such as vacuum fluctuations.

VI. NUCLEON-NUCLEUS SCATTERING

We would like to relate the meson-nucleon field theory to two nucleon scattering in free space and to nucleon-nucleus scattering. Relativistic Brueckner Bethe Goldstone (RBBG) theory is introduced to solve the relativistic two-nucleon problem in the medium.[12] This allows one to calculate correlation corrections to the MFT results. Clearly, two nucleons will be anticorrelated at short distances because of the repulsive vector exchange.

Space does not allow us to describe RBBG results for the binding energy of nuclear matter (see the talk at this summer school by R. Machleidt). Basically the correlations increase the binding energy (indeed, one has to work very hard in order to calculate this accurately) but they have little effect on the *individual* strong scalar and vector optical potentials.

RBBG theory can also be used to describe nucleon-nucleus scattering. In this section, relativistic optical potentials are calculated at energies near 200 MeV for elastic proton scattering from closed shell nuclei. The 200 MeV energy region is interesting for several reasons. First, as a bridge between high energies, where a simple impulse approximation is valid, and nuclear structure calculations at very low energies. Second, some nonrelativistic optical models show non Wood-Saxon shapes for the potential. These interesting shapes (see Fig. 19) can be explained by quadratic potential terms from a second order reduction of the Dirac equation. Finally, there exists new spin rotation function (Q) data from IUCF[13] for scattering from C^{12}, O^{16}, Ca^{40} and Ca^{48} at 200 MeV. Thus, with A and $d\sigma/d\Omega$ already measured, one can compare theory to a complete set of elastic proton scattering observables.

We begin by identifying the optical potential with the self-energy calculated in relativistic Brueckner theory. Although this program has not yet been carried out completely the present calculations still contain important physics beyond the original relativistic impulse approximation (RIA).

First, the original RIA representation of the NN amplitudes involve ambiguities which become increasingly serious as the energy is decreased. To resolve these ambiguities, a simple direct plus exchange model of the NN amplitudes[14] is employed which allows the use of a pseudo*vector* ($\not{q}\gamma_5$ rather than simple γ_5) pi-N coupling. Next, medium modifications from Pauli blocking are included by using nuclear matter calculations in a local density approximation.

We describe the formalism for these calculations and then show results which quantitatively reproduce all measured spin observables (both A and Q) for closed shell nuclei. Finally, these calculations are compared to the relativistic results of Tjon and Wallace[15] and the nonrelativistic ones of Rikus and von Geramb.[16]

Work is underway to develop a relativistic Brueckner formalism to include NN correlations.[12] Here a reaction matrix Γ (or G matrix) is introduced which sums the ladder diagrams between two nucleons in the medium (Feynman diagrams in which two nucleons exchange any number of mesons but only one at a time). The reaction matrix satisfies a relativistic Brueckner Bethe Goldstone (RBBG) equation which is a generalization of the nonrelativistic result. This equation can be solved in the NN center of velocity frame and the results expressed as five Lorentz invariants (see below) which are easy to transform to the nuclear rest frame. Summing this interaction over the occupied states of the target (in its rest frame) gives the self-energy or optical potential (U) which contains complex Lorentz scalar and vector pieces.

$$U = \sum_{2}^{\text{occupied}} \langle 12|\Gamma|12 - 21\rangle \tag{6.1}$$

Here, 1 is the projectile state and 2 is an occupied target state ($|12 - 21\rangle$ indicates an antisymmetrized matrix element).

There are two effects of the medium on the calculation of Γ. First, the Pauli principle limits the scattering to intermediate states above the Fermi sea. This Pauli blocking is included as described below. Second, the nucleons move in an average potential (given self-consistently by the real part of Eq. (6.1). Such binding energy corrections change the free relationship between momentum and energy *and* enhance the lower components of the Dirac spinors.

We would like to examine these effects step by step so we include Pauli blocking but not binding energy corrections in these first calculations. Binding energy corrections are not expected to be large near 200 MeV because the real part of the nonrelativistic optical potential is small at this energy.

However, binding E corrections may be larger at lower energies and will be included in future calculations.

In general, Γ is a very complicated $4 \times 4 \quad \times \quad 4 \times 4$ Dirac matrix in the spinor space of the two nucleons. In principle, a complete solution of the RBBG equation would determine this complex 256 component matrix, (Tjon and Wallace have calculated the complete matrix for one model but only in the zero density limit[15]). It is much easier to simply solve the RBBG equation for its positive energy matrix elements and make a model for the full matrix structure. The simple model also gives insight into the form of Γ which is lacking in the full calculation. Furthermore, since the positive energy matrix elements of Γ should reproduce the exp. NN amplitudes at zero density, we will take these from data and only calculate the ratio of Pauli blocked to free NN interactions.

To construct an optical potential in a relativistic "$t\rho$" approximation, one must make important assumptions regarding the "off-shell" behavior of the NN amplitudes. First, Dirac spinor matrix elements of an operator $\hat{F}$ are equated to experimental amplitudes. The standard representation of $\hat{F}$ is

$$\hat{F} = \sum_{i=s}^{t} F_i(q, E)\lambda_1^i \, \lambda_2^i$$

i	λ^i
s (Scalar)	1
v (Vector)	γ^μ
p (Pseudovector)	γ^5
a (Axial-vector)	$\gamma^5\gamma^\mu$
t (Tensor)	$\sigma^{\mu\nu}$

(6.2)

Here λ_1^i is a Dirac operator in the spinor space of particle one and the Lorentz invariant amplitudes F_i are functions of energy E and momentum transfer q. We emphasize, only the free spinor matrix elements of Eq. (6.2) are determined by NN data while the full "off-shell" operator is needed to construct an optical potential.

To resolve ambiguities in the operator $\hat{F}$, a simple model which includes the direct and exchange first Born contributions from a number of mesons was developed.[14] This model divides the Lorentz invariant amplitudes into direct $\tilde{t}_i^D$ and exchange $\tilde{t}_i^X$ contributions,

$$F_i(q) = \tilde{t}_i^D(q) + \tilde{t}^X(Q) \tag{6.3}$$

Here, the exchange momentum transfer Q is for scattering angle $\pi - \theta$. The exchange parts of the scalar and vector invariants $\tilde{t}_s^X$, $\tilde{t}_v^X$ have large contributions from one pion exchange. As discussed in [14], it is important to evaluate these using a pseudo*vector* ($\not{q}\gamma_5$) invariant in place of the pseudo*scalar* γ_5 in Eq. (6.2). This serves to decrease the effects of $\tilde{t}_s^X$ and $\tilde{t}_v^X$ substantially, at low energies. Note, the original RIA implicitly uses a pseudoscalar pi-N coupling.

The scalar $i = s$ and vector $i = v$ optical potentials U_i are calculated by folding t with ρ and making a local density approximation for the exchange term.

$$U_i(r) = -\frac{4\pi i p}{M} \int d^3s \left[t_i^D(s)\rho_i(\vec{r}+\vec{s}) + j_0(ps)t_i^X(s)\rho(\vec{r}+\frac{1}{2}\vec{s})\rho_{\text{off}}(k_f s) \right] \tag{6.4}$$

Here, p is the projectile momentum, $j_0(ps)$ a local density approximation to the projectile wavefunction and the off diagonal density matrix has been approximated by its nuclear matter value $\rho_{\text{off}}(ks) = 3/(sk)^3\,(\sin(sk) - sk\cos(sk))$.[17] The baryon and scalar densities have been taken from Section IV[2] and have not been varied. Finally, t^D and t^X are fourier transforms of $\tilde{t}^D$, $\tilde{t}^X$.

It remains to include medium modifications from Pauli blocking. Nuclear matter calculations of the lowest order relativistic Brueckner optical potentials have been performed both with and without the Pauli operator. These calculations started with the HEA one-boson-exchange potential and do not include binding E corrections. From the *ratio* of two nuclear matter calculations we define Pauli blocking correction factors a_i which can be applied to finite nuclei in a local density approximation. Eq. (6.4) is corrected

$$U_i(r) \Leftarrow \left\{ 1 - a_i(E) \left[\frac{\rho_b(r)}{\rho_0} \right]^{2/3} \right\} U_i(r) \tag{6.5}$$

Here $\rho_0 = 0.19\ \text{fm}^{-3}$ and the k_F^2 density dependence is based on simple phase space arguments. At $E = 200$ MeV, the nuclear matter calculations give

	Scalar		Vector	
	Real	Imag	Real	Imag
a_i	-0.01	0.10	0.06	0.21

(6.6)

The calculations can be summarized as follows. Ref. [14] is used for the t matrix which is folded with mean field densities from Section IV[2] and then multiplied by the Pauli blocking correction of Eqs. (6.5) and (6.6).

We now present results. Fig. 18 shows the scalar and vector optical potentials for Ca^{40} at 200 MeV. These agree well with the phenomenological

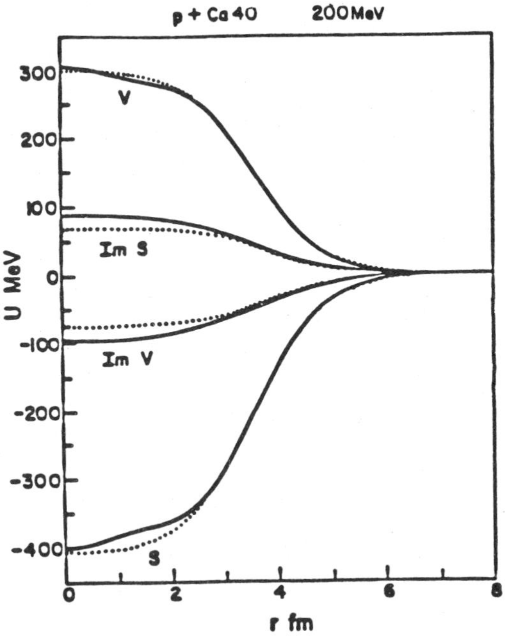

Fig. 18: Optical potentials, Eqs. (6.4–6.5), for Ca^{40} at 200 MeV (solid). The dotted curves are phenomenological fits to elastic data.[3,19]

Dirac Wood-Saxon fits to the elastic data. Fig. 19 shows the Schrödinger-like equivalent potential for Pb^{208}. Compared to the nonrelativistic Paris G matrix calculation of [16], we have a non-Wood-Saxon shape for the central and a much stronger real spin-orbit potential. Fig. 20 shows the cross section for Ca^{40}. The use of pseudoscalar NN amplitudes leads to too high a cross section, while omitting the Pauli blocking corrections also overestimates sigma and gives too much structure. The original RIA corresponds to both omitting Pauli blocking and using psuedoscalar amplitudes and gives a curve above all those those in Fig. 20. A full folding of the density dependence of the interaction instead of the simple Eq. (6.5) may improve the structure in the cross section, although the Pauli blocked pseudovector calculation is not seriously in error. The analyzing power A (both TRIUMF and IUCF data) and the new *preliminary* IUCF spin rotation function Q data[13] are shown for Ca^{40}, O^{16} and Ca^{48} in Figs. 21 – 23. The calculations agree very well with data if Pauli blocking is included and pseusovector amplitudes are used.

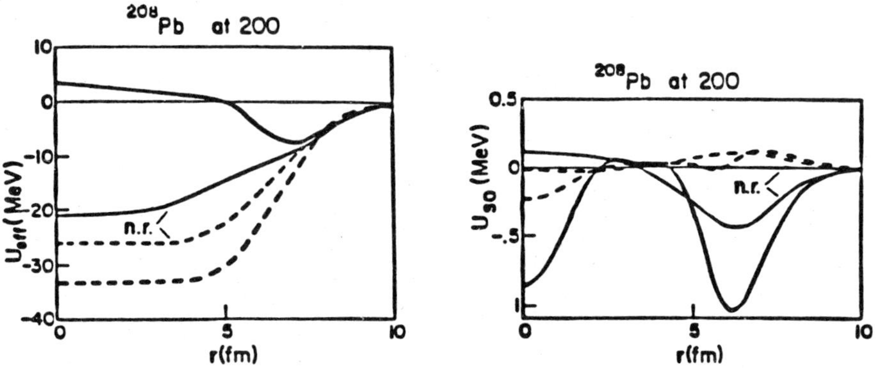

Fig. 19: Central (left) and spin-orbit (right) Schrödinger equivalent optical potentials for Pb^{208}. The real part is solid and the imaginary is dashed. The nonrelativistic results of [16] are marked n.r.

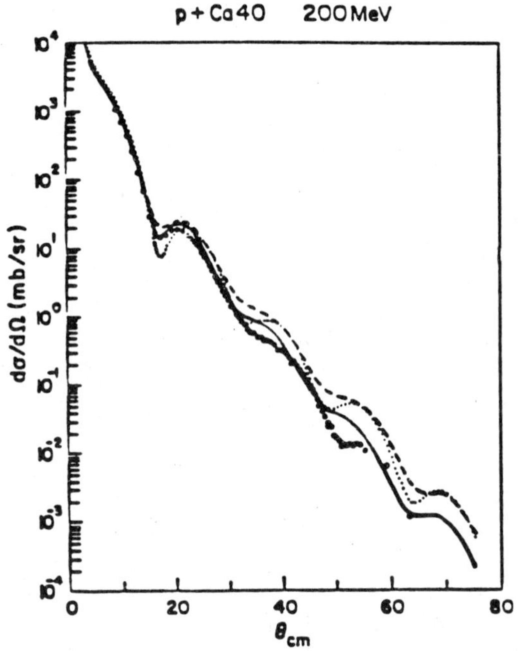

Fig. 20: Cross section for Ca^{40} (solid). The dashed curve uses pseudoscalar rather than pseudovector NN amplitudes and the dotted curve omits Pauli blocking.

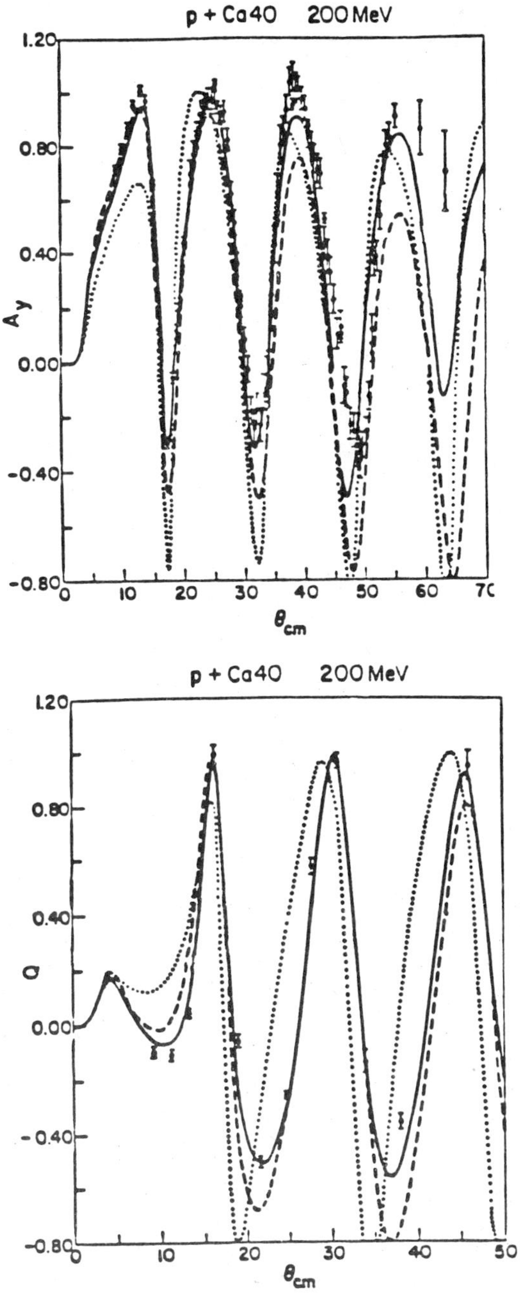

Fig. 21: Spin observables for Ca^{40}, see fig. 20.

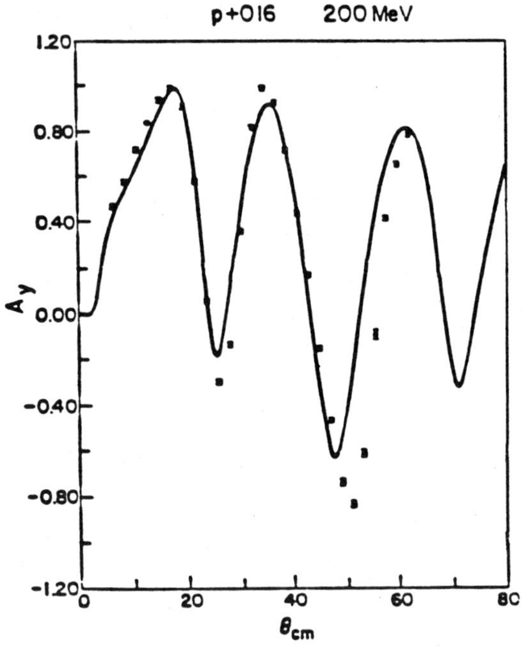

Fig. 22:
Spin observables for O^{16}.

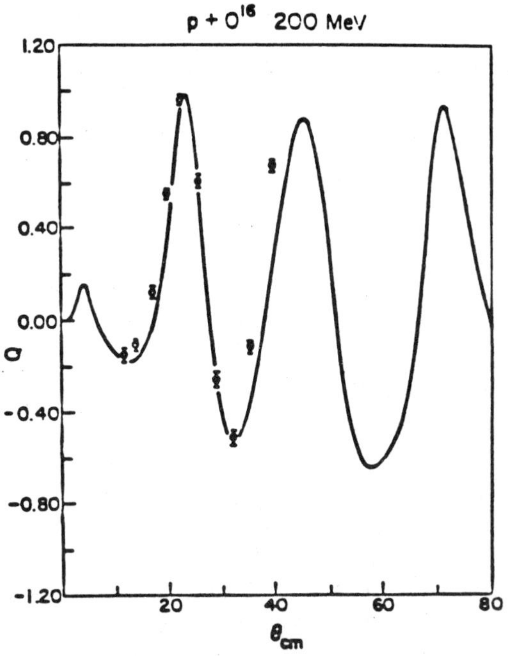

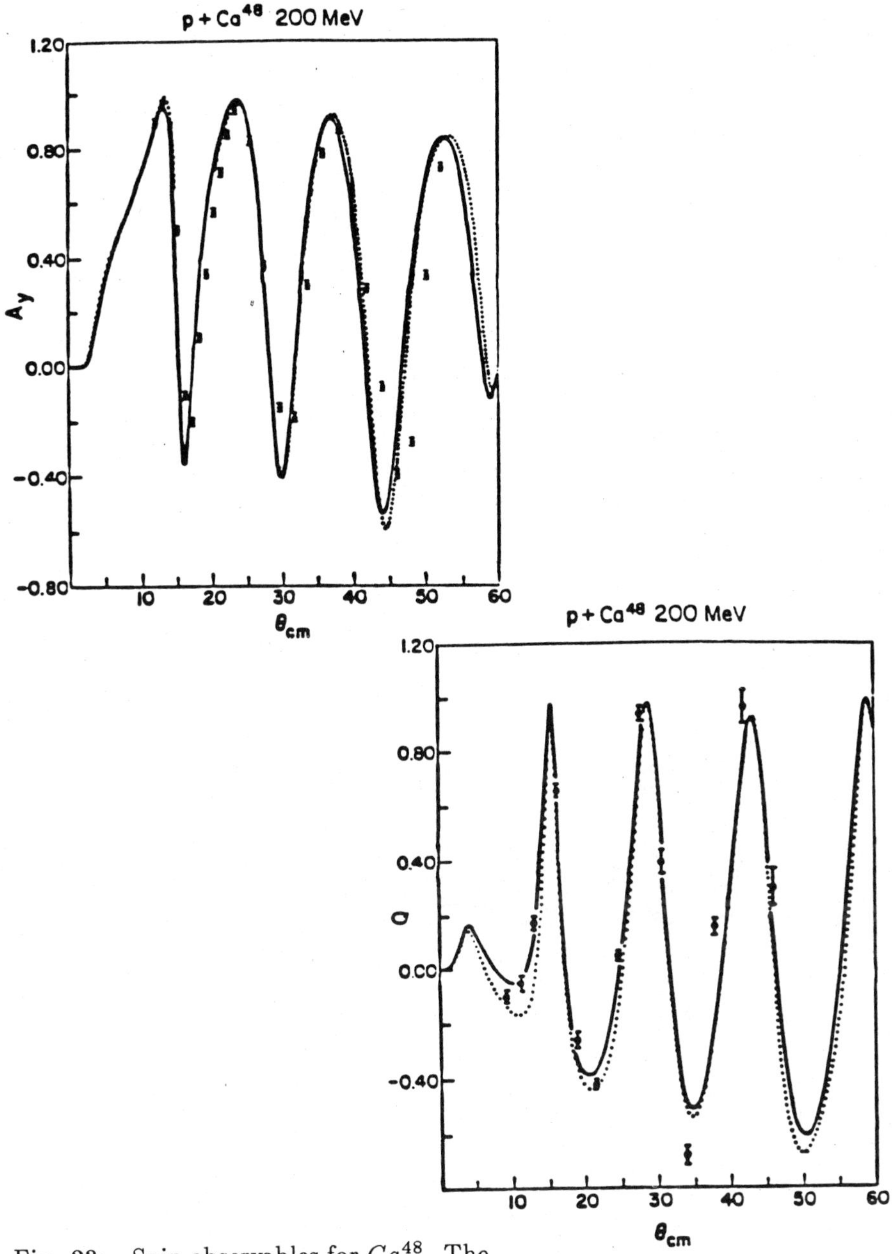

Fig. 23: Spin observables for Ca^{48}. The dotted curve includes a small Lorentz tensor potential.

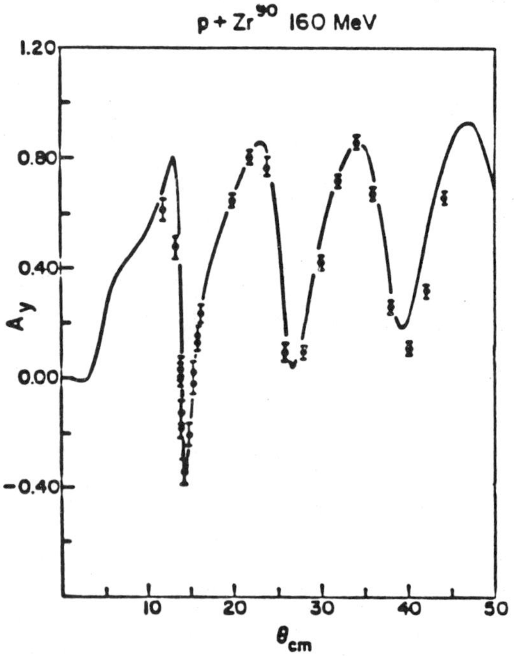

Fig. 24: Analyzing power in Zr^{90} at 160 MeV.

Our calculations essentially agree with Tjon and Wallace[15] for the pseudovector NN amplitudes and for the spin observables in Ca^{40}. However, these authors don't include Pauli blocking which may explain their overestimation of the cross section.

The effect of the Lorentz tensor potential is seen in Fig. 23 to be very small. All of the other calculations shown omit this term.

For the heavier nuclei Q has not yet been measured but A is shown in Figs. 24 – 25 for Zr^{90} and Pb^{208}. For Pb, the nonrelativistic calculation of [16] is seen to underestimate the depth of the minima in A. The relativistic results are very good for both nuclei.

The energy dependence of various corrections is examined with calculations for Pb^{208} at 400 MeV. At this energy there is almost no difference between pseudovector and pseudoscalar NN amplitudes. In addition, Pauli

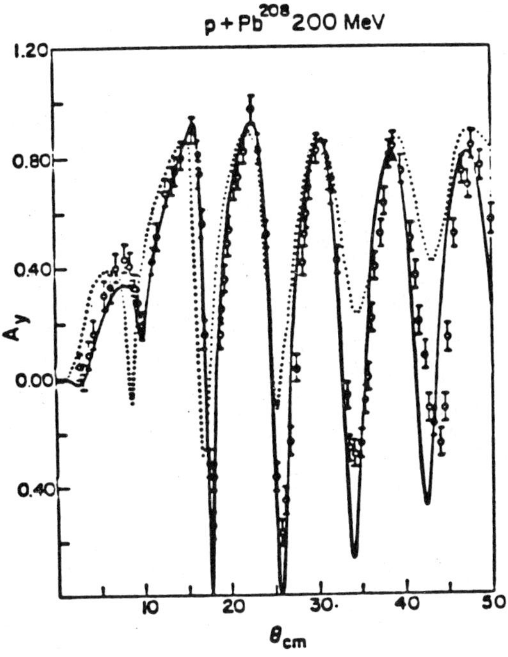

Fig. 25: Analyzing power in Pb^{208}. The nonrelativistic results[16] are dotted.

blocking has only a small effect. Fig. 26 shows the analyzing power in Pb^{208}. Calculations including Pauli blocking (solid) have only slightly larger maxima than free calculations (dotted). Thus, the effects of Pauli blocking while not absent are much smaller than in some nonrelativistic calculations. As the energy increases above 400 MeV, the calculations reduce to the original RIA (which provides an excellent description of data at these energies.)

In conclusion, relativistic optical potentials have been calculated for closed shell nuclei at energies near 200 MeV. These calculations go beyond the RIA by resolving ambiguities in the relativistic NN amplitudes and including Pauli blocking.

The calculations quantitatively reproduce all measured elastic spin observables (both A and Q) for closed shell nuclei. It remains to be seen if this good description is unique. Further nonrelativistic work to compare with

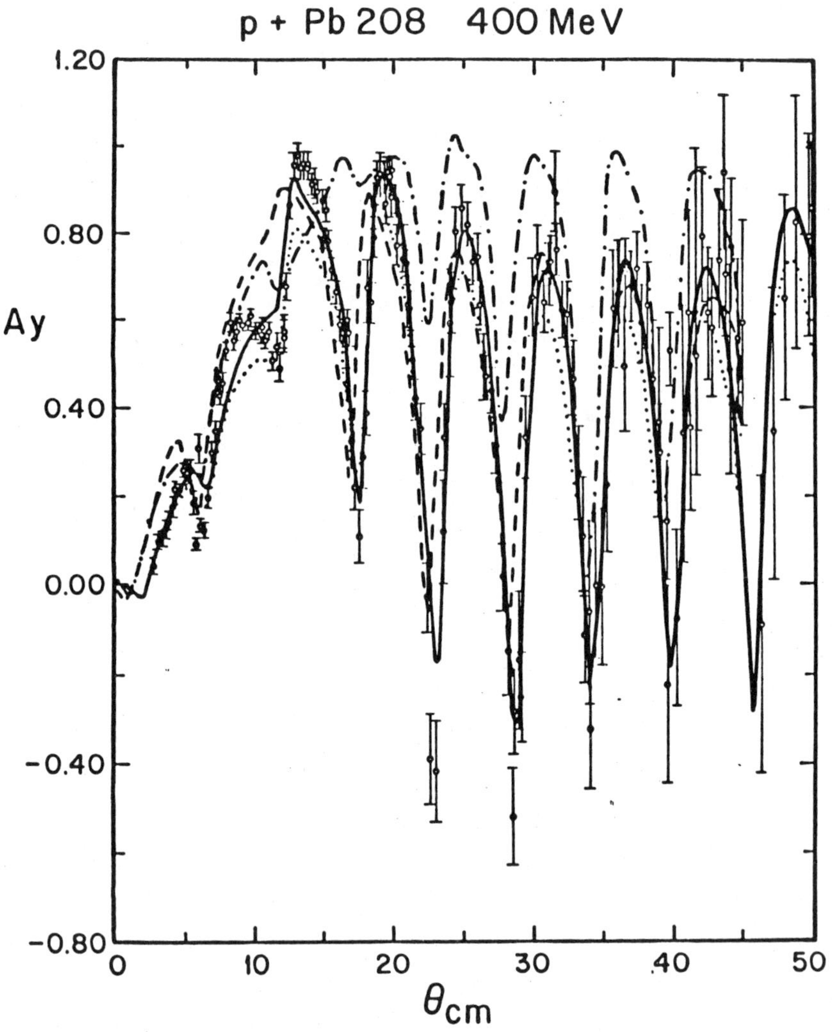

Fig. 26: Analyzing power in Pb^{208} at 400 MeV (TRIUMF data). Curves with Pauli blocking are rel. (solid) and nonrel.[16] (dashed). Curves without medium modifications are rel. (dotted) and nonrel.[6] (dot-dashed).

this relativistic approach would be very useful. Also, measuring Q for Pb or Zr might be useful to help examine the role of Pauli blocking in nuclei with larger interiors.

VII. SUMMARY

In this series of lectures we have introduced a model meson-baryon field theory to describe strong scalar and vector potentials in nuclei. The model was solved in a mean field approximation to yield densities, wavefunctions and spectra for closed shell nuclei. The strong potentials shift the nucleon's mass from M to M^* which may be as small as $M/2$. Furthermore, the lower components of the Dirac wavefunctions are significantly enhanced.

Section 5 discussed vacuum fluctuations from shifts in the mass of the Dirac sea nucleons. It is important to consider all of the relativistic effects of strong potentials not just the easy-to-calculate changes in the positive energy wavefunctions. Given that the potentials are strong, there is no reason vacuum fluctuations need be small. Furthermore, the difference between scalar and baryon densities may be sensitive to vacuum effects.

Section 6 has discussed proton nucleus elastic scattering. Here the Dirac equation provides a good description of spin observables. In the future, it will be important to look for relativistic effects on nuclear structure with electron and inelastic proton scattering.

ACKNOWLEDGEMENTS

The calculations of Section 6 were done with D. Murdock. The kind hospitality of the University of Colorado Nuclear Physics Laboratory (where this was written) is gratefully appreciated.

REFERENCES

1. J. D. Walecka, *Ann. Phys.* (NY) **83**, 491 (1983).
2. C. J. Horowitz and B. D. Serot, *Nucl. Phys.* **A368**, 503 (1981).
3. L. G. Arnold, *et al.*, *Phys. Rev.* **C23**, 1449 (1981); B. C. Clark, *et al.*, Proceedings of "The Interaction Between Medium Energy Nucleons in Nuclei," H. O. Meyer, ed., (Bloomington, IA) (1982).
4. J. A. McNeil, J. Shepard and S. J. Wallace, *Phys. Rev. Lett.* **50**, 1439 (1983); J. Shephard, *et al.*, *Phys. Rev. Lett.* **50**, 1443 (1983).
5. B. D. Serot and J. D. Walecka, "The Relativistic Nuclealr Many-Body Problem," to be published in *Advances in Nuclear Physics*, J. Negele and E. Vogt, eds.
6. L. Ray and G. W. Hoffman, *Phys. Rev.* **C31**, 538 (1985).

7. J. D. Bjorken and S. D. Drell, "Relativistic Quantum Fields," (McGraw-Hill, NY) (1965).
8. S. Weinberg, "Gravitation and Cosmology", (Wiley, NY) (1972).
9. J. J. Sakurai, "Advanced Quantum Mechanics," (Addison-Wesley, Reading, MA) (1967).
10. C. J. Horowitz and B. D. Serot, *Phys. Lett.* **140B**, 181 (1984).
11. S. A. Chin, *Ann. Phys.* **108**, 301 (1977).
12. C. J. Horowitz and B. D. Serot, *Phys. Lett.* **137B**, 287 (1984).
13. E. Stevenson, Proceedings of "Dirac Approaches to Nuclear Physics," Los Alamos New Mexico (1985) and private communication (1985).
14. C. J. Horowitz, *Phys. Rev.* **C31**, 1340 (1985).
15. J. A. Tjon and S. J. Wallace, University of Maryland preprint 85-052 (1985).
16. L. Rikus and H. V. von Geramb, *Nucl. Phys.* **A426**, 469 (1984).
17. F. A. Brieva and J. R. Rook, *Nucl. Phys.* **A291**, 317 (1977).
18. L. Ray, *Phys. Rev.* **C19**, 1855 (1979).
19. B. C. Clark, private communication (1985).

ANTIPROTON PHYSICS AT INTERMEDIATE ENERGY

George E. Walker
Indiana University
Bloomington, IN 47405

ABSTRACT

The basic character of antinucleon-nucleon ($\bar{N}N$) and antinucleon-nucleus ($\bar{N}A$) interactions is reviewed. The overlap and contrast with nucleon-nucleon (NN) and nucleon-nucleus (NA) interactions is discussed. The usefulness of the antinucleon as a probe for elucidating the strong force and the nuclear many-body system is stressed. Exotic reactions leading possibly to hypernuclei, quark-gluon plasmas and studies of charmed meson systems are included in the antiproton physics qualitatively summarized.

INTRODUCTION

In this qualitiative overview of antinucleon ($\bar{N}$) physics the discussion is divided into the following topics:

Basic Properties of the Antinucleon

$\bar{N}N$ Interactions

$\bar{N}$-Nucleus Elastic and Inelastic Scattering

$\bar{N}N$ and $\bar{N}A$ "Exotic" Reactions

In the process of putting together these lectures I have drawn heavily on the work of others, including several excellent reviews of aspects of antiproton physics.[1-5]

As we approach the subject of antiproton-nucleus interactions it will be useful to keep in mind those areas where the $\bar{p}$ is expected to yield similar and complementary information on the nucleus compared to kaon-, pion-, proton-, electron-nucleus scattering. Each projectile has its own characteristic spin and isospin dependent nuclear interaction, nuclear mean free path (see figure 1), and energy and momentum transfer projectile-target nucleon-interaction dependence. In addition strong interaction conservation laws involving, for example, baryon number, strangeness, and G parity may play an important role in differentiating various projectile-nucleus interactions.

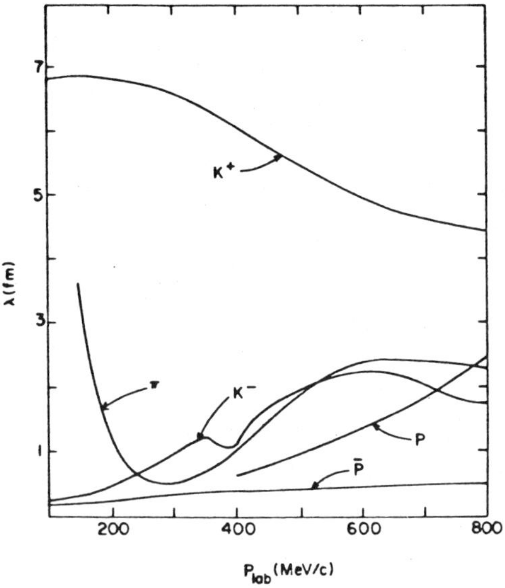

FIG. 1. Mean free path λ of various hadrons in nuclei as a function of lab momentum P_{lab}. The nuclear matter estimate $\lambda=(\rho\sigma_{AV})^{-1}$ is used, where $\rho=0.16\ fm^{-3}$ is the density and $\sigma_{AV}=(\sigma_p+\sigma_n)/2$ is the average of elementary hadron-proton and hadron-neutron cross sections σ_p and σ_n. For incident protons (P), $\sigma_{AV}(b)=\sigma_{np}\approx 8.5/T_{lab}(MeV)$ is adopted, where T_{lab} is the lab kinetic energy. For antiprotons ($\overline{P}$), $\sigma_{\overline{p}n}\approx\sigma_{\overline{p}d}-\sigma_{\overline{p}p}$ and $\sigma^{\overline{P}}_{AV}(mb)\approx 64.5+39/P_{lab}(GeV)$ is assumed. The elementary cross sections have not been Fermi averaged, which would tend to smooth out the rapid energy dependence of λ for pions (π) and for K^- near 400 MeV/c [the Y*(1520) resonance]. The proton results do not include the important effect of Pauli blocking. Figure taken from Ref. 6.

BASIC PROPERTIES

It may be useful to remind ourselves of the history of the concept of the antiparticle as a hole in the Dirac Sea (vacuum). The Dirac equation has positive and negative energy solutions, i.e.

$$(\not{p} - m\epsilon_r)\psi = 0 \quad (\text{where } \epsilon_r = \pm 1) \tag{1}$$

with the minus sign being associated with negative energy solutions. Stability apparently requires a filled Dirac sea of negative energy solutions. One can make a one-to-one correspondence between negative energy solutions and antiparticle solutions of the Dirac equation. In fact the antiparticle states are associated with holes in the filled Dirac Sea (absence of a negative energy nucleon). The quantum numbers of the hole state are taken relative to the vacuum. Since the nucleon is a positive energy particle and the anti-nucleon is a hole (missing negative energy nucleon) one can imagine a situation where the positive energy nucleon emits energy (in the form of a boson quanta i.e. π or γ) and falls into the negative energy hole - thus destroying the hole. This process is called $\bar{N}N$ pair annihilation and is accompanied by the emission of energy (in the form of photons, pions, kaons, $\bar{\Lambda}\Lambda$ etc.). Similarly there can be $\bar{N}N$ pair creation induced by photons, pions, etc.

Some of the relevant quantum numbers for describing a Hadron are intrinsic spin (J), parity (P), baryon number (B), charge (Q), strangeness (S), hypercharge (Y), isospin and third component (I, I_3), where

$$S = Y - B \tag{2a}$$

$$Q = I_3 + \frac{1}{2}Y = I_3 + \frac{1}{2}(B+S) \ . \tag{2b}$$

If a particle has quantum numbers J, P, B, I, I_3, Y, S then the corresponding antiparticle has quantum numbers J, -P, -B, $I, -I_3$, -Y, -S. Thus since the proton has quantum numbers B=1, Q=1, S=0, Y=1, $I=I_3=\frac{1}{2}$ then the antiproton has B=-1, Q=-1, S=0, Y=-1, $I=\frac{1}{2}$, $I_3=-\frac{1}{2}$. Note the magnetic moment of the antinucleon is opposite in sign to that of the nucleon. Since the intrinsic parity of the nucleon and antinucleon have opposite sign, the $\overline{N}N$ system in a relative S state has <u>odd</u> parity.

Because the $\overline{N}N$ system has baryon number zero there are many more important reaction channels below $P_{lab} \lesssim$ 10 GeV/c than for the NN system. These include $\overline{p}p \rightarrow$ pions, kaons, charmed mesons, $\overline{\Lambda}\Lambda$ and $\overline{\Lambda}_c\Lambda_c$. The pair annihilation into several mesons or strange particles may provide the doorway to the study of nuclear quark gluon plasmas.[7,8] The $\overline{p}p \rightarrow \overline{\Lambda}\Lambda$ reaction leads, in the nuclear environment, to the study of hypernuclei [hyponuclei] via the reaction $(p,\overline{\Lambda})$ $[p,\Lambda]$. The $\overline{p}p \rightarrow \overline{c}c$ reaction allows the study of charmed meson production in a wide variety of angular momentum, pair and charge conjugation channels - not just the $J^{PC}=1^{--}$ available from e^+e^- annihilation.

THE $\overline{N}N$ INTERACTION

The one boson exchange (OBE) part of the $\overline{N}N$ interaction is related to the OBE part of the NN interaction by G parity (charge conjugation followed by a 180° rotation in isospin space). In what follows we describe these t channel exchanges in a non-relativistic potential model. The basic idea is that one describes the long range real part of the $\overline{N}N$ interaction via one meson exchange Feynman diagrams. One then makes a non-relativistic reduction and defines the resulting expression as an $\overline{N}N$ potential. The resulting potential is related to the OBE NN potential by changing the sign of the terms arising from the exchange of odd G parity mesons (π,ω,δ). In what follows we briefly list the results for $\overline{N}N$ and NN OBE potentials and discuss some of the main features and differences for the two systems. A more complete discussion is found in references 9 and 10.

We consider the exchange of scalar (δ,ϵ), pseudoscalar (π,η), and vector (ρ,ω) mesons. [Isoscalar mesons are listed second in the parenthesis]. The resulting non-relativistic potentials can be written in the form

$$V_{NN} = \sum_i V^i(r) \tag{3a}$$

$$V_{\overline{N}N}(r) = \sum_i (-1)^{G^i} V_i(r) \tag{3b}$$

where (lower sign for $\overline{N}N$ includes $(-1)^{G^i}$ factor[9,10])

$$V_\pi(r) = \pm\, \tau_1\cdot\tau_2\left[(\sigma_1\cdot\sigma_2)V_\sigma^\pi(r) + S_{12}V_T^\pi(r)\right] \tag{4a}$$

$$V_\eta(r) = \sigma_1\cdot\sigma_2 V_\sigma^\eta(r) + S_{12}V_T^\eta(r) \tag{4b}$$

$$V_\rho(r) = \tau_1\cdot\tau_2\left[V_o^\rho(r) + \sigma_1\cdot\sigma_2 V_\sigma^\rho(r) - S_{12}V_T^\rho(r) - L\cdot S V_{LS}^\rho(r) + Q_{12}V_{LS2}^\rho(r)\right] \tag{4c}$$

$$V_\omega(r) = \pm\left[V_o^\omega(r) + \sigma_1\cdot\sigma_2 V_\sigma^\omega(r) - S_{12}V_T^\omega(r) - L\cdot S V_{LS}^\omega(r) + Q_{12}V_{LS2}^\omega(r)\right] \tag{4d}$$

$$V_\delta(r) = \mp\, \tau_1\cdot\tau_2\left[V_o^\delta(r) + L\cdot S V_{LS}^\delta(r) + Q_{12}V_{LS2}^\delta(r)\right] \tag{4e}$$

$$V_\epsilon(r) = -\left[V_o^\epsilon(r) + L\cdot S V_{LS}^\epsilon(r) + Q_{12}V_{LS2}^\epsilon(r)\right] \tag{4f}$$

where

$$L\cdot S = (r_1-r_2) \times (p_1-p_2) \cdot (\sigma_1+\sigma_2) \tag{5a}$$

$$S_{12} = 3(\sigma_1\cdot\hat{r})(\sigma_2\cdot\hat{r}) - \sigma_1\cdot\sigma_2 \tag{5b}$$

$$Q_{12} = \frac{1}{2}\left[\sigma_1\cdot L\,\sigma_2\cdot L + \sigma_2\cdot L\,\sigma_1\cdot L\right] \tag{5c}$$

and the radial functions V(r) are positive.[9,10] Some of the gross features of the $\overline{N}N$ and NN systems, such as the behavior of phase shifts and the ordering of bound levels can be understood by examining the behavior of the different potential "coherences" associated with the two systems.

There are two types of coherences that have been

discussed by Buck et al.[10]

a) When all contributions (i.e. V_i due to all mesons) to a particular component $V_i=(V_o,V_\sigma,V_{LS},V_T,V_{LS2})$ have the same sign.

b) When all components ($\Sigma_i V_i$) of the potential generated by mesons in the same nonet (for example π,η form one nonet) have the same sign.

In reference 10 some coherences in the NN system are listed. For example:

1) The isotriplet spin-spin potential, $\Sigma_j V_\sigma^j$, is attractive (repulsive) for singlet even (triplet odd) waves. (coherence type a)

2) Isospin-spin triplet spin-orbit potentials are attractive (repulsive) for J=L+1 (J=L,L-1). (coherence type a) (Note the notation $^{2T+1,2S+1}L_J$ is used for Partial Waves)

3) For isospin-spin triplet states (coherence type b)

   i) pseudoscalar mesons give a repulsive contribution for J=L
   ii) vector mesons give a repulsive contribution for J=L-1
   iii) scalar mesons give an attractive contribution for J=L+1

An example of the effect of coherences in the NN system results in the understanding of the sign change in the $^{33}P_o$ phase shift as a function of energy.[10] This results because the <u>long range</u> pion exchange potential is attractive but the <u>short range</u> spin-spin and spin-orbit potentials are repulsive (1 and 2 above). The vector meson exchange force is also repulsive (3ii above).

For the $\bar{N}N$ channel it is also useful to consider the coherences obtained due to t channel meson exchange even though the interaction also contains an important annihilation piece. [This procedure has also been found useful in qualitatively understanding KN interactions in cases where one might believe that single t channel exchange is a naive model for the resulting transition matrix.]

The coherences in the $\bar{N}N$ system have been discussed in ref. 10. We list some results below.

1) The spin independent potential, V_o, is attractive for ρ, ω, δ, and ϵ exchange in the isosinglet channel.
2) The tensor potential is attractive (repulsive) for π, η, ρ, and ω exchange in the isosinglet channel for J=L (J=L±1).
3) There are nonet coherences in the isosinglet-spin triplet channels where:

   i) the vector meson (ρ,ω) interaction is attractive for J=L-1
   ii) the pseudoscalar meson (π,η) interaction is repulsive for J=L
   iii) the scalar meson (δ,ϵ) interaction is attractive for J=L+1.

Note that the combination of 2) and 3)i above imply that the $\bar{N}N$ $^{13}P_o$ channel interaction should be especially attractive. This implication is substantiated by more detailed calculations as can be seen from fig. 2 where the $\bar{N}N$ system with these quantum numbers is predicted to be especially tightly bound.[4]

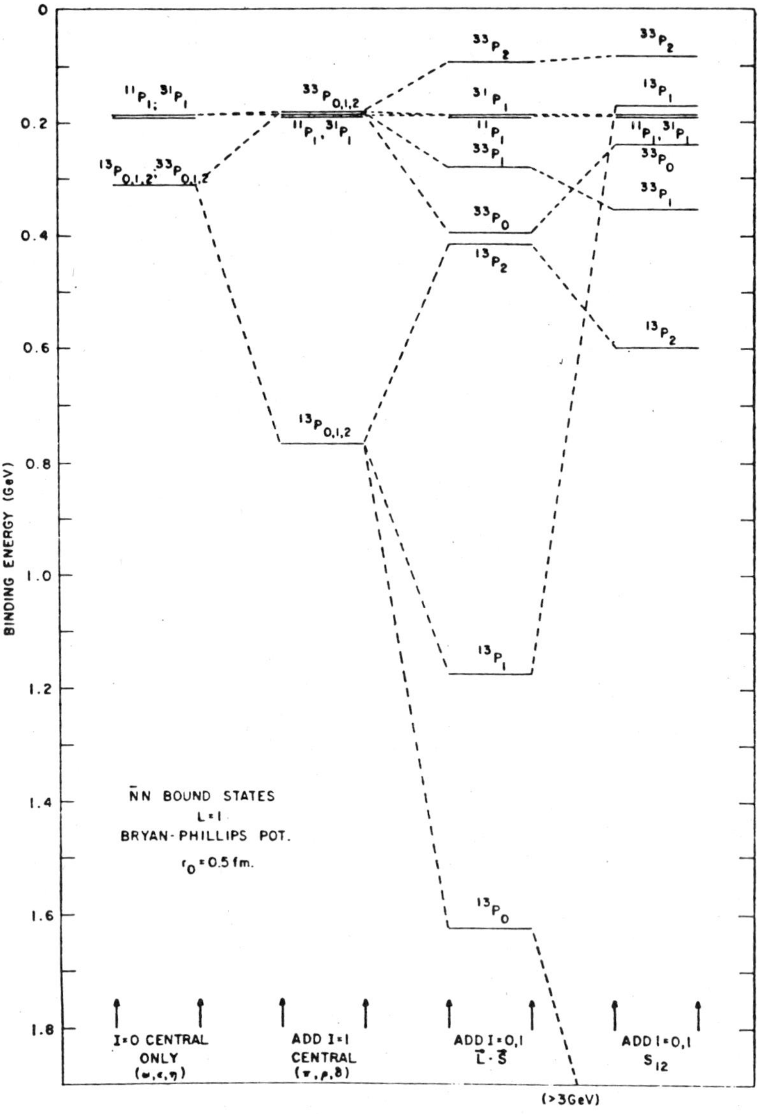

FIG. 2. The $\bar{N}N$ spectrum of L=1 states obtained for the Bryan Phillips Model [11] by successively introducing isoscalar Wigner and spin-spin potentials (col. 1), isovector ones (col. 2), spin-orbit terms (col. 3) and diagonal tensor forces (col. 4). A value r_o=0.5 fm of the cutoff was chosen so that all the P-states are bound; this spectrum is only of illustrative significance. Figure taken from Ref. 10.

In comparing the NN and $\overline{N}N$ t channel meson exchange potentials one finds that the NN interaction has a very important spin-orbit contribution (L·S) while for the $\overline{N}N$ system the tensor (S_{12}) and quadratic spin-orbit potential (V_{LS2}) play an important role due to coherences. In fact, even with a spin independent short range annihilation potential included (see later discussion), theoretical predictions show that the elastic $\overline{p}p$ polarization is determined by the tensor and quadratic spin-orbit contributions with a negligible sensitivity to the V_σ and V_{LS} terms.[4] The polarization transfer is predicted to be considerably enhanced if vector meson exchange (see 3i above) is included with the pion exchange interaction.[4]

Since the $\overline{N}N$ system has baryon number zero there are strongly coupled annihilation channels (missing in the NN, B=2 system) that play an important role in determining the $\overline{N}N$ interaction, and the elastic channel bound or resonant state properties. The real part of the $\overline{N}N$ "optical potential" may be mainly due to the OBE contributions discussed above, especially at long and intermediate range. However at short range the "hard core" in phenomenological NN potentials and the presence of the annihilation potential means that the G parity arguments used above cannot be usefully applied to relate the NN and $\overline{N}N$ interaction. For the $\overline{N}N$ annihilation potential we must build new models often with recourse to a bag picture and introducing quark and gluon degrees of freedom. Before discussing representative models of the annihilation potential we very briefly mention the situation for resonance structure in the $\overline{N}N$ system.

There is conflicting evidence for the existence of a resonance in the $\overline{N}N$ system at $\simeq$ 1940 MeV (called the S meson).[12] There is some evidence for a resonant bump 2.02 GeV/$_c{}^2$ in the production experiments $\overline{p}p \to p_{fast}$ $\overline{n}\pi^+\pi^-\pi^-$, $\overline{p}p \to \pi^+_{fast}$ $\overline{p}n\pi^+\pi^-$ and $\gamma p \to p\overline{p}p$.[12] An important challenge for models of the $\overline{N}N$ interaction and of the $\overline{N}N$ system is to predict the bound system energy levels and branching ratios for various annihilation channels. We shall return to this point later in the discussion.

There have been many models suggested for the annihilation piece of the $\bar{N}N$ interaction. Boundary condition black disk type models provide an understanding of total cross sections; however, one needs to have more detailed models to predict spin and isospin dependences, angular distributions and branching ratios to reaction channels.

There have been several optical model parametrizations of the $\bar{N}N$ annihilation potential.[4] For example Bryan and Phillips used a form for the $\bar{N}N$ potential given by:[11]

$$V_{\bar{N}N} = \underset{\text{t channel}}{V^t_{OBE}} + \underset{-(V_o+iW_o)/(1+e^{(r-R)/a})}{V_{ann}(r) + iW(r)} \tag{6}$$

with

$$V_o = 0,\ W_o = 62\ \text{GeV},\ R = 0,\ a = 1/6\ \text{fm}.$$

Dover and Richard[13] also used a Saxon-Woods form for the annihilation potential with V^t_{OBE} being given by the Paris potential.[14] Dover-Richard give different parametrizations denoted, for example, by I and II below.

I) $V_o = 21$ GeV, $W_o = 20$ GeV, $R = 0.$ $a = .2$fm

II) $V_o = 500$ MeV, $W_o = 500$ MeV, $R = .8$fm $a = .2$fm

These potentials fit the $\bar{p}p \rightarrow \bar{n}n$ data but underestimate the back angle elastic scattering data.[15] Although potentials I and II above have different depth constants, V_o, they are comparable in the region $\simeq$ 1fm. At short distances the differences in the potentials are irrelevant because of the very strong absorption.

The Paris group of Cote et al[16] obtained good fits to the available differential cross section and polarization data by using the following expression for the annihilation potential:

$$W(r) = \left\{ g_c(1+f_c E) + g_{ss}(1+f_{ss}E)\sigma_1\cdot\sigma_2 + g_T S_{12} + \frac{g_{LS}}{4m^2} L\cdot S \frac{1}{r}\frac{\delta}{\delta r} \right\} \frac{K_o(2mr)}{r} \quad (7)$$

where the coefficients, g_i, are isospin dependent. The radial function, K_o, is a modified Bessel function of range $\frac{1}{2m} \approx .1\text{fm}$. A version of the Paris Potential[17] is used for the real part of the potential. The W above has a strong energy dependence. The strong spin dependence (which will be of considerable importance in the study of $\bar{N}$-nucleus inelastic scattering) can be exhibited by the ratios of W^{IS} in the S-wave channel:[4]

(I ≡ isospin, S ≡ spin of the $\bar{N}N$ system)

$$W^{00}:W^{10}:W^{01}:W^{11} \longrightarrow \begin{array}{ll} 1:0.81:0.11:0.073 & (E=0\text{ MeV}) \\ 0.92:1.\ \ :0.15:0.035 & (E=100\text{MeV}). \end{array} \quad (8)$$

There have been other attempts to use macroscopic models to fit the $\bar{N}N$ data. Timmers et al[18,4] use a coupled channels model to parametrize the loss of flux from the elastic channel. The off-diagonal annihilation potentials are spin independent which is in contrast to the single channel model of Cote et al.[16] It is interesting that both models do a respectable job of fitting the existing data but give considerably different predictions, for example, for the larger angle polarization observable. There exist other models of the types listed above but all these studies are limited in that a) there is insufficient $\bar{N}N$ data to truly test the suggested forms and b) these "macroscopic" phenomenological models cannot provide detailed individual reaction channel predictions in their present form (for example, leading to strange mesons in the final state).

More microscopic models based on QCD, confined quarks and gluons (bag type models) provide the possibility of a richer set of reaction channel predictions. Some suggestive diagrams at the quark level are shown in figs. 3 and 4.

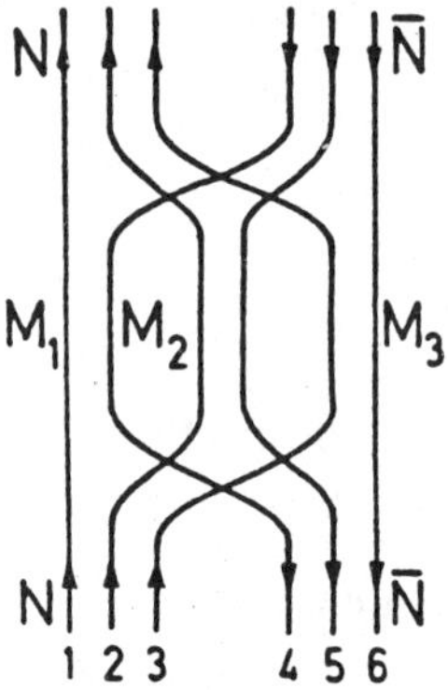

FIG. 3. A quark rearrangement graph leading to a three meson intermediate state is shown. Figure taken from Ref. 4.

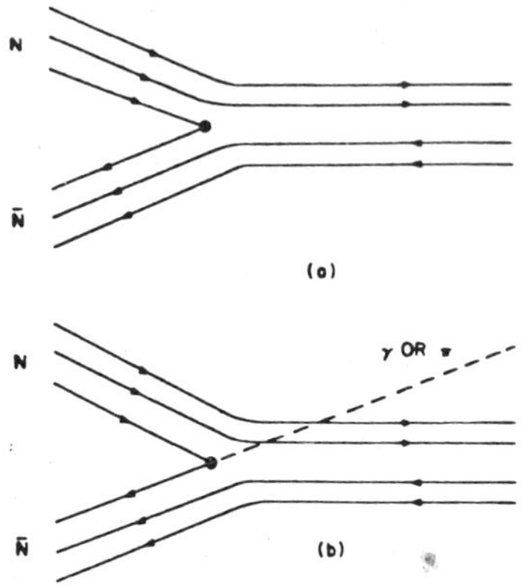

FIG. 4. Mechanism for direct $\bar{N}N$ coupling to $q^2\bar{q}^2$ bag states ($^3P_0$ model) is shown in (a); one of the processes for populating $q^2\bar{q}^2$ states via γ or π emission is shown in (b). Figure taken from Ref. 4.

In figure 3 we show how the $\bar{N}N$ system can be rearranged so that the intermediate "compound" system may be regarded as a multipion $3(\bar{q}q)$ system. These intermediate states may provide the doorway to final channels involving mesons, photons and "baryonium." In figure 4 we illustrate how quark-antiquark ($\bar{q}q$) annihilation [accompanied perhaps by photon or meson emission (fig. 4b)] may lead to final $\bar{q}^2q^2$ states. Such baryonium states that are mesonic-like but weakly coupled to final mesons have been studied in bag models and their existence and properties would provide fundamental information on modern microscopic theories of strong interactions.[19] Note in fig. 4a the $\bar{q}q$ annihilation is accompanied by an internal gluon (not shown). In the process allowing for γ or π emission (fig. 4b), γ or π production followed by $\bar{q}q$ annihilation into an internal gluon should also be considered.

In the various models available the detailed geometry (especially at short distances) is probably not crucial. Most models adopt an overlap-absorption radius that allows a certain degree of flexibility. More important are the detailed predictions associated with branching ratios to reaction channels and the predicted spin-isospin channel strength ratios associated with elastic and charge exchange $\overline{N}N$ scattering. These predictions should not only be experimentally verifiable in the future for the $\overline{N}N$ system but can be tested, as discussed below, in $\overline{N}$-nucleus reactions where one can use the nucleus as a spin-isospin filter.

Models that obtain the $\overline{N}N$ annihilation interaction via $\overline{N}N \to \overline{q}^2 q^2$ + gluon in the never-come-back approximation yield good fits to total elastic and charge exchange cross sections.[20] The adjustable parameters are the bag radius R and α_s^*, an effective QCD coupling constant. An important point is that the α_s^* that allows a good fit to the data is about an order of magnitude greater than that suggested by perturbative QCD. Thus the effective coupling constant is probably mocking up the effects of more complicated multistep processes. Therefore the predicted spin-isospin ratios based on a first order process (with an <u>effective</u> overall coupling constant) may not be very meaningful. Nevertheless it is interesting that at least one such model does not yield a strong spin dependence for W^{IS} giving:[4,21]

$$W^{00}: W^{10}: W^{01}: W^{11} = 0.18: 0.65: 1: 0.49 \quad (9)$$

for the non-tensor part of W(r).

More recently Myhrer et al[22] have considered models which allow for multiquark-antiquark annihilations into several gluons adopting the chiral bag model. Reasonable fits to the $\overline{p}p \to \overline{p}p$ and $\overline{p}p \to \overline{n}n$ angular distributions are obtained. Not surprisingly the results are insensitive to form factor changes at short distances. It will be interesting to extend and apply this model to calculate specific annihilation channels.

In addition to $\overline{q}q$ annihilation one may consider models[23,24] where the quarks rearrange as shown in fig. 3. The three $\bar{q}q$ intermediate states may be viewed as 3 meson virtual states which may then decay into the final physical mesons observed in a particular reaction channel. Coupled channels models[24] including Δ (3/2, 3/2) isobar states also exist for these rearrangement models. The predicted spin-isospin dependence of W^{IS} tends to be a function of the microscopic reaction mechanism but is generally weaker than that obtained by Cote et al.[16] The general lack of data (in various spin-isospin channels) and the number of uncertain parameters associated with the reaction mechanism are major challenges currently. This is a very important area for future experimental and theoretical research.

In what follows we turn to $\bar{N}$-nucleus interactions with the twin goals of learning about the $\bar{N}N$ reaction in the many-body environment and using the $\bar{N}$ to probe selected spin-isospin modes of nuclei.

$\bar{N}$-NUCLEUS ELASTIC AND INELASTIC SCATTERING

The LEAR (low energy antiproton ring) facility at CERN is responsible for the considerable recent interest in antinucleon-nucleon and antinucleon-nucleus reactions among nuclear physicists. The recent and anticipated data from LEAR should be very important in confronting some of the models associated with $\bar{N}N$ annihilation discussed in the last section. If we consider the new $\bar{N}$-nucleus data, it will be interesting to compare it with $(\bar{p},\bar{p}')$ predictions using the nucleus as a spin-isospin filter to isolate particular pieces of the $\bar{N}N$ transition operator.

Using results of multiple scattering theory[25] it is clear that an important piece of theoretical input is the transition matrix operator, $t_{\bar{N}N}$, or scattering amplitude, M, for free $\bar{N}N$ scattering expressed in a form that can be used in the many-body problem using the impulse approximation. The transition matrix, $t_{\bar{N}N}$, is related to M via

$$t_{\overline{N}N} = -\frac{4\pi (hc)^2}{E_{CM}} M(E_{CM},\theta) \equiv \eta M \tag{10}$$

where

$$M_{\overline{N}N}(E_{CM},\theta) = A + B\vec{\sigma}_{\overline{N}}\cdot\hat{n}\ \vec{\sigma}_N\cdot\hat{n} + C(\vec{\sigma}_{\overline{N}} + \vec{\sigma}_N)\cdot\hat{n}$$
$$+ E\vec{\sigma}_{\overline{N}}\cdot\hat{q}\ \vec{\sigma}_N\cdot\hat{q} + F\vec{\sigma}_{\overline{N}}\cdot\hat{Q}\ \sigma_N\cdot\hat{Q}\ . \tag{11}$$

In eq. (11) A, B, C, E, F are functions of θ, E_{CM} and isospin. If $\vec{k}$ and $\vec{k}'$ are the initial and final momentum, respectively in the center-of-mass frame then the vectors $\vec{q}$, $\vec{Q}$ and $\hat{n}$ can be written

$$\vec{q} = \vec{k} - \vec{k}' \qquad \text{(momentum transfer)} \tag{12a}$$
$$\vec{Q} = \vec{k} + \vec{k}' \tag{12b}$$
$$\hat{n} = \frac{\vec{k} \times \vec{k}'}{|\vec{k} \times \vec{k}'|}\ . \qquad \text{(normal to scattering plane)} \tag{12c}$$

Using spin singlet (P_S) and triplet (P_T) projection operators the scattering amplitude, M, may be written in the form[2]

$$M = A'P_S + B'P_T + C'(\sigma_{\overline{N}}+\sigma_N)\cdot\hat{n} + E'S_{12}(q)$$
$$+ F'S_{12}(Q) \tag{13}$$

where A', B', C', E', F' are functions of θ, E_{CM} and isospin and are related to A→F by the relations

$$A' = A - (B+E+F) \qquad E' = \frac{E-B}{3}$$
$$B' = A + \frac{(B+E+F)}{3} \qquad F' = \frac{F-B}{3} \tag{14}$$
$$C' = C\ .$$

One can transform M to a $\overline{N}N$ transition matrix, $t_{\overline{N}N}$, by using eq. (10). It is useful to make the spin and isospin dependence explicit and therefore $t_{\overline{N}N}$ is written in the form

$$t = t_o^C + t_\sigma^C \vec{\sigma}_{\overline{N}}\cdot\vec{\sigma}_N + t_\tau^C \vec{\tau}_{\overline{N}}\cdot\vec{\tau}_N + t_{\sigma\tau}^C \vec{\sigma}_{\overline{N}}\cdot\vec{\sigma}_N\, \vec{\tau}_{\overline{N}}\cdot\vec{\tau}_N$$
$$+ (t_o^{LS} + t_\tau^{LS} \vec{\tau}_{\overline{N}}\cdot\vec{\tau}_N)\, \vec{L}\cdot\vec{S} + (t_o^T + t_\tau^T \vec{\tau}_{\overline{N}}\cdot\vec{\tau}_N)\, S_{12}(q)$$
$$+ (t_o^{TQ} + t_\tau^{TQ} \vec{\tau}_{\overline{N}}\cdot\vec{\tau}_N)\, S_{12}(Q) \tag{15}$$

where the t_i are functions of the C.M. energy and the momentum transfer q. The $\overline{N}N$ central t matrix components having a subscript τ (σ) are isovector (spin-vector) operators in the nuclear target space for ($\overline{N},\overline{N}'$) reactions on nuclei. A quadratic spin-orbit contribution is contained in the last term in eq. (15). Note this component is a function of $\vec{Q}=\vec{k} + \vec{k}'$.

A topic of interest in studying nuclear currents is the investigation of nuclear longitudinal ($\sigma_N\cdot q$) and transverse ($\sigma_N\times q$) spin densities. It turns out that electrons and pions are sensitive to nuclear transverse spin densities. If we consider a simple static model plane wave impulse approximation of the nucleon- and antinucleon-nucleus inelastic response then the differential cross section can be written:[26]

$$\frac{d\sigma}{d\Omega} \propto |M_0 V^0|^2 + |M_1 V^t|^2 \quad \text{(natural parity excitations, } \Delta\pi=(-1)^J) \tag{16a}$$

$$\frac{d\sigma}{d\Omega} \propto |X_\ell V^\ell|^2 + |X_t V^t|^2 \quad \text{(unnatural parity excitations, } \Delta\pi=(-1)^{J+1}) \tag{16b}$$

where M_0, M_1, X_ℓ, and X_t are tensors defined in ref. 26. The longitudinal coupling, $\vec{\sigma}\cdot\vec{q}$, (transverse coupling, $\vec{\sigma}\times\vec{q}$) is denoted by V^ℓ (V^t). In terms of the coefficients given in eq. (11), one obtains:[2]

$$V^0(q) = \eta(A^2+C^2)^{1/2} \tag{17a}$$

$$V^\ell(q) = \eta E \tag{17b}$$

$$V^t(q) = \frac{\eta}{\sqrt{2}}(C^2 + B^2 + F^2)^{1/2} . \tag{17c}$$

The longitudinal and transverse interactions are

plotted in fig. 5.[2] The $\bar{N}N$ interaction used[28] to obtain this figure is discussed in more detail later in this section. Note that near q=0 the longitudinal and transverse couplings for $\bar{N}N$ $\Delta T=0$ transitions are stronger than the corresponding NN $\Delta T=0$ transitions. The range of the isovector $\bar{N}N$ interaction V_{τ}^{ℓ} is found[2] to be quite a bit longer than that of the transverse $\bar{N}N$ interaction, V_{τ}^{t}. To the extent that the transition operator range has some resemblance to the range of an underlying meson exchange potential one can understand why V_{τ}^{ℓ} (associated with pion exchange) would be considerably longer ranged than V_{τ}^{t} (associated with ρ-meson exchange). For a probe such as the $\bar{N}$ which is strongly absorbed in the nucleus this implies that for $\Delta T=1$ $(\bar{p},\bar{n})$ transitions, longitudinal spin excitations should dominate the nuclear response.[2] One uses the charge exchange reaction to filter out the $\Delta T=0$ transverse spin response states. Note that a plane wave model which does not take into account strong absorption will not emphasize the importance of the range of the transition potential.

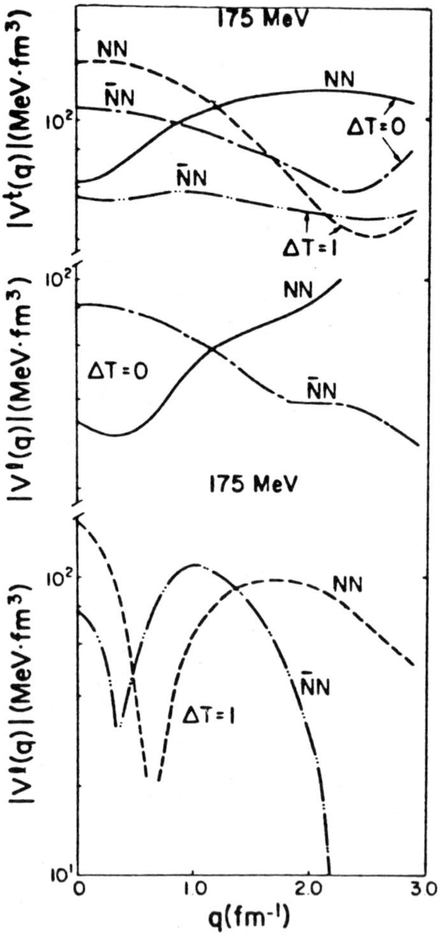

FIG. 5. Momentum transfer dependence of the longitudinal and transverse NN (direct + exchange) and $\bar{N}N$ t-matrix interactions at T_p=175 MeV. The $\bar{N}N$ [NN] t matrix is discussed in Ref. 28 [29]. Figure taken from Ref. 2.

We now consider two different $\bar{N}N$ potential models that were discussed earlier in our summary of models of the $\bar{N}N$ interaction. We shall use these two potential models to generate $t_{\bar{N}N}$ and study the differences obtained and the implications for studies of antinucleon-nucleus inelastic scattering. The main difference between the two models is that model I adopts the spin independent form for the complex annihilation potential given by eq. (6) with the parametrization set of Dover and Richard denoted by I (see below eq. 6). While model II uses the strongly spin dependent form for the annihilation potential given by Cote et al[16] (see eqs. 7 and 8). Both models I and II use versions of the Paris potential for the real t channel meson exchange part of the $\bar{N}N$ interaction (see Ref. 28 for more discussion of models I and II).

The potentials of models I and II are converted to yield the $\bar{N}N$ transition matrix $t_{\bar{N}N}$ by solving the nonrelativistic Schrodinger equation for nonidentical fermions.[28] It is useful to compare the various components of the $\bar{N}N$ t matrix obtained in models I and II with each other and with the corresponding terms in t_{NN} for the nucleon-nucleon t matrix given by Love and Franey.[29]

The energy dependence of components of the central t are shown in fig. 6. The energy dependence and magnitude of t_o^c (not shown) are very similar in the two models with both yielding a $|t_o^c|$ which is about an order of magnitude larger than the $t_{\sigma\tau}^c$ component for model I. A significant contrast between the two models occurs for the t_σ^c term, which is quite large (second only to t_o^c) in model II and the least important term in model I. The term t_σ^c is largely responsible for the excitation of low spin $\Delta T=0$, non-normal parity states at low q. If model II is qualitatively correct, $(\bar{N},\bar{N}')$ inelastic scattering should be valuable for exciting isoscalar spin-flip resonances. By contrast, if model I is qualitatively correct, t_σ^c will, as in the case of NN scattering, be so small, that the study of such states would be very difficult. The strong excitation of low J non-normal parity isoscalar states in $(\bar{N},\bar{N}')$ in model II results from the strong spin-dependence of the imaginary part of the $\bar{N}N$ annihilation potential.

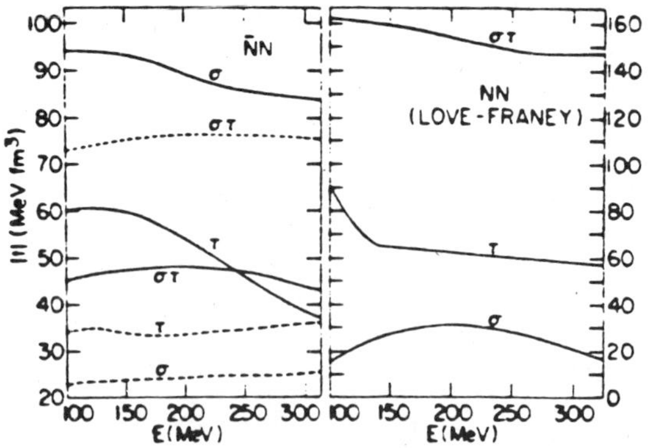

FIG. 6. The magnitudes of various spin-isospin components of the central part t^c of the $\bar{N}N$ t matrix are shown on the left as a function of laboratory kinetic energy E for fixed momentum transfer q=0. The solid curves correspond to model II for the $\bar{N}N$ potential, while the dashed curves refer to model I. On the right, we display the corresponding components of the NN t matrix for comparison. The NN curves are taken from Love and Franey (ref. 29). This figure is taken from ref. 28.

For small angle inelastic $\bar{N}$ scattering, the central parts t^c_σ ($\Delta T=0$) and $t^c_{\sigma\tau}$ ($\Delta T=1$) provide the most important contributions to the unnatural parity transition amplitudes. In this region, Models I and II yield significantly different values for the quantity $|t^c_\sigma/t^c_{\sigma\tau}|^2$. For example at $\theta=0°$, for energies $100 \le E \le 300$ MeV, we find[28,30]

$$R_t = \left|\frac{t^c_\sigma}{t^c_{\sigma\tau}}\right|^2 \approx \begin{matrix} 1/9 \text{ (Model I)} \\ 4 \text{ (Model II)} \end{matrix} \quad . \tag{18}$$

We shall discuss inelastic scattering in more detail in a later section.

As mentioned earlier the role of distortion (particularly absorption) is quite important for $\bar{N}$-nucleus reactions. (Thus the ranges of different terms in $t_{\overline{N}N}$ play an important role and the relative strengths of $t_{\overline{N}N}$ components at zero momentum transfer are insufficient for predicting the spin-isospin nature of the nuclear response.) In what follows we follow the discussion of Love et al[2] of the distortion function for antinucleon- and nucleon-nucleus interactions.

The distorted wave impulse approximation (DWIA) expression for the antinucleon-nucleus T matrix is given by

$$T^{DW}_{f_i} = \int d\vec{r}_{\overline{N}}\; \chi^{(-)^*}(\vec{k}',\vec{r}_{\overline{N}}) \left\langle J_f M_f\, S'_{z_{\overline{N}}} \,\middle|\, \sum_{i=1}^{A} t_{\overline{N}i} \,\middle|\, J_i M_i S_{z_{\overline{N}}} \right\rangle \chi^{(+)}(\vec{k},\vec{r}_{\overline{N}}) \tag{19}$$

where the $\chi^{\pm}$ are the antinucleon distorted waves, $t_{\overline{N}i}$ is the antinucleon-ith nucleon transition matrix given by eq. (15), and $J_i, M_i (J_f M_f)$ denotes the initial (final) target spin and z projection respectively. The symbols $S_{z_{\overline{N}}}$ $(S'_{z_{\overline{N}}})$ represent the initial (final) antinucleon spin projection and $\vec{k}$ $(\vec{k}')$ is the initial (final) relative antinucleon-nucleus momentum in the many body C.M. system. In writing the DWIA amplitude in the simple form above a local form for the transition operator has been assumed. Suppressing discrete spin (and isospin) indices the DWIA amplitude can be written in the schematic form[2]

$$T^{DW}_{fi} = \int d\vec{r}_{\overline{N}}\; D(\vec{k},\vec{k}',\vec{r}_{\overline{N}})\; U(r_{\overline{N}}) \tag{20}$$

where U is a product of the nuclear transition density and $t_{\overline{N}N}$ and D is given by

$$D(\vec{k},\vec{k}',\vec{r}_{\overline{N}}) = \chi^{(-)^*}(\vec{k}',\vec{r}_{\overline{N}})\; \chi^{(+)}(\vec{k},\vec{r}_{\overline{N}}) \;. \tag{21}$$

Love et al[2] study the angle-average, $\int d\Omega_{\overline{N}}$, of eq. (21) which is given by

$$D(r_{\bar{N}},\vec{k}\cdot\vec{k}') = \sum_L (2L+1) \frac{\chi_L(k,r_{\bar{N}})\chi_L(k',r_{\bar{N}})}{k\,k'\,r_{\bar{N}}^2} P_L(\hat{k}\cdot\hat{k}'). \tag{22}$$

This distortion term is the only contributor for monopole transitions. It can also yield some intuition regarding the role of distortions in $\bar{N}N$ scattering. In figure 7 the results of Love et al[2] are shown for $D(r_{\bar{N}},k^2)$. The comparison for protons and antiprotons on ^{12}C and ^{208}Pb demonstrates the large absorption effect for the antiproton.

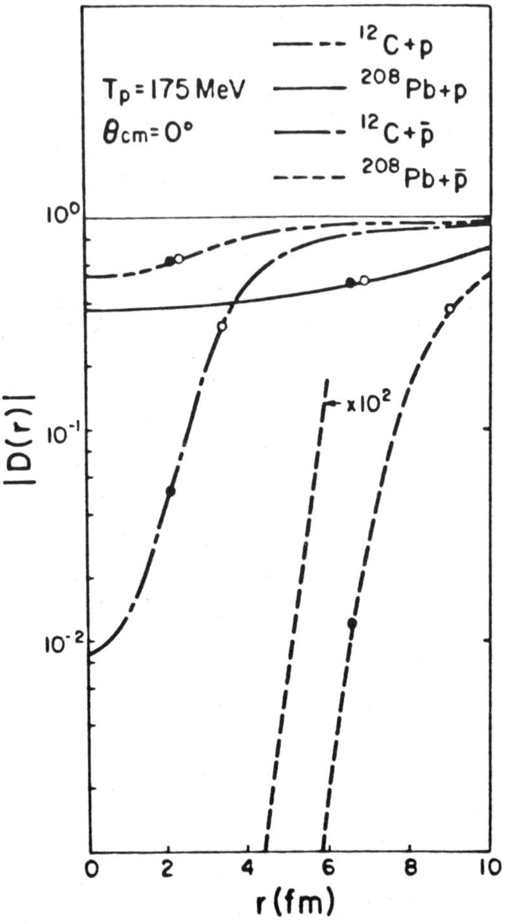

FIG. 7. Monopole distortion functions for 175 MeV protons and antiprotons on ^{12}C and ^{208}Pb. Figure taken from Ref. 2.

It may also be instructive to study the localization in angular momentum, ℓ, associated with $\bar{N}$-nucleus scattering. This can be obtained by studying the absorption coefficients $\eta_\ell \equiv |S_\ell|^2$ of the S matrix elements, S_ℓ, as a function of ℓ. Again following Love et al[2] in Table 1 the relevant quantities are given for studying the angular momentum "windows" associated with proton- and anti-proton-nucleus elastic scattering. The quantities given in the table are defined as follows:

a) $T_{lab} \longrightarrow$ Laboratory kinetic energy of the projectile
b) $k_{CM} \longrightarrow$ wavenumber in the projectile-target center-of-mass
c) $\sigma_{ABS} \longrightarrow \pi(R_{SA}+\frac{1}{k_{cm}})^2$, ($R_{SA} \longrightarrow$ strong absorption radius)
d) $r_i \equiv \frac{\ell_i}{k_{cm}}$ where ℓ_i is the angular momentum where $\eta_\ell = i$
e) $\Delta r \equiv r_{0.9} - r_{0.1}$, $\Delta\ell \equiv \ell_{0.9} - \ell_{0.1}$
f) $\ell_g \longrightarrow$ angular momentum where wave is half absorbed, $\eta_{\ell g}=0.5$.

A value of $\eta_\ell = 0$ (1) indicates total (no) absorption.

Note, from Table 1,[2] for similar targets and energy the antiproton elastic scattering is more localized in ℓ ($\Delta\ell$) and r (Δr) than proton elastic scattering. In terms of r space localization 800 MeV proton scattering is somewhat similar to 175 $\bar{p}$ scattering. However the localization in angular momentum is more dramatic for the antiproton in these cases.

TABLE I

COMPARISON OF ANTIPROTON AND PROTON ELASTIC SCATTERING FROM ^{12}C and ^{208}Pb

SYSTEM	T_{Lab} (MeV)	k_{cm} (fm^{-1})	σ_{ABS} (mb)	R_{SA} (fm)	$r_{0.9}$ (fm)	$r_{0.1}$ (fm)	Δr (fm)	ℓ_g	$\ell_{0.9}$	$\ell_{0.1}$	$\Delta\ell$
$\bar{p}$+C	175	2.76	489	3.6	4.6	2.8	1.8	10	13	8	5
p+C	175	2.76	229	2.3	4.0	0.0	4.0	6	11	0	11
p+C	800	6.46	282	2.8	3.9	1.8	2.1	18	25	12	13
$\bar{p}$+Pb	175	3.02	2710	9.0	10.2	8.0	2.2	27	31	24	7
p+Pb	175	3.02	1629	6.9	8.8	4.8	4.0	21	26.5	14.5	12
p+Pb	800	7.35	1878	7.6	8.8	6.5	2.3	55	65	48	17

Table I is from Ref. 2.

We now turn to a comparison between the (recent) data from LEAR and predictions based on the DWIA. In figure 8 the results of 180 MeV $\bar{p}+^{12}C$ elastic scattering is given. The theoretical curve is generated by using $U=\rho t_{\bar{N}N}$. The two theoretical models shown correspond to potentials discussed earlier (DRII→Model I, Paris→Model II). Both models yield good fits to the elastic scattering data. The process is characterized by strong absorption and r and ℓ space localization. It is interesting that purely phenomenological optical potential fits to the data[31,32] and the ρt approximation give an optical potential which is strongly absorptive and weakly attractive in the nuclear surface region.[1,32] The deep "diffractive" minuma exhibited in fig. 8 are characteristic of strong absorption.

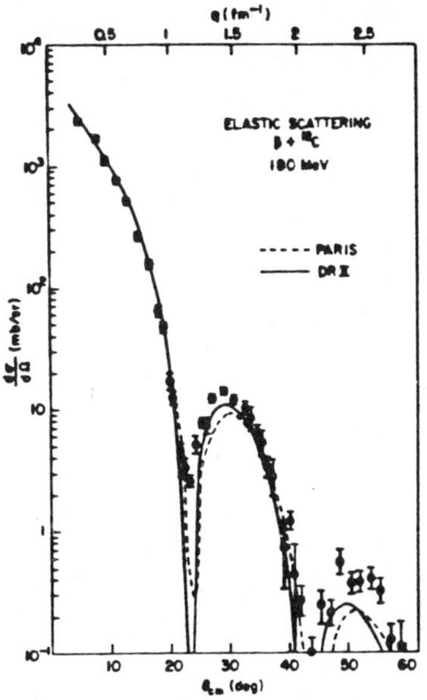

FIG. 8. Angular distribution for $\bar{p}$ elastic scattering from ^{12}C at 180 MeV. The data points are taken from Garreta et al.[31], while the solid and dashed curves represent calculations using the "$\rho t_{\bar{N}N}$" approximation to the $\bar{p}$ optical potential. Figure taken from Ref. 1.

The piece of the $t_{\bar{N}N}$ matrix contributing to the elastic scattering above is the piece denoted by t_o^c earlier. Since this very strong piece is also the dominant contributor to $\Delta T=0$ normal parity transitions ($\pi=(-1)^{J_f}$) in $(\bar{p},\bar{p}')$, we expect such transitions to also show characteristics of strong absorption. In figures 9 and 10 we show the recent data of Garreta et al[31,32] for excitation of the 2^+(4.4 MeV) and 3^-(9.6 MeV) T=0 levels for 180 MeV $\bar{p}$ inelastic scattering. The 185 MeV (p,p') data[33] is also exhibited. The comparison with DWIA model II predictions shows the theory yields good agreement with experiment. (The results for model I are very similar[1] since the crucial input is the value of t_o^c - i.e. differences in the spin-isospin dependent pieces of $t_{\overline{N}N}$ are unimportant for these $\Delta T=0$ normal parity (non-spin flip transitions). Note that the input nuclear transition densities for these states have been scaled[1] to fit the experimental (e,e') longitudinal form factor. Note also that angular distributions for the 2^+ and 3^- transitions obey the Blair Phase rule[34] [$\pi/2$ relative phase with elastic scattering] for diffraction patterns associated with strong absorption. Medium corrections to the free space $t_{\overline{N}N}$ amplitudes may be decreased because the inelastic scattering is localized in the nuclear surface.[1]

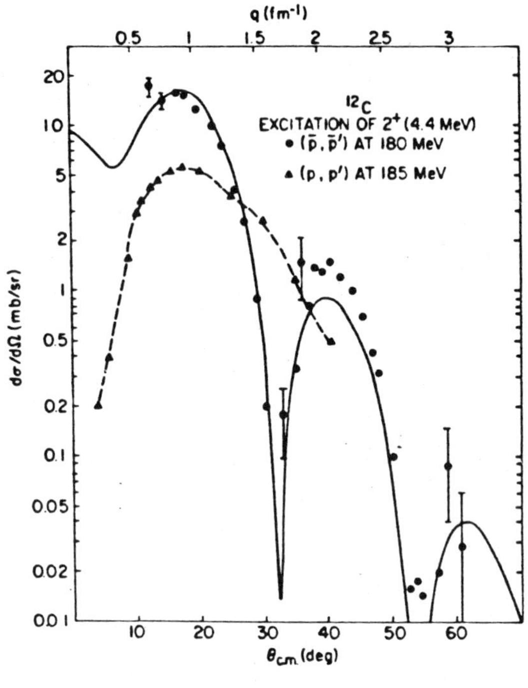

FIG. 9. Excitation of the 2^+ (4.4 MeV, T=0) state by p and $\overline{p}$ inelastic scattering at 180 MeV. The $\overline{p}$ and p data are from Refs. 32 and 33 respectively, while the solid curve is a DWIA calculation given in Ref. 1. Figure taken from Ref. 1.

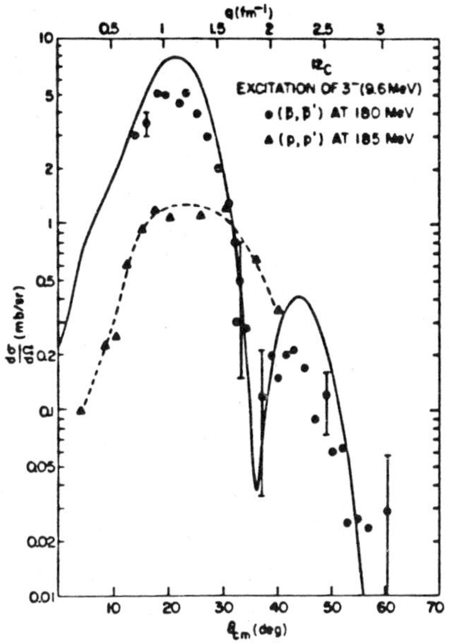

FIG. 10. Excitation of the 3^- (9.6 MeV, T=0) state in ^{12}C by $\overline{p}$ and p at 180 MeV; data and curves as in Fig. 9. Figure taken from Ref. 1.

We have earlier mentioned (see eq. (18) and the accompanying discussion) that the ratio of $\Delta T=0$ and 1 spin-flip transitions, having similar nuclear transition densities could be very helpful in selecting between proposed models of the $\overline{N}N$ interaction. Because of the different ranges associated with, for example, t_σ^C and $t_{\sigma\tau}^C$ and the strong surface localization associated with ($\overline{p}$,$\overline{p}$') reactions it is necessary to carry out distorted wave calculations before making comparison with experiment. (The ratio R_t given by eq. 18 is only suggestive and should not be used directly to confront experimental data.) We define the quantity $R_J(\theta)$ as follows:

$$R_J(\theta) = \left| \frac{d\sigma}{d\Omega} (J^\pi, T=0) \right| \Big/ \left| \frac{d\sigma}{d\Omega} (J^\pi, T=1) \right| \qquad (23)$$

Detailed DWIA calculations for ^{12}C ($\overline{p}$,$\overline{p}$') $^{12}C^*$

have been performed to obtain the ratios $R(\theta)$ for the 1^+ doublet (T=0, 12.7 MeV and T=1, 15.1 MeV) and 2^- doublet (T=0, 18.3 MeV and T=1, 19.4 MeV).[30] The results are given in Table 2 for models I and II. For angles ≤5° where t_σ^c and $t_{\sigma\tau}^c$ dominate the transition amplitudes, the differences between the predictions of models I and II for $R_1(\theta)$ result from the different predictions of the 12.7 MeV state cross section. Similar results are obtained for the ratio $R_2(\theta)$ associated with the 2^- doublet mentioned above.[30] Note that there are significant differences in the ratios predicted (model II, which is based on a spin-dependent absorption potential yields a much larger ratio at forward angles). However, the quantitative results are significantly different than those obtained from eq. (18) due to the different ranges of t_σ and $t_{\sigma\tau}$.

TABLE II

θ_{cm}(deg)	$R(\theta)$ [II]	$R(\theta)$ [I]	$d\sigma/d\Omega$ (15.1, II) (mb/sr)	$d\sigma/d\Omega$ (15.1, I) (mb/sr)
0	0.44	0.05	0.33	0.45
5	0.60	0.11	0.20	0.26
10	0.38	0.15	0.20	0.20
15	0.30	0.16	0.14	0.14

Antiproton inelastic cross sections and ratios for excitation of the 1^+, T=0 (12.71 MeV) and 1^+, T=1 (15.1 MeV) states in ^{12}C for $E_{\bar{p}}^{Lab}$ = 175 MeV. The ratio of cross sections, $R_1(\theta)$, is defined by eq. (23). The two models for the transition operator are denoted I and II and are discussed in the text. The model II based on a spin dependent absorption potential yields a much larger ratio R at forward angles. Table taken from Ref. 30.

Comparison with experiment for these doublet ratios has not yet been possible because of the presence of other strongly excited states near-by and the coarse energy resolution of these first generation experiments.[1] In light nuclei excited states tend to more widely separated so $(\bar{p},\bar{p}')$ exclusive studies should be concentrated in this region. For example, Dover and Millener[1] have mentioned $^{6}Li(\bar{p},\bar{p}')^{6}Li^{*}$ as an attractive reaction to study. One method of eliminating unwanted $\Delta T=0$ states would be to consider the $(\bar{p},\bar{n})$ reaction. This reaction also has the advantage (for $N\neq Z$ targets) of allowing one to study T giant resonances without the background of strongly excited lower isospin states. In this latter regard the $(\bar{p},\bar{n})$ reaction acts like the (n,p) reaction. The $(\bar{p},\bar{n})$ reaction may also be a useful complementary way to (p,n) for studying Gamow-Teller resonances in N=Z nuclei. Note that since the $(\bar{p},\bar{n})$ $\pi+\rho$ excitation potential has a different relative sign ($V_{\rho}+V_{\pi}$ is coherent for the $\bar{N}N\to\Delta\bar{N}$ transition potential[1]) than for the (p,n) reaction the role of Δ's in the excitation process might be enhanced. This is of importance because the quenching of Gamow-Teller strength seen in (p,n) may be attributable to Δ particle-N hole components in the many-body nuclear wavefunction.

Of course spin-observables such as polarization ± analyzing power are predicted to be useful in selecting between models of the excitation operator $t_{\bar{N}N}$ and distinguishing between various spin-isospin excitation modes.[28,30] This spin observable data, which will be very challenging to obtain, is of considerable importance.

EXOTIC REACTIONS

Because the $\bar{p}p$ system has baryon number zero there are interesting reaction channels such as $\bar{p}p\to\bar{\Lambda}\Lambda$ or $K^{+}K^{-}$ which are unavailable in nucleon-nucleon interactions. It is of interest to study these reactions both for $\bar{p}$ beams on free nucleons and for $\bar{p}$'s on nuclei. As mentioned at the beginning of these lectures, the $(\bar{p},\bar{\Lambda})$, $[\bar{p},\Lambda]$ reaction on nuclei leads to a final hypernucleus with S=-1 [hyponucleus with S=+1]. These are important to study because it allows one to see how the nucleus is perturbed by the introduction of an impurity with strangeness non-zero (the S or $\overline{S}$ quark). On the other hand the elementary $\bar{p}p\to\bar{\Lambda}\Lambda$ reaction has been the focus of both theoretical[35,36] and experimental[37] investigation recently. Tabakin and Eisenstein[35] adopt a model for the reaction $\bar{p}p\to\bar{\Lambda}\Lambda$ involving the exchange of K (pseudoscalar), K^* (vector), and K^{**} (tensor) mesons. Initial and final state interactions are handled approximately using the Sopkovich approximation.[38] This results in a form for the partial wave amplitude M^J given by

$$M^J(\bar{p}p\to\bar{\Lambda}\Lambda) = \sqrt{S^J_{\bar{\Lambda}\Lambda}}\; T^J \sqrt{S^J_{\bar{p}p}} \qquad (24)$$

where $S^J_{\bar{\Lambda}\Lambda}$ ($S^J_{\bar{p}p}$) is the final (initial) state S matrix for the Jth partial wave. The main effect of the inclusion of the S matrix terms is to damp out low partial waves. Since little is known about $S_{\bar{\Lambda}\Lambda}$ it is adjusted to fit the sparse existing $\bar{\Lambda}\Lambda$ information. The spin dependence of the initial and final state interactions as well as the one kaon exchange reaction mechanism play an important role in determining the polarization.[35] Interestingly the K^* and K^{**}, which yield important spin effects, are found to be more important than the K in contributing to the process. The agreement between theory and the meager existing data (for the angular distribution and polarization) is reasonable. The forward peaking of the data apparently is caused by the damping of the $^1S_0$ amplitude which is caused in the model of ref. 35 by the final state $\bar{\Lambda}\Lambda$ interaction. Dillig and Frankenberg[36] note that crossing symmetry connects the exchange piece of the ΛN interaction and the $\bar{N}N\to\bar{\Lambda}\Lambda$ process. Since there is a large momentum mismatch ($q\geq600$ MeV/c) in the $\bar{p}p\to\bar{\Lambda}\Lambda$ reaction the connection with the exchange part of the ΛN interaction means that one has an additional handle on the short range piece of this term potentially important for the level structure of hypernuclei. Estimates given in ref. 36 suggest that initial state

$\bar{p}p$ interactions should be much more important than the final state $\bar{\Lambda}\Lambda$ interactions. More experimental and theoretical work is expected and needed for elucidating this interesting reaction. Clearly the elementary $\bar{p}p \to \bar{\Lambda}\Lambda$ process in the nuclear many-body environment should be the subject of future investigation.

The possibility of partial quark deconfinement or the formation of a quark-gluon plasma in nuclei at high baryon density and/or temperature has led to several suggestions for obtaining this fundamental new form of matter. One suggestion is to use relativistic heavy ions to obtain the appropriate baryon density speculated to be necessary for the phase change to the quark-gluon plasma. Another possibility[7] is to use antiproton annihilation in nuclei to obtain a high energy density (~ 1 GeV/fm^3) and/or average kinetic energy (temperature) (> 180 MeV). Gibbs and Strottman[7] point out that for higher energy antiprotons (P_{lab} ~ 6 GeV/c) the annihilation cross section is lower than for antiprotons currently available at LEAR and thus the annihilation is not constrained to occur at the nuclear surface. This means that the pions produced in the annihilation can be absorbed in the nucleus and thus more efficiently raise the nuclear energy density. Since the mesons are well focused the increased energy density is more localized. It turns out that theoretical models predict that a large part of the energy available in the original $\bar{N}N$ system is transferred to target nucleons.[7] Thus this technique for causing hadronic matter to undergo a phase transition to a quark-gluon plasma appears deserving of serious consideration.

A question which naturally arises in the study of the phase transition discussed above is - What would be the signal of such a transition or how would one know that a quark-gluon plasma had been formed? Derreth et al[8] have used statistical arguments to predict the abundance of strange particles produced in $\bar{p}$-nucleus annihilations. They suggest that since their predictions are an upper limit based on a canonical ensemble picture assuming a hadronic model of the system that the observation of substantially more strange particles produced in $\bar{p}A$ experiments might be a signal of a quark-gluon plasma.

FINAL REMARKS

There is considerable interest in the new data available from LEAR on the $\bar{N}N$ and $\bar{N}A$ systems. But this data will just scratch the surface of the fundamentally important physics that can be investigated using an intense, variable energy $\bar{p}$ beam having excellent energy resolution. Thus an appropriate facility dedicated to producing a high quality variable energy $\bar{p}$ beam in the low energy range (0 to a few GeV) appears to be of high priority. Such a facility could be used to study $\bar{N}N$ reactions leading to a wide variety of final channels in both free space and the nuclear medium. Exotic states of matter (Hypernuclei, Hyponuclei, Charmed Nuclei, Quark-gluon Plasmas) could be studied using such a facility.

For theorists, the current data should provide an opportunity for studying fundamental strong interaction processes at either the quark-gluon or hadron levels and then comparing model predictions with experimental results over a wide range of strongly interacting systems.

ACKNOWLEDGEMENTS

I would like to thank C. B. Dover, W. G. Love, and D. J. Millener for permission to use figures and tables from their recent review talks (see refs. 1, 2, 4). This work was supported in part by the National Science Foundation.

REFERENCES

1. C. B. Dover and D. J. Millener, in Antinucleon- and Nucleon-Nucleus Interactions, G. E. Walker, C. D. Goodman and C. Olmer, eds. (to be published by Plenum, New York), conference proceedings-Telluride, Colorado, March 18-21, 1985.

2. W. G. Love, Amir Klein, and M.A. Franey, in Antinucleon- and Nucleon-Nucleus Interactions, op. cit., Ref. 1.

3. D. Garreta, in Antinucleon- and Nucleon-Nucleus Interactions, op. cit., Ref. 1.

4. C. B. Dover, Nucl. Phys. A416, 313c (1984).

5. E. Predazzi, "Lectures on Review of Nucleon-Antinucleon Physics" (Institut de Physique Nucleaire, Universite Claude Bernard, Lyon 1, LYCEN 8053, 1980).

6. C. B. Dover and G. E. Walker, Phys. Rev. C19, 1393 (1979).

7. W. R. Gibbs and D. Strottman in Antinucleon- and Nucleon-Nucleus Interactions, op. cit., Ref. 1.

8. C. Derreth, W. Greiner, H.-Th. Elze, J. Rafelski, Phys. Rev. C31, 1360 (1985).

9. M. Nagels, T. Rijken, and J. J. de Swart, Phys. Rev. D12, 744 (1975).

10. W. W. Buck, C. B. Dover, and J. M. Richard, Ann. Physics 121, 47 (1979).

11. R. A. Bryan and R.J.N. Phillips, Nucl. Phys. B5, 201 (1968).

12. See Ref. 4 for a summary discussion and references to the relevant experiments.

13. C. B. Dover and J. M. Richard, Phys. Rev. C21, 1466 (1980).

14. M. Lacombe et al., Phys. Rev. D12, 1495 (1975).

15. M. Alston-Garnjost et al., Phys. Rev. Lett. 43, 1901 (1979).

16. J. Cote et al., Phys. Rev. Lett. 48, 1319 (1982).

17. M. Lacombe et al., Phys. Rev. C21, 861 (1980)

18. P. H. Timmers, W. A. van der Sander, and J. J. de Swart, Nijmegen preprints THEF-NYM-83.06, 83.07; Phys. Rev. D29, 1928 (1984).

19. R. L. Jaffe, Phys. Rev. D17, 1445 (1978).

20. R. A. Freedman et al., Phys. Rev. D23 1103 (1981); M. A. Alberg et al., Phys. Rev. D27, 536 (1983).

21. A. Faessler et al., University of Tubingen preprint (1982). (Result and reference quoted and taken from Ref. 4.)

22. F. Myhrer, "Microscopic Approaches to $\overline{N}N$ Annihilation Potentials," presented at Int. Conf. on Hadronic Probes and Nuclear Interactions, Phoenix, AZ, March 1985, and APS Spring Meeting, Crystal City, VA, April 1985 (with R. Tegen and T. Mizutani).

23. A. M. Green, J. A. Miskanen and J. M. Richard, Phys. Lett. 121B, 101 (1983). See Ref. 24.

24. A. M. Green and J. A. Niskanen, Helsinki preprint HU-TFT-83-27 (1983). (Results and references 23 and 24 taken from Ref. 4.)

25. A. K. Kerman, H. McManus and R. M. Thaler, Ann. Phys. 8, 551 (1959).

26. J. M. Moss in Spin Excitations in Nuclei, eds. F. Petrovitch et al., (Plenum, New York, 1984) p. 355.

27. W. G. Love, M. A. Franey and F. Petrovitch in Spin Excitations in Nuclei, op.cit., Ref. 26, p. 205.

28. C. B. Dover, M. E. Sainio, and G. E. Walker, Phys. Rev. C28, 2368 (1983).

29. W. G. Love and M. A. Franey, Phys. Rev. C24, 1073 (1981); (E) C27, 438 (1983); M. A. Franey and W. G. Love, Phys. Rev. C31, 488 (1985).

30. C. B. Dover, M. A. Franey, W. G. Love, M. E. Sainio and G. E. Walker, Phys. Lett. 143B, 45 (1984).

31. D. Garreta et al., Phys. Lett. 135B, 266 (1984); (E) 139B, 464 (1984).

32. D. Garreta et al., Phys. Lett. 149B, 64 (1984).

33. D. Hasselgren et al., Nucl. Phys. 69, 81 (1965).

34. J. S. Blair, Lectures in Theoretical Physics Vol. VIII-C (eds. P. D. Kunz, D. A. Lind, and W. E. Brittin, Univ. of Colorado, Boulder 1966).

35. F. Tabakin and R. A. Eisenstein, Phys. Rev. C31, 1857 (1985).

36. M. Dillig and R. V. Frankenberg in Antinucleon- and Nucleon-Nucleus Interactions, op. cit., Ref. 1.

37. P. D. Barnes et al., proposal CERN/PSCC/81-40 (1981).; P. D. Barnes, in Antinucleon- and Nucleon-Nucleus Interactions, op. cit., Ref. 1.

38. N. J. Sopkovich, Nuovo Cimento 26, 186 (1962).

DEVELOPMENT OF THE DIRAC SCATTERING APPROACH

B. C. Clark
Ohio State University
Columbus, OH 43065

ABSTRACT

The phenomenological success of the Dirac optical model approach is reviewed. A discussion of the characteristic features of the second order equation is presented. A comparison of the usual Lorentz scalar - Lorentz vector (SV) model with other types of models is given. Applications of the SV model to proton, neutron and antiproton scattering at a number of energies and for a range of targets show that the results exhibit in general, very systematic behaviour. The use of the Kemmer-Duffin-Petiau formulation for spin zero or spin one projectiles is presented.

INTRODUCTION

It is now generally recognized that the Dirac equation is a viable alternative to the usual Schrödinger equation approach for analysing nucleon-nucleus scattering data. This was not the case several years ago. This paper reviews the development of Dirac phenomenology, stressing how closely this development has been tied to experiment. In fact, the first use of the Dirac equation in analysing pA elastic scattering suffered from the lack of spin observables.[1] It was found that either Lorentz scalar or Lorentz four-vector optical potentials could be used to fit the data. This situation changed completely when p-^{4}He elastic cross section (σ) and analysing power (A_y) experiments were performed at the ZGS shortly before its demise.[2] The availability of both σ and A_y put a critical constraint on the Dirac approach. Some new feature was necessary if the large spin observables were to be reproduced.

It became clear, as I shall discuss in more detail later, that there were two ways to proceed. One, which more nearly resembled the nonrelativisitic approach, involved introducing a tensor optical potential which, in the second order Dirac equation, produces a spin-orbit term. The tensor in combination with either a Lorentz scalar or a Lorentz vector potential would then form the basis of the model. The second approach, which was motivated by meson exchange models of the nucleon-nucleon force, was to use large canceling Lorentz scalar and vector potentials to obtain the required spin orbit enhancement. This is the approach we took.[3] It is safe to say that if we had not been able to get good fits to the p-^{4}He data much of the subject matter of this talk might well be missing.

This paper is organized as follows. In Section 2 the second order Dirac equation for local, time independent interactions is given. In addition some ambiguities in the choice of Lorentz character of the optical potentials are discussed. Section 3 presents applications of Dirac phenomenology to proton, neutron and anti-proton nucleus scattering. Section 4 introduces an application of the Kemmer-Duffin-Petiau[4-6] wave equation, which is similar in form to the Dirac equation, to meson-nucleus scattering. The last section contains conclusions and suggestions for future work.

THE DIRAC EQUATION

In order to better appreciate the choice of the scalar-vector (SV) model usually employed in Dirac phenomonology it is instructive to consider the Dirac equation in its second order form. As discussed by Miller,[7] the Dirac equation containing local potentials may be written

$$\{\vec{\alpha}\cdot\vec{p} + \beta\,[m + S(\vec{r}) + \gamma^{\mu}U_{v\mu}(\vec{r}) + \gamma^{5}U_{ps}(\vec{r})$$

$$+ \gamma^{\mu}\gamma^{5}U_{a\mu}(\vec{r}) + \sigma^{\mu\nu}U_{t\mu\nu}(\vec{r})]\}\ \psi(\vec{r}) = E\psi(\vec{r}) \quad . \tag{1}$$

In the case that potentials are spherically symmetric, Hama[8] has given an expression for the second order Dirac equation. For spherical symmetry the potentials take the form

$$S(\vec{r}) = S(r),$$

$$\gamma^{\mu}U_{v\mu}(\vec{r}) = \gamma^{o}U_{v}^{o}(r) - \vec{\gamma}\cdot\hat{r}U_{v}^{r}(r) \ ,$$

$$\gamma^{5}U_{ps}(\vec{r}) = \gamma^{5}U_{ps}(r) \quad ,$$

$$\gamma^{\mu}\gamma^{5}U_{a\mu}(\vec{r}) = \beta\gamma^{5}U_{a}^{o}(r) + \gamma^{5}\vec{\gamma}\cdot\hat{r}U_{a}^{r}(r),$$

and

$$\sigma^{\mu\nu}U_{t\mu\nu}(\vec{r}) = -\gamma^{o}\vec{\gamma}\cdot\hat{r}U_t^r(r) = \beta i\vec{\alpha}\cdot\hat{r}\beta U_t(r) \quad .$$

Equation (1) becomes,

$$\{\vec{\alpha}\cdot\vec{p} + \beta(m + S) - (E - U_v^o - V_c) - \beta\vec{\gamma}\cdot\hat{r}U_v^r$$
$$+ \beta\gamma^5 U_{ps} + \gamma^5 U_a^o + \beta\gamma^5\vec{\gamma}\cdot\hat{r}U_a^r + i\vec{\alpha}\cdot\hat{r}\beta U_t\}\ \psi(\vec{r}) = 0 \quad , \tag{2}$$

where the notation of Bjorken and Drell[9] is used for the γ-matrices and V_c is the static Coulomb potential for charged particle scattering. Equation (2) may be rewritten as two coupled equations for the upper (ψ_u) and the lower (ψ_ℓ) components of $\psi(\vec{r})$. Solving for ψ_ℓ in terms of ψ_u gives

$$\psi_\ell(\vec{r}) = \frac{1}{(E + m)\ A + (\vec{\sigma}\cdot\hat{r})U_a^r}\ [\vec{\sigma}\cdot\vec{p} - \vec{\sigma}\cdot\hat{r}U_v^r - U_{ps} + i(\vec{\sigma}\cdot\hat{r})U_t + U_a^o]\ \psi_u(\vec{r}) \tag{3}$$

where

$$A(r) = (m + S + E - U_v^o - V_c)/(m + E) \quad . \tag{4}$$

The equation for $\psi_u(\vec{r})$ is

$$\{(E - U_v^o - V_c + (\vec{\sigma}\cdot\hat{r})U_a^r)^2 - (m + S)^2 - Q(r)\}\psi_u(\vec{r})=0, \tag{5}$$

where

$$Q(r) = [(E + m)\ A + (\vec{\sigma}\cdot\hat{r})U_a^r][\vec{\sigma}\cdot\vec{p} - \vec{\sigma}\cdot\hat{r}U_v^r + U_{ps} - i(\vec{\sigma}\cdot\hat{r})U_t + U_a^o] \tag{6}$$

$$\times \frac{1}{[(E + m)\ A + (\vec{\sigma}\cdot\hat{r})U_a^r]}\ [\vec{\sigma}\cdot\vec{p} - \vec{\sigma}\cdot\hat{r}U_v^r - U_{ps} + i(\vec{\sigma}\cdot\hat{r})U_t + U_a^o].$$

Using the identity

$$\frac{1}{(E + m)\ A + (\vec{\sigma}\cdot\hat{r})U_a^r} \equiv \frac{1}{(E + m)^2\ A^2 - U_a^{r2}}[(E + m)\ A - (\vec{\sigma}\cdot\hat{r})U_a^r], \tag{7}$$

Hama[8] has shown that Q(r) can be written

$$
\begin{aligned}
Q(r) = \{ & \vec{p}^{\,2} + 2iX^2 \left(B\frac{\partial B}{\partial r} - U_a^r \frac{\partial U_a^r}{\partial r}\right) \left(-B^2(U_v^r - iU_t) + U_a^{r^2} U_{ps}\right) \\
& + iX\frac{\partial B}{\partial r}[B(U_v^r - iU_t) - U_a^r U_{ps}] + \frac{2iB^2 X}{r}(U_v^r - iU_t) + i\left(\frac{\partial U_v^r}{\partial r} - i\frac{\partial U_t}{\partial r}\right) \\
& - \frac{2i}{r} X U_a^r B U_{ps} + iX\frac{\partial U_a^r}{\partial r}[BU_{ps} + U_a^r(U_v^r - iU_t)] + U_v^{r^2} + U_t^2 - U_{ps}^2 \\
& + U_{ps} U_a^{\circ}\} + \left[-\frac{X}{r}\left(B\frac{\partial B}{\partial r} - U_a^r \frac{\partial U_a^r}{\partial r}\right) + \frac{X}{r} 2iBU_a^r (U_a^{\circ} - U_{ps})\right. \\
& \left. + \frac{X}{r} 2iU_a^{r^2} (U_v^r - iU_t) + \frac{2}{r}U_t\right](\vec{\sigma}\cdot\vec{L}) + \left[\frac{iX}{r}\left(B\frac{\partial B}{\partial r} - U_a^r \frac{\partial U_a^r}{\partial r}\right)\right. \qquad (8) \\
& \left. - \frac{2iXU_a^{r^2}}{r^2} - \frac{2U_v^r}{r}\right](\vec{r}\cdot\vec{p}) - \left[\frac{2XU_a^{r^2}}{r}(U_a^{\circ} - U_{ps}) + \frac{2BU_a^r X}{r}(U_v - iU_t)\right] \\
& (\vec{\sigma}\cdot\vec{r})(\vec{r}\cdot\vec{p}) + \left[\frac{2iXU_a^r B}{r} + iX\left(B\frac{\partial U_a^r}{\partial r} - U_a^r \frac{\partial B}{\partial r}\right) + 2XU_a^r B(U_v^r - iU_t)\right. \\
& \left. + 2XB^2 U_a^{\circ} - 2XU_a^{r^2} U_{ps}\right](\vec{\sigma}\cdot\vec{p}) + \frac{2U_a^{r^2} X}{r^2}\vec{L}^2 - \frac{2iU_a^{r^2} X}{r^2}(\vec{\sigma}\cdot\vec{L})(\vec{r}\cdot\vec{p}) \\
& + 2iB\frac{U_a^r}{r}X(\vec{\sigma},\ \vec{L})(\vec{\sigma}\cdot\vec{p}) + \\
& + \left\{iU_a^{\circ} X\left(B\frac{\partial B}{\partial r} - U_a^r \frac{\partial U_a^r}{\partial r}\right) + \frac{2iXU_a^{r^2}}{r}(U_a^{\circ} - U_{ps}) - i\left(\frac{\partial U_a^{\circ}}{\partial r} - \frac{\partial U_{ps}}{\partial r}\right)\right. \\
& + 2iX^2\left(B\frac{\partial B}{\partial r} - U_a^r \frac{\partial U_a^r}{\partial r}\right)[-B^2 U_{ps} + U_a^{r^2}(U_v^r - iU_t)] \\
& + iX\frac{\partial B}{\partial r}[BU_{ps} + U_a^r(U_v^r - iU_t)] - iX\frac{\partial U_a^r}{\partial r}[B(U_v^r - iU_t) + U_a^r U_{ps}] \\
& \left. + \frac{2iXU_a^r B}{r}(U_v^r - iU_t) + 2iU_{ps}U_t - U_v^r U_a^{\circ} - iU_a^{\circ} U_t\right\}(\vec{\sigma}\cdot\hat{r}),
\end{aligned}
$$

where

$$B = (E + m)A,$$

and

$$X = \frac{1}{B^2 - U_a^{r^2}} .$$

This complicated expression for the second order equation serves to illustrate the advantage of working with Eq. (2) rather than Eq. (5).

If we assume that parity is a good quantum number of the system, pseudoscalar and axial vector potentials are zero. In this case the second order Dirac equation may be written

$$\begin{aligned} \{\nabla^2 &+ (E - U_v^{\circ} - V_c)^2 - (m + S)^2 - U_v^{r^2} - T^2 \\ &+ [\frac{1}{rA}\frac{\partial A}{\partial r} - 2\frac{T}{r}](\vec{\sigma}\cdot\vec{L}) - \frac{2}{r}(iU_v^r + T) \\ &+ \frac{1}{A}\frac{\partial A}{\partial r}(iU_v^r + T) + \frac{1}{r}[(\vec{r}\cdot\vec{p})(U_v^r - iT)] \\ &- [i\frac{1}{rA}\frac{\partial A}{\partial r} - 2\frac{U_v^r}{r}](\vec{r}\cdot\vec{p})\}\,\psi_u(\vec{r}) = 0 , \end{aligned} \tag{9}$$

where

$$T(r) \equiv U_t(r) + U_{AM}(r) . \tag{10}$$

Here $U_{AM}(r)$ is the potential due to the interaction of the anomalous magnetic moment of the projectile with the Coulomb field of the target nucleus, and is given by

$$U_{AM}(r) = \frac{\kappa}{2m}\frac{\partial}{\partial r}V_c(r) , \tag{11}$$

where κ is the anomalous magnetic moment of the projectile. ($\kappa = 1.79$ for protons and $\kappa = -1.91$ for neutrons.) From Eq. (9) we see that the tensor term contributes to both the spin-orbit term and the central term. The three-vector part of U_v contributes to both the $(r\cdot p)$ or Darwin term and the central terms.

To remove the first derivative terms, we let

$$\psi_u(\vec{r}) = K(r)\,\phi(\vec{r}) \tag{12}$$

with $K(r) \to 1$ as $r \to \infty$. Direct substitution of (12) into (9) gives

$$\frac{\partial}{\partial r}K(r) = \frac{1}{2}[\frac{1}{A}\frac{\partial A}{\partial r} + 2iU_v^r]K(r) , \tag{13}$$

or

$$K(r) = A^{1/2} \exp \int i\, U_v^r(r)\, dr . \tag{14}$$

Using Eq. (13), we may write what has been termed the Schrödinger equivalent equation i.e. a second order equation which contains central and spin-orbit terms as,

$$\{\nabla^2 + (E - U_v^\circ - V_c)^2 - (m + S)^2 - T^2$$

$$+ \frac{T}{A}\frac{\partial A}{\partial r} - 2\frac{T}{r} - \frac{\partial T}{\partial r} - \frac{3}{4A^2}\left(\frac{\partial A}{\partial r}\right)^2$$

$$+ \frac{1}{2r^2 A}\frac{\partial}{\partial r}\left(r^2 \frac{\partial A}{\partial r}\right) + \left(\frac{1}{rA}\frac{\partial A}{\partial r} - 2\frac{T}{r}\right)(\vec{\sigma}\cdot\vec{L})\}\phi(\vec{r}) = 0 \quad . \tag{15}$$

Notice that the three-vector part of the vector potential does not appear. The tensor potential contributes in a complicated way to the central potential and also contributes to the spin-orbit term. In addition, there are cross terms between T and derivatives of U_v°, V_c and S. One may define what are called Schrödinger equivalent central, spin-orbit and Darwin potentials. They are ($U_o = U_v^\circ$)

$$U_{cent} = \frac{1}{2E}(2EU_o + 2mS - U_o^2 + S^2 - 2V_c U_o + U_t^2 \tag{16}$$

$$+ 2U_t U_{AM} - \frac{(U_t + U_{AM})}{A}\left(\frac{\partial A}{\partial r}\right) + \frac{2U_t}{r} + \frac{\partial U_t}{\partial r} + 2E\, U_{Darwin}) \quad ,$$

$$U_{Darwin} = \frac{1}{2E}\left[-\frac{1}{2r^2 A}\frac{\partial}{\partial r}\left(r^2\frac{\partial A}{\partial r}\right) + \frac{3}{4A^2}\left(\frac{\partial A}{\partial r}\right)^2\right] \quad , \tag{17}$$

and

$$U_{so} = \frac{1}{2E}\left[-\frac{1}{rA}\left(\frac{\partial A}{\partial r}\right) + \frac{2}{r}(U_t + U_{AM})\right] \quad , \tag{18}$$

for a Schrödinger equivalent equation given by

$$[\vec{p}^{\,2} + 2E(U_{cent} + U_{so}\,\vec{\sigma}\cdot\vec{L})]\phi(\vec{r})$$

$$= [(E - V_c)^2 - m^2 - \frac{2U_{AM}}{r} - \frac{\partial U_{AM}}{\partial r} - U_{AM}^2]\phi(\vec{r}) \quad . \tag{19}$$

Thus, for parity conserving potentials, it is clear that the spin orbit term depends on S, U_o and U_t.

It is interesting to consider under what conditions Eq. (2) can be simplified. As already mentioned for systems where parity is assumed to be a good quantum number the pseudoscalar and axial vector potentials are eliminated. Further, the space part of the Lorentz vector potential may, if it is spherically symmetric, be removed by a gauge transformation

$$\psi(\vec{r}) = G(r)\phi(\vec{r}) \quad , \tag{20}$$

where

$$G(\vec{r}) = \exp i \int_0^1 U_v^r (tr)\, r dt. \qquad (21)$$

Under these conditions we can simplify Eq. (2) to the following form

$$\{\vec{\alpha}\cdot\vec{p} + \beta(m + S) - (E - V) - i\, \beta\vec{\alpha}\cdot\hat{r}\, T\}\phi(\vec{r}) = 0 \quad . \qquad (22)$$

Here V is the time-like Lorentz vector and contains the static Coulomb potential as well as the nuclear potential and T is the sum of U_{AM} and U_t.

It can be shown that in order to obtain agreement with pA data at least two of the three potentials in Eq. (22) must appear. In fact, a transformation of the form,

$$\phi(\vec{r}) = e^{i\gamma^o F(r)}\psi(\vec{r}) \qquad (23)$$

where $F(r) \to 1$ as $r \to \infty$

yields for $\psi(\vec{r})$

$$[(E - V)\gamma^o \cos 2F - (m + S) \cos 2F + i(E - V) \sin 2F$$

$$- i(m + S)\gamma^o \sin 2F + i\gamma^o\vec{\gamma}\cdot\hat{r}T + \gamma^o\vec{\gamma}\cdot\hat{r}\,\frac{\partial F}{\partial r} - \vec{\gamma}\cdot\vec{p}]\psi(\vec{r}) = 0 \; , \qquad (24)$$

showing that by proper choice of F(r) one can change from a SV model to an equivalent ST or VT or SVT model.[10]

For the ST model we eliminate the nuclear vector term by requiring that

$$(E - V - V_c) \cos 2F - i(m + S) \sin 2F = (E - V_c), \qquad (25)$$

where V_c is the Coulomb potential. This produces a new scalar potential S'

$$m + S' = (m + S) \cos 2F - i(E - V - V_c) \sin 2F \quad , \qquad (26)$$

and a new tensor potential T'

$$T' = T - i\,\frac{\partial F}{\partial r} \quad . \qquad (27)$$

To remove the scalar potential and obtain VT model, we require

$$(m + S) \cos 2F - i(E - V - V_c) \sin 2F = m. \qquad (28)$$

This produces a new vector potential V'

$$E - V' - V_c = (E - V - V_c) \cos 2F - i(m + S) \sin 2F \qquad (29)$$

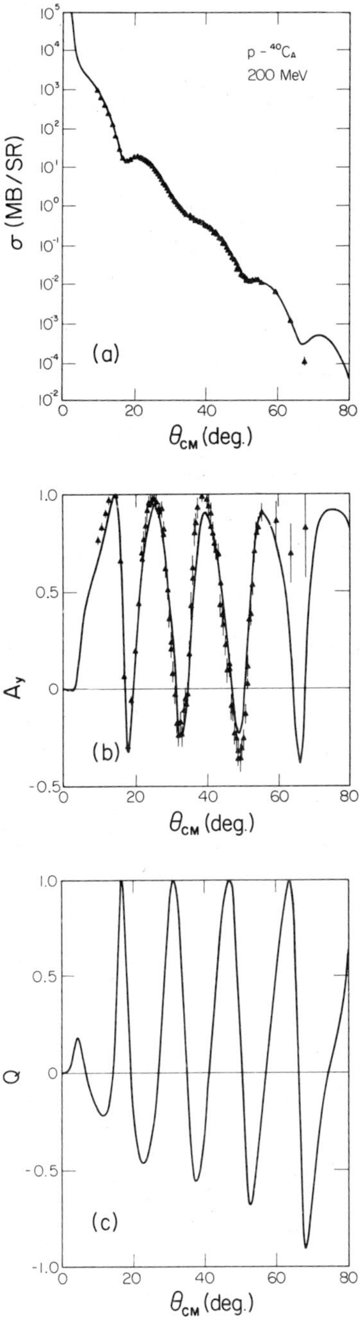

FIG. 1. Elastic scattering observables for p-^{40}Ca at 200 MeV. The calculation shown is from the ST model obtained from using the transformation Eq. (23) on a SV model 12-parameter fit to the data of Ref. 11.

and a new tensor T'

$$T' = T - i \frac{\partial F}{\partial r} . \qquad (30)$$

The transformation Eq. (23) has been used to obtain equivalent ST and VT potentials from fits obtained using a SV model. Figure 1 shows the results obtained for p-^{40}Ca at 200 MeV[10] using the equivalent scalar and tensor potentials from Eqs. (26) and (27). Identical results are obtained using the equivalent VT potentials of Eqs. (29) and (30). The transformation has also been used to obtain ST and VT models from our fit to p-^{40}Ca and 497.5 Mev.[10] In both cases the original SV model contained twelve parameters, the same number as in the Schrödinger phenomenology. The form factors used were two parameter Fermi shapes. Figure 2 shows the scalar and vector potentials obtained from fitting the p-^{40}Ca data. The corresponding second order Dirac equation central and spin orbit potentials are shown in Fig. 3. Figure 4 shows the tensor potentials and the vector potentials obtained using the transformation. The central and spin orbit potentials obtained from these potentials are the same as those shown in Fig. 3. Thus, there exists a set of equivalent

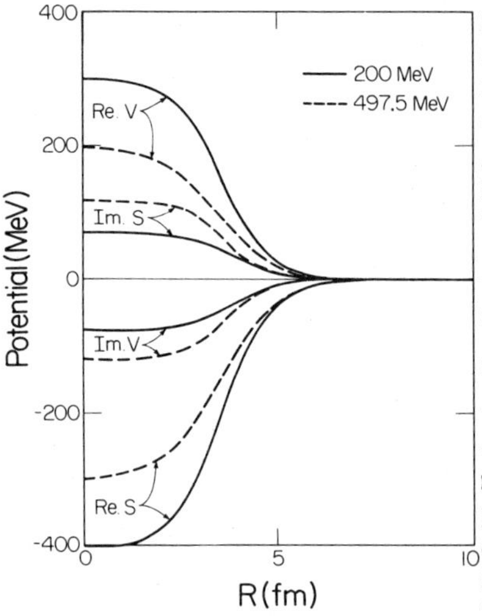

FIG. 2. Real and imaginary vector and scalar potentials obtained from 12-parameter SV fits to the data of Refs. 11 and 12.

potentials and an ambiguity in their Lorentz character.

There is, however, strong theoretical evidence from relativistic nuclear structure[13-14] and relativistic impulse approximation[15-18] calculations favoring the SV model. That it is a preferable phenomonological choice is apparent. It is a more local representation, the strengths of the scalar and vector potentials vary smoothly with energy and are almost constant from 20 to 200 MeV.[19] The strengths of the real scalar or vector potentials in the ST or VT models must vary rapidly with energy in order to change from attraction at low energies to

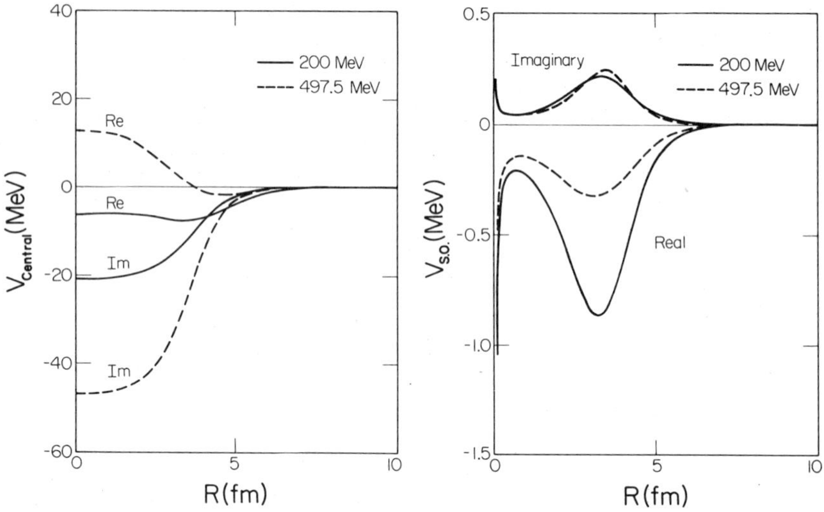

FIG. 3. Real and imaginary central and spin orbit potentials corresponding to the potentials of Fig. 2.

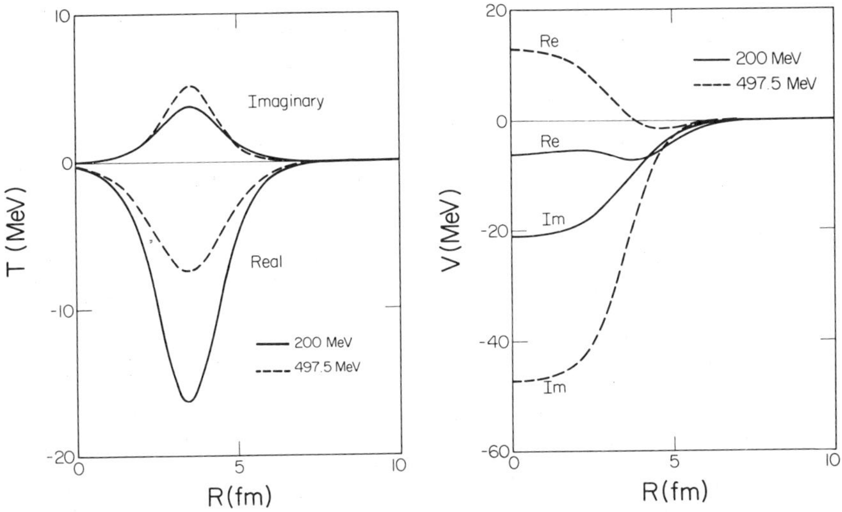

FIG. 4. Real and imaginary vector and tensor potentials obtained from the potentials shown in Fig. 2.

repulsion at higher energies. The geometries of the real scalar and vector potentials resemble the nuclear density;[19] however, the corresponding geometry in the ST or VT model has, as can be seen from Fig. 4, a much more complicated shape in the transition energy region. In fact, we have been unable to use the ST or VT modles with simple geometries to obtained good fits to the p-^{40}Ca data at 200 MeV.[10]

Applications of Dirac Phenomenology

There are several features required of any phenomenological optical model analysis. It should give good fits to the data, the number of parameters should be held to a minimum, it should be systematic and it should make contact with theory. In this section I will discuss how well the usual SV model of Dirac phenomenology meets these tests.

The scalar and vector optical potentials are written

$$V = V_R f_V(r) + i V_I g_V(r) \tag{31}$$

$$S = S_R f_S(r) + i S_I g_S(r) \tag{32}$$

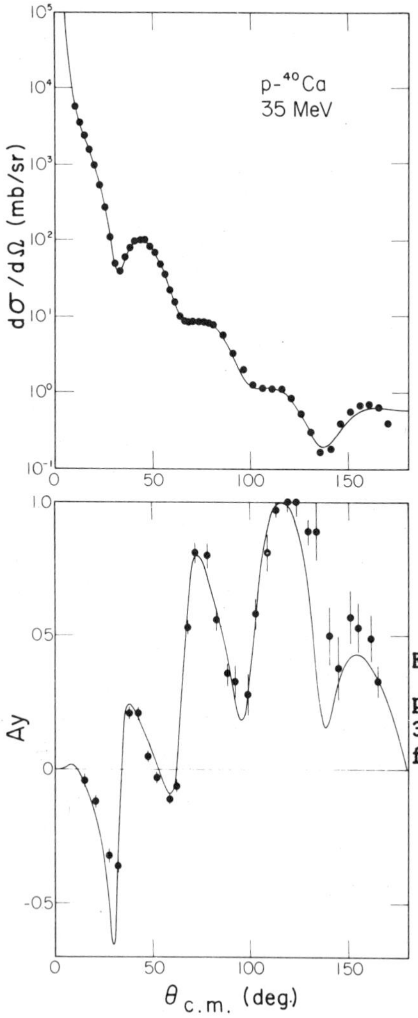

FIG. 5. Cross sections and analyzing powers for the best fit to p-^{40}Ca at 35 MeV. The cross section data are from Ref. 20 and the A_y data from Ref. 21.

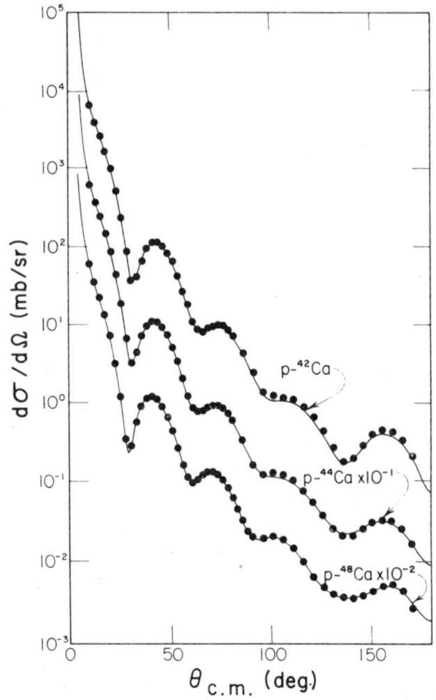

FIG. 6. Best cross sections for p-^{42}Ca, p^{44}Ca and p-^{48}Ca at 35 MeV. The cross section data are from Ref. 20.

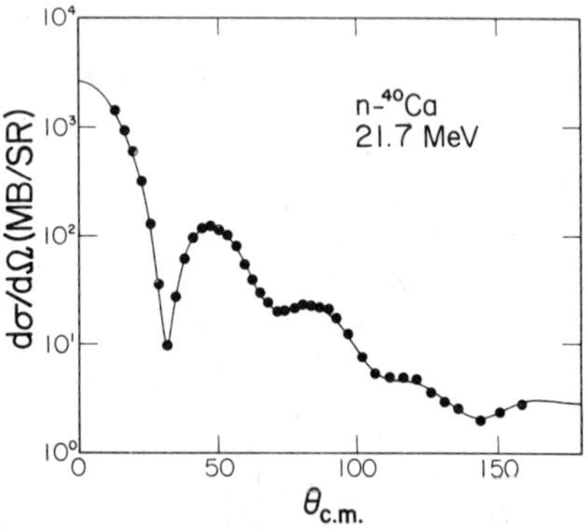

FIG. 7. Best fit cross sections for n-^{40}Ca at 21.7 MeV. The data are from Ref. 22.

where f and g are often taken to be 2-parameter Fermi shapes. Thus, the model can contain 12 parameters the same number used in most nonrelativistic analyses. This model has been used to analyse pA data in the energy region $10 \leq T_p \leq 1000$ MeV.

At low energies the analysis is hampered by the lack of high quality spin measurements. In spite of that a number of interesting contrasts between the Dirac and Schrödinger approaches emerge. In both cases quite reasonable fits to cross section data can be obtained, however, in the Dirac case this can be done with fewer parameters. Particularly striking is the agreement with large angle data as shown in Fig. 5 for p-^{40}Ca at 35 MeV and Fig. 6 for p-^{42}Ca, p-^{44}Ca and p-^{48}Ca at the same energy. Nonrelativistic analyses usually require an additional term in the optical potential which depends on the orbital angular momentum to obtained such agreement at large angles.[20] Further, in every case the spin observables were better reproduced in the Dirac case.[20,22] A similar situation also holds for neutron scattering. The Dirac fit shown in Fig. 7 can be reproduced in a nonrelativistic analysis with the addition of more parameters.[22]

At higher energies where spin observables are usually available we find that the SV model is superior to the usual Schrödinger approach in reproducing experiment, expecially with regard to the spin observables. The classic example is afforded by p-^{40}Ca at 497.5 MeV where the complete set, cross section, analysing power and spin rotation function measurements were available.[23] In this case a comparison of Schrödinger and Dirac optical model calculations, where both used simple Fermi shape (or derivative Fermi shape) for the geometries, clearly showed the ability of the Dirac model to reproduce the spin observables while the Schrödinger calculation failed. In this phenomenological study we concluded that the additional spin observable produced little change in the real potential parameters but did influence the imaginary potentials. Recent studies by Kobos et al.[24] have stressed the importance of the role of spin observables in removing possible ambiguities in the imaginary potentials. Applications to other targets have shown that the relativistic optical model is capable of fitting the data quite well. We shown in Fig. 8 our results for p-^{208}Pb at 800 MeV. The disagreement at large angles.may be due to coupled channel effects.[25,26]

Over the years a number of systematic features of the relativistic optical model have emerged. Perhaps the first

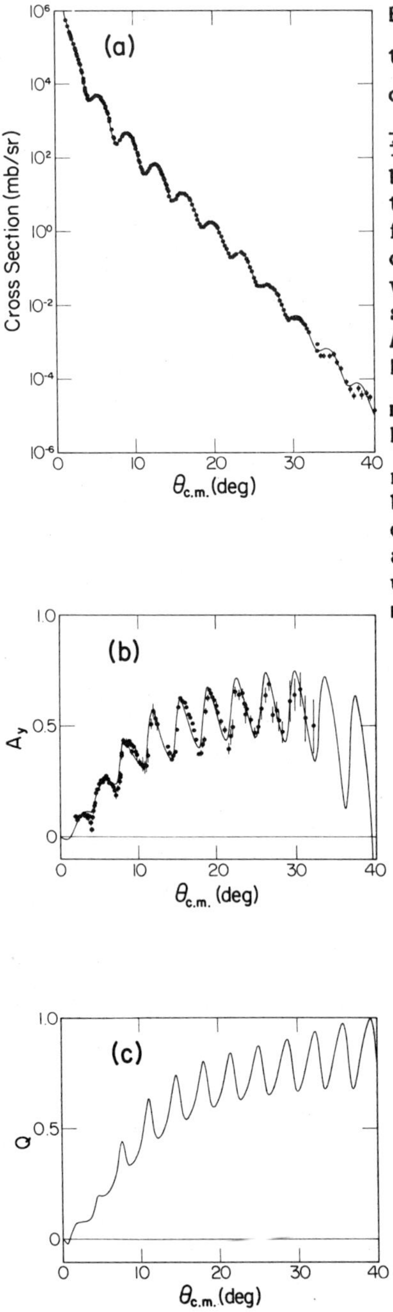

FIG. 8. Dirac SV optical model fit to p-^{208}Pb 800 MeV σ and A_y data of Ref. 12 and the predicted Q.

indication that the results could be understood in terms of relativistic mean field theory came from the observation that the ratio of the volumn integrals of the real vector to real scalar potentials showed a linear energy dependence. At zero energy this ratio, shown in Fig. 9, agrees with the Walecka model[27] and with relativistic Hartree calculations of finite nuclei.[28,29] Other systematic behaviour, such as the energy dependence of the effective central and spin-orbit potentials can be understood in terms of relativistic mean field theory at low energies

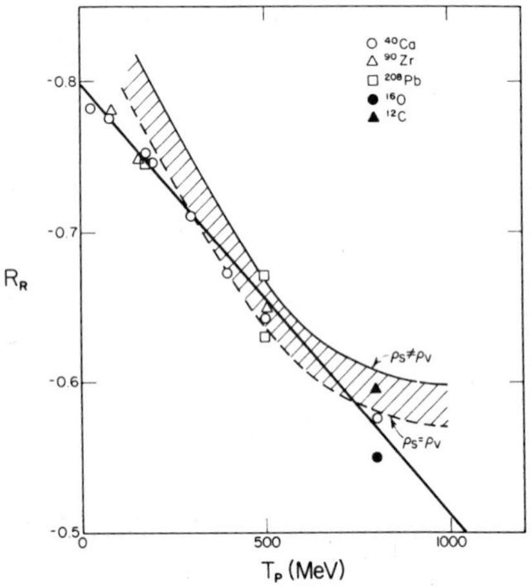

FIG. 9. The ratio, R_R, determined from fitting p-A data. The value at zero agrees well with relativistic mean field theory. The solid line is a least square fit.

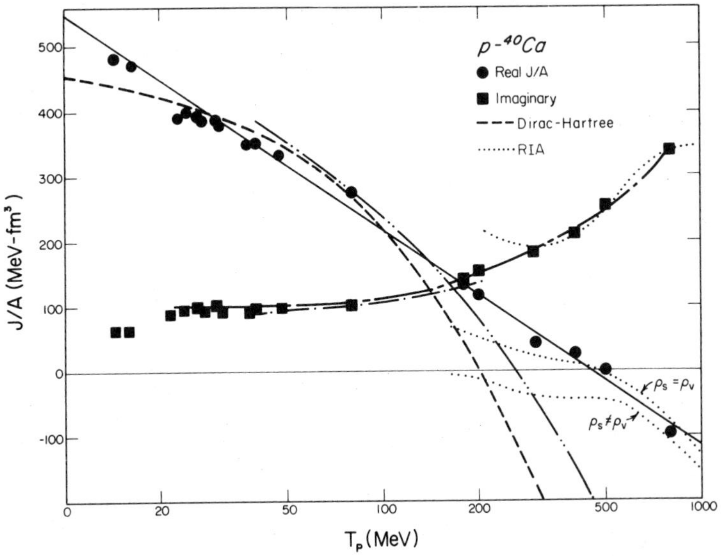

FIG. 10. The real and imaginary volume integrals of the effective effective central potential. The dash-dot-dot line is calculation of Ref. 30, the dash-dot show the results of Ref. 31.

and the RIA at higher energies, see Figs. 10 and 11. In fact, a simple folding model, described in Ref. 19, based on relativistic Hartree calculations[29] has been used to obtain the geometries of the real scalar and vector potentials. These potentials have been used to obtain good fits to p-^{40}Ca data from 20 to 1000 MeV.[19] The real and imaginary effective central and spin-orbit potentials obtained are shown in Figs. 12-14.

It now appears that improvements in the RIA which include the effects of exchange[32,33] and Pauli blocking[33] produce potentials which agree quite well with the phenomenological results. In Fig. 15 the strengths of the real scalar and vector optical potentials are compared with the calculations of Tjon and Wallace.[32] Even at energies as low as 200 MeV the RIA optical potentials calculated by Horowitz[33] are very close to the phenomenological results as is shown in Fig. 16. While much work remains to be done a consistant picture is emerging and the presence of the large cancelling nuclear potentials,

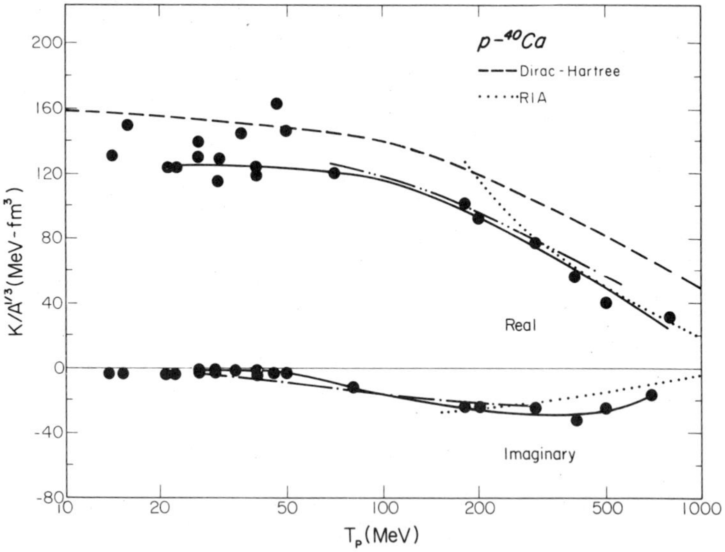

FIG. 11. The real and imaginary volume integrals of the spin-orbit potential. The dash-dot-dot and dash-dot are as in Fig. 10.

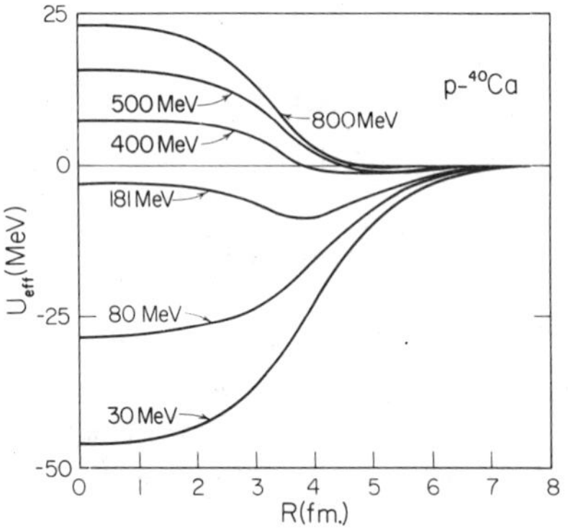

FIG. 12. Effective real central potentials from the from the analysis of p-^{40}Ca data.

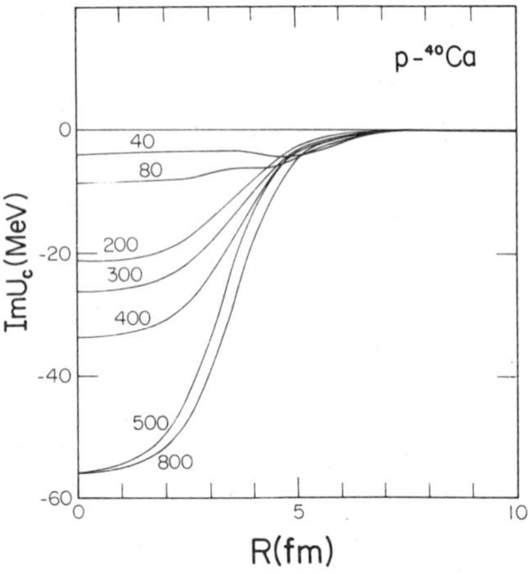

Fig. 13. Same as Fig. 12 for imaginary central potential.

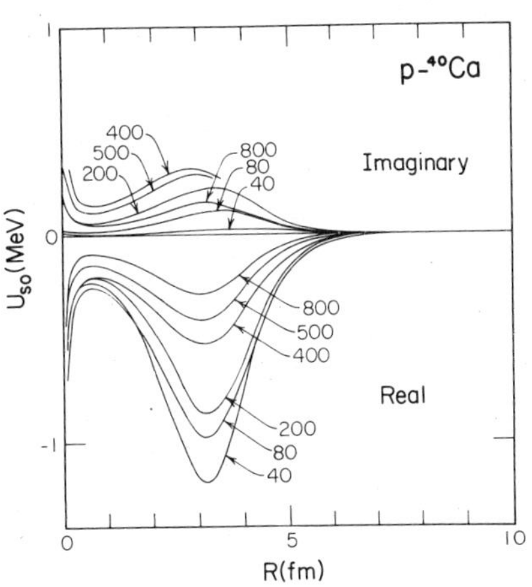

FIG. 14. Same as Fig. 12 for the real and imaginary spin-orbit potential.

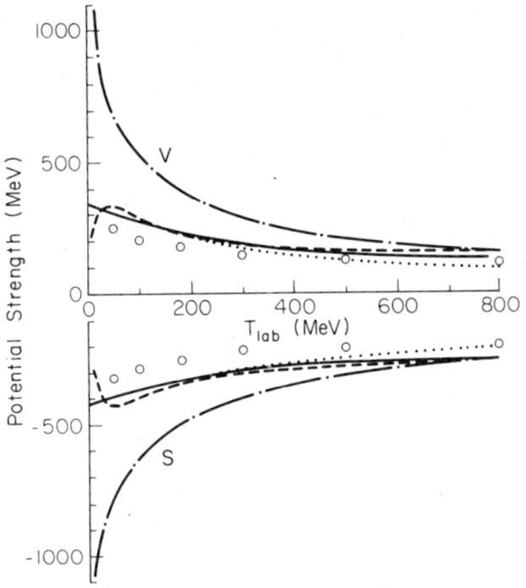

FIG. 15. The solid line is the results from Dirac Optical model fits to pA data. The dash line gives the RIA without exchange while the other lines and open circles are from the calculations of Ref. 32 which include exchange.

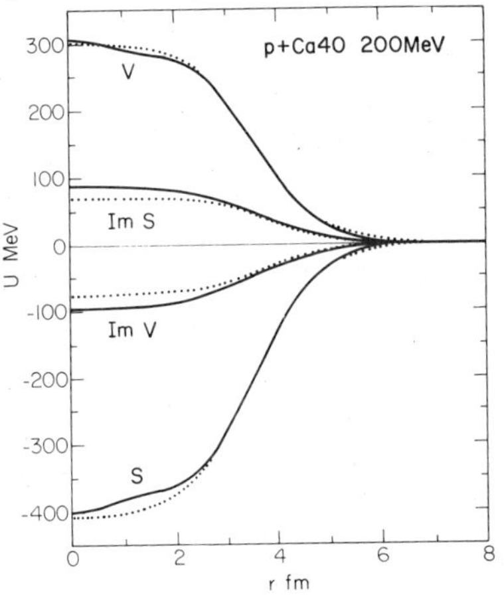

FIG. 16. The dotted lines are the results of a SV Dirac fit to the data while the solid line is the result of Ref. 33 which contains both Pauli blocking and exchange effects.

inherent in the SV optical model, seems to be a required freature of both the phenomenological and theoretical treatments.

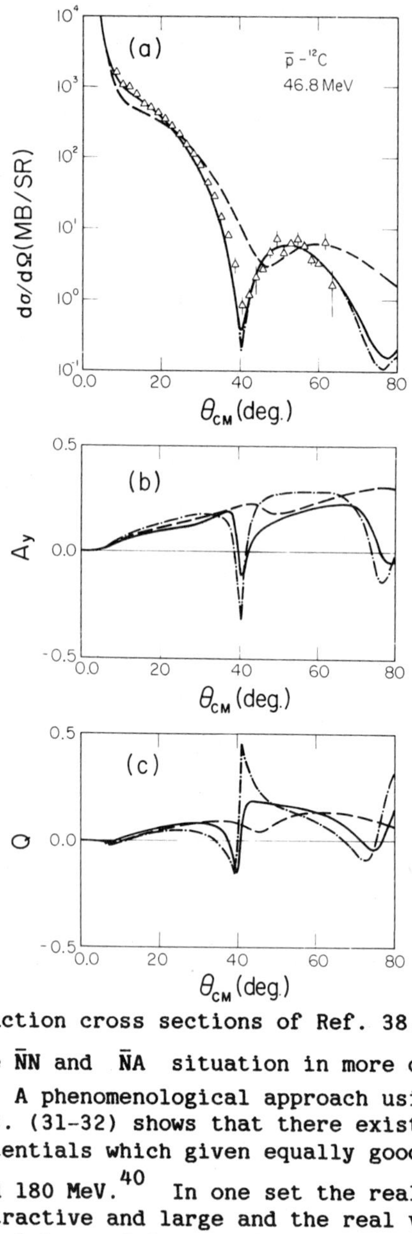

FIG. 17. Calculated differential cross sections for $\bar{p}$-^{12}C scattering at 46.8 MeV. The data are from Ref. 35. The smooth curve is the RIA calculation, the dashed curve is the RIA calculation neglecting the range of the $\bar{N}N$ amplitudes, the dash-dot curve shows the nonrelativistic impulse approximation.

In the case of antiproton-nucleus scattering the situation is not as clear. The recent $\bar{p}$-^{12}C measurement at 47 MeV can be reproduced quite well by both the RIA and nonrelativistic impulse approximation (NRIA) calculations as shown in Fig. 17.[34] The results are dominated by the long range of the imaginary potentials in both cases. Similar agreement between RIA and NRIA calculations occurs at higher energies.[36,37] Figure 18 shows the good agreement between the RIA calculations and the measured reaction cross sections of Ref. 38. George Walker discusses the $\bar{N}N$ and $\bar{N}A$ situation in more detail in these proceedings.

A phenomenological approach using the SV model given by Eqs. (31-32) shows that there exist several types of optical potentials which given equally good fits to the data at both 47 and 180 MeV.[40] In one set the real scalar potential is attractive and large and the real vector potentials is repulsive and large. In set two, both real potentials are

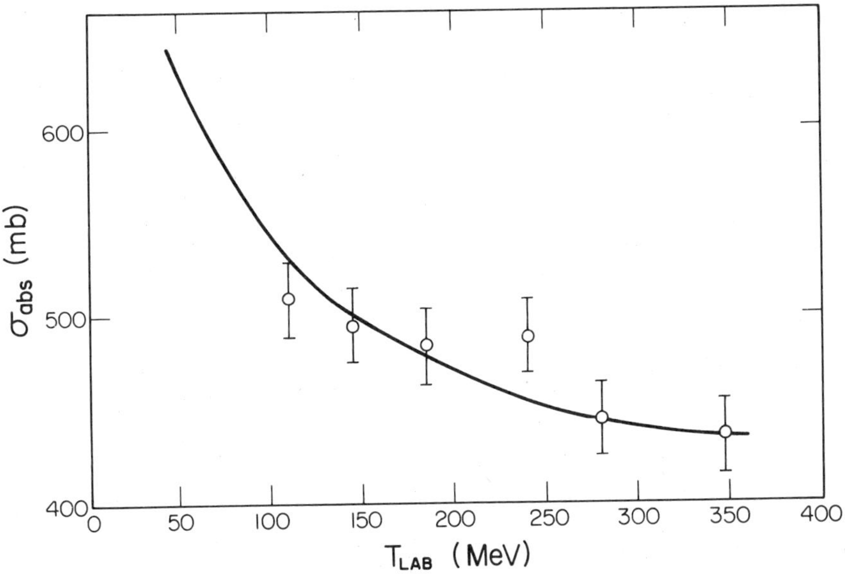

FIG. 18. Total reaction cross sections for $\bar{p}$-^{12}C calculated using the RIA with Paris $\bar{N}N$ amplitudes.[39] The data are from Ref. 38.

large and attractive. The spin orbit potentials are very different in the two cases and the predicted spin observables differ greatly as shown in Fig. 19 for $\bar{p}$-^{12}C at 180 MeV. It appears that the situation with respect to real potential strengths would be clarified if the spin observables were measured. Further, such measurements could well give information regarding the $\bar{N}N$ t-matrix.

Meson Scattering in the Kemmer-Duffin-Petiau Formalism

As relativistic dynamics are the subject of this workshop it seems appropriate to consider other relativistic equations. In this section I will discuss an application of the Kemmer-Duffin-Petiau[4-6] (KDP) equation to meson nucleus scattering. The equation, like the Dirac equation, is linear in the energy. This feature facilitates the construction of RIA optical potentials in a manner analogous the proton-nucleus case. The KDP formalism has been used by Fischbach, Nieto, Primakoff and Scott in treating meason decays.[42] An historical

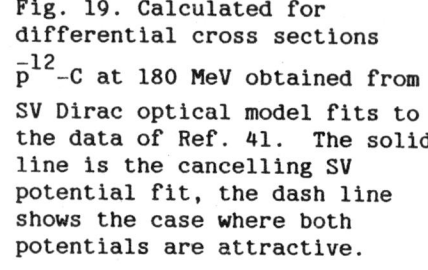

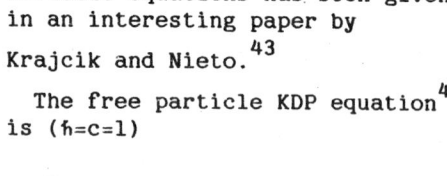

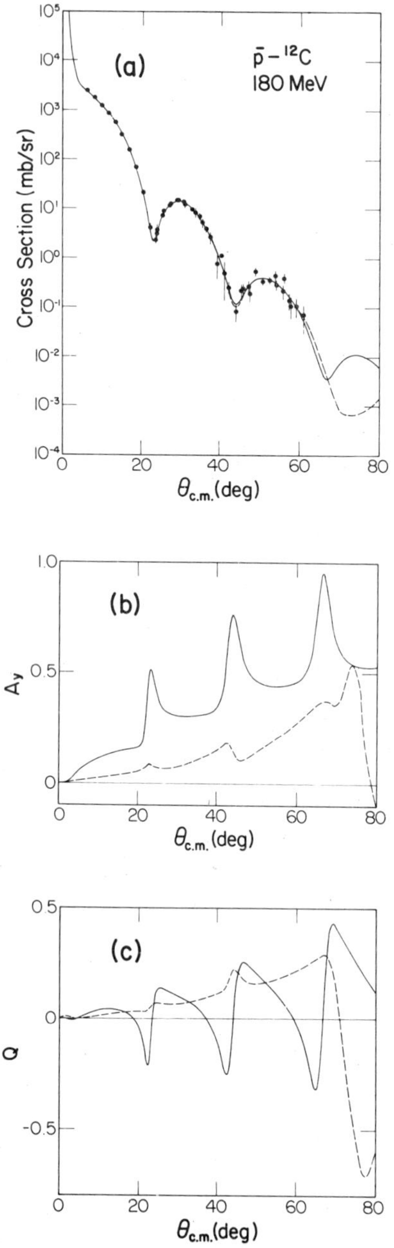

Fig. 19. Calculated for differential cross sections $\bar{p}$-^{12}C at 180 MeV obtained from SV Dirac optical model fits to the data of Ref. 41. The solid line is the cancelling SV potential fit, the dash line shows the case where both potentials are attractive.

review of first-order relativistic equations has been given in an interesting paper by Krajcik and Nieto.[43]

The free particle KDP equation[4] is ($\hbar$=c=1)

$$(i\beta^{\mu}\partial_{\mu} - m)\,\phi = 0, \tag{33}$$

where μ = 0, 1, 2, 3, m is the mass parameter and the β^{μ} obey[44]

$$\beta^{\mu}\beta^{\nu}\beta^{\lambda}+\beta^{\lambda}\beta^{\nu}\beta^{\mu}=g^{\mu\nu}\beta^{\lambda}+g^{\lambda\nu}\beta^{\mu}. \tag{34}$$

The algebra generated by the four β^{μ}'s has three irreducible representations of dimension one, five and ten. The five dimensional representation yields a spin operator whose eigenvalues are zero, the ten dimensional case corresponds to spin one and the one dimensional case is trivial. The first component of the five dimensional Kemmer wave function for the spin zero case satisfies the Klein-Gordon equation for massive particles.

In order to apply the KDP formalism to meson-nucleus scattering we must introduce interactions in Eq. (33). If one writes

$$(i\beta^{\mu}\partial_{\mu}-m-U)\phi=0 \tag{35}$$

the most general form for U contains two scalar, two vector and two tensor terms.[44] We omit the tensors to avoid noncausal effects.[44] For the spin zero case the scalar operators are the unit operator I and the 5 x 5 operator P whose elements are all zeros except the (1,1) element; thus P acts as a projection operator onto the first component of ϕ. The vector operators are β^{μ} and $\tilde{\beta}^{\mu}=P\beta - \beta^{\mu}P$. The form for U is

$$U = U_s I + U_s^1 P + \beta^{\mu} U_v + \beta^{\mu} P U_v^1 . \qquad (36)$$

The last two terms may also be written as $\beta^{\mu} U_v + P\beta^{\mu} U_v^1$.

In order to construct impulse approximation optical potentials consistent with Eq. (36) we need an invariant form for the meson- nucleon t matrix. The choice for the invariant form used here is

$$t = I_N I t_s + I_N P t_s^1 + \gamma_{\mu} \beta^{\mu} t_v + \gamma_{\mu} \beta^{\mu} P t_v^1 , \qquad (37)$$

where I_N and γ_{μ} are the unit and Dirac γ-matrices for the nucleon. As in the nucleon-nucleus RIA we equate the matrix elements of the empirical c.m. two-body scattering amplitude,

$$F(q) = f(q) + \vec{\sigma} \cdot \vec{n}\, g(q), \qquad (38)$$

taken between Pauli spinors for the nucleon with the matrix elements of the invariant t matrix between Dirac and Kemmer free particle spinors. The scattering amplitude and the invariant t matrix are related by a (2x4) matrix, thus certain choices need to be made in order to proceed. We have limited the t-matrix to only two of the four possible terms in Eq. (37). Thus, we consider models with two scalars, two vectors or a vector-scalar mixture. For each choice a transformation matrix K relates t and F. For example, for a scalar-vector mixture

$$\begin{pmatrix} t_s \\ t_v \end{pmatrix} = - \frac{2\pi\sqrt{s}K^{-1}}{Mm} \begin{pmatrix} f \\ g \end{pmatrix} , \qquad (39)$$

where $\sqrt{s}$ is the total meson-nucleon energy and m(M) is the meson (nucleon) mass.

The combination of two scalars or two vectors produces a matrix K with zero determinant, thus, we consider t to consist of a scalar and a vector amplitude. There are, however, several choices for the form of t depending on whether the operator P is in both terms (case 1), in the scalar only (case 2), in the vector only (case 3) or in neither (case 4). The forms $\gamma_\mu P\beta^\mu t_v$ and $\gamma_\mu\beta^\mu P t_v$ produce identical results. Note that t_s and t_v depend on both f and g, thus, even in the impulse approximation, the scalar and vector potentials contain contributions from f and g. The usual first-order nonrelativistic calculation only contains contributions from f[45].

The invariant amplitudes for each of the four cases are used to construct optical potentials for use in Eq. (35). The optical potentials for spin zero targets are given by

$$U_{s,v} = \sum_{i=p,n} \int \frac{dq^3}{(2\pi)^3} e^{i\vec{q}\cdot\vec{r}}\, t^i_{s,v}(q)\, \rho^i_{s,v}(q), \tag{40}$$

where $\rho_{s,v}(q)$ are the Fourier transforms of the relativistic Hartree densities of Horowitz and Serot[29]. The KDP equation for meson-nucleus scattering may now be written as,

$$[i\beta^\mu\partial_\mu - A_\mu\beta^\mu - U_j - m]\,\phi = 0, \tag{41}$$

where $j = 1, 2, 3, 4$ for the four cases used. The electromagnetic potential A_μ has been added by minimal substitution. We take A_μ as the static Coulomb potential obtained from the empirical charge distribution. In addition, the space-like components of U_v do not contribute for spin zero targets.

The KDP elastic cross sections are obtained by solving the second order equation obtained for the first component of the KDP wave function. For conserved current interactions, such as the EM interaction, this second order equation is identical to the KG equation for EM interactions.[46] Here, however, a different second-order equation results for each case. They are:

Case 1 $$[(E-U_c-U_v)(E-U_c) - m(m+U_s) + \nabla^2]\,\phi_1 = 0; \tag{42}$$

Case 2 $$[(E-U_c-U_v)^2 - m(m+U_s) + \nabla^2]\,\phi_1 = 0; \tag{43}$$

Case 3 $$[(E-U_c-U_v)(E-U_c) - (m+U_s)^2 + \nabla^2 - \vec{U}_D\,\vec{\nabla}]\,\phi_1 = 0; \tag{44}$$

Case 4 $$[(E-U_c-U_v)^2 - (m+U_s)^2 + \nabla^2 - \vec{U}_D\cdot\vec{\nabla}]\,\phi_1 = 0; \tag{45}$$

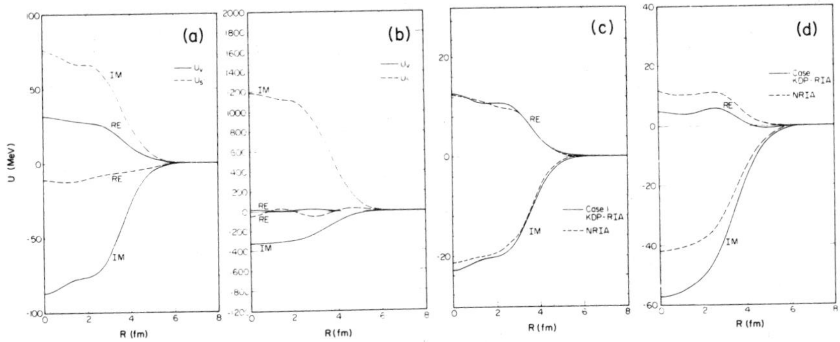

FIG. 20. Calculated KDP-RIA scalar and vector potentials for case 1. The potentials $K^+ - {}^{40}Ca$ at 800 MeV/c are shown in (20a); those for $\pi^+ - {}^{40}Ca$ at 800 MeV/c are shown in (20b). The effective central potentials for the cases shown in Figs. (20a) and (20b) are given in (20c) and (20d) (solid curves). The NRIA potentials are the dashed curves.

where

$$\vec{U}_D = \frac{1}{m+U_s} \vec{\nabla} U_s . \tag{46}$$

The non-local Darwin term may be replaced by an equivalent local term using a wave function transformation, just as in the second order Dirac equation.

As a first application of the formalism we considered elastic scattering of 800 MeV/c beams of K^+, π^+ and π^- from a ^{40}Ca target. The Martin amplitudes[47] were used for the K^+ and the E100 solutions from Arndt[48] for $\pi^\pm$ in constructing the invariant t matrices. The elastic cross sections were obtained by solving the appropriate Klein-Gordon equation[49]. We considered all four cases and found that cases 3 and 4 gave poorer agreement with $\pi^\pm$ data than did cases 1 and 2. At this energy the optical potentials for cases 1 and 2 are almost identical for a given reaction and it is not possible to choose between them on the basis of experiment. This is not true for $\pi^\pm$ scattering at lower energies where case 1 produces substantially better agreement with experiment. The results for K^+ were very similar for all four cases. Figures (20a)

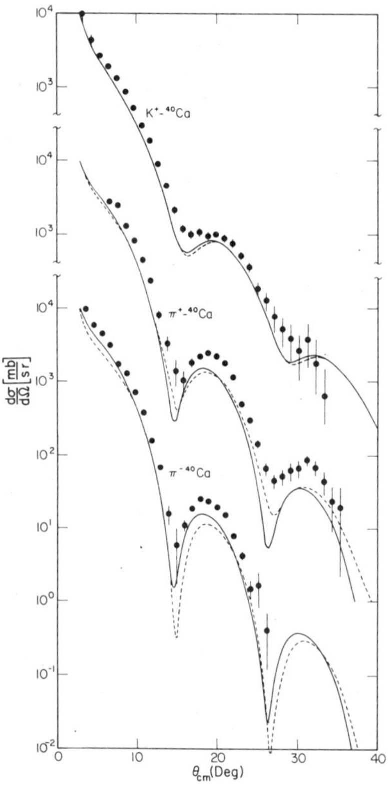

FIG. 21. Calculated KDP-RIA cross sections (solid curves) for case 1 potentials for K^+-^{40}Ca and $\pi^{\pm}$-^{40}Ca at 800 MeV/c. The data are from Refs. 50 and 51. The NRIA calculations are given by the dashed curve.

and (20b) show the case 1 optical potentials for K^+-^{40}Ca and π^+-^{40}Ca at 800 MeV/c. The scalar and vector potentials are large and tend to cancel just as in the pA RIA. The π^--^{40}Ca potentials are almost identical to the π^+-^{40}Ca potentials. Figures (20c) and (20d) show the effective central potentials for these two cases.

Figure (21) shows the calculated K^+-^{40}Ca cross section for case 1 potentials, along with the measured cross sections from Ref. 50. For the kaon all four choices produce essentially identical cross sections. The calculated cross sections for $\pi^{\pm}$ for cases 1 and 2 are also very similar, however cases 3 and 4 are in much poorer agreement with experiment. Also shown in Fig. (21) are the $\pi^{\pm}$ case 1 results along with the measured cross sections from Ref. 51. The dashed curves in Fig. (21) are nonrelativistic impulse approximation (NRIA) calculations using the Schrödinger equation with relativistic kinematics[45] and the same input amplitudes and densities (vector densities only). The K^+- nucleon spin-dependent amplitude g is very small at this energy. Because of this one would expect KDP and NRIA calculations to be quite similar, and this is the case. For the $\pi^{\pm}$ nucleon amplitude, g is not small, and the differences between KDP and NRIA are larger, especially at small angles for the π^- case. In addition, the presence of the cross term U_cU_v in the KDP appears to improve the agreement with experiment at small angles.

It is not clear at this time whether this new treatment of meson-nucleus scattering using KDP formalism will prove advantageous. It is, however, a different approach and one which has a number of intersting possibilities. The formalism is currently being considered for descriptions of inelastic meson-nucleus scattering.[52]

CONCLUSIONS

Over the past few years interest in relativistic treatments of nuclear reactions and nuclear structure has grown tremendously. I have considered only one aspect of the problem, that of the relativistic optical model. Other presentations at this workshop will review the various relativistic treatments of finite nuclei; nuclear matter and nuclear reactions; both elastic and inelastic treatments.

The systematic behaviour of the SV relativistic optical model has been verified by analyses of a large body of data. The general features are: 1) the ratio R_R varies linearly with energy almost independent of target nuclei; 2) the strengths of the real scalar and vector potentials exhibit a smooth energy variation; 3) the geometries of the real scalar and vector potentials resemble the nuclear density, they do not exhibit anomalous shapes; 4) Dirac phenomenology seems to be better able to obtain good fits to spin observables than are obtained in the usual Schrödinger approach; 5) the effective central and spin-orbit potentials very systematically with energy. The phenomenological results can be understood in terms of relativistic nuclear structure calculations and the relativistic impulse approximation.

In conclusion, there are strong indications, both phenomenological and theoretical, that nuclei contain large cancelling scalar and vector fields. The central problem is to definitely establish (or refute) their existence experimentally.

ACKNOWLEDGEMENTS

The author thanks her collaborators and colleagues for their many contributions to various aspects of this work. They include E.D. Cooper, S. Hama, C.J. Horowitz, G.W. Hoffmann, G. Kälbermann, R. Kozack, J.A. McNeil, R.L. Mercer, L. Ray, B.D. Serot, C.M. Shakin, D.A. Sparrow and K. Stricker-Bauer. The author also thanks C.B. Dover, M. Lacombe, R. Vinh Mau and M.E. Sainio for providing the $\bar{N}N$ amplitudes. This work was supported by the National Science Foundation, Grant Phy-8306268.

REFERENCES

1. B.C. Clark, R.L. Mercer, D.G. Ravenhall and A.M. Saperstein, Phys. Rev. C7, 466 (1973).

2. R. Klem, G. Igo, R. Talaga, A Wriekat, H. Courant, K. Eisweiler, T. Joyce, H. Kagen, Y. Makdisi, M. Marshak, B. Mossberg, E. Peterson, K. Ruddick and T. Walsh, Phys. Lett. 70B, 155 (1977); Phys. Rev. Lett. 38, 1272 (1977).

3. L.G. Arnold, B.C. Clark, R.L. Mercer, D.G. Ravenhall, and A.M. Saperstein, Phys. Rev. C19, 917 (1979).

4. N. Kemmer, Proc. Roy. Soc. (London) A173, 91 (1939).

5. R.J. Duffin, Phys. Rev. 54, 1114 (1938).

6. G. Petiau, Acad. Roy. Belg. Cl. Sci. Mom. Collect 8 16, No. 2 (1936).

7. L.D. Miller, Ann. Phys. (N.Y.) 91, 40 (1975).

8. S. Hama, Ph.D. Dissertation, The Ohio State University (1984) (unpublished).

9. J.D. Bjorken and S.D. Drell, Relativistic Quantum Mechanics (McGraw-Hill Book Co., New York, 1964).

10. B.C. Clark, S. Hama, S.G. Kälbermann, E.D. Cooper and R.L. Mercer, Phys. Rev. C31, 694 (1985); C31, 1975 (1985).

11. D. Hutcheon et al., private communication and IUCF Tech. Rep. 1983.

12. G.W. Hoffmann et al., Phys. Rev. Lett. 47, 1436 (1981) and Phys. Rev. C24, 541 (1981); A. Rahbar et al., Phys. Rev. Lett. 47, 1811 (1981); and L. Ray et al., Phys. Rev. C23, 828 (1981).

13. M.R. Anastasio, L.S. Celenza, W.S. Pong and C.M. Shakin, Physics Reports 100, 327 (1983), and references therein.

14. B.D. Serot and J. D. Walecka in Advances in Nuclear Physics, edited by J.W. Negele and Erich Vogt (Plenum, New York in press), Vol. 16, and references therein.

15. J.A. McNeil, J. Shepard and S.J. Wallace, Phys. Rev. Lett. 50, 1439 (1983).

16. J. Shepard, J.A. McNeil and S.J. Wallace, Phys. Rev. Lett. 50, 1443 (1983).

17. B.C. Clark, S. Hama, R.L. Mercer, L. Ray and B.D. Serot, Phys. Rev. Lett. 50, 1644 (1983) and B.C. Clark, S. Hama, R.L. Mercer, L. Ray, G.W. Hoffmann and B.D. Serot, Phys. Rev. C28 Rapid Communications, 1421 (1983).

18. M.V. Hynes, A. Picklesimer, P.C. Tandy and R.M. Thaler, Phys. Rev. Lett. 52, 978 (1984).

19. B.C. Clark, S. Hama and R.L. Mercer,. AIP Conf. Proc. No. 97: The Interaction Between Medium Energy Nucleons in Nuclei-1982, edited by H. O. Meyer (A.I.P. Press, New York, 1983), p. 260.

20. R.H. McCamis, T.N. Nasr, J. Birchall, N.E. Davison, W.T.H. van Oers, P.J.T. Verheijen, R.F. Carlson, A.J. Cox, B.C. Clark, S. Hama, E.D. Cooper and R.L. Mercer, submitted for publication 1985.

21. K.H. Bray, K.S. Jayaraman, G.A. Moss, W.T.H. van Oers, D.O. Wells and Y.I. Wu, Nucl. Phys. A167, 57 (1971).

22. R. Alarcon and J. Rapaport private communication (1985).

23. B.C. Clark, R.L. Mercer and P. Schwandt, Phys. Lett. 122B, 211 (1983).

24. A.M. Kobos, E.D. Cooper, J.R. Rook and W. Haider, Nucl. Phys. A435, 677 (1985).

25. L. Ray and G.W. Hoffmann, Phys. Rev. C30, 1593 (1984).

26. R.D. Amado and D.A. Sparrow, Phys. Rev. C29, 932 (1984).

27. J.D. Walecka, Ann. Phys. (N.Y.) 83, 491 (1974).

28. L.D. Miller and A.E.S. Green, Phys. Rev. C5, 241 (1972).

29. C.J. Horowitz and B.D. Serot, Nucl. Phys. A368, 503 (1981).

30. M. Jaminon, private communication (1983).

31. C.J. Horowitz, Phys. Lett. 117B, 153 (1982), Nucl. Phys. A412, 228 (1984).

32. J.A. Tjon and S.J. Wallace, University of Maryland preprint ORO-5126-244 (1985), and S.J. Wallace, Workshop on Dirac Approaches to Nuclear Physics, Los Alamos, NM,(1985) (to be published).

33. C.J. Horowitz, Phys. Rev. C31, 1340 (1985); Workshop on Dirac Approaches to Nuclear Physics, Los Alamos, NM, 1985 (to be published) and International Conference on Antinuclear-and Nucleon-Nucleus Interactions Telluride, Collorado, (1985) (to be published).

34. B.C. Clark, S. Hama, J.A. McNeil, R.L. Mercer, L. Ray, B.D. Serot, D.A. Sparrow and K. Stricker-Bauer, Phys. Rev. Lett. 53, 1423 (1984).

35. D. Garreta, P. Birien, G. Bruge, A. Chaumeaux, D.M. Drake, S. Janovin, D. Legrand, M.C. Mallet-Lemaire, B. Mayer, J. Pain, J.C. Peng, M. Berrada, J.P. Bocquet, E. Monnand, J. Mougey, P. Perrin, A. Erell,. J. Lichtenstadt and A.I. Yavin, Phys. Lett. 135B 266 (1984).

36. D.A. Sparrow, Workshop on Dirac Approaches in Nuclear Physics, Los Alamos, NM, (1985) (to be published).

37. A. Picklesimer, P.C. Tandy and J. A. Tjon, preprint, 1985.

38. K. Nakamura et al. Phys. Rev. Lett. 52, 731 (1984).

39. J. Côte et al. Phys. Rev. Lett. 48, 1319 (1982).

40. R. Kozack, B.C. Clark, S. Hama and R.L. Mercer, Bul. Am. Phys. Soc. 1985.

41. D.Garreta, P. Birien, G. Bruge, A. Chaumeaux, D.M. Drake, S. Janouin, D. Legrand, M.C. Lemaire, B. Mayer, J. Pain, J.C. Peng, M. Berrada, J.P. Bocquett, E. Monnand, J. Mougey, P. Perrin, E. Aslanides, O. Bing, J. Lichtenstadt and A.I. Yavin, Phys. Lett. 149B, 64 (1984).

42. E. Fischbach, M.M. Nieto, H. Primakoff and C.K. Scott, Phys. Rev. D9, 2183 (1974) and references therein.

43. R.A. Krajcik and M.M. Nieto, Am. J. Phys. 45, 818 (1977).

44. R.F. Guertin and T.L. Wilson, Phys. ReV. D15, 1518 (1977).

45. W.R. Coker, J.D. Lumpe and L. Ray. Rev. C31, 1412 (1985). The NRIA calcuations here do not include the A-1/A factor.

46. E. Fishbach, M.M. Nieto and C.K. Scott, J. Math. Phys. 14, 1760 (1973), and Phys. Rev. D7, 207 (1973).

47. B.R. Martin, Nucl. Phys. B94, 413 (1975).

48. R. A. Arndt, Scattering Analysis Interactive Dial-in Program and Private Communication.

49. The calculations are done by partial wave decomposition of the Klein-Gordon equation using Program KGPRG; B. C. Clark, OSU Technical Report, 1974. The program is available from the author.

50. D. Marlow et al., Phys. Rev. C25, 2619 (1982).

51. D. Marlow et al., Phys. Rev. C30, 1662 (1984).

52. C.-Y. Cheung and J.R. Shepard private communication and J.R. Shepard in the AIP Conference Proceedings of the International Conference on Hadronic Probes and Nuclear Interactions, Tempe, Arizona (1985), (to be published).

IMPLICATIONS OF DIRAC NUCLEAR STRUCTURE FOR ELECTRON SCATTERING

E. R. Siciliano
University of Georgia
Athens, GA 30602

ABSTRACT

The use of electron scattering to distinguish between Dirac and Schrödinger bound state single-particle wave functions is discussed within the one-photon-exchange impulse approximation.

INTRODUCTION

Historically, most of our detailed knowledge of nuclear structure has come from electron-nucleus scattering. For example, elastic and inelastic scattering of electrons from nuclei have been used to determine the spatial distributions of nuclear charge and current densities.[1] The main reason why electron scattering is a useful tool for studying nuclear structure is that to a large degree of certainty, the electron-nucleus interaction is known (and, at least in principle, any corrections can be calculated in a well-defined convergent scheme). Furthermore, the electron-nucleus interaction is relatively weak (as compared to the intra-nuclear forces), which enables the electron to probe the target without greatly disturbing it.

A useful model of atomic nuclei is the shell model. In this model, a nucleus consists of nucleons mutually bound together by some self-consistent effective interaction, and the individual nucleon's wave function can be obtained by solving a one-body equation of motion in which this effective interaction enters. The conventional non-relativistic approach to this model adopts the Schrödinger equation as the appropriate one-body equation of motion. The strength of the effective Schrödinger interaction is around 50 MeV. As we have heard at this workshop, there is emerging a competing relativistic approach in which the Dirac equation is the appropriate equation of motion. In this Dirac approach, the effective interaction consists of two Lorentz invariant potentials of strengths around 400 MeV. The question I will attempt to answer in this talk is how can electron scattering be used to distinguish between these two different approaches?

The outline of my talk is as follows: I begin with a brief recapitulation of the general formulas that result from the one-photon-exchange formulation of electron-nucleus scattering. Next, I choose an impulse model for the nuclear current operator. This current operator is to be evaluated between single-particle wave functions. I then discuss the general features of Dirac single-particle wave functions and how they essentially differ from Schrödinger wave functions. Guided by this discussion, specific transitions that may be sensitive to the different wave functions are suggested and some model calculations are shown. I conclude by summarizing the main points of this talk.

GENERAL RESULTS FROM ELECTRON SCATTERING

The formulation and eventual calculation of electron-nucleus scattering basically consists of three steps:
(1) A general formula is derived for the differential cross section for unpolarized electron-nucleus scattering using Quantum Electro Dynamics (QED) at the one-photon-exchange level. Throughout the derivation, the nuclear information (beyond the Mott or point-charge cross section) is represented by the four-current $J^\mu = (\rho, \vec{J})$,
(2) Without choosing a specific nuclear model, a multipole decomposition of J^μ is performed and general selection rules are deduced.
(3) Finally, a specific model for J^μ is chosen and numerical results are obtained. It is interesting to note that all of the results from steps (1) and (2) obtain for any nuclear model, provided the J^μ is Lorentz and time-reversal invariant, and satisfies current and parity conservation. These results are summarized in the following discussion.

The differential cross section for electron-nucleus scattering for the case of unpolarized electrons and unorientated nuclei can be written as[1]

$$\frac{d\sigma}{d\Omega}_{fi} = 4\pi\sigma_{Mott} \left\{ \left(\frac{q_\mu^4}{q^4}\right) F_L^2(q) + \left[\frac{1}{2}\left(\frac{q_\mu^2}{q^2}\right) + \tan^2\left(\frac{\Theta}{2}\right) \right] F_T^2(q) \right\} \tag{1}$$

where the longitudinal and transverse form factors have the multipole expansions

$$F_L^2(q) = (2J_i + 1)^{-1} \sum_{J=0} |\langle J_f || \hat{M}_J(q) || J_i \rangle|^2 \, \eta(J_f J J_i) \tag{2}$$

and

$$F_T^2(q) = (2J_i+1)^{-1} \sum_{J=1} \left\{ |\langle J_f || \hat{T}_J^{el}(q) || J_i \rangle|^2 + |\langle J_f || \hat{T}_J^{mag}(q) || J_i \rangle|^2 \right\} \eta\,(J_f\, J\, J_i) \tag{3}$$

In these expressions $q_\mu(q)$ is the four (three)-momentum transfer, Θ is the scattering angle, $J_i(J_f)$ is the initial (final) spin of the nucleus, and η restricts the sum over J such that the triangle condition is satisfied, i.e.

$$\eta\,(j_1 j_2 j_3) = \begin{cases} 1, & j_1+j_2 \geq j_3 \geq |j_1-j_2| \\ 0, & \text{otherwise} \end{cases}$$

For simplicity, I have taken the electron mass equal to zero and I have ignored target recoil. These restrictions may be relaxed; however, the resulting formula becomes unnecessarily cluttered. The form factors depend upon the nucleus through the nuclear current

$$J^\mu(\vec{q}) = \int d\vec{x}\, e^{-i\vec{q}\cdot\vec{x}} \langle f|\, \hat{J}^\mu(\vec{x})\, |i\rangle = (\rho(\vec{q}), \vec{J}(\vec{q})), \tag{4}$$

where the intrinsic nuclear states are characterized by angular momentum, parity, and isospin, i.e. $|i\rangle = |J_i^\pi, J_{zi}; T_i, T_{zi}\rangle$ and $|f\rangle = |J_f^\pi, J_{zf}; T_f, T_{zf}\rangle$. The multipole operators are defined in terms of the nuclear charge-current operator.

$$\hat{M}_{JM}(q) \equiv \int d\vec{x}\, j_J(qx)\, Y_{JM}(\Omega_x)\, \hat{\rho}(\vec{x}) \tag{5}$$

$$\hat{T}_{JM}^{el}(q) \equiv \frac{1}{q} \int d\vec{x} \left\{ \vec{\nabla}_x \left[j_J(qx) \vec{Y}_{JJ1}^m(\Omega_x) \right] \right\} \cdot \vec{J}(\vec{x}), \tag{6}$$

$$\hat{T}_{JM}^{mag}(q) \equiv \int d\vec{x} \left\{ j_J(qx)\, \vec{Y}_{JJ1}^m(\Omega_x) \right\} \cdot \vec{J}(\vec{x}). \tag{7}$$

In these expressions, j_ℓ is a spherical Bessel function, $Y_{\ell m}$ is a spherical harmonic, and $\vec{Y}_{J\ell}^m$ is a vector spherical harmonic. From considering the parity of these operators, it can be shown that $\vec{M}$ and $\vec{T}^{el}$ contribute to natural parity transitions, whereas $\hat{T}^{mag}$ contributes to unnatural parity transitions. Considering time reversal invariance, it can also be shown that the reduced matrix elements have the following properties.

$$< J_f || \hat{M}_J || J_i > = (-)^{J_i - J_f + J} < J_i || \hat{M}_J || J_f > \quad (8)$$

and for either $\hat{T}^{el}$ or $\hat{T}^{mag}$

$$< J_f || \hat{T}_J || J_i > = (-)^{J_i - J_f + J + 1} < J_i || \hat{T}_J || J_f > . \quad (9)$$

These properties give us a very nice selection rule for elastic scattering ($J_i = J_f$): only even multipoles of M_J and odd multipoles of $\hat{T}_J^{mag}$ will contribute to elastic scattering.

NUCLEAR CHARGE-CURRENT OPERATOR

Since we do not have a complete relativistic quantum field theory of nuclear structure for which a consistent, conserved electromagnetic current can be identified, I need to make a model for the operator. The obvious first guess is the impulse model. That is, I shall assume that the exchanged photon couples to the nucleons as if they were free, and I shall neglect meson exchange currents and other many-body effects that could change the properties of the free-space nucleon current operator.

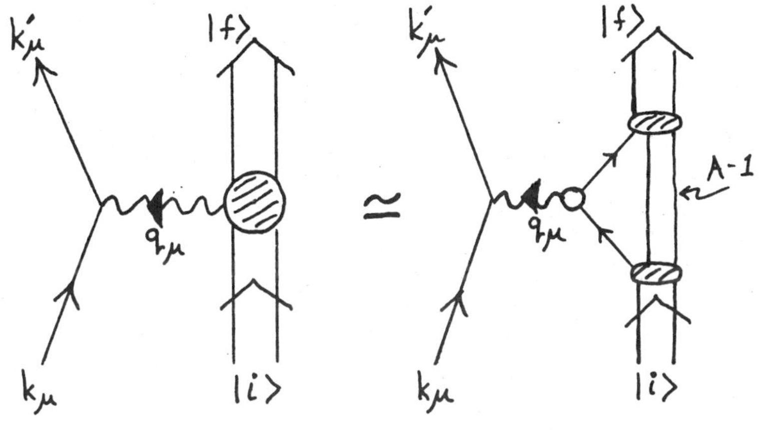

FIG. 1. Impulse approximation for nuclear current operator. The notation is k_μ (k'_μ) for the electron's initial (final) four-momentum, $|i>$ ($|f>$) for the initial (final) nuclear state.

As indicated in Fig. 1, the nuclear charge-current operator in this model is a one-body operator on the nuclear space,

$$\hat{J}^{\mu}(\vec{x}) = \sum_{i=1}^{A} \hat{J}_i^{\mu}(\vec{x}) \quad . \tag{10}$$

If the nucleon were a structureless (bare) Dirac particle, the single-nucleon current operator would then be

$$\hat{J}_i^{\mu}(\vec{x}) = e(i)\, \gamma^{\mu}(i)\, \delta(\vec{x} - \vec{x}_i), \tag{11}$$

where $e(j) = [1+\tau_3(j)]/2$ projects onto protons since the third component of the nucleon isospin operator gives an eigenvalue of +1(-1) for protons (neutrons). We know Eq. (11) must be modified, however, because this simple model predicts the wrong magnetic moments for the proton and neutron, i.e. $\mu_p = 1.0\ \mu_B$ and $\mu_n = 0$ $(\mu_B = \frac{e}{2m_p})$, where their actual values are $\mu_p = 2.79\mu_B$ and $\mu_n = -1.91\mu_B$. In terms of an "anomalous" magnetic moment κ, the experimental results can be expressed as

$$\mu_p = (1+\kappa_p)\,\mu_B \quad \text{and} \quad \mu_n = \kappa_n\,\mu_B, \tag{12}$$

for $\kappa_p = 1.79$ and $\kappa_n = -1.91$. Theoretically, it is possible to account for these anomalous magnetic moments by including the influence of the strong interactions and considering standard radiative corrections to the vertex.[2] Diagrams indicating some of these corrections are shown in Fig. 2. The upshot of these corrections is to write for the single-nucleon operator

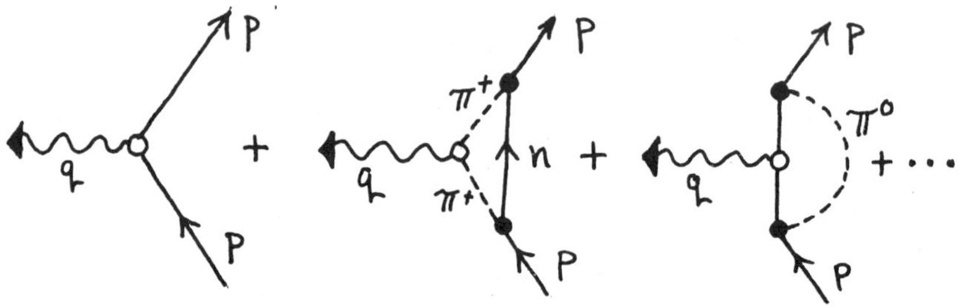

FIG. 2. Standard radiative corrections to the photon-nucleon vertex.

$$\hat{j}_j^{\mu}(q^2) = e(j)\,\gamma^{\mu}(j)\,F_1(q^2) + \frac{i\kappa(j)}{2m_N}\,\sigma^{\mu\nu}(j)\,q_{\nu}\,F_2(q^2), \qquad (13a)$$

where $\kappa = \kappa_p\,(1+\tau_3)\tfrac{1}{2} + \kappa_n\,(1-\tau_3)\tfrac{1}{2}$ and $\sigma^{\mu\nu} = i\,[\gamma^{\mu}, \gamma^{\nu}]\tfrac{1}{2}$. Because I am interested in separating the Dirac single-particle large and small component dependence, it is convenient to define the auxiliary matrices $\Gamma_1 = \begin{pmatrix}1 & 0\\ 0 & 1\end{pmatrix}$, $\Gamma_2 = \begin{pmatrix}1 & 0\\ 0 & -1\end{pmatrix}$, $\Gamma_3 = \begin{pmatrix}0 & 1\\ 1 & 0\end{pmatrix}$, and $\Gamma_4 = \begin{pmatrix}0 & 1\\ -1 & 0\end{pmatrix}$. By taking the nucleon form factors to be approximately equal, $F_1(q^2) \simeq F_2(q^2) \simeq f(q^2)$, where $f(q^2) = [1 + q_{\mu}^2/(855\ \mathrm{MeV})^2]^{-1}$; the above single-nucleon current operator may be written as

$$\hat{j}_j^{\,0}\,(q^2) = f(q^2)\left\{ e(j)\,\Gamma_2(j) + \frac{\kappa(j)}{2m_N}\,\vec{\sigma}(j)\cdot\vec{q}\;\Gamma_3(j)\right\},$$

$$\hat{\vec{J}}_j\,(q^2) = f(q^2)\left\{ e(j)\,\vec{\sigma}\,(j)\,\Gamma_4(j) + \frac{\kappa(j)}{2m_N}\left[\omega\vec{\sigma}(j)\,\Gamma_3(j)\right.\right.$$

$$\left.\left. + i\vec{\sigma}\,(j)\times\vec{q}\;\;\Gamma_1(j)\right]\right\} \qquad (13b)$$

These operators may be expressed in the coordinate-space representation by making the substitutions: $1 \to \delta\,(\vec{x}-\vec{x}_j)$, $\vec{\sigma}(j)\cdot\vec{q} \to -i\,\vec{\nabla}_x \cdot \vec{\sigma}(j)\,\delta\,(\vec{x}-\vec{x}_j)$, and $\vec{\sigma}\,(j)\times\vec{q} \to i\vec{\nabla}_x \times \vec{\sigma}\,(j)\,\delta(\vec{x}-\vec{x}_j)$. Apart from the single nucleon form factor f(q2), which will multiply all matrix elements, Eq. (13) then yields the total nuclear density and current operators

$$\hat{\rho}\,(\vec{x}) = \sum_{j=1}^{A}\left\{ e(j)\;\;\Gamma_2(j) - \frac{\kappa(j)}{2m_N}\,\vec{\nabla}\cdot\vec{\sigma}\,(j)\,\Gamma_3\,(j)\right\}\sigma(\vec{x}-\vec{x}_j)$$

and

$$\hat{\mathrm{J}}\,(\vec{x}) = \sum_{j=1}^{A}\left\{ e(j)\,\vec{\sigma}(j)\,\Gamma_4(j) + \frac{\kappa(j)}{2m_N}\left[\omega\vec{\sigma}\,(j)\,\Gamma_3(j)\right.\right.$$

$$\left.\left. - \vec{\nabla}\times\vec{\sigma}\,(j)\,\Gamma_1(j)\right]\right\}\;\delta(\vec{x}-\vec{x}_j). \qquad (14)$$

It is instructive to compare these results to the standard nonrelativistic results (on two-component Pauli space),[3]

$$\hat{\rho}_{NR}(\vec{x}) = \sum_{j=1}^{A} e(j)\ \delta(\vec{x}-\vec{x}_j) \tag{15}$$

and

$$\hat{\vec{J}}_{NR}(\vec{x}) = \sum_{j=1}^{A} \left\{ \frac{e(j)}{2im_N} \left[\vec{\nabla}_j\ \delta(\vec{x}-\vec{x}_j) + \delta(\vec{x}-\vec{x}_j)\ \vec{\nabla}_j \right] \right\} + \vec{\nabla} \times \vec{\mu}(\vec{x}),$$

where

$$\hat{\vec{\mu}}(\vec{x}) = \sum_{j=1}^{A} \frac{\left[e(j)\mu_B + \kappa(j) \right]}{2m_N}\ \vec{\sigma}(j)\ \delta(\vec{x}-\vec{x}_j).$$

The form for $\hat{J}^{\mu}$ given by Eq. (14) is consistent with minimal coupling, and leads to a conserved current if the Dirac potentials which bind the nucleons are real and the same for initial and final nucleon states. It is interesting to note that in the usual non-relativistic formulation using Schrödinger wave functions, the transition current is not conserved when the wave functions are generated from velocity dependent potentials such as the spin-orbit interaction. The problem of current conservation as applied to the model nuclear wave functions used in most calculations is generally not a concern - and it should be! How large are the effects of violating current conversation? The answer is not known at this time, but in the Dirac model, if things are calculated consistently, one may have an easier way to generate a self-consistent conserved transition current.

DIRAC-HARTREE SINGLE-PARTICLE WAVE FUNCTIONS

Thus far, I have discussed the nuclear current operator, which is to be evaluated between bound Dirac single-particle wave functions. Now I discuss the single-particle wave functions, $\Psi_\alpha(\vec{r})$ for $\alpha = n,\ell,j,m$, which are assumed to be solutions of the time-independent Dirac equation,

$$\left\{ \gamma^\circ E_\alpha - \vec{\gamma}\cdot\vec{p} - \underset{\sim}{1}\, m_N - \underset{\sim}{1}\, U_s - \gamma^\circ U_0 \right\} \Psi_\alpha(\vec{r}) = 0. \tag{16}$$

Here, I use the Bjorken and Drell metric and definition of the Dirac matrices. I also assume natural units ($\hbar = c = 1$) and I explicitly show the Dirac identity matrix ($\underset{\sim}{1}$). The potentials U_S and U_0 are assumed real, central potentials. Next, we divide Ψ_α into upper and lower components,

$$\Psi_\alpha(\vec{r}) = \frac{1}{r} \begin{pmatrix} G_\alpha(r) \\ iF_\alpha(r)\ \vec{\sigma}\cdot\hat{r} \end{pmatrix} \mathcal{Y}_\alpha(r), \tag{17}$$

where the angular function $\mathcal{Y}_\alpha(\hat{r})$ results from coupling a Pauli spin or $(\chi_{\frac{1}{2}})$ to a spherical harmonic $(Y_{\ell\mu})$, i.e.

$$\mathcal{Y}_\alpha(\hat{r}) = \left[Y_\ell(\hat{r}) \times \chi_{\frac{1}{2}}\right]_{jm} \tag{18}$$

By inserting Eq. (17) into (16), we obtain the coupled radial equations

$$\left[E - m_N - (U_s + U_o)\right] G_\alpha(r) = -\left\{\frac{d}{dr} + \frac{K_\alpha}{r}\right\} F_\alpha(r)$$

$$\left[E + m_N + (U_s - U_o)\right] F_\alpha(r) = \left\{\frac{d}{dr} + \frac{K_\alpha}{r}\right\} G_\alpha(r) \quad , \tag{19}$$

where $$K_\alpha = \begin{cases} -(\ell+1), & j=\ell+\frac{1}{2} \\ +\ell, & j=\ell-\frac{1}{2} \end{cases} \quad .$$

From these equations, F_α can be related to G_α, i.e.

$$F_\alpha(r) = \frac{1}{\left[E + m_N + (U_s(r) - U_o(r))\right]}\left\{\frac{d}{dr} + \frac{K_\alpha}{r}\right\} G_\alpha^{(NR)}(r) \tag{21}$$

To a good approximation, the Relativistic and Non-Relativistic upper components can be equated, $G_\alpha \simeq G_\alpha^{(NR)}$, and we obtain the approximate relation for the lower components,

$$F_\alpha(r) \simeq \left[\frac{E + m_N}{E + m_N + (U_s - U_o)}\right] F_\alpha^{(NR)}(r) \quad . \tag{22}$$

Now by inserting some typical values for the scalar and vector potentials, $U_s(r{=}o) \simeq -450$ MeV and $U_o(r{=}o) \simeq +350$ MeV, we see inside nuclei $F_\alpha \simeq 1.7 F_\alpha^{(NR)}$. Thus, we see that one of the main implications of using the Dirac-Hartree approach is an enhancement of the small components. This enhancement is a direct consequence of the large and cancelling potentials, which characterize all of the relativistic models and are directly related to the magnitude of the nucleon-nucleus (N-A) spin-orbit interaction. The enhancement of the small components also implies an enhance of the N-A spin-orbit interaction in the nuclear interior, which has been shown to be partly responsible for the success of the Dirac Relativistic Impulse Approximation.[4]

In Figs. (3a) and (3b), I show typical Dirac single-particle wave functions and densities for ^{16}O. These results are from the work of Bouyssy, Marcos, and Mathiot.[5] The scalar, vector, and tensor Dirac densities are obtained by summing over different combinations of single-particle wave functions, i.e.

$$\rho_{s(v)}(r) = \frac{1}{4\pi} \sum_{\alpha} (2j+1)\,(G_{\alpha}^{2} \;(\mp)\; F_{\alpha}^{2})\, r^{-2},$$

$$\rho_{T} = \frac{1}{2\pi} \sum_{\alpha}(2j+1)\, F_{\alpha}\, G_{\alpha}\, r^{-2} \qquad (23)$$

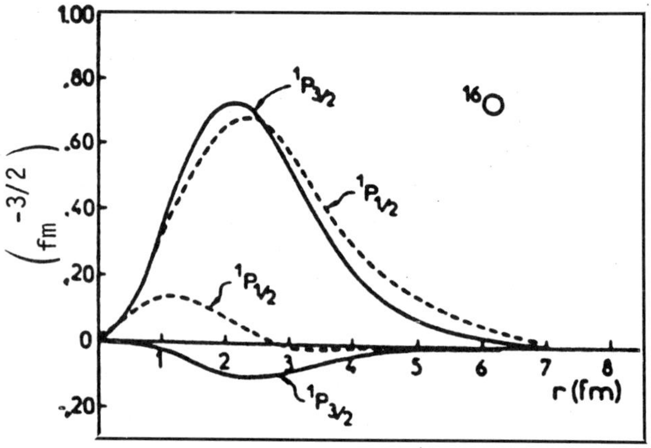

FIG. 3a. Dirac single-particle wave functions for ^{16}O. The upper curves are for the large components (G) and the lower curves are for the small components (F). The figure is from Ref. (5).

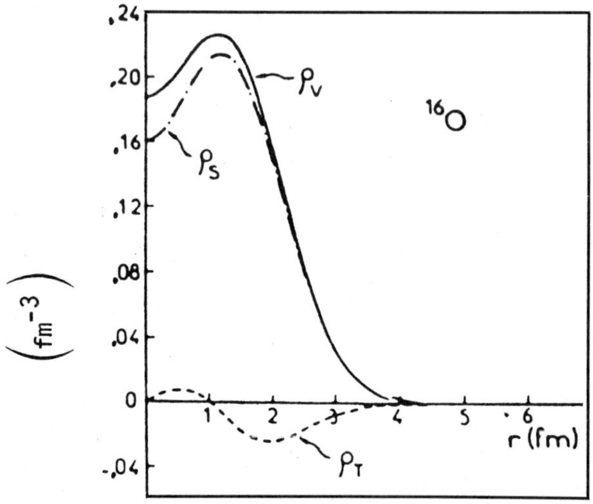

FIG. 3b. Dirac scalar, vector, and tensor densities from the single-particle wave functions of Ref. (5).

The difference between vector and scalar densities indicates the contribution from the small components, and from Fig. (3b), we see that $\rho_V - \rho_S$ is greatest at r=0.

Now getting back to the evaluation of the nuclear current operator given by Eq.(14), we expect the largest difference between the relativistic and non-relativistic models to occur for matrix elements linear in the small components: those involving Γ_3 and Γ_4. Two of these terms are proportional to $(2m_N)^{-1}$, and thus are generally negligible. The third term is proportional to $\vec{\sigma}$ and contributes to the transverse form factor. The transverse form factor also involves the $\vec{\sigma} \times \vec{q}$ term, which connects large to large to large components and is not expected to be strongly altered by the use of Dirac wave functions. Because the $\vec{\sigma} \times \vec{q}$ term competes with the $\vec{\sigma}$ term, we need some other means of filtering out the $\vec{\sigma} \times \vec{q}$ contribution to transverse amplitudes. One way is to use isospin selectivity. Because the $\vec{\sigma} \times \vec{q}$ term is dependent on κ and the $\vec{\sigma}$ term is not, we can use the strong isospin dependence of κ, i.e.

$$\frac{\kappa_{T=1}}{\kappa_{T=0}} = \frac{\kappa_p - \kappa_n}{\kappa_p + \kappa_n} = \frac{3.70}{-.12} = -31,$$

to de-emphasize the $\vec{\sigma} \times \vec{q}$ contribution. We then reach the conclusion that the use of Dirac wave functions are likely to be most pronounced for T=0 (isoscalar) transverse amplitudes where the $\vec{\sigma} \times \vec{q}$ term is suppressed. Based upon this conclusion, I will next discuss two specific cases where electron scattering may be sensitive to the use of Dirac single-particle wave functions.

CANDIDATES FOR OBSERVING DIFFERENCES BETWEEN SCHRODINGER AND DIRAC APPROACHES

The idea is to find electron scattering processes which are dominated by the isoscalar part of the three-vector nuclear current $\hat{\vec{J}}$. Since $\vec{J}$ contributes to the transverse multipole operators [Eqs. (6) and (7)], the processes we seek involve transverse form factors. Experimentally, transverse form factors are observed by measuring cross sections near $\theta \approx 180^\circ$, so that the $\tan^2(\theta/2)$ kinematic factor multiplying F_T^2 in Eq. (1) can be used to suppress the longitudinal form factor. One such process is elastic magnetic scattering.[6] This process involves $\theta \approx 180^\circ$ elastic scattering ($|f\rangle = |i\rangle$) from $J_i \neq 0$ nuclei. The name derives from the fact that only odd multipoles of the transverse magnetic form factor contribute to the cross section [as mentioned following Eq. (7)].

As an example of a model relativistic calculation of elastic magnetic scattering, I quote an article by B. D. Serot,[7] in which ^{209}Bi was treated as a ^{208}Pb core plus a 1h9/2 valence proton. In this model, the core scalar and vector potentials result from a self-consistent Dirac-Hartree calculation of ^{208}Pb. These potentials are then used to determine the valence wave function. The results for the valence proton wave function are shown in Fig. (4).

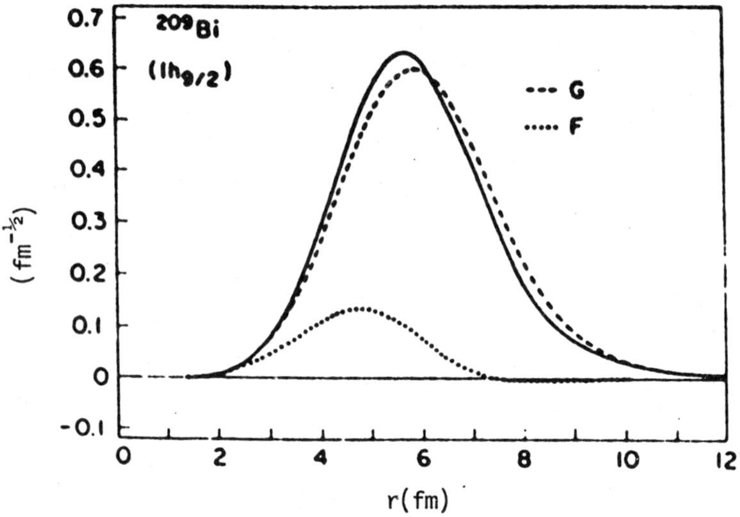

FIG. 4. Wave functions for valence proton in ^{209}Bi. The Dirac large (small) components are shown by the dashed (dotted) curves. An equivalent Schrödinger solution is shown by the solid curve. The figure is from Ref. (7).

The corresponding plane wave transverse form factors are shown in Fig. (5). Note that for $q \gtrsim 2fm^{-1}$, the Dirac and Schrödinger models yield similar form factors. This results from dominance of the M9 multipole for these q values, and the fact that the nucleon's magnetic moment dominates this multipole. At $q < 2fm^{-1}$, we see there are significant differences between the models, with the Dirac results approximately a factor of two larger than the Schrödinger results. Unfortunately, the Dirac calculations go in the wrong direction with respect to the trend of data. [It would be useful to have more data in the low q region.] This problem is apparently related to the problem of calculated magnetic

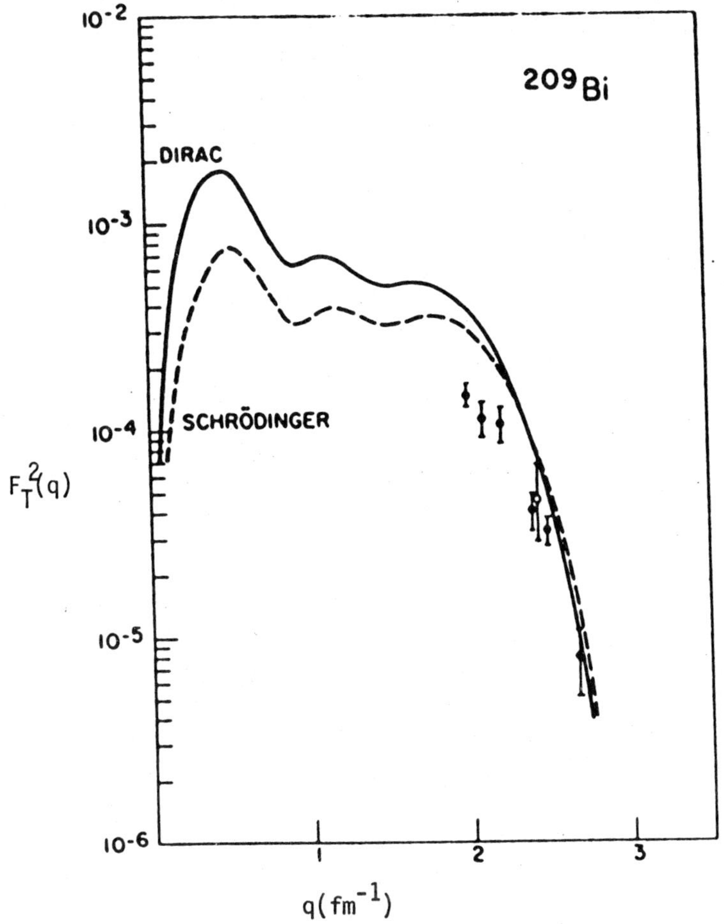

FIG. 5. Dirac (solid) and Schrödinger (dashed) transverse magnetic form factors from Ref. (6). The data are from Ref. (8).

moments, which are also too large for the Dirac scalar plus vector model.[5] Although these are model calculations, they clearly suggest that elastic magnetic scattering may be a sensitive testing place for relativistic nuclear structure effects, and perhaps in the future more data will exist for such comparisons.

Another way of selecting the isoscalar part of the nuclear current is to use inelastic scattering to specific nuclear final states. For examples of such calculations, I quote an article by J. R. Shepard et. al,[7] in which Dirac single-particle wave functions were used to calculate the excitation of 2^+, T = 0 (4.44 MeV); 1^+, T = 0 (12.71 MeV); and for contrast the 1^+, T = 1 (15.11 MeV) states in ^{12}C. The single-particle wave functions were determined by solving the Dirac equation with real scalar and vector potentials of a Saxon-Woods form. The strengths and geometries were chosen to be qualitatively consistent with low-energy Dirac proton-nucleus scattering, to reproduce the single-particle binding energies, and to be consistent with elastic electron scattering. Because of the time of these calculations there were no Dirac shell-model admixture coefficients, non-relativistic one-body spectroscopic amplitudes were used. Of course, this is <u>not</u> a <u>consistent</u> relativistic treatment of the relevant nuclear structure and caution must be exercised in drawing conclusions with respect to data. Dirac shell-model calculations are in progress, however, and in the near future consistent calculations will be possible.[10]

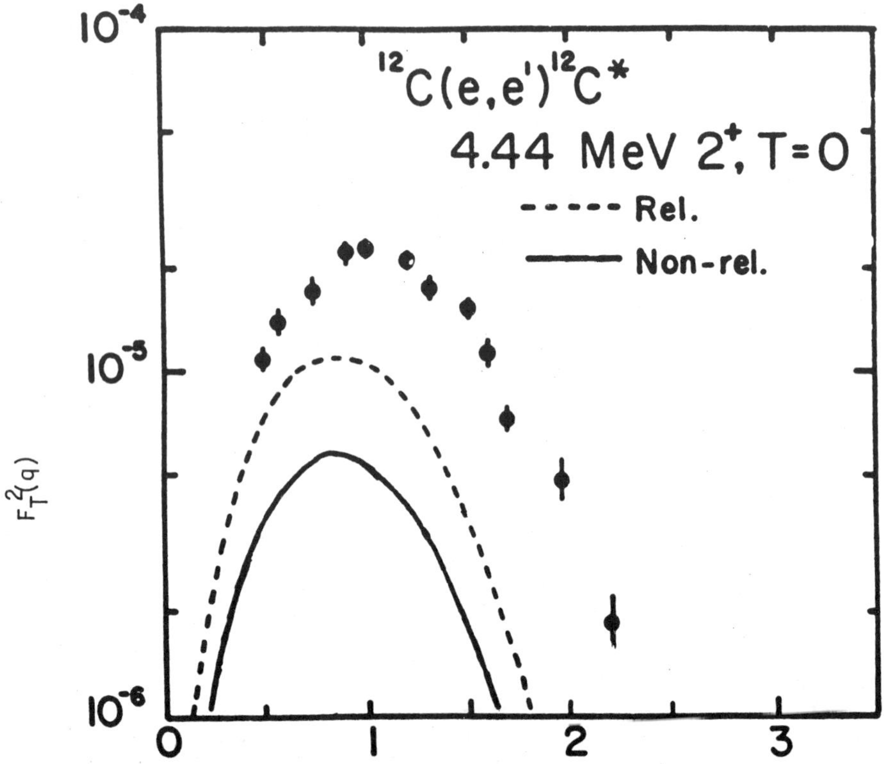

FIG. 6. Dirac (dashed) and Schrödinger (solid) form factors from Ref. (9). The data are from Ref. (13).

In Fig. (6), calculations of the transverse form factors for the 2^+, T = 0 (4.44 MeV) state in ^{12}C are shown. The non-relativistic shell-model amplitudes for this state were taken from Ref. (11). We see the form factor that results from Dirac bound-state wave functions is about a factor of two larger than the one resulting from Schrödinger wave functions. This is a direct consequence of the enhancement of the Dirac lower components and the dominance of the $\vec{\sigma}$ term for this transverse isoscalar transition.

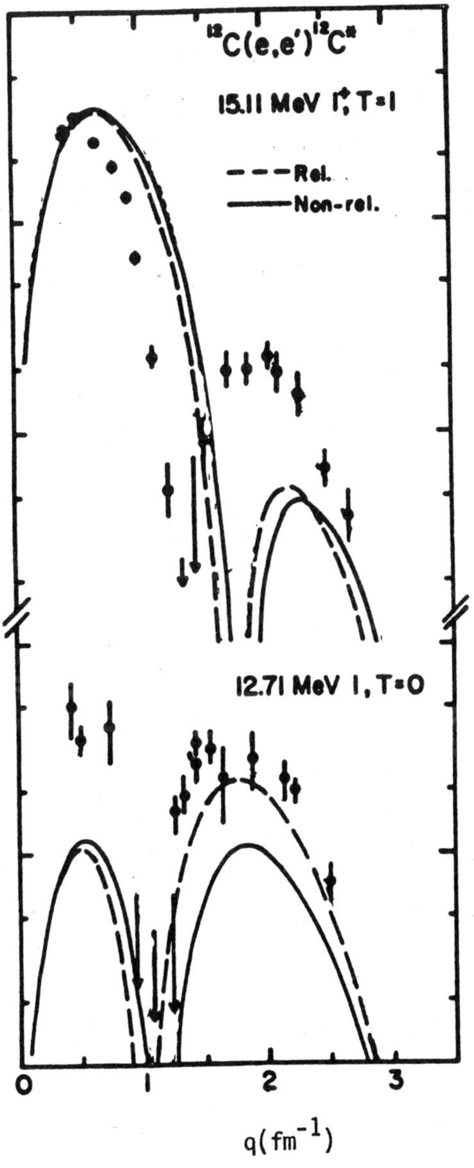

FIG. 7. Dirac (dashed) and Schrödinger (solid) form factors from Ref. (9). The data are from Ref. (13).

In Fig. (7), calculations of the transverse form factors for the 1^+, T = 0 (12.71 MeV) and 1^+, T = 1 (15.11 MeV) states in ^{12}C are shown. The non-relativistic shell-model amplitudes from Ref. (12) were used for these states. We see that there is very little difference between the use of Dirac and Schrödinger bound states for the 1^+, T = 1 states, as we expected due to the dominance of the $\vec{\sigma} \times \vec{q}$ term in the current. For the 1^+, T = 0 state, however, we see a factor of 2 enhancement around $q \sim 1.7\ fm^{-1}$. This is again a direct consequence of the Dirac lower component enhancement, because for this region of momentum transfer the $\vec{\sigma}$ term dominates the transition amplitude.

CONCLUSIONS

In this talk, I have discussed certain electron scattering processes that are expected to be sensitive to a unique feature of the Dirac-Hartree approach to nuclear structure: enhanced lower components of the bound Dirac single-particle wave functions. By considering the impulse model for the nuclear charge-current operator, it was deduced that transverse isoscalar processes which are dominated by the $\vec{\sigma}$ term of the current operator would be most likely to show differences between Schrödinger and Dirac bound nucleon wave functions. Model calculations of elastic magnetic scattering and specific inelastic transitions were shown to exhibit the expected differences. Before definite conclusions can be reached, however, consistent Dirac shell-model wave functions are needed. Work along these lines are in progress.[10]

Another issue was raised concerning the Dirac model for nuclear magnetic moments. Because the strong interactions are believed to cause the radiative corrections leading to the free nucleon's anomalous magnetic moment, it is possible that strong nuclear medium effects may renormalize the impulse model for the current. This is an important problem and needs to be addressed.

REFERENCES

1. For more discussion see the lectures by J. Dubach in these proceedings.

2. See, for example, J. D. Bjorken and S. D. Drell, Relativistic Quantum Mechanics, p. 244, McGraw-Hill, New York (1964).

3. Actually, certain relativistic corrections to the non-relativistic charge density given in Eq. (15) have been found necessary to understand the neutron contributions to extracted charge densities. For example, see the discussion of the Darwin-Foldy terms for ^{48}Ca vs. ^{40}Ca in the review article by J. Friar and J. Negele appearing in Advances in Nuclear Physics (M. Baranger and E. Vogt, eds.), Vol. 8, Plenum Press, New York (1979).

4. J. R. Shepard, J. A. McNeil, and S. J. Wallace, Phys. Rev. Lett. 50, 1443 (1984).

5. A. Bouyssy, S. Marcos, and J. F. Mathoit, Nucl. Phys. A415, 497 (1984).

6. For an up-to-date review on elastic magnetic scattering see the article by T. W. Donnelly and I. Sick, Rev. Mod. Phys. 56, 431 (1984).

7. B. D. Serot, Phys. Lett. 107B, 263 (1981).

8. P. K. A. deWitt Huberts, Proc. Conf. Modern Trends in elastic electron scattering (IKO, Amsterdam, 1978).

9. J. R. Shepard, E. Rost, E. R. Siciliano, and J. A. McNeil, Phys. Rev. C 29, 2243 (1984).

10. See the lectures by Charles Price and Michael Price in these proceedings.

11. K. Amos and I. Morrison, Phys. Rev. C19, 2108 (1979).

12. T.-S. H. Lee and D. Kurath, Phys. Rev. C 21, 293 (1980).

13. J. B. Flanz et al., Phys. Rev. Lett. 43 1922 (1979).

A RELATIVISTIC SCHEMATIC MODEL OF NUCLEAR STRUCTURE IN LIGHT NUCLEI

Charles E. Price
Indiana University
Bloomington, IN 47405

ABSTRACT

A relativistic schematic model is discussed incorporating a Lorentz invariant zero range interaction. The model is investigated using the Tamm-Dancoff and Random Phase approximations including both direct and exchange terms. Energy spectra and electromagnetic form factors in ^{16}O are presented and, the qualitative features associated with relativistic effects are outlined.

INTRODUCTION

Recently there has been considerable interest in the application of Dirac Phenomenology to the prediction of nuclear transition rates.[1,2] In order to make such predictions a relativistic theory of the nuclear response is required. This theory should include relativistic effects both in the transition operators and in the nuclear structure. For example, the effects of using a relativistic model to obtain the configuration mixed nuclear structure wave functions might negate (or enhance) the effects of using relativistic transition operators. Many applications require a relativistic model of the nuclear linear response.

For this reason we have developed a relativistic schematic model of nuclear structure using both the Tamm-Dancoff (TDA) and Random Phase (RPA) Approximations. Earlier work on non-relativistic schematic models[3] has demonstrated their utility in understanding the collectivity of certain states whose energies are considerably shifted from the unperturbed particle hole energies. In addition, a schematic model may be a simple way to determine a good starting point for a more complete calculation of relativistic nuclear structure.

RELATIVISTIC RPA EQUATIONS

The response of a system to a perturbation of the form

$$\hat{H}^{ex}(t) = \int \hat{\Psi}^{\dagger}(x)\Gamma^{ex}\hat{\Psi}(x)\hat{\Phi}(x)d^3x \quad , \qquad (1)$$

where $\hat{\Psi}$ is a nucleon field operator, $\hat{\Phi}$ is a boson field operator, and Γ represents a Lorentz coupling, is governed by the density correlation function, or polarization propagator.[I] In a finite system, the polarization propagator is defined by

$$i\Pi(x,x') = \langle\Psi_0|T\left[\hat{\Psi}^{\dagger}(x)\hat{\Psi}(x)\hat{\Psi}^{\dagger}(x')\hat{\Psi}(x')\right]|\Psi_0\rangle \qquad (2)$$

where T denotes the usual time ordered product. This propagator can be evaluated to a given order in $\hat{H}^{int}$, the interaction Hamiltonian, which is given by

$$\hat{H}^{int} = \int \hat{\Psi}^{\dagger}(x)\Gamma^{int}\hat{\Psi}(x)\hat{\Phi}(x)d^3x$$

where $\hat{\Phi}$ is now a meson field operator, by transforming the fields into the interaction representation and using Wick's Theorem. Then, the lowest order contribution is

$$i\Pi^{0}_{\lambda\mu;\alpha\beta}(\omega) = -(2\pi)^{-2}\int d\omega_1 \int d\omega_2 2\pi\delta(\omega-\omega_1+\omega_2) \qquad (3)$$

$$iG^{0}_{\lambda\alpha}(\omega_1)\,iG^{0}_{\beta\mu}(\omega_2)$$

where

$$G^{0}_{\lambda\alpha}(\omega)=\delta_{\lambda\alpha}\left[\frac{\theta(\alpha-F)}{\omega-\omega_\alpha+i\eta} + \frac{\theta(F-\alpha)}{\omega-\omega_\alpha-i\eta}\right] \qquad (4)$$

is the Green's function which arises from the contraction of two nucleon fields. The theta functions in Eq. (4) project α above or below the occupied levels

---

[I] For more detail, see A.L. Fetter and J.D. Walecka, Quantum Theory of Many Particle Systems (McGraw-Hill, New York, 1971) pp. 558-566.

and the factors of $\pm i\eta$ are included to insure proper boundary conditions. The next term in the polarization propagator is second order in $\hat{H}^{int}$ and has the form

$$i\Pi^{(2)}_{\lambda\mu;\alpha\beta}(\omega) = \frac{-i}{\hbar} \sum_{\eta\nu\rho\sigma} \frac{d\omega_1}{2\pi} \cdots \frac{d\omega_5}{2\pi} \frac{d^3k}{(2\pi)^3} \tag{5}$$

$$\langle\rho\sigma| \frac{e^{i\vec{k}\cdot(\vec{x}_1-\vec{x}_2)}}{\vec{k}^2-\omega_5^2-\mu^2+i\epsilon} |\eta\nu\rangle \Big[-iG^O_{\eta\alpha}(\omega_1)\, iG^O_{\lambda\rho}(\omega_3)$$

$$iG^O_{\nu\mu}(\omega_4)\, iG^O_{\beta\sigma}(\omega_2)\, 2\pi\delta(\omega_1-\omega_3-\omega_5)\, 2\pi\delta(\omega_4+\omega_5-\omega_2)$$

$$+iG^O_{\eta\alpha}(\omega_1)\, iG^O_{\beta\rho}(\omega_2)\, iG^O_{\nu\mu}(\omega_4)\, iG^O_{\lambda\sigma}(\omega_3)\, 2\pi\delta(\omega_1-\omega_2-\omega_5)$$

$$2\pi\delta(\omega_4-\omega_5-\omega_3)\Big]\, 2\pi\delta(\omega-\omega_1-\omega_2)$$

Here, the difference between the relativistic and non-relativistic derivations appears as an explicit frequency dependence in the factor,

$$\frac{e^{i\vec{k}\cdot(\vec{x}_1-\vec{x}_2)}}{\vec{k}^2-\omega_5^2-\mu^2+i\epsilon} . \tag{6}$$

This factor arises from the contraction of two meson field operators and without the frequency dependence in the intermediate propagator would lead to a simple Yukawa form for the spatial dependence of the interaction. After performing the integrals over the delta functions in Eq. (5), this factor has the form

$$\frac{e^{i\vec{k}\cdot(\vec{x}_1-\vec{x}_2)}}{\vec{k}^2-(\epsilon_p-\epsilon_h)^2-\mu^2+i\epsilon} . \tag{7}$$

Thus, the retardation correction is on the order of the particle-hole energy squared divided by the meson mass squared, and so, to lowest order, this correction can be neglected. With this minor approximation, the relativistic polarization propagator is identical to the non-relativistic polarization propagator, and therefore leads to the familiar RPA equations:

$$(E_o+\epsilon_a-\epsilon_b-E_n)\varphi_{JT}^{(n)}(ab)+\Sigma_{lm}\Big[v_{ab;lm}^{JT}\varphi_{JT}^{(n)}(lm) \qquad (8)$$

$$+u_{ab;lm}^{JT}\varphi_{JT}^{(n)}(lm)\Big]=0 \quad ,$$

$$(E_o-(\epsilon_a-\epsilon_b)-E_n)\varphi_{JT}^{(n)}(ab) - \Sigma_{lm}[v_{ab;lm}^{JT\,*}\varphi_{JT}^{(n)}(lm) \qquad (9)$$

$$+u_{ab;lm}^{JT\,*}\varphi_{JT}^{(n)}(lm)\Big] = 0$$

where

$$v_{ab;lm}^{JT} = -\sum_{J'T'} (2J'+1)\ (2T'+1) \begin{Bmatrix} j_m j_a J' \\ j_b j_l J \end{Bmatrix} \qquad (10)$$

$$\begin{Bmatrix} 1/2 & 1/2 & T' \\ 1/2 & 1/2 & T \end{Bmatrix} [\langle lbJ'T'|V|amJ'T'\rangle+(-1)^{T'+j_a+j_m+J'}$$

$$\langle lbJ'T'|V|maJ'T'\rangle]$$

and

$$u_{ab;lm}^{JT} = (-1)^{1-T+j_m-j_l+J}\ v_{ab;ml}^{JT} \qquad (11)$$

In Eq. (8) and Eq. (9), E_o is the energy of the ground state, E_n is the energy of the excited state, $(\epsilon_a - \epsilon_b)$ is the energy of the pure particle-hole states, $\varphi_{JT}^{(n)}$ (ab) are the admixture coefficients of pure particle-hole states which come from creating a particle-hole pair in the ground state, and $\varphi_{JT}^{(n)}$ (ab) are the admixture coefficients of pure particle-hole states which come from destroying a particle-hole pair in the ground state. Equations (10) and (11) express the particle-hole interaction in terms of the particle-particle matrix elements, where V is the particle-particle interaction.

SCHEMATIC MODEL

In the schematic model, we use a Lorentz invariant, zero-range interaction

$$V = \delta^4(x-x')\,[g_s^2+g_v^2(\gamma^\mu)^1(\gamma_\mu)^2+g_{\rho s}(\gamma^5)^1(\gamma^5)^2 \quad (12)$$

$$+g_{av}^2(\gamma^5\gamma^\mu)^1\ (\gamma^5\gamma_\mu)^2+g_T^2\ (\sigma^{\mu\nu})^1(\sigma^{\mu\nu})^2]$$

with

$$g_i^2 = g_i^2(1)+g_i^2(2)\tau_1\cdot\tau_2\ .$$

This interaction contains the five Lorentz invariants and an explicit isospin dependence. The Dirac delta function is obtained by allowing the meson mass in the intermediate propagator of Eq. (6) to go to infinity. In this limit the retardation correction discussed above is identically zero. The interaction in Eq. (12) contains ten independent coupling constants which can be treated as parameters to be determined by fits to experimental results or can be determined from some more fundamental theory of the effective meson-nucleon Lagrangian. The crucial schematic model assumption is that the radial integrals arising from matrix elements of the particle-particle interaction are independent of the specific particle-hole configuration, and thus can be removed from the sums over pure particle-hole states which appear in Eq. (8) and (9). With this assumption, the particle-hole interaction (Eq. 10) takes on the following separable form.

$$v^{JT}_{ab;lm} = -\left\{Av^{JT}_{ab}v^{JT}_{lm}+B\bar{v}^{JT}_{ab}\bar{v}^{JT}_{lm}\right\} \quad (13)$$

where

$$v^{JT}_{ab} = (-1)^{j_a+1/2}\,[(2j_a+1)(2j_b+1)]^{1/2} \quad (14)$$

$$\begin{pmatrix} j_a & j_b & J \\ -1/2 & 1/2 & 0 \end{pmatrix}\ ;$$

$$\bar{v}^{JT}_{ab} = [(2j_a+1)(2j_b+1)]^{1/2} \begin{pmatrix} j_a & j_b & J \\ 1/2 & 1/2 & -1 \end{pmatrix} . \qquad (15)$$

In Eq. (13), A and B are complicated functions of the ten coupling constants and contain a simple phase dependence on the total angular momentum, J, and total isospin, T.[3]

For simplicity, the following discussion will be restricted to the TDA equations which can be obtained from Eq. (8) and (10) by setting ϕ equal to zero; however, the important features of the discussion apply equally well in the RPA. With the separable form of the particle-hole interaction the TDA equations become

$$(E_n - E_o - (\epsilon_a - \epsilon_b))\psi^{(n)}_{JT}(ab) = - Av^{JT}_{ab}\Sigma_{lm} v^{JT}_{lm}\psi^{(n)}_{JT}(lm) \qquad (16)$$

$$-B\bar{v}^{JT}_{ab}\Sigma_{lm}\bar{v}^{JT}_{lm}\psi^{(n)}_{JT}(lm).$$

The sums over the admixture coefficients are constants and can be eliminated to give the following secular equation,

$$1 + A\,\Sigma_{ab} \frac{(v^{JT}_{ab})^2}{(E^* - \epsilon_{ab})} + B\,\Sigma_{ab} \frac{(\bar{v}^{JT}_{ab})^2}{(E^* - \epsilon_{ab})} \qquad (17)$$

$$+ \frac{AB}{2} \Sigma_{ablm} \frac{(v^{JT}_{ab}\bar{v}^{JT}_{lm} - v^{JT}_{lm}\bar{v}^{JT}_{ab})^2}{(E^* - \epsilon_{ab})(E^* - \epsilon_{lm})} = 0$$

where $E^* \equiv E_n - E_o$ and $\epsilon_{ab} \equiv \epsilon_a - \epsilon_b$.

In a non-relativistic schematic model, when the exchange term is neglected, and L-S coupling is used, the secular equation contains only the first two terms in Eq. (17). Such a secular equation can easily be solved graphically as in Fig. 1, in which the vertical lines are the unperturbed energy values and the dots mark the solutions for the energy eigenvalves.

The important feature is that the energy eigenvalves are trapped between adjacent unperturbed energies. This implies that all of the energy values are shifted in the same direction relative to the unperturbed values. Figure 2 shows a similar graph for the full secular equation, Eq. (17). Here, both because the exchange term is retained and because Dirac spinors require j-j coupling the energy eigenvalues are no longer trapped. In this case, it is possible to have more than one eigenvalue between adjacent unperturbed levels. It is also possible to shift some of the levels to higher energies and others to lower energies, which did not occur for non-relativistic models when exchange was neglected.

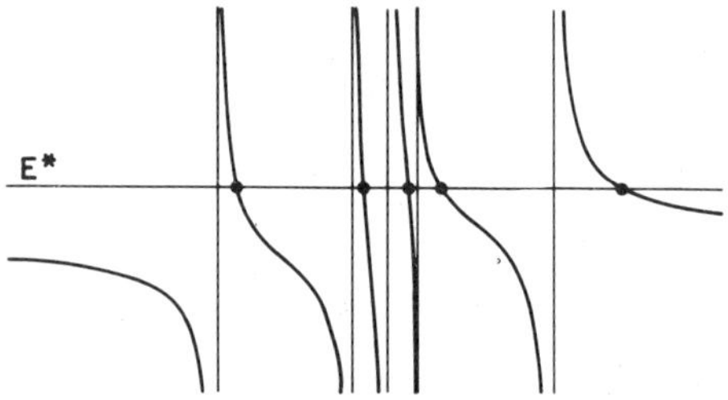

FIG. 1. Graphical solution of non-relativistic secular equation for J=T=1. Vertical lines are the unperturbed energy levels. Dots mark the energy eigenvalues.

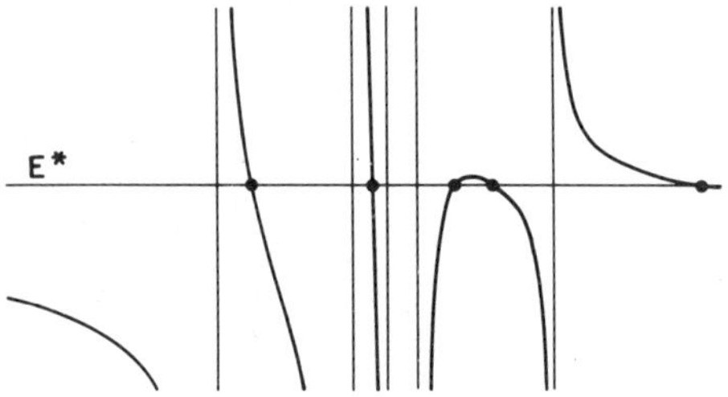

FIG. 2. Graphical solution of full secular equation for J=T=1.

ENERGY LEVELS

In order to fit the energy spectrum, it is sufficient to vary the constants A and B in Eq. (17) for each independent choice of J and T. Once A and B have been determined for each valve of J and T, the coupling constants are obtained by inverting a simple matrix.

One possible fit to the energy spectrum of ^{16}O (RPA) is shown in Fig. 3, where the stars (*) indicate the levels that were used to fix the parameters.

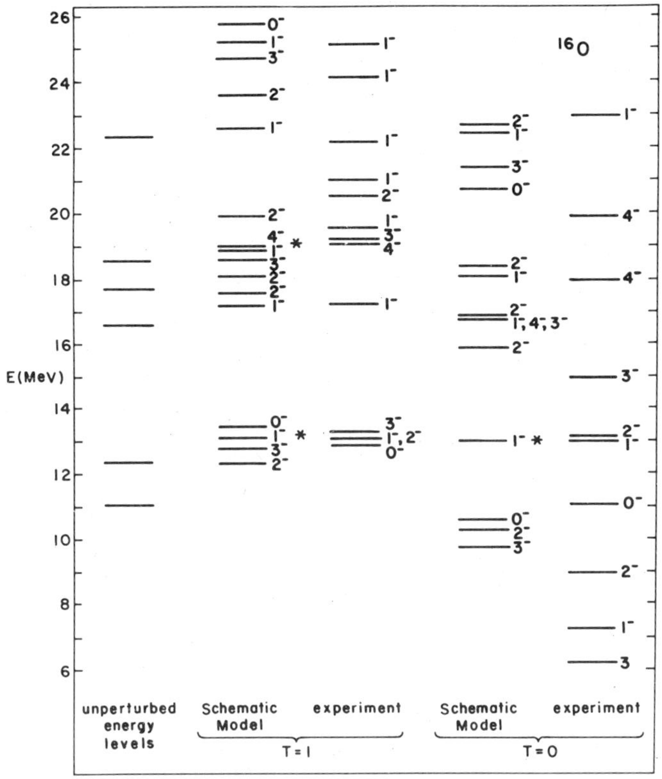

FIG. 3. Comparison of experimental energy spectrum and schematic model predictions for both T=1 and T=0 states in ^{16}O.

Qualitatively, the schematic model provides a reasonable description of the experimental energy levels, with the possible exception of the low-lying T=0 states. Unfortunately, using all possible information from particle-hole energies in ^{16}O, particle-particle energies in ^{6}Li and the electromagnetic form factors it is only possible to place three conditions on the ten coupling constants. This is because in this schematic model the appropriate equations contain only three independent combinations of the ten model parameters. So, in order to obtain a parameter set, it is necessary to specify seven of the ten parameters. Not surprisingly, this leads to a wide variety of parameter sets that give exactly the same predictions for the energy spectrum and electromagnetic transition rates. Therefore, this model is not useful for motivating a unique or reasonable parameter set to be used as a starting point in a more complete calculation.

ELECTROMAGNETIC FORM FACTORS

Using a relativistic form of the nucleon current operator,

$$j^{\mu} = \bar{\Psi}\,\gamma^{\mu}\,{}^{1}/_{2}\,(1+\tau_{3})\,\Psi + \frac{1}{2M}\,\partial_{\nu}(\bar{\Psi}\,\lambda\,\sigma^{\mu\nu}\Psi) \tag{18}$$

where λ is the anomalous moment of the nucleon, it is possible to derive a relativistic electromagnetic transition multipole expansion in terms of the following multipole operators

$$\hat{M}_{JM}(q) = \int d^{3}x\; j_{J}(qx)\; Y_{J}^{M}(\Omega_{x})\; j^{o}(x)\;, \tag{19}$$

$$\hat{T}_{JM}(q) = \int d^{3}x[\tfrac{1}{q}\,\vec{\nabla}\mathrm{x}(j_{J}(qx)\vec{Y}^{M}_{JJ1}(\Omega_{x}))]\cdot\vec{j}(x)\;, \tag{20}$$

and

$$\hat{T}^{\,mag}_{JM}(q) = \int d^{3}x[j_{J}(qx)\vec{Y}^{\,M}_{JJ1}(\Omega_{x})]\cdot\vec{j}(x) \tag{21}$$

where q is the momentum transfer and $\vec{Y}^{M}_{JJ1}$ is a vector spherical harmonic.[1] The electromagnetic cross section for electron scattering from a nucleus can be written[5]

$$\frac{d\sigma}{d\Omega} = 4\pi\sigma_{M}\left[1 + 2\,\frac{E_{e}}{M_{T}}\,\sin^{2}(\theta/2)\right]^{-1}$$

$$\times\left[\left(\frac{q_{\mu}^{2}}{q^{2}}\right)^{2} F_{L}^{2}(q) + \frac{q_{\mu}^{2}}{2q^{2}} + \tan^{2}(\theta/2)\right] F_{T}^{2}(q) \tag{22}$$

where σ_m is the Mott cross section, E_e is the incident electron energy, θ is the scattering angle and q_μ is the four-momentum transfer. $F_L^2(q)$ and $F_T^2(q)$ are the longitudinal and transverse form facters respectively, and are defined in terms of reduced matrix elements of the multipole operators,

$$F_L^2(q) = f_{sn}^2(q_\mu^2) g_{cm}^2(q^2) \sum_J |(\Psi_J || \hat{M}_J || \Psi_o)|^2 \quad (23)$$

$$F_T^2(q) = f_{sn}^2(q_\mu)^2 g_{cm}^2(q^2) \sum_J \left\{ |(\Psi_J || \hat{T}^{el}_J || \Psi_o)|^2 + |(\Psi_J || \hat{T}^{mag}_J || \Psi_o)|^2 \right\} \quad (24)$$

where f_{sn} is a single nucleon form factor and g_{cm} is a standard center of mass correction. Finally, it is convenient to define the quantity[6]

$$F^2(q) = \frac{F_L^2(q)}{{}^1/_2 + \tan^2 {}^\theta/_2} + F_T^2(q) \ , \quad (25)$$

which, for states dominated by the transverse multipoles, is relatively angle-independent.

Using the schematic model assumption that the radial integrals are independent of the specific particle-hole configurations, it is possible to estimate these form factors[3]. It is important to point out that for form factors this assumption is not well motivated, and as a result, the predicted form factors can be used only as a guide to some of the qualitative features that may be seen in a full calculation.

In Figure 4a we compare the predicted T=1 spectrum in ^{16}O to the electron scattering data of Sick et.al.[7]

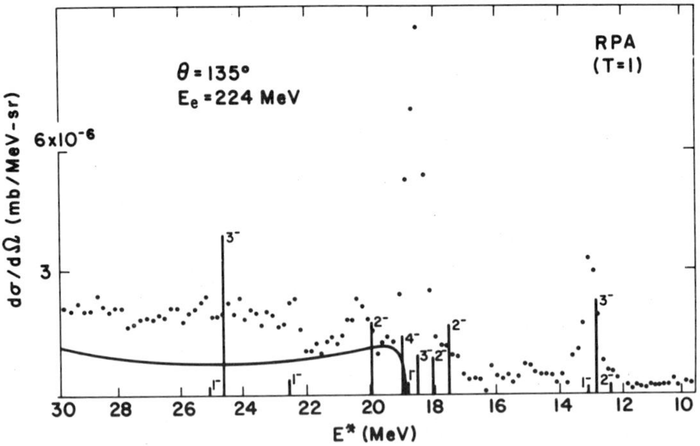

FIG. 4a. Experimental differential cross section for inelastic electron scattering at an incident electron energy of 224 MeV and a scattering angle of 135°, taken from Ref. 7. The vertical spikes are the schematic model predictions in the RPA. Only T=1 states are shown.

The major features of the data are the strongly excited complexes in the 13 and 19 MeV regions. These complexes are reasonably well reproduced in the schematic model. This model predicts a group of three states near 13 MeV; a 1^- state at 13.09 MeV, a 2^- state at 12.32 MeV and a 3^- state at 12.74 MeV. The 3^- state is dominant for this incident electron energy. In the 19 MeV region, the schematic model predicts a group of 5 states; a 1^- state at 18.79 MeV, 2^- states at 18.02 and 19.93 MeV, a 3^- state at 18.54 MeV and a 4^- state at 18.98 MeV. The strength is fairly evenly distributed among these states.

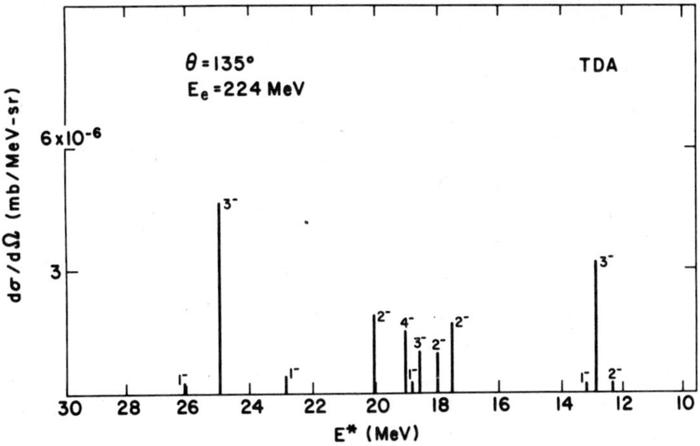

FIG 4b. The schematic model predictions in the TDA.

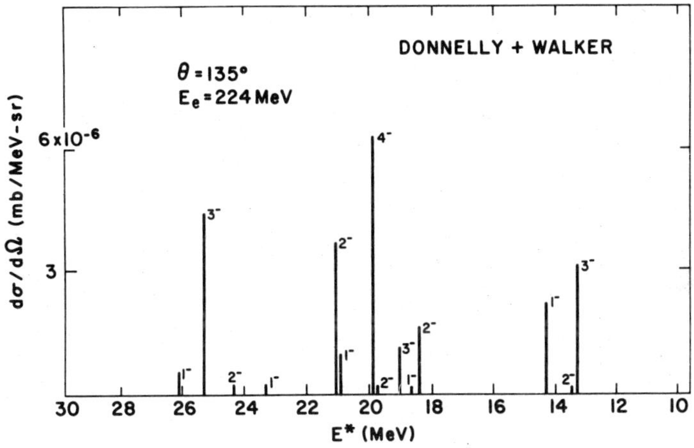

FIG. 4c. The non-relativistic calculations (TDA) of Ref. 5.

Figure 4b shows the predicted spectrum in the schematic model using the Tamm-Dancoff approximation. Clearly, the TDA results are nearly equal to the RPA predictions of Fig. 4a. In general, the TDA gives slightly larger values for the cross sections, and the high-lying states are shifted to slightly higher energies. This relationship between the TDA and RPA calculations is in agreement with non-relativistic calculations. For comparison, Fig. 4c shows the non-relativistic calculated spectrum of Donnelly and Walker.[5] The most striking feature of this comparison is the large reduction in the predicted cross sections in the schematic model, particularly for states in the region of the 19 MeV complex. Unfortunately, it is not clear whether this reduction is due to relativistic effects or is simply an artifact of the schematic model assumptions. Clearly, this is an interesting area for investigation in a more complete calculation.[8]

Figure 5 shows the predicted spectrum (RPA) for ^{16}O as a function of momentum transfer.

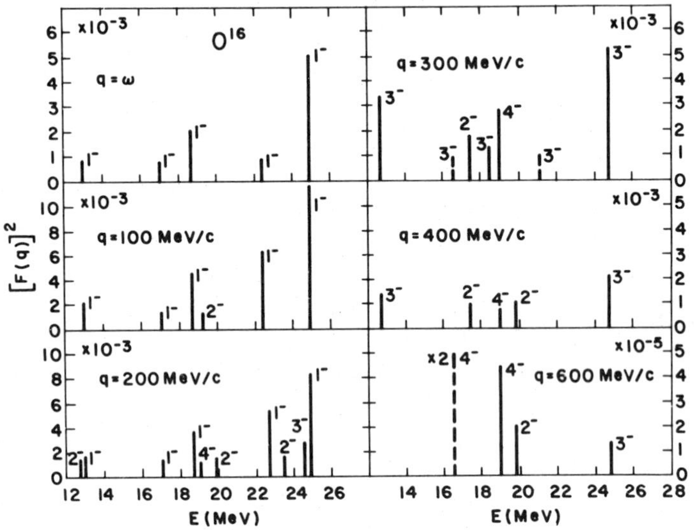

FIG. 5. The schematic model predictions for the form factor $F^2(q^2)$ (defined in Eq. 19) as a function of momentum transfer, q. The solid lines are T=1 states and the dashed lines are T=0 states.

As expected from classical angular momentum arguments, low spin states dominate the spectrum near q=0 and high spin states dominate for large q. In addition to the T=1 states, there are three significant T=0 states (dashed lines) in the predicted spectra. The enhancement of these states relative to non-relativistic calculations may be a signal of the importance of the lower components of the nuclear wave functions as predicted by Shepard et.al.[1] It is particularly interesting that the 4^- T=0 state dominates the spectrum at 600 MeV/c, since non-relativistically this state is always smaller than the 4^- T=1 state by about a factor of thirty. Again, these enhancements must be investigated in a more complete calculation.[8]

Finally, Figs. (6) and (7) show the predicted form factors for the various complexes in the electron scattering data.[7,9,10,11,12]

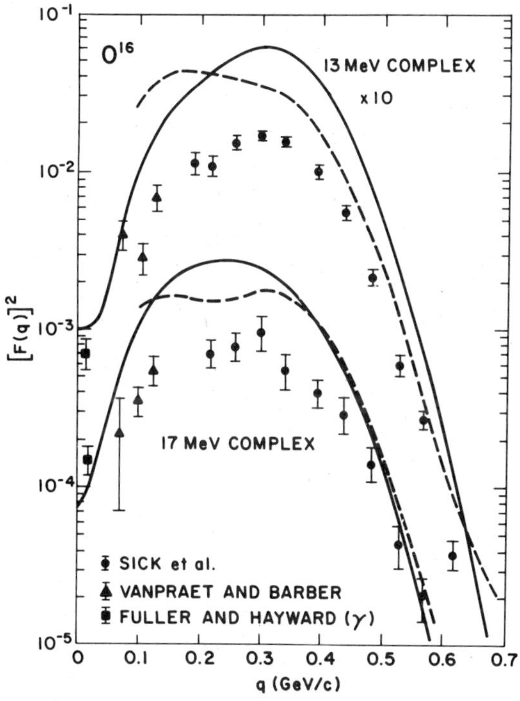

FIG. 6. $F^2(q^2)$ vs. q for the 13 and 17 MeV complexes in ^{16}O. The points with error bars are the experimental results and the solid curves are the non-relativistic

predictions of Ref. 5. The dashed curves are the schematic model predictions.

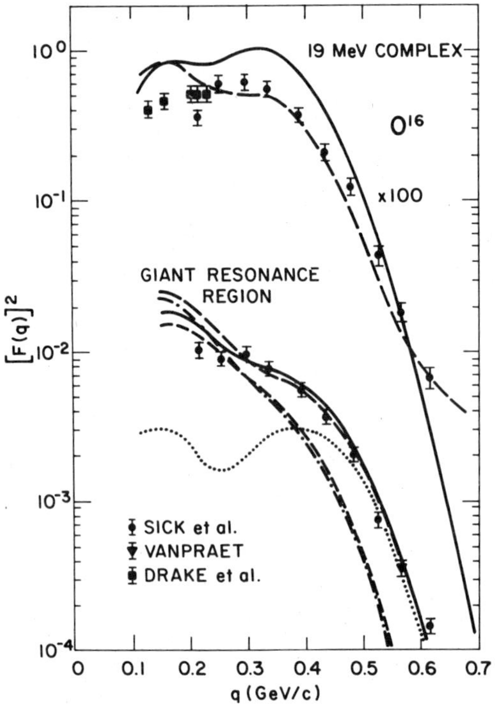

FIG. 7. $F^2(q^2)$ vs. q for the 19 MeV complex and the giant resonance region. The points with error bars are the experimental results. The solid curves are the non-relativistic results of Ref. 5, and the long-dashed curves are the schematic model predictions. In the giant resonance region the quasielastic contribution (dotted curve) was added to the model calculations (short-dashed--non-relativistic, dot-dashed--schematic) to obtain the full predictions.

The non-relativistic calculations of Donnelly and Walker[6] (solid curves) are shown for comparison. The schematic model results are somewhat different than the non-relativistic predictions, although some of this difference is undoubtably due to the schematic model assumptions. In general the schematic model predictions require approximately the same overall reduction factors as the non-relativistic calculations in order to be in agreement with the data. Although, it is not shown on either figure, the TDA and RPA schematic model predictions are essentially equal.

SUMMARY AND CONCLUSION

In conclusion, the relativistic schematic model developed here gives qualitative agreement with experimental particle-hole energy levels and reasonable predictions for electromagnetic form factors that are not radically different from previous non-relativistic predictions. Unfortunately, this model is not a viable method for determining the necessary coupling constants of the Lorentz invariant interaction potential; and, thus, cannot provide detailed input to a more sophisticated calculation of relativistic nuclear structure. However, the schematic model is very useful for identifying interesting areas for further investigation. For example, this model predicts a reduction of the states in the 19 MeV complex in ^{16}O, and an enhancement of certain T=0 states at high momentum transfer, specifically, the 4^- T=0 state at 600 MeV/c and the 3^- T=0 states at 300 MeV/c.

ACKNOWLEDGEMENTS

I wish to thank Professor G.E. Walker for numerous discussions and debates on this topic. I also wish to thank Michael Price for his help in checking the calculations. This work was supported in part by the National Science Foundation.

REFERENCES

1. J. R. Shepard, E. Rost, E. R. Siciliano, and J. A. McNeil, Phys. Rev. C29 (6), 2243 (1984).

2. B. C. Clark, S. Hama, R. L. Mercer, L. Ray, and B. D. Serot, Phys. Rev. Lett. 50 , 1644 (1983).

3. C. E. Price and G. E. Walker, "A Schematic Model Calculation of Relativistic Effects in Nuclear Structure," to be submitted for publication.

4. G. E. Brown and M. Bolsterli, Phys. Rev. Lett. 3, 472 (1959).

5. T. deForest Jr. and J. D. Walecka, Adv. in Phys. 15, 1 (1966).

6. T. W. Donnelly and G. E. Walker, Ann. of Phys. 60 (1), 209 (1970).

7. I. Sick, E. B. Hughes, T. W. Donnelly, J. D. Walecka, and G. E. Walker, Phys. Rev. Lett. 23, 1117 (1969).

8. M. W. Price, these proceedings; M. W. Price and G. E. Walker, "Studies of Nuclear Structure and the Electromagnetic Response Function using a Finite Range Dirac Residual Interaction" to be submitted for publication. (In fact, these calculations are in closer agreement with non-relativistic results.)

9. G. J. Vanpraet and W. C. Barber, Nucl. Phys. 79, 550 (1966).

10. E. G. Fuller and E. Hayward, in Nuclear Reactions (North-Holland Publishing Co., Amsterdam, The Netherlands, 1962), Vol.II, p. 113.

11. G. J. VanPraet, Nucl. Phys. 74, 219 (1965).

12. T. E. Drake, E. L. Tomusiak, and H. S. Caplan, Nucl. Phys., A118, 138, (1968).

STUDIES OF NUCLEAR STRUCTURE USING A FINITE RANGE DIRAC INTERACTION

Michael W. Price
Indiana University
Bloomington, IN 47405

ABSTRACT

Various parameterizations of a Dirac particle-hole interaction are employed to predict particle-hole four-component wave function amplitudes and excited state energy levels of ^{16}O in the RPA and TDA approximations. The predicted energy levels are compared with the experimental energy levels of ^{16}O. Comparison is also made between calculated relativistic form factors and nonrelativistic form factors for (e,e′) scattering from ^{16}O.

INTRODUCTION

The successes of relativistic nuclear dynamics in providing an accurate theoretical account of spin observables and differential cross-sections in nucleon-nucleus scattering have necessitated the existence of relativistic nuclear wave functions. An important class of such wave functions are those constructed from an appropriate linearized particle-hole (p-h) theory incorporating a Dirac p-h interaction and reduced in a Dirac single-particle basis. Heretofore, for lack of a better alternative, one has been forced to use nonrelativistic single-particle admixture amplitudes as input to Dirac phenomenological calculations of inelastic scattering utilizing *relativistic* transition operators. The possible dangers of such inconsistencies between relativistic nuclear scattering and nonrelativistic nuclear structure have been alluded to elsewhere in this workshop[1,2].

In an effort to help remove the inconsistencies between nuclear scattering and nuclear structure, we employ several parameterizations of a Dirac p-h interaction in

Tamm-Dancoff approximation (TDA) and random-phase approximation (RPA) calculations of Dirac p-h admixtures and excited state energy levels. Comparison is made
a) between predicted and experimental energy levels of excited states of ^{16}O and b) between nonrelativistic form factors for inelastic electron scattering to various levels in ^{16}O, and corresponding form factors obtained by adopting Dirac p-h admixtures and relativistic generalizations of electromagnetic transition operators.

THE TDA AND RPA METHODS IN A DIRAC SINGLE-PARTICLE BASIS

Following the notation of Fetter and Walecka[3], we consider a transition operator, in the RPA, such that the nth nuclear excited state $|\Psi^{(n)}_{JT}\rangle$, can be reached either by creating or destroying a p-h configuration from the ground state, $|\Psi^{(0)}\rangle$. Such a procedure includes, for example, the possibility of 2p-2h components in the ground state. In terms of p-h creation and annihilation operators $\hat{\zeta}^{\dagger}_{\alpha\beta}$ and $\hat{\zeta}_{\alpha\beta}$, respectively, this is equivalent to the assumption that the nth nuclear excited state can be written,

$$|\Psi^{(n)}_{JT}\rangle = \sum_{ab} \psi^{(n)\star}_{JT}(ab)\hat{\zeta}^{\dagger}(abJT)|\Psi^{(0)}\rangle + \sum_{ab}\phi^{(n)}_{JT}(ab)\hat{\zeta}(abJT)|\Psi^{(0)}\rangle , \tag{1}$$

where

$$\hat{\zeta}^{\dagger}(abJT) = \sum_{m_\alpha m_\beta} \sum_{t_{3_\alpha} t_{3_\beta}} \langle j_\alpha m_\alpha j_\beta m_\beta | j_\alpha j_\beta JM\rangle \times \langle \tfrac{1}{2} t_{3_\alpha} \tfrac{1}{2} t_{3_\beta} | \tfrac{1}{2} \tfrac{1}{2} TM_T\rangle \hat{\zeta}^{\dagger}_{\alpha\beta} , \tag{2}$$

and

$$\hat{\zeta}^{\dagger}_{\alpha\beta} = a^{\dagger}_{\alpha} b^{\dagger}_{\beta} . \tag{3}$$

In Eq. (3), a_α creates a particle state and b_β creates a hole state:

$$a^\dagger_\gamma \mid 0\rangle = \mid j_\gamma m_\gamma\rangle \qquad \gamma > \text{Fermi level} \qquad (4)$$

$$b^\dagger_\gamma \mid 0\rangle = \mid j_\gamma m_\gamma\rangle \qquad \gamma < \text{Fermi level.} \qquad (5)$$

The operator $\hat{\zeta}^\dagger(abJT)$, in Eq. (2), therefore creates a jj-coupled p-h state of total angular momentum J, and isospin T, from the particle state a, and the hole state b. In the RPA, the admixture coefficients $\psi^{(n)}_{JT}(ab)$ and $\phi^{(n)}_{JT}(ab)$ are determined by the coupled set of equations:

$$\left\{ [E_0 + (\epsilon_a - \epsilon_b)] - E_n \right\} \psi^{(n)}_{JT}(ab) \qquad (6)$$

$$+ \sum_{lm} [v^{JT}_{ab;lm} \, \psi^{(n)}_{JT}(lm) + u^{JT}_{ab;lm} \, \phi^{(n)}_{JT}(lm)] = 0$$

$$\left\{ [E_0 - (\epsilon_a - \epsilon_b)] - E_n \right\} \phi^{(n)}_{JT}(ab) \qquad (7)$$

$$-\sum_{lm} [v^{JT^\star}_{ab;lm} \, \phi^{(n)}_{JT}(lm) + u^{JT^\star}_{ab;lm} \, \psi^{(n)}_{JT}(lm)] = 0$$

where, ϵ_a and ϵ_b are single-particle energies of the particle and hole, respectively. The RPA equations employed here have the same form as in the nonrelativistic case and have been motivated elsewhere[3,4,5,6].

The p-h matrix elements $u^{JT}_{ab;lm}$ and $v^{JT}_{ab;lm}$ are given by [3,4],

$$v^{JT}_{ab;lm} = - \sum_{J'T'} (2J' + 1)\ (2T' + 1) \tag{8}$$

$$\times \begin{Bmatrix} j_m & j_a & J' \\ j_b & j_l & J \end{Bmatrix} \begin{Bmatrix} \tfrac{1}{2} & \tfrac{1}{2} & T' \\ \tfrac{1}{2} & \tfrac{1}{2} & T \end{Bmatrix} [\langle lbJ'T'|V|\, amJ'T'\rangle$$

$$- (-1)^{\frac{1}{2} + \frac{1}{2} + T'} (-1)^{j_a + j_m + J'} \langle lbJ'T'|V|maJ'T'\rangle]$$

and

$$u^{JT}_{ab;lm} = (-1)^{\frac{1}{2} - \frac{1}{2} - T} (-1)^{j_m - j_l - J}\ v^{JT}_{ab;ml} \quad . \tag{9}$$

The equations, thus far, have undergone a basis reduction in total angular momentum and isospin space since J and T are assumed to be good quantum numbers. The two-particle matrix elements in the direct and exchange terms are taken between jj-coupled Dirac-Hartree [7,8] single-particle spinors.

A somewhat cruder approximation is the TDA wherein the nuclear ground state is assumed to be a closed-shell (i.e., $|\Psi(0)\rangle = |0\rangle$ and $a_\alpha|0\rangle = b_\alpha|0\rangle = 0$) and one constructs a nuclear excited state by creating p-h states from this inert core,

$$|\Psi^{(n)}_{JT}\rangle = \sum_{ab} \psi^{(n)\star}_{JT}(ab)\ \hat{\zeta}^{\dagger}(abJT)\,|0\rangle \quad . \tag{10}$$

The admixture amplitudes, $\psi^{(n)}_{JT}(ab)$,can be determined from the TDA equations,[3,4] which result from setting ϕ=0 in the RPA equations.

The usual procedure in either approximation is to adopt a form for the p-h interaction. After evaluating the p-h matrix elements and inserting numerical values (see below) for the single-particle and hole energies, one obtains a matrix eigenvalue problem for the nuclear excited state energies and amplitudes $\psi^{(n)}_{JT}$ and $\phi^{(n)}_{JT}$ in the RPA and $\psi^{(n)}_{JT}$ in the TDA.

The single-particle and hole energies can either be obtained from nature by using binding energies in neighboring nuclei, or perhaps more consistently from the Hartree-Fock single-particle binding energies. A relativistic Hartree-Fock theory does not exist to date, therefore we choose, when adopting Dirac-Hartree single-particle wave functions, to use binding energies obtained from neighboring nuclei.

The form adopted for the p-h interaction, as in the nonrelativistic case, is based largely on one's prejudices regarding the nucleon-nucleon interaction in the nuclear medium. One can (and we think should) for example, regard the p-h interaction as an iterated interaction in the nuclear medium (a G-matrix[†]). On the other hand, one could adopt the approximation that higher-order iterations of the meson-fermion interaction Hamiltonian are unimportant so that a lowest-order one boson exchange "potential" would suffice. We discuss both viewpoints below.

PARAMETERIZATIONS OF THE PARTICLE-HOLE INTERACTION

The p-h interactions considered in this paper can each be cast in a general form consisting of a sum of five Lorentz invariants. Each invariant is multiplied by a finite range

---

†If a relativistic form for the G-matrix is not available then one can study a parameterized effective form.

interaction of the Yukawa type and dependent on the total isospin of the state through a factor $(A + B\,\vec{\tau}_1 \cdot \vec{\tau}_2)$,

$$V(1,2) = V_S \frac{e^{-\mu_S r_{12}}}{\mu_S r_{12}} (A_S + B_S\,\vec{\tau}_1 \cdot \vec{\tau}_2)$$

$$+ V_V \frac{e^{-\mu_V r_{12}}}{\mu_V r_{12}} (A_V + B_V\,\vec{\tau}_1 \cdot \vec{\tau}_2)\, \gamma^{\mu} \gamma_{\mu}$$

$$+ V_{PS} \frac{e^{-\mu_{ps} r_{12}}}{\mu_{ps} r_{12}} (A_{PS} + B_{PS}\,\vec{\tau}_1 \cdot \vec{\tau}_2)\, \gamma^5 \gamma_5$$

$$+ V_{AV} \frac{e^{-\mu_{av} r_{12}}}{\mu_{av} r_{12}} (A_{AV} + B_{AV}\,\vec{\tau}_1 \cdot \vec{\tau}_2)\, \gamma^5 \gamma^{\mu} \gamma_5 \gamma_{\mu}$$

$$+ V_T \frac{e^{-\mu_t r_{12}}}{\mu_t r_{12}} (A_T + B_T\,\vec{\tau}_1 \cdot \vec{\tau}_2)\, \sigma^{\mu\nu} \sigma_{\mu\nu} \,. \qquad (11)$$

In Eq. (11), the conventions and notations for the γ-matrix operators are those of reference 9. In what follows, we consider parameterizations based on a) QHD-I,[8] a mean field theory involving neutral scalar and vector mesons (σ and ω); b) a Finite-Hartree[7,8] model involving σ, ω and ρ exchange; and finally c) a chi-squared parameterization of the strengths of all five bilinear forms with assumed values for the ranges. The parameters used in each case are shown in Table I. All strengths are in MeV and ranges in F^{-1}. Values for coupling constants are shown where applicable in parentheses. In the parameterizations based on Finite-Hartree, the Compton wave number of the ρ (3.90 F^{-1}) is taken to be equal to that of the ω (3.97 F^{-1}).

In c) above, the chi-squared parameterization consisted of minimizing χ^2 defined as,

$$\chi^2 = \frac{1}{N} \sum_{i=1}^{N} (E_i^{exp} - E_i^{theo})^2 \quad , \tag{12}$$

TABLE I

Particle-Hole Interaction Parameters.

	QHD-I	Finite-Hartree	χ^2-Parameterization
V_S (MeV)	-3,799 (91.64)	-4,544 (109.6)	-535.7
A_S	1	1	.5789
B_S	0	0	-.4211
μ_s (F^{-1})	2.64	2.64	2.64
V_V	8,491 (136.2)	12,870	4,246
A_V	1	.9224 (190.4)	.9100
B_V	0	.0776 (65.23)	.08997
μ_v	3.97	3.97	3.97
V_{PS}	---	---	-426.3
A_{PS}	---	---	.9419
B_{PS}	---	---	.0581
μ_{ps}	---	---	.699
V_{AV}	---	---	-7,810
A_{AV}	---	---	.7119
B_{AV}	---	---	.2881
μ_{av}	---	---	5.00
V_T	---	---	-1,086
A_T	---	---	.9916
B_T	---	---	.0084
μ_t	---	---	5.00

where E_i^{exp} is the ith chosen experimental energy level of ^{16}O and E_i^{theo} is the corresponding energy eigenvalue computed in the TDA by independently varying the strengths (both isospin dependent and independent) of each term in the invariant p-h interaction Eq. (11). The selected experimental energy levels[10] and the resulting TDA energy eigenvalues are shown in Table II. The value of χ^2 corresponding to the parameter set shown in Table I and the energy eigenvalues in Table II is .526 MeV^2.

TABLE II

Selected experimental energy levels of ^{16}O for χ^2- parameterization.

J^π	T	E^{exp} (MeV)	E^{theo} (MeV)
3^-	0	6.13043	6.37
0^-	0	10.957	11.1
0^-	1	12.797	12.4
2^-	1	12.9685	10.8
1^-	1	13.090	12.8
3^-	1	13.258	13.1
1^-	1	17.1	17.1
4^-	0	17.788	17.6
9^-	1	18.975	18.7
3^-	1	19.206	19.2

ELECTROMAGENTIC TRANSITION OPERATORS

In addition to predicting energy levels of nuclear excited states, the admixture amplitudes resulting from a p-h model calculation allow a prediction of form factors for inelastic scattering to these states. Inelastic electron scattering, for example, is an excellent process for elucidating nuclear structure. The electromagnetic one-body operator $\hat{J}^{\mu}(x)$ (Eq. (19)), links the nuclear ground state to the excited states studied in the TDA and RPA. As discussed in this workshop[11] and elsewhere,[12,13] the differential cross-section for inelastic electron scattering may be written,

$$\frac{d\sigma}{d\Omega} = 4\pi\sigma_M \left[1 + (2\epsilon \, \mathrm{Sin}^2\theta/2)/M_T\right]^{-1} F^2 \tag{13}$$

where, $F^2 \equiv (q_\mu^2/q^2)^2 F_L^2 +$

$$\left[\frac{1}{2}(q_\mu^2/q^2) + \mathrm{Tan}^2\theta/2\right] F_T^2 \tag{14}$$

$$F_L^2 = (2J_i+1)^{-1} \sum_{J=0}^{\infty} |\langle J_f || \hat{M}_J(q) || J_i\rangle|^2 \tag{15}$$

$$F_T^2 = (2J_i+1)^{-1} \sum_{J=1}^{\infty} \left\{ |\langle J_f || \hat{T}_J^{el}(q) || J_i\rangle|^2 + |\langle J_f || \hat{T}_J^{mag}(q) || J_i\rangle|^2 \right\}. \tag{16}$$

The incident energy of the electron is ϵ, the scattering angle is θ, and the mass of the target is M_T. The Mott cross-section is

$$\sigma_M = \left[\alpha \operatorname{Cos}\frac{\theta}{2} / \left(2\,\epsilon \operatorname{Sin}^2\frac{\theta}{2}\right)\right]^2 . \tag{17}$$

This result, from reference 13, uses the convention $q_\mu = (\vec{q}, -i\,\omega)$ $(\omega = E_f - E_i)$. It can be shown [3] that an arbitrary irreducible one-body tensor operator $\hat{T}_{JM}$ of rank J in the nuclear Hilbert space has reduced matrix elements between initial and final states given by,

$$\langle J_f || \hat{T}_J || J_i \rangle = \sum_{ab} \Big\{ \langle a || T_J || b \rangle \, \psi_J(ab) + (-1)^{j_b - j_a - J} \langle b || T_J || a \rangle \, \phi_J(ab) \Big\} . \tag{18}$$

In the present calculation, the multipole operators $M_J(q)$, $T_J^{el}(q)$ and $T_J^{mag}(q)$ operate in the space of single-particle Dirac spinors. These operators may be derived [8,14,15] from a nuclear current in a manner analogous to the nonrelativistic treatment[12].

From a nuclear current [8,14] including the anomalous magnetic moment λ,

$$\hat{J}^{\mu}(x) = \hat{\bar{\psi}}(x)\,\gamma^{\mu}\,\frac{1}{2}(1+\tau_3)\,\hat{\psi}(x) + \partial_\nu\left(\hat{\bar{\psi}}(x)\,\frac{\lambda}{2M}\,\sigma^{\mu\nu}\,\hat{\psi}(x)\right) , \tag{19}$$

where the $\psi(x)$ are Heisenberg field operators, and the general expressions for the multipole operators,

$$\hat{M}_{JM}(q) \equiv \int d\vec{x}\; j_J(qx)\, Y_{JM}(\hat{x})\, \hat{J}_0(\vec{x}) \tag{20}$$

$$\hat{T}^{el}_{JM}(q) \equiv \frac{1}{q}\int d\vec{x}\left\{\vec{\nabla} \times j_J(qx)\,\vec{Y}^{M}_{JJ1}(\hat{x})\right\}\cdot$$

$$\hat{\vec{J}}(\vec{x}) \tag{21}$$

$$\hat{T}^{mag}_{JM}(q) \equiv \int d\vec{x}\; j_J(qx)\,\vec{Y}^{M}_{JJ1}(\hat{x})\cdot\hat{\vec{J}}(\vec{x}), \tag{22}$$

one obtains the following forms for the Dirac multipole operators[14,15],

$$T^{el}_{JM}(q) = \begin{pmatrix} q\lambda'\Sigma^{M}_{J}(q\vec{x}) & i(Q+\lambda'\omega)\,\Sigma'^{M}_{J}(q\vec{x}) \\ i(Q-\lambda'\omega)\,\Sigma'^{M}_{J}(q\vec{x}) & -q\lambda'\Sigma^{M}_{J}(q\vec{x}) \end{pmatrix} \tag{23}$$

$$T^{mag}_{JM}(q) = \begin{pmatrix} iq\lambda'\Sigma'^{M}_{J}(q\vec{x}) & (Q+\lambda'\omega)\,\Sigma^{M}_{J}(q\vec{x}) \\ (Q-\lambda'\omega)\Sigma^{M}_{J}(q\vec{x}) & -i\,q\lambda'\,\Sigma'^{M}_{J}(q\vec{x}) \end{pmatrix} \tag{24}$$

$$M_{JM}(q) = \begin{pmatrix} QM^{M}_{J}(q\vec{x}) & -i\lambda' q\Sigma''^{M}_{J}(q\vec{x}) \\ i\lambda' q\Sigma''^{M}_{J}(q\vec{x}) & QM^{M}_{J}(q\vec{x}) \end{pmatrix}. \tag{25}$$

In Eqs. (23)-(25), $\lambda' \equiv \lambda/2M$ and the following abbreviated notations [16] have been used:

$$Q \equiv \frac{1}{2}(1+\tau_3) \tag{26}$$

$$\omega \equiv E_f - E_i \tag{27}$$

$$M^M_J(q\vec{x}) \equiv j_J(qx)\, Y_{JM}(\hat{x}) \tag{28}$$

$$\vec{M}^M_{JJ}(q\vec{x}) \equiv j_J(qx)\, \vec{Y}^M_{JJ1}(\hat{x}) \tag{29}$$

$$\Sigma^M_J(q\vec{x}) \equiv \vec{M}^M_{JJ}(q\vec{x}) . \vec{\sigma} \tag{30}$$

$$\Sigma'^M_J(q\vec{x}) \equiv -i\left\{\frac{1}{q}\vec{\nabla}\times\vec{M}^M_{JJ}(q\vec{x})\right\}.\vec{\sigma} \tag{31}$$

$$\Sigma''^M_J(q\vec{x}) \equiv \frac{1}{q}\vec{\nabla}M^M_J(q\vec{x})\cdot\vec{\sigma} . \tag{32}$$

It is important to note that transition operators such as the Dirac multipole operators in Eqs. (23) - (25), (or for example, the weak interaction transition operator involving additional γ_5 matrix structure) will include some contributions to the total transition strength, that arise from matrix elements connecting upper and lower Dirac spinors.

In the present results, we plot for the sake of comparison, $\mathcal{F}^2(q)$, defined as

$$\mathcal{F}^2(q) \equiv \frac{F(q)^2}{[1+\mathrm{Tan}^2\theta/2]} \tag{33}$$

We also include in $F^2(q)$, (see Eq. (14)), single-nucleon form factors [12,17], $f^2_{SN}(q^2_\mu)$, and center-of-mass correction factors [12,17], $g^2_{CM}(q^2)$:

$$f_{SN}(q^2_\mu) = [1+q^2_\mu/q^2_N]^{-2},\ q_N = 855\ \frac{\mathrm{MeV}}{c} \tag{34}$$

$$g_{CM}(q^2) = e^{y/A} \ , \ y = (\tfrac{1}{2}bq)^2 \tag{35}$$

In Eq. (35), A is the atomic mass number and b is the oscillator constant of a nonrelativistic shell model. We choose [17] b=1.77 F for ^{16}O.

RESULTS AND CONCLUSIONS

Energy Levels

The calculated energy levels representing "1 $\hbar\omega$" transitions from the ground state to T=1 and T=0 excited states in ^{16}O are shown in Figs. 1-6. In each figure, the unperturbed p-h levels are shown to the far left and the experimental levels believed to contain important p-h amplitudes are shown to the far right. The unperturbed levels are those determined from binding energies in neighboring nuclei unless otherwise specified (for example in Figs. 5 and 6, unperturbed levels are determined from a relativistic Hartree calculation).

Possible sources of error in the fitting of energy levels include attempts to fit levels in the TDA or RPA that are not p-h in nature; or the χ^2-fitting of such levels in the case where a G-matrix is being parameterized. Another uncertainty associated with a χ^2-parameterization of the G-matrix is the assignment of ranges in each Lorentz invariant term. Our calculations indicate that some p-h levels are greatly affected by the choice of range (for a fixed strength) particularly in the Lorentz scalar and vector terms. This needs to be systematically studied. As discussed earlier there is also some uncertainty associated with the source of the single-particle and hole energies. Two possible sources are binding energies in neighboring nuclei with one more or one less neutron or the computed single-particle binding energies in a Finite-Hartree calculation. The results of RPA calculations using each approach are shown and discussed below. One remaining uncertainty arises from the treatment of the continuum states. Our calculations, in fact, use a basis restricted to "1 $\hbar\omega$" transitions so that transitions to continuum states are prohibited, resulting in the production of fewer states in a given spectrum. The effect of continuum transitions is discussed below.

In Figs. 1 and 2, we show the RPA levels corresponding to parameterizations of the p-h interaction based on a) QHD-I parameters; b) Finite-Hartree model parameters; and c) parameters resulting from a minimization of chi-squared (Eq. (12)) calculated in the TDA. The parameter values in each case are shown in Table I.

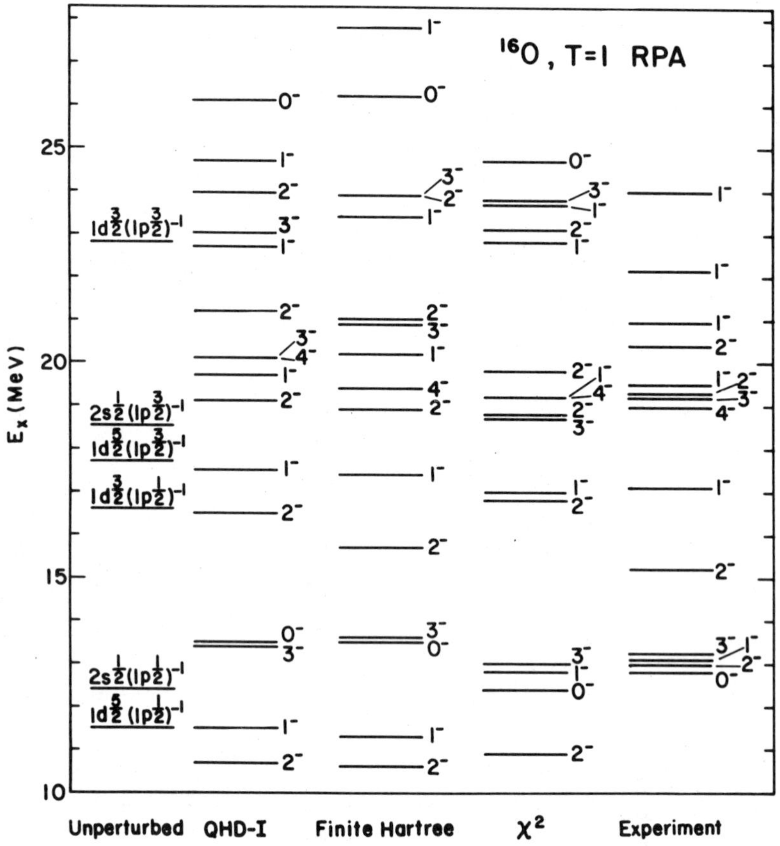

Fig. 1. T=1 RPA energy levels of ^{16}O.

The T=1 levels resulting from a QHD-I parameterization (Fig. 1) that are within roughly 1 MeV of the experimental

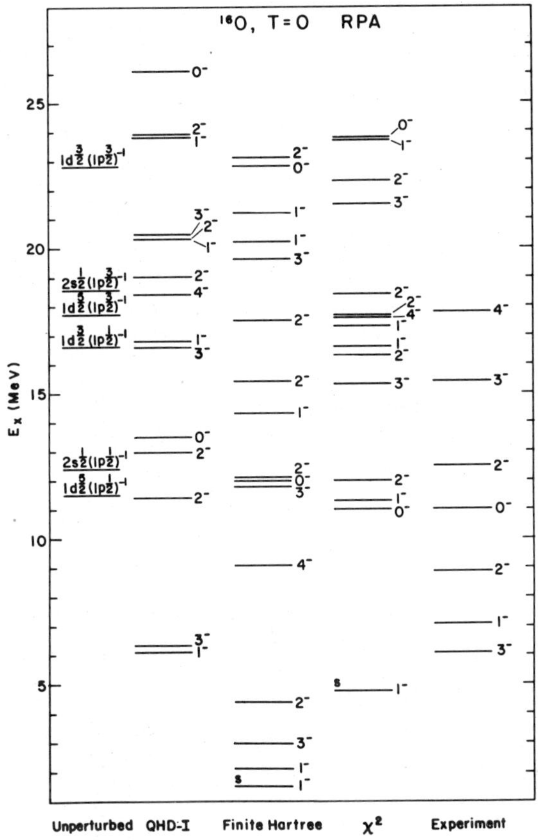

Fig. 2. T=0 RPA energy levels of ^{16}O. Spurious levels are labeled with an S.

levels include the 1^- dipole† state at 22.7 MeV; the 1^- dipole spin-flip† state at 24.7 MeV; the 2^- state at 21.2 MeV; the 3^- and 4^- states at 20.1 MeV; and finally in this energy region, we find a 1^- state at 19.7 MeV and a 2^- state at 19.1 MeV. The QHD-I model residual interaction is also successful in predicting within 1 MeV of experiment the 1^- state at 17.5 MeV and the lowest-lying 0^- and 3^- states at about 13.5 MeV. The 2^- state seen experimentally at 15.2 MeV is placed by the QHD-I parameterized RPA at 16.5 MeV making it slightly more than 1 MeV too high in excitation energy. The lowest-lying 1^- and 2^- levels, on the other hand, are pushed too far down to agree with experiment. Although these lowest 1^- and 2^- levels have the right ordering, they are approximately 2 MeV lower than their experimental values each at $E_x \approx 13$ MeV. The highest-lying 3^- state has $E_x = 23.1$ MeV and the highest-lying 0^- state is predicted at about 26 MeV.

The Finite-Hartree parameters improve the T=1 levels below 18 MeV relative to the QHD-I results by interchanging the 0^- and 3^- levels at $E_x \approx 13.5$ MeV making the 3^- highest, and by predicting a 2^- level at $E_x = 15.7$ MeV. The remaining levels below 18 MeV have values that lie within .2 MeV of the corresponding QHD-I predictions. Hence the lowest-lying 1^- and 2^- states are still roughly 2 MeV too low. The nonnormal parity states in the 19-22 MeV range are shifted slightly downward relative to the QHD-I RPA results giving rise to a noticeable improvement for the 4^-

---

†Here and in what follows, we identify the 1^- states in the giant resonance region as dipole or dipole spin-flip according to the relative magnitudes for each state of the form factor squared at low momentum transfer. We thus identify, as in the nonrelativistic theory[17], the 1^- state giving the <u>dominant</u> form factor squared at low q as the <u>dipole</u> state.

state which is then within .4 MeV of the experimental level at 19 MeV. The normal parity states in this energy range are shifted upward relative to the QHD-I predictions worsening the fit to the experimental 3^- level at 19.2 MeV by nearly 1 MeV and the fit to the experimental 1^- level at 19.5 MeV by .5 MeV. Of the states above 22 MeV, the 0^- and 2^- are predicted by the Finite-Hartree parameterization to be at virtually the same levels of excitation as found in the QHD-I case; the highest-lying 3^- state is predicted at about 24 MeV.

The fits to the 1^- dipole and dipole spin-flip levels are worsened with the Finite-Hartree parameterization relative to that from QHD-I. The most dramatic difference appears in the dipole spin-flip state predicted at 27.8 MeV, nearly 3 MeV higher still than that from QHD-I. The experimental 1^- dipole state at 22.2 MeV is predicted to have $E_x = 23.4$ MeV, slightly more than 1 MeV too high in excitation energy.

The use of parameters determined from a minimization of χ^2 (Eq. (12)) calculated in the TDA, results in a greatly improved RPA T=1 spectrum of ^{16}O. With the χ^2-parameterization, the clustering of levels into complexes at 13, 17, and 19 MeV begins to take place. Moreover, the orderings of the levels within the clusters are reasonably well determined, especially in the 13 MeV complex. With the exception of the two lowest 2^- states, all levels fall easily within 1 MeV of experiment, where it exists. The χ^2-parameterized p-h interaction places the lowest 2^- level at $E_x \approx 11$ MeV and the second lowest 2^- at $E_x \approx 17$ MeV. These excitation energies are 2 MeV too low and too high, respectively, when compared to experiment (see Fig. 1). This is a serious flaw of this particular parameterization and is also found in the TDA calculation. At the same time predictions of the 1^- dipole and dipole spin-flip levels are greatly improved relative to the QHD-I and Finite-Hartree

RPA predictions. The dipole state is predicted at 22.8 MeV and the dipole spin-flip at 23.7 MeV to be compared with the experimental values of 22.3 MeV and 24 MeV, respectively.

The calculated T=0 RPA levels of ^{16}O are shown in Fig. 2. The lowest-lying 1^-, T=0 state is a spurious level corresponding to nuclear center-of-mass motion, which in general mixes with the remaining levels. It has been shown[18] nonrelativistically that such a spurious level should have a zero energy eigenvalue in a self-consistent RPA calculation derived from a Hartree-Fock (HF) formalism. A self-consistent RPA calculation such as this would of course utilize the computed HF single-particle binding energies and wave functions as well as a residual p-h interaction that results from the full HF Hamiltonian. It should be noted that this result of zero energy excitation for the spurious 1^-, T=0 holds only for the RPA. The present calculations use p-h interactions that are at best based on a relativistic Hartree formalism. Furthermore, except where comparison is made to other work, we choose to use single-particle binding energies from neighboring nuclei instead of using the relativistic Hartree single-particle binding energies.

Using a QHD-I parameterization of the residual p-h interaction, we in fact find an imaginary energy eigenvalue, $E_x = 4.9i$ for the lowest 1^-, T=0 state. The second lowest 1^- and lowest 3^- levels are roughly within 1 MeV of experiment. The lowest 0^- level, experimentally at 11 MeV, is predicted to lie at about 13.5 MeV. Similarly in this energy range, the lowest 2^- level is predicted to have an excitation energy that is about 2 MeV higher than the experimental result at about 9 MeV. The next 2^- level, on the other hand, is within .5 MeV of the experimental 2^- level at 12.5 MeV. The prediction for the 3^- level at 15.4 MeV is roughly 1 MeV too high at 16.6 MeV, while that for the 4^- level is much closer to experiment at 18.4 MeV. Note that the agreement found for the 4^- level from a QHD-I parameterization is destroyed by the introduction of the charged ρ meson as included in the Finite-Hartree case.

The utilization of Finite-Hartree parameters results in placing the spurious level at 1.5 MeV in contrast to the QHD-I imaginary result. The next 1^- level which should lie at about 7 MeV is predicted 5 MeV too low at 2 MeV. A spectrum which is in fact lower overall than that found from QHD-I parameters appears to be the trend for the remaining Finite-Hartree T=0 levels. The two lowest 3^- levels, for example, drop roughly 4 MeV from the QHD-I values resulting in a much less acceptable spectrum when compared to experiment. The most dramatic case of lowered excitation is that for the 4^- state predicted to be about 9 MeV lower than the QHD-I value placing it at about 9 MeV. This calculated 4^- level excitation energy therefore misses the experimental level by nearly 9 MeV. In a similar manner, the lowest 2^- state found experimentally at 8.9 MeV is calculated to lie at 4.4 MeV. Actually, the only T=0 levels which are in better agreement with experiment in the Finite-Hartree case as compared to the QHD-I case are the 2^- and lowest-lying 0^- levels each predicted at about 12 MeV.

The χ^2-parameterization results in a spurious 1^- level at 4.8 MeV and the 1^- state seen at 7.1 MeV is predicted to lie at 11.3 MeV. Thus each is about 4 MeV too high in excitation energy. Similarly, the entire 2^- spectrum appears to be at an unsatisfactorily high excitation energy. For example, the first and second lowest-lying 2^- levels each require a 3 MeV shift downward in order to agree with experiment. On the other hand, the χ^2-parameterization does reproduce the 4^- state near 17.8 MeV, the 3^- at 15.4 MeV and the lowest-lying 0^- at 11.0 MeV. The lowest-lying 3^- state seen at 6 MeV is pushed down to an imaginary energy eigenvalue (E_x = 1.9i) with this parameterization. A result such as this is not surprising since the RPA calculation utilizes in this instance parameters resulting from a minimization of χ^2 computed in the TDA. The failure of the fits to these low-lying collective 1^- and 3^- levels appears to indicate the need for an RPA calculation based on parameters resulting from an RPA χ^2-minimization. One subject of current research is therefore a minimization of χ^2 computed in the RPA.

The greatest differences between predicted TDA and RPA levels occur for the 1^- and 3^- T=0 levels. In Figs. 3 and 4, we show the 1^- and 3^- T=1 and T=0 levels, respectively, for both TDA and RPA calculations. The parameterization

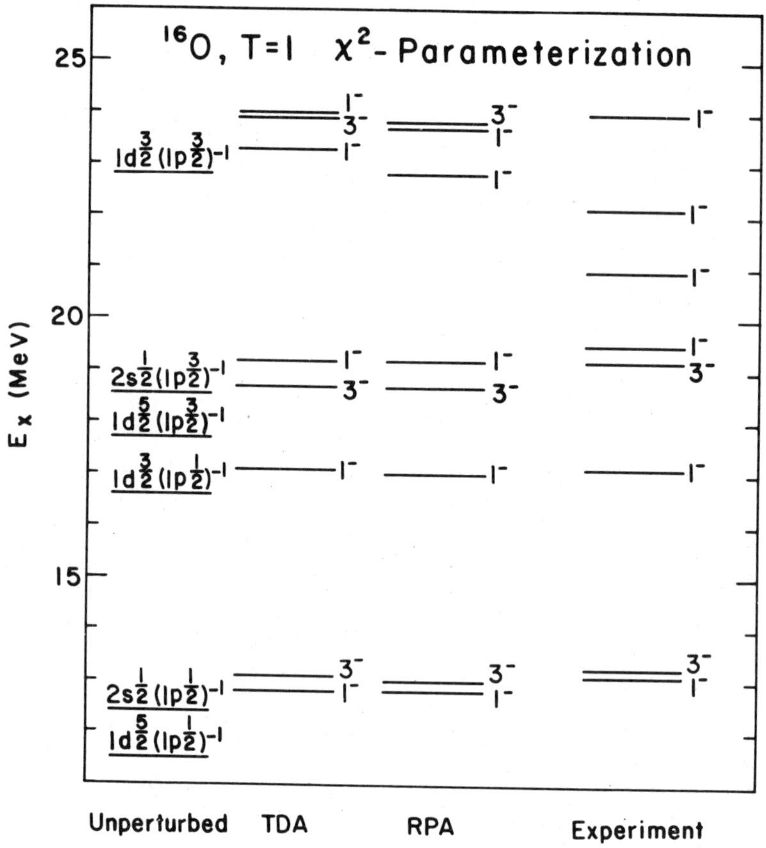

Fig. 3. Comparison of T=1 TDA and RPA energy levels of ^{16}O. Each calculation uses a TDA χ^2-parameterized p-h interaction.

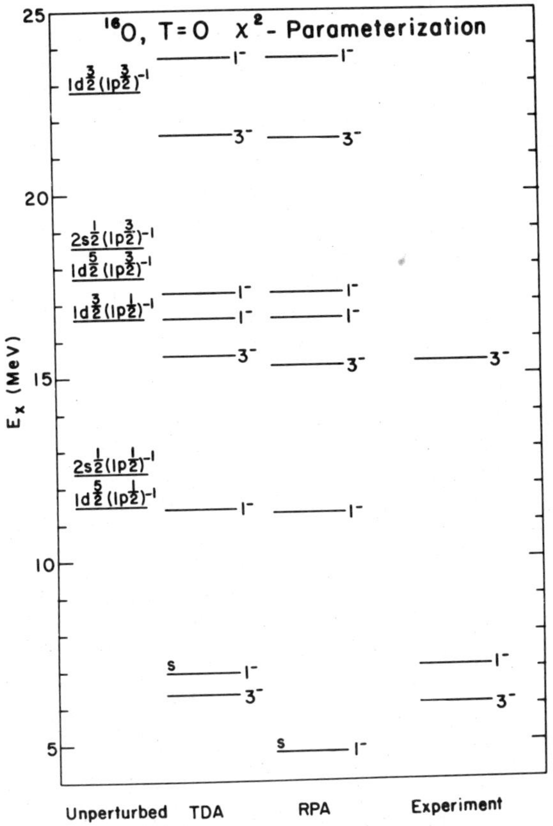

Fig. 4. Comparison of T=0 TDA and RPA energy levels of ^{16}O. Each calculation uses a TDA χ^2-parameterized p-h interaction. Spurious levels are labeled with an S.

throughout is that resulting from the minimization of χ^2 computed in the TDA. The T=1 spectra (Fig. 3) are nearly identical, with the RPA levels slightly lower in excitation than the corresponding TDA levels (as in the nonrelativistic treatment). This effect is most pronounced for the 1^- dipole and dipole spin-flip states. The largest difference (.5 MeV) is in fact that between the TDA dipole and the RPA dipole levels. The T=0 levels are shown in Fig. 4. With the exception of the lowest-lying 1^- and 3^- levels, the TDA and

RPA levels bear the same close relation as in the T=1 case. The spurious 1^- level (not fitted in χ^2) falls at roughly 5 MeV in the RPA. The RPA is therefore a slight improvement over the TDA for this level and this parameterization. The lowest-lying collective 3^- was included in the χ^2-fitting and thus was, in the TDA, within .3 MeV of experiment. In the RPA, we find an imaginary eigenvalue for this level, $E_x = 1.9i$, which represents the largest difference between TDA and RPA.

In Figs. 5 and 6, we compare our QHD-I parameterized RPA levels with those from a similar calculation by R.J. Furnstahl[6]. The present calculation uses the same unperturbed levels as derived from the relativistic-Hartree single-particle binding energies in the Furnstahl calculation. These unperturbed levels differ from the corresponding levels from neighboring nuclei by as much as 2.5 MeV. It is also important to note that the two calculations differ greatly in the size of single-particle basis used. The present calculation uses a "1 $\hbar\omega$" truncated basis whereas the Furnstahl calculation allows for transitions to the continuum. Two manifestations of the inclusion of continuum states is the overall compression of the levels as can be seen in the Furnstahl levels of Figs. 5 and 6, and the capability of producing many more states than from a "1 $\hbar\omega$" truncated basis.

The inclusion of continuum states does indeed cause significant changes in excitation energies for specific states that one would expect to be sensitive to the treatment of basis. Among these, for this parameterization, are the T=1 giant dipole state (at 20.5 MeV in the Furnstahl results and at about 23 MeV in the present work) and the lowest-lying collective 3^- and spurious 1^-, T=0 levels.

Particle-Hole Admixture Amplitudes and (e,e′) Form Factors

In Table III, we show as a representative sample, TDA admixture amplitudes for 1^-, T=1 states of ^{16}O at calculated excitation energies of 13, 22 and 24 MeV. The p-h interaction parameterization is from the χ^2-minimization. Also

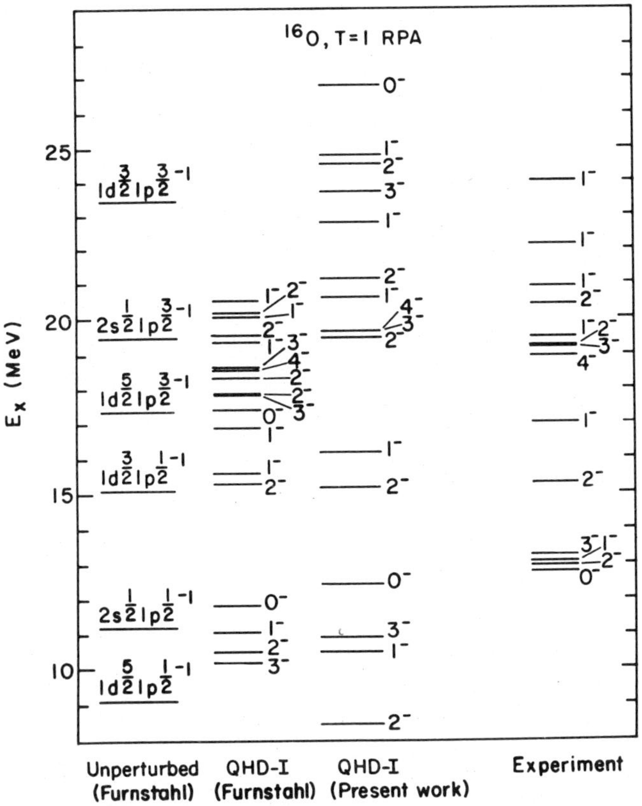

Fig. 5. Comparison with the T=1 RPA ^{16}O energy levels of reference 6.

in Table III are the corresponding nonrelativistic TDA results of Donnelly and Walker[17], and Gillet and Vinh-Mau[19]. We find that the degree of mixing is preserved in the relativistic calculation i.e., important p-h configurations are the same in both the relativistic and the nonrelativistic treatments. Any discrepancies that occur are among the less important p-h configurations. Note, however, that discrepancies among these less prominent p-h configurations arise in some cases between the two nonrelativistic TDA calculated admixtures as well.

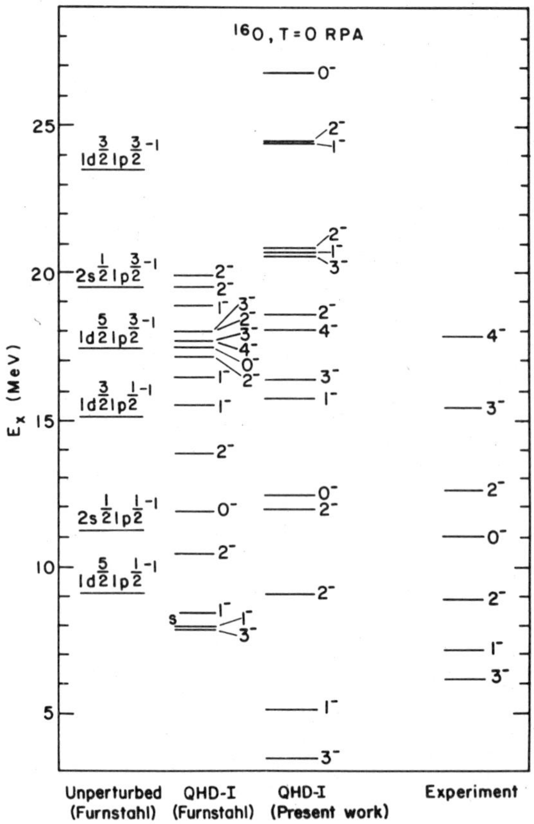

Fig. 6. Comparison with the T=0 RPA ^{16}O energy levels of reference 6. Spurious levels are labeled with and S.

The calculated RPA form factors squared, Eq. (33), for (e,e′) scattering are plotted as a function of q in Figs. 7 and 8 (the solid lines). These form factor predictions result from the χ^2-parameterized RPA admixtures and energy levels. The form factors squared for transitions to the 13 MeV and 17 MeV complexes are shown in Fig. 7 and for transitions to the 19 MeV complex and giant resonance region in Fig. 8. The constituents of each complex and their calculated energies are shown in Table IV.

Table III

TDA admixture amplitudes for $J^{\pi} = 1^{-}$, T=1 states of ^{16}O at E_x = 13,22 and 24 MeV.

	Particle-Hole Configuration	Present Work (lsj coupling)	Donnelly & Walker† (lsj coupling)	Gillet & Vinh-Mau* (lsj coupling)
E_x = 13 MeV	$1d\frac{5}{2}(1p\frac{3}{2})^{-1}$	.044	-.106	.096
	$1d\frac{3}{2}(1p\frac{3}{2})^{-1}$	.002	-.026	.020
	$1d\frac{3}{2}(1p\frac{1}{2})^{-1}$	-.004	.008	-.008
	$2s\frac{1}{2}(1p\frac{3}{2})^{-1}$	-.028	-.057	-.026
	$2s\frac{1}{2}(1p\frac{1}{2})^{-1}$	.999	.992	.995
E_x = 22 MeV	$1d\frac{5}{2}(1p\frac{3}{2})^{-1}$	.921	.882	.880
	$1d\frac{3}{2}(1p\frac{3}{2})^{-1}$	.310	.181	.259
	$1d\frac{3}{2}(1p\frac{1}{2})^{-1}$	.205	.329	-.345
	$2s\frac{1}{2}(1p\frac{3}{2})^{-1}$	-.110	.261	.180
	$2s\frac{1}{2}(1p\frac{1}{2})^{-1}$	-.043	.111	-.088
E_x = 24 MeV	$1d\frac{5}{2}(1p\frac{3}{2})^{-1}$	-.260	-.051	-.145
	$1d\frac{3}{2}(1p\frac{3}{2})^{-1}$	.939	.949	.943
	$1d\frac{3}{2}(1p\frac{1}{2})^{-1}$	-.217	-.273	.270
	$2s\frac{1}{2}(1p\frac{3}{2})^{-1}$	.053	-.149	-.131
	$2s\frac{1}{2}(1p\frac{1}{2})^{-1}$	.010	.013	-.006

†Reference 17

*Reference 19

Table IV

Constituents of the T=1 Complexes in Figs. 7 and 8.

		E^{theo}
13 Mev Complex:	1^-	12.8 (MeV)
	2^-	10.9
	3^-	13.0
17 MeV Complex:	1^-	17.0
	2^-	16.8
19 MeV Complex:	1^-	19.2
	2^-	18.8
	2^-	19.8
	3^-	18.7
	4^-	19.2
Giant Resonance:	1^-	22.8
	1^-	23.7
	2^-	23.1
	3^-	23.8

We include in both figures the nonrelativistic TDA results of Donnelly and Walker[17] (dashed lines) for comparison. The points carrying error bars are experimental data which was taken at $\theta = 135^o$. Both theory and experiment in the 13 MeV and 19 MeV complexes have been multiplied by 10 and 100, respectively for graphical reasons.

We first note that the reduction factors required nonrelativistically to reproduce experiment are still needed for our relativistic results. These factors are roughly the same for the 13 and 17 MeV complexes (factors of 3 and 2, respectively) as well as the 19 MeV complex (a factor of 2).

No reduction appears to be needed for the giant resonance region as was the case nonrelativistically.

At low q, the RPA predictions of the form factor squared are consistently enhanced realtive to the nonrelativistic TDA results. It is also interesting to note that after the appropriate reductions, the RPA curves for the 13 and 17 MeV complexes are roughly as much above the photon points (q=ω) as the TDA curves are below the photon points. For the giant resonance region, the RPA predicted photon point is about a factor of 2 larger than experiment.

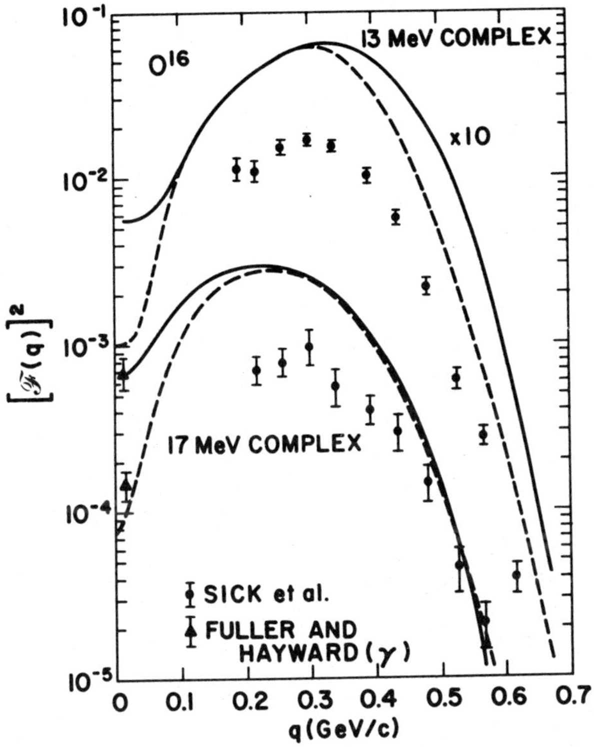

Fig. 7. $\mathcal{F}^2$(q) vs. q for (e,e′) transitions to the T=1 13 MeV and 17 MeV complexes of ^{16}O. The points carrying error bars are the experimental data (references 20 and 21). The solid lines are the RPA predictions of the present work. The dashed lines are the nonrelativistic TDA predictions of reference 17.

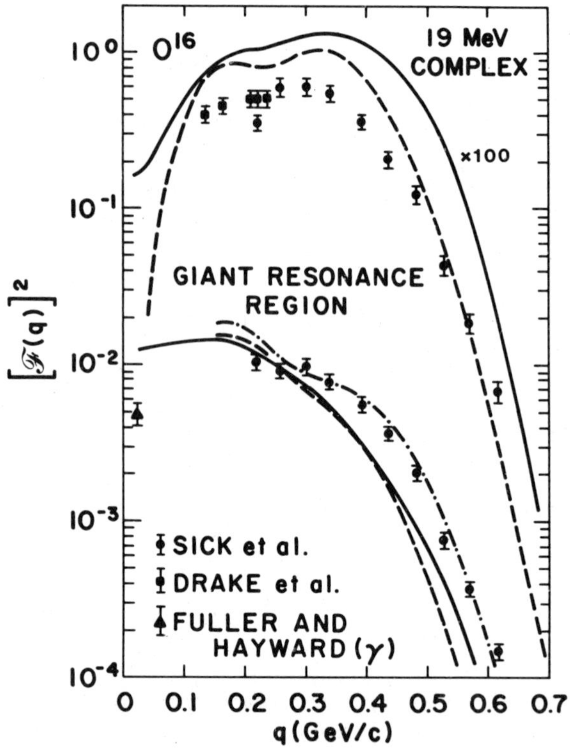

Fig. 8. $\mathscr{F}^2(q)$ vs. q for (e,e′) induced transitions to the T=1, 19 MeV complex and giant resonance region. The points carrying error bars are the experimental data (references 20-22). The Solid lines are the RPA predictions of the present work. The dashed lines are the nonrelativistic TDA predictions of reference 17. The dot-dashed line is the discrete TDA (dashed line of the giant resonance region) plus the nonrelativistic quasi-elastic contribution. (reference 17).

In most cases, at intermediate values of q the RPA and TDA curves tend to coincide while at high q the RPA curves are again enhanced relative to the TDA curves. One exception, however, is the 17 MeV complex where the RPA curve and TDA curve remain coincident until about 500 MeV/c after which the RPA curve falls below the TDA curve.

Summary and Conclusions

While the calculated RPA excited state energy levels resulting from QHD-I and Finite-Hartree parameterized p-h interactions reproduce some experimental levels, there are several that are consistently not fitted. Some examples are the lowest 1^- and 2^-, T=1 states and the lowest 2^-, T=0 state. In addition, a completely consistent RPA calculation would place the spurious 1^-, T=0 level at 0 MeV excitation. Our calculations fail to achieve this. Also the Finite-Hartree parameterization leads to a 4^-, T=0 level that is about 9 MeV too low when compared to experiment. We note that difficulty in fitting this particular level has been encountered elsewhere[6] for this parameterization. These difficulties suggest that a p-h interaction based on σ, ω and ρ exchange treated in lowest order is not adequate to calculate nuclear excited state energy levels.

The χ^2-parameterization leads to reasonable T=1 spectra when compared to experiment with slight improvements needed in the fits to the 2^- levels. The lowest-lying T=0 levels are unsuccessfully fitted by this TDA χ^2-parameterization. An RPA χ^2-parameterization (with ranges systematically studied) needs to be performed and is currently being completed[23]. Inclusion of the continuum needs to be investigated as well.

The relativistic (e,e′) form factors are interesting because of the enhancements at low and high q. We have shown that reduction factors are still required to reproduce experiment and are about the same size as found nonrelativistically. Similarly the relativistic p-h admixture amplitudes appear not to have changed much from the nonrelativistic results.

Acknowledgements

The author would like to thank G.E. Walker for his advice and encouragement. The author also would like to thank Charles Price for his help with the numerical checks of the calculations.

References

1. E.R. Siciliano, "Implications of Dirac Nuclear Structure on Electron Scattering," Los Alamos Workshop on Relativistic Dynamics and Quark-Nuclear Physics, Los Alamos, NM, (1985) (to be published in the workshop proceedings).

2. J.R. Shepard, "Nonrelativistic and Relativistic Approaches to Inelastic Proton Scattering," Los Alamos Workshop on Relativistic Dynamics and Quark-Nuclear Physics, Los Alamos, NM, (1985) (to be published in the workshop proceedings).

3. A.L. Fetter and J.D. Walecka, Quantum Theory of Many-Particle Systems, (McGraw-Hill, Inc., New York, 1971).

4. C.E. Price, "A Relativistic Schematic Model of Nuclear Structure in Light Nuclei," Los Alamos Workshop on Relativisitic Dynamics and Quark-Nuclear Physics, Los Alamos, NM, (1985) (to be published in the workshop proceedings).

5. C.E. Price and G.E. Walker, (to be submitted for publication).

6. R.J. Furnstahl, "Particle-Hole Excitations in ^{16}O in a Relativistic Field Theory of Nuclei," (Stanford University preprint).

7. C.J. Horowitz and B.D. Serot, Nucl. Phys. A 368 (1981) 503.

8. B.D. Serot and J.D. Walecka in Advances in Nuclear Physics edited by J.W. Negele and E. Vogt (Plenum, New York in press).

9. J.D. Bjorken and S.D. Drell, Relativistic Quantum Mechanics (McGraw-Hill, Inc., New York, 1964).

10. F. Ajzenberg-Selove, Nucl. Phys. A375 (1982) 1.

11. J.F. Dubach, "Introduction to QED and Electronuclear Scattering," Los Alamos Workshop on Relativistic Dynamics and Quark-Nuclear Physics, Los Alamos, NM, (1985) (to be published in the workshop proceedings).

12. T.deForest Jr. and J.D. Walecka in Advances in Physics 15, 1 (1966).

13. T.W. Donnelly and J.D. Walecka in Ann. Rev. Nucl. Sci. 25 (1975) 329.

14. B.D. Serot, Phys. Lett., 107B (1981) 263.

15. J.R. Shepard, E. Rost, E.R. Siciliano and J.A. McNeil Phys. Rev., C29 (1984) 2243.

16. T.W. Donnelly and W.C. Haxton, AT. Data and Nucl. Data Tables 23, (1979) 103-176.

17. T.W. Donnelly and G.E. Walker, Ann of Phys. 60 (1970) 209.

18. D.J. Thouless, Nucl. Phys. 22 (1961) 78.

19. V. Gillet and N. Vinh-Mau, Nucl. Phys. 54 (1964) 321.

20. I. Sick et al., Phys. Rev. Lett. 23 (1969) 1117.

21. E.G. Fuller and E. Hayward, "Nuclear Reactions", P.M. Endt and P.B. Smith, eds., Vol. II, p.113 North-Holland, Amsterdam, 1962.

22. T.F. Drake et al., Nucl. Phys. A118 (1968) 138.

23. M.W. Price and G.E. Walker, (to be submitted for publication).

RELATIVISTIC AND NONRELATIVISTIC DYNAMICS IN THE (e,e′p) REACTION

J. W. Van Orden
University of Maryland
College Park, MD 20742

ABSTRACT

Calculations of the $(\vec{e},e'p)$ reaction using both Dirac and Schrödinger dynamics in the DWIA are presented. Relativistic effects are most clearly manifest in a 20% suppression of the longitudinal response function R_L due to the Dirac final state interaction. A rough measure of the violation of current conservation is presented and the origins of this problem are discussed.

INTRODUCTION

The ejection of protons from nuclei by electron scattering, the (e,e'p) reaction, is shown schematically in Fig. 1, where k,s_e and k',s'_e are the four-momenta and spins of the incident and scattered electrons. The target nucleus has four-momentum P and intrinsic quantum numbers represented by i, while the four-momentum and intrinsic quantum numbers of the residual nucleus are P' and f. The ejected proton has four-momentum p' and spin s'. The four-momentum transferred to the nucleus by the scattered electron is q.

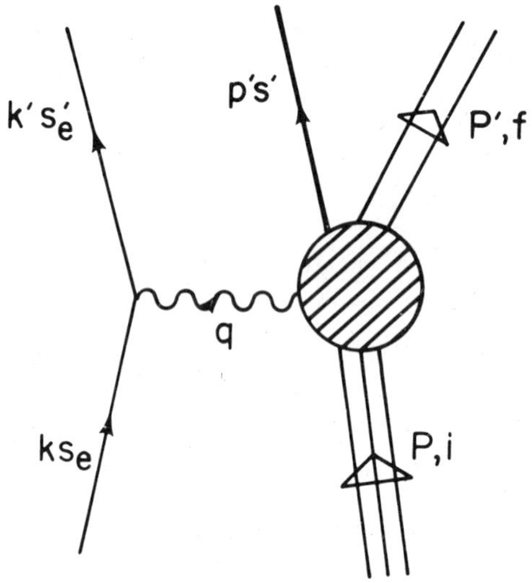

FIG. 1. Schematic diagram of the (e,e'p) reaction.

One motivation for examining the (e,e'p) reaction in the quasielastic region is provided by the inclusive quasielastic (e,e') reaction. Figure 2[1] shows the double-differential cross section for 60° scattering of 500 MeV electrons from ^{12}C in the

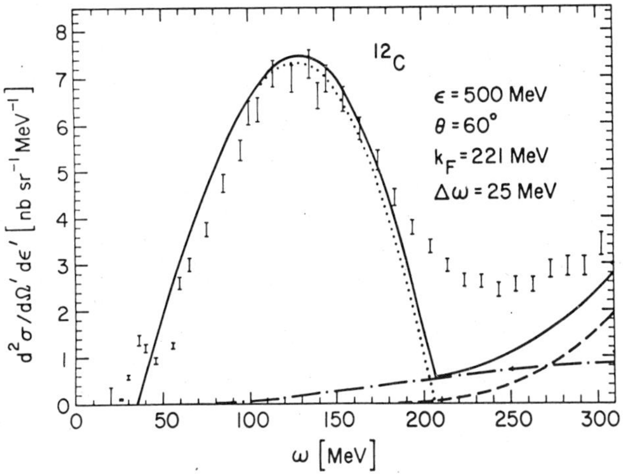

FIG. 2. Differential cross section for quasielastic scattering of 500 MeV electrons from ^{12}C at 60° [Ref. 1].

quasielastic region. The dotted line represents the simplest possible model of quasielastic scattering. This is the Fermi gas model where the nucleus is modeled as a noninteracting gas of nucleons. The quasielastic scattering in this model is described by elastic scattering of the electron from the individual nucleons in the gas which results in the nucleon being left above the filled Fermi sea. There are two adjustable parameters in this model; the Fermi momentum k_F which is the magnitude of the momentum of the highest filled plane wave state in the gas, and an energy shift $\Delta\omega$ which simulates the binding energy of nucleons in the nuclear medium. As shown in Fig. 2, it is possible to obtain a good representation of the cross section data in the region of the quasielastic peak.

The differential cross section for quasielastic (e,e') has the form[2]

$$\frac{d^2\sigma}{d\Omega_{k'} d\epsilon_{k'}} = \left(\frac{d\sigma}{d\Omega_{k'}}\right)_{MOTT} \left\{\frac{q^4}{\underset{\sim}{q}^4} R_L + \left(\tan^2\frac{\theta}{2} - \frac{1}{2}\frac{q^2}{\underset{\sim}{q}^2}\right) R_T\right\} , \tag{1}$$

where $\epsilon_{k'}$ is the energy of the scattered electron, θ is the electron scattering angle, $q^2 = \omega^2 - \tilde{q}^2$, ω is the energy transfer, and

$$\left(\frac{d\sigma}{d\Omega_{k'}}\right)_{MOTT} = \frac{\alpha^2 \cos^2 \theta/2}{4|\tilde{k}|^2 \sin^4 \theta/2} \qquad (2)$$

is the Mott cross section for electron scattering from an infinitely heavy point "nucleon". All of the information about the response of the nuclear target is contained in the longitudinal and transverse response functions, R_L and R_T. It is possible to extract these two response functions from cross section data by means of a Rosenbluth separation. The longitudinal and transverse response functions for ^{12}C at a three-momentum transfer of 400 MeV/c are shown in Figs. 3 and 4[3]

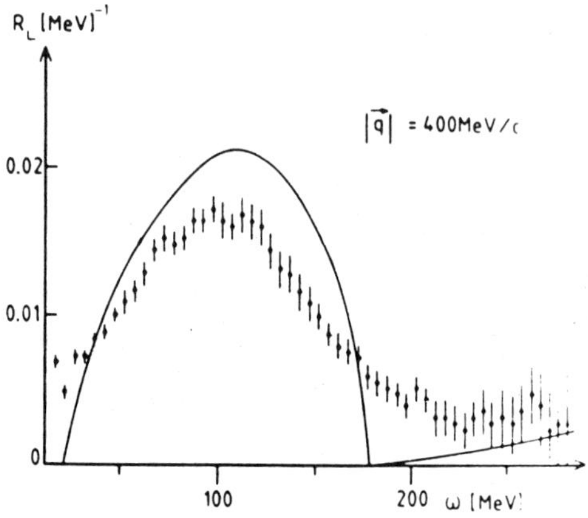

FIG. 3. Longitudinal response function for ^{12}C at 400 MeV/c momentum transfer [Ref. 3].

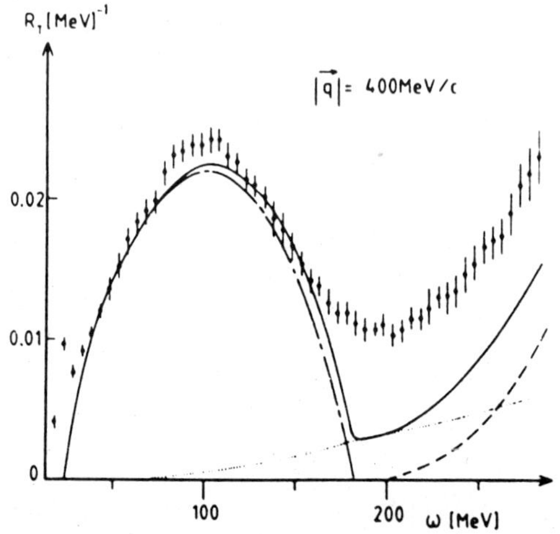

FIG. 4. Transverse response function for ^{12}C at 400 MeV/c momentum transfer [Ref. 3].

as a function of energy transfer. In both cases the solid line represents the same Fermi gas model results as in Fig. 2 for reference. Here, although the size of the calculated peak in R_T is in rough agreement with the data, the calculated peak in R_L is 20% larger than the data. In reality the situation is somewhat worse because the single-nucleon form factors used in this calculation are not correct. Using form factors consistent with experiment further increases the size of the calculated R_L. More sophisticated conventional calculations of quasielastic (e,e') are also incapable of simultaneously describing the transverse and longitudinal response functions. At this time, all calculations which describe the suppression of R_L use a prescription for, or simple model of, medium modifications to the single nucleon current.[4-7] One obvious way to deal with various theoretical descriptions of this suppression of R_L is to examine the various exclusive processes, such as (e,e'p), which contribute to the inclusive (e,e') reaction.

The traditional motivation for studying the quasielastic (e,e'p) reaction relies on the assumption that this process is truly quasielastic. That is, that the electron scatters incoherently from the individual protons in the target, ejecting the proton from the nucleus and that the ejected proton experiences no further interaction with the residual nucleus. This is represented schematically by Fig. 5. By measuring the four-momentum of the ejected proton for a known four-momentum transfer q, it is possible to determine the four-momentum of the struck proton in the nucleus. Figure 6[8] shows the missing energy spectrum for quasielastic (e,e'p) on ^{16}O. The noteworthy feature of this spectrum is the existence of sharp $1/2^-$ and $3/2^-$ peaks corresponding to the ejection of protons from the $1p_{1/2}$ and $1p_{3/2}$ shells. By varying the three-momentum transfer while adjusting the energy transfer such that the cross section remains on one of these two peaks, the three-momentum distributions of protons in the $1p_{1/2}$ and $1p_{3/2}$ shell can be determined. Of course, this analysis is dependent upon the reaction being

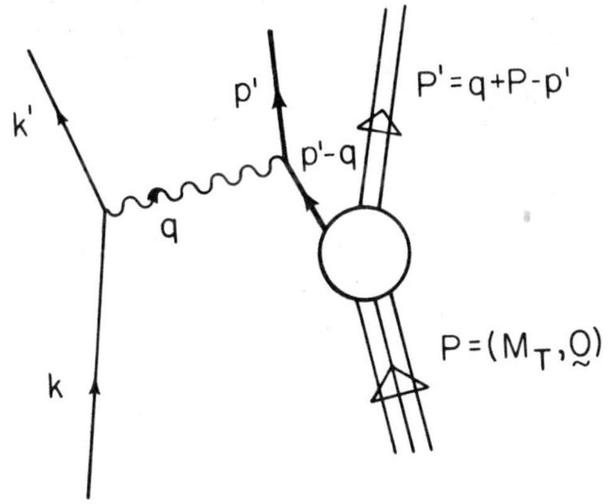

FIG. 5. Schematic diagram representing "truly" quasi-elastic (e,e'p).

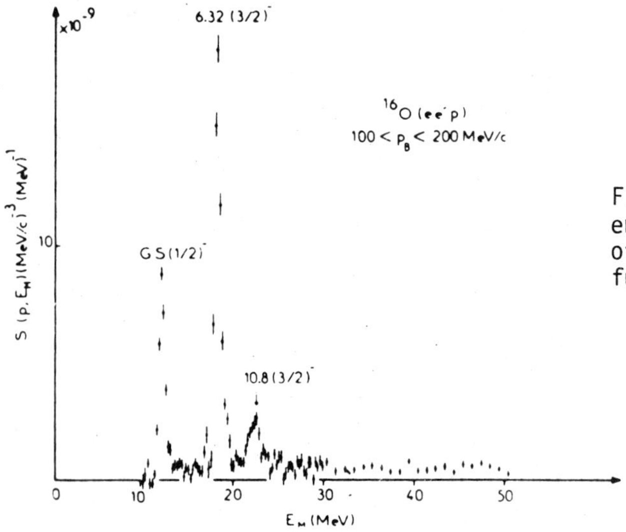

FIG. 6. Missing energy distribution of the ^{16}O spectral function.

truly quasielastic. In reality the nucleus does not simply consist of independent nucleons moving in a mean field potential since there are nucleon-nucleon correlations and virtual charged mesons are exchanged. Any ejected proton will also interact strongly with the residual nucleus. Therefore, it is not really possible to extract accurately nuclear ground state properties from the (e,e'p) reaction unless final state interaction, meson exchange and correlation effects are well understood.

ELECTRON SCATTERING FORMALISM

In the one-photon-exchange approximation, a general form of the (e,e'p) cross section can be determined by means of fairly simple and straightforward arguments. For the purposes of this workshop it is useful to outline the derivation of the (e,e'p) cross section. This derivation is generalized to include the possibility of having a polarized incident electron beam.[9] This reaction for polarized incident electrons is denoted by ($\vec{e}$,e'p).

Using Eq. (B.1) from Appendix B of Bjorken and Drell,[10] the differential cross section for ($\vec{e}$,e'p) can be written as

$$d\sigma = \frac{m_e}{|\underset{\sim}{k}|} \frac{m_e d^3k'}{(2\pi)^3 \varepsilon_{k'}} \frac{m d^3p'}{(2\pi)^3 E_{p'}} \sum_{s_e'} \sum_{s'} \sum_{f} \bar{\sum_{i}} \int \frac{d^3P'}{(2\pi)^3} \Phi_f(P')$$

$$\times (2\pi)^4 \delta^4(k-k'+P-P'-p')$$

$$\times |\bar{u}(k's_e')\gamma_\mu u(ks_e) \frac{e^2}{q^2} \langle p's'(-),\psi_{fP'}|\hat{J}^\mu(q)|\psi_{iP}\rangle|^2 , \quad (3)$$

where m_e and m are the electron and nucleon masses, and

$E_{p'} = \sqrt{\underset{\sim}{p}'^2+m^2}$. The electron spinors and gamma matrices follow the standard conventions.[10] $\Phi_f(P')$ is a density of states factor appropriate for the intrinsic spin of the residual nucleus. Since the recoil momentum of the residual nucleus is nonrelativistic, $\Phi_f(P') \cong 1$. The electron charge is e and $\hat{J}^\mu(q)$ is the electromagnetic current operator for the nuclear system. The incoming (spherical) wave boundary condition is appropriate for the continuum part of the final state which is indicated by the (-) in Eq. (3). The notation $\bar{\sum}_i$ indicates an average over the initial angular momentum projections of the target.

It is convenient to express Eq. (3) in terms of an electron response tensor,

$$\eta_{\mu\nu} = m_e^2 \sum_{s_e'} \bar{u}(ks_e)\gamma_\mu u(k's_e')\bar{u}(k's_e')\gamma_\nu u(ks_e) , \quad (4)$$

and a nuclear response tensor,

$$W^{\mu\nu} = \sum_{s'} \sum_{f} \bar{\sum_{i}} \int \frac{d^3P'}{(2\pi)^3} \Phi_f(P')(2\pi)^3 \delta^4(q+P-P'-p')$$

$$\times \langle \psi_{iP}|\hat{J}^{\mu\dagger}(q)|p's'(-),\psi_{fP'}\rangle\langle p's'(-),\psi_{fP'}|\hat{J}^\nu(q)|\psi_{iP}\rangle$$

$$= W^{\mu\nu S} + W^{\mu\nu A} . \quad (5)$$

The nuclear tensor can be written in terms of the sum of pieces symmetric ($W^{\mu\nu S}$) and antisymmetric ($W^{\mu\nu A}$) under interchange of the indices μ and ν.

In the extreme relativistic limit (ERL), where the electron mass can be neglected, the electron response tensor can be written as

$$\eta_{\mu\nu} \underset{ERL}{\cong} \frac{1}{4}\, tr[\gamma_\mu \not{k}' \;\gamma_\nu (1+h\gamma_5)\not{k}]$$

$$= (k'_\mu k_\nu + k_\mu k'_\nu - k'\cdot k g_{\mu\nu}) + i\varepsilon_{\mu\nu\lambda\kappa}\, k'^\lambda k^\kappa \tag{6a}$$

$$= \eta^S_{\mu\nu} + \eta^A_{\mu\nu} , \tag{6b}$$

where h is the helicity of the incoming electron and the electron response tensor has been written as the sum of a symmetric ($\eta^S_{\mu\nu}$) and an antisymmetric ($\eta^A_{\mu\nu}$) part. These parts can be identified easily from Eq. (6a). Note that the last term in Eq. (6a), which is the only antisymmetric term, is also the only term dependent on the electron helicity. In terms of the electron and nuclear response tensors Eq. (3) can be rewritten as

$$d\sigma = \frac{1}{|\underset{\sim}{k}|} \frac{d^3k'}{(2\pi)^3\varepsilon_{k'}} \frac{md^3p'}{(2\pi)^3E_{p'}} \frac{e^4}{q^4} \left(\eta^S_{\mu\nu} W^{\mu\nu S} + \eta^A_{\mu\nu} W^{\mu\nu A}\right) , \tag{7}$$

where use has been made of the property that the contraction of a symmetric and an antisymmetric tensor is zero.

In order to arrive at a general form of the cross section, it is necessary to construct a general form for the nuclear response tensor $W^{\mu\nu}$. This can be done by using two general properties of this second-rank Lorentz tensor. The first of these properties is that $W^{\mu\nu}$ must have the same transformation properties under parity as the direct product of two true vectors. This is the result of the fact that the electromagnetic interaction conserves parity and therefore the electromagnetic current must be a true vector under parity transformation. The second property of $W^{\mu\nu}$ is the direct result of the conservation of electromagnetic current as embodied in the continuity equation. In momentum space the continuity equation can be written as

$$q_\mu \langle p's'(-), \psi_{fP'} | \hat{J}^\mu(q) | \psi_{iP} \rangle = 0 . \tag{8}$$

This requires that the nuclear response tensor has the property

$$q_\mu W^{\mu\nu} = q_\nu W^{\mu\nu} = 0 . \tag{9}$$

The Lorentz tensor $W^{\mu\nu}$ must be constructed from the three independent 4-vectors q, p', and P, and the four independent Lorentz scalars q^2, $q\cdot p'$, $q\cdot P$ and $p'\cdot P$ which can be constructed from them. The nuclear response tensor can then be written as the sum of five independent terms each of which has the proper parity and satisfies Eq. (9).

$$W^{\mu\nu} = W_1 G^{\mu\nu} + W_2 V_i^\mu V_i^\nu + W_3 V_f^\mu V_f^\nu + W_4(V_i^\mu V_f^\nu + V_f^\mu V_i^\nu) + W_5(V_i^\mu V_f^\nu - V_f^\mu V_i^\nu) \tag{10}$$

where the coefficients W_i are functions of the four independent Lorentz scalars and

$$G^{\mu\nu} = g^{\mu\nu} - \frac{q^\mu q^\nu}{q^2} ,$$

$$V_i^\mu = \frac{1}{M_T} (P^\mu - \frac{P \cdot q}{q^2} q^\mu) ,$$

$$V_f^\mu = p'^\mu - \frac{P' \cdot q}{q^2} q^\mu . \tag{11}$$

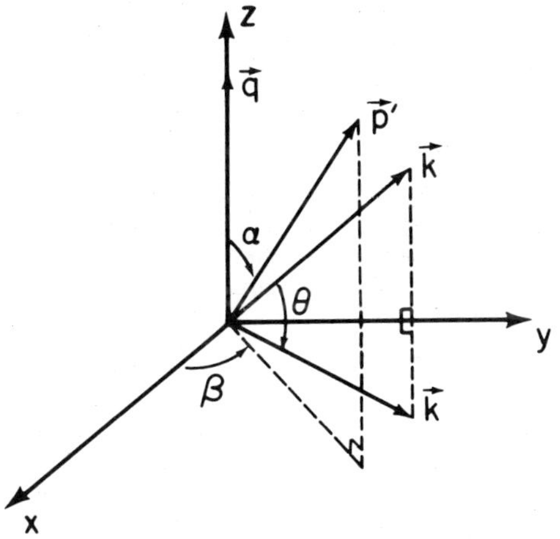

FIG. 7. Coordinate axes used to define the angles α and β.

Choosing the coordinate system shown in Fig. 7 and using Eq. (10) the cross section for the $(\vec{e},e'p)$ reaction can be written as

$$\left(\frac{d\sigma}{d\varepsilon_{k'}d\Omega_{k'}d\Omega_{p'}}\right) = \frac{m|\underset{\sim}{p}'|}{(2\pi)^3}\left(\frac{d\sigma}{d\Omega_{k'}}\right)_{MOTT}$$

$$\times \left\{\frac{q^4}{\underset{\sim}{q}^4} R_L + \left(\tan^2\frac{\theta}{2} - \frac{q^2}{2q^2}\right) R_T - \frac{q^2}{2q^2}\cos 2\beta\, R_{TT}\right.$$

$$\left. + \frac{q^2}{\underset{\sim}{q}^2}\left(\tan^2\frac{\theta}{2} - \frac{q^2}{\underset{\sim}{q}^2}\right)^{1/2} \sin\beta\, R_{LT} - h\,\frac{q^2}{\underset{\sim}{q}^2}\tan\frac{\theta}{2}\cos\beta\, R_{LT'}\right\} \tag{12}$$

The five response functions R_L, R_T, R_{TT}, R_{LT} and $R_{LT'}$ contain all of the information about the nuclear transition currents and are defined as

$$\begin{aligned} R_L &= [W^{00}] , \\ R_T &= [W^{11} + W^{22}] = [W^{++} + W^{--}] , \\ \cos 2\beta\, R_{TT} &= [W^{22} - W^{11}] = 2\mathrm{Re}[W^{+-}] , \\ \sin\beta\, R_{LT} &= [W^{02} + W^{20}] = -\sqrt{2}\,\mathrm{Im}[W^{0+} + W^{0-}] \\ \cos\beta\, R_{LT'} &= \sqrt{2}\,\mathrm{Im}[W^{0-} - W^{0+}] \end{aligned} \tag{13}$$

where the + and - superscripts refer to the spherical basis defined by the unit vectors,

$$\hat{e}^{\pm} = \mp \frac{1}{\sqrt{2}} (\hat{e}^1 \pm i\hat{e}^2) , \tag{14}$$

and the square brackets are defined to mean

$$[W^{\mu\nu}] = \int_{line} dE_{p'}\, W^{\mu\nu} .$$

Current conservation (Eq. (8)) can be used to eliminate the component of the three-vector current parallel to the momentum transfer in favor of the charge density. This has been used in deriving Eq. (12). Thus, longitudinally polarized virtual photons couple to the nuclear transition charge density J^0, while virtual photons with helicity ±1 couple to the two components $J^{\pm}$ of the three-vector transition current density transverse to the direction of the photon. Since all three independent components of the four-vector current can exist for a given transition, response functions which involve the interference of the various current components are present. Remembering the definition of $W^{\mu\nu}$ (Eq. (5)) and using Eq. (13), the response function R_L is determined entirely by the transition charge density. The response function R_T is given by the sum of the squares of the two transverse components of the transition current density. This corresponds to summing over the photon helicities ±1. The response function R_{TT} is the result of interference between the two different transverse components of the nuclear electromagnetic current. The response functions R_{LT} and $R_{LT'}$ are the result of intereference between the transition charge density and the two transverse components of the current. The azimuthal angular dependence in Eq. (12) arises as a result of the helicity of the absorbed virtual photon. In inclusive scattering, where terms involving β do not appear due to the azimuthal angle integration, it is only possible to differentiate between the transverse and longitudinal polarizations of the virtual photon, not between the ±1 helicity states of the photon. A short discussion of the extraction of the five response functions from the cross section and of what may be learned by doing so can be found in Ref. 11.

From the definition of $W^{\mu\nu}$ in Eq. (5) and the cross section expression given by Eqs. (14) and (15), it is clear that the cross section can be obtained if the matrix element,

$$J^{\mu} = \langle p's'(-),\psi_{fP'}|\hat{J}^{\mu}(q)|\psi_{iP}\rangle \tag{16}$$

can be calculated. This requires knowledge of the exact many-body electromagnetic current operator $\hat{J}^{\mu}(q)$ for the nuclear system as well as the exact many-body wave functions for the target nucleus, the final ejected nucleon, and the residual nucleus. Since it is not possible to obtain the exact wave functions or the full many-body electromagnetic current operator, this matrix element is usually calculated using wave functions from some calculable model and some approximation to the current operator. For example, in the case of (e,e'p) the bound states are usually described in terms of shell model wave functions $|\psi_i^{SM}\rangle$ and the final state is described in terms of optical model wave functions $|\psi_{p's'}^{(-)},f\rangle_{op}$.

In principle, the exact wave functions can always be written in terms of the model wave functions using Moller-type operators which "evolve" the model wave functions into the exact wave functions. That is,[11]

$$|\psi_{iP}\rangle = \Omega_B|\psi_i^{SM}\rangle\ , \tag{17a}$$

$$|p's'(-),\psi_{fP}\rangle = \Omega_S|\psi_{p's'}^{(-)},f\rangle_{op}\ , \tag{17b}$$

where Ω_B and Ω_S are Moller-type operators for the bound and scattering states. Using Eq. (17), Eq. (16) can be rewritten as

$$J^{\mu} = {}_{op}\langle\psi^{(-)}_{p's'},f|\Omega_S^{\dagger}\hat{J}^{\mu}(q)\Omega_B|\psi_i^{SM}\rangle \ . \tag{18}$$

In writing Eq. (18) the wave functions have been simplified at the expense of introducing the extremely complicated operators Ω_B and Ω_S. Equation (18) is still exact so all of the complexity of the original problem has simply been shifted into the creation of an effective current operator

$$\hat{J}^{\mu}_{eff} = \Omega_S^{\dagger}\,\hat{J}^{\mu}(q)\Omega_B \ , \tag{19}$$

which can not be calculated exactly. It is therefore necessary to approximate this effective current operator. The approximation which is usually used in calculating the (e,e'p) reaction is the distorted wave impulse approximation (DWIA) which corresponds to replacing the omega operators by the unit operator, $\hat{J}^{\mu}(q)$ by the free one-body current operator $\hat{J}^{(1)\mu}(q)$. This approximation is done entirely for calculational convenience and is made without any regard for the constraint of current conservation Eq. (8), and, as is shown here, has serious consequences for the maintenance of this constraint.

RELATIVISTIC DWIA CALCULATION OF ($\vec{e}$,e'p)

It is interesting to consider a DWIA calculation of the ($\vec{e}$,e'p) reaction using wave functions which are solutions to the four-component Dirac equation. The motivation for such a calculation is the success of Dirac-based calculations of elastic proton scattering[12-14] and nuclear ground state densities.[15,16]. As a first step in a program to obtain a consistent description of the ($\vec{e}$,e'p) reaction in a Dirac-based approach, the DWIA calculation has the attraction of providing a relatively simple context for the study of possible relativistic effects. It should be kept in mind that any conclusions derived from this calculation must be tempered by the knowledge that the violation of current conservation is inherent in the DWIA approach.

In momentum space the DWIA current matrix element, neglecting target recoil, can be written as[11]

$$J^{\mu} = \int \frac{d^3p}{(2\pi)^3}\,\bar{\psi}^{(-)}_{p's'}(\underset{\sim}{p})\ J^{(1)\mu}(q)\ \psi_i^{DH}(\underset{\sim}{p}-\underset{\sim}{q}) \ . \tag{20}$$

In this expression the bound state wave function is chosen to be a Dirac-Hartree wave function.[15,16] The one-body current operator is the usual free Dirac current operator

$$J^{(1)\mu}(q) = \gamma^0 \hat{J}^{(1)\mu}(q) = F_1(q^2)\gamma^\mu + \frac{F_2(q^2)}{2m} i\sigma^{\mu\nu}q_\nu \ . \qquad (21)$$

The wave function for the ejected proton is generated by the momentum-space Dirac optical model code WIZARD.[14,17] Note that in Eq. (20) the Dirac conjugate wave function $\bar{\psi} = \psi^\dagger\gamma^0$ is used for consistency with Eq. (21). Details concerning construction of the Dirac distorted wave can be found in Ref. 11.

The relationship of the Dirac matrix element, Eq. (20), to the more usual DWIA matrix element calculated with Schrodinger wave functions can be seen by recalling a simple derivation of the usual one-body nonrelativistic current operator. This derivation consists of the following steps:

(i) Evaluate the Dirac current operator, Eq. (21), between free positive energy Dirac spinors $\bar{u}(p's')$ and $u(ps)$.

(ii) Remove the Pauli spinors contained in the Dirac spinors and multiply the remnants of the Dirac spinors with the gamma matrices to obtain the equivalent operator in the Pauli spin space.

(iii) Expand this operator in powers of the momenta divided by the nucleon mass, keeping only the leading orders in this expansion.

(iv) Fourier transform to obtain an operator in coordinate space.

(v) Evaluate this operator between Schrödinger wave functions.

With the exception of steps (iii) and (iv), this is equivalent to using Dirac wave functions in Eq. (20) which have the ratio of upper to lower components fixed at the positive energy plane wave value for each momentum. This is the sense in which some of the calculations presented here are said to be nonrelativistic. In these nonrelativistic calculations, the Fourier amplitudes of the positive energy spinors for the scattering states are obtained from a nonrelativistic calculation while for the bound states the Dirac wave functions are projected onto positive energy spinors and renormalized.

RESULTS

The calculations presented here[11] are performed for the case of 135 MeV protons ejected from ^{16}O. The final proton energy is chosen to be within the energy range attainable by currently available experimental facilities. In Figs. 8 through 18, relativistic and nonrelativistic calculations of the five response functions are compared. Figures 8 through 12 show the response functions for the ejection of a 135 MeV proton from the $1p_{1/2}$ shell of ^{16}O. Figures 13 through 18 are for the ejection of a 135 MeV proton from the $1p_{3/2}$ shell of ^{16}O. The response functions are plotted as a function of the magnitude of the recoil momentum of the residual nucleus $|p'-q|$ at a constant momentum transfer $|\underset{\sim}{q}| = 2.64\ fm^{-1}$. The behaviõr of the five response functions as functions of $|\underset{\sim}{p}-\underset{\sim}{q}|$ and $|\underset{\sim}{q}|$ have been fully examined and the figures shown are representative slices of the complete three-dimensional graphs.

The solid line in each figure represents a completely relativistic calculation of the response function using Dirac distorted waves for the ejected proton and Dirac-Hartree[15,16] wave functions for the bound state. Results based on wave functions from Refs. 15 and 16 give essentially the same results. The dot-dashed line represents a nonrelativistic calculation, in the sense described earlier, while the dashed line is the undistorted calculation using the Dirac-Hartree bound state and a Dirac plane wave for the ejected proton. Calculations have also been performed using a nonrelativistic bound state wave function and a relativistic scattering wave function. In all cases, the results of this calculation differ from the fully relativistic calculation by only one or two percent. Thus, the curves shown for the relativistic calculation are also representative of these "semi-relativistic" calculations. Figures 8 through 12 also contain a calculation labeled "on-shell" which is represented by a short-dashed line. In this calculation only the pole part of the propagator in the Moller operator is retained; this corresponds to keeping only the on-energy-shell part of the T-matrix in the Moller operator. The purpose of this calculation is to provide a rough measure of the sensitivity of the response functions to the off-shell parts of the nucleon-nucleus T-matrix in the final state interaction.

Figure 8 shows the longitudinal response function R_L for the $1p_{1/2}$ shell. Here the relativistic calculation is 54% as large as the plane wave result, showing the effect of the loss of flux into other reaction channels. The nonrelativistic R_L is about 20% larger than the relativistic one. We find this to be almost entirely the result of the difference between the nonrelativistic and relativistic distorted wave function for the ejected nucleon; the Dirac-Hartree shell-model wave function for the ejected nucleon; the Dirac-Hartree shell model wave function accounts for only a small part of this difference. The "on-shell" calculation is, coincidentally, almost identical to the nonrelativistic calculation in this case, so that the sensitivity to off-shell effects is also roughly 20%.

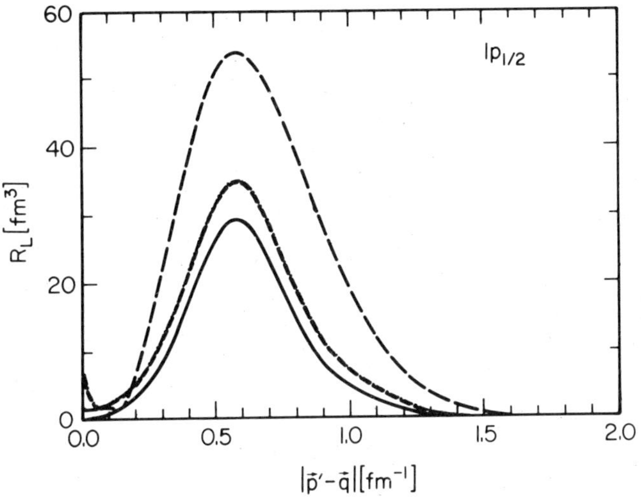

FIG. 8. Longitudinal response function, R_L, as a function of recoil momentum $|\underset{\sim}{p}'-\underset{\sim}{q}|$ for knockout of a 135 MeV proton from the $1p_{1/2}$ shell of ^{16}O.

Figure 9 shows the transverse response function R_T for the $1p_{1/2}$ shell. As shown in this figure, the relativistic R_T is 68% of the plane wave result. The nonrelativistic R_T is 1% smaller than the relativistic value, indicative of little dependence of this response function on relativistic versus nonrelativistic dynamics. The "on-shell" calculation is very close in size to both of the distorted wave calculations, except for small recoil momenta, so that there appears to be little off-shell sensitivity in this response function. It is interesting to contrast the differences in the various calculations of R_T to those of R_L. While the differences between the relativistic and nonrelativistic calculations of R_T are small, the calculation of R_L which includes relativistic effects is suppressed by a substantial amount relative to the nonrelativistic result. This is the type of change which is necessary to account for the suppression of the longitudinal response function in inclusive (e,e'). Note that the effect need not be of the same magnitude in (e,e') as it is in the present ($\vec{e}$,e'p) reaction which includes only one reaction channel. The interesting feature of this suppression is that it is due to dynamical relativistic effects in the final state interaction. This can be seen by noting that the relativistic and

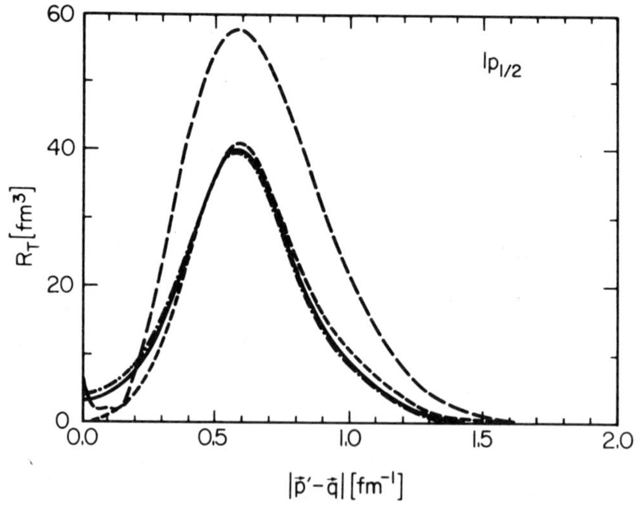

FIG. 9. Transverse response function, R_T, as a function of recoil momentum $|\underset{\sim}{p}'-\underset{\sim}{q}|$ for knockout of a 135 MeV proton from the $1p_{1/2}$ shell of ^{16}O.

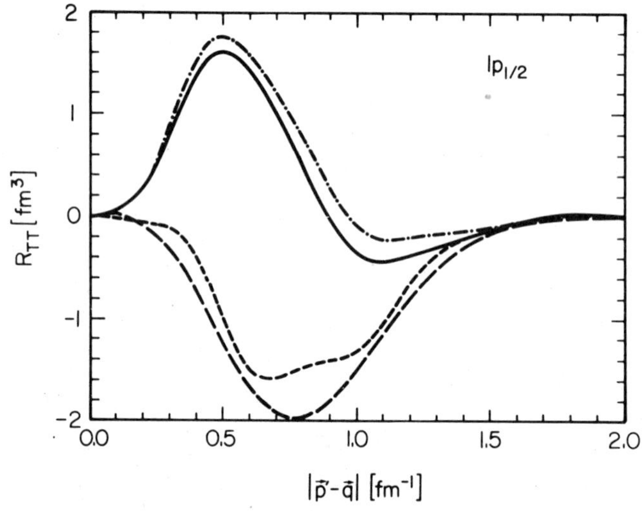

FIG. 10. Transverse-transverse response function, as a function of recoil momentum $|\underset{\sim}{p}'-\underset{\sim}{q}|$ R_{TT}, for knockout of a 135 MeV proton from $1p_{1/2}$ shell of ^{16}O.

"semi-relativistic" calculations, which differ only by whether the bound state wave function is relativistic or nonrelativistic, are essentially the same. Thus, the relativistic bound state effects are small in both cases.

Figure 10 shows the transverse-transverse interference response function R_{TT} for the $1p_{1/2}$ shell. Notice that this response function is small compared to R_L and R_T, being only about 6% as large as R_L. The magnitude of the relativistic calculation is about 20% less than the plane wave result but is of opposite sign. The relativistic and nonrelativistic R_{TT} values differ by about 10%. The "on-shell" calculation differs both in sign and shape from the other two distorted wave calculations so that this response function seems to be especially sensitive to off-shell effects. Unfortunately, this response function is sufficiently small relative to the other response functions that it will be difficult to extract experimentally.

Figure 11 shows the longitudinal-transverse interference response function R_{LT} for the $1p_{1/2}$ shell. The relativistic calculation is only about 40% as large as the plane wave, showing a greater sensitivity to final state interaction effects than either R_L or R_T. On the other hand, the relativistic and nonrelativistic calculations differ by about 10%, showing

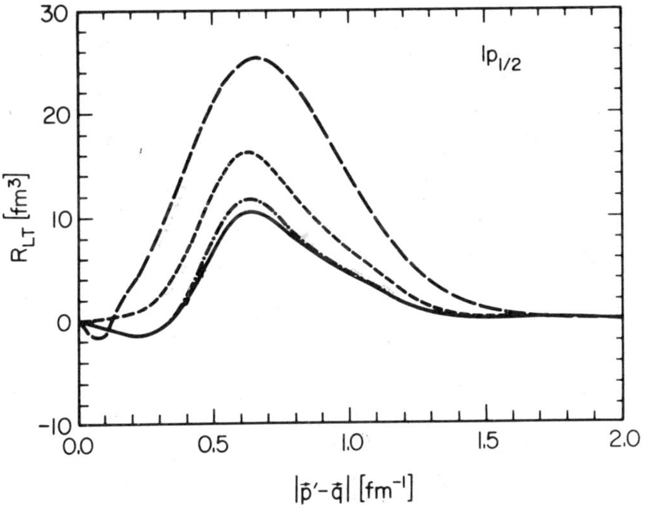

FIG. 11. Longitudinal-transverse response function, R_{LT}, as a function of recoil momentum $|p'-q|$ for knockout of a 135 MeV proton from the $1p_{1/2}$ shell of ^{16}O.

relativistic dynamic sensitivity intermediate between R_L and R_T. The "on-shell" calculation is about 55% larger than the relativistic result suggesting that this response function is quite sensitive to the off-shell character of the final state interaction. R_{LT} is about half as large as R_L and so should be experimentally observable with good accuracy.

Figure 12 shows the helicity dependent longitudinal-transverse interference response function $R_{LT'}$ for the $1p_{1/2}$ shell. This response function can only be measured using a polarized electron beam. The plane wave value for $R_{LT'}$ is identically zero for all recoil momenta since this response function can only be nonzero if the ejected proton experiences final state interactions with the residual nucleus. Therefore, $R_{LT'}$ exists only as a result of the final state interaction. Consistent with this statement, the relativistic and nonrelativistic distorted wave results are nonzero and differ by about 10%, indicating dependence on relativistic effects similar to that of R_{LT}. The "on-shell" calculation differs considerably from the two distorted wave calculations. This suggests that the interference structure functions may be, in general, quite sensitive to off-shell effects. Since $R_{LT'}$ is 55% of R_L it is quite possible that an

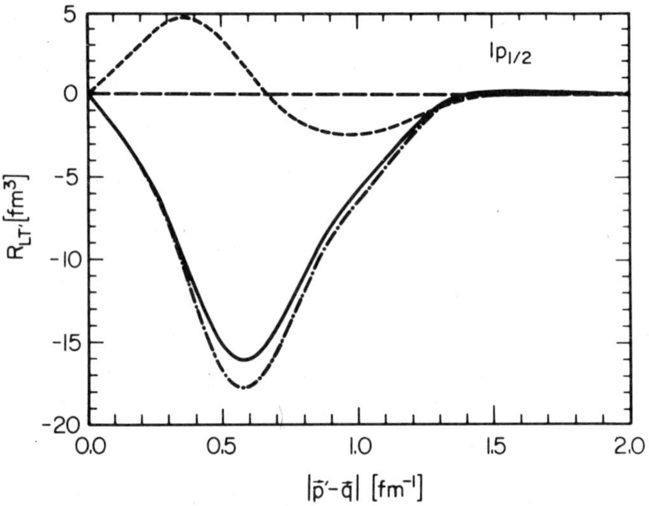

FIG. 12. Helicity dependent longitudinal-transverse response function, $R_{LT'}$, as a function of recoil momentum $|p'-q|$ for knockout of a 135 MeV proton from the $1p_{1/2}$ shell of ^{16}O.

intense polarized electron beam will allow $R_{LT'}$ to be extracted from the cross section data. However, this can be accomplished only with an out-of-plane spectrometer.

Figures 13 through 17 show that the results for the $1p_{3/2}$ shell are similar to those of the $1p_{1/2}$ shell. There is, however, an interesting difference between the calculations in the two shells. Distorted wave calculations, both relativistic and nonrelativistic, of R_{TT} and $R_{LT'}$ in the $1p_{1/2}$ shell produce opposite signs relative to their analogous calculations in the $1p_{3/2}$ shell. However, the sign of R_{LT} is the same in both shells. The existing cross section[18] data indicates that R_{LT} changes sign when going from one shell to the other. Nonrelativistic calculations[19] have been able to reproduce this feature by adjusting the imaginary part of the spin-orbit contribution to the optical potential. This freedom is not available within the context of our microscopic treatment of the proton distorted waves. However, recent work has suggested that a more complex approach based on meson theory is needed at low proton energy to produce a good description of proton-nucleus elastic scattering.[20,21]. It is possible that these improvements to the optical potential at low energy would remove this deficiency in the present calculations. An implementation of the improved

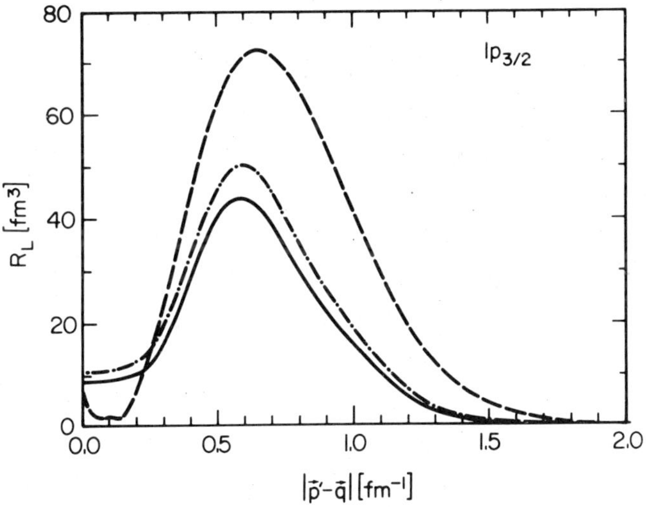

FIG. 13. Longitudinal response function, R_L, as a function of recoil momentum $|p'-q|$ for knockout of a 135 MeV proton from the $1p_{3/2}$ shell of ^{16}O.

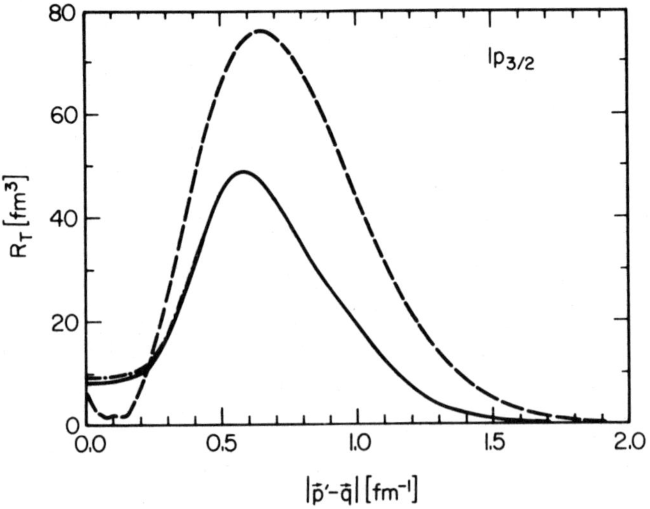

FIG. 14: Transverse response function, R_T, as a function of recoil momentum $|p'-q|$ for knockout of a 135 MeV proton from the $1p_{3/2}$ shell of ^{16}O.

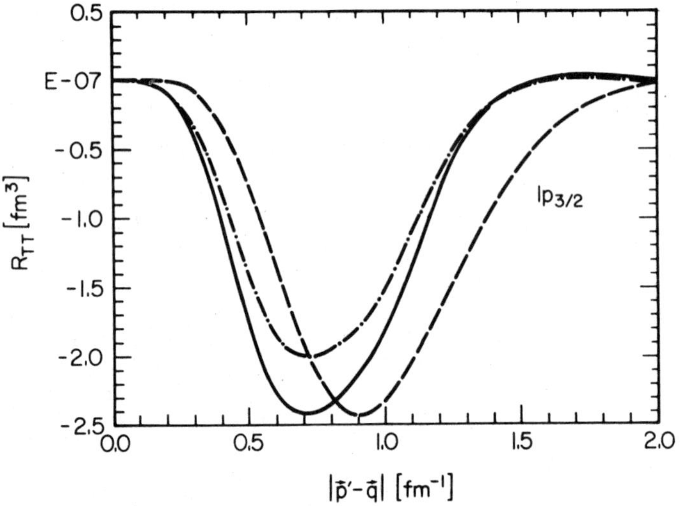

FIG. 15. Transverse-transverse response function, as a function of recoil momentum $|p'-q|$ R_{TT}, for knockout of a 135 MeV proton from the $1p_{3/2}$ shell of ^{16}O.

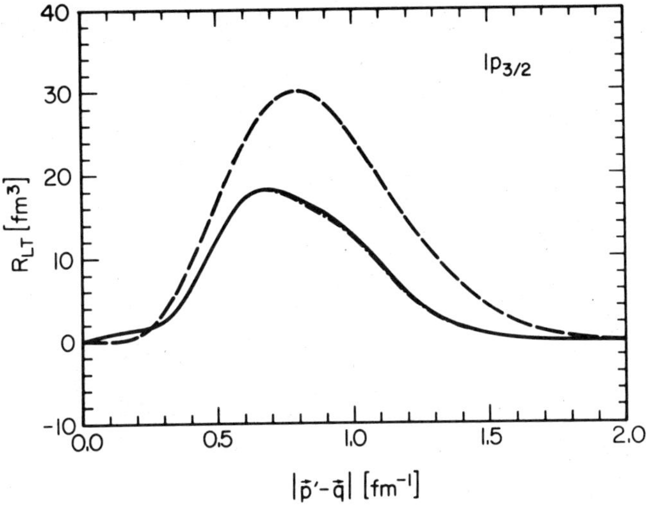

FIG. 16. Longitudinal-transverse response function, R_{LT}, as a function of recoil momentum $|p'-q|$ for knockout of a 135 MeV proton from the $1p_{3/2}$ shell of ^{16}O.

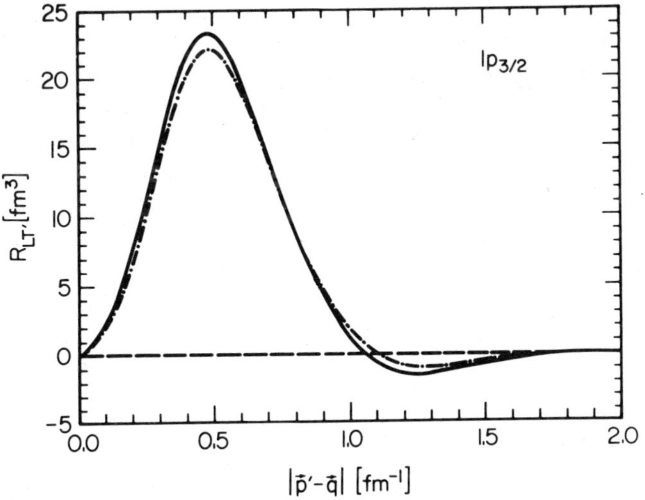

FIG. 17. Helicity dependent longitudinal transverse response function, $R_{LT'}$, as a function of recoil momentum $|p'-q|$ for knockout of a 135 MeV proton from the $1p_{3/2}$ shell of ^{16}O.

proton distortion for momentum-space calculations is being developed and this is expected to lead to an improved analysis in the near future.

Another, possibly related, problem is current conservation. In deriving the expression for the $(\vec{e},e'p)$ cross section, it is assumed that the electromagnetic current is conserved, which implies that

$$J^0 = \underset{\sim}{q} \cdot \underset{\sim}{J}/\omega = |\underset{\sim}{q}| J_L/\omega \ , \qquad (22)$$

where J_L is the longitudinal component of the three-vector current. We employed this constraint in deriving the cross section (12) to write R_L entirely in terms of J^0. If (22) is satisfied the same longitudinal response function R_L must be obtained by using (22) to eliminate J^0 in favor of J_L in the expression for R_L. Thus, as a rough test of the effect of current nonconservation, the calculations of the longitudinal response function based on the two different alternatives (J^0 or J_L) can be compared. For clarity, the response function calculated on the basis of J_L is referred to as $\tilde{R}_L$. Figures 18 and 19 show the quantity

$$C_{TEST} = \frac{R_L - \tilde{R}_L}{R_L + \tilde{R}_L} \qquad (23)$$

for both the $1p_{1/2}$ and $1p_{3/2}$ shells. The line types in these figures correspond to those used in the figures displaying the response functions. Clearly, the constraint of current conservation requires C_{TEST} to be zero for all recoil momenta. The present calculations indicate violations of current conservation by as much as 40% by this measure. This is not unexpected due to the orthogonality problems which show up in DWIA calculations of all types. Moreover, all DWIA calculations omit two-body currents and meson exchange currents. However, it may be worth noting that the effects of the violation of current conservation for the relativistic calculations in the region of the peaks in the response functions is 10% or less (by this measure) so that the calculations may be more reliable than expected a priori. In principle, it is possible to restore current conservation in a dynamically meaningful way, but this is difficult to achieve in practice. While it is important to determine whether or not such an extension will introduce any new features in the response functions, there is no reason to expect it to result in any qualitative changes to the results presented here.

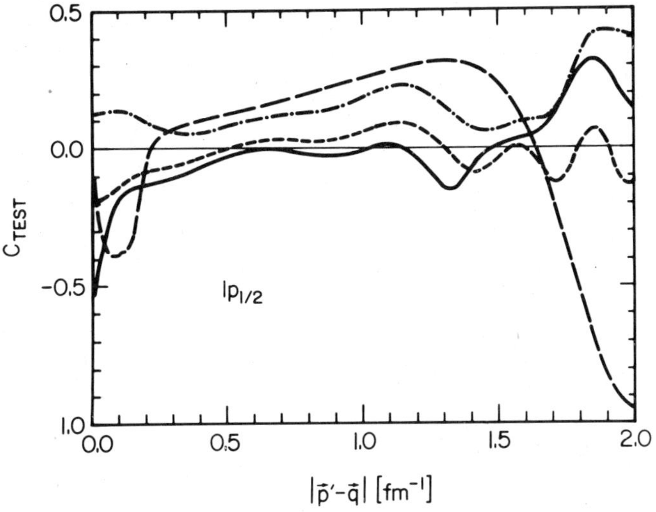

FIG. 18. C_{TEST} as a function of the recoil momentum for the $1p_{1/2}$ shell of ^{16}O. See text.

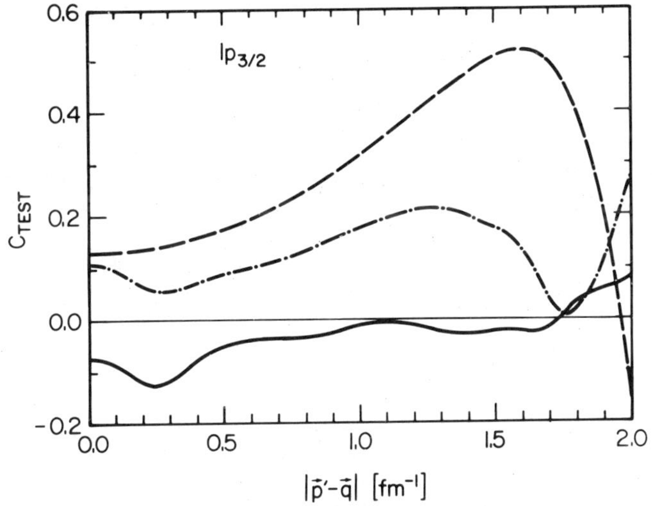

FIG. 19. C_{TEST} as a function of the recoil momentum for the $1p_{3/2}$ shell of ^{16}O. See text.

CONCLUSIONS

The results of the relativistic DWIA calculation of (e,e'p) can be summarized by the following conclusions:

(i) The longitudinal response function R_L is suppressed by about 20% relative to the transverse response function R_T due to relativistic final state effects.
(ii) Other relativistic effects are on the order of 1% to 10%.
(iii) The interference response functions R_{TT}, R_{LT} and $R_{LT'}$ are sensitive to off-shell effects and to details of the optical potential.

These conclusions should be taken as tentative due to the lack of current conservation inherent in DWIA calculations of this process. The inadequacy of the impulse approximation optical potential used here is manifest by the lack of a change in sign of R_{LT} in going from the $1p_{1/2}$ to $1p_{3/2}$ shells.

ACKNOWLEDGEMENTS

The support of the U. S. Department of Energy for this research is gratefully acknowledged.

REFERENCES

1. J. W. Van Orden and T. W. Donnelly, Ann. Phys. (N.Y.) 131, 451 (1980).

2. T. W. Donnelly and J. D. Walecka, Ann. Rev. Nucl. Sci. 25, 329 (1975).

3. P. Barreau, et al., Nucl. Phys. A402, 515 (1983).

4. R. Rosenfelder, Ann. Phys. 128, 188 (1980).

5. J. V. Noble, Phys. Rev. Lett. 46, 412 (1981).

6. G. Do Dang and Pham Van Thieu, Phys. Rev. C28, 1845 (1983); G. Do Dang and N. Van Giai, Phys. Rev. C30, 731 (1984).

7. L. S. Celenza, A. Harindranath, C. M. Shakin and A. Rostenthal, "Quark Effects in Nuclear Longitudinal Response Functions," preprint BCINT 84/111/132; L. S. Celenza, A. Harindranath and C. M. Shakin, "Quark Effects in Deep-Inelastic Electron Scattering from ^{12}C," preprint BCINT 84/121/134.

8. M. Bernheim, et al., Nucl. Phys. A375, 381 (1982).

9. T. W. Donnelly, in Nuclear and Subnuclear Degrees of Freedom and Lepton Scattering (Erice Lectures, Erice, Italy, 1984); T. W. Donnelly, in Proceedings of Workshop on Perspectives in Nuclear Physics at Intermediate Energies, S. Boffi, C. Ciof degli Atti and M. M. Giannini, eds. (World Scientific, Trieste, 1984).

10. J. D. Bjorken and S. D. Drell, Relativistic Quantum Mechanics (McGraw-Hill, New York, 1964).

11. A. Picklesimer, J. W. Van Orden and S. J. Wallace, "Final State Interactions and Relativistic Effects in the ($\vec{e}$,e'p) Reaction," University of Maryland preprint ORO 5126-249.

12. J. A. McNeil, J. R. Shepard and S. J. Wallace, Phys. Rev. Lett. 50, 1439 (1983); J. R. Shepard, J. A. McNeil and S. J. Wallace, Phys. Rev. Lett. 50, 1443 (1983).

13. B. C. Clark, S. Hama, R. L. Mercer, L. Ray and B. D. Serot, Phys. Rev. Lett. 50, 1644 (1983).

14. M. V. Hynes, A. Picklesimer, P. C. Tandy and R. M. Thaler, Phys. Rev. Lett. 52, 978 (1984); Phys. Rev. C31, 1438 (1985).

15. C. J. Horowitz and B. D. Serot, Nucl. Phys. A368, 503 (1981).

16. L. D. Miller, Phys. Rev. C14, 706 (1976).

17. A. Picklesimer, P. C. Tandy, R. M. Thaler and D. H. Wolfe, Phys. Rev. C29, 1582 (1984); Phys. Rev. C30, 1861 (1984).

18. M. Bernheim, et al., Nucl. Phys. A375, 381 (1982).

19. S. Boffi, C. Giusti and F. D. Pacati, Nucl. Phys. A386, 599 (1982).

20. S. J. Wallace, in Proceedings of the LAMPF Workshop on Dirac Approaches to Nuclear Physics, January 1985, to be published.

21. J. A. Tjon and S. J. Wallace, Phys. Rev. Lett. 54, 1357 (1985); "Meson Theoretical Basis for Dirac Impulse Approximation," University of Maryland preprint ORO 5126-244, to be published.

A DYNAMICAL BASIS FOR NN AMPLITUDES AND DIRAC OPTICAL POTENTIALS

Stephen J. Wallace
University of Maryland
College Park, MD 20742

ABSTRACT

Key points which motivated the relativistic impulse approximation for the Dirac optical potential are reviewed. A formal definition of the impulse approximation is presented in terms of the four-point function for NN scattering and the nuclear ground state wave functions. The first implementation of the Dirac optical potential, which was based only upon free NN scattering data, is shown to follow from four approximations to the general definition. A more satisfactory implementation of the Dirac optical potential, particularly for low proton energy, follows from a meson exchange model of NN scattering. A complete set of Lorentz-invariant NN amplitudes is calculated from the meson model. Preliminary results for elastic scattering of protons by ^{40}Ca are presented.

INTRODUCTION

Development of a relativistic impulse approximation for analyses of proton-nucleus scattering data was originally motivated by some empirical facts regarding quasi-free scattering and by a simple model of relativistic multiple scattering. These motivational points are briefly reviewed because they have led to successful connections between NN scattering and N-Nucleus scattering.[1-3] A formal definition is suggested for the Dirac impulse approximation optical potential which is independent of specific dynamical models. Corrections to this relativistic potential require the development of a more comprehensive theory. Present analyses necessarily involve approximations even within the impulse framework. Nevertheless there is a well-defined starting point.

Basic ingredients of the optical potential are NN amplitudes and nuclear wave functions. The first attempt to implement the formal definition of the impulse approximation potential can be understood in terms of four approximations. It is now clear that some of these approximations were inadequate at low energy, for example, with regard to the constraints which were assumed to be sufficient to characterize the relativistic NN amplitudes.[4,5]

In general, the construction of an optical potential requires the fully off-shell NN interaction. Virtual pair effects have been emphasized by Hynes et al. to be the main new ingredient of the Dirac approach.[6] Peter Tandy has reviewed the Dirac approach in its simplest formulation in which a Dirac potential is inferred from a nonrelativistic potential.[7] For elastic scattering, this is, in practice, a good approximation to the original formulation based on NN amplitudes and therefore has similar limitations. As Peter has emphasized, the separation of effects which are included in "nonrelativistic" treatments from the new virtual pair effects shows that the latter are instrumental to the Dirac successes.

A serious question is thereby raised. What is the dynamical basis for predicting the virtual pair effects? A second question, intimately related to the first, has been raised by Adams and Bleszynski.[8] This concerns off-shell ambiguities in the representation of relativistic NN amplitudes. Many representations in terms of Dirac covariants are equivalent for positive energy matrix elements, however, they lead to different predictions for negative energy matrix elements. The latter are essential for controlling the virtual pair effects in proton-nucleus scattering. The question is how to unambiguously characterize the Lorentz-invariant NN amplitudes.

In order to answer such questions, a dynamical basis is needed for the relativistic NN amplitude.[9,10] A meson theoretical approach to this problem has been formulated by J. A. Tjon and myself over the course of the last year and a half which builds upon earlier work by Tjon and collaborators on the NN interaction.[11-15] The main features of the impulse approximation as constrained by a meson theoretical foundation are discussed in the several recent papers.[9,10,16] These results, although incomplete, are reviewed in the latter half of this article.

MOTIVATIONS: QUASI-FREE DOMINANCE AND RELATIVISTIC POTENTIAL SCATTERING

Experimental data in the intermediate energy region provide direct evidence that quasi-free NN collisions are the predominant reaction mechanism. Reference 17 reviews developments in proton-nucleus scattering in the pre-Dirac impulse approximation era, i.e., prior to the time when Dirac potentials were calculated from NN amplitudes.

The "fundamental" approach based on solving the many-body Schrodinger equation is not practical nor has it led to useful insights or approximations. Multiple scattering theories developed in the 1950's provide the basis for essentially all analyses.[18-20] The formulation of a t-matrix for elastic scattering illustrates some of the main points. Let $|0\rangle$ represent the nuclear ground state and assume the projectile plus nucleus to be governed by a Schrödinger hamiltonian with two-body interactions. Then elastic scattering of protons involves matrix elements of an operator T_{00} defined in terms of an optical potential U and the Lippmann-Schwinger equation,

$$T_{00} = U + U\,G_0\,T_{00}\ , \tag{1}$$

where

$$U = \langle 0|\sum_i \tau_i^Q + \sum_{i\neq j} \tau_i^Q\,G^Q\,\tau_j^Q + \ldots|0\rangle\ , \tag{2}$$

in which $Q \equiv 1 - |0\rangle\langle 0|$ projects off the nuclear ground state, and τ_i^Q is a generalization of the NN t-matrix.[18] In terms of the projectile-nucleon potential, v_i, the projectile kinetic energy, K, and the nuclear hamiltonian, H_A, τ_i^Q is defined by the integral equation

$$\tau_i^Q = v_i + v_i\,G^Q\,\tau_i^Q\ , \tag{3}$$

where

$$G^Q = \frac{Q}{E-K-H_A}\ . \tag{4}$$

Equation (2) shows the multiple scattering structure with single-scattering terms plus double-scattering terms, and so on. To the extent that quasi-free processes dominate, the τ_i should be closely related to free NN t-matrices which are obtained by letting $Q \to 0$ and $H_A \to K_i$, the struck nucleon kinetic energy, in Eqs. (3) and (4). Moreover the organization of Eq. (2) is meaningful in that single-scattering terms are predominant and double-scattering terms represent a correction of secondary importance according to available estimates. Perhaps the nicest organization of the optical potential is that due to Kerman, McManus and Thaler in which U is expressed in terms of τ operators without any projection operators, Q.[20] In any event, the first term of the theory admits of a simple impulse approximation,

$$U_{IA} = \langle 0|\sum_i t_i|0\rangle\ , \tag{5}$$

where t_i is the free NN t-matrix obtained from τ_i as discussed above. This is found to provide a reasonable starting point for theoretical analyses of intermediate energy elastic scattering. Correspondingly, one describes inelastic scattering using the distorted wave impulse approximation (DWIA). The free t-matrix is sandwiched between initial and final nuclear states and distorted wave initial and final states for the projectile. This procedure is qualitatively successful with regard to describing the main features of elastic and inelastic scattering of protons. Details are generally thought to be understandable in terms of corrections to the impulse approximation, although no systematic theoretical calculations have been performed.

The ideas of quasi-free multiple scattering also lead to a successful understanding of total cross sections, reaction cross sections and inclusive proton-nucleus spectra. Figure 1 shows a doubly differential cross section obtained at Los Alamos by scattering 800 MeV protons from carbon.[21] At a fixed scattering angle of 8°, the inelastically scattered protons are seen to give direct evidence for the quasi-free reactions in the form of two broad maxima. The maximum centered at 1.4 GeV/c is due to quasi-free NN → N'N scattering and the yet broader maximum near

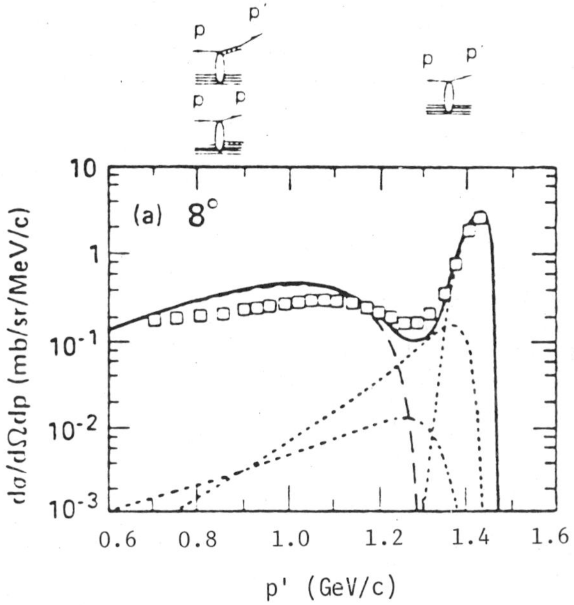

FIG. 1. Inclusive cross section for p=1.46 GeV/c protons scattered by ^{12}C. Peak at p'=1.4 GeV/c is due to quasi-free pN→p'N scattering. Broad peak at p'=1.0 GeV/c is due to quasi-free $pN \to N\Delta \to p'N\pi$ scattering. Solid line shows theoretical calculation of R. Smith. Dotted lines show single, double and triple scattering components. Dashed line shows Δ-production component.

1.0 GeV/c is due to quasi-free NN → N'Nπ reactions. One observes the energy of N' in each case. These data can be understood as the consequence of multiple quasi-free processes consistent with the dominance of the elastic scattering optical potential by the impulse approximation, Eq. (5). The theoretical line in Fig. 1 comes from a recent Ph.D. thesis by R. Smith.[22]

The weight of empirical evidence at intermediate energies suggests that quasi-free NN interactions are the predominant mechanism in proton-nucleus scattering. One therefore has an empirical basis for the multiple scattering organization of the theory, whether it be nonrelativistic or relativistic.

RELATIVISTIC POTENTIAL SCATTERING

A schematic model for relativistic multiple scattering illustrates some of the features one expects in a more general theory.[23] Consider a Dirac wave scattering from N fixed potentials located at positions $\vec{r}_1, \vec{r}_2, \ldots, \vec{r}_N$. If just one fixed potential were present, the potential $\hat{v}(\vec{r}-\vec{r}_i)$ would be chosen to reproduce NN scattering. In this sense, the potential is representative of the meson exchange interaction between two nucleons. When N fixed potentials are present, the Dirac equation is solved with the superposition of N potentials,

$$[E\gamma^0 - \vec{\gamma}\cdot\vec{p} - m - \sum_i \hat{v}(\vec{r}-\vec{r}_i)]\psi(\vec{r},\vec{r}_T) = 0 \ , \tag{6}$$

where $\vec{r}_T$ denotes the collection $\{\vec{r}_1, \vec{r}_2, \ldots, \vec{r}_N\}$. The wave function depends parametrically upon the positions of the potentials.

Scattering of the Dirac wave may be described by Dirac plane-wave matrix elements of a transition operator, $\hat{T}_F$. In the following, a carat is used to denote an operator on Dirac states. If one follows the same development which leads to the nonrelativistic multiple scattering expansion, $\hat{T}_F$ can be expressed as an infinite series which begins as follows:

$$\hat{T}_F = \sum_i \hat{t}_i + \sum_{i\neq j} \hat{t}_i \frac{1}{\not{p}-m+i\eta} \hat{t}_j + \ldots \ , \tag{7}$$

where the two-body scattering operator $\hat{t}_i$ is defined by the relativistic integral equation,

$$\hat{t}_i = \hat{v}_i + \hat{v}_i \frac{1}{\not{p}-m+i\eta} \hat{t}_i \ . \tag{8}$$

As noted above, the potentials $\hat{v}_i$ may be chosen so that plane-wave matrix elements of $\hat{t}_i$ will reproduce NN scattering (in the laboratory frame since the potentials are fixed). The interpretation of $\hat{T}_F$ relevant for nucleon-nucleus scattering is that of the t-matrix for sudden passage of the Dirac incident wave. Locations of the potentials (target nucleons) are fixed in the sense that there is too little time for the particles to recoil appreciably during the passage of the Dirac wave. The subscript, F, stands for flash photograph more appropriately than for fixed-scatters.

Elastic scattering can be defined in terms of the probabiltiy amplitude $\psi_o(\vec{r}_1,\vec{r}_2,\ldots,\vec{r}_N)$, for finding N particles in the ground state. Applying T_F to ψ_o defines the probability amplitude for scattering. Finally calculating the overlap with ψ_o as final probability amplitude, one is led to the elastic scattering t-matrix,

$$\hat{T}_{oo} = \langle 0|\hat{T}_F|0\rangle \ . \tag{9}$$

All averaging over the nuclear ground state wave function is now finished. Dirac plane wave matrix elements of $\hat{T}_{oo}$ must be calculated to obtain the scattering amplitude for the projectile. Just as in the nonrelativistic theory, elastic scattering can be described in terms of an optical potential, $\hat{U}$ introduced as follows:

$$\hat{T}_{oo} = \hat{U} + \hat{U}\frac{1}{\not{p}-m+i\eta}\hat{T}_{oo} \ , \tag{10}$$

$$\hat{U} = \langle\psi_o|\sum_i \hat{t}_i^Q + \sum_{i\neq j}\hat{t}_i^Q\frac{1}{\not{p}-m+i\eta}\hat{t}_j^Q + \ldots|\psi_o\rangle \ , \tag{11}$$

where Q is once again a projection off the ground state, and $\hat{t}_i^Q$ is defined as in Eq. (8) except with Q inserted as in Eqs. (4) and (5).

Many obstacles arise in attempts to derive the multiple scattering structure from a relativistic quantum field theory. Therefore, Eqs. (6)-(11) represent a schematic model. Nevertheless we believe that the multiple scattering structure is essential since empirical evidence suggests quasi-free dominance. Indeed one expects the impulse approximation to be the correct starting point on physical grounds. The mathematics does not provide any such guide. In this schematic relativistic model, the impulse approximation optical potential is

$$\hat{U}_{IA} = \langle 0|\sum_i \hat{t}_i|0\rangle \ . \tag{12}$$

Provided that $\hat{t}_i$ is representative of the free NN t-matrix, this is the same idea as Eq. (5). Similarly, one may discuss inelastic scattering within a DWIA framework in this schematic relativistic model. The simple structure discussed here played a guiding role in development of the Dirac impulse approximation of Ref. 1.

FORMAL DEFINITION OF DIRAC IMPULSE OPTICAL POTENTIAL

Formidable difficulties stand in the way of attempts to define a complete and satisfactory "relativistic theory" for nuclear physics. For example, quantum field theory has yet to explain the NN interaction. In place of a fundamental theory, one generally relies upon cutoff meson theory.[24] Basic meson baryon couplings are specified in this approach, however, form factors involving cutoff masses are prescribed at vertices of Feynman graphs in order to eliminate high momentum contributions. Ultimately the cutoff parameters must be justified from a more fundamental basis, e.g., quantum chromodynamics. The main assumptions of cutoff meson theory are (i) that this will be possible and (ii) that the meson-baryon degrees of freedom are the most important ingredients for nuclear physics.

Independent of the details of the underlying theory, it is possible to define the Dirac impulse approximation optical potential in terms of two basic ingredients: NN amplitudes and relativistic nuclear wave functions. The definition is

$$\hat{U}_{IA} \equiv \langle\psi_A|\sum_{i=1}^{A} \hat{\mathcal{M}}_i|\psi_A\rangle \ , \tag{13}$$

where $\hat{\mathcal{M}}$ represents the sum of all Feynman diagrams which can contribute to NN → NN. Thus $\hat{\mathcal{M}}$ is the Feynman four-point function for nucleon-nucleon scattering. It is the relativistic analog of the t-matrix. The wave function $|\psi_A\rangle$ represents the baryon number equal to A relativistic nuclear ground state. The total four-momentum of this nuclear ground state must be characterized by $(M_A,\vec{0})$ in the rest frame, where M_A is the rest mass. If the complete ground state is replaced by its dominant Fock-space component, ψ_{AN}, which corresponds to A nucleons, the formal definition of the impulse optical potential is

$$\hat{U}_{IA} = \sum_i \int \frac{d^4p_1}{(2\pi)^4} \cdots \int \frac{d^4p_{A-1}}{(2\pi)^4} \bar{\psi}_{AN}(p_1,\ldots,p_i,\ldots)\hat{\mathcal{M}}_i(p,p_i\to p',p_i')$$
$$\times\ \psi_{AN}(p_1,\ldots,p_i',\ldots) \ . \tag{14}$$

The potential depends on four-momenta p and p' which describe the initial and final state of the projectile nucleon. Internal momenta must be integrated over and ψ_{AN} depends upon A-1 independent internal four-momenta. The Bethe-Salpeter vertex function provides an example of ψ_{AN} for the two-particle case.[25]

Given an effective Lagrangian, one can, in principle, proceed to construct the NN four-point function and a nuclear ground state. However, approximations seem inescapable and Eq. (13) is expected to serve as a formal definition of what one aims to calculate in the same sense as Eq. (5) serves nonrelativistically. It is worth noting that exact nuclear ground state wave functions are not known in either case! The main point is that a sensible formal definition is possible. Of course, it remains to develop suitable corrections to the impulse approximation. Equation (13) provides only a starting point.

CONSTRAINTS ON $\hat{\mathcal{M}}$ FROM NN SCATTERING EXPERIMENTS

Scattering experiments can in principle determine the positive-energy, on-mass-shell matrix elements of $\hat{\mathcal{M}}$. Helicity amplitudes provide a convenient basis for discussing the experimental constraints on $\hat{\mathcal{M}}$.[5,26] These are defined in the nucleon-nucleon cm frame by Dirac matrix elements as follows:

$$\phi^{(++,++)}_{\lambda_1'\lambda_2',\lambda_1\lambda_2} \equiv \bar{u}^{(+)}_{\lambda_1'}(\vec{p}\,')\bar{u}^{(+)}_{-\lambda_2'}(-\vec{p}\,')\hat{\mathcal{M}}\, u^{(+)}_{\lambda_1}(\vec{p})u^{(+)}_{-\lambda_2}(-\vec{p}) \;/\; \kappa \qquad (15)$$

where $u^{(+)}_{\lambda}(p) = N_p \begin{pmatrix} 1 \\ \frac{2\lambda p}{E_p+m} \end{pmatrix} \chi_\lambda$ is a positive energy Dirac spinor for helicity eigenvalue λ. The product Dirac states therefore have helicity eigenvalues as follows,

$$\tfrac{1}{2}(\vec{\sigma}_1+\vec{\sigma}_2)\cdot\hat{p}\; u^{(+)}_{\lambda_1}(\vec{p})u^{(+)}_{-\lambda_2'}(-\vec{p}) = (\lambda_1-\lambda_2)u^{(+)}_{\lambda_1}(\vec{p})u^{(+)}_{-\lambda_2'}(-\vec{p}) \;. \qquad (16)$$

The divisor, κ, in Eq. (15) is a kinematic factor necessary to relate the Feynman amplitude to scattering amplitudes. It is defined by

$$\kappa = 2ip\left(\frac{2\pi\sqrt{s}}{m^2}\right) = \frac{4\pi i\; P_{lab}}{m} \;, \qquad (17)$$

where p is the cm momentum, $s = (p_1+p_2)^2$, and P_{lab} is the laboratory frame momentum. Table I indicates five helicity amplitudes, ϕ_1 to ϕ_5, which are linearly independent when parity invariance, time-reversal invariance and charge symmetry are constraints. These helicity amplitudes are "known" from phase-shift analyses of experimental data.[27] Currently the pp phase shifts are well determined up to 800 MeV. The pn phase shifts are known up to 500 MeV and experiments at Los Alamos are extending this knowledge towards 800 MeV.

Consistent with parity invariance, time-reversal invariance and charge symmetry, the relativistic amplitude may also be characterized in terms of five Lorentz invariant amplitudes and associated covariants. One such Lorentz invariant representation is in terms of Fermi covariants;[4,5]

$$\hat{\mathcal{M}} = \kappa \sum_{n=1}^{5} F_n \mathcal{K}_n , \tag{18a}$$

where $\mathcal{K}_n$ represents the five Fermi covariants,

$$\mathcal{K}_1 = S \equiv 1 \tag{18b}$$

$$\mathcal{K}_2 = V \equiv \gamma_1 \cdot \gamma_2 \tag{18c}$$

$$\mathcal{K}_3 = T \equiv \sigma_1^{\mu\nu} \sigma_{2\mu\nu} \tag{18d}$$

$$\mathcal{K}_4 = P \equiv \gamma_1^5 \gamma_2^5 \tag{18e}$$

$$\mathcal{K}_5 = A \equiv \gamma_1^5 \gamma_1^\mu \gamma_2^5 \gamma_{2\mu} . \tag{18f}$$

Thus the amplitudes are scalar (F_1), vector (F_2), tensor (F_3), pseudoscalar (F_4), and axial vector (F_5). More covariants could be added, but only five can be determined from NN data. The five Fermi covariants were originally chosen by Goldberger, Grisaru, MacDowell and Wong (GGMW).[5] Other covariants were discussed by Goldberger, Nambu and Oehme (GNO),[28] but they were found to be less satisfactory with regard to analytic properties. As pointed out by Adams and Bleszynski, the choice between GGMW and GNO covariants leads to some arbitrariness since either set is capable of describing the NN data.[8] Two reasons in favor of the GGMW choice are that the invariant amplitudes are free of kinematic singularities at $\theta = 0$ or π and that the one-meson exchange contributions can be naturally written in terms of the Fermi covariants.[5]

The experimental constraints on invariant amplitudes may be derived by inserting Eq. (18) into Eq. (15). The result is the 5×5 matrix equation,

TABLE I

HELICITY AMPLITUDES

ϕ_n	ϕ_1	ϕ_2	ϕ_3	ϕ_4	ϕ_5
$\lambda_1'\lambda_2', \lambda_1\lambda_2$	$\frac{1}{2}\frac{1}{2}, \frac{1}{2}\frac{1}{2}$	$\frac{1}{2}\frac{1}{2}, -\frac{1}{2}-\frac{1}{2}$	$\frac{1}{2}-\frac{1}{2}, \frac{1}{2}-\frac{1}{2}$	$\frac{1}{2}-\frac{1}{2}, -\frac{1}{2}\frac{1}{2}$	$\frac{1}{2}\frac{1}{2}, \frac{1}{2}-\frac{1}{2}$

$$\phi_n = A_{nm} F_m , \tag{19}$$

where

$$A_{nm} \equiv \bar{u}^{(+)}_{\lambda_1'}(\vec{p}\,')\bar{u}^{(+)}_{-\lambda_2'}(-\vec{p}\,')\,\mathcal{K}_m\, u^{(+)}_{\lambda_1}(\vec{p})u^{(+)}_{-\lambda_2}(-\vec{p}) . \tag{20}$$

Subscript n refers to the helicity labels as defined in Table I. Matrix A depends only on kinematic quantities such as $z = \cos\theta$, where θ is the scattering angle, the nucleon mass, m, the relative cm momentum, p, and the energy, $E = \sqrt{p^2+m^2}$. As shown by GGMW[5] (see also Ref. 10), the invariant amplitudes can be directly expressed in terms of the five helicity amplitudes. This amounts to an explicit construction of the inverse of matrix A. The result is

$$\begin{bmatrix} F_1 \\ F_2 \\ F_3 \\ F_4 \\ F_5 \end{bmatrix} = (16ip)^{-1} \begin{bmatrix} 1 & 6 & 4 & 4 & 1 \\ 1 & 0 & -2 & 2 & -1 \\ -1/2 & 1 & 0 & 0 & -1/2 \\ 1 & 6 & -4 & -4 & 1 \\ 1 & 0 & 2 & -2 & -1 \end{bmatrix} \begin{bmatrix} G_1 \\ G_2 \\ G_3 \\ G_4 \\ G_5 \end{bmatrix} \tag{21}$$

$$\begin{bmatrix} G_1 \\ G_2 \\ G_3 \\ G_4 \\ G_5 \end{bmatrix} = \begin{bmatrix} \frac{m^2}{E^2} & -\frac{m^2}{E^2} & \frac{m^2-p^2z}{E^2} & -\frac{m^2+p^2z}{E^2} & -z \\ 0 & 0 & -1 & -1 & -\frac{E^2}{p^2} \\ 0 & 0 & -1 & 1 & 0 \\ 0 & 0 & 1 & 1 & \frac{m^2}{p^2} \\ -\frac{m^2}{p^2} & -\frac{m^2}{p^2} & -z & -z & -\frac{(E^2+m^2)z}{p^2} \end{bmatrix} \begin{bmatrix} \phi_1 \\ \phi_2 \\ \frac{m^2}{p^2}\frac{\phi_3}{1+z} \\ \frac{m^2}{p^2}\frac{\phi_4}{1-z} \\ \frac{2m\phi_5}{E\sin\theta} \end{bmatrix} \tag{22}$$

A noteworthy point is that helicity amplitude ϕ_3 behaves as $1+\cos\theta$ as $\theta \to \pi$, ϕ_4 behaves as $1-\cos\theta$ as $\theta \to 0$, and ϕ_5 behaves as $\sin\theta$ as $\theta \to 0$ or π. These kinematic zeroes are divided out in Eq. (22) which means that the Lorentz-invariant amplitudes remain finite at the points $\theta = 0$ or π as one expects from meson exchange dynamics. This is the origin of the statement that the Fermi amplitudes are free of kinematic singularities.

Equations (21)-(22) provide an elegant way to express the invariant amplitudes in terms of known helicity amplitudes. An equivalent determination of the invariant amplitudes is given in Ref. 4 starting from Wolfenstein amplitudes.[17] The answers are the same. Determination of the Lorentz-invariant NN amplitudes from phase shifts was the crucial step leading to the first implementation of Eq. (13).[4]

FIRST IMPLEMENTATION OF $\hat{U}_{IA}$

Four approximations were used in the first attempts to implement Eq. (13).[1-3]

(1) Assume the five-term on-mass-shell representation of Eq. (18) to be an adequate operator form of $\hat{\mathcal{M}}$. This corresponds to using $\hat{\mathcal{M}}$ just as one would use the corresponding meson exchange terms which are embedded in the invariant amplitudes.

(2) Assume relativistic Hartree wave functions give an adequate description of the nuclear ground state.[29,30]

These two assumptions lead to the approximate form

$$\hat{U}_{IA} \simeq \sum_{\alpha} \int \frac{d^3p_1}{(2\pi)^3} \bar{\psi}_\alpha(\vec{p}_1 + \tfrac{1}{2}\vec{q})\hat{\mathcal{M}}(\vec{p},\vec{p}_1 - \tfrac{1}{2}\vec{q} \to \vec{p}-\vec{q},\vec{p}_1 + \tfrac{1}{2}\vec{q}) \times \psi_\alpha(\vec{p}_1 - \tfrac{1}{2}\vec{q}) , \quad (23)$$

where $\psi_\alpha(\vec{p})$ is the Fourier transform of a Dirac-Hartree wave function for occupied state α. Calculations are usually based on the wave functions of Ref. 30.

(3) Following techniques commonly used in the nonrelativistic impulse approximation, assume that a factorization approximation is adequate.[31]

This consists in expanding $\hat{\mathcal{M}}$ in (23) about the point $\vec{p}_1 = 0$, which is the most likely value due to rapid falloff of nuclear wave functions with increasing momentum. The expansion term linear in $\vec{p}_1$ vanishes due to symmetry of the product of nuclear wave functions under $\vec{p}_1 \to -\vec{p}_1$. In this sense, $\vec{p}_1 = 0$ is an optimal factorization point. From the factorization assumption one obtains a form analogous to the nonrelativistic "tρ" optical potential,

$$\hat{U}_{IA}(\vec{p},\vec{q}) \simeq \frac{1}{4} \mathrm{Tr}_2\{\hat{\mathcal{M}}(\vec{p},-\frac{1}{2}\vec{q} \to \vec{p}-q, \frac{1}{2}\vec{q})\hat{\rho}(q)\} . \tag{24}$$

Only the contribution due to $\hat{\mathcal{M}}$ at $\vec{p}_1=0$ is retained. This is referred to as the "Breit frame" amplitude.[17]. The nuclear density operator is defined in the Hartree model by

$$\hat{\rho}(q) \equiv 2 \sum_\alpha \int \frac{d^3p_1}{(2\pi)^3} \psi_\alpha (\vec{p}_1 - \frac{1}{2}\vec{q})\, \bar{\psi}_\alpha(\vec{p}_1 + \frac{1}{2}\vec{q})$$

$$= \rho_S(q) + \gamma_2^0\, \rho_V(q) - \frac{\vec{\alpha}_2\cdot\vec{q}}{2m}\, \rho_T(q) , \tag{25}$$

where the latter form follows for closed shell nuclei. Thus there are scalar, vector and tensor contributions to the density. The three densities ρ_S, ρ_V and ρ_T are all comparable in size, however, the factor $\vec{q}/(2m)$ generally suppresses tensor contributions.

(4) In order to localize the nucleon interchange contributions, assume $\hat{\mathcal{M}}$ to be a function of $\vec{q}^2$ only.

Nucleon interchange contributions are nonlocal and they are inherent due to the Pauli principle. However, they mainly contribute when the NN amplitudes are evaluated at backward scattering angles. Thus momentum transfer q must be comparable to projectile momentum. The density $\hat{\rho}(q)$ tends to suppress the nucleon interchange contributions for energies such as 500-800 MeV. Exchange localization amounts to taking $\hat{\mathcal{M}}$ in Eq. (24) to depend only on q^2 through the on-mass-shell relation, $\cos\theta = 1-q^2/(4p^2)$, where p is the cm frame momentum. These four assumptions lead to an optical potential of the form

$$\hat{U}(q) = S(q) + \gamma^0\, V(q) - \frac{\vec{\alpha}\cdot\vec{q}}{m}\, T(q) , \tag{26}$$

where

$$S(q) = \kappa\, F_1(q)\, \rho_S(q) , \tag{27}$$

$$V(q) = \kappa\, F_2(q)\, \rho_V(q) , \tag{28}$$

$$T(q) = \kappa\, F_3(q)\, \rho_T(q) . \tag{29}$$

Note that $\kappa F_1(q)$ contains the scalar meson exchange contribution to the Feynman amplitude, $-g_S^2/(m_S^2+q^2)$, and may be expected to be attractive. $\kappa F_2(q)$ contains the vector meson exchange contribution, $g_V^2/(m_V^2+q^2)$, and may be expected to be repulsive. Determination of the invariant amplitudes directly from the NN phase shifts using Eqs. (21)-(22) bears out these expectations.[1,4]

When the potential is Fourier transformed to coordinate space and inserted in the Dirac equation, one has

$$[E\gamma^0 - \vec{\gamma}\cdot\vec{p} - m - S(r) - \gamma^0 V(r) - i\vec{\alpha}\cdot\hat{r}\,\frac{T'(r)}{m}]\psi = 0 \qquad (30)$$

where $T'(r) = (d/dr)T(r)$. Figure 2 shows results for scattering of 500 MeV protons by ^{40}Ca based on this parameter-free determination of the Dirac potential.[3] The tensor term is omitted but it does not matter much. Also shown are experimental data and the corresponding nonrelativistic impulse approximation results.[32]

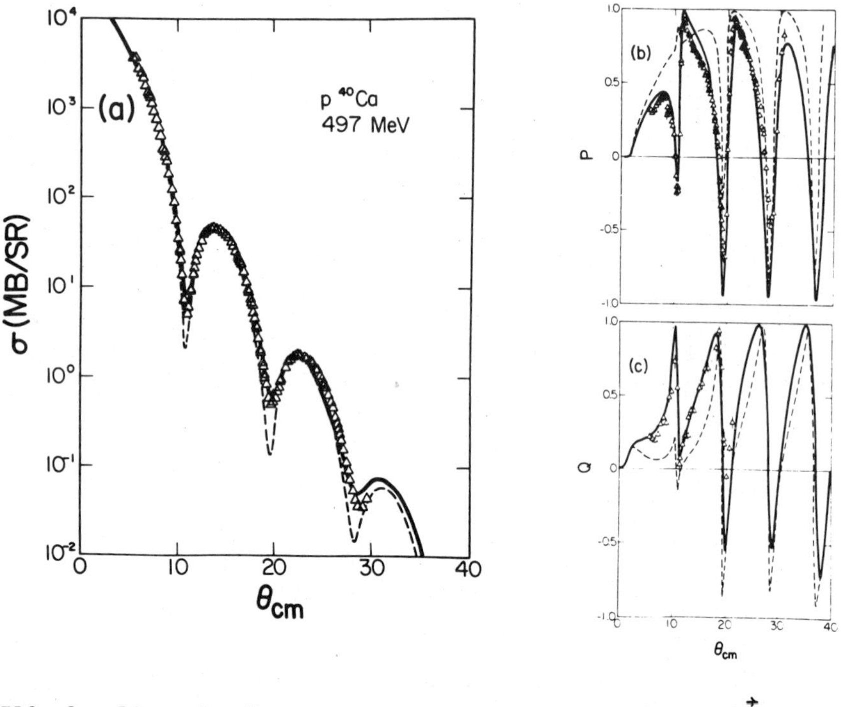

FIG. 2. Dirac impulse approximation results for 497 MeV $\vec{p}+^{40}$Ca elastic scattering are shown by solid lines. Nonrelativistic impulse approximation results are shown by dashed lines.

Much of the current interest in Dirac approaches stems from the exceptional agreement of these first calculations with experimental data, and the clear disagreement of nonrelativistic calculations with the spin observables.

The main ingredient of the Dirac success can be understood in terms of the virtual NN pair contribution which is implicit in solutions of the Dirac equation.[6] Peter Tandy's lectures have established the basic point that a completely equivalent way to write the Dirac equation is in terms of positive and negative energy basis states of the free Dirac equation.[7] For the positive energy component of the wave function one obtains (spin labels are suppressed)

$$(E - E_p - U^{++} - U_{pair})\psi^{+} = 0 \tag{31}$$

where

$$U_{pair} \equiv U^{+-}[E + E_p - U^{--}]^{-1}\, U^{-+} \tag{32}$$

and

$$U^{\rho'\rho} \equiv \bar{u}^{(\rho')}(\vec{p}\,')\,\hat{U}\,u^{(\rho)}(\vec{p})\,, \qquad (\rho,\rho' = \pm)\,. \tag{33}$$

Each of the potentials U^{++}, U^{+-}, etc., is a 2×2 matrix in the Pauli spin space.

To a good approximation, U^{++} is the "nonrelativistic" potential. More precisely, one may define a no-pair potential which, in the impulse approximation, contains purely positive energy NN matrix elements for both scattering and bound particles. Due to Eqs. (15) and (27)-(29), it follows that the no-pair potential is equal to the nonrelativistic impulse potential. Since the use of Hartree wave functions[30] in (25)-(29) does not lead to large differences from the use of positive-energy projected wave functions, $U_{no\text{-}pair} \simeq U^{++}$. One has the result

$$U^{++} = \bar{u}^{(+)}_{S'}(\vec{p}-\vec{q}) \begin{bmatrix} S+V & -\dfrac{\vec{\sigma}\cdot\vec{q}}{m}\,T \\ -\dfrac{\vec{\sigma}\cdot\vec{q}}{m}\,T & S-V \end{bmatrix} u^{(+)}_{S}(\vec{p})\,. \tag{34}$$

Provided the three densities of Eq. (25) are taken equal to $\rho_V(q)$, the baryon density, the Dirac matrix element in (33) produces

$$U^{++} \approx \chi^{+}_{S'}[A + iC\,\vec{\sigma}\cdot\vec{q}\times\hat{p}]\chi_S\,\rho_V(q) \tag{35}$$

where A and C are the usual scalar and spin-flip Wolfenstein amplitudes.[17] The constraints from positive energy NN scattering guarantee that the no-pair potential will agree with the nonrelativistic one.

A crude but useful approximation for U_{pair} is the following[16]

$$U_{pair} = \bar{u}_{S'}^{(+)}(\vec{p}\,') \begin{bmatrix} 0 & 0 \\ 0 & \frac{(S-V)^2}{E+m} \end{bmatrix} u_S^{(+)}(\vec{p}) \ . \tag{36}$$

One now sees that if the tensor term is neglected, Wolfenstein amplitudes A and C (which are known) essentially determine S and V. Once this is done, the pair potential is predicted as shown by (36). The latter step means that the NN amplitudes F_1 and F_2 of Eqs. (27)-(28) are being treated in the same fashion as relativistic meson exchange to predict negative energy couplings in the sense of Eq. (32). Remarkably, the NN amplitude representation of Eq. (13) seems to do this in just the way required to fit the proton-^{40}Ca data at 500 MeV. See Ref. 33 for calculations on other nuclei and at other energies.

Success of the impulse approach at 500 MeV does not carry over to the low energy region. Particularly at energies of 200 MeV or lower one sees that the scalar and vector potential strengths become too large.[10] This is illustrated in Fig. 3 where potential

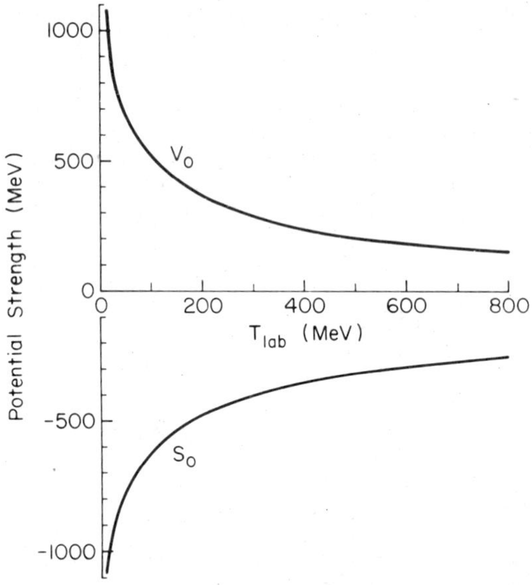

FIG. 3. Strengths of real scalar (S_0) and vector (V_0) potentials in nuclear matter (ρ=0.16 fm^{-3}) based on impulse approximation.

strengths for constant nuclear density $\rho_V = \rho_S = 0.16\ fm^{-3}$ are shown. Below 200 MeV, real parts of scalar and vector potentials become much too large and one sees from Eq. (36) that U_{pair} will be correspondingly large.

The low energy problem can be traced to one-pion exchange contributions associated with nucleon interchange.[23] In the representation of Eq. (18), all one-pion exchange is forced to be pseudoscalar or interchange pseudoscalar (due to the Pauli principle). If one calculates the one-pion exchange contribution, it turns out to be entirely due to the nucleon interchange process because direct pion exchange vanishes due to parity conservation. Figure 4 shows the one-pion exchange contributions to the optical potential for two cases--pseudoscalar πN coupling, γ^5, and pseudovector πN coupling, $\gamma^5 \not{q}/(2m)$. These are equivalent for positive energy states but they lead to very different contributions to U^{+-} and U^{-+} potentials which enter U_{pair}. The contributions to Dirac scalar and vector potentials are also very different. Pseudo-vector πN coupling eliminates the rapid energy dependence which is also seen to characterize the full NN amplitudes in the representation of Eq. (18).

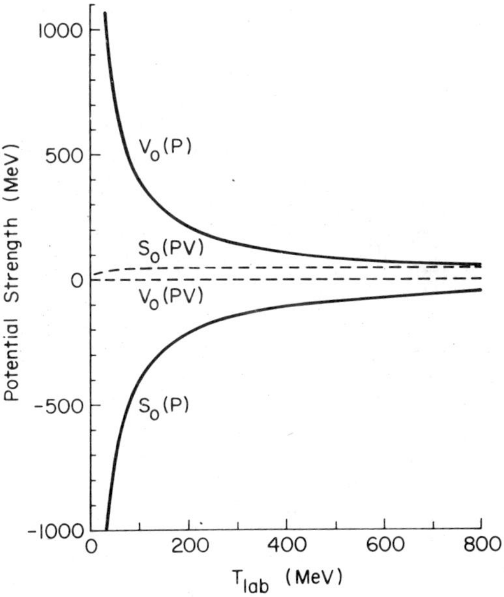

FIG. 4. One-pion exchange contributions to impulse approximation potential strengths, S_0 and V_0. P indicates pseudoscalar πN coupling and PV indicates pseudovector πN coupling.

FIG. 5. Impulse approximation potential strengths (see text).

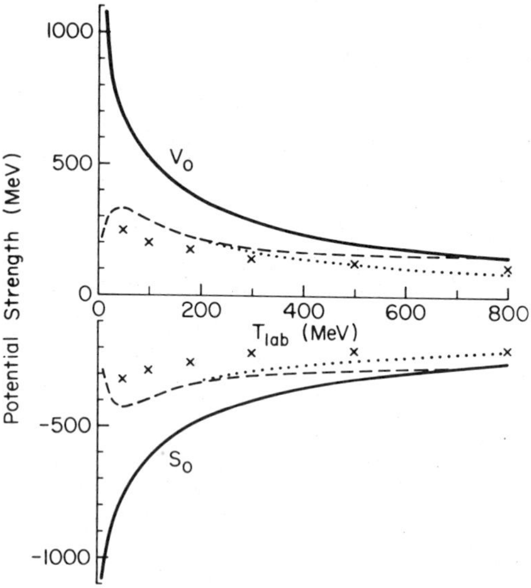

Meson theory succeeds in describing the NN scattering data only when a pair suppression mechanism such as pseudovector πN coupling is used. Pseudoscalar coupling predicts much too large pair couplings at low energy.[11] Now we see that exactly the same thing happens in the Dirac impulse approximation in its original form. The problem is evident only at low energy since the nucleon interchange process is otherwise suppressed. Figure 5 shows what happens if a new representation is used for the NN covariants in place of Eq. (18) such that pionic contributions fall into pseudovector terms.[10] This cannot be done in a unique way. Dashed and dotted lines in Fig. 5 show two possible ways of replacing pseudoscalar covariants by pseudovector ones. In each pseudovector case the potential strengths are predicted to be much smaller and the result is in rough accord with Dirac phenomenology. The x's will be explained further on.

Equivalence of pseudoscalar and pseudovector couplings for positive energy, on-mass-shell matrix elements stems from the identity,

$$\bar{u}_{1'}^{(+)}\bar{u}_{2'}^{(+)}[P-PV]u_1^{(+)}u_2^{(+)} = 0 , \tag{37}$$

where

$$P \equiv \gamma_1^5 \gamma_2^5 , \tag{38}$$

$$PV \equiv \gamma_1^5 \gamma_1 \cdot \frac{(p_1 - p_1')}{2m} \gamma_2^5 \gamma_2 \cdot \frac{(p_2 - p_2')}{2m} , \tag{39}$$

which is known as the equivalence theorem.[34] Differences between pseudoscalar and pseudovector contributions exactly vanish. This means that one may add an arbitrary amount of P-PV to the NN amplitude and still satisfy the positive energy constraints (15). If $\mathcal{M}_1$ is an amplitude which obeys Eq. (15), then

$$\hat{\mathcal{M}}_2 = \hat{\mathcal{M}}_1 + \lambda F_\pi (PV-P) \tag{40}$$

also obeys Eq. (15) for any constant λ!. Figure 5 illustrates the range of possibilities which are obtained with λ = 0 (pseudoscalar) and λ = 1 (pseudovector). The positive energy NN data clearly are necessary but not sufficient input to determine the relativistic NN amplitude needed in the impulse approximation. Indeed, U_{pair} is strongly dependent upon components of $\hat{\mathcal{M}}$ which are unconstrained by NN data, although the simple predictions do succeed at 500 MeV and above.

Analysis of this problem shows that 44 amplitudes are independent in a general Lorentz invariant representation of $\hat{\mathcal{M}}$ on-mass-shell.[35] Five amplitudes can be constrained by positive energy data but 39 others remain unconstrained. I believe the inescapable conclusion to be that one must have a dynamical basis for the NN amplitude in order to define the impulse approximation. For the present, this means we must consider a meson theoretical dynamics.

CONSTRAINTS ON $\hat{\mathcal{M}}$ BASED ON MESON EXCHANGE DYNAMICS

Dr. Machleidt has discussed the meson exchange approach to NN scattering. The model is formulated in terms of an effective Lagrangian which specifies meson-baryon vertices. In addition, cutoff parameters are introduced and one chooses a finite set of meson exchange diagrams to develop the "potential". Unitarity is enforced by solving an integral equation. The approach is relativistic in nature.

Work by J. A. Tjon and collaborators has extended the usual meson exchange dynamics in several ways.[11-15] The relativistic Bethe-Salpeter equation is solved numerically and channel couplings to NΔ and ΔΔ states are introduced in order to describe inelasticity which becomes important at energies above pion production threshold.[14,15] As a result the NN phase shifts are reproduced reasonably well in the 0 to 1000 MeV region. This is rather remarkable progress. One consistent set of meson-baryon couplings and cutoff parameters describe all the essential features of NN scattering over a wide energy region. Moreover, the dynamical equations used predict not only the positive energy matrix elements but also a complete set of negative energy matrix elements. Therefore one has at hand a model capable of providing a dynamical basis for $\hat{\mathcal{M}}$, $\hat{U}_{IA}$ and the implicit pair couplings of the Dirac equation. Currently, no other approach exists to give a general description of $\hat{\mathcal{M}}$.

Figure 6 illustrates the dynamical ingredients of the van Faassen-Tjon model for NN scattering.[15] Coupled integral equations are set up in a Bethe-Salpeter framework for N and Δ baryons with exchange of π, ε, ρ, ω, and η mesons, where ε is a scalar meson with mass and coupling constant essentially the same as the σ meson of one-boson exchange models. The πN coupling is chosen to be pseudovector. Further details may be found in Ref. 15.

As shown by van Faassen and Tjon, the full Bethe-Salpeter dynamics is approximated reasonably well by a three-dimensional quasi-potential equation provided some coupling constants are changed by 10-20%.[15] The quasi-potential formalism is less demanding computationally and therefore has been used in the collaboration of Tjon and myself to calculate meson exchange predictions for $\hat{\mathcal{M}}$.[9,10]

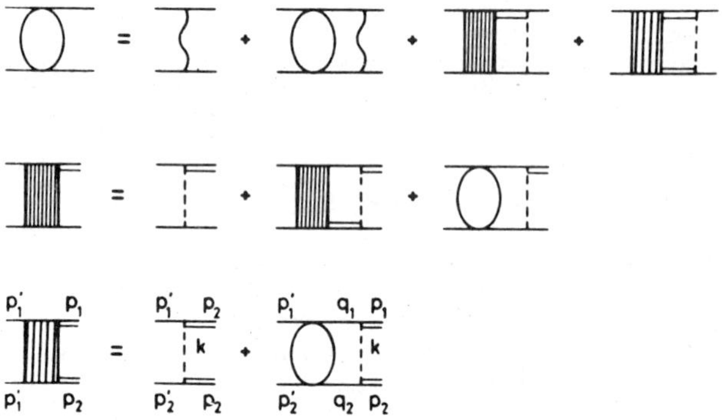

FIG. 6. Diagrams for coupled NN, NΔ, ΔΔ integral equations solved by van Faassen and Tjon. Nucleon lines are shown by ——— and Δ lines are shown by ═══.

Our analysis is a straightforward extension of the analysis leading from Eq. (15) to Eq. (22). The NN integral equations are solved in the cm frame where a partial wave expansion can be performed. Partial wave t-matrix elements of all the needed helicity amplitudes are determined. In the general case, assuming only parity invariance, the helicity amplitudes are[36]

$$\phi^{(\rho_1'\rho_2',\rho_1\rho_2)}_{\lambda_1'\lambda_2',\lambda_1\lambda_2} = \bar{u}^{(\rho_1')}_{\lambda_1'}(\vec{p}\,')\bar{u}^{(\rho_2')}_{-\lambda_2'}(-\vec{p}\,')\hat{\mathcal{M}}\,u^{(\rho_1)}_{\lambda_1}(\vec{p})u^{(\rho_2)}_{-\lambda_2}(-\vec{p}\,')\;/\;\kappa \tag{41}$$

where each $\rho_i = \pm$ denotes the sign of energy, e.g., $p^0 = \rho\sqrt{\vec{p}^{\,2}+m^2}$. There are 256 helicity amplitudes (2^4 energies $\times$ 2^4 helicities). Parity invariance reduces the number of independent amplitudes by half to 128. Partial wave expansions are summed to calculate each of the 128 helicity amplitudes. Due to time-reversal invariance and charge symmetry, only 44 of the amplitudes are actually independent.[35]

The optical potential is generally needed in the proton-nucleus cm frame where NN scattering amplitudes depend on more than one angle and thus partial wave expansions are impractical. Therefore it is necessary to construct the amplitudes in the NN cm frame and then to boost to the N-nucleus cm frame. This generally involves Wigner rotations, except at $\theta=0$. A more convenient procedure is to introduce Lorentz-invariant amplitudes which take the same form in all frames. Moreover, the invariant amplitudes can be defined so that the representation of Eq. (18) is naturally embedded.

In Ref. 35 a general Lorentz-invariant representation of the NN scattering amplitude is developed in the following form:

$$\begin{aligned}\hat{F} = {} & \hat{F}^{11} + \Lambda^{(-)}_{2'}\hat{F}^{12} + \hat{F}^{13}\Lambda^{(-)}_{2} + \Lambda^{(-)}_{2'}\hat{F}^{14}\Lambda^{(-)}_{2} \\ & + \Lambda^{(-)}_{1'}[\hat{F}^{21} + \Lambda^{(-)}_{2'}\hat{F}^{22} + \hat{F}^{23}\Lambda^{(-)}_{2} + \Lambda^{(-)}_{2'}\hat{F}^{24}\Lambda^{(-)}_{2}] \\ & + [\hat{F}^{31} + \Lambda^{(-)}_{2'}\hat{F}^{32} + \hat{F}^{33}\Lambda^{(-)}_{2} + \Lambda^{(-)}_{2'}\hat{F}^{34}\Lambda^{(-)}_{2}]\Lambda^{(-)}_{1} \\ & + \Lambda^{(-)}_{1'}[\hat{F}^{41} + \Lambda^{(-)}_{2'}\hat{F}^{42} + \hat{F}^{43}\Lambda^{(-)}_{2} + \Lambda^{(-)}_{2'}\hat{F}^{44}\Lambda^{(-)}_{2}]\Lambda^{(-)}_{1}\;.\end{aligned} \tag{42}$$

There are sixteen $\hat{F}^{ij}$. The first superscript of $\hat{F}^{ij}$ is the class index, i = 1 to 4, which refers to the asssociated negative energy projectors for particle 1 (projectile) and the second superscript of $\hat{F}^{ij}$ is the subclass index, j = 1 to 4, which refers to the associated negative energy projectors for particle 2 (target). The form of (42) is chosen so that the representation of Eq. (18) is naturally embedded as $\hat{F}^{11}$. The use of negative energy projection operators guarantees that positive energy matrix elements, $(++|\hat{F}|++)$, involve only the $\hat{F}^{11}$ term. Furthermore, the representation contains all possible negative energy components consistent with parity invariance.

In the complete representation of (42), the class 1 terms (first row) determine potential U^{++}. The coupling potential U^{-+} is determined by class 1 and class 2 terms (first and second rows), and coupling potential U^{+-} is determined by class 1 and class 3 terms. The U^{--} potential involves contributions from all four classes. The subclass 1 terms (left column) enter proportional to the full density (ρ_S or ρ_V) and subclass 2, 3 and 4 amplitudes which involve $\Lambda_2^{(-)}$ projection operators enter the potentials proportional to the lower component density ρ_ℓ, where $\rho_\ell = 1/2(\rho_V - \rho_S)$ is typically a few percent of ρ_V.

Each $\hat{F}^{ij}$ in (42) is expanded in an overcomplete set of nine covariants:

$$\hat{F}^{ij} = \sum_{n=1}^{9} F_n^{ij} \mathcal{K}_n , \tag{43}$$

where

$$\mathcal{K}_n = \{S,V,T,P,A,\gamma_2\cdot Q_1,\gamma_1\cdot Q_2,P\gamma_2\cdot Q_1,P\gamma_1\cdot Q_2\} . \tag{44}$$

Here $Q_1^\mu = (p_1+p_1')^\mu/(2m)$, $Q_2^\mu = (p_2+p_2')/(2m)$, and p_1 and p_2 are initial momenta, p_1' and p_2' are final momenta. There exists one linear relation among the helicity basis matrix elements of the nine $\mathcal{K}_n$ which reduces them to eight linearly independent covariants. Equation (42) is capable of representing a general parity conserving NN interaction since only half of the $4^4 = 256$ matrix elements of $\hat{F}$ are independent when parity is conserved.

Invariant amplitudes F_n^{ij} are related to the helicity amplitudes as defined by Eq. (41) through a set of kinematic matrices, for each of the 16 combinations of i and j values. Equation (21) illustrates how the matrices are determined to connect invariant amplitudes to helicity amplitudes. Using the meson theoretical helicity amplitudes and these matrices, one simply solves for the invariant amplitudes in complete analogy with the analysis of positive energy amplitudes except that it is done numerically rather than analytically. The result is the first determination of a complete set of Lorentz-invariant NN amplitudes.

As noted above, the $\hat{F}^{11}$ part of Eq. (42) contains all the information about positive energy scattering. Therefore the five invariant amplitudes F_n^{11}, n = 1 to 5, are exactly as determined by Eqs. (21)-(22). Amplitudes F_n^{11}, n = 6 to 9, vanish due to symmetries. If all the other invariant amplitudes were zero, the original impulse approximation based on Eq. (18) would be reproduced. In order to guarantee their accuracy, the F_n^{11} amplitudes are constrained by experimental phase shifts in our analysis. Thus the meson theory is used to obtain all the other amplitudes which are unconstrained by NN data. These other amplitudes turn out to be far from negligible as may be appreciated by the fact that they contain the difference between pseudoscalar and pseudovector πN coupling.

Figure 5 shows the results for real potential strengths based on the meson exchange model as x. There is rough accord with calculations shown by dash and dot lines in Fig. 5 which are based upon the replacement of pseudoscalar covariants by pseudovector ones in simple 5 or 6 term representations based on positive energy data. Our results are therefore consistent with the interpretation that the most important difference between $\mathcal{M}$ of Eq. (18) and $\hat{\mathcal{M}}$ based upon meson exchange dynamics lies in the consistent treatment of pseudovector πN coupling.

IMPULSE OPTICAL POTENTIAL BASED ON MESON THEORY

In order to determine the meson theoretical optical potential, the analysis given before is reconsidered from the viewpoint of replacing assumption 1 by

1') Assume the 128 term representation of Eqs. (42)-(43) to be an adequate operator form for $\hat{\mathcal{M}} \equiv \kappa F$ (see Eq. (17)).

Assumptions 2 and 3 of the previous discussion are retained. Therefore we need a trace of the meson theoretical $\hat{\mathcal{M}}$ just as in Eq. (24). This leads to the following expansion for the optical potential

$$\hat{U} = \hat{U}^1 + \Lambda_{1'}^{(-)} \hat{U}^2 + \hat{U}^3 \Lambda_1^{(-)} + \Lambda_{1'}^{(-)} \hat{U}^4 \Lambda_1^{(-)} , \tag{45}$$

where each of the four terms contains scalar, vector and tensor contributions,

$$\hat{U}^i = S^i + \gamma^0 V^i - \frac{\vec{\alpha}\cdot\vec{q}}{m} T^i \; . \tag{46}$$

Each of the scalar, vector and tensor contributions in Eq. (46) is defined in terms of NN amplitudes from Eq. (42) and scalar, vector and tensor densities from Eq. (25). The potentials are nonlocal functions of $\vec{p}$ and $\vec{q}$. Details of these constructions are not too revealing and therefore they are omitted.

The general structure is defined by Eqs. (45) and (46). This structure is most appropriate to use in a momentum-space formulation of Dirac scattering. Calculations along these lines are planned but have not yet been performed.

Initial calculations have been performed in coordinate space and this involves use of assumption 4 regarding localization of nucleon interchange contributions. For low incident proton energy, nucleon interchange contributions are important and therefore the results of calculations using a localization scheme are only reliable at small scattering angles. Improvements to the localization scheme have been discussed by Horowitz at this workshop[37] following the nonrelativistic technique of Brieva and Rook, and based upon Yukawa fits to NN amplitudes. Improved calculations using this technique are planned for the near future.

If the NN amplitudes are localized, and the negative energy projection operators are approximated as follows,

$$\Lambda^-(\vec{p}) \simeq (-E\gamma^0 + \vec{\gamma}\cdot\vec{p} + m)/(2m) \; , \tag{47}$$

where E is the incident proton energy (in place of $E_p = \sqrt{\vec{p}^2+\vec{m}^2}$), then we find that the optical potential can be transformed to coordinate space.[9] The result is

$$\hat{U}(\vec{r}) = S(r) + \gamma^0 V(r) - \{C(r), \frac{\not{p}-m}{2m}\}_+$$
$$- i\vec{\alpha}\cdot\hat{r}\,\frac{T'(r)}{m} - [S_{LS}(r) + \gamma^0 V_{LS}(r)]\vec{\sigma}\cdot\vec{L} \; , \tag{48}$$

where $\not{p} \equiv E\gamma^0 - \vec{\gamma}\cdot\vec{p}$ and $\vec{p} = (\hbar/i)\vec{\nabla}$. This potential contains six of the eight possible terms which were shown in Ref. 38 to be admissible parity-invariant components of a Dirac optical potential. The two missing terms vanish due to time-reversal invariance in our analysis. All terms in the potential received contributions from NN amplitudes which are not constrained by positive energy data. Nevertheless, U^{++} in Eq. (31) is essentially the same as in nonrelativistic analyses and in the original formulation of the impulse approximation. U_{pair} changes in ways which are consistent with meson theory and is much smaller at low proton energy.

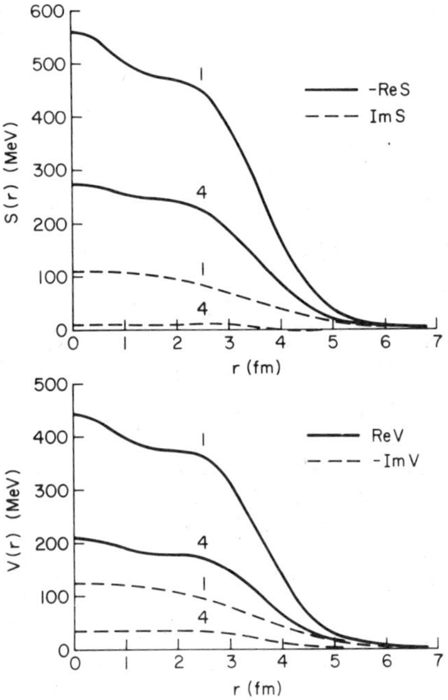

FIG. 7. Scalar and vector potentials for 181 MeV $p+^{40}Ca$ based on original impulse approximation, 1, and meson theoretical impulse approximation, 4.

Figure 7 shows results for 181 MeV protons scattering from ^{40}Ca.[9] The real and imaginary scalar and vector potentials are shown based on the original impulse approximation, Eqs. (27)-(29), with label 1, meaning that only F_n^{11} amplitudes from one class in Eq. (42) are used. The meson theoretical results based on using amplitudes from four classes, F_n^{11}, F_n^{21}, F_n^{31} and F_n^{41}, are shown with label 4. The meson theoretical calculation reduces the real parts of scalar and vector potentials by about 50% compared to the original impulse approximation predictions. Table II illustrates that the scalar and vector potentials at r=0 have rather modest

TABLE II

SCALAR AND VECTOR POTENTIALS AT r=0

T_{lab}(MeV)	S(r=0) (MeV)	V(r=0) (MeV)
181	-275+i12	212-i36
500	-219+i38	150-i75
800	-214+i36	138-i98

energy dependence in the meson theoretical approach. Results for scattering observables σ and A_y at 181 MeV are shown in Fig. 8.

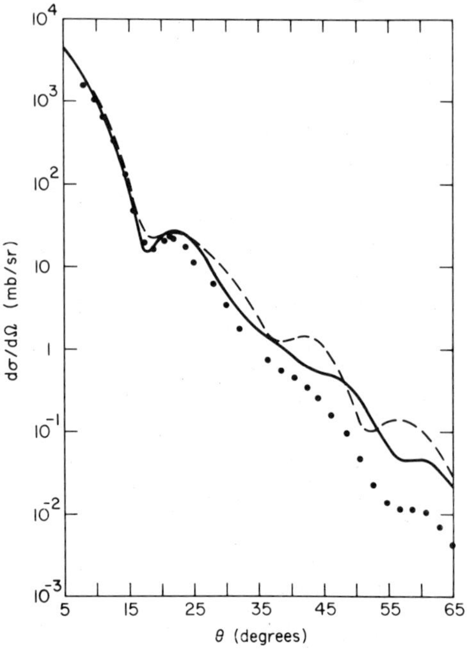

(a)

FIG. 8. Cross section (a) and analyzing power (b) for 181 MeV $\vec{p}+{}^{40}Ca$ elastic scattering. Solid line shows meson-theoretical optical-potential result omitting terms which vanish when nuclear wave functions are proportional to free Dirac spinors. Dash line shows original impulse approximation result.

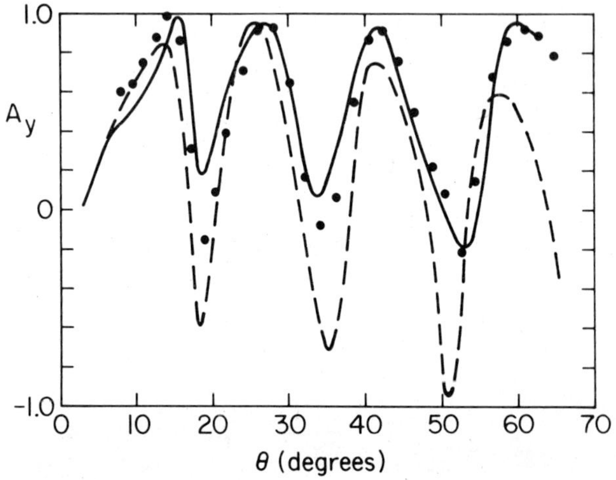

(b)

Experimental points are from Ref. 39. The meson theoretical approach (solid lines) is a modest improvement over the original impulse approximation (dash lines). Figure 9 shows a more complete calculation based on all 16 classes of NN amplitudes from Eq. (42) in comparison with very recent experimental data of E. J. Stephenson and collaborators,[40] on the third observable, Q, for 200 MeV $\vec{p}+{}^{40}Ca$ scattering. The agreement here is very encouraging.

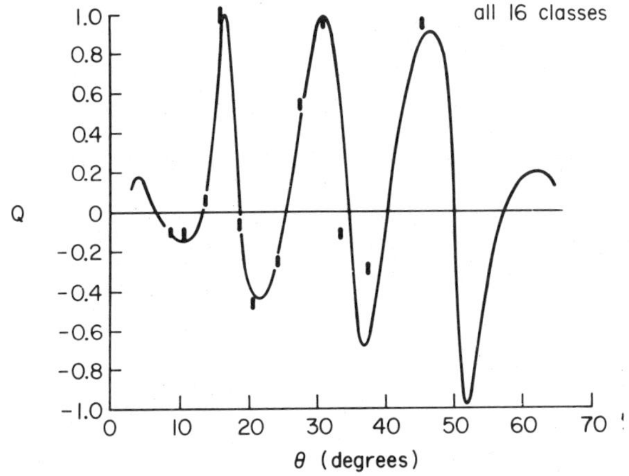

FIG. 9. Spin rotation function Q for 200 MeV $\vec{p}+{}^{40}Ca$ elastic scattering. Solid line shows meson theoretical result including all NN invariant amplitude components.

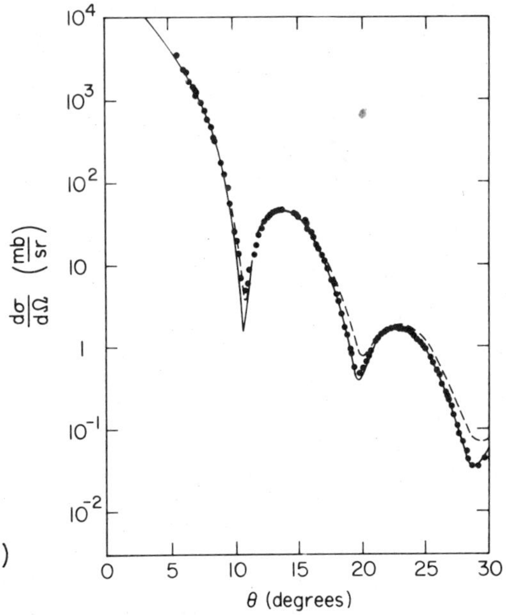

(a)

FIG. 10. Cross section (a), analyzing power (b), and spin rotation function (c) for 497 MeV $\vec{p}+{}^{40}Ca$. Solid lines show results based on meson theoretical potential omitting terms which vanish when nuclear wave functions are proportional to free Dirac spinors. Dash line shows results based on original impulse approximation.

(b)

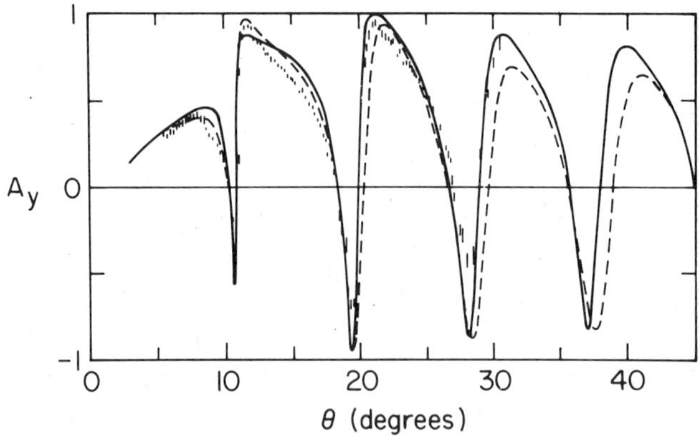

(c)

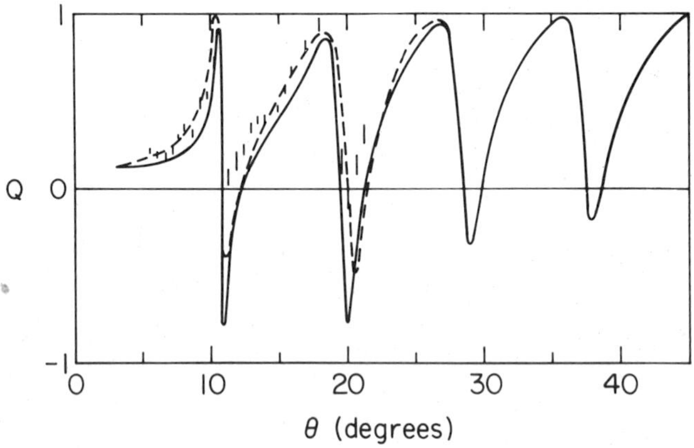

Figure 10 shows results for 500 MeV $\vec{p}+^{40}Ca$ based only on the dominant four classes of NN amplitudes (those not involving $\overset{(-)}{\Lambda}_2$ or $\overset{(-)}{\Lambda}_{2'}$). The differential cross section is in very nice agreement with the experimental data[33,41] and this is an improvement over the original impulse approximation shown by the dashed line. Analyzing power and Q results are shown in Figs. 10b and 10c. Again the results are reasonable. The available calculations are neither systematic nor complete, but they do indicate that the meson theoretical approach is capable of explaining experimental results.

SUMMARY

Empirical evidence in proton-nucleus interactions suggests that quasi-free NN scatterings are the predominant reaction mechanism. Theoretical developments are organized accordingly to display a multiple scattering structure. The impulse approximation provides a good starting point for theoretical analysis.

In a relativistic framework, the impulse approximation can be defined in terms of all contributions to the free NN→NN scattering which are collectively denoted by $\hat{\mathcal{M}}$, the Feynman four-point function. Moreover, the relativistic nuclear ground state function is needed. These ingredients are combined to obtain the impulse approximation optical potential. Recent developments in relativistic treatments of proton-nucleus scattering can be understood as various approximations to the formally defined optical potential.

First attempts to construct the optical potential used only the constraints which follow from positive energy NN scattering data.[1-3] A five-term Lorentz-invariant representation of free NN scattering data based on the Fermi covariants naturally displayed large scalar attraction and large vector repulsion in accord with one-boson exchange models of the nuclear force.[4] Moreover, the predictions for the Dirac optical potential gave a very successful description of 500 MeV proton scattering by ^{40}Ca with no free parameters.[2,3]

Further analyses established that the virtual N$\bar{N}$ pair contributions, which are implicit in use of the Dirac equation, are instrumental to the success.[6] The virtual pair contribution is the essential difference from nonrelativistic analyses. At low proton energy, the impulse approach based on the Fermi covariants produces much too large pair contributions because pseudoscalar πN coupling is implicit.[23] This problem and a related ambiguity in the representation of NN amplitudes can only be satisfactorily resolved by developing a dynamical basis for the NN interaction.

Conventional meson theory of the nuclear force is used to construct a complete set of Lorentz-invariant amplitudes for NN scattering.[35] One consistent set of meson-baryon coupling constants and cutoff masses provides a reasonable description of NN data from 0 to 1000 MeV.[30] This same model is adopted as the fundamental ingredient of the optical potential in order to provide a dynamical approach free of ambiguity with regard to choices of covariants in the NN representation.[9,10] Virtual pair contributions are thereby given a dynamical foundation.

Calculations for proton-nucleus scattering demonstrate reasonable, even encouraging, agreement with experimental data. More complete calculations are in progress in order to establish whether the dynamical approach leads to systematic improvement in the description of data. It already seems clear that the original impulse approximation was not misleading at proton energies of 500-800 MeV. However it is essential to adopt a dynamical basis in order to obtain satisfactory results at low energy. The meson exchange model described consistently incorporates pseudovector πN coupling which is a key feature missing in the original impulse approximation.

ACKNOWLEDGEMENT

This work was supported under U. S. Department of Energy contract DE-AS05-76-ER05126.

REFERENCES

1. J. A. McNeil, J. R. Shepard and S. J. Wallace, Phys. Rev. Lett. 50, 1439 (1983).

2. J. R. Shepard, J. A. McNeil and S. J. Wallace, Phys. Rev. Lett. 50, 1443 (1983).

3. B. C. Clark, S. Hama, R. L. Mercer, L. Ray and B. D. Serot, Phys. Rev. Lett. 50, 1644 (1983).

4. J. A. McNeil, L. Ray and S. J. Wallace, Phys. Rev. C27, 2123 (1983).

5. M. L. Goldberger, M. T. Grisaru, S. W. MacDowell and D. Y. Wong, Phys. Rev. 120, 2250 (1960).

6. M. V. Hynes, A. Picklesimer, P. C. Tandy and R. M. Thaler, Phys. Rev. Lett 52, 978 (1984).

7. P. C. Tandy, this proceedings.

8. D. Adams and M. Bleszynski, Phys. Lett. 136B, 10 (1984).

9. J. A. Tjon and S. J. Wallace, Phys. Rev. Lett. 54, 1357 (1985).

10. J. A. Tjon and S. J. Wallace, Phys. Rev. C (in press).

11. J. Fleischer and J. A. Tjon, Phys. Rev. D21, 87 (1980).

12. M. J. Zuilhof and J. A. Tjon, Phys. Rev. C24, 736 (1981).

13. J. A. Tjon and E. E. van Faassen, Phys. Lett. 120B, 39 (1983).

14. E. E. van Faassen and J. A. Tjon, Phys. Rev. C28, 2354 (1983).

15. E. E. van Faassen and J. A. Tjon, Phys. Rev. C30, 285 (1984).

16. S. J. Wallace, in Proceedings of LAMPF Workshop on Dirac Approaches to Nuclear Physics, Conference Proceedings LA-10438-C (Los Alamos National Laboratory, Los Alamos, New Mexico, 1985).

17. S. J. Wallace in Advances in Nuclear Physics, edited by J. W. Negele and E. Vogt, Vol. 12 (Plenum, New York, 1981), p. 135.

18. M. L. Goldberger and K. M. Watson, "Collision Theory" (Wiley, New York, 1964).

19. R. J. Glauber, in Lectures in Theoretical Physics, edited by W. E. Britten and L. G. Dunham, Vol. I (Interscience, New York, 1959), p. 315.

20. A. Kerman, H. McManus and R. M. Thaler, Ann. Phys. (N.Y.) 8, 551 (1959).

21. J. A. McGill, G. W. Hoffmann, M. L. Bartlett, R. W. Fergerson, E. C. Milner, R. E. Chrien, R. J. Sutter, T. Kozlowski and R. L. Stearns, Phys. Rev. C29, 204 (1984); J. A. McGill, C. Glaushausser, K. Jones, S. K. Nanda, M. Bartlett, R. Fergerson, J. A. Marshall, E. C. Milner and G. W. Hoffmann, Phys. Lett. 134B, 157 (1984).

22. R. A. Smith, "Spin Observables in Inelastic Proton-Nucleus Scattering at Intermediate Energy," Ph.D. dissertation, University of Maryland (1984).

23. S. J. Wallace, in Proceedings of the Seventeenth LAMPF Users Group Meeting, Conference Proceedings LA-10080-C (Los Alamos National Laboratory, Los Alamos, 1984).

24. R. Machleidt, this proceedings.

25. N. Nakanishi, Suppl. Prog. Theor. Phys. (Japan) 43, 1 (1969).

26. M. Jacob and G. C. Wick, Ann. Phys. (N.Y.) 7, 404 (1959).

27. R. A. Arndt, L. D. Roper, R. A. Bryan, R. B. Clark, B. J. VerWest and P. Signell, Phys. Rev. D28, 97 (1983).

28. M. L. Goldberger, Y. Nambu and R. Oehme, Ann. Phys. 2, 226 (1957).

29. L. D. Miller, Phys. Rev. C14, 706 (1976).

30. C. J. Horowitz and B. D. Serot, Nucl. Phys. A368, 503 (1981).

31. A. Picklesimer, P. C. Tandy, R. M. Thaler and D. Wolfe, Phys. Rev. C29, 1582 (1984).

32. L. Ray and G. W. Hoffmann, Phys. Rev. C31, 538 (1985).

33. G. W. Hoffmann, L. Ray, M. L. Bartlett, R. Fergerson, J. McGill, E. C. Milner, K. K. Seth, D. Barlow, M. Bosko, S. Iverson, M. Kaletka, A. Saha and D. Smith, Phys. Rev. Lett. 47, 1436 (1981).

34. J. Hamilton, Theory of Elementary Particles (Oxford Univ. Press, London, 1959).

35. J. A. Tjon and S. J. Wallace, University of Maryland technical report OR05126-246 (1985), "General Lorentz-Invariant Representation of NN Scattering Amplitudes," submitted for publication.

36. J. J. Kubis, Phys. Rev. D6, 547 (1972).

37. C. J. Horowitz, this proceedings.

38. L. S. Celenza and C. M. Shakin, Phys. Rev. C28, 1256 (1983).

39. L. G. Arnold, B. C. Clark, R. L. Mercer and P. Schwandt, Phys. Rev. C23, 1949 (1981).

40. E. J. Stephenson, in Proceedings of LAMPF Workshop on Dirac Approaches to Nuclear Physics, Conference Proceedings LA-10438-C (Los Alamos National Laboratory, Los Alamos, New Mexico, 1985).

41. A. Rahbar, B. Aas, E. Bleszynski, M. Bleszynski, M. Haji-Saeid, G. J. Igo, F. Irom, G. Pauletta, A. T. M. Wang, J. McClelland, J. F. Amann, T. A. Carey, W. D. Cornelius, M. Barlett, G. W. Hoffmann, C. Glashausser, S. Nanda and M. M. Gazzaly, Phys. Rev. Lett. 47, 1811 (1981).

RELATIVISTIC vs NONRELATIVISTIC MODELS OF PROTON-NUCLEUS INELASTIC SCATTERING

James R. Shepard
University of Colorado
Boulder, CO 80309

ABSTRACT

Relativistic and non-relativistic models of proton-nucleus inelastic scattering based on the impulse approximation are compared and contrasted. In the version of the impulse approximation we consider, a class of terms not present in the non-relativistic formulation are found to arise naturally in the relativistic treatment. These terms, when expressed as matrix elements of two-component wavefunctions, are found to involve the non-local current operator. Possible experimental signatures of these terms as well as of the strong scalar and time-like vector potentials which are present in relativistic models of nuclear dynamics are examined.

GENERAL PROPERTIES OF INELASTIC AMPLITUDES

Proton-nuleus inelastic scattering is the process by which a nuclear level other than the ground state is excited by proton scattering. The form of the transition amplitude is restricted by the requirements of invariance under space-inversion (parity), rotations, and time-reversal. We may define a right-handed coordinate system in terms of the projectile initial and final momenta according to Figure 1. (For simplicity, we assume that the overall Q-value of the reaction is zero.) A triad of orthogonal unit vectors is $\{\hat{x},\hat{y},\hat{z}\} = \{\hat{n},\hat{p},\hat{q}\}$. We quantize along the $\hat{z} = \hat{q}$ direction. Assuming the initial nuclear state to have spin-parity $J_i^{\pi_i} = 0^+$, we may write the inelastic scattering amplitude to a final state of spin-parity J^{π} and spin projection M as

$$T_M = \langle J^\pi, M ; \vec{p}_f, m_f | \mathcal{T} | 0^+ ; \vec{p}_i, m_i \rangle \tag{1}$$

where we have explicitly included reference to the initial and final spin projections of the projectile, m_i and m_f, respectively.

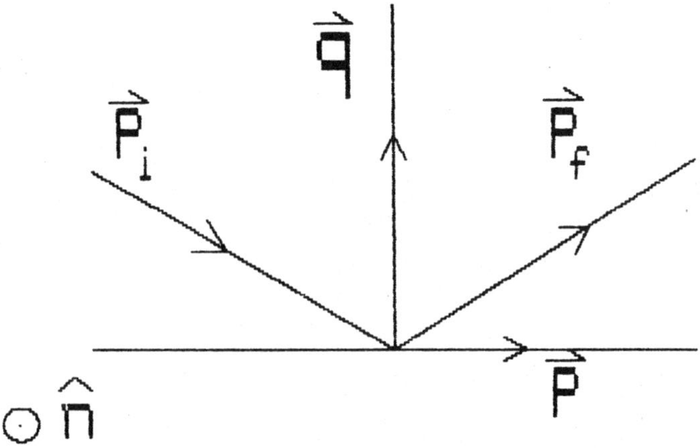

FIG. 1 Coordinate system for inelastic scattering.

This transition amplitude may be expressed in terms of operators acting in the spin-space of the projectile; i.e., in terms of the Pauli spin matrices of the projectile. We define components of the Pauli spin operator as follows:

$$\sigma_n = \vec{\sigma}\cdot\hat{n}, \quad \sigma_p = \vec{\sigma}\cdot\hat{p}, \text{ and } \sigma_q = \vec{\sigma}\cdot\hat{q}.$$

It is convenient to identify the final nuclear states as having either natural ($\pi = (-1)^J$) or unnatural ($\pi = -(-1)^J$) parity. Assuming the interaction of Eq.1 to be invariant with respect to the parity transformation and rotations, we find [1] that the transition amplitude must have the general form given in Table I, where we use the definitions $T_M^{(+)} = i/\sqrt{2}\,(T_M + T_{-M})$ and $T_M^{(-)} = -1/\sqrt{2}\,(T_M - T_{-M})$, assuming $M \geq 1$.

TABLE I

FORM OF THE INELASTIC AMPLITUDE

Natural Parity	Unnatural Parity
$T_0 = A_0 + B_0 \sigma_n$	$T_0 = C_0 \sigma_p + D_0 \sigma_q$
$T_M^{(+)} = A_M + B_M \sigma_n$	$T_M^{(+)} = C_M \sigma_p + D_M \sigma_q$
$T_M^{(-)} = C_M \sigma_p + D_M \sigma_q$	$T_M^{(-)} = A_M + B_M \sigma_n$

Having established the general form of the transition amplitude, we now indicate how it is related to experimentally measureable quantities. Modern intermediate energy facilities like LAMPF, TRIUMF, and IUCF are (or will soon be) able to deliver proton beams polarized in any direction and furthermore can measure the polarization state of the proton after scattering. This permits the measurement of so-called spin-transfer observables in addition to the more usual cross sections and analyzing powers. In order to understand what these spin-transfer quantities are and how they are related to the amplitudes of Table I, we must first consider the spin density matrix for a spin-1/2 particle. We have

$$\rho(\vec{P}) = \tfrac{1}{2}\left(\mathbb{1} + \vec{P}\cdot\vec{\sigma}\right) \tag{2}$$

where $\vec{P}$ is the polarization of the prepared system. Clearly, ρ is an operator in the spin space of the particles making up the system. The expectation value of a spin operator, $\mathcal{O}$, for this system is given by

$$\langle \mathcal{O} \rangle = \mathrm{Tr}\, \mathcal{O}\rho / \mathrm{Tr}\,\rho . \tag{3a}$$

For example, the polarization in the $\hat{y}$-direction is given by

$$\langle \sigma_y \rangle = \tfrac{1}{2}\, \mathrm{Tr}\, \sigma_y \left(\mathbb{1} + \vec{P}\cdot\vec{\sigma}\right) = P_y \tag{3b}$$

where we have made use of the following properties of the Pauli spin matrices:

$$\sigma_i \sigma_j = i\varepsilon_{ijk}\sigma_k + \delta_{ij}\mathbb{1}, \quad \mathrm{Tr}\,\sigma_i = 0, \text{ and } \mathrm{Tr}\,\mathbb{1} = 2. \tag{3c}$$

For a prepared system (i.e., a proton beam) described by the density matrix ρ_i, the density matrix after scattering is given by

$$\rho_f = \sum_M T_M \rho_i T_M^{+} \tag{4}$$

where the T_M are the transition amplitudes of Eq.1 and the incoherent sum over M arises from the fact that we do not observe the M substate of the residual nucleus. Then, for example, the proton polarization in the $\hat{j}$-direction after scattering is given by

$$\begin{aligned} P_j(\text{final}) &= \mathrm{Tr}\, \sigma_j \rho_f / \mathrm{Tr}\, \rho_f \\ &= \mathrm{Tr} \sum_M \sigma_j T_M \rho_i T_M^{+} \Big/ \mathrm{Tr} \sum_M T_M \rho_i T_M^{+} . \end{aligned} \tag{5}$$

By preparing the beam so that it is polarized in one of the directions specified by the unit vectors $\{\hat{n},\hat{p},\hat{q}\}$ and then measuring its polarization in one of these directions after scattering it is possible to measure the spin-transfer observables, Dij, defined by

$$I D_{ij} = \tfrac{1}{2} \mathrm{Tr} \sum_M \sigma_i T_M \sigma_j T_M^{+} \tag{6}$$

where $i = \{0, n, p, q\}$ and $\sigma_i = \{\mathbb{1}, \sigma_n, \sigma_p, \sigma_q\}$

and where $I = \tfrac{1}{2} \mathrm{Tr} \sum_M T_M T_M^{+}$.

It is then straightforward to establish the relationships between the Dij's and the elements of the scattering amplitude presented in Table I. These relationships are given in Table II.

We note that D_{on} corresponds to the analyzing power, Ay, and D_{no} to the polarization, P. Furthermore, Dqp is the spin rotation parameter, Q, as originally defined by Glauber and Osland [2]. It is convenient to define the spin-sum and spin-difference functions, Σ_s and Δ_s, respectively, as

$$\Sigma_s = (D_{no} + D_{on}) + i\,(D_{qp} - D_{pq}) = (P + A_y) + i\,(Q + B) \tag{7a}$$

$$\Delta_s = (D_{qp} + D_{pq}) + i\,(D_{no} - D_{on}) = (Q - B) + i\,(P - A_y) \tag{7b}$$

where we have also defined B=-Dpq.

Having established the general properties of the

TABLE II

Relationships Between Elements of Inelastic Amplitudes and Experimental Observables

$$I(1 + D_{nn} + D_{pp} + D_{qq}) = 4 \sum_M{}' |A_M|^2$$

$$I(1 + D_{nn} - D_{pp} - D_{qq}) = 4 \sum_M{}' |B_M|^2$$

$$I(1 - D_{nn} + D_{pp} - D_{qq}) = 4 \sum_M{}' |C_M|^2$$

$$I(1 - D_{nn} - D_{pp} + D_{qq}) = 4 \sum_M{}' |D_M|^2$$

$$I\left[(D_{no} + D_{on}) + i(D_{qp} - D_{pq})\right] = 4 \sum_M{}' A_M B_M^*$$

$$I\left[(D_{pq} + D_{qp}) + i(D_{no} - D_{on})\right] = 4 \sum_M{}' C_M^* D_M$$

inelastic amplitudes and their relation to experimental quantities, we now consider specific reaction models and examine the amplitudes and observables they give rise to.

THE DISTORTED WAVE IMPULSE APPROXIMATION (DWIA) FOR (p,p')

We begin by considering a simple model of proton-nucleus elastic scattering based on the schematic diagrams of Figure 2.

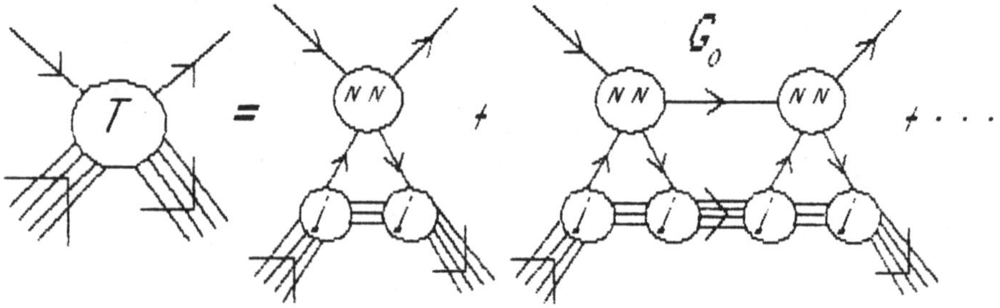

FIG. 2 Schematic multiple-scattering model for proton-nucleus elastic scattering.

Mathematically, these diagrams yield a full elastic scattering t-matrix, T_E , which is related to the single-scattering t-matrix $T_E^{\,0}$ by

$$T_E = T_E^{\,0} + T_E^{\,0} G_0 T_E^{\,0} + \cdots = T_E^{\,0} + T_E^{\,0} G_0 T_E \tag{8}$$

where G_0 is the free projectile propagator. We may use Eqn.8 to write the full elastic scattering wavefunction, ψ , as

$$\psi = \phi + G_0 T_E^{\,0} \psi \tag{9a}$$

or, inverting,

$$\left(G_0^{-1} - T_E^{\,0} \right) \psi = 0 \tag{9b}$$

where ϕ is the free projectile wavefunction. Using the properties the free propagator, we may rewrite Eqn.9b as

$$(H_0 + T_E^0)\psi = E\psi \tag{10}$$

where H_0 is the free projectile Hamiltonian. This is the differential wave equation for elastic scattering and the single-scattering t-matrix, T_E^0 (evaluated between the nuclear ground state wavefunctions), is identified as the optical potential. If we choose a Schroedinger propagator for the projectile, Eq.10 is a Schroedinger equation, whereas a Dirac propagator yields a Dirac equation.

We may now generalize this simple model as is indicated in Figure 3 to account for inelastic scattering.

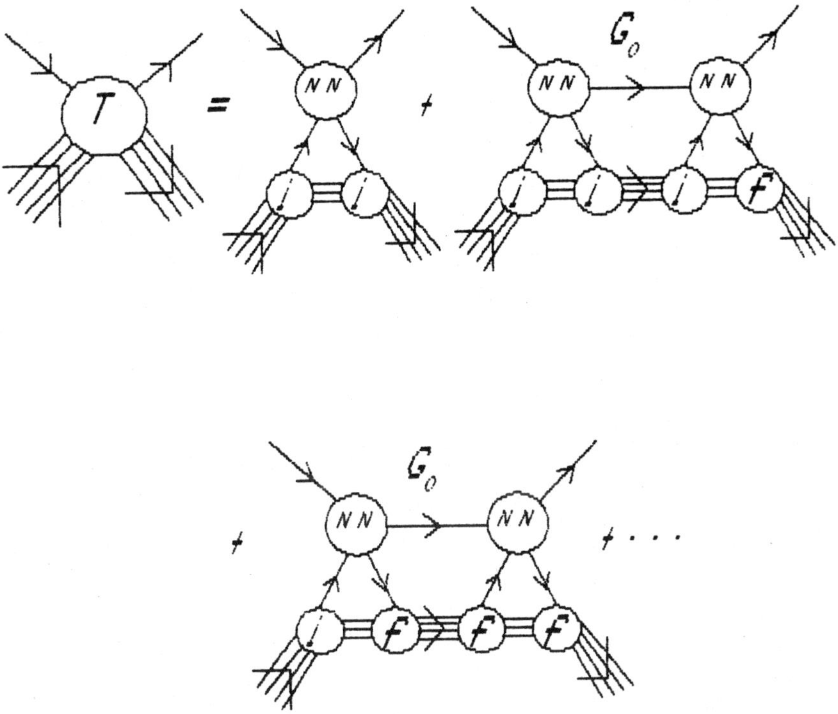

FIG. 3 Schematic multiple-scattering model of proton-nucleus inelastic scattering.

As indicated in the figure, we assume that there is but a single inelastic event with the remaining multiple scatterings being interpreted as elastic events in either the initial or final channel. This identification is accomplished formally as follows: The full inelastic t-matrix, T_X, may be expressed in terms of the single-scattering elastic and inelastic t-matrices, T_E° and T_X°, respectively, as

$$T_X = T_X^{\circ} + T_E^{\circ} G_0 T_X^{\circ} + T_X^{\circ} G_0 T_E^{\circ} + \cdots \tag{11a}$$

$$= (1 + T_E^{\circ} G_0 + T_E^{\circ} G_0 T_E^{\circ} G_0 + \cdots) T_X^{\circ} (1 + G_0 T_E^{\circ} + G_0 T_E^{\circ} G_0 T_E^{\circ} + \cdots)$$

The inelastic amplitude, T_{fi}, may then be written as

$$T_{fi} = \langle \phi_f | (1 + T_E^{\circ} G_0 + \cdots) T_X^{\circ} (1 + G_0 T_E^{\circ} + \cdots) | \phi_i \rangle \tag{11b}$$

$$= \langle \psi_f | T_X^{\circ} | \psi_i \rangle$$

where the ψ's are the elastic scattering projectile wavefunctions or "distorted waves" of Eqn.10. When the single-scattering t-matrices are constructed from the free nucleon-nucleon t-matrices, t_{NN}, as will be discussed below, the T_{fi} of Eqn.11b constitutes the Distorted Wave Impulse Approximation or DWIA for proton-nucleus inelastic scattering.

THE NON-RELATIVISTIC DWIA FOR (p,p')

The non-relativistic DWIA inelastic amplitude may be written as (where the distorted waves are now ϕ's)

$$T_{JM}^{NR} = \phi_f^{(-)\dagger}(\rho)\, \Phi_f^{\dagger}(t)\, \tilde{t}_{NN}^{NR}\, \phi_i^{(+)}(\rho)\, \Phi_i(t) \tag{12}$$

where ϕ and Φ are two-component wavefunctions for the projectile and target, respectively, and where the NN t-matrix, $\tilde{t}_{NN}^{NR}$, is an operator in the space of such two-component wavefunctions; e.g.,

$$\tilde{t}_{NN}^{NR} = a + b\vec{\sigma}_1 \cdot \vec{\sigma}_2 + iqc(\vec{\sigma}_1 + \vec{\sigma}_2)\cdot\hat{n} + d\vec{\sigma}_1\cdot\vec{q}\,\vec{\sigma}_2\cdot\vec{q} + e\vec{\sigma}_1\cdot\hat{p}\,\vec{\sigma}_2\cdot\hat{p} \tag{13}$$

Consistent with the definition of the impulse approximation, the coefficients {a,b,c,d,e} are constrained by the requirement that the matrix elements of $\tilde{t}_{NN}^{NR}$ for free nucleons reproduce the measured NN amplitudes. That is, $\tilde{t}_{NN}^{NR}$ is fixed on-shell by experiment.

In order to get a feeling for the physics contained in the (p,p') amplitude of Eqn.12, we will make the approximation that the projectile wavefunctions are plane waves. This results in a great mathematical simplification and most of the intresting physics is retained. (Note that here and in the development which follows, we do not explicitly treat the antisymmetrization of the projectile and target nucleons. That is, knock-on exchange processes are only included implicitly.) In this Plane Wave Impulse Approximation (PWIA), Eq.12 becomes

$$T_{JM}^{NR} \longrightarrow \langle JM | \tilde{t}_{NN}^{NR} e^{-i\vec{q}\cdot\vec{r}} | 00 \rangle \tag{14}$$

where T_{JM}^{NR} is now an operator in the spin-space of the projectile. We now re-express the NN t-matrix of Eq.13 as

$$\tilde{t}_{NN}^{NR} = \mathcal{A} + \vec{\mathcal{B}}\cdot\vec{\sigma}(target) \tag{15a}$$

$$\text{where} \quad \mathcal{A} = a + iqc\,\sigma_n(proj.) \tag{15b}$$

$$\text{and} \quad \vec{\mathcal{B}} = b\,\vec{\sigma}(proj.) + iqc\,\hat{n} + d\,\vec{\sigma}(proj.)\cdot\vec{q}\,\vec{q} + e\,\vec{\sigma}(proj.)\cdot\hat{p}\,\hat{p} \tag{15c}$$

which causes Eq.14 to become

$$T_{JM}^{NR} = \mathcal{A}\langle JM | e^{-i\vec{q}\cdot\vec{r}} | 00 \rangle + \vec{\mathcal{B}}\cdot\langle JM | \vec{\sigma} e^{-i\vec{q}\cdot\vec{r}} | 00 \rangle \tag{16}$$

Note that, in Eq.16, all nuclear structure dependence resides in the nuclear matrix elements,

$$\rho_J = \langle J0 | e^{-i\vec{q}\cdot\vec{r}} | 00 \rangle$$

$$\text{and} \quad \Sigma_{J,M} = \langle J,M | \sigma_M e^{-i\vec{q}\cdot\vec{r}} | 00 \rangle \tag{17a}$$

while the projectile dependence appears only in the quantities $\mathcal{A}$ and $\vec{\mathcal{B}}$.

We may use standard angular momentum techniques to obtain the plane wave form of T_{JM}^{NR} for, e.g., unnatural parity transitions:

$$T_{JM=0}^{NR} = \vec{\mathcal{B}}\cdot\hat{q}\,\Sigma_{J,0}\,; \quad T_{JM=1}^{(+)} = \vec{\mathcal{B}}\cdot\hat{p}\,\Sigma_{J,1}\,; \quad T_{JM=1}^{(-)} = \vec{\mathcal{B}}\cdot\hat{n}\,\Sigma_{J,1}\,. \tag{17b}$$

These results may be compared with the general form of

the inelastic amplitude given in Table I. Such a comparison yields the identifications presented in Table III where the results for natural parity transitions also appear.

Observables arising in this model can readily be obtained using the expressions shown in Table II. For example, we find, for unnatural parity transitions,

$$\tfrac{1}{4} I(1+D_{nn}+D_{pp}+D_{qq}) = \sum_M |A_M|^2 = |A_n|^2 = |iqc\Sigma'_{J,1}|^2 \tag{18}$$

Note however, that in this non-relativistic model, the spin-difference function is identically zero for both natural and unnatural parity transitions. Nevertheless, we know [4,5] that both P-Ay and Q-B are substantially different from zero for 12C(p,p') to the first two 1+ levels. The physics responsible for the non-zero spin-difference function in these cases has clearly been omitted from our simple model.

TABLE III

Non-Relativistic Plane Wave Inelastic Amplitudes

Natural Parity	Unnatural Parity
$A_q = a\rho_J$; $B_q = iqc\rho_J$	$C_q = 0$; $D_q = (b+q^2d)\Sigma'_{J,0}$
$A_p = qc\Sigma'_{J,1}$; $B_p = -ib\Sigma'_{J,1}$	$C_p = (b+e)\Sigma'_{J,1}$; $D_p = 0$
$C_n = i(b+e)\Sigma'_{J,1}$; $D_n = 0$	$A_n = iqc\Sigma'_{J,1}$; $B_n = b\Sigma'_{J,1}$

THE RELATIVISTIC DWIA FOR (p,p')

The relativistic analogue of the inelastic transition amplitude of Eq.12 is

$$T^R_{JM} = \psi_f^{\dagger(-)}(\rho)\, \Psi_f^{\dagger}(t)\, \gamma^0(\rho)\gamma^0(t)\, \tilde{t}^R_{NN}\, \psi_i^{(+)}(\rho)\, \Psi_i(t) \tag{19}$$

where ψ and Ψ are now four-component wavefunctions for the projectile and target, γ^0 is the usual Dirac matrix and the NN t-matrix is an operator in the space of four-component wave functions, a common form of which is

$$\tilde{t}_{NN}^{R} = t_S + t_V \gamma(1)\cdot\gamma(2) + [t_P + t_A \gamma(1)\cdot\gamma(2)]\gamma^5(1)\gamma^5(2) + t_T \sigma^{\mu\nu}(1)\sigma_{\mu\nu}(2). \tag{20}$$

The impulse approximation now implies that the matrix-elements of $\tilde{t}_{NN}^{R}$ for free, positive-energy nucleons give the measured NN amplitudes.

We again make the plane wave approximation by letting the projectile wavefunctions become

$$\psi \rightarrow u(p)\, e^{-ik\cdot x}$$

where

$$u(p) = \sqrt{\frac{E+m}{2m}} \begin{pmatrix} 1 \\ \dfrac{\vec{\sigma}\cdot\vec{k}}{E+m} \end{pmatrix} \tag{21}$$

is the usual free Dirac spinor. The inelastic amplitude (after the trivial time integration is performed) is then

$$T_{JM}^{R} = u_f^{\dagger}(p) \langle \widetilde{JM} | \gamma^0(p)\gamma^0(t)\, \tilde{t}_{NN}^{R}\, e^{-i\vec{q}\cdot\vec{r}} | \widetilde{00} \rangle \tag{22}$$

where the tildes over the bra and ket indicate a four-component wavefunction. In analogy with Eq.15, we rewrite $\tilde{t}_{NN}^{R}$ as

$$\gamma^0(1)\gamma^0(2)\, \tilde{t}_{NN}^{R} = \sum_{j=1}^{4} (f_j + \vec{\sigma}_1\cdot\vec{\sigma}_2 g_j)\, \Gamma_j(1)\Gamma_j(2) \tag{23a}$$

where the Γ-matrices act only in the space of the components of the wavefunctions and are defined by

$$\Gamma_1 = \begin{pmatrix} 1 & 0 \\ 0 & 1 \end{pmatrix},\ \Gamma_2 = \begin{pmatrix} 1 & 0 \\ 0 & -1 \end{pmatrix},\ \Gamma_3 = \begin{pmatrix} 0 & 1 \\ 1 & 0 \end{pmatrix},\ \Gamma_4 = \begin{pmatrix} 0 & 1 \\ -1 & 0 \end{pmatrix} \tag{23b}$$

while the $\{f_j\}$ and $\{g_j\}$ are related to the $\{t_j\}$ of Eq. 20 by

$$\begin{array}{llll} f_1 = t_V & f_2 = t_S & f_3 = t_A & f_4 = t_P \\ g_1 = -t_A & g_2 = 2t_T & g_3 = -t_V & g_4 = 2t_T \end{array} \tag{24}$$

This finally allows us to write T_{JM}^{R} in a form analogous to T_{JM}^{NR} as given in Eq.16; i.e.,

$$T_{JM}^{R} = \sum_{j=1}^{4} \left\{ a_j \langle \widetilde{JM} | \Gamma_j e^{-i\vec{q}\cdot\vec{r}} | \widetilde{00} \rangle + \vec{B}_j \cdot \langle \widetilde{JM} | \vec{\sigma}\Gamma_j e^{-i\vec{q}\cdot\vec{r}} | \widetilde{00} \rangle \right\} \tag{25a}$$

where $a_j = u_f^{+}(p)\,\Gamma_j\,u_i(p)$ (25b)

and $\vec{B}_j = u_f^{+}(p)\,\vec{\sigma}\,\Gamma_j\,u_i(p).$ (25c)

In evaluating the nuclear matrix elements, we must generalize slightly the techniques used to deal with the non-relativistic nuclear matrix elements of Eqs. 16 and 17a. This is easily accomplished by remembering that upper and lower components of four-component bound-state wavefunctions have opposite parities. For nuclear matrix elements involving Γ_1 or Γ_2 and therefore connecting upper-to-upper and lower-to-lower components, the usual non-relativistic selection rules apply. For example, we find

$$\rho_J^j = \langle \widetilde{J0} | \Gamma_j e^{-i\vec{q}\cdot\vec{r}} | \widetilde{00} \rangle ;\ j = 1,2 \tag{26a}$$

is non-zero only for natural parity transitions while

$$\Sigma_{J,0}^{j} = \langle \widetilde{J0} | \sigma_0 \Gamma_j e^{-i\vec{q}\cdot\vec{r}} | \widetilde{00} \rangle ;\ j = 1,2 \tag{26b}$$

is non-zero only for unnatural parity transitions. Since Γ_3 and Γ_4 connect upper-to-lower and lower-to-upper components, these selection rules are just reversed for j= 3 and 4.

It is now straightforward [1], although tedious to construct a table of inelastic amplitudes for the relativistic PWIA which is analogous to Table III for the non-relativistic PWIA. Such a table is given in reference 1 and a modified version of it is presented here in Table IV.

Without having yet defined all of the symbols used in Table IV, we may observe that, in contrast with the non-relativistic amplitude of Table III, all of the amplitudes allowed by general invariance principles can be present. In particular, we see that non-zero spin-difference functions are now possible in our relativistic PWIA.

We now address some of the considerations which went into the construction of Table IV. Comparison of the relativistic inelastic amplitude of Eq.25a with its non-relativistic counterpart of Eq.16 suggests that there are four times as many nuclear matrix-elements (or

TABLE IV

FRIA Plane Wave Inelastic Amplitudes

Natural Parity	Unnatural Parity
$A_q = a\rho_J + \frac{iq}{m} \cdot t_T \cdot \langle \vec{\sigma} \times \vec{j} \rangle_0$	$C_q = p/m \cdot t_A \cdot \langle \vec{\sigma} \cdot \vec{j} \rangle$
$B_q = iqc\rho_J - \frac{2p}{m} \cdot t_T \cdot \langle \vec{\sigma} \times \vec{j} \rangle_0$	$D_q = (b + q^2 d)\, \Sigma'_{J,0}$
$A_p = qc\, \Sigma'_{J,1} - p/m \cdot t_V \cdot \langle \vec{j} \rangle_1$	$C_p = (b+e)\, \Sigma'_{J,1} - q/2m \cdot t_V \cdot \langle \vec{j} \rangle_1$
$B_p = -ib\, \Sigma'_{J,1} + \frac{iq}{2m} \cdot t_V \cdot \langle \vec{j} \rangle_1$	$D_p = -\frac{2ip}{m} \cdot t_T \cdot \langle \vec{\sigma} \times \vec{j} \rangle_1$
$C_n = i(b+e)\, \Sigma'_{J,1} - \frac{iq}{2m} \cdot t_V \cdot \langle \vec{j} \rangle_1$	$A_n = iqc\, \Sigma'_{J,1} - \frac{ip}{m} \cdot t_V \cdot \langle \vec{j} \rangle_1$
$D_n = \frac{2p}{m} \cdot t_T \cdot \langle \vec{\sigma} \times \vec{j} \rangle_1$	$B_n = b\, \Sigma'_{J,1} - \frac{q}{2m} \cdot t_V \cdot \langle \vec{j} \rangle_1$

nuclear transition densities) entering in the former as in the latter. While this is true in a strict sense, many of the relativistic nuclear matrix elements are closely related to the non-relativistic ones. We may obtain these relationships by considering a specific model for the four-component nuclear wavefunctions. In analogy with the free Dirac spinor for the projectile (Eq.21), we can define the following bound-state Dirac wavefunction:

$$\psi_F(\vec{r}) = N \begin{pmatrix} 1 \\ \frac{-i\vec{\sigma}\cdot\vec{\nabla}}{E+m} \end{pmatrix} \phi(\vec{r}) \tag{27}$$

where N is a normalization constant and $\phi(\vec{r})$ is the usual two-component (or Schrpedinger) bound-state wavefunction. Since ψ_F approximately satisfies the free Dirac equation, we refer to the relativistic impulse approximation which employs this prescription for the bound-state wave functions as the Free Relativistic Impulse Approximation or FRIA. It is then straightforward to show that, in FRIA , we have, e.g.,

$$\begin{aligned} \rho_J^1 &= \langle \widetilde{J0} | \Gamma_1 e^{-i\vec{q}\cdot\vec{r}} | \widetilde{00} \rangle \\ &\longrightarrow \langle J0 | \left(1 + \frac{\vec{\sigma}\cdot\vec{\nabla}_f\, \vec{\sigma}\cdot\vec{\nabla}_i}{(E+m)^2} \right) e^{-i\vec{q}\cdot\vec{r}} | 00 \rangle \\ &\simeq \langle J0 | e^{-i\vec{q}\cdot\vec{r}} | 00 \rangle = \rho_J , \end{aligned} \tag{28a}$$

where we have, in effect, ignored the lower-to-lower component combination. Similarly, we find

$$\rho_J^2 \simeq \rho_J \quad \text{and} \quad \Sigma^1_{J,M} ,\ \Sigma^2_{J,M} \simeq \Sigma_{J,M} \tag{28b}$$

where we have used

$$\rho_J^i \equiv \langle \widetilde{J0} | \Gamma_i e^{-i\vec{q}\cdot\vec{r}} | \widetilde{00} \rangle \ \text{and}\ \Sigma^i_{JM} \equiv \langle \widetilde{JM} | \sigma_M \Gamma_i e^{-i\vec{q}\cdot\vec{r}} | \widetilde{00} \rangle .$$

The correspondances for the remaining relativistic nuclear matrix-elements are somewhat more difficult to establish. For example, we have

$$\begin{aligned} \rho_J^4 &= \langle \widetilde{J0} | \begin{pmatrix} 0 & 1 \\ -1 & 0 \end{pmatrix} e^{-i\vec{q}\cdot\vec{r}} | \widetilde{00} \rangle \\ &\longrightarrow \langle J0 | (-i/2m)\, \vec{\sigma}\cdot(\vec{\nabla}_f + \vec{\nabla}_i)\, e^{-i\vec{q}\cdot\vec{r}} | 00 \rangle \\ &= \langle J0 | \vec{\sigma}\cdot\vec{q}/2m\; e^{-i\vec{q}\cdot\vec{r}} | 00 \rangle = q/2m\, \Sigma_{J,0} . \end{aligned} \tag{29}$$

In contrast, we write

$$\rho_J^3 = \langle \tilde{J}0 | \begin{pmatrix} 0 & 1 \\ 1 & 0 \end{pmatrix} e^{-i\vec{q}\cdot\vec{r}} | \tilde{0}\tilde{0} \rangle \tag{30}$$
$$\rightarrow \langle J0 | (i/2m)\, \vec{\sigma}\cdot(\vec{\nabla}_f - \vec{\nabla}_i)\, e^{-i\vec{q}\cdot\vec{r}} | 00 \rangle$$
$$= \langle J0 | \vec{\sigma}\cdot\vec{j}\, e^{-i\vec{q}\cdot\vec{r}} | 00 \rangle$$

where $\vec{j} = \frac{i(\vec{\nabla}_f - \vec{\nabla}_i)}{2m} = \frac{\vec{p}_f + \vec{p}_i}{2m}$ is the usual non-relativistic probability or convection current operator. We see that ρ_J^3 has no counterpart in the non-relativistic PWIA presented earlier and that it is one of the new nuclear matrix-elements arising in the FRIA which permits a non-zero spin-difference function (since the quantity Cq in Table IV is proportional to it).

Similar proceedures allow the re-expression of the remaining relativistic nuclear matrix-elements and we obtain

$$\Sigma^{13}_{J,0} \rightarrow 0 \quad \text{(by current conservation)} \tag{31a}$$
$$\Sigma^{13}_{J,1} \rightarrow q/2m\, \Sigma^{1}_{J,1} + \langle \vec{j} \rangle_1 \tag{31b}$$
$$\Sigma^{14}_{J,0} \rightarrow q/2m\, \rho_J - \langle \vec{\sigma}\times\vec{j} \rangle_0 \tag{31c}$$
$$\Sigma^{14}_{J,1} \rightarrow -i \langle \vec{\sigma}\times\vec{j} \rangle_1 \tag{31d}$$

where $$\langle \vec{j} \rangle_1 = \langle J,1 | (\vec{j})_1\, e^{-i\vec{q}\cdot\vec{r}} | 00 \rangle \tag{31e}$$

and $$\langle \vec{\sigma}\times\vec{j} \rangle_M = \langle J,M | (\vec{\sigma}\times\vec{j})_M\, e^{-i\vec{q}\cdot\vec{r}} | 00 \rangle . \tag{31f}$$

The relations of Eqs.28-31 coupled with other relations [6] between the {a,b,c,d,e} of Eq.13 and the α_j and β_j of Eq.25 give rise to the form of the FRIA inelastic amplitude presented in Table IV. Referring to Tables III and IV, we observe that all the amplitudes of the non-relativistic PWIA are also present relativistically. Furthermore, all terms appearing only in the relativistic formulation involve the nuclear current operator, $\vec{j}$, either by itself or in combination with the Pauli spin operator, $\vec{\sigma}$, as in $\vec{\sigma}\cdot\vec{j}$ or $\vec{\sigma}\times\vec{j}$. The latter two are referred to [7] as "composite current" operators. These three nuclear operators, in contrast to all the others giving rise to the matrix-elements of Tables III and IV, are non-local by virtue of the gradients they contain. Also as, mentioned earlier, their presence is required for non-zero spin-difference functions, at least for the impulse approximations considered here.

In trying to understand the significance of these

current operators, we may recall that the convection current, at least, appears in standard microscopic treatments of inelastic electron scattering [8]. Since we usually think of the nuclear part of this treatment as being non-relativistic, we are led to ask whether the presence of currents is a fundamental difference between relativistic and non-relativistic models of (p,p'). The answer is, "No!." Similar non-localities arise in extended non-relativistic impulse approximations [7] when, for instance, knock-on exchange processes are treated explicitly. Therefore, we may, if we wish, think of the FRIA as a particular version of an extended non-relativistic impulse approximation. This suggestion is strongly reinforced by the realization [1] that the standard microscopic treatments of (e,e') referred to above are, in effect, just the FRIA outlined above with the free electromagnetic current operator substituted for t^R_{NN} in Eq.19 (and, of course, with free electron plane waves replacing the proton waves).

While the presence of currents is not a fundamental difference between the two approaches, it appears in some cases to be a practical difference. That is, the non-localities appearing in the form of currents in the relativistic impulse approximation are quite different from those emerging due to explicit knock-on exchange contributions in standard non-relativistic treatments of (p,p') [7]. These differences are best examined by considering observables which are sensitive to current contributions. Such observables - for FRIA - are readily identified using Tables II and IV. For example, we find, as mentioned above, that the spin-difference function depends crucially on the composite currents; it is identically zero without them. In Figure 4, we compare the 150 MeV 12C(p,p')12C*(12.71 MeV 1+ T=0) P-Ay data of Carey et al. [4] with relativistic (labelled DRIA [9]) and non-relativistic (labelled DW81 [10]) calculations. In these latter calculations, the NN amplitudes of Franey and Love [11] based on the Arndt SP84 phase-shifts [12] are employed. The relativistic calculations use a relativistic parametrization of the NN amplitudes based on Arndt's SM84 [12] phase-shift solution. Both sets of calculations make use of impulse approximation optical potentials to generate the distorted waves. The nuclear structure amplitudes of Lee and Kurath [13] were used, except that, where indicated, the LS=10 amplitude was dropped. This modification has been shown [14,15] to result in a global improvement in the agreement between intermediate energy (p,p') data and relativistic impulse approximation calculations for this particular transition. Note that P-Ay in Figure 4 is substantially non-zero for the non-relativistic calculations only because knock-on exchange effects are treated explicitly [7]. It is apparent that the relativistic calculations describe the data significantly better than the non-relativistic ones, especially when the LS=10 amplitude is dropped. A similar comparison with complementary Q-B data (labelled as Ds'1 + Dl's) [5] is made in Figure 5. Again the superiority of the relativistic calculations is demonstrated. On the basis of these comparisons, we conclude that the physical non-localities responsible for the spin-difference function for this transition are better accounted for by the composite-currents arising in lowest order in the relativistic impulse approximation than by those coming from the knock-on exchange contributions present in the extended non-relativistic impulse approximation. However, it should be noted that neither type of calculation gives a satisfactory description of the P-Ay data [4] for the 12C(p,p')12C*(15.11 MeV 1+ T=1) transition at 150 MeV, suggesting that neither model embodies the the relevant physics in this case. The origin of this problem is unknown at present. Further investigation of the differences between the non-localities occuring in relativistic and non-relativistic calculations such as those discussed here would be very desireable. Unfortunately, the nearly complete lack of relevant (p,p') spin-transfer data makes this impossible at present.

We may now ask if there are any fundamental differences between relativistic and non-relativistic impulse approximations. The answer to this question is, "Yes!", if, in our expression for the relativistic inelastic amplitude T^{R}_{JM} (Eq.19), we use continuum and bound-state wavefunctions consistent with modern relativistic models of nuclear dynamics such as the impulse approximation for elastic scattering discussed above and Dirac-Hartree models [16] of nuclear ground-states. In the FRIA discussed above, in addition to the assumption of plane waves for the projectile, we employed bound-state four-component wavefunctions of the type defined in Eq.27 which approximately satisfy the free Dirac equation; i.e.,

$$(i\gamma\cdot\partial - m)\psi_F = 0 \qquad (32)$$

In contrast, in the relativistic models just discussed, the wavefunctions satisfy a Dirac equation like

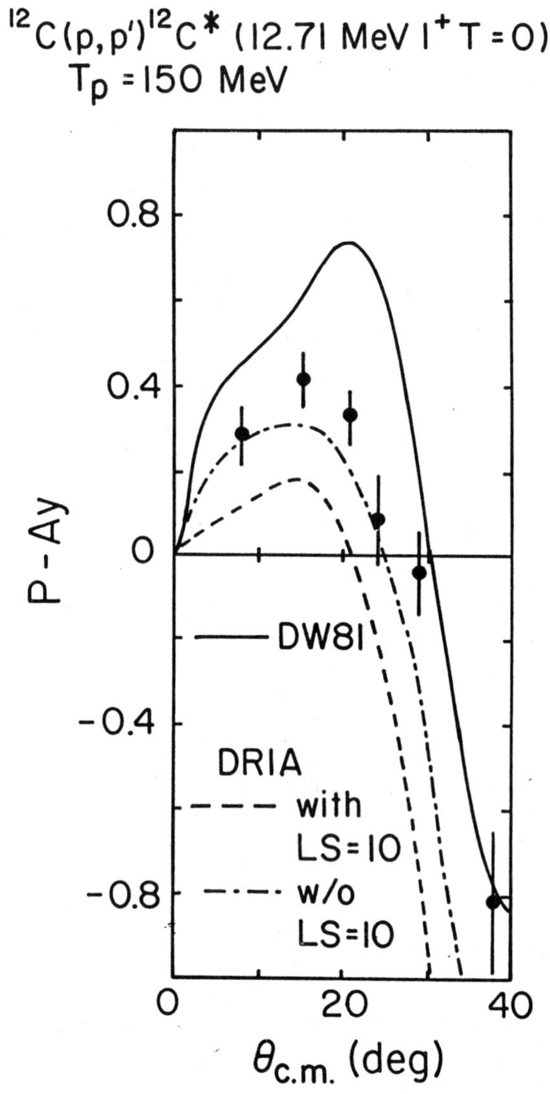

FIG. 4 The P-Ay data of Ref. 4 are compared with non-relativistic (DW81) and relativistic (DRIA) impulse approximation calculations. See text for details.

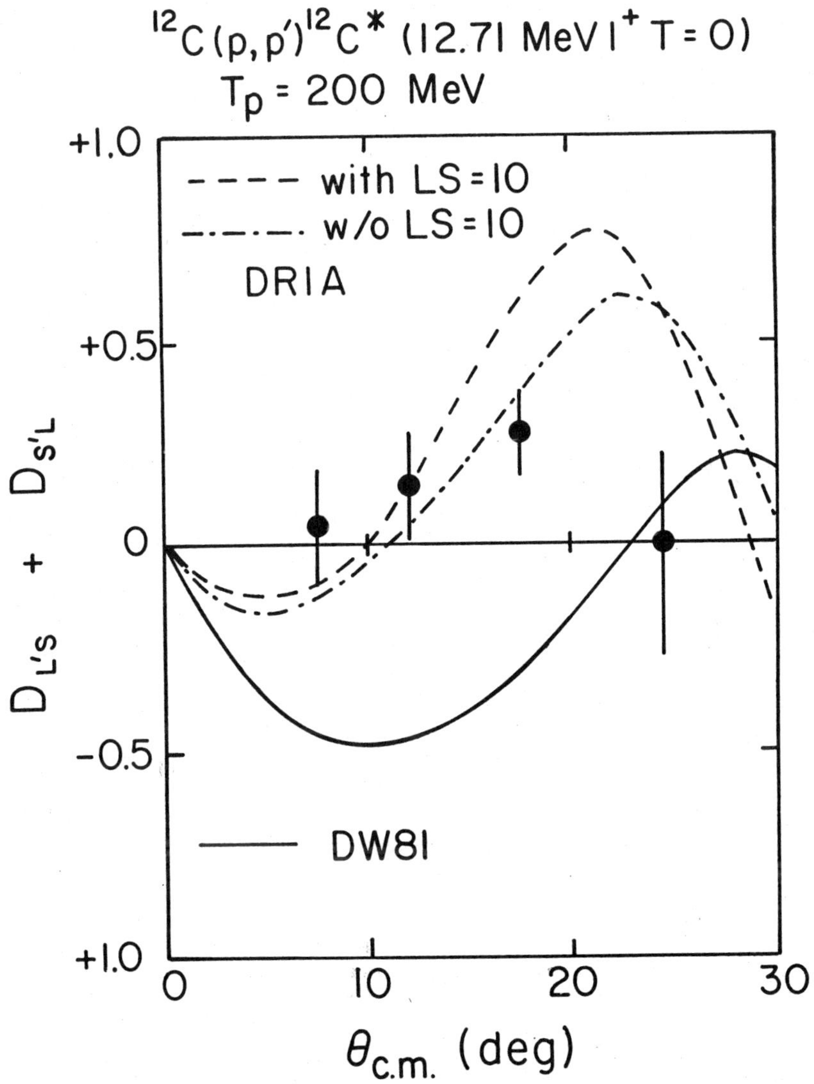

FIG. 5 The Dl's + Ds'l (=Q-B) data of Ref. 5 are are compared with non-relativistic (DW81) and relativistic (DRIA) impulse approximation calculations. See text for details.

$$\left[i\gamma\cdot\partial - m - S(r) - \gamma^0 V(r) \right] \psi_D = 0 \tag{33}$$

where S and V are the strong scalar and time-like vector potentials characteristic of such models. For the bound-states, for example, we can write ψ_D in a form similar to the expression in Eq.27 which defines ψ_F :

$$\psi_D(\vec{r}) = N \begin{pmatrix} 1 \\ \frac{-i\mathcal{S}(r)\vec{\sigma}\cdot\vec{\nabla}}{E+m} \end{pmatrix} \mathcal{S}^{-\frac{1}{2}}(r)\, \phi(\vec{r}) \tag{34a}$$

$$\text{where } \mathcal{S}(r) \equiv (E+m)/(E+m-V(r)+S(r)) \tag{34b}$$

and where $\phi(\vec{r})$ is the solution of some effective Schroedinger equation. Note that, for $\mathcal{S}(r) \rightarrow 1$ everywhere, $\psi_D \rightarrow \psi_F$, provided $\phi(\vec{r})$ is the same in Eqs.7 and 34. However, S and V are typically +450 and -350 MeV, respectively, in the nuclear interior. Their algebraic difference is then appreciable even on the scale set by E+m and $\mathcal{S}(r)$ can be as large as 1.75 in this region. Examination of Eq.34a reveals that the effect of the strong potentials is then to enhance the lower components of ψ_D . We refer to the version of the relativistic impusle approximation which employs ψ_D for the continuum and bound-states - and thereby incorporates the effects of the strong scalar and time-like vector potentials - as the Dynamical Relativistic Impulse Approximation or DRIA. We can consider the possibility of having values of $\mathcal{S}(r)$ in Eq. 34 different from unity as constituting a fundamental difference between the DRIA and the FRIA or any extended non-relativistic impulse approximation.

We can now attempt to identify possible experimental signatures of the differences between the DRIA and the FRIA; i.e., effects arising from the enhanced lower components in the former. This is most readily accomplished by using Eq.34 to rewrite our original expression (Eq.19) for the relativistic inelastic transition amplitude, T^R_{JM} . We may re-express Eq.34a as

$$\psi_D(\vec{r}) = N\, \tilde{u}(S,V)\, \phi(\vec{r}) \tag{35a}$$

where we have defined a new operator,

$$\hat{u}(S,V) = \begin{pmatrix} 1 \\ \frac{-i\mathcal{S}(r)\vec{\sigma}\cdot\vec{\nabla}}{E+m} \end{pmatrix} \mathcal{S}^{-\frac{1}{2}}(r)\,. \tag{35b}$$

We then have

$$T^{R}_{JM} = \phi^{+(-)}_{f}(\rho)\, \Phi^{+}_{f}(t)\, \tilde{t}^{eff}_{NN}\, \phi^{(+)}_{i}(\rho)\, \Phi_{i}(t) \tag{37a}$$

where

$$\tilde{t}^{eff}_{NN} = \hat{u}^{+}_{f}(\rho;S,V)\, \hat{u}^{+}_{f}(t;S,V)\, \gamma^{0}(\rho)\, \gamma^{0}(t)\, \tilde{t}^{R}_{NN}\, \tilde{u}_{i}(\rho;S,V)\, \tilde{u}_{i}(t;S,V). \tag{37b}$$

Note that Eq.37 has been written so as to facilitate comparison with the original expression for the non-relativistic inelastic transition amplitude, TfiNR, of Eq. 12. We in fact observe that, if $S,V \rightarrow 0$ everywhere or $\Delta(r) \rightarrow 1$ everywhere, we have $\tilde{t}^{eff}_{NN} \rightarrow \tilde{t}^{NR}_{NN}$. We may therefore say that the DRIA vs. FRIA difference lies in the effective density dependence introduced in the $\tilde{t}^{eff}_{NN}$ appearing in Eq.37 by the strong potentials S(r) and V(r) present in the operators u(i;S,V). Such density dependence, in the effective spin-orbit interaction, was noted in the original [17] relativistic impulse approximation treatment of elastic scattering where is was found to be essential for the excellent agreement between the calculations and experiment. In order to identify potential effects in inelastic scattering, we observe that the nuclear currents are linear in the lower components and will therefore be enhanced in the DRIA. We thus focus on observable combinations which are maximally sensitive to either convection- or composite-currents. For example, in both DRIA and FRIA, we find, by referring to Tables II and IV, that for unnatural parity transitions, the combination

$$\tfrac{1}{4} I (1 + D_{nn} + D_{pp} + D_{qq}) = \sum_{M} |A_M|^2 = |A_n|^2 \tag{38}$$

$$= \left| iqc\, \Sigma_{J,1} - \frac{ip}{m} \cdot t_V \cdot \langle \vec{j} \rangle_1 \right|^2$$

is maximally sensitive to the convection current, $\vec{j}$. In Figure 6, we compare DRIA and FRIA calculations of this observable with the 500 MeV 12C(p,p')12C*(12.71 MeV 1+ T=0) data of reference 18. The theoretical input is the same as for the 150 and 200 MeV calculations described earlier. Note that omission of the LS=10 nuclear structure amplitude appears to be required by the data. As anticipated, enhancement of the lower components in the DRIA results in the enhancement of this observable.

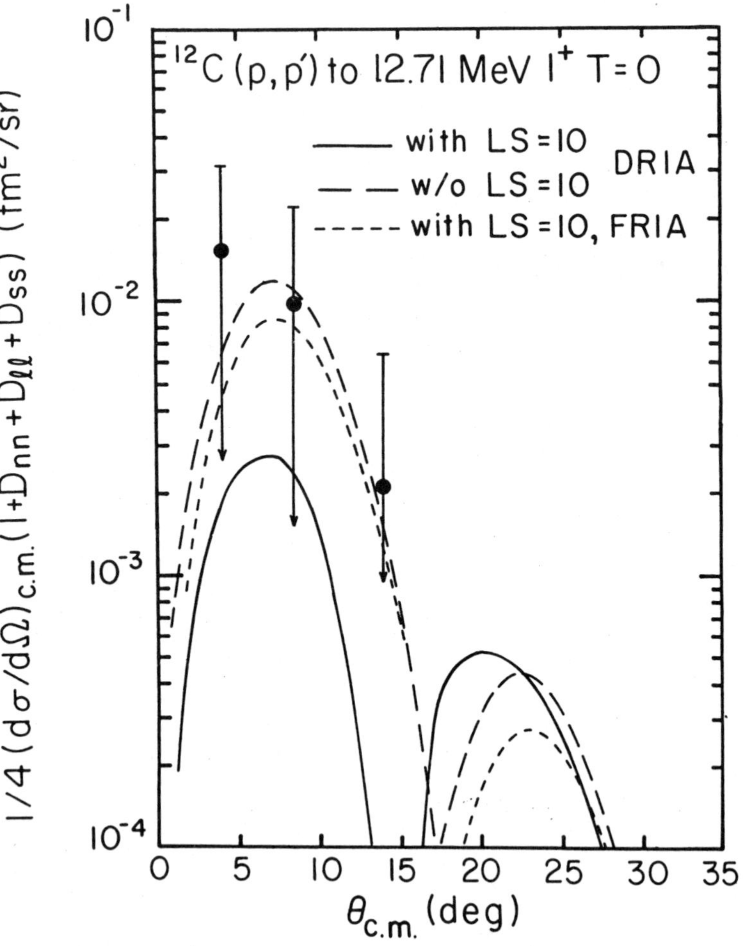

FIG. 6 The data of Ref. 18 are compared with FRIA and DRIA calculations for the observable combination equivalent to 1/4 I(1+Dnn+Dpp+Dqq). See text for details.

However, the magnitude of the enhancement is quite small, not only relative to the error bars of the data, but also relative to theoretical uncertainties such as the nuclear structure input. A similar situation is revealed in Figure 7 where calculations are compared with data for the observable which is maximally sensitive to the composite-current, $\vec{\sigma}\cdot\vec{j}$. The $\vec{\sigma}\cdot\vec{j}$ contribution to this observable combination is large as is demonstrated by the curve labelled LS=11→ 0 which effectively omits it. Again the anticipated enhancement in the DRIA result is observed, but again it is small compared to other uncertainties and cannot provide definitive evidence regarding the existence of the enhancement in nature.

The DRIA enhancement of the spin-orbit interaction referred to earlier in regard to elastic scattering suggests that inelastic scattering observable combinations sensitive to the spin-orbit may also be of intrest. For instance, it has long been known [19] that the cross section of the 90Zr(p,p')90Zr (1st 8+)level in the 150 to 200 MeV range is dominated by the spin-orbit contribution. We find that the DRIA calculatios of these cross sections are 1.6 to 2 times larger than FRIA results. Unfortunately, nuclear structure uncertainties again preclude a definitive confrontation with data [19] for the present.

As we did earlier for the 1+ transitions in 12C, we can try to use the information in Tables II and IV to identify specific observable combinations which are maximally sensitive to the elements of the inelastic transition amplitude we wish to study. For example, we know that, for natural parity, collective, isoscalar transitions, the spin-independent nuclear transition density (Eqs.17c and 26a) is dominant, implying that, of all of the terms in Table IV, only

$$A_q = a\rho_J \quad \text{and} \quad B_q = iqc\rho_J$$

are important. Enhancement of the lower components of the projectile and target wavefunctions should alter the relationship between these two terms of the transition amplitude. Table II tells us that we may emphasize them separately in the observable combinations

$$\tfrac{1}{2} I(D_{nn} + D_{pp}) \simeq |A_q|^2 \tag{39a}$$

$$\text{and} \quad \tfrac{1}{2} I(D_{nn} - D_{pp}) \simeq |B_q|^2. \tag{39b}$$

We consider the specific natural parity

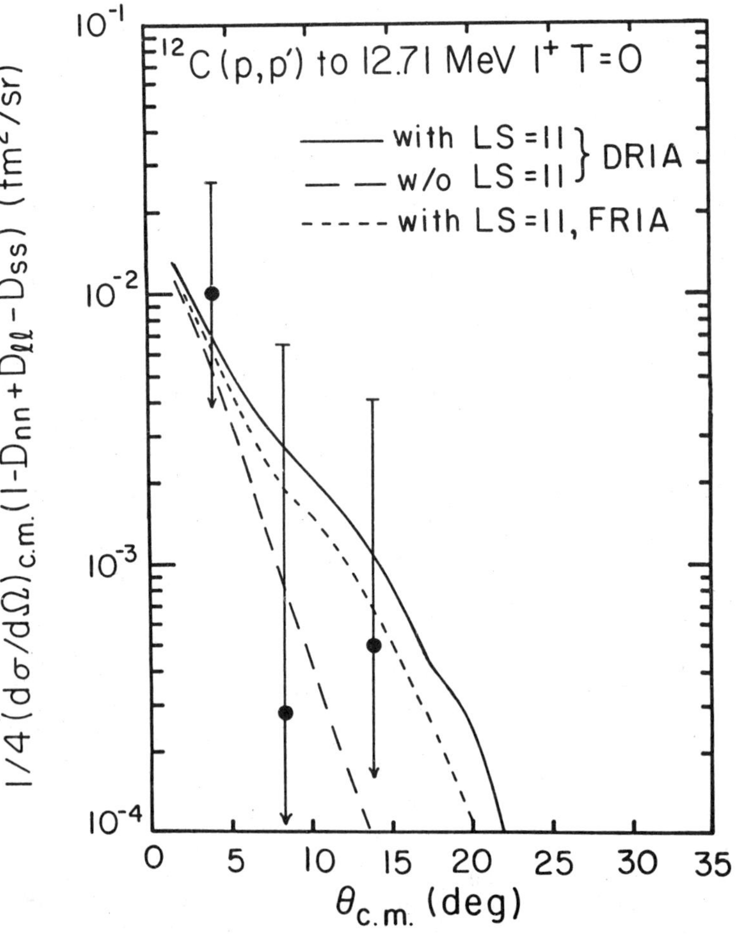

FIG. 7 The data of Ref. 18 are compared with FRIA and DRIA calculations for the observable combination equivalent to 1/4 I(1-Dnn+Dpp-Dqq). See text for details.

collective, isoscalar transition 28Si(p,p')28Si* (9.70 MeV 5- T=0) at 500 MeV. Here the nuclear structure input can be tightly constrained by (e,e') data for the same transtition and our fit to the (e,e') data [20] is shown in Figure 8. Note that DRIA and FRIA results are identical here [21]. Inelastic (p,p') calculations using identical nuclear structure input are compared with cross section data [22] in Figure 9. We see that the DRIA cross sections are uniformly enhanced compared to the FRIA results, as we might have anticipated for a transition known [22] to have very large spin-orbit contributions. We also observe that the DRIA calculations are in better agreement with the data. While this result is quite intresting by itself, it would be of obvious value to find an observable combination with a stronger signature of the enhancement of the lower components. In Figures 10 and 11, we show DRIA and FRIA calculations of the observable combinations of Eqs.39a and 39b, respectively. The one involving Dnn-Dpp is nearly the same for both calculations while the one involving Dnn+Dpp is enhanced in DRIA by a factor somewhat larger than that which was shown in Figure 9 for the cross section alone. These observable combinations have recently been measured at the HRS at LAMPF [23] and comparison with the calculations of Figures 10 and 11 are likely to be very intresting. It is clear quite generally that future tests of (p,p') reaction models - including but not limited to those discussed here - would be greatly improved by having complete sets of spin-transfer data in addition to the cross sections and (sometimes) analyzing powers which are typically available at present.

SUMMARY

We have examined the general structure of the (p,p') inelastic transition amplitude and have established its relationship to experimental observables. We have considered specific relativistic and non-relativistic impulse approximation models of the transition amplitude. We have found that a new class of terms involving nuclear currents arise naturally in lowest order in the relativistic formulation but that similar terms also appear in higher order in extended non-relativistic treatments. However, effects originating with the enhancement of the lower components of the projectile and target wavefunctions caused by

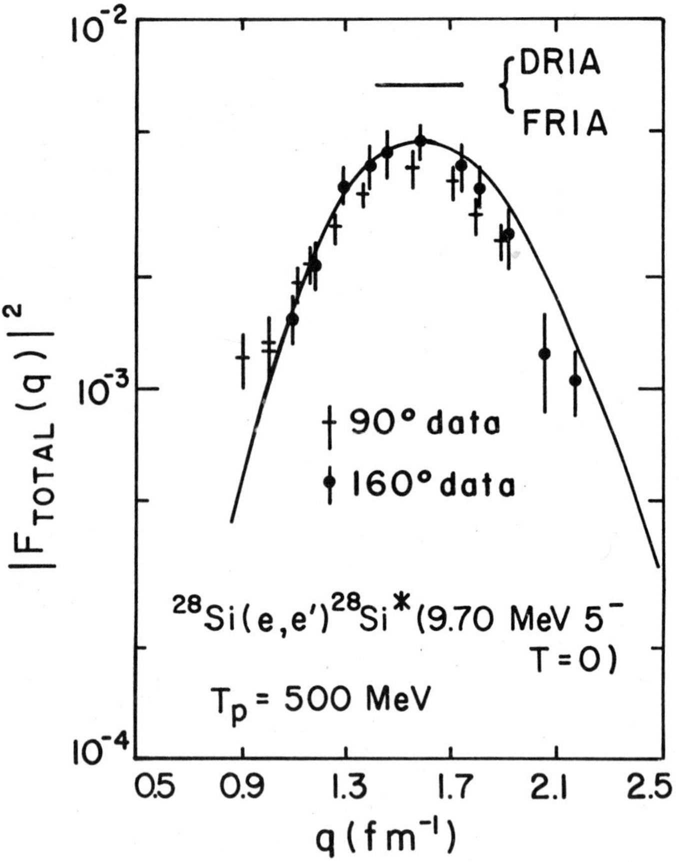

FIG. 8 Relativistic impulse approximation (e,e′) fits to the data of Ref. 20 are shown. The FRIA and DRIA results differ negligibly.

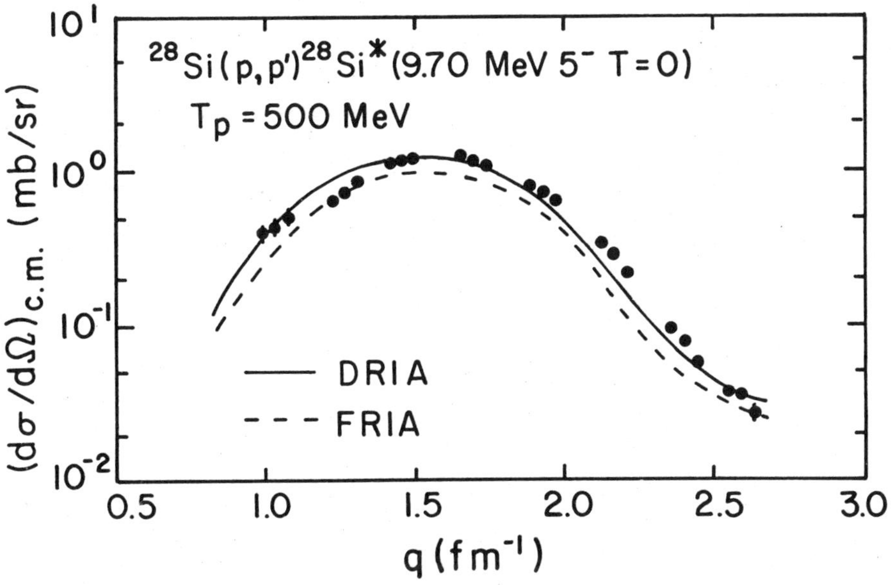

FIG. 9 The (p,p') data of Ref. 22 are compared with FRIA and DRIA calculations which use the same nuclear structure input as the (e,e') calculations of Fig. 8.

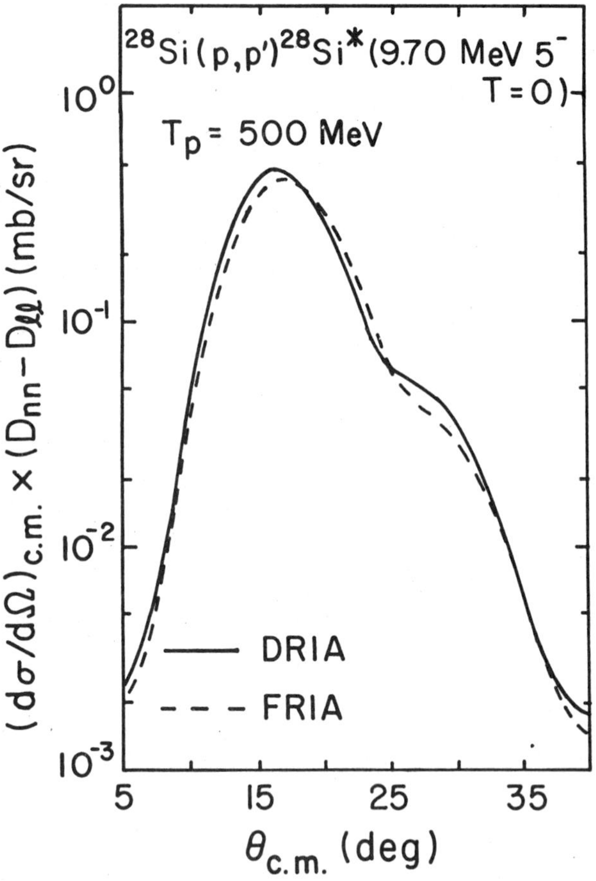

FIG. 10 FRIA and DRIA calculations of the observable combination related to $|B_q|^2$ are shown. See text for details.

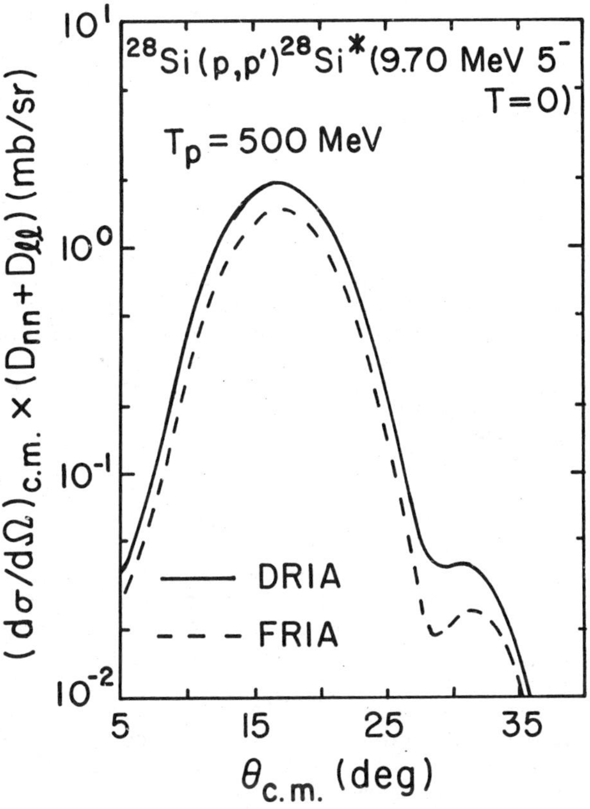

FIG. 11 FRIA and DRIA calculations for the observable combination related to $|A_q|^2$ are shown. See text for details.

strong scalar and time-like vector potentials were not found to have any obvious counterparts in non-relativistic models. Possible experimental signatures of these enhancements were examined in some detail, but no definitive comparisons with experiment were made.

FUTURE DIRECTIONS

The relativistic model presented here is obviously quite primitive. A refinement of clear importance is the explicit treatment of knock-on exchange processes. This has already been done in approximate ways for elastic scattering [24-26] and appears to be important for bombarding energies below about 200 MeV. The explicit treatment of knock-on exchange for (p,p') represents a formidable computational challenge, but explicitly anti-symmetrized relativistic representations of the NN amplitude already exist [24-26] and the development of an inelastic computer code which can make use of them is in its final stages [27].

At energies below 200-300 MeV, the impulse approximation itself fails as effects of the nuclear medium arising principally from modification of the phase space available for NN scattering due to the Pauli principle become increasingly important. Such effects should be included in the relativistic treatment of (p,p') in this energy range.

The relativistic calculations presented here for 12C(p,p') have made use of nuclear structure information obtained from non-relativistic models. This inconsistency can be removed in some cases by determining the important nuclear structure quantities phenomenologically as was done for the 5- transition in 28Si. However, the situation really calls for the development of relativistic models of nuclear structure, a process which is now only in its beginning stages [28-30].

Relativistic models of inelastic processes involving probes other than the nucleon have been or are being developed [15,21]. The complementarity of measurements with these various probes should be exploited maximally to reduce inevitable theoretical uncertainties. The similarity of the relativistic formulations for these processes is especially conducive to such a program [15].

Experimentally, as already emphasized above,

measurements of complete sets of spin-transfer observables for certain carefully selected (p,p') transitions would certainly warrent the time and expense they would require. The experimental capabilities already exist and it seems certain that such data will be available within a few years. It is also likely that such data will suggest various measurements with complementary probes and vice-versa. In any case, it seems probable that - in a few years - our picture of intermediate-energy proton-nucleus inelastic scattering will be quite different from the prevalent ones of today.

This work was supported in part by the U.S.D.O.E.

REFERENCES

1. J.R.Shepard, E.Rost, and J.A.McNeil, Univ. of Colo. Preprint NPL-993 and to be published.
2. R.J.Glauber and P.Osland, Phys.Lett. 80B, 401 (1979)
3. D.A.Sparrow, J.Piekarewicz, E.Rost, J.R.Shepard, J.A.McNeil,T.A.Carey, and J.B.McClelland, Phys. Rev.Lett. 54, 2207 (1985)
4. T.A.Carey et al., Phys.Rev.Lett. 49, 266 (1982)
5. C.Olmer, Invited talk, International Conference on Antinucleon- and Nucleon-Nucleus Interactions, Telluride, CO, March, 1985, and to be published.
6. See, e.g., J.A.McNeil, L.Ray, and S.J.Wallace, Phys.Rev. C27, 2123 (1983)
7. W.G.Love and J.R.Comfort, Phys.Rev. C29, 2135 (1984) and W.G.Love, in Proc. of LAMPF Workshop on Dirac Approaches to Nuclear Physics, J.R. Shepard, C.-Y.Cheung, R.L.Boudrie, eds., LANL Report LA-10438-C, p.220 (1985)
8. T.W.Donnelly and J.D.Walecka, Annu.Rev.Nucl.Sci. 25, 329 (1975)
9. E.Rost and J.R.Shepard, Computer Code "DRIA", unpublished; see also J.R.Shepard, E.Rost, and J.Piekarewicz, Phys.Rev. C30, 1604 (1984)
10. J.R.Comfort, Computer Code DW81, modification of DW70, R.Schaeffer and J.Raynal, unpublished; J.Raynal, Nucl.Phys. A97, 572 (1967)
11. M.A.Franey and W.G.Love, Phys.Rev. C31, 488 (1985)
12. R.A.Arndt, Program SAID, unpublished.
13. T.-S.H.Lee and D.Kurath, Phys.Rev. C21, 293 (1980)
14. J.R.Shepard, in Proc. of LAMPF Workshop Dirac Approaches to Nuclear Physics, J.R.Shepard, C.-Y.Cheung, and R.L.Boudrie, eds., LANL Report LA-10438-C, p.300 (1985)
15. J.R.Shepard, Invited talk, International Conference on Antinucleon- and Nucleon-Nucleus Interactions, Telluride, CO, March, 1985, and to be published.
16. B.D.Serot and J.D.Walecka, in Advances in Nuclear Physics, eds. J.W.Negele and E.Vogt, (Plenum, New York, to be published.)
17. J.R.Shepard, J.A.McNeil, and S.J.Wallace, Phys.Rev.Lett. 50, 1443 (1983)

18. J.B.McClelland et al., Phys.Rev.Lett. 52, 98 (1984)
19. W.G.Love, in Spin Excitations in Nuclei, F.Petrovich et al., eds., Plenum, N.Y., p.205 (1984)
20. S.Yen et al., Phys.Rev. C27, 1939 (1983)
21. J.R.Shepard, E.Rost, E.R.Siciliano, and J.A. McNeil, Phys.Rev. C29, 2243 (1984)
22. N.Hintz el al., Phys.Rev. C30, 1976 (1984)
23. C.Glashausser, private communication.
24. C.J.Horowitz, Phys.Rev. C31, 1340 (1985)
25. C.J.Horowitz, Contribution to this conference.
26. S.J.Wallace, Contribution to this conference.
27. E.Rost and J.R.Shepard, Computer Code "DREX", unpublished.
28. R.Furnstahl, in Proc. of LAMPF Workshop on Dirac Approaches to Nuclear Physics, J.R. Shepard, C.-Y.Cheung, and R.L.Boudrie, eds., LANL Report LA-10438-C, p.377 (1985)
29. C.E.Price, Contribution to this conference.
30. M.W.Price, Contribution to this conference.

RECENT EXPERIMENTAL RESULTS IN (e,e′p)

Robert W. Lourie
Massachusetts Institute of Technology
Cambridge, MA 02139

ABSTRACT

The results of recent (e,e'p) measurements on ^{12}C at the M.I.T.–Bates Linear Accelerator will be presented. Data were taken at the quasielastic peak, in the dip region and at the peak of the Δ–resonance. The physics of these regions, as revealed by both previous inclusive (e,e′) experiments and the current coincidence experiments will be discussed. The questions to be addressed include the enhanced transverse response in the quasielastic peak, the excess strength in the dip region and the reaction process at the Δ–resonance peak.

INTRODUCTION

The electromagnetic probe measures the response of the nucleus to real or virtual photons. Since the electromagnetic interaction is weak it is usually sufficient to consider the process to lowest order, i.e. the exchange of a single photon. Figure 1a shows the Feynman diagram for inclusive electron scattering and fig. 1b shows the exclusive (e,e′p) process. The exchanged virtual photon, unlike a real one, may posess both longitudinal and transverse polarization components (with respect to $\vec{q}$). Lorentz covariance, parity and current conservation restrict the nuclear electromagnetic response to two independent functions of $\vec{q}$ and ω. The doubly–differential cross section for the inclusive (e,e′) process when no polarizations are observed is[1]

$$\frac{d^2\sigma}{d\Omega_e d\omega} = \sigma_M \left[v_L R_L(\vec{q},\omega) + v_T R_T(\vec{q},\omega)\right] \quad , \qquad (1.1)$$

where

$$v_L = \left(\frac{q_\mu}{\vec{q}}\right)^4 \quad ,$$

$$v_T = \frac{1}{2}\left(\frac{q_\mu}{\vec{q}}\right)^2 + tan^2(\frac{\theta_e}{2}) \quad ,$$

and

$$\sigma_M = \frac{\alpha^2 cos^2(\theta_e/2)}{4E_0^2 sin^4(\theta_e/2)}$$

is the Mott cross section for scattering from a structureless particle. The electron kinematics are as follows: E_0 is the incident energy, θ_e is the electron scattering angle, $\vec{q} = \vec{k} - \vec{k}'$ is the 3-momentum transfer, $\omega = E_0 - E_f$ is the energy transfer and $q_\mu^2 = \vec{q}^2 - \omega^2$ is the 4-momentum transfer. R_L and R_T are the nuclear response functions to longitudinal and transverse photons respectively. All the nuclear structure information is contained in these response functions. One sees that by keeping $\vec{q}$ and ω fixed and making measurements at different E_0 and θ_e one can separate R_L and R_T (Rosenbluth separation).

When, in addition, one detects an outgoing nucleon in coincidence two additional response functions are present. One of them results from an interference between the longitudinal and transverse components of the nuclear electromagnetic current and the other from interference between the two transverse components. The kinematics for (e,e′p) are shown in fig. 2. The coincidence cross section is given by[2]

$$\begin{aligned}\frac{d^4\sigma}{d\Omega_e d\Omega_p dE_p d\omega} = \sigma_M[&v_L W_L(\vec{q},\omega,\theta_p,E_p) + v_T W_T(\vec{q},\omega,\theta_p,E_p) \\ &+ v_{LT} W_{LT}(\vec{q},\omega,\theta_p,E_p)\cos(\phi_p) \\ &+ v_{TT} W_{TT}(\vec{q},\omega,\theta_p,E_p)\cos(2\phi_p)]\end{aligned} \quad (1.2)$$

where v_L and v_T are given above,

$$v_{LT} = \frac{1}{\sqrt{2}}\left(\frac{q_\mu}{\vec{q}}\right)^2 \sqrt{(\frac{q_\mu}{\vec{q}})^2 + tan^2(\frac{\theta_e}{2})} \quad ,$$

and

$$v_{TT} = \frac{1}{2}\left(\frac{q_\mu}{\vec{q}}\right)^2 \quad .$$

The coincidence experiments at Bates were performed under the condition of parallel kinematics, i.e. $\vec{p}_F$ parallel to $\vec{q}$. This choice of kinematics eliminates the two interference terms W_{TT} and W_{LT}, reducing ambiguity in interpreting the results. To determine W_{TT} and W_{LT} requires measurement of protons out of the electron scattering plane, a difficult task that has not yet been attempted.

A schematic spectrum of electrons scattered from a "generic" nucleus is shown in fig. 3. Four distinct regions are present: 1) at the lowest energy losses discrete states and Giant Resonances are excited, 2) at kinematics approximately corresponding to scattering from a free nucleon the quasielastic

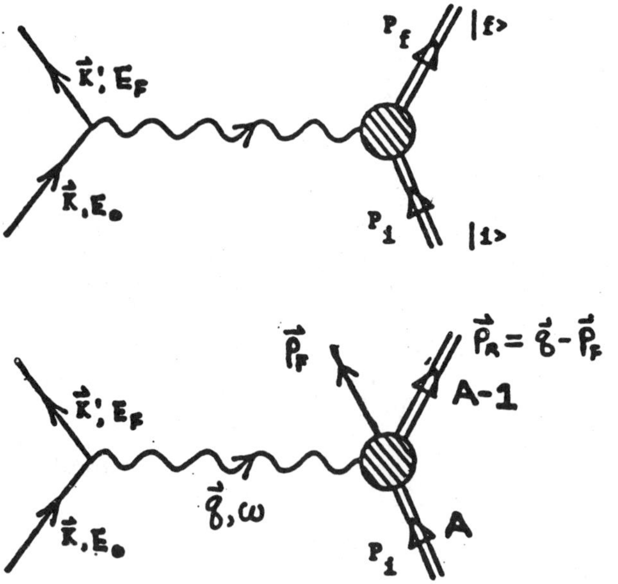

FIG. 1. Feynman diagrams for a) (e,e′) and b) (e,e′p) in the one–photon exchange approximation.

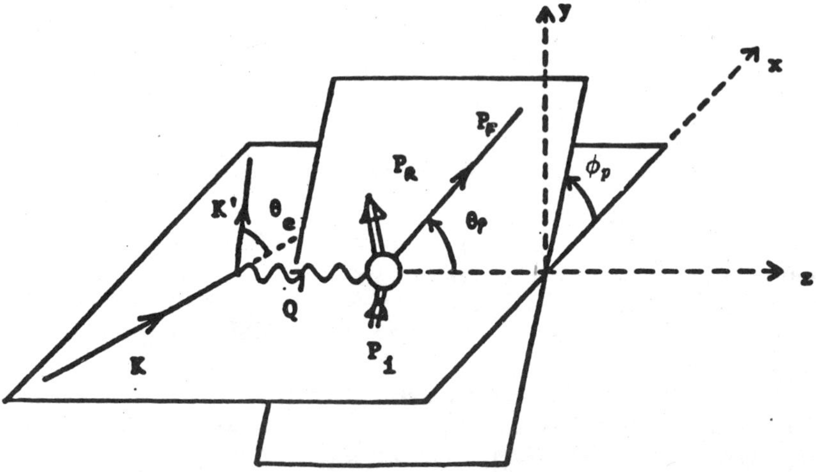

FIG. 2. Kinematics and coordinates for (e,e′p).

(QE) peak appears, 3) above the QE peak is the so–called dip region and 4) there is the Δ–resonance peak. I will discuss the physics of the later three regions in some detail.

THE QUASIELASTIC PEAK

The quasielastic peak arises from the knockout of quasifree nucleons. If the nucleon were truly free the peak would appear at $\omega = \vec{q}^{\,2}/2M$ where M is the nucleon mass. In actuality the peak is at somewhat higher (20–30 MeV) excitation since the nucleons are bound and may interact with other nucleons on their way out of the nucleus. The peak is also Doppler broadened by the nucleon's Fermi motion. The reaction process in the QE region is expected to be dominantly one–body in character, providing oppurtunity for observing the properties of single nucleons in the nuclear medium.

The impulse approximation (IA) to QE scattering assumes that the electron scatters incoherently from individual nucleons. This is expected to be valid provided $|\vec{q}| >> R^{-1}$ and $\omega >> \epsilon_B$ where R and ϵ_B are typical nuclear dimensions and binding energies respectively. When these conditions are met the quasielastic data should exhibit y–scaling[3]. Then the response functions no longer depend on $\vec{q}$ and ω independently but rather are functions of the single variable y. When relativistic kinematics are used

$$y = -\frac{|\vec{q}|}{2M} + \frac{\omega}{|\vec{q}|}\sqrt{\left(\frac{\vec{q}}{q_\mu}\right)^2\left(1+\frac{q_\mu^2}{2M^2}\right)} \quad . \tag{1.3}$$

Since it is assumed that all the energy and momentum are absorbed on a single nucleon the conservation equations may be solved for k_1, the component of the nucleon's initial momentum parallel to $\vec{q}$. The result is that $k_1 = My$. So if y-scaling is valid the response functions only depend on the longitudinal component of the nucleon's initial momentum. It is assumed that $\vec{q}$ is large enough that the convection current contribution to the transverse response is negligible. If the reaction process is not the quasielastic knockout of objects with the mass and form factor of the nucleon the data will not scale.

The quasielastic response functions may be defined in terms of dimensionless (reduced) response functions as

$$R_L(\vec{q},\omega) = \frac{A}{|\vec{q}|}G_E^2(q_\mu^2)\left(\frac{\vec{q}}{q_\mu}\right)^2 f_L(\vec{q},\omega) \quad , \tag{1.4}$$

$$R_T(\vec{q},\omega) = \frac{A}{|\vec{q}|}G_M^2(q_\mu^2)\left(\frac{q_\mu^2}{2M^2}\right) f_T(\vec{q},\omega) \quad , \tag{1.5}$$

where

$$G_E^2(q_\mu^2) = ZG_{E,proton}^2(q_\mu^2) + NG_{E,neutron}^2(q_\mu^2) \quad ,$$
$$G_M^2(q_\mu^2) = ZG_{M,proton}^2(q_\mu^2) + NG_{M,neutron}^2(q_\mu^2)$$

and the G_Es and G_Ms are the electric and magnetic form factors of the proton and neutron. If the data y–scale then the reduced nuclear structure functions depend on y only:

$$f_L(y) = f_T(y) = \pi M \int_{My}^{\infty} n(k)kdk \quad , \tag{1.6}$$

where $n(k)$ is the (spherically symmetric) nucleon momentum distribution. For an N=Z nucleus like ^{12}C $n(k)$ is expected to be the same for protons and neutrons.

The separated ^{12}C(e,e′) data from Saclay[4] have been analyzed[5] for y–scaling behavior. Figure 4 shows the $f(y)$ derived from this data assuming that the nucleons have the free mass and form factor. The data are plotted versus the relativistic y–scaling variable. In the region of the QE peak both the longitudinal and transverse data are $|\vec{q}|$ independent, i.e. lie on a universal curve. However, this universal curve is different, at least in magnitude, for the transverse data than for the longitudinal data. The longitudinal response is expected to be the best measure of the one–body part of the reaction process since the other processes that may contribute (meson exchange currents (MECs), low energy–loss tail of the Δ) are, at least to lowest order, entirely transverse. Indeed, one can see that the dip and Δ regions do not scale in the transverse data. One is forced to conclude from this analysis that the nucleon is modified in the medium and that the modification is considerably more pronounced in the transverse response. Recent theoretical work[6,7] indicate that the presence of relativistic scalar and vector fields may produce just such a modification. In particular the effective mass that results when a scalar field is present has been shown[5] to enhance the transverse response R_T over the longitudinal R_L. The (e,e′) measurements with their scaling behavior indicate that the basic reaction process is indeed quasielastic nucleon knockout though perhaps of a modified nucleon.

A more stringent test of the reaction process is provided by (e,e′p) experiments. In the quasielastic region the coincidence cross section is usually assumed[8] to factor into the product of the off–shell e–p cross section times the spectral function $S(\vec{p}, \epsilon_m)$:

$$\frac{d^4\sigma}{d\Omega_e d\Omega_p d\omega dT_p} = K\sigma_{ep} S(\vec{p}, \epsilon_m) \tag{1.7}$$

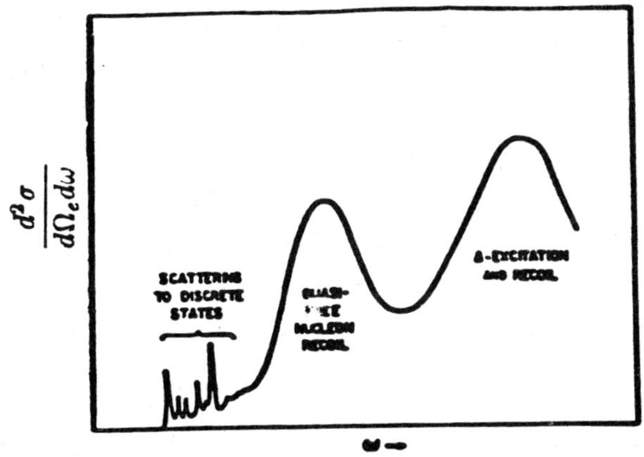

FIG. 3. Schematic spectrum of electron scattering from a nucleus.

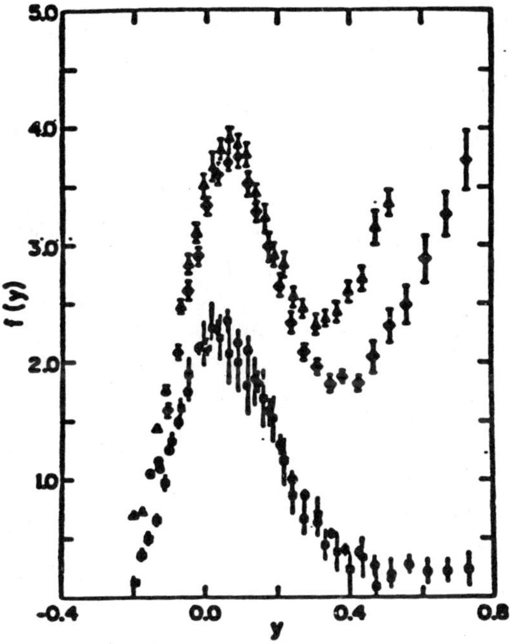

FIG. 4. Reduced longitudinal and transverse response functions (f_L and f_T) for ^{12}C (e,e′). The longitudinal(transverse) data at 400 MeV/c are shown by circles(diamonds) and at 500 MeV/c by squares(triangles).

where K is a kinematical factor. In the shell model the spectral function, which is the joint probability of finding the proton with initial momentum $\vec{p}$ and removal (separation) energy ϵ_m, is

$$S(\vec{p},\epsilon_m) = \sum_{\alpha} N_\alpha |\phi_\alpha(\vec{p})|^2 \delta(\epsilon_m - \epsilon_\alpha) \quad , \tag{1.8}$$

where N_α is the occupation number of the shell model orbit α, ϵ_α is the shell binding energy and $\phi_\alpha(\vec{p})$ is the momentum–space wavefunction. It is the single–particle momentum distributions and the occupation numbers that one hopes to determine. It must be noted that distortion effects and final state interactions somewhat complicate this simple picture. The classic quasielastic (e,e′p) results of Mougey *et al.*[8] are shown in fig. 5. The spectrum is plotted versus missing energy, $\epsilon_m = \omega - T_p - \vec{p_R}^2/2M_{A-1}$ where T_p is the kinetic energy of the detected proton. The p–shell shows up at ϵ_m=18 MeV, the binding energy for a $1p_{3/2}$ proton. The broad structure around 35 MeV results from the knockout of the more tightly bound $1s_{1/2}$ protons. Allowing for the removal energy the missing energy is the excitation energy of the residual A-1 system. The momentum distribution of the protons in the p and s–shell regions is shown in fig. 6. They have the characteristic shapes of l=1 and l=0 wavefunctions respectively. These data have both longitudinal and transverse contributions but show no evidence for the presence of any reaction process other than single nucleon knockout. The experimentally determined occupation numbers of 1 and 2.5 ,corrected for distortion effects, for the the $1s_{1/2}$ and $1p_{3/2}$ orbitals differ considerably from the shell model values of 2 and 4. The origin of this reduction remains an open question.

At Bates we have attempted the first separation of the coincidence response into its longitudinal and transverse components. The measurements were made at ω = 120 MeV, $|\vec{q}|$ = 400 MeV/c. Data were taken at E_0 = 446 MeV, θ_e = 60 deg. and E_0 = 291 MeV, θ_e = 120 deg. It must be noted that these data are preliminary. The missing energy spectra for the two energy/angle combinations are shown in fig. 7. As in the Saclay data the p and s–shells are observed. There is an indication that the s–shell in the 120 deg. data is enhanced compared to the forward angle data. The preliminary separated data show this effect even more strongly but the separations are extremely sensitive to the overall normalization of the data which are not as yet completely determined. However, a relatively normalization independent quantity is the ratio of the integrated p to s–shell strength. $(p/s)_{LONG.}/(p/s)_{TRANS}$ is greater than unity on the order of 50%, again indicative of a transverse enhancement primarily of the s–shell. Neither the forward or backward angle data indicate anything other than p and s–shell knockout. The tentative conclusion is that the one–body nucleon knockout

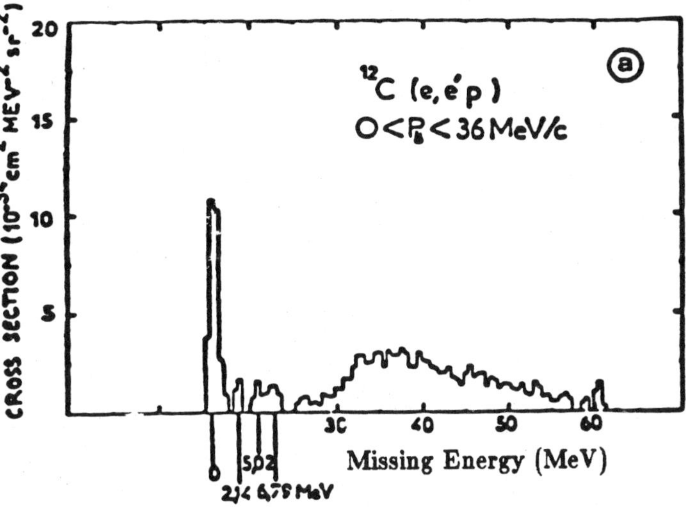

FIG. 5. Missing energy spectrum at the quasielastic peak of ^{12}C measured at Saclay[8].

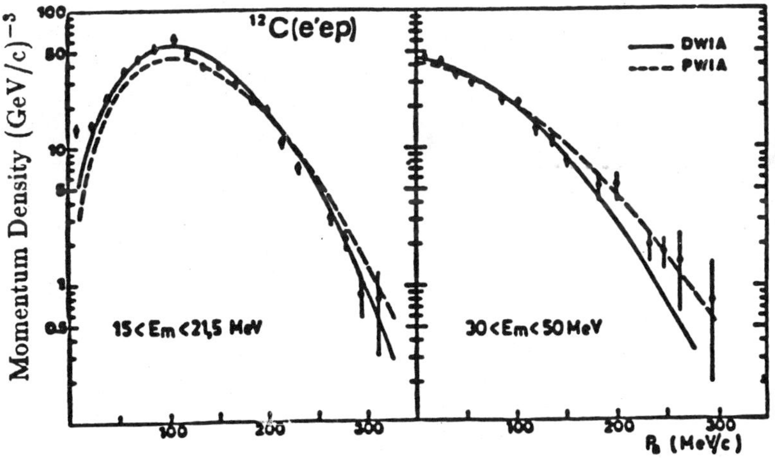

FIG. 6. $1p_{3/2}$ and $1s_{1/2}$ momentum distributions for ^{12}C from the Saclay[8] experiment..

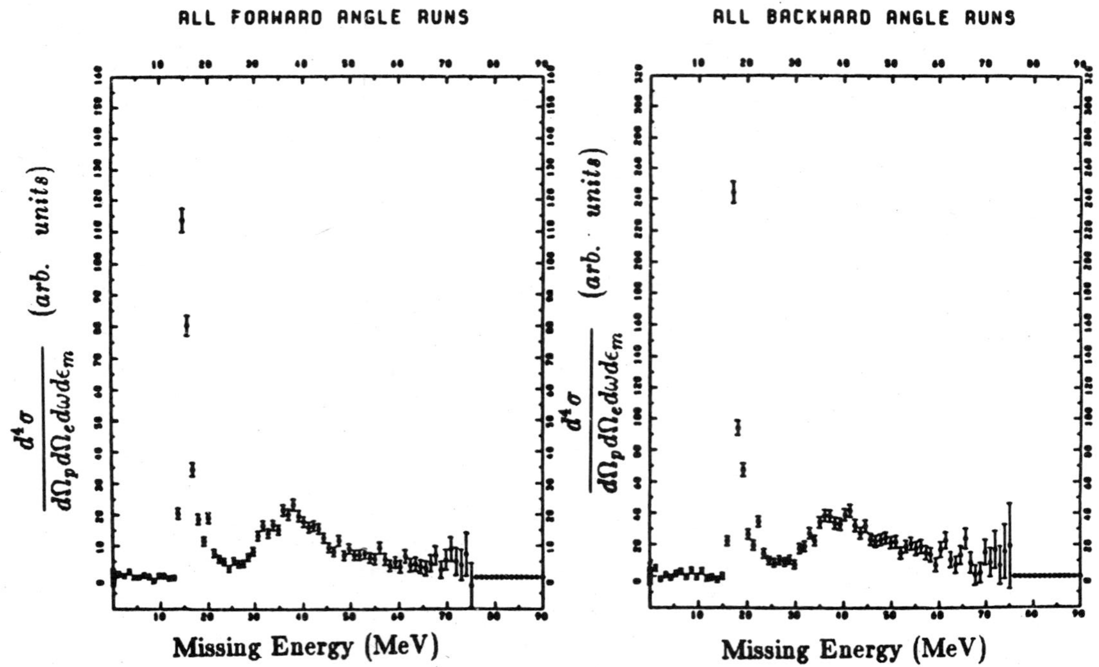

FIG. 7. Preliminary missing energy spectra from the Bates ^{12}C (e,e′p) quasielastic experiment: a) E_0 = 446 MeV, θ_e = υ0 deg., b) E_0 = 291 MeV, θ_e = 120 deg.

response is modified in the transverse channel as the y–scaling analysis of the (e,e′) data would imply.

THE DIP REGION

Intermediate between the QE and Δ peaks is the dip region. Figure 8 shows R_L and R_T for ^{12}C at $|\vec{q}| = 400$ MeV/c. The origin of the excess strength observed in this region by inclusive (e,e′) experiments has been a mystery in electromagnetic nuclear physics for several years. All attempts to predict the cross section, based on the known reaction processes of quasielastic nucleon knockout, quasifree Δ production, MECs and coherent pion production have been insufficient to explain the data.[9] In particular the dominant one–body process, quasielastic scattering, provides only 20–30% of the experimental cross section in ^{12}C at $\vec{q}$=400 MeV/c and ω= 200 MeV. Even with the inclusion of a large 2–body contribution, calculated by Laget[10] using a phenomeonological quasi–deuteron model, the data still lie considerably above the sum of the calculated contributions. The various reaction processes in this region will lead to different distributions of missing energy in the (e,e′p) reaction. The first coincidence experiment at Bates was made in the dip region ($\omega = 200$ MeV, $|\vec{q}| = 400$ MeV/c). These data are shown in fig. 9. Several features are immediately obvious. There is a pronounced p–shell peak from the one–body part of the reaction and then a strong, nearly uniform population of the continuum out to the largest missing energies we are able to measure. There is an indication of s–shell strength above the continuum in the region of 30–50 MeV missing energy. The p–shell strength is roughly that expected from the Saclay[8] momentum distributions and the ratio of p to s–shell strength is also consistent with the Saclay results if the continuum under the s–shell is assumed flat. The integrated continuum yield is about 10 times that of the s–shell and 4 times that of the p–shell. It is clear from these results that a process other than QE knockout is present and indeed dominates. The uniformity of the distribution suggests some sort of two–body effect.

We completed another measurement in the dip region last Fall. The electron energy loss was increased to 275 MeV in order to further suppress the QE contribution. A preliminary spectrum from this experiment is shown in fig. 10. Unfortunately, time limitations prevented us from achieving the desired statistical accurracy so these data are shown in 20 MeV wide bins. There is a rise in the cross section at about $\epsilon_m = 60$ MeV which is the threshold for 2–body knockout. The distribution is uniform until $\epsilon_m = 160$ MeV where real pion production becomes possible. Here the cross section rises sharply.

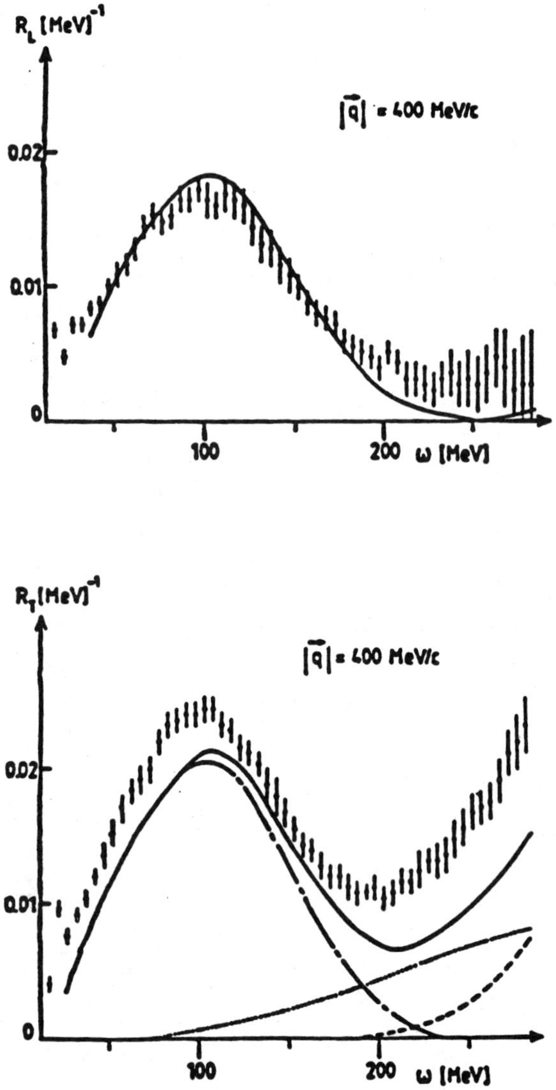

FIG. 8. The inclusive (e,e′) separated response functions in ^{12}C at $|\vec{q}|$ = 400 MeV/c from ref. 5.

FIG. 9. Missing energy spectrum in the dip region (ω = 200 MeV) of ^{12}C measured at Bates.

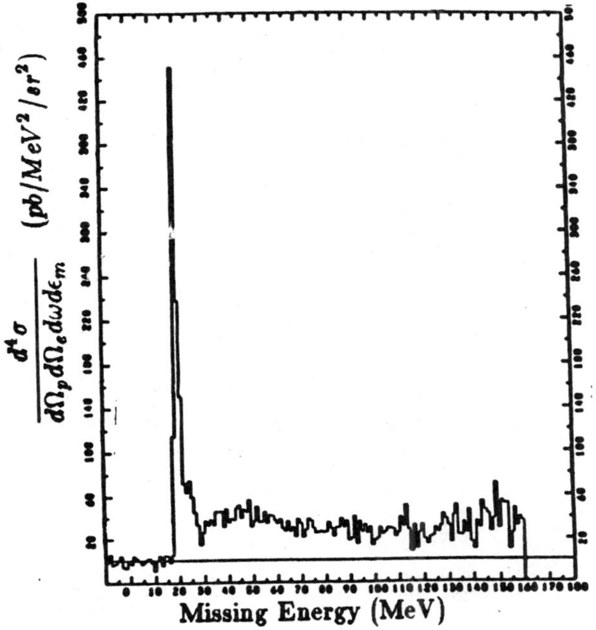

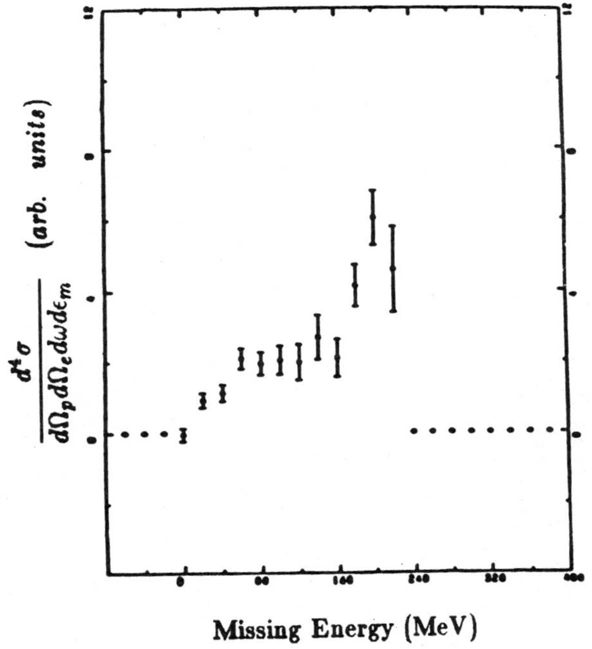

FIG. 10. Preliminary missing energy spectrum in the dip region (ω = 275 MeV) of ^{12}C measured at Bates.

The two dip region measurements indicate the contribution of at least three processes to the reaction: 1) QE nucleon knockout with about the strenght expected, 2) real pion production and 3) the process responsible for the uniform population of missing energies in the region above the 2–nucleon knockout threshold and below pion threshold. It is this third process that is responsible for most of the strength in the dip at $\omega = 200$ MeV. Theoretical support is urgently needed to elucidate the nature of this process.

THE Δ-RESONANCE REGION

The Δ peak results from the quasifree excitation of a nucleon to its first excited state. 300 MeV is required to excite the Δ. The Δ has spin 3/2 and isospin 3/2. In modern language it is thought of as a quark spin–flip transition. Pions and both real and virtual photons may excite a nucleon to a Δ. The electroexcitation of the Δ in several nuclei[11] is shown in fig. 11. The Δ peak is broadened and slighly displaced in energy compared to the free nucleon. In addition the area under the peak in a complex nucleus is about 35% larger than the area under the ^{1}H(e,e′)Δ peak. The Δ is clearly modified inside a nucleus, perhaps via the N–Δ interaction and the large number of final states, both with and without a real pion present, that are available.

The distribution of protons emerging from a nucleus after Δ excitation can give information on the possible reaction processes. We have measured the missing energy spectrum in parallel kinematics at $\omega = 382$ MeV and $|\vec{q}| = 473$ MeV/c, i.e. at the peak of the Δ. The preliminary spectrum is shown in fig. 12. There is a substantial amount of cross section below pion threshold, indicating that a process that produces a nucleon but no real pion in the final state is important. Above pion threshold the cross section increases sharply and remains roughly constant. The cross section below threshold may result from a two–nucleon knockout mechanism, perhaps via the NΔ—> NN process that is only possible in a nucleus. Pion reabsorption is also expected to play a role.

From an experimental point of view an interesting comparision is shown in fig. 13 where tagged real photon data[12] is displayed along with our Δ–region (e,e′p) results. The qualitative similarity of the structures observed is obvious. One can think of (e,e′p) as providing a tagged source of virtual photons. The (e,e′p) cross section can then be written as

$$d\sigma(e,e'p) = \Gamma_V \, d\sigma(\gamma_V, p) \tag{1.9}$$

where

$$\Gamma_V = \frac{\alpha^2}{4\pi^2}\frac{k}{q_\mu^2}\frac{E_0-\omega}{E_0}\frac{2}{1-\epsilon}$$

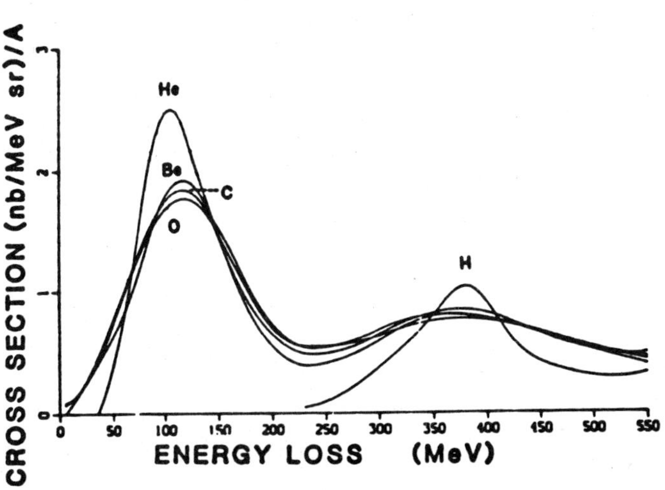

FIG. 11. **Inclusive (e,e′) from a variety of nuclei in the Δ-resonance region from ref. 11.**

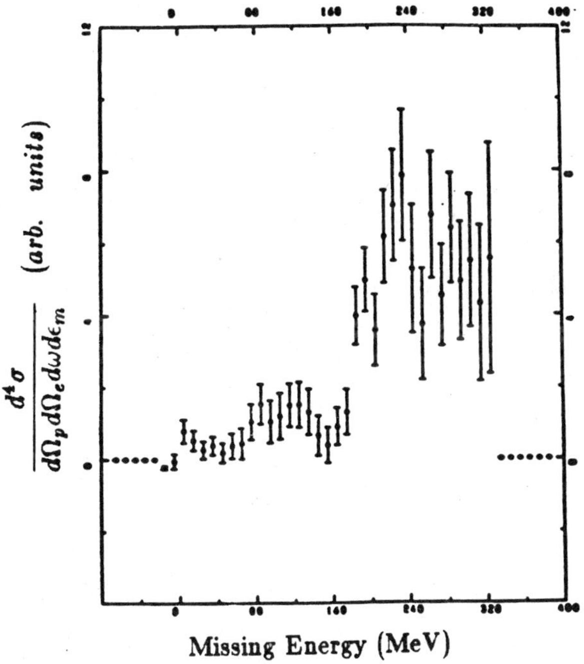

FIG. 12. Preliminary missing energy spectrum in the Δ region ($\omega = 383$ MeV) of ^{12}C measured at Bates.

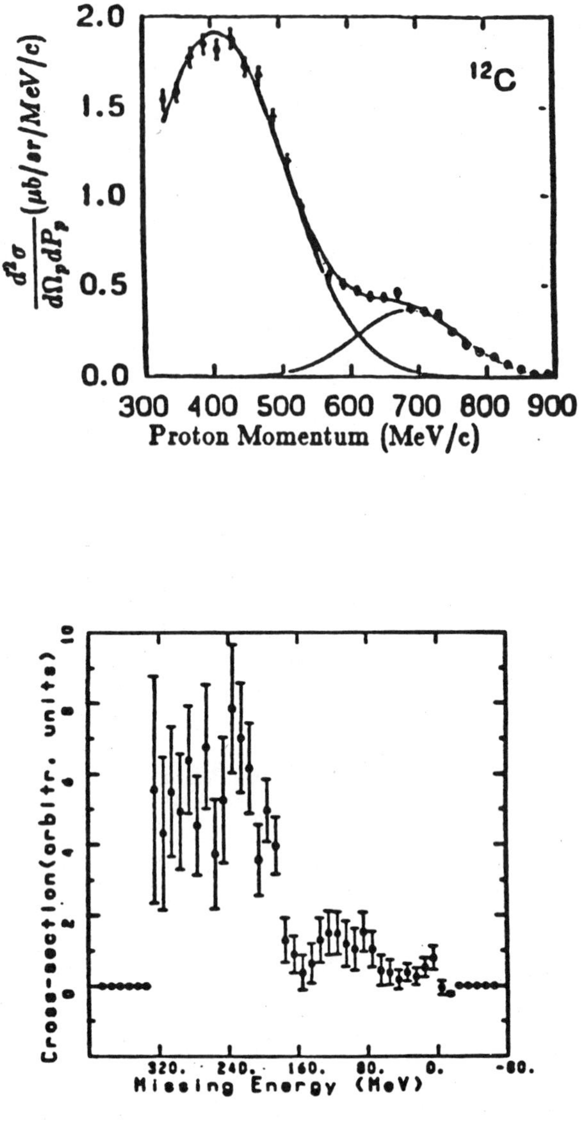

FIG. 13. Comparision of spectra in the Δ region produced by a) tagged real photons[12] and b) virtual photons from the Bates (e,e′p) experiment.

is the virtual photon flux, $\sigma(\gamma_V, p)$ is the virtual photoabsorption cross section, $k = (W^2 - M^2)/2M$ is the energy of a real photon which creates an excited nucleon of invariant mass W from a nucleon of mass M and

$$\epsilon = \left(1 + \frac{2\vec{q}^2}{q_\mu^2} tan^2(\theta_e/2)\right)^{-1}$$

is the longitudinal polarization of the virtual photon. The number of virtual photons per unit time is

$$N_V = \frac{< I >}{e} \Gamma_V \, (\Delta\Omega\Delta\omega)$$

where $< I >$ is the average current, $\Delta\Omega\Delta\omega$ is the solid angle–energy acceptance of the electron spectrometer and e is the electron charge. For our conditions ($E_0 = 650$ MeV, $\theta_e = 40$ deg., $< I > = 5\ \mu$A, $\Delta\Omega = 20$ msr. and $\Delta\omega = 70$ MeV) we get about $7 \mathrm{x} 10^7$ photons/sec. with 4 MeV resolution. The tagged real photon rate is around 10^5/sec. with 20 MeV resolution. The (e,e'p) reaction is clearly a very competitive technique for this type of work.

SUMMARY

The current coincidence program at the Bates accelerator is designed to elucidate the nature of the (e,e'p) reaction mechanism. Evidence for a transverse enhancement of the quasielastic response, particularly in the s–shell region, has been seen. A new reaction process has been observed in the dip region. This process is not one–body in character and represents the dominant contribution to proton knockout from the dip region. Two processes have been observed contributing to proton knockout from the Δ–resonance region. One of them occurs below pion threshold and is evidence that even at the peak of the Δ processes other than $\Delta \longrightarrow N + \pi$ are important. The investigation of the $\vec{q}$ and A dependencies of the processes we have observed is the focus of our current and planned experiments. Measurement of ^{12}C(e,e'p) at the quasielastic peak with $\vec{q} = 800$ MeV/c has just been completed. The light nuclei ^{3}He and ^{4}He will be studied in the near future.

ACKNOWLEDGMENTS

I would like to thank Paul Ulmer and Hossain Baghaei for allowing their data to be presented prior to publication. Prof. W. Bertozzi, Dr. J. M. Finn and Dr. T. W. Donnelly are to be thanked for many enligtening discussions. I acknowledge the support of the Fannie and John Hertz Foundation. This work was supported in part by the U. S. Department of Energy (D.O.E.) under contract DE–AC02–76ER03069.

REFERENCES

[1] T. deForest Jr. and J. D. Walecka, Adv. in Phys. **15**, 1 (1966).

[2] T. W. Donnelly, Symmetries in Nuclear Structure, ed. by K. Abrahams, K. Allaart and A. E. L. Dieperink, Plenum Publishing Copr.,(1983).

[3] G. B. West, Phys. Rep. **18C**, 269 (1975); Y. Kowazoe, G. Takeda and H. Matsuzaki, Prog. Theo. Phys., **54**, 1394, (1975).

[4] P. Barreau *et. al.*, Nucl. Phys. **A402**, 515 (1983).

[5] J. M. Finn, R. W. Lourie and B. H. Cottman, Phys. Rev. **C29**, 2230 (1984).

[6] G. Do Dang and Pham Van Thieu, Phys. Rev. **C28**, 1845 (1983).

[7] L. S. Celenza, A. Harindranath and C. M. Shakin, submitted to Phys. Rev. C.

[8] J. Mougey *et al.*, Nucl. Phys. **A262**, 461 (1976).

[9] J. W. Van Orden and T. W. Donnelly, Ann. Phys. (N.Y.) **131**, 451 (1980); T. W. Donnelly, J. W. Van Orden, T. deForest Jr. and W. C. Herman, Phys. Lett. **76B**, 393, (1978).

[10] J. M. Laget, From Collective States to Quarks in Nuclei, Lecture Notes in Physics, vol. 137, ed. by H. Arenhövel and A. M. Sarais, Springer, Berlin (1981).

[11] J. S. O'Connell *et al.*, Phys. Rev. Lett. **53**, 1627 (1984).

[12] S. Homma *et al.*, Phys. Rev. Lett. **53**, 2536 (1984).

ANTIPROTONS ARE ANOTHER MATTER

Michael V. Hynes
(Collaborators: Richard J. Hughes and Alan Picklesimer)
Los Alamos National Laboratory
Los Alamos, NM 87545

ABSTRACT

The advent of the new beam cooling techniques and their application to antiproton production has already made possible major advances in high energy physics. These same techniques offer uniquely exciting possibilities for ultralow energy physics. Through a combination of deceleration stages, antiprotons produced at several GeV (where the production cross section is at a maximum) can be made available for experiments at thermal velocities. High precision measurements of the antiproton mass and magnetic moment can be performed. Comparison of these measurements with those for the proton will test the CPT invariance of internal baryon dynamics at an unprecedented level. In addition the gravitational constant for antimatter can be measured for the first time, and to high accuracy. Each of these measurements will provide very important information on the dynamical symmetry between matter and antimatter in our universe. Antiprotons at thermal velocities will also make these fundamental particles available for experiments in condensed matter and atomic physics. The recent speculation that antiprotons may form metastable states in some forms of normal matter could open many new avenues of basic and applied research.

I. INTRODUCTION TO ANTIPARTICLES

The concept of antiparticles began with the work of P.A.M. Dirac in the early 1930's on the dynamics of electrons.[1] This work for the first time wedded the, then new, quantum mechanics with Einstein's relativistic kinematics. The need for this advance arose from atomic physics where it had recently been estimated that the electrons in an atom are moving in their orbits with velocities near the velocity of light. Dirac's new relativistic theory of electrons was an enormous breakthrough and explained a host of observed phenomena in an elegant and fundamental way. However, the new theory predicted the existence of a new particle in nature that was in every way the mirror image or anti-particle counterpart of the electron. In the mid-1930's the anti-electron, the positron, was discovered.[2]

The tremendous success of the Dirac theory and its experimentally confirmed prediction of the existence of an antiparticle for the electron touched off widespread speculation that the existence of antiparticles was a fundamental symmetry of nature. All particles have an opposite, an antiparticle. For protons there are antiprotons. For neutrons there are antineutrons. One can see that with positrons, antiprotons and antineutrons, you have all the ingredients needed to make anti-atoms. Thus, it was speculated that there could exist a whole periodic table of anti-elements identical in every way to the familiar elements except that they are constructed of antiparticles. Soon the term antimatter was coined.

Although the existence of the antiproton was predicted in the 1930's it was not until 1955 that its existence was experimentally observed. Chamberlain, Segre, and coworkers[3] at the Lawrence Berkeley Laboratory had labored since the late 1940's to build a proton particle accelerator with enough energy to produce antiprotons. They knew exactly what they were after and tailored the accelerator design for the production of antiprotons. Their discovery of this new antiparticle rocked the world of physics. Chamberlain and Segre were awarded the Nobel Prize in physics for this observation. The award cited specifically the experimental confirmation of the particle-antiparticle symmetry in nature. This work opened the door for cosmologists and astronomers to ask in earnest if there were antimatter in our universe and stimulated a host of other investigations.

II. ANTIPROTON PRODUCTION, COLLECTION, AND COOLING

A. Fundamental Production Process

The basic reaction that produces antiprotons ($\bar{p}$) is illustrated in Fig. II-1. In this figure, a high-energy proton with momentum P_0 is incident on a nucleon at rest in, for instance, a liquid hydrogen (LH_2) target. Such an initial state can reach a multitude of possible final states (channels) ranging from simple elastic scattering to multiple pion and kaon production, depending on the incident beam momentum. However, let us consider only the final state that produces $\bar{p}$'s. To conserve baryon number and charge, antiprotons are produced as part of a $p\bar{p}$ pair, as shown in Fig. II-1 (where the scattered incident proton and recoiling target nucleon are also shown). The minimum beam momentum required for this reaction is P_0 = 6.5 GeV/c (5.6 GeV kinetic energy) whereas the production probability increases rapidly with increasing P_0. Typical $\bar{p}$ production facilities for basic research use incident beam energies in excess of 20 GeV. Usually these facilities use production targets of beryllium (Be), carbon (C), or tungsten (W) instead of LH_2. This simplifies the production system structure and leads to slightly different kinematic properties of the distribution of $\bar{p}$'s emerging from the target. There have been a great number of measurements of $\bar{p}$ production from nuclear targets.[4-15] Although there is little fundamental understanding of the production process, several

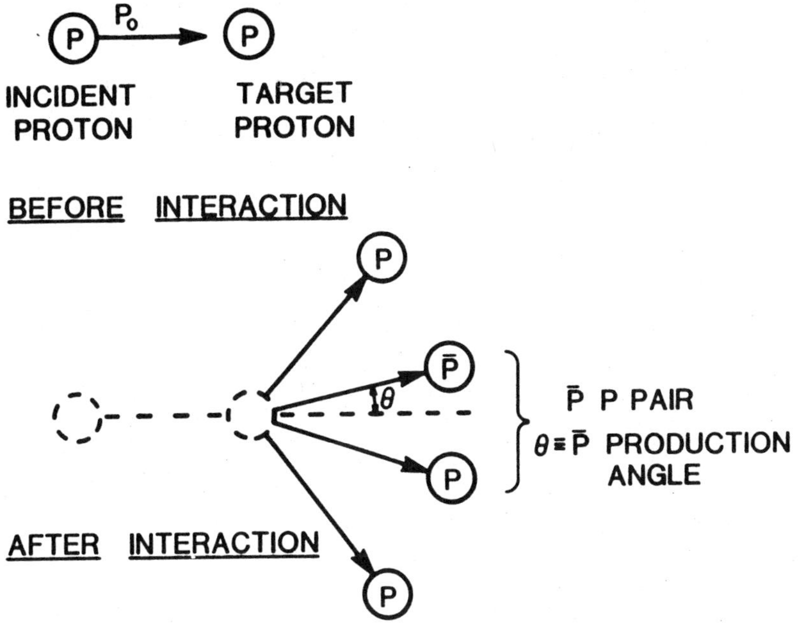

Fig. II-1. The basic interaction leading to $\bar{p}$ production in pp interactions. Note the definition of the $\bar{p}$ production angle (see text also).

empirically derived production cross-section formulations describe the available data.[16,17]

A modern $\bar{p}$ production system is shown schematically in Fig. II-2. A high-energy proton beam from an accelerator is transported through a series of magnets to the $\bar{p}$ production target. The proton beam transport magnets allow for a certain amount of adjustment in the beam focus. The production target is usually of heavy metal about an interaction length long in the beam direction. An interaction length target is defined as a target that has a nuclear density corresponding to an e-fold

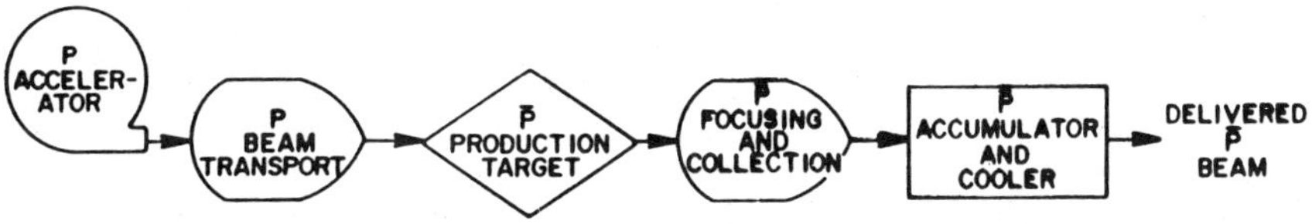

Fig. II-2. Block diagram of modern $\bar{p}$ production facility showing the major elements involved.

attenuation of the incident beam because of inelastic nuclear reactions. The antiprotons emerging from the target are dispersed in production angle (see Fig. II-1) and momentum. In another terminology, the $\bar{p}$'s fill a large phase space with a low phase-space density. Thus the production target has a large emittance. The $\bar{p}$ beam transport usually selects a small part of the $\bar{p}$ phase space for further use and thus has an acceptance in momentum and production angle that is small compared to the total amount that is available from the production target. The beam emerging from the $\bar{p}$ beam transport system is very often simply made available for experiments. In this case, filters designed to remove other negative particles such as pions and kaons in the beam are included in the transport system. This type of $\bar{p}$ beam is available at the Brookhaven National Laboratory (BNL)[18] and at KEK in Japan.[19,20] Another approach to filtering, proposed at BNL is the so-called time-separated beam (TSB).[21] Here the $\bar{p}$ beam transport system ranges from 1 to 2 km in length, over which distance many of the pions and kaons will decay or be separated spacially from the $\bar{p}$'s due to the differential velocity.

The new facilities at CERN and FNAL, however, have introduced a new "cooling" technology to antiproton beams. In these systems the $\bar{p}$'s from the beam transport are directed into a large ring structure where rf elements[22] and/or a co-rotating electron beam[23] are used to compress the phase-space density of the antiprotons. Thus the beam emittance is made smaller while the phase-space density increases, as shown schematically in Fig. II-3 (Ref. 24). At CERN and FNAL the cooled antiproton beam is injected into a much larger synchrotron where the $\bar{p}$'s are accelerated and used for high-energy $\bar{p}p$ research. Also at CERN there is a Low Energy Antiproton Ring (LEAR) facility where the $\bar{p}$'s can be decelerated to very low energies (5 to 20 MeV).

B. Antiproton Production Cross Section

The antiproton production cross section is the basic ingredient in designing the $\bar{p}$ collection and cooling systems. Empirical fits to the data have been available since 1967 with the work of Sanford and Wang (SW).[16] The more recent work of Hojvat and van Ginneken (HG)[17] incorporates all of the new data in the fit and agrees with the earlier SW results on beryllium for proton energies less than about 30 GeV.

The $\bar{p}$ production cross section peaks near 0° (in production angle) and increases with increasing incident proton energy,[17] as shown in Fig. II-4. The momentum of the $\bar{p}$'s at which the cross section is maximum also increases with increasing incident beam energy. Tungsten is the target material for the cross sections shown in Fig. II-4. Although these cross sections seem quite large, they fall very quickly with increasing production angle as shown in Fig. II-5. In this figure the cross sections for $\bar{p}$ production from 30.9-GeV protons on beryllium are shown for the production angles indicated. By 12° the cross section has fallen about an order of magnitude. Also, from these figures it is

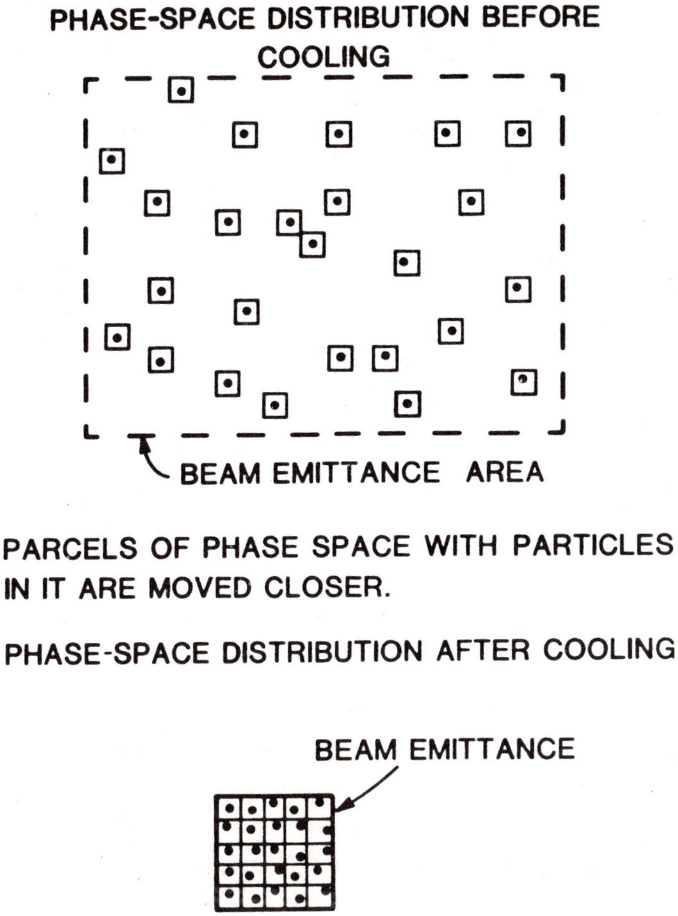

Fig. II-3. Phase-space distributions of $\bar{p}$ beam before and after cooling.

evident that the momentum spread of the emerging $\bar{p}$'s is quite broad and increasing with increasing incident energy.

Typically the focusing and collection systems accept about 2 to 4 degrees in production angle and about 4% in momentum spread ($\Delta p/p$). Thus, only a small fraction of the antiprotons produced fit into the acceptance of the rest of the system. The total number of antiprotons produced per interacting proton, defined as the multiplicity, is a large and rapidly increasing function of incident energy, as shown for both lead and hydrogen targets in Fig. II-6. The multiplicity shown in the figure results from the integration of the HG cross sections.[17] At 40 GeV and above, the

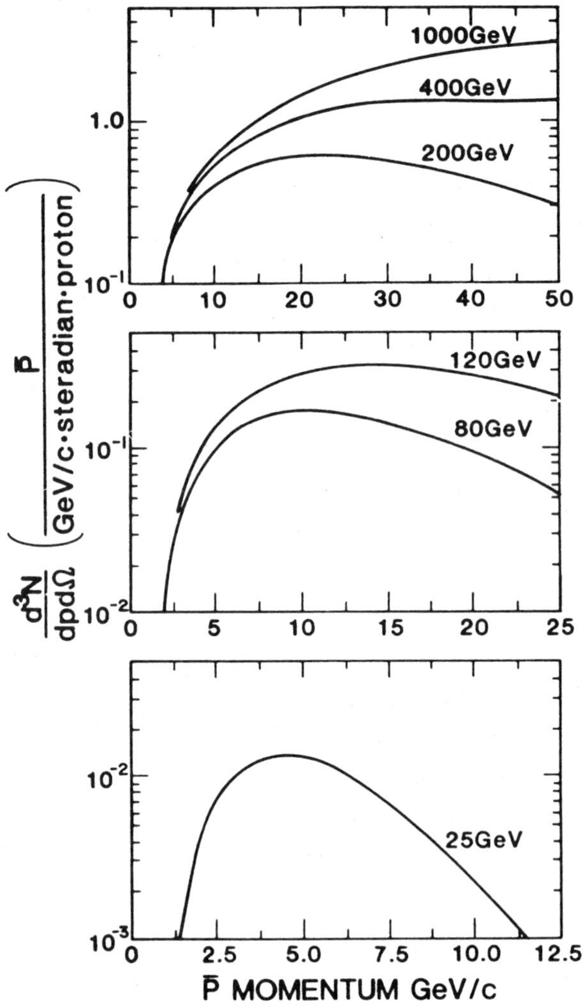

Fig. II-4. Differential $\bar{p}$ production cross section for tungsten near 0°. The different beam energies are indicated. (See Ref. 17.)

multiplicity is several per cent, getting to 10% near 120 GeV (for lead). From these considerations, it is evident that large-acceptance $\bar{p}$ transport systems can achieve significant increases in $\bar{p}$ flux.

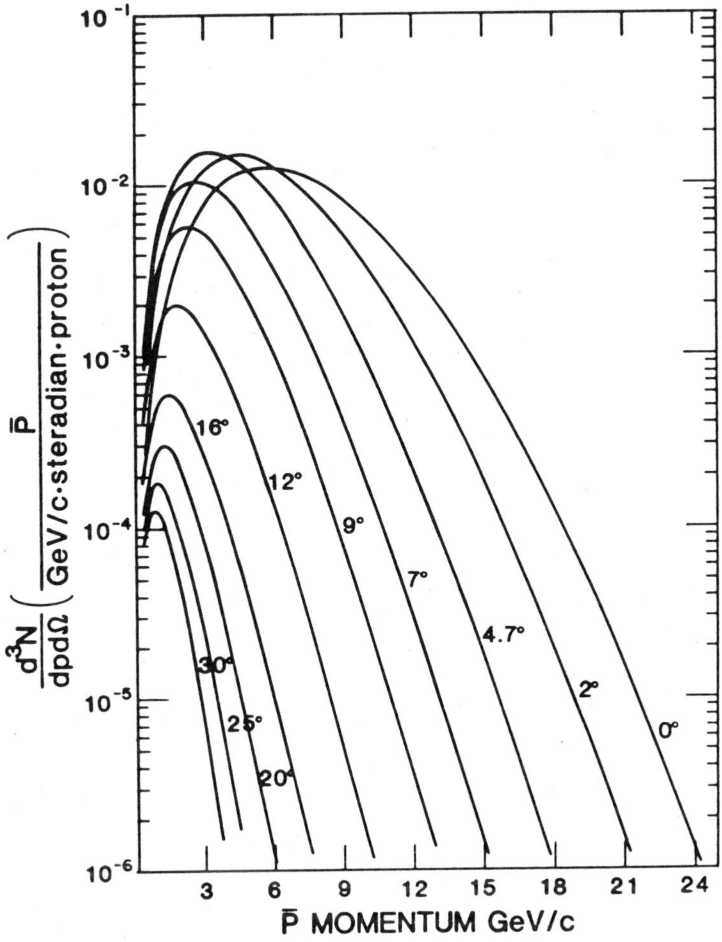

Fig. II-5. Differential $\bar{p}$ production cross section for 30.9-GeV protons on beryllium. Each contour is for the production angle indicated. (See Ref. 16.)

C. The Production Target

Because most $\bar{p}$ beam transport systems image the production beam spot onto the acceptance of the antiproton accumulator/cooler, a high premium is placed on the transverse and longitudinal size of the $\bar{p}$ production spot. The density of $\bar{p}$'s in the transverse phase-space acceptance of the collection system increases with decreasing transverse proton beam size until secondary $\bar{p}$ production and multiple scattering effects begin to

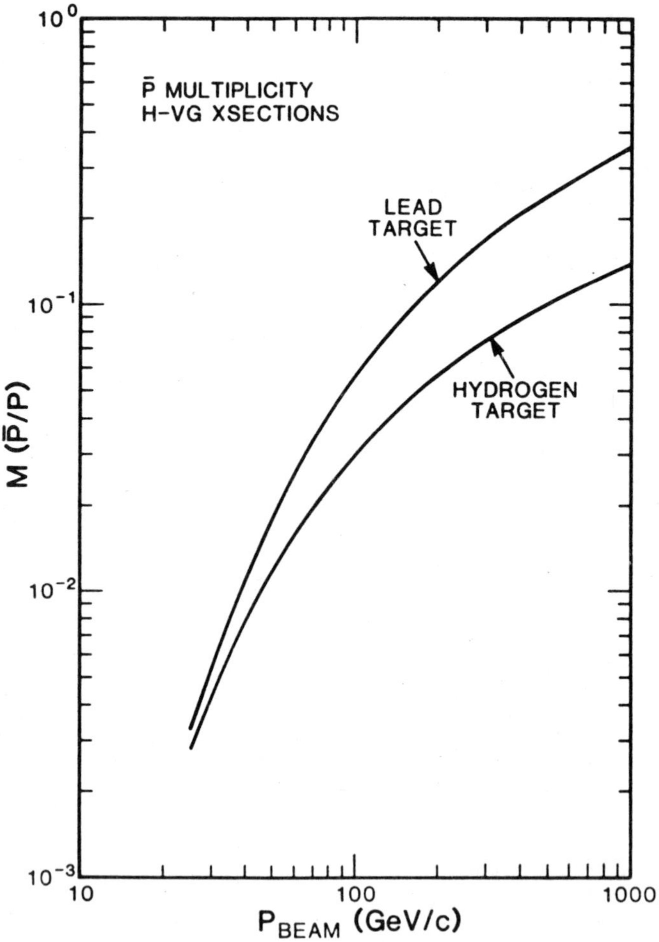

Fig. II-6. The $\bar{p}$ multiplicity as a function of incident energy for both lead and hydrogen targets.

dominate. In addition, the use of short high-density targets, such as tungsten, makes the longitudinal size of the production spot small, which is important to offset the depth-of-focus restriction of most short focal length imaging systems.[17] A combination of $\bar{p}$ absorption in the production target and increased longitudinal spot size tends to limit the lengths of tungsten targets to less than an interaction length (~10 cm). For the FNAL system, $\bar{p}$ transport is maximized at 6 to 7 cm (Ref. 17).

A high-energy and -intensity pulsed proton beam passing through an interaction length target will load the target material with a relatively large amount of energy in a time short compared with thermal diffusion times. Energy densities above 200 J/g are expected to result in density depletion due to shock wave formation and eventual target failure. A tungsten target less than an interaction length long will reach this energy deposition point when the number of protons per square millimeter is 2 to 2.5 $\times 10^{13}$ (Ref. 25). Thus for solid targets, such as tungsten, a limit to the transverse beam spot size is imposed by the mechanical properties of the material. This limit will naturally be different for lead or copper targets. The nominal energy deposition for the FNAL and CERN designs is less than the 200 J/g limit.

Several methods have been proposed to improve on this targeting problem. Sweeping the p and $\bar{p}$ beams through the target to outrun the shock waves or rapidly moving the target have been proposed.[26] Liquid targets (for example, mercury) are also possibilities.

D. Focusing and Collection Systems

Antiprotons are produced in the target generally in the forward direction (small production angles) but with a large momentum spread that increases with increasing incident beam energy. At 25 GeV, 95% of the $\bar{p}$'s are produced forward of 10° but with momenta ranging from 2 to 10 GeV/c. At 80 GeV, the same percentage of $\bar{p}$'s are inside a cone of 5° but with momenta ranging from 2 to 40 GeV/c. The concentration of antiprotons near zero degrees makes the solid angle requirement of a focusing and collection system rather modest; 100 msr generally will be adequate. However, the very broad momentum spread makes collection of most of the antiprotons through conventional focusing systems very difficult.

Conventional systems in use today for antiprotons accept ~4% in $\Delta p/p$ (Ref. 17) and use lithium lens[27-29] or magnetic horn[30,17] focusing elements to image the production beam spot on the acceptance plane of the accumulator/cooler. These devices are specially designed pulsed magnetic elements with short focal lengths and with axially symmetric focusing properties unlike quadrupoles. Very high field gradients, 1000 to 1500 T/m, are used to pull the diverging particles back to a focus. Typical repetition rates are ~0.5 Hz although some lithium lens systems have been operated at 15 Hz. Both the CERN and FNAL $\bar{p}$ collection system designs originally used quadrupoles as the focusing elements.[30,31] However, these were abandoned in the design stage in favor of a lithium lens or horn because larger acceptances were possible.

A schematic cross section and end-view of a lithium lens is shown in Fig. II-7. The assembly is mounted directly on the secondary of an air core transformer to keep the system inductance and resistance low. The lens itself is a cylindrical lithium conductor excited by a unipolar current pulse. This pulse is typically about 50 μs long at 150 kA (Ref. 27). The skin depth in

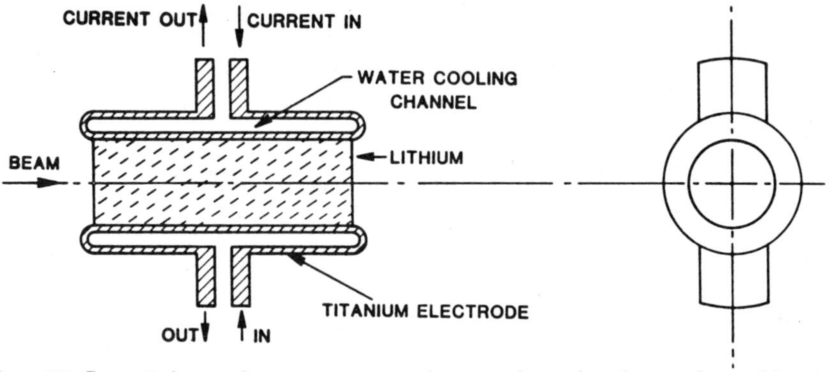

Fig. II-7. Schematic cross section and end view of a lithium lens. The current feed path is also shown.

lithium is about 2 mm for these pulses, which results in a nearly uniform current distribution for cylinders of about 5-mm diameter. The resulting azimuthal magnetic field rises linearly with distance from the symmetry axis to a maximum value of about 100 kG at the inside edge of the electrode. With the beam passing directly through the lithium, this field configuration is exactly that required for axially symmetric focusing. In typical designs, about 10% of the beam is lost due to nuclear interactions.[27]

Although the concept is simple, the fabrication of the device is complicated by the thermal expansion of the lithium due to joule heating. The electrode and most of the rest of the lens assembly (not shown in Fig. II-7) is made of titanium to withstand the 400 kg/cm^2 pressure normally encountered. A schematic view of the FNAL lithium lens collector system is shown in Fig. II-8.

The antiproton facility at CERN uses a magnetic horn[30] for its focusing and collection system. This device is schematically shown in Fig. II-9. The exterior walls of the horn are usually

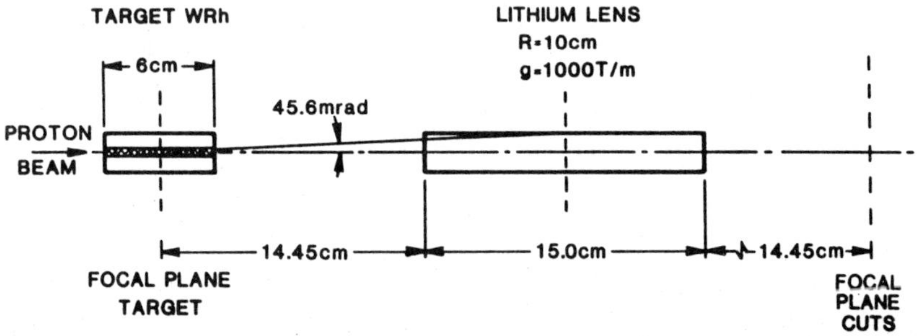

Fig. II-8. Schematic view of the FNAL lithium lens collector system.

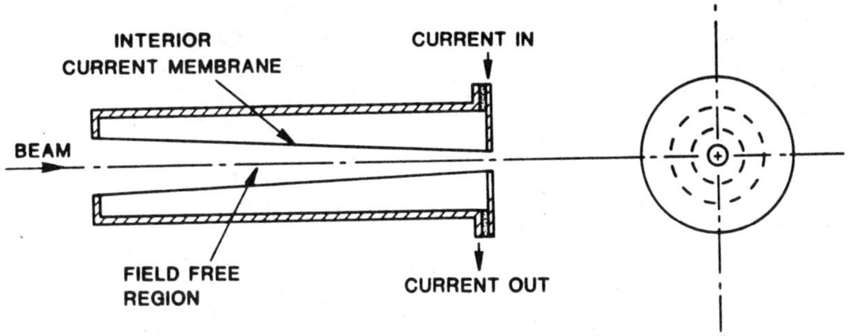

Fig. II-9. Schematic cross section and end view of a magnetic horn collector. The current feed path is also shown.

quite thick for mechanical stability, whereas the interior walls are quite thin to allow for particle passage with low interaction probability. The current path shown in the figure will generate an axially symmetric azimuthal B-field, which is zero inside the throat of the horn and falling as 1/r for radial distances larger than the distance to the inside current membrane. This field configuration also will provide axially symmetric focusing with the incident beam along the symmetry axis.

The CERN horn is excited by a unipolar 150-kA current pulse about 20 μsec wide at 0.5 Hz. The joule heating in the thin inner membrane as well as the compressional force on it from the magnetic field require that this membrane be about 0.5 mm thick. Particle losses due to interactions in the horn are usually quite low. A schematic view of the CERN horn collector system is shown in Fig. II-10.

The emittance of the collection and focusing system must be matched to the acceptance of the accumulator/cooler structure. The mission of this structure is to compress the large phase space of the antiprotons emerging from the collection system to a very small, high-density volume suitable for injection into conventional accelerator or storage devices (see Fig. II-3). The CERN (Ref. 30) and FNAL (Ref. 31) systems are designed to inject

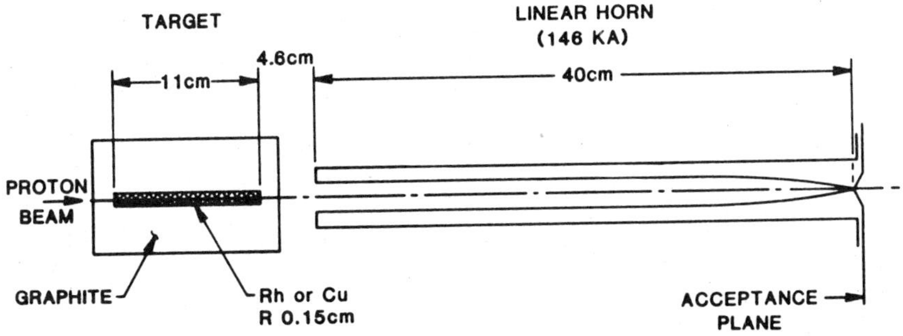

Fig. II-10. Schematic view of the CERN horn collector system.

into a large synchrotron ring that will accelerate the antiprotons to very high energies (>200 GeV). At CERN there is, in addition, a facility (LEAR) for low-energy antiproton storage.

The concept of beam cooling or phase-space compression would appear on the surface to violate Liouville's theorem.[32,33] This theorem, however, strictly applies to continuous distribution in phase space whereas the beams used in accelerators and storage rings are truly distributions of points. Thus, in a typical beam, there are large volumes of phase space that are simply empty, as shown in Fig. II-3. Liouville's theorem is not violated when the point distribution emerging from the collection system is compressed to fill these voids.

The term "cooling" implies that the beam's quality can be characterized by a temperature (T). At best there is only an informal analogy to the usual definition of temperature via the Maxwell-Boltzmann distribution. Most cooled beams can neither be characterized by a single T nor can the temperature distribution in the beam be independent of position. Nevertheless, the concept of an effective beam temperature is a useful parameter describing beam quality and makes more intuitive the operation of phase-space compression techniques.

Two temperatures are crucial to understanding cooling techniques. The longitudinal temperature (T_L) characterizes the distribution of velocities in the beam along the dimension parallel to the beam momentum. The transverse temperature (T_T) characterizes the distribution of velocities in the beam along the dimension transverse to the beam momentum. These temperatures are very difficult to change with techniques other than those used in beam cooling.

Two types of transverse temperature cooling are in use, electron and stochastic. In 1966 Budker[34] discussed the concept of electron cooling in storage rings, crediting O'Neil[35] for independent discovery. In this technique an electron beam moves parallel to and mixes with a heavy particle beam, such as antiprotons, at the same longitudinal velocity, as shown in Fig. II-11. The electron beam, which overlaps the antiproton beam for a straight section in the $\bar{p}$ storage ring, has a lower transverse temperature than does the antiproton beam. As the two beams come to thermal equilibrium through Coulomb interactions, the longitudinal and transverse temperatures of the antiproton beam will be reduced or cooled. Changes in longitudinal temperature are easily trimmed with rf; the transverse cooling, however, is crucial.

In 1972 Van Der Meer[36] proposed that cooling could be effected by an electronic feedback system that sensed the mean lateral position of the beam in a short section of the storage ring. In an opposite section of the ring a transverse kick is applied to the beam in the direction that reduces the mean lateral position as shown in Fig. II-12. This technique is termed stochastic cooling because it works on the statistical fluctuations of the beam.

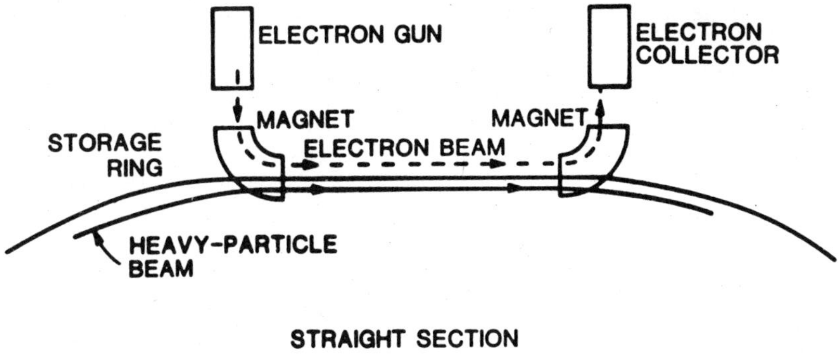

Fig. II-11. Schematic view of electron cooling[37] (see text also).

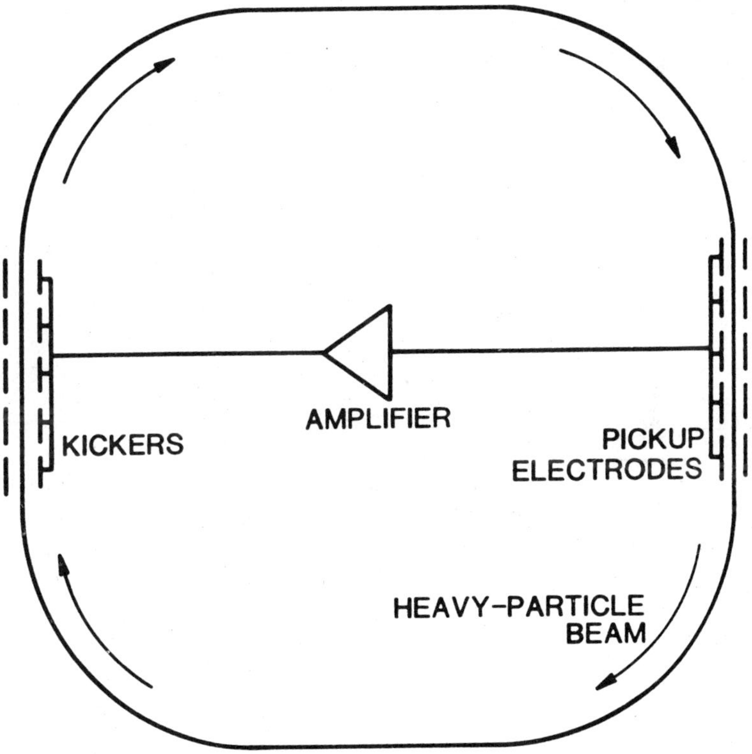

Fig. II-12. Schematic view of stochastic cooling[37] (see text also).

Electron cooling is very efficient at antiproton energies of a few hundred MeV or less but falls off rapidly at higher energies and at high antiproton beam currents. Stochastic cooling is not as efficient at low energies but does not fall as rapidly as electron cooling with increasing energy. Depending upon the antiproton beam energy, electron, stochastic, or a combination of these two cooling techniques is used.

III. Low Energy Antiprotons at CERN

A. The CERN Antiproton Complex

Several very complete reviews of the CERN antiproton complex, shown in Fig. III-1, are available.[30,38]

In brief, the operation starts with a 26 GeV beam of protons from the Proton Synchrotron (PS) focusing onto a copper production target. Antiprotons (among other particles) emerge from the target with a variety of angles and momenta. The production cross section has a maximum at an antiproton momentum between 3 and 4 GeV/c. Consequently, the Antiproton Accumulator (AA) is tuned to a central momentum of 3.57 GeV/c with a $\Delta p/p$ acceptance of $\pm$ 7.5 $\times$ 10^{-3} and an angular acceptance of 100 $\pi\cdot$mm$\cdot$mr. The antiprotons are produced in 500 ns bursts every 2.4 s. For every burst of 10^{13}, 26 GeV protons from the PS, 2 $\times$ 10^{7} antiprotons are injected into the AA. On injection the $\bar{p}$'s are broadly spread out in both longitudinal (along the beam direction) and transverse momentum. The cooling systems in the AA reduce this momentum spread and thus compress the antiprotons into a smaller volume of phase space. Each new burst of $\bar{p}$'s starts on the tail of this phase space distribution and is eventually stacked into the central high density core. Stacks of $\bar{p}$'s are accumulated over a period of days. Eventually accumulation is stopped and small slices of the

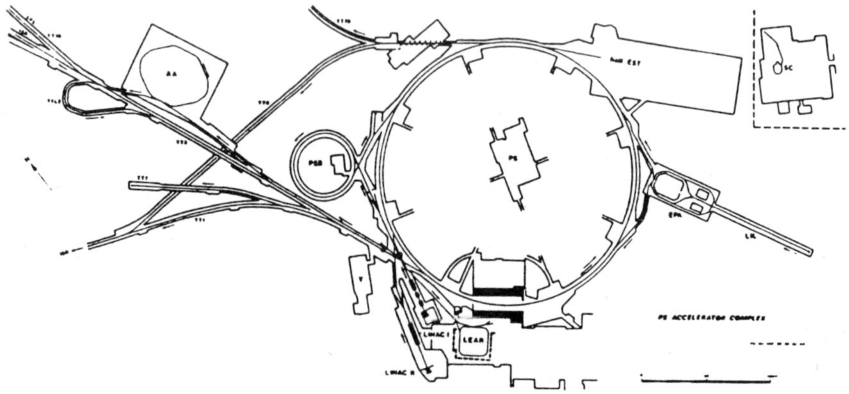

Fig. III-1. The CERN antiproton complex.

core are peeled off into an ejection orbit in the AA. The ejected beam is then transported back to the PS where it is either accelerated up to 26 GeV/c for ejection to the SPS or decelerated to 600 MeV/c for transfer to LEAR.

B. The LEAR Facility

Typically the Low Energy Antiproton Ring (LEAR) receives a pulse of 10^9 $\bar{p}$'s every hour. In LEAR these $\bar{p}$'s can be accelerated up to 2 GeV/c (1.3 GeV kinetic energy) or decelerated to about 100 MeV/c (5 MeV kinetic energy). The general layout of LEAR is shown in Fig. III-2. The ring is almost square in shape allowing for long straight sections where equipment can be installed. The beam

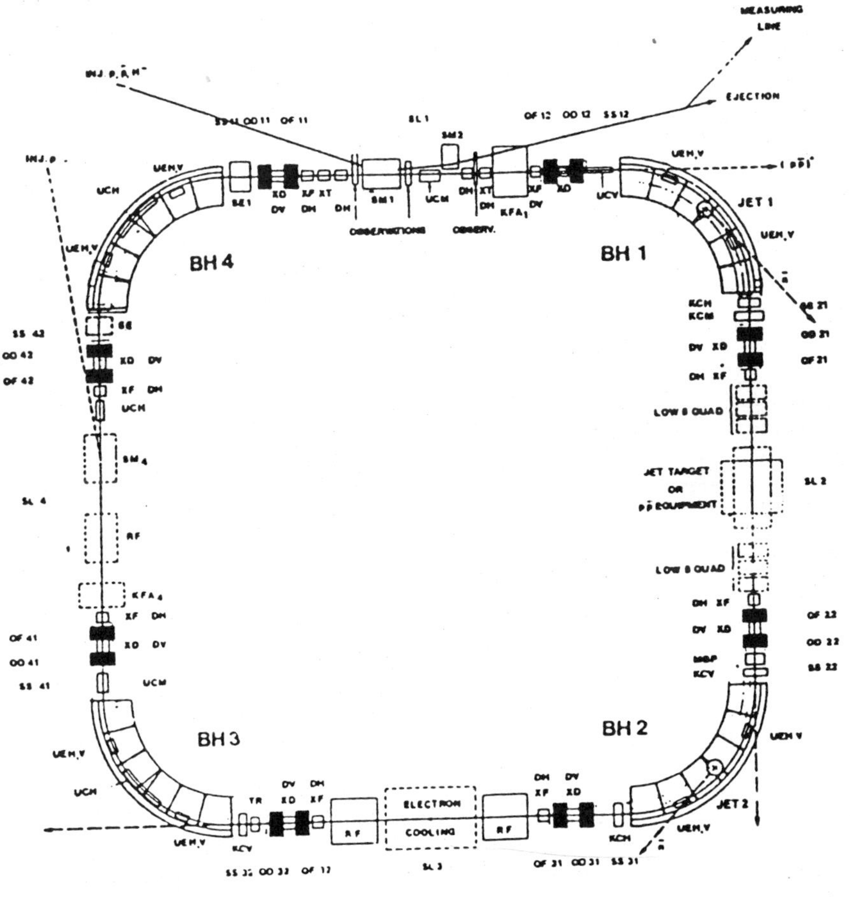

Fig. III-2. LEAR general lay out.

TABLE III-1
BASIC LEAR PARAMETERS

Momentum (kinetic energy) range	0.1-2 GeV/c (5.3 MeV-1.3 GeV)
Injection momentum (kinetic Energy)	0.6 GeV/c (175.4 MeV)
Circumference	78.54 m
Typical cycle	10^9 $\bar{p}$ injected every 10^3 s
Typical extracted beam	10^6 $\bar{p}$'s per second
Typical spill length	900 s
Long straight sections	4 of 8 m length each
Short straight sections	8 of 1 m length each
Bending magnets:	
No., arc length, field at 2 GeV/c	4, 6.55 m, 1.6 T
Quadrupoles:	
No., magnetic length, max gradient	16, 0.5 m, 12 T/m
Focusing structure	4 superperiods, BoDFOFDoB
Betatron wave number	Q_H~2.3, Q_V~2.7
Momentum compaction	$\alpha = \gamma_{tr}^{-2} = 4.8 \times 10^{-3}$
Aperture limitations	$a_H = \pm 70$ mm, $a_V = \pm 29$ mm
Acceptances	$\varepsilon_H = 240\ \pi\cdot$mm$\cdot$mr
	$\varepsilon_V = 48\ \pi\cdot$mm$\cdot$mr
	$\Delta p/p = \pm 1.1\%$
Vacuum system design pressure	$10^{-11} - 10^{-12}$
Bake-out temperature	300° C
Pump down time	40 h
RF system frequency range (h = 1)	0.4 - 3.5 MHz
Peak voltage per turn	12 kV

height is 1.66 m above the floor of the hall. The machine parameters are listed in Table III-1.

The injection and ejection process use the same septum magnet located in SL1 (see Fig. III-2). For injection the beam receives its final deflection by a fast ferrite kicker located in SL4. For ejection a thin electrostatic septum in SS11 is used to deflect the beam across the common magnetic septum after which the final deflection is delivered by a thick septum magnet. The ejection system provides a long spill at a high duty cycle so that continuous beam is available between two antiproton refills from the AA. The system is based on the normal betatron resonance ejection technique but uses an RF noise or stochastic driver[39] to slowly diffuse the particles towards the resonant momentum that leads to extraction.

Stochastic cooling systems for several energies have been included in the ring design. In addition an electron cooling system is due to be installed and will be used for momenta below 300 MeV/c. Extracted beam emittances (unnormalized) for the low momentum operation anticipated for our program of experiments are summarized in Table III-2. These emittances are determined from the cooling efficiency and various beam-blow-up mechanisms (such as intra-beam scattering). Because the stochastic ejection system peels off the horizontal and longitudinal emittances, the

TABLE III-2

LEAR LOW MOMENTUM EMITTANCES

Momentum (MeV/c)	100	50
Kinetic energy (MeV)	5	1.3
Unnormalized emittances (π·mm·mrad)		
ε_H	1	2
ε_V	3	6
$\Delta p/p$	0.15×10^{-3}	0.3×10^{-3}

properties of the extracted beam can be far better than those of the circulating beam.

C. The Los Alamos Ultra-Cold Antiproton Program

The Los Alamos National Laboratory, in conjunction with experimental groups from Genoa, NASA, Pisa, Rice, and Texas A&M and theoretical groups from Case-Western, Kent State, and Texas A&M again has commenced a program to construct a thermal source of antiprotons. The motivation for this work was to measure the gravitational mass of antiprotons, or antimatter, originally discussed by Goldman and Nieto.[40] The basic approach in this experiment involves a time of flight (TOF) measurement where the antiproton is launched either up or down in a gravitational field. Such an approach is in analogy to that used by Witteborn and Fairbank[41] for the electron measurement. In order to bring the TOF distance and timing into an experimentally tractable range, very low antiproton energies (velocities) are required. Indeed these energies are better described in terms of degrees Kelvin rather than in electron volts. The scale for these energies and how they relate to the physics and experimental equipment used at other energies is displayed in Fig. III-3. The thermometer shown in the figure is calibrated in electron volts down the left hand side with the corresponding temperature in degrees Kelvin on the right. The physics of interest is schematically indicated on the right hand side whereas on the left, the experimental equipment used is listed. The realm of high energy physics is somewhat arbitrarily indicated to start at 10^{14} °K (10^{10} eV) and extend upward without limit thru 10^{16} °K (10^{12} eV). The SPS for instance operates at ~ 2.7×10^{15} °K (2.5×10^{11} eV). Normal LEAR operations span the temperatures between 1.5×10^{13} - 5.8×10^{10} °K (1.3 GeV - 5 MeV). On the upper end of this range the $\bar{\Lambda}\Lambda$ experiment runs with all the stop experiments at the lower end. As shown in the figure, antiproton temperatures in the range of 1 - 10 °K are required for the gravity experiment. In order to achieve this antiproton temperature we are proposing to build a Radio Frequency Quadrupole (RFQ) decelerator and ion trap system. Such a system will extend the normal LEAR operating temperatures by ten orders of magnitude. Such an extension of antiproton

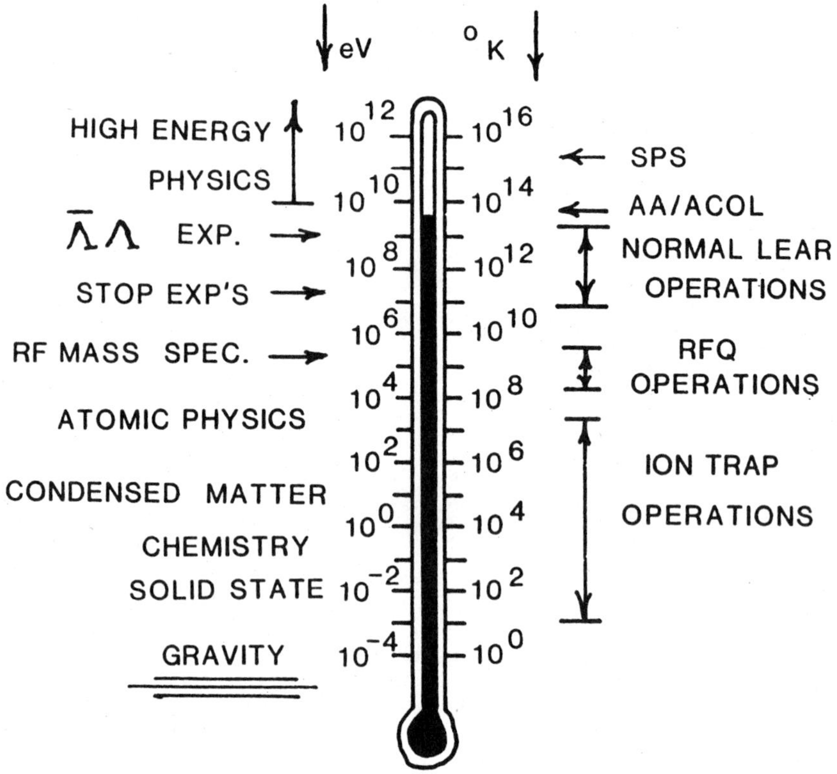

Fig. III-3. The antiproton thermometer shown is calibrated in electron volts down the left hand side, with the corresponding temperature in degrees Kelvin down the right hand side. The physics of interest along with the equipment used in each regime is also shown.

temperatures opens very exciting possibilities in atomic physics, condensed matter, chemistry and solid state.

The overall system we are planning is displayed schematically in Fig. III-4. A bunched beam of antiprotons is extracted from LEAR at 5 MeV or lower. The bunching cavity required in the LEAR ring would be matched to the operating frequency of the RFQ. The insertion of such a cavity and the techniques for both slow and fast extraction have been discussed recently by Lefevre and Mohl.[42] Several RFQ designs have been generated at Los Alamos for both 2 and 5 MeV injection energies and 200 and 425 MHz operating frequencies.[43] All designs have 99% or better particle transmission characteristics with very low emittance growth (<10%) for a 100 keV output energy. The lowest energy possible with RFQ

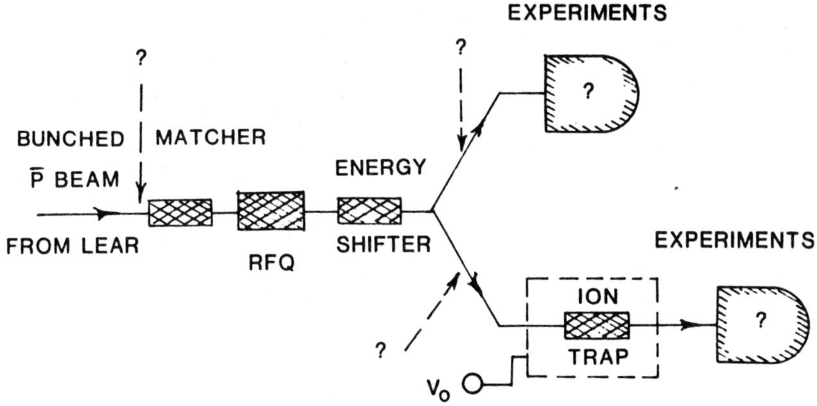

Fig. III-4. Overall system for low energy/low temperature operation (see text also).

technology is still under study. The 100 keV output energy was chosen to illustrate the use of an energy shifting cavity located after the RFQ. This device allows for a continuous adjustment of the beam energy by as much as ± 80 keV. Thus beams between 20 - 180 keV can be made available. A debunching cavity follows the energy shifter at a suitable drift length to once again rotate the output phase ellipse to achieve the best energy resolution. For example, at 20 keV the beam energy spread would be ± 0.5 keV. As shown in the figure, the beam now can be used directly for experiments. The RF mass spectrometer experiment which will measure the $\bar{p}$ inertial mass to 1×10^{-9} could use this beam. Many of the stop experiments could eliminate their degrader foils using this new beam as well. On the other hand, the beam can be transported to our planned ion trap for further energy reduction. Using an ion trap in such a capacity (and indeed for the gravity measurement) has been discussed by our collaborators from Pisa and Genoa.[44]

In our ion trap scheme, we envision using a large, possibly cylindrical, ion trap[45] for catching and cooling the beam from the RFQ to electron volt energies. Such a trap is shown schematically in cross section in Fig. III-5. The entire trap system will most likely be operated at an elevated voltage to take advantage of some amount of D. C. deceleration (V_0 in Figs. III-4 and III-5). As the beam burst approaches the trap, the front endcap electrode (see Fig. III-5) is run at the elevated ground potential (V_0), while the rear endcap electrode is operated at a total voltage greater than the beam energy. As the beam burst enters the trap it will turn around at the rear endcap electrode, at which point the voltage on the front endcap electrode is quickly brought to the same voltage as the rear electrode. The beam burst will now also turn around at the front electrode and is thus trapped. Fast

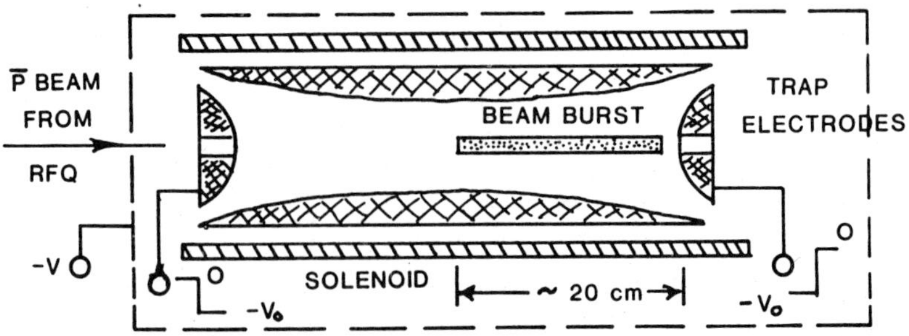

Fig. III-5. High energy (20 keV) bulk catching and cooling trap shown in schematic cross section (see text also).

rise time, high voltage pulsers have been built at Los Alamos for other applications.[46] Which designs can be readily adapted to an ion trap application. The trap assembly can now be lowered in voltage adiabatically to normal ground potential. Additional cooling using resistive damping of image currents[47] and electron cooling[48] can be used to further reduce the antiproton temperature. Subsequently, the antiprotons will be moved to an "harmonic trap" at cryogenic temperatures for further cooling. This whole process may take several minutes after which another burst can be caught, cooled and added to the previous cooled burst(s) in the trap. In a fast extraction mode 10^7 to 10^8 $\bar{p}$'s could be in each burst. Thus in a day 10^{10} $\bar{p}$'s could be stored in the trap at low temperatures, ready for experiments. The cold antiprotons can be launched, one or several at a time and be transported to the experiment at hand or experiments using the antiprotons in the trap as a target can be performed.

Two ion trap limitations need to be considered in designing experiments for such a system. Firstly, the number density of antiprotons is limited by a space-charge-like effect which leads to radial deconfinement in the trap and subsequent $\bar{p}$ loss.[49] Because the magnetic field is the source of radial confinement in the traps we envision, the limit to the number density is a function only of this field as shown in Fig. III-6. The fields we plan using in our traps will be between 5 and 6 Tesla where 10^{10} $\bar{p}$'s/cm^3 can be safely stored. Secondly, collisions with residual gas molecules in the trap will lead to a fixed loss rate. The experimental execution time combined with this loss rate sets the vacuum required in trap. To estimate this loss rate, the cross sections from Ref. (50) for $\bar{p}$ + H annihilation can be used. The results of this estimate are shown in Fig. III-7, where contours of constant annihilation rate are plotted as a function of antiproton temperature and residual gas pressure. Pressures better than 10^{-15} Torr and temperatures of about 10°K are required to keep the loss rate below 10^{-6} sec^{-1}. This vacuum requirement may seem extreme but the storage trap we envision will be at

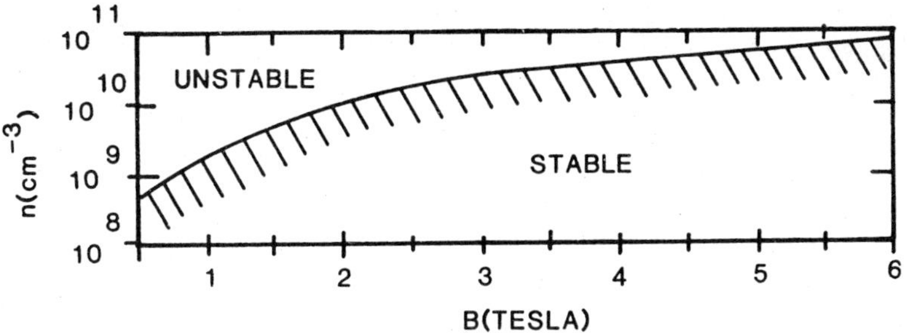

Fig. III-6. Stability limit for antiproton density in a Penning trap as a function of the applied magnetic field.

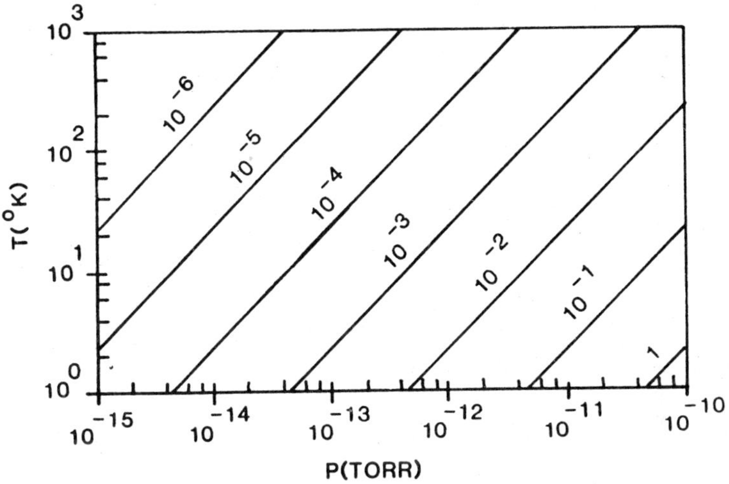

Fig. III-7. Contours of constant annihilation rate as a function of antiproton temperature and residual gas pressure. The cross sections are taken from Ref. 50.

liquid helium temperatures or better (< 4°K). At such cryogenic temperatures residual gas molecules are simply frozen out on the walls of the cooler container. The only thing left is helium vapor which can be readily pumped using a super cooled adsorption surface. Vacuums considerably better than 10^{-15} Torr can be obtained using cryogenic techniques. Loss rates can thus be enormously reduced.

Originally our interest in low temperature antiprotons arose from our plans for a gravitational mass experiment. In designing a system to achieve an antiproton temperature appropriate for this measurement it became obvious that many other experimental possibilities could be opened in atomic physics, chemical physics, condensed matter and solid state.

In atomic physics for example, many of the experiments planned for ELENA could be carried out using the ion trap as a source or a target of thermal antiprotons. One can envision ejecting the thermal antiprotons down a long vacuum isolation line into a gas cell or polarized gas jet at very low pressure. To have one interaction length of gas will require approximately 0.3 Torr pressures (assuming $\sigma = 10^{-16}$ cm^2). Such low pressures are completely free of Stark mixing effects and one can imagine now using the Stark mixing effects as a possible probe of chemical and molecular dynamics as one increases the pressures involved. There are also some very exciting possibilities for obtaining a polarized beam of antiprotons using the ion trap as a source.

In condensed matter physics, ions have been used as a probe of superfluid dynamics.[51] Antiprotons are a unique charged ion with very different dynamics than normal matter ions. Their interactions in a superfluid may reveal some interesting surprises. This last speculation is driven by the fact that atomic scale barriers of ~ 1.0 eV and ~ 1 Angstrom lead to very small barrier penetration coefficients. The results for a barrier penetration calculation for antiprotons are shown in Fig. III-8. In the figure the transmission coefficient (T) is plotted versus the barrier width (a) in Angstroms for several barrier heights (V_0). The antiproton energies are also indicated. With such small transmission probabilities on an atomic scale there may exist meta-stable states of antiprotons in normal matter. The principle reason for the small transmission coefficients for antiprotons is due to their mass. Thus negative muons or other negatively charged particles (not electrons) or lower mass will not exhibit the meta-stable state behavior.

IV. The Gravity Experiment

A. Introduction

Theoretical approaches to gravitational theory abound whereas experiments in gravity can be counted on one hand.[52-55] A fundamental measurement in experimental gravity that has not been executed to date is the measurement of the gravitational force on antimatter. Although several erroneous arguments[56,57] have been advanced as indirect evidence for the equivalence of the gravitational mass of particle and antiparticle, there is no direct measure of the equivalence. We propose to directly test this equivalence in the baryon sector using protons and antiprotons.[58]

Conventional quantum field theories are known to possess CPT invariance.[59] This symmetry systematizes the relationship between matter and antimatter, first introduced by Dirac.[60] The most

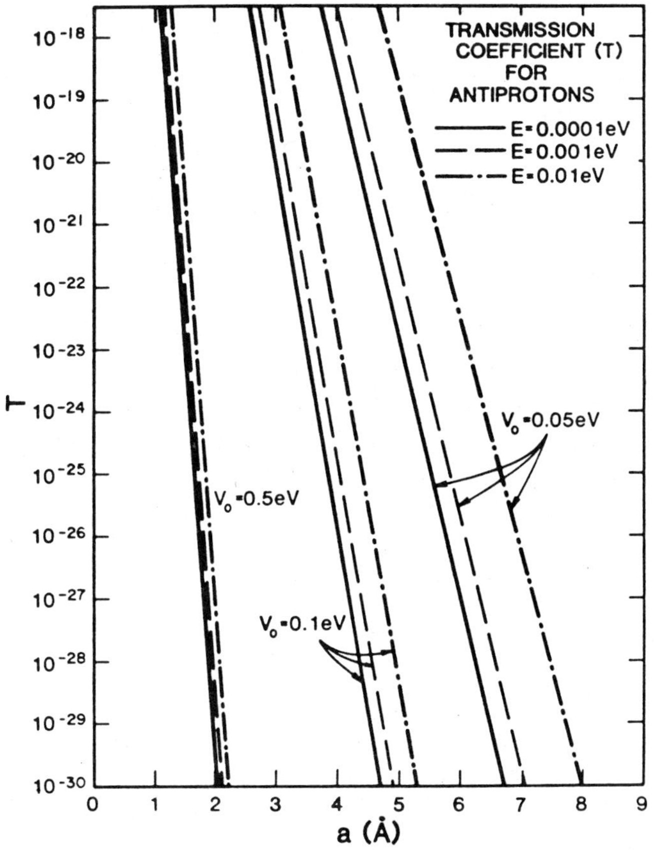

Fig. III-8. Antiproton barrier penetration coefficient plotted versus barrier width. The barrier height (V_0) and antiproton energy are also shown.

familiar consequences of CPT symmetry are that for each particle there is an antiparticle with the same

- Inertial mass
- Total lifetime
- Magnetic moment (with opposite sign).

One of the assumptions underlying the CPT theorem is that space-time is flat. Thus, it is not obvious that a quantum theory of gravity would possess CPT symmetry. Moreover, Wald[61] and others[62] have shown that the violations of quantum mechanics induced by black-holes would lead to the failure of CPT symmetry, while Geroch[63] and Kiskis[64] have shown that the topology of space-time has a direct bearing on the existence of C, P, and T symmetries in quantum field theory.

Experimentally, CPT symmetry has only been tested where gravitational effects can ignored. Tests of this symmetry using the inertial mass, lifetime, and magnetic moment are available for a number of particle-antiparticle pairs.[65] However, it should be noted that these tests are normalized to either an electro-weak or a strong interaction scale as appropriate. Because of the extreme weakness of the gravitational interaction we would not expect a CPT violating effect to be apparent in these tests.

The principle of equivalence[52,66] has been only tested for normal matter. CPT symmetry cannot be invoked to argue that antimatter also obeys the equivalence principle because it is not known if the CPT theorem holds in a gravitational field.

At the present time, there is neither direct nor indirect experimental or theoretical determination of the gravitational properties of antimatter. The measurement of the gravitational mass ot the antiproton will provide information on the basic symmetry between matter and antimatter in our universe at a level well beyond that which is currently available. Due to the fact that the experiment is gravitational in nature it will also provide a test of such symmetries in a context where recent theoretical work has indicated a mechanism for possible matter-antimatter asymmetries. It is no exaggeration to note that the observation of such an asymmetry would profoundly affect the way we both view and attempt to describe our physical universe. In the event that complete symmetry between protons and antiprotons is observed, the experiment will provide essential input to future theoretical attempts to unify gravitation with the other fundamental interactions.

B. Experimental Considerations

An apparatus for measuring the gravitational acceleration of electrons and positrons has been designed and tested by Witteborn and Fairbank (WF).[54] This experiment basically consisted of directing thermal electrons up a shielded copper drift tube at 4.2 K. As the experiment was underway, Schiff and Barnhill[67] pointed out that due to the electrons in the copper "sagging" under the influence of gravity, there should be an electric field acceleration on the electron equal and opposite to the acceleration due to gravity. Thus an electron in the drift tube would experience an effective gravitational force of zero, while a positron would experience an effective force twice the gravitational one. In general, in this apparatus a charged particle should experience an effective gravitational acceleration

$$g_{eff} = g(1 - \frac{Qm_e}{eM})$$

where Q(M) is the charge (mass) of the drifting particle, m_e is the electron mass and g is the normal gravitational acceleration. The WF experiment found that for the electron, g_{eff} was less than 0.09g. Despite difficulties which have delayed the program, this group continues to consider efforts to go on to test positrons.

To muddy the waters further, it was later observed that if in addition to the sagging electrons, the ions sag in the gravitational field then the WF experiment should see an electric force ~2000 times larger and in the opposite direction[68] to that predicted by Schiff and Barnhill. This new effect was observed at room temperature[69] but appears to go away with decreasing temperature so that at 4.2 K, the Schiff-Barnhill effect will be dominant.[70] These competing effects are not thoroughly understood[71,72] but it is conjectured that at low temperatures frozen residual gas deposits on the walls of the drift tube shield the effect of the ions. This effect must be understood for the analysis of our experiment.

For electrons or positrons the Schiff-Barnhill effect is a 100% correction to the gravitational acceleration. In total contrast this effect is a 0.05% correction if protons or antiprotons are used. Such a vanishingly small correction could even be calibrated out of the measuring apparatus if very high precision were required. This observation then leads us to propose to measure g for protons and antiprotons.

Although the mass of the proton and antiproton resolves the above criticisms of the WF experiment, the resolution of an additional criticism due to not measuring both the upward and downward gravitational effects in the same apparatus will require a different approach to the experimental design. At the core of this additional criticism is the fact that stray electric and magnetic fields can easily overwhelm the gravitational forces. To appreciate this one need only observe that

$$\left|\frac{m_e g}{e}\right| = 5.6 \times 10^{-11} \text{ V/m}$$

and

$$\left|\frac{m_p g}{e}\right| = 1.03 \times 10^{-7} \text{ V/m} \quad .$$

Despite the fact that the above field strengths are ~2000 times easier to control with antiprotons than with electrons, they are still so small that it is easy to imagine numerous effects that could generate them. In the WF experiment the following sources of stray fields were anticipated:

- The "patch" effect
- Thompson emf
- Thermal fluctuations
- Spatial and temporal variations in the magnetic field
- Mechanical alignment
- Residual gas collisions.

Despite the care exercised by these workers they still had problems with the Schiff-Barnhill effect and with the ion-sag controversy.

Let us consider what the time of flight (TOF) spectrum for antiprotons launched upward would be under ideal circumstances. In general, for a particle with initial velocity v_0, the time of flight either up or down a drift tube of length L will be

$$t = \int_0^L \frac{dz}{[v_0^2 \pm 2gz]^{1/2}}$$

$$\approx \frac{L}{v_0} (1 \pm \frac{1}{2} \frac{gL}{v_0^2}) \equiv \frac{L}{v_0}(1 \pm \delta)$$

For antiprotons launched up a drift tube, the TOF spectrum will be spanned by one of three possible features; normal gravity, no gravity, or antigravity. These possibilities are displayed in Fig. IV-1 for vertical flight paths in the gravitational field. If there were no gravitational acceleration for antiprotons the TOF would form a sharp peak (under ideal conditions) at TOF = L/v_0 as shown in the figure. If the gravitational acceleration for antiprotons were the normal one, then the TOF up would be longer by $\delta L/v_0$ and the TOF down would be just as much shorter (δ is defined in the previous equation). The opposite would be true for antigravity.

Under actual measuring conditions the sharp spikes in the TOF spectrum, illustrated in Fig. IV-1, would appear as broadened peaks due to thermal fluctuations and experimental uncertainties. In order to resolve the normal gravity peak location from that for no gravity (or any position in between) the width of this peak must be kept small compared to the separation $\delta L/v_0$. The quantities L, v_0 and g will set the scale for this separation. Because we wish to work with sensibly dimensioned equipment, we choose L = 1m. The initial velocity v_0, will be set by the temperature of the stored antiprotons. The value of $\delta L/v_0$ for several antiproton temperatures is tabulated in Table IV-1. For antiprotons at 100 K, the separation between the no gravity and normal gravity possibilities is ~ 1.3×10^{-9} s. Timing resolution much better than 5×10^{-10} s would be quite difficult to achieve and the TOF measurement at 100 K would not be possible (in a 1m drift length). The antiprotons in our traps, however, will be at 4 K, where the gravity-no gravity separation will be 1.6×10^{-7} s. This separation is easily resolvable.

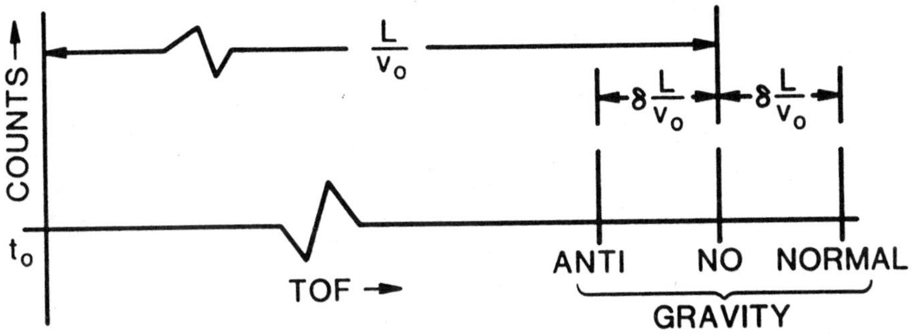

Fig. IV-1. TOF spectrum for vertical launch of the antiproton. The three possibilities spanning the region of interest are shown (see text also).

TABLE IV-1

ANTIPROTON TOF PARAMETERS

T(K)	v_0(m/s)	L/v_0(s)	$\delta \frac{L}{v_0}$(s)
1	157.4	6.35×10^{-3}	1.26×10^{-6}
4	314.9	3.18×10^{-3}	1.57×10^{-7}
10	497.9	2.01×10^{-3}	3.97×10^{-8}
50	1113.3	8.98×10^{-4}	3.55×10^{-9}
100	1574.5	6.35×10^{-4}	1.26×10^{-9}

The TOF resolution we can expect in our apparatus will depend upon:

- The velocity distribution of $\bar{p}$'s in the traps at launch time.
- The jitter in the launch and arrival pulses.
- The uncertainty in the launch position in the traps.

All of these factors can be controlled by moving the antiprotons, one at a time, into the compensated measuring trap before their launch. There the $\bar{p}$ motion can be well characterized. We estimate that a timing resolution of less than a nanosecond is possible. This compares very well with $\delta L/v_0$ = 157 ns for 4 K antiprotons. Our estimate of the timing resolution indicates that a gravity measurement is quite feasible, although considerable refinement of the experimental design is clearly required.

The availability of H^- ions in addition to protons as a calibration particle is a fortunate circumstance in measuring the gravitational mass for antiprotons. The mass of the H^- differs from that of the proton (or antiproton) by ~ 1 part in 10^3, but it has the same charge as the antiproton. Thus any stray electric fields will perturb the H^- measurements in almost exactly the same way as they will for the antiproton. This enormously improves the reliability and the precision of the calibration procedure.

V. Antiprotons in Your Future

Since their discovery, the rate of antiproton production has increased by an order of magnitude every 2.5 years (on the average). This trend line is shown in Fig. V-1 where the relevant physics and detector technology are indicated as well.[73-90] The slope of this trend line is limited by funding and the available accelerator and magnet technology. The Low Energy Antiproton Ring (LEAR) facility,[83-86] which recently came on-line at CERN, fits clearly on the trajectory as does a proposed upgrade of the Los Alamos Meson Physics Facility LAMPF II.[88,89] The early part of this trend line was driven by the advent of the zero gradient synchrotron (ZGS)[78] at the Argonne National Laboratory and the alternating gradient synchrotron (AGS)[77] at the Brookhaven National Laboratory. In fact, most antiproton production in this era actually exceeds the trend line which is drawn on a conservative trajectory. The present and future production rates will be driven by a new technology, stochastic and electron cooling. The facility at the Fermi National Accelerator Laboratory (FNAL)[87] scheduled to come on-line in 1985 is already considerably above the trend line. In addition, a practical antiproton factory, using existing magnet and accelerator technology, could be built by 1990 and would produce 100 to 1000 times more antiprotons than the conservative LAMPF II proposal. This possible facility is still above the trend line, which shows that the limits of the new cooling technology are not properly indicated. Actual limits are considerably higher. Nevertheless,

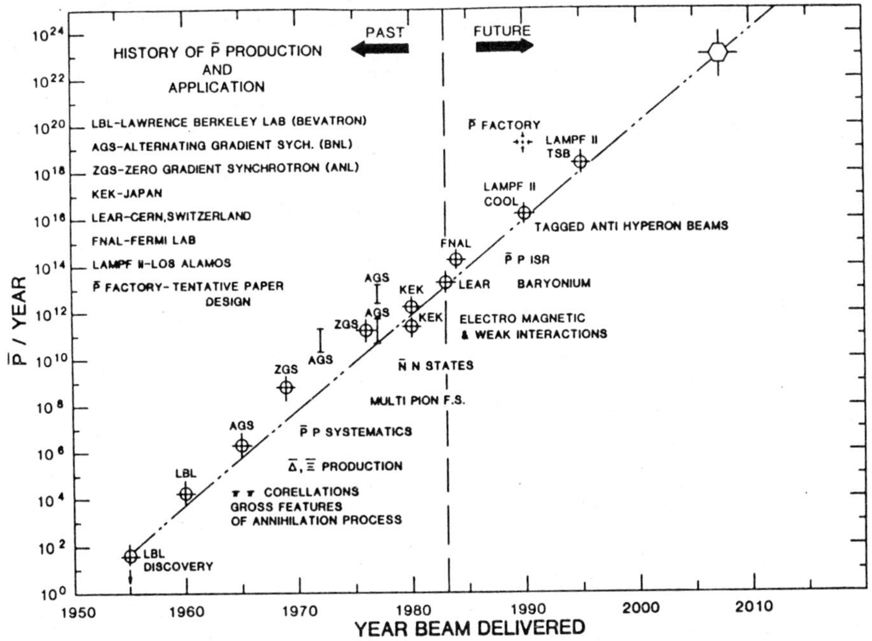

Fig. V-1. Annual antiproton production versus year for most high-energy physics facilities around the world. The circled points represent the published flux value; the vertical bar indicates the range of fluxes cited in the literature (Refs. 73-90). The point labeled $\bar{p}$ factory represents a practical design using existing magnet and accelerator technology. The physics of interest for each era is also noted.

if the conservative trend line is followed, the annual production of antiprotons will exceed a gram by the year 2010.

In order to maintain the trends in production and cooling, very significant progress needs to be made in accelerator and accumulator/cooling technology. Increasing the production and cooling rates is a high priority task at antiproton facilities around the world. Rapid progress in these areas can be expected in the short term.

In the light of these observations, let us consider whether or not there will be an antiproton in your future in physics. Most of you gathered here today at the last session of this Summer School will be well along in your careers by 1990. But to be all inclusive let use base our estimate of the likelyhood on the year 1995. According to the trend line, by that year roughly 10^{18} antiprotons will be produced annually at some facility, hopefully

LAMPFII or a variant thereof, in the United States. At that time let's estimate that there will be some 10^4 physicists involved in some research relating to antiprotons. That's a substantial fraction of the basic research community in the U.S., but as pointed out in this presentation, antiprotons have exciting research possibilities for physicists in atomic and condensed matter as well as for those in nuclear and particle areas (and no doubt in areas we have not yet dreamed of). In any event, this means that there will be some 10^{14} antiprotons available yearly for each of these researchers including very possibly yourselves. Now, 10^{14} antiprotons is about a nanogram of material. Doesn't sound like much, but considering that it represents 10^5 hours of beam time at the current LEAR facility, I'm certain that you'll recognize that it is indeed very likely that there will be an antiproton in your future.

REFERENCES

1. P. A. M. Dirac, Proc. Roy. Soc. A126, 360 (1930); A133, 60 (1931).

2. C. D. Anderson, Phys. Rev. 43, 491 (1933); 44, 406 (1933).

3. O. Chamberlain, E. Segre, C. Wiegand, and T. Ypsilantis, Phys. Rev. 100, 1947 (1955).

4. C. W. Akerlof, S. G. Crabb, J. L. Day, N. P. Johnson, P. Kalbach, A. D. Krisch, M. T. Lin, et al., Phys. Rev. D3, 645 (1971).

5. J. V. Allaby, F. Binon, and A. N. Diddens, CERN report 70-12 (1970).

6. B. Alper, H. Boggild, P. Booth, F. Bulos, L. J. Carroll, D. Damgaard, G. von Dardel, et al., Nucl. Phys. B100, 237 (1975).

7. J. Baider, R. Bland, J. C. Chollet, T. Devlin, J. M. Gaillard, J. LeFrancois, B. Merkel, et al., Phys. Lett. 39B, 414 (1972).

8. L. M. Barkov, Serpukhov report IHEP-79-92 (1979).

9. W. Bozzoli, A. Bussiere, G. Giacomelli, E. Lesquoy, R. Meunier, L. Moscoso, A. Muller, et al., Nucl. Phys. B140, 271 (1978).

10. D. Dekkers, J. A. Geibel, R. Mermod, G. Weber, T. R. Willitts, K. Winter, B. Jordan, et al., Phys. Rev. 137B, 962 (1965).

11. A. N. Diddens, W. Galbraith, E. Lillethun, G. Manning, A. G. Parham, A. E. Taylor, T. G. Walker, and A. M. Wetherwell, Nuovo Cimento 31, 961 (1964).

12. T. Eichten, D. Haidt, J. B. M. Pattison, W. Venus, H. W. Wachsmuth, O. Worz, T. W. Jones, et al., Nucl. Phys. B44, 333 (1972).

13. H. J. Frisch, N. D. Giokaris, J. M. Green, C. Grosso-Pilcher, M. D. Mestayer, L. Schachinger, M. J. Shochet, et al., Phys. Rev. Lett. 44, 511 (1980).

14. K. Guettler, B. G. Duff, M. N. Prentice, S. J. Sharrock, W. M. Gibson, A. Duane, H. Newman, et al., Nucl. Phys. B116, 77 (1976).

15. M. A. J. Botje, Thesis, Utrecht (1980).

16. B. R. Sanford and C. L. Wang, Brookhaven National Laboratory report BNL-11479.

17. C. Hojvat and A. Van Ginneken, Nucl. Instrum. Methods 206, 67 (1983).

18. G. M. Bunce, Brookhaven National Laboratory report BNL-50874 (1978).

19. H. Hirabayashi, S. Kurokawa, A. Yamamoto, and A. Kusumegi, IEEE Trans. Nucl. Sci. NS-26, 3191 (1979).

20. A. Yamamoto, H. Ikeda, S. Kurokawa, M. Takasaki, M. Taino, Y. Zuzuki, H. Ishii, et al., Nucl. Instrum. Methods 203, 35 (1982).

21. H. Brown, AGS/EPS Technical Note (1980).

22. S. Van Der Meer, CERN report, CERN/ISR-PO/72-31.

23. G. I. Budker, At. Ener. 22, 346 (1967).

24. C. Rubbia, CERN report EP77-2.

25. A. Van Ginneken and S. Mohkov, "High Intensity Targeting Workshop", University of Wisconsin report (FNAL 1980).

26. V. Krievien and F. E. Mills, "High Intensity Targeting Workshop", FNAL report (1979).

27. B. F. Bayanov, T. N. Petrov, G. I. Silvestrov, J. A. Maclachlan, and G. L. Nicholls, Nucl. Instrum. Methods 190, 9 (1981).

28. T. A. Vsevolozhskaya, M. A. Lyubimova, and G. I. Silvestrov, Zh. Tekh. Fiz., (Eng. Trans.) 20, 1556 (1975).

29. B. F. Bayanov and G. I. Sil'vestrov, Zh. Tekh. Fiz. (Eng. Trans.) 23, 94 (1978).

30. "Design Study $\bar{p}p$ Facility", CERN report, CERN/PS/AA 78-3.

31. D. Cline, P. McIntyre, F. Mills, and C. Rubbia, FNAL report FNAL TM-689.

32. J. Liouville, J. Math. Pures. Appl. 2, 16 (1837).

33. A. G. Ruggerio, Lawrence Berkeley Laboratory report LBL-7574 (1978).

34. G. I. Budker, Brookhaven National Laboratory report BNL-TR-635.

35. G. K. O'Neil, Phys. Rev. 102, 1418 (1956).

36. S. Van Der Meer, CERN-ISR internal report, August 1972.

37. F. P. Cole and F. E. Mills, Ann. Rev. Nucl. Part. Sci. 31, 295 (1981).

38. S. Van Der Meer, Particle Accelerator Conference, Washington, D.C. (1981).

39. S. Van Der Meer, CERN Report, CERN/PS/AA 78-6 (1978).

40. T. Goldman and M. M. Nieto, Phys. Lett. 112B, 437 (1982).

41. F. C. Witteborn and W. M. Fairbank, Phys. Rev. Lett. 19, 1049(1967); Rev. Sci. Instrum. 48, 1(1977).

42. P. Lefevre and D. Mohl, PS/LEA/PL/DM/MCO.

43. J. Billen, K. R. Crandall, T. P. Wangler, and M. Weiss, Proc. of the Third LEAR Workshop, Tignes-Savoir, France, January 1985 (to be published).

44. N. Beverini, et al., CERN/PSCC/P68; N. Beverini, "Physics at LEAR with Low-Energy Cooled Antiprotons," U. Gastaldi and R. Klapisch eds., Plenum Press, New York, 1984; G. Torelli, Proc. 5th Europ. Symp. on Nucleon-Antinucleon Interactions; G. Torelli, $\bar{p}$-LEAR- Note 16-Karlsruhe Workshop 1979.

45. R. F.Bonner, et al., Int. Jour. Mass. Spec. and Ion Phys. 24, 255 (1977).

46. G. Krausse, "Third IEEE Pulsed Power Conference," Albuquerque, NM, 1981; "16th Power Modulator Symposium," Washington, D. C., 1983.

47. H. G. Dehmelt, Adv. in At. and Mol. Phys. 3 (1967) 53.

48. W. Kells, et al., Fermilab-Conf.-84/68-E.

49. D. J. Wineland, et al., Adv. in At. and Mol. Phys. (to be published).

50. L. Bracci, et al., Phys. Lett. 85B, 280 (1979).

51. M. L. Ott-Rowland, et al., Phys. Rev. Lett. 49, 1708(1982).

52. R. V. Eötvös, D. Pekár, and E. Fekete, Ann. Phys. (Leipzig) 68, 11(1922); P. G. Roll, R. Krotov, and R. H. Dicke, Ann. Phys. (NY) 26, 442(1962).

53. R. V. Pound, and G. A. Rebka, Jr., Phys. Rev.Lett. 4, 337(1960); R. V. Pound and J. L. N. Snider, Phys. Rev. 140, B788(1965).

54. F. C. Witteborn and W. M. Fairbank, Phys. Rev. Lett. 19, 1049(1967); Nature 220, 436(1968); Rev. Sci. Inst. 48, 1(1977).

55. R. Collella, A. W. Overhauser, and S. A. Werner, Phys. Rev. Lett. 34, 1472(1975).

56. L. I. Schiff, Proc. Nat. Aca. Sci. 45, 69(1959); Phys. Rev. Lett. 1, 254(1958).

57. M. L. Good, Phys. Rev. 121, 311(1961); see also W. Thirring in Essays in Physics, G. K. T. Conn and G. N. Fowler, eds., Academic Press, (New York) 1972, Vol. 4, P. 125.

58. T. Goldman and M. M. Nieto, Phys. Lett. 112B, 437(1982).

59. J. Luders, Det. Kon. Dan. Vid. Sel. Mat. Fys. Medd. 28, no. 5(1954); J. S. Bell, Proc. Roy. Soc. (London) A231, 479(1955).

60. P. A. M. Dirac, Proc. Roy. Soc. (London) A126, 360(1930).

61. R. M. Wald, Phys. Rev. D21, 2742(1980).

62. D. N. Page, Phys. Rev. Lett. 44, 301(1980); Gen. Rel. Grav. 14, 299(1982).

63. R. Geroch, in General Relativity and Cosmology, Proc. of the Int. Sch. of Phys. "Enrico Fermi", Course 47, B. K. Sachs, ed., Academic Press (New York) 1971.

64. J. Kiskis, Phys. Rev. D17, 3196(1978).

65. Particle Data Group, Rev. Mod. Phys. 56, (1984) p. s97.

66. V. B. Braginskii and V. I. Panov, Sov. Phys. JETP 34, 463(1972).

67. L. I. Schiff and M. V. Barnhill, Phys. Rev. 151, 1067(1966).

68. A. J. Dessler, F. C. Michel, H. E. Rorschach, and G. T. Trammell, Phys. Rev. 168, 737(1968); F. C. Michel, Phys. Rev. Lett. 21, 104(1968); C. Herring, Phys. Rev. 171, 1361(1968).

69. P. P. Craig, Phys. Rev. Lett. 22, 700(1969).

70. J. M. Lockhart, F. C. Witteborn, and W. M. Fairbank, Phys. Rev. Lett. 38, 1220(1977).

71. A. R. Hutson, Phys. Rev. B17, 1934(1978).

72. R. S. Hanni, and J. M. J. Madey, Phys. Rev. B17, 1976(1978).

73. O. Chamberlain, E. Segre, C. Wiegand, and T. Ypsilantis, Phys. Rev. 100, 1947 (1955).

74. L. E. Agnew, T. Elioff, W. B. Fowler, R. L. Lander, W. M. Powell, E. Segre, H. M. Steiner, et al., Phys. Rev. 118, 1371 (1960).

75. C. A. Steinberger, N. Gelfand, U. Nauenberg, M. Nussbaum, J. Schultz, J. Steinberger, H. Brugger, and L. Kirsch, NEVIS Laboratory report NEVIS-126 (1964).

76. N. Barash, P. Franzini, L. Kirsch, D. Miller, J. Steinberger, T. H. Tan, R. Plano, and P. Yaeger, Phys. Rev. 139, B1659 (1965).

77. G. M. Bunce, Brookhaven National Laboratory report BNL-50874 (1978).

78. "Zero Gradient Synchrotron Users Manual," Argonne National Laboratory (1969).

79. E. Colton, Nucl. Instrum. Methods 138, 599 (1976).

80. H. Hirabayashi, S. Kurokawa, A. Yamamoto, and A. Kusumegi, IEEE Trans. Nucl. Sci. NS-26, 3191 (1979).

81. A. Yamamoto, H. Ikeda, S. Kurokawa, M. Takasaki, M. Taino, Y. Zuzuki, H. Ishii, et al., Nucl. Instrum. Methods 203, 35 (1982).

82. S. Kurokawa and H. Hirabayashi, Nucl. Instrum. Methods 212, 91 (1983).

83. "Design Study $\bar{p}p$ Facility," CERN report, CERN/PS/AA 78-3.

84. U. Gastaldi, K. Kilian, and D. Mohl, Proc. 10th Int. Conf. on High-Energy Acc., Protvino (IHEP, Serpukov, 1977).

85. U. Gastaldi, Proc. 4th European Antiproton Conf., A. Fridman, Ed., Barr, France (CNRS, Paris, 1979).

86. W. Hardt, 5th European Symposium on Nucleon Antinucleon Interactions, Bressanone, Italy (Instituo Nazionale De Fisica Nuclear, Padova, 1980).

87. E. R. Gray, D. E. Johnson, F. R. Huson, F. E. Mills, L. C. Teng, G. S. Tool, P. M. McIntyre, et al., IEEE Trans. Nucl. Sci. NS-24, 1854 (1977).

88. F. E. Mills, Proc. Second LAMPF II Workshop, Los Alamos, New Mexico, July 19-22, 1982, Los Alamos National Laboratory report LA-9572-C, Vol. II (November 1982).

89. T. E. Kalogeropoulos, Proc. Workshop Nucl. Part. Phys. at Energies up to 31 GeV: New and Future Aspects, Los Alamos National Laboratory report LA-8775-C, Vol. II (March 1981) and Los Alamos National Laboratory report LA-9572-C, Vol. II (November 1982).

90. H. Brown, AGS/EPS Technical Note (1980).

PART II

QUARK-NUCLEAR PHYSICS

DEEP INELASTIC SCATTERING WITH APPLICATION TO NUCLEAR TARGETS *

R. L. Jaffe
Massachusetts Institute of Technology
Cambridge, MA 02139

MOTTO

"Looking for the quarks in the nucleus is like looking for the Mafia in Sicily: Everyone knows they're there, but its hard to find the evidence."

Anonymous

PREFACE

These lectures are addressed to a very specific audience: graduate students and young researchers in theoretical nuclear physics. They contain very little that cannot be found elsewhere in the vast literature on the subject, but they summarize a particular viewpoint which is both straightforward and fairly correct. My intention is to provide a reasonably thorough introduction to inclusive, inelastic electron scattering and enough advanced material to excite the reader to go on by him/herself. The reader is assumed to have some familiarity with relativistic quantum mechanics and with the more elementary side of quantum field theory. Some previous exposure to QCD – quarks, color, gluons, $SU(3)$ and the like – will help. But, previous exposure to renormalization theory, the renormalization group and the operator product expansion is not assumed.

I have tried to avoid the temptation to oversimplify – the subject has its subtleties and it is not possible to do good work in the field without understanding them. More simplified treatments can be found in books by Feynman (Photon – Hadron Interactions, (W. A. Benjamin, New York (1972)) and by Close (An Introduction to Quarks and Hadrons (Academic Press, New York, 1979)). Equally, I have tried to avoid too much formalism. I hope the power of the more advanced methods developed in §5 and

* This work is supported in part through funds provided by the U.S. Department of Energy (D.O.E.) under contract #DE-AC02-76ER03069.

applied in §6 will encourage the reader to pursue the subject more formally in the future. More advanced treatments can be found in the lecture notes of D. J. Gross (in Methods in Field Theory, Proc. of 1975 Les Houches Summer School, (ed., R. Balian and J. Zinn-Justin, North-Holland, Amsterdam) and C. H. Llewellyn Smith (Topics in Quantum Chronomdynamics, in Quantum Flavordynamics Quantum Chromodynamics and Unified Theories (ed., K. T. Mahanthappa and J. Randa, Plenun Press, New York, 1980) in the recent books by C. Itzykson and J. Zuber (Introduction to Quantum Field Theory, McGraw-Hill, New York, 1980) and T. P. Cheng and L. F. Li Gauge Theories of Elementary Particle Physics, (Clarendon Press, Oxford, 1984).

In the interest of time, I have had to eliminate all mention of inclusive, inelastic scattering of neutrinos and other related processes such as electron-positron annihilation and Drell-Yan production of lepton pairs. These too can be found treated in detail elsewhere in the literature.

Finally I would like to thank Frank Close, Chris Llewellyn-Smith, Dick Roberts and especially Graham Ross for collaboration on many of the ideas presented here. I would also like to thank Gerry Garvey and Mikkel Johnson for making it possible for me to attend the school, and Roger Gilson for preparing the manuscript.

§0 INTRODUCTION

The quark description of hadrons is now universally accepted, but its implications for nuclear physics are anything but clear. Nuclear structure and low energy nuclear reactions are well-described by a variety of semi-phenomenological theories which leave little room for insight from thinking about quark and gluon dynamics. The reason is obvious: the energies involved in most nuclear phenomena are so low compared to the natural scale of quark dynamics that quark and gluon degrees of freedom are effectively frozen into nucleon and meson "quasiparticles". To support this, consider first that the string tension in QCD is ~ 1 GeV/fm, so it costs ~ 1 GeV to separate a quark from two others (or from an antiquark) by an additional fermi beyond their equilibrium separation; and second, that the $N-\Delta$ mass difference is 300 MeV, so it costs ~ 300 MeV to flip a quark's spin relative to the two others which together with it form a nucleon. Both energies are much larger than typical nuclear energy scales. Thus Δ's are relegated to a relatively minor role in the description of nuclei, and quarks (as distinct from nucleons and mesons) are even less important.

It is not even clear, as a matter of principle, how to establish the necessity of a fundamentally "quarkic" description of nuclei. Because color is confined, it is almost always possible to find an entirely equivalent hadronic description of some quark-dynamical process. For example, there has been much interest in isolating so-called "hidden color" components in the deuteron wave function. These are of the form $\left[(q^3)^{\mathbf{8}} - (q^3)^{\mathbf{8}}\right]^{\mathbf{1}}$: although the entire six quark system is a color singlet, each group of three quarks is coupled to a color octet. Such admixtures have been invoked, for example, to resolve discrepancies in models of deuteron photodisintegration. It is trivial to show, however, that any such configuration can be rewritten, after a change of coupling transformation, as a sum of configurations in which each set of three quarks is coupled to a color singlet, *i.e.*, an ordinary baryon. [The proof is merely to note that ϵ_{abc} is the only covariant invariant tensor in SU(3).] So any hidden color state can be equally well represented as a sum over conventional baryon-baryon configurations, and any phenomenon ascribed to hidden color could be explained equivalently by a sufficiently clever theorist who had never heard of quarks and color. The quark theorist might triumph in the end if his description were simpler and more economical: one "hidden color" state might do as well as some complex superposition of baryon resonances. But this is not the "smoking gun" enthusiasts are seeking.

Similar observations could have been made about low energy hadron physics twenty years ago, and one therefore wonders how the quark description of hadrons was established in the first place, which brings me to the subject of these lectures. Of course, quarks were recognized as a convenient bookkeeping device as early as 1964, but the need for a quark dynamics for hadrons was not compelling until a set of experiments was performed at SLAC in the late 1960's. High energy electrons were scattered inelastically from nucleons in close analogy to the α-particle scattering experiments performed by Rutherford in the early 1900's. In both cases, the experimenters were surprised to find that the scattering cross section remained large even at high momentum transfer indicating the presence of point-like scattering centers within the target. Further studies with electrons and then with neutrinos established the spin, charge and baryon number of the point-like objects within hadrons and showed they were quarks. It was also recognized that these experiments not only detect quarks but directly measure their distribution within the target baryon.

During the early 1970's, it became clear that only a non-Abelian gauge field theory, quantum chronodynamics (QCD), could explain the qualitative features of quark dynamics. One of its triumphs was the quantitative prediction of the details of inelastic electron scattering in the very high momentum transfer, or deep inelastic limit. Despite the complexity of QCD, the simple Rutherford picture remains valid: the scattering electron measures the quark distribution in the target.

These lectures explore the use of deep inelastic electron scattering as a probe of the quark distribution in nuclei. For more than a decade it apparently was thought that the quark distribution in a nucleus was simply given by the quark distribution in so many neutrons and protons, corrected for the fact that they are in motion within the nucleus (Fermi motion). Little attention was paid to quark distributions in nuclei. It came as a surprise to particle and nuclear theorists alike when the first careful comparison of a nuclear target (iron) with a nucleon (actually a deuteron), presented at the 1982 Paris Conference by the European Muon Collaboration (EMC), showed a $\sim$ 15% difference in a region where Fermi motion effects are thought to be negligible. The "EMC effect" as it is known, was immediately confirmed by experiments at SLAC and elsewhere. Theorists quickly presented a variety of explanations of the EMC effect. Now, three years later, there are several well-developed schools of thought, much controversy and only a little agreement among partisans about the origin of the effect. Most explanations are based on a "convolution model" formalism in which one supposes the nucleus contains, in addition to nucleons, some small admixture of more exotic constituents (pions, multiquark bags, Δ's, α clusters...), and then adds up their quark distributions. Unfortunately, this approach has a hitch: the assumption that quark distributions in constituents add incoherently is not justified and in many cases probably wrong. Another approach, known as "rescaling" suggested by the scale transformation properties of QCD, avoids the questionable assumption of convolution models by dealing directly and solely with the quark distribution of the nucleus. The result is simple and striking: the EMC effect shows that the typical length scale associated with quark propagation in the nuclear ground state is longer than the corresponding length scale in the nucleon. For a particle physicst, one of the most interesting features of this approach is its implications for QCD at finite density. Nuclei provide samples of quark/nuclear matter at a variety of mean densities (as A increases, the surface to volume ratio goes to zero so the mean density grows). The A dependence of the EMC effect closely reflects the variations in mean nuclear density, indicating that the quark length scale in quark/nuclear matter grows with density.

In these lectures, I have tried to avoid detailed analysis of models of the EMC effect, although critics of the "rescaling" approach, to which I am devoted, will point out that I present it in considerable detail. In fact, much of the rescaling analysis is model independent and essential for a modern education in the subject. In §1, I introduce the kinematic variables in coordinate and momentum space. The whole discussion is set in the laboratory or target rest frame. Certain general tools like dispersion relations are reviewed and summarized there. In §2, I present the parton model from a somewhat unfamiliar point-of-view, one which experts will recognize as imitating the more formal operator product expansion analysis used in QCD. The reason for this approach is to keep as close to coordinate space as possible since considerable insight into the parton model and the EMC effect comes from an easy fluency between coordinate and momentum space. In §3, I summarize the data, as far as I know it, and use the parton model to interpret it. §4, is dedicated to convolution models. Since so much work has been based on these models, I have tried to present them in detail, illustrated by the case in which the constituents are the nucleons themselves (Fermi motion), and let the reader judge their utility for himself. In §5, I return to fundamentals. QCD changed our understanding of inelastic scattering and corrected and extended the parton model. I have tried to introduce the QCD analysis with as little excess formalism as possible, though for a real working knowledge the reader will have to learn more about gauge field theory and the renormalization group elsewhere. Finally, in §6, I use the powerful methods developed in §5 to help give a new way of looking at the EMC effect, and draw some surprisingly simple conclusions.

§1 KINEMATICS AND OTHER GENERALITIES

1.1 Structure Functions

We are interested in the process $eA \to e'X$ where A is a nucleus, the proton and neutron being important special cases, and X is an unobserved hadronic final state. The electron and nucleus are assumed unpolarized, although polarization dependent effects can be handled in the same fashion. The process is known as inelastic electron scattering or inclusive electroproduction. To lowest order in α, the process is described by one photon exchange (Fig. 1):[1]

$$A \propto \overline{u}(k')\gamma^{\mu}u(k)\frac{1}{q^2}\langle X|J_{\mu}(0)|p\rangle \quad , \tag{1.1}$$

where $J_{\mu}(0)$ is the hadronic electromagnetic current operator. The differential cross-section for scattering in which X is not observed is proportional to $\sum_X |A|^2(2\pi)^4\delta^4(p+q-p_X)$, or

$$d\sigma \propto \ell^{\mu\nu}W_{\mu\nu} \quad , \tag{1.2}$$

where

$$\ell^{\mu\nu} = \frac{1}{2} Tr \; k\!\!\!/'\gamma^\mu k\!\!\!/\gamma^\nu = 2(k'^\mu k^\nu + k'^\nu k^\mu - g^{\mu\nu} k \cdot k') \tag{1.3}$$

(we ignore the electron mass), and

$$W_{\mu\nu} \equiv \frac{1}{4\pi} \sum_X (2\pi)^4 \delta^4(p + q - p_X) \langle p|J_\mu(0)|X\rangle\langle X|J_\nu(0)|p\rangle \quad . \tag{1.4}$$

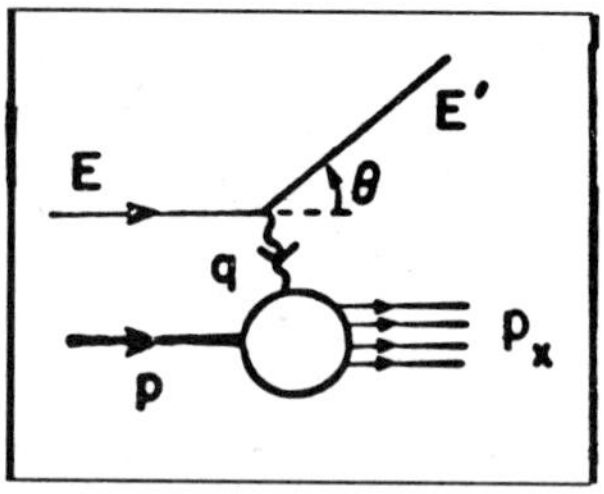

Fig. 1: Inclusive inelastic electron scattering via one photon exchange. E, E' and θ are defined in the target rest frame.

$W_{\mu\nu}$ contains all reference to hadronic states. It is represented graphically in Fig. 2.

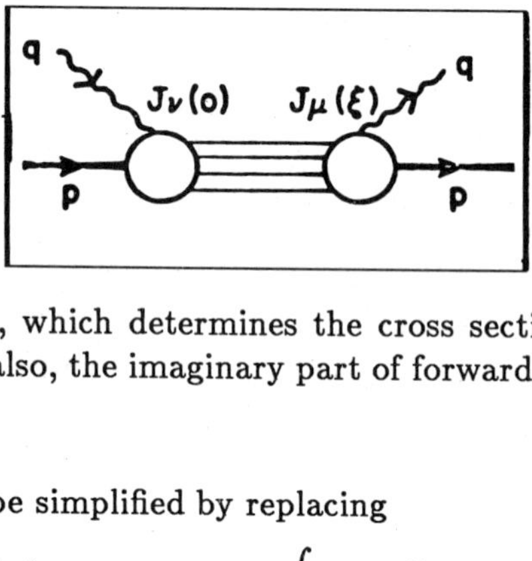

Fig. 2: $W_{\mu\nu}$, which determines the cross section for electron scattering and also, the imaginary part of forward, virtual Compton scattering.

Eq. (1.4) may be simplified by replacing

$$(2\pi)^4 \delta^4(p + q - p_X) \equiv \int d^4\xi \; e^{i(p+q-p_X)\cdot\xi} \quad , \tag{1.5}$$

translating $J_\mu(0)$ to the space-time point ξ, and using completeness $(\sum_X |X\rangle\langle X| = 1)$:

$$W_{\mu\nu} = \frac{1}{4\pi} \int d^4\xi \; e^{iq\cdot\xi} \langle p|J_\mu(\xi)J_\nu(0)|p\rangle_c \quad . \tag{1.6}$$

The subscript c on the matrix element denotes "connected" and ensures that vacuum transitions of the form $\langle 0|J_\mu(\xi)J_\nu(0)|0\rangle\langle p|p\rangle$ are excluded. The current product in Eq. (1.6) can be replaced by a commutator

$$W_{\mu\nu} = \frac{1}{4\pi}\int d^4\xi\ e^{iq\cdot\xi}\langle p|\left[J_\mu(\xi),\ J_\nu(0)\right]|p\rangle_c \tag{1.7}$$

because the term we have subtracted vanishes for stable targets if $q^0 > 0$. In quantum field theory, it is most convenient to deal with time-ordered products of operators. Thus,

$$T_{\mu\nu} = i\int d^4\xi\ e^{iq\cdot\xi}\langle p|T\left(J_\mu(\xi)J_\nu(0)\right)|p\rangle_c \tag{1.8}$$

is the amplitude for forward scattering of a virtual photon of momentum q from a hardonic target of momentum p (Fig. 3).[2]

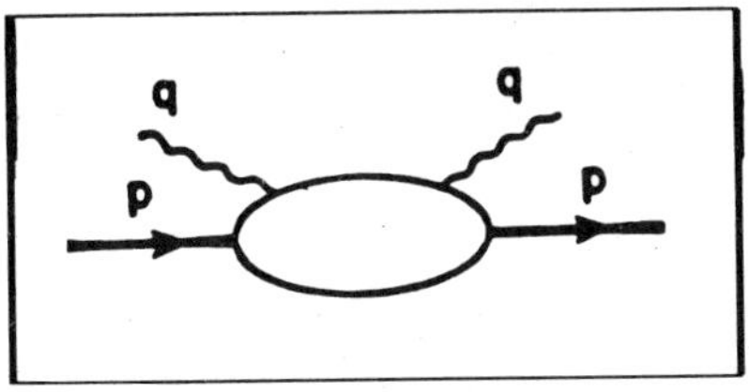

Fig. 3: Forward virtual Compton scattering.

It is easily seen that $W_{\mu\nu}$ is the imaginary part of $T_{\mu\nu}$ (when q^0 is taken to have a small positive imaginary part)

$$W_{\mu\nu} = \frac{1}{2\pi} Im\ T_{\mu\nu}\ . \tag{1.9}$$

So, inclusive electroproduction is intimately related to virtual Compton scattering.

$W_{\mu\nu}$ (and $T_{\mu\nu}$) can be decomposed in terms of a pair of Lorentz invariant "structure functions" $W_{1,2}$ ($T_{1,2}$):

$$\begin{aligned} W_{\mu\nu} = &-\left(g_{\mu\nu} - \frac{q_\mu q_\nu}{q^2}\right)W_1(q^2,\nu) \\ &+ \frac{1}{M_T^2}\left(p_\mu - \frac{M_T\nu}{q^2}q_\mu\right)\left(p_\nu - \frac{M_T\nu}{q^2}q_\nu\right)W_2(q^2,\nu) \end{aligned} \tag{1.10}$$

and likewise for $T_{\mu\nu}$. $W_{1,2}$ are functions of the Lorentz invariants q^2 and $p \cdot q = M_T \nu$ ($q^2 = -4EE' \sin^2 \frac{\theta}{2}$ and $\nu = E - E'$ in the target rest frame) where M_T is the target mass. No other terms are allowed by Lorentz invariance, current conservation ($q^\mu W_{\mu\nu} = W_{\mu\nu} q^\nu = 0$) and parity. The structure functions depend on the squared four momentum transfer $q^2 \equiv -Q^2 = \nu^2 - \vec{q}^{\,2}$ which is spacelike (so $Q^2 > 0$) and the energy transfer in the laboratory, $\nu = q^0$, which is positive. The squared mass of the final hadronic state, X, is $(p+q)^2 = M_T^2 + 2M_T\nu - Q^2 \equiv W^2$ and is greater than M_T^2. Thus, the "scaling" variable, $x_T \equiv Q^2/2M_T\nu$ must be between 0 and 1. Often experimentalists and theorists prefer to use a uniform scaling variable $x = x_N = Q^2/2M\nu$ ($M \equiv M_N$ is the nucleon mass) for all targets. The reason for this is that the structure functions of different nuclei look to first order like the sum of structure functions for A independent nucleons. They are therefore very small for $x > 1$ regardless of A. I will continue to use the variable intrinsic to the target (x_T) except where it is necessary to convert to the conventional definition for comparison between targets.

Bjorken suggested[3] that in the limit of large Q^2 at fixed x (now known as the Bjorken or "deep" inelastic limit) W_1 and $\frac{\nu}{M_T} W_2$ should become functions of x_T alone:

$$\begin{aligned} \lim_{Bj} W_1(q^2, \nu) &= F_1(x_T) \\ \lim_{Bj} \frac{\nu}{M_T} W_2(q^2, \nu) &= F_2(x_T) \quad , \end{aligned} \tag{1.11}$$

which is approximately verified by experiment. This phenomenon is known as "Bjorken scaling" or just "scaling" for short. The kinematic range of inclusive scattering is shown in Fig. 4. Note that the $Q^2 \to 0$ limit gives photoproduction.

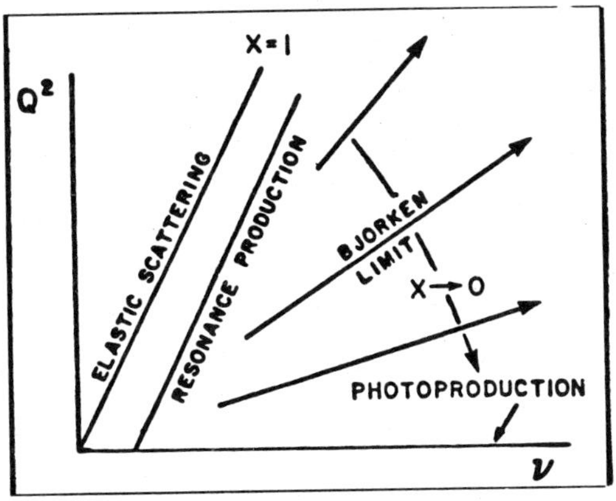

Fig. 4: Kinematic variables for inelastic electron scattering.

1.2 Dispersion Relations

The virtual forward Compton amplitude, $T_{\mu\nu}$, possesses simple properties when regarded as an analytic function of ν at fixed q^2. These are special cases of the general results of dispersion theory, most of which are forgotten.[4] Since I will need these properties in the subsequent chapters, I will review them here. For pedagogical simplicity I will ignore spin and analyze a hypothetical virtual "Compton" amplitude $T(q^2,\nu)$ for a scalar "photon" defined by

$$T(q^2,\nu) \equiv i\int d^4\xi\, e^{iq\cdot\xi}\langle p|T\big(J(\xi)J(0)\big)|p\rangle_c \quad . \tag{1.12}$$

The generalization to the physically interesting case of $T_{\mu\nu}$ will be quoted at the end.

$T(q^2,\nu)$ is a real analytic function of ν at fixed q^2,

$$T(q^2,\nu^*) = T^*(q^2,\nu) \quad , \tag{1.13}$$

and is crossing symmetric,

$$T(q^2,\nu) = T(q^2,-\nu) \quad . \tag{1.14}$$

The fundamental assumption of dispersion theory is that scattering amplitudes are analytic except at values of the kinematic variables which allow intermediate states to be physical (*i.e.*, on shell). This can be proven to all orders of perturbation theory,[5] but must be regarded as an assumption in QCD where the quanta of the perturbation theory (quarks and gluons) are not physical states. When $\nu \geq -q^2/2M_T$, the virtual photon-target system can form a physical hadronic intermediate state so $T(q^2, \nu)$ has a cut along the positive real-ν axis. More precisely, $T(q^2, \nu)$ has a pole at $\nu = -q^2/2M_T$ corresponding to elastic scattering ($ep \rightarrow e'p'$) and a cut beginning at pion production threshold. $T(q^2, \nu)$ also has a cut on the negative real axis corresponding to the "crossed" process $p \rightarrow$"γ"$+x$ which is physically allowed when $\nu \leq -|q^2|/2M_T$. The discontinuity across the right hand cut is

$$\text{disc}\, T(q^2, \nu) = 2i\, Im\, T(q^2, \nu) = 4\pi i\, W(q^2, \nu) \quad . \tag{1.15}$$

These analytic properties are summarized in Fig. 5.

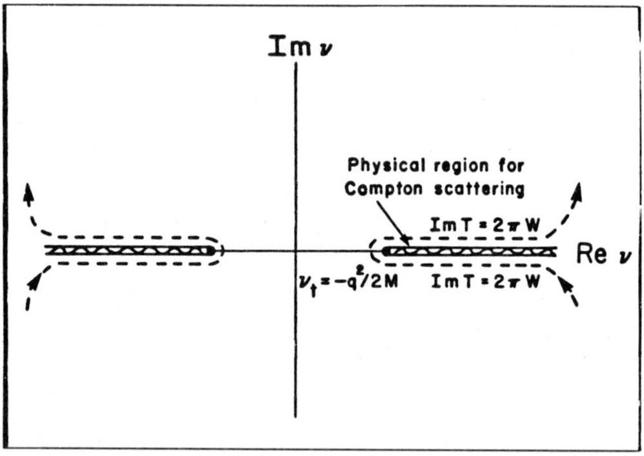

Fig. 5: The complex ν-plane. The contour shown figures in the derivation of dispersion relations.

Using Cauchy's theorm on the contour shown in Fig. 5, it is possible to derive a "dispersion relation" for $T(q^2, \nu)$,

$$T(q^2, \nu) = 4 \int_{-q^2/2M}^{\infty} \frac{d\nu' \nu'}{\nu'^2 - \nu^2} W(q^2, \nu') \quad . \tag{1.16}$$

Here, ν is complex (ν' is real) and the physical Compton amplitude is obtained by letting it approach the positive real axis from above, $\nu \rightarrow \nu_R + i\epsilon$. If $W(q^2, \nu')$ falls too slowly as $\nu' \rightarrow \infty$ then the integral may not converge.

In that case one can derive a weaker, "subtracted" dispersion relation. Formally, take Eq. (1.16) for two different choices of ν and subtract, *e.g.*

$$T(q^2,\nu) = T(q^2,0) + 4\nu \int_{-q^2/2M}^{\infty} \frac{d\nu'}{\nu'(\nu'^2-\nu^2)} W(q^2,\nu') \ . \tag{1.17}$$

The integral is better behaved at large ν'. This procedure can be continued as far as necessary and works as long as $W(q^2,\nu)$ is polynomial bounded as $\nu \to \infty$, which we will assume (see below). Modulo subtractions, the real part of T can be calculated if W is known, thus dispersion relations can be (and have been) verified experimentally.

It is convenient to change variables to $\omega \equiv -2M_T\nu/q^2$ in Eq. (1.16),

$$T(q^2,\omega) = 4\int_1^{\infty} \frac{\omega' d\omega'}{\omega'^2-\omega^2} W(q^2,\omega') \ . \tag{1.18}$$

Note that $T(q^2,\omega)$ is analytic in the circle of radius 1 about $\omega = 0$ and may therefore be expanded in a Taylor series for $|\omega| < 1$:

$$T(q^2,\omega) = 4\sum_{\substack{n \\ \text{even}}} M^n(q^2)\omega^n \tag{1.19}$$

with

$$\begin{aligned} M^n(q^2) &= \int_1^{\infty} d\omega' \, \omega'^{-n-1} W(q^2,\omega') \\ &= \int_0^1 dx \, x^{n-1} W(q^2,x) \ . \end{aligned} \tag{1.20}$$

The physically interesting case of $T_{\mu\nu}$ is summarized by dispersion relations for the two invariant amplitudes $T_{1,2}(q^2,\omega)$. $T_1(q^2,\omega)$ requires subtraction,

$$\begin{aligned} T_1(q^2,\omega) &= T_1(q^2,0) + 4\omega^2 \int d\omega' \frac{W_1(q^2,\omega')}{\omega'(\omega'^{\,2}-\omega^2)} \\ \frac{\nu T_2}{M_T}(q^2,\omega) &= 4\omega \int_1^{\infty} d\omega' \frac{\frac{\nu W_2}{M_T}(q^2,\omega')}{\omega'^2-\omega^2} \ . \end{aligned} \tag{1.21}$$

1.3 Light Cone Coordinates and the Target Rest Frame

Most introductory treatments of deep inelastic lepton scattering are formulated in the "infinite momentum frame" where the target is boosted to some arbitrarily large momentum P_∞. I find this approach both unnecessary and misleading and prefer instead to work in the target rest frame, where I suppose nuclear physicists also feel most at home. Of course, the physics is frame independent, at least if one is careful enough.[6] In the target rest frame the Bjorken limit takes on a particularly simple form. We choose the negative z-axis to lie along the virtual photon direction:

$$q = \left(\nu,\ 0,\ 0, -\sqrt{\nu^2 + Q^2}\right) \ . \tag{1.22}$$

As $Q^2 \to \infty$ with x fixed, $Q^2/\nu^2 \to 0$ so

$$q \to (\nu,\ 0,\ 0,\ -\nu - M_T x_T) \ . \tag{1.23}$$

The significance of this form is most transparant if we introduce "light-cone" coordinates,

$$q^\pm \equiv \frac{1}{\sqrt{2}}(q^0 \pm q^3) \ . \tag{1.24}$$

[Note: $a \cdot b = a^+ b^- + a^- b^+ - \vec{a}_\perp \cdot \vec{b}_\perp$, $a^2 = 2a^+ a^- - \vec{a}_\perp^2$, so $g^{+-} = g^{-+} = 1$ and $a_\mp = a^\pm$. To avoid confusion, I will stick to contravariant indices.] In the Bjorken limit $q^- \to \infty$ but $q^+ \to -M_T x_T/\sqrt{2}$, *i.e.*, q^+ remains finite. Note $M_T x_T = Mx = Q^2/2\nu$, so the limiting value of q^+ is independent of M_T.

We will frequently discuss the important distance scales which contribute to electroproduction. The distance referred to is the space-time separation, ξ_λ, between the points at which the currents J_μ and J_ν act. Note that ξ_λ is only defined in the Compton amplitude, $T_{\mu\nu}$, or its imaginary part $W_{\mu\nu}$, not in the electroproduction amplitude A itself (Eq. 1.1). Since ξ and q appear as conjugate variables in the definition of $W_{\mu\nu}$ and $q \cdot \xi = q^+ \xi^- + q^- \xi^+$, $q^- \to \infty$ forces $\xi^+ \to 0$, but $q^+ = -Mx/\sqrt{2}$ requires only $|\xi^-| < \sqrt{2}/Mx$. The first of these relations, $\xi^+ \to 0$, follows from general theorems on Fourier transforms.[7] Let

$$\tilde{f}(q^-) = \int d\xi^+ \, e^{iq^- \xi^+} f(\xi^+) \ . \tag{1.25}$$

If $f(\xi^+)$ is smooth (infinitely differentiable everywhere) and well-behaved as $|\xi^+| \to \infty$, then $\tilde{f}(q^-)$ vanishes faster than any power of q^- as $q^- \to \infty$. But, if $f(\xi^+)$ has singularities then the $q^- \to \infty$ limit is dominated by the behavior of $f(\xi^+)$ near the singularities. Suppose, for example, $f(\xi^+) = 0$ for $\xi^+ \le 0$, then as $q^- \to \infty$

$$\tilde{f}(q^-) \sim \frac{i}{q^-} f(0) - \left(\frac{i}{q^-}\right)^2 f'(0) \ + \ldots , \tag{1.26}$$

as easily obtained from integration by parts. The integrand in Eq. (1.7) defining $W_{\mu\nu}$ is singular at $\xi^+ = 0$ since it vanishes when $\xi^2 < 0$, so the $q^- \to \infty$ limit is dominated by $\xi^+ \approx 0$. The second relation, $|\xi^-| < \sqrt{2}/Mx$, is more subtle. After all the dust settles (in §2) the structure functions will be given by Fourier transforms in $Mx/\sqrt{2}$ of a smooth function of ξ^-. The Fourier transform, however, is only conditionally convergent at large ξ^-. It diverges like $1/\sqrt{x}$ or $1/x$ (for "valence" or "ocean" quarks, see §2) as $x \to 0$, so the oscillation in the exponential acts as a large ξ^- cutoff. The larger $\sqrt{2}/Mx$, the larger range of ξ^- contributes to the structure function. For a more complete discussion with examples see the Heidelberg talk by Llewellyn Smith.[8]

Since the commutator in Eq. (1.7) is causal, $\xi^2 = 2\xi^+\xi^- - \vec{\xi}_\perp^{\,2}$ is positive. Hence, $\xi^+ \to 0$ with ξ^- finite requires $\vec{\xi}_\perp \to 0$: All components of ξ^λ except ξ^- vanish in the Bjorken limit. Thus, deep inelastic scattering is not, as is often incorrectly remarked, a short distance ($\xi_\lambda \to 0$) phenomenon; it is instead a light-cone ($\xi^2 \to 0$) dominated process. This distinction is crucial to understanding the dynamics and yields considerable insight into nuclear effects in inclusive electron scattering. Together $\xi^+ \to 0$ and $|\xi^-| < \sqrt{2}/Mx$ imply $|\xi^0| < 1/Mx$ and $|\xi^3| < 1/Mx$. $W_{\mu\nu}$ measures a current-current correlation function in the target ground state (*c.f.*, Eq. 1.7). In the Bjorken limit, the correlation probed becomes light-like, but may extend to very large spatial distances (and times) in the small x limit. The relation between x and $|\xi^3|$ will be crucial to our analysis of nuclear effects in leptoproduction.

1.4 Alternative Structure Functions

It is often convenient to use different structure functions to describe $W_{\mu\nu}$. A particularly important pair are W_L and W_T defined by

$$\begin{aligned} W_T &= W_1 \\ W_L &= \left(1 + \frac{\nu^2}{Q^2}\right) W_2 - W_1 \ . \end{aligned} \tag{1.27}$$

W_T and W_L arise when one considers the polarization of the virtual photon, ϵ_μ. The leptonic current produces a flux of virtual photons which may be transverse,

$$\begin{aligned} \epsilon^\mu_{T_1} &= (0,\ 0,\ 1,\ 0) \\ \epsilon^\mu_{T_2} &= (0,\ 1,\ 0,\ 0) \ , \end{aligned} \tag{1.28}$$

or longitudinal

$$\epsilon_L^\mu = \frac{1}{\sqrt{Q^2}}\left(\sqrt{\nu^2+Q^2},\ 0,\ 0,\ -\nu\right) \ , \tag{1.29}$$

in the target rest frame in which q^μ is given by Eq. (1.21). Note $\epsilon \cdot q = 0$ and $\epsilon_{Tj}^2 = -\epsilon_L^2 = -1$. W_T and W_L are the components of $W_{\mu\nu}$ which couple to ϵ_T and ϵ_L, respectively:

$$\begin{aligned} W_T &\equiv \epsilon_{Tj}^\mu W_{\mu\nu} \epsilon_{Tj}^\nu \\ W_L &\equiv \epsilon_L^\mu W_{\mu\nu} \epsilon_L^\nu \ . \end{aligned} \tag{1.30}$$

They are proportional to the total cross sections for absorption of a transverse or longitudinal polarized virtual photon, respectively,

$$\sigma_{T,L}(q^2,\nu) = \frac{4\pi^2\alpha}{k} W_{T,L}(q^2,\nu) \tag{1.31}$$

where $k = (W^2 - M_T^2)/2M_T$. σ_L and σ_T are positive, so $W_1 \geq 0$ and $(1+\nu^2/Q^2)W_2 \geq W_1$. As $q^2 \to 0$, σ_L vanishes and σ_T approaches the photoproduction cross section for real photons.

Experimentalists measure W_1 and W_2 by comparing cross sections at fixed q^2 and ν, but different values of E, E' and θ:

$$\frac{d^2\sigma}{dE'd\Omega} = \frac{4\alpha^2 E'^2 \cos^2\theta/2}{q^4}\left[W_2(q^2,\nu) + 2W_1(q^2,\nu)\tan^2\frac{\theta}{2}\right] \ . \tag{1.32}$$

In practice, it is convenient to write the differential cross section in terms of σ_L, σ_T and a parameter ϵ:

$$\epsilon^{-1} \equiv 1 + 2\left(1 + \frac{\nu^2}{Q^2}\right)\tan^2\frac{\theta}{2} \tag{1.33}$$

$$\frac{d^2\sigma}{dE'd\Omega} = \frac{\alpha}{4\pi^2 Q^2}\frac{kE'}{E}\left(\frac{2}{1-\epsilon}\right)\left[\sigma_T(q^2,\nu) + \epsilon\sigma_L(q^2,\nu)\right] \tag{1.34}$$

Apart from kinematic factors, $d^2\sigma$ is a linear function of ϵ. A linear fit to the data yields $\sigma_L + \sigma_T \propto \nu W_2$ as the intercept at $\epsilon = 1$, and $R \equiv \sigma_L/\sigma_T$ as the slope. σ_L/σ_T is difficult to measure. Data sets from different spectrometer settings and different beam energies must be combined, which introduces systematic uncertainties. At high beam energies $\epsilon \approx 1$ so the experiments are not sensitive to R.

§2 THE PARTON MODEL

Anyone who studies inelastic electron scattering should begin with the parton model of Bjorken and Feynman. There are many fine sources from which to learn it;[1-6] one or more should be studied in conjunction with these lectures. I, too, will describe the parton model, but quickly and maintaining as much contact with coordinate space as possible. If you have never seen the parton model before, the derivation presented here will appear difficult and rather formal. The parton model is often misused. To learn how not to misuse it one must approach the model more formally, *e.g*, via the operator product expansion (OPE). Anyone who intends to do research is this field is strongly advised to study the OPE and the renormalization group in QCD beforehand,[1] but I will only touch briefly on them.

It is conventional to motivate the parton model by arguing that in some sense interactions can be ignored near the light cone. There is no realistic theory in which this is true, though in QCD it is approximately true. In §5, we will develop a more precise language for such matters. Here, I will only state the assumptions which lead to the model.

The first assumption of the parton model is that the current J_μ couples to quarks (as opposed to fundamental scalars, etc.). Then the contributions to the forward virtual Compton amplitude can be classified by the flow and interactions of quark lines. The second assumption is that at large values of Q^2 the currents, but not the states, may be treated as in free field theory. Thus, final state interactions (Fig. 6b) and vertex corrections (Fig. 6c) are ignored.

This leaves the one and two particle contributions shown in Figs. 6a and 6d, respectively. Methods similar to those we shall apply to Fig. 6a show that the contributions of Fig. 6d vanish faster by a power of Q^2, so I will ignore them henceforth. This leaves only Fig. 6a: the elastic and incoherent scattering of each quark in the target, *i.e.*, "quasi-elastic" scattering. It is important to keep in mind the place of the parton model in QCD. It is valid modulo logarithms: quantities which scale, *i.e.*, become functions of x-alone in the parton model, will be modulated by powers of $ln\ Q^2$ when QCD interactions are included. Quantities which vanish like a power of Q^2 in the parton model may vanish only like a power of $ln\ Q^2$ in QCD.

The structure function, $W_{\mu\nu}$, in the parton model is obtained by placing the intermediate state in Fig. 6a on-shell. The struck quark and the remnants of the target appear separately as physical intermediate states as shown in Fig. 7.

This, of course, is wrong – quarks are confined by non-perturbative

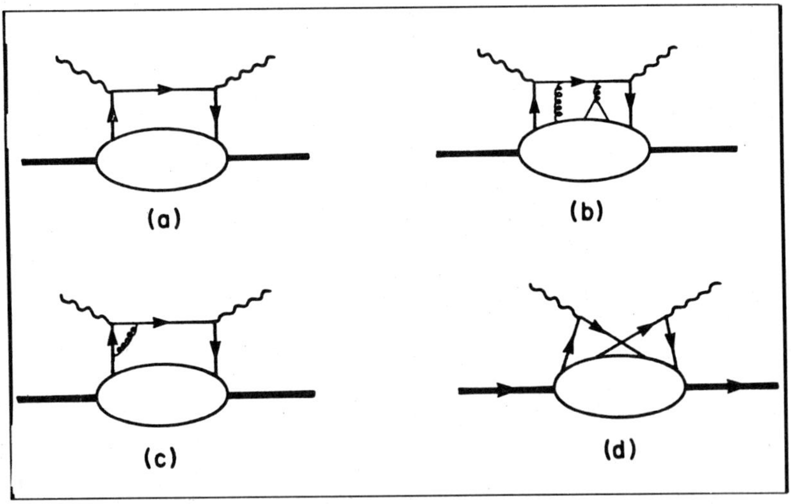

Fig. 6: Some contributions to forward virtual Compton scattering in QCD: a) The parton model diagram; b) final state interactions, c) vertex corrections; d) interference.

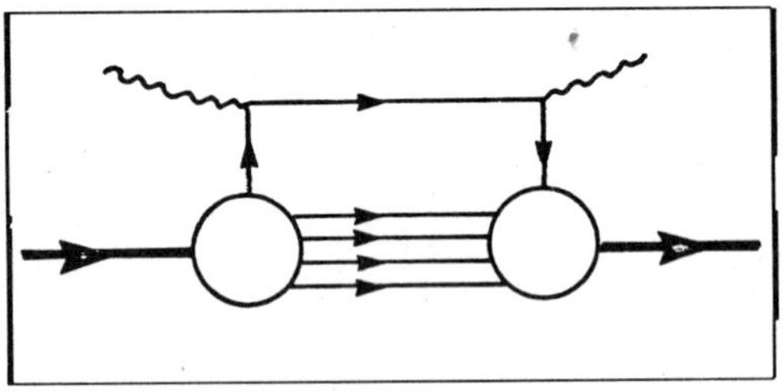

Fig. 7: Parton model for $W_{\mu\nu}$.

effects in QCD. It is assumed that the processes which neutralize quark quantum numbers do not affect the dominant terms at large Q^2. The justification for this is that those non-perturbative, confining effects in QCD which have been studied, while strong, appear to vanish rapidly with Q^2. Until the non-perturbative aspects of QCD are better understood this will remain a major, though reasonable, assumption.

2.1 The Derivation of the Parton Model

The current $J_\mu(\xi)$ reduces to $\overline{\psi}(\xi)\mathcal{Q}\gamma_\mu\psi(\xi)$ in a free quark model, where $\mathcal{Q}$ is the quark charge matrix: $\mathcal{Q} = \text{diag}\left(2/3,\ -1/3,\ -1/3,\ldots\right)$ for $u,\ d,\ s,\ldots$ The current commutator in $W_{\mu\nu}$ reduces to[2]

$$\left[J_\mu(\xi),\ J_\nu(0)\right] = \overline{\psi}(\xi)\gamma_\mu\ S(\xi)\gamma_\nu\mathcal{Q}^2\psi(0) - \overline{\psi}(0)\gamma_\nu\ S(-\xi)\gamma_\mu\mathcal{Q}^2\psi(\xi) \quad (2.1)$$

where $S(\xi)$ is the anticommutator function[1.1]

$$S(\xi) = \{\psi(\xi), \overline{\psi}(0)\} = -\gamma^\rho \partial/\partial\xi^\rho \Delta(\xi) = S(-\xi) \quad (2.2)$$

and

$$\Delta(\xi) = \frac{1}{2\pi}\delta(\xi^2)\epsilon(\xi^0) + \ldots \quad (2.3)$$

The additional terms in $\Delta(\xi)$ vanish for zero quark mass. Since quark masses generate only $0(m^2/Q^2)$ corrections in the Bjorken limit, we ignore them from now on.

The product of operators in Eq. (2.1) is singular when $\xi_\lambda \to 0$, but the singularity is a C-number – namely the term removed by normal ordering – so it doesn't contribute to a connected matrix element. Thus,

$$\begin{aligned} \langle p|\overline{\psi}(\xi)\psi(0)|p\rangle_c &= \langle p| : \overline{\psi}(\xi)\psi(0) : |p\rangle \\ &= -\langle p|\psi(0)\overline{\psi}(\xi)|p\rangle_c \end{aligned} \quad (2.4)$$

where the second equality is a consequence of normal ordering. The Lorentz structure of Eq. (2.1) is simplified by the identity

$$\begin{aligned} \gamma_\mu\gamma_\rho\gamma_\nu &\equiv S_{\mu\rho\nu\sigma}\gamma^\sigma - i\epsilon_{\mu\rho\nu\sigma}\gamma^\sigma\gamma^5 \\ S_{\mu\rho\nu\sigma} &\equiv g_{\mu\rho}g_{\nu\sigma} + g_{\mu\sigma}g_{\nu\rho} - g_{\mu\nu}g_{\rho\sigma} \end{aligned} \quad . \quad (2.5)$$

The ϵ-term does not contribute because it is impossible to construct a pseudovector from ξ_λ and p_λ. Putting together the pieces, we find

$$\begin{aligned} \lim_{Bj} W_{\mu\nu} = -\frac{S_{\mu\rho\nu\sigma}}{8\pi^2}\int d^4\xi\ e^{iq\cdot\xi}\left[\frac{\partial}{\partial\xi_\rho}\delta(\xi^2)\epsilon(\xi^0)\right]\times \\ \times\langle p|\overline{\psi}(\xi)\gamma^\sigma\mathcal{Q}^2\psi(0) - \overline{\psi}(0)\gamma^\sigma\mathcal{Q}^2\psi(\xi)|p\rangle_c\ . \end{aligned} \quad (2.6)$$

The "$\lim_{Bj}$" reminds us that the parton model assumptions which went into Eq. (2.6) are only (approximately) valid as $Q^2 \to \infty$ at fixed x. Integrating by parts and introducing light-cone coordinates in the target rest frame

$$\begin{aligned} \lim_{Bj} W_{\mu\nu} = \lim_{q^-\to\infty} \frac{S_{\mu\rho\nu\sigma}iq^\rho}{8\pi^2}\int d\xi^+ d\xi^- d^2\xi_\perp\ e^{iq^+\xi^- + iq^-\xi^+}\times \\ \times\ \delta(2\xi^+\xi^- - \vec{\xi}_\perp^{\,2})\epsilon(\xi^+ + \xi^-)\times \\ \times\ \langle p|\overline{\psi}(\xi)\gamma^\sigma\mathcal{Q}^2\psi(0) - \overline{\psi}(0)\gamma^\sigma\mathcal{Q}^2\psi(\xi)|p\rangle_c\ . \end{aligned} \quad (2.7)$$

We have dropped the term in which $\partial/\partial\xi_\rho$ acts on the matrix element since it generates at most a factor p^ρ or $\xi^\rho\mu^2$ (μ^2 is some mass characteristic of the target) both of which are negligible with respect to q^ρ in the Bjorken limit. [Note ξ^ρ Fourier transforms into q^ρ/q^2.]

The form of $S_{\mu\rho\nu\sigma}$ implies that the coefficient of $-g_{\mu\nu}$ in $W_{\mu\nu}$ equals half the trace of $W_{\mu\nu}$, or referring to Eqs. (1.10) and (1.11),

$$F_1 = \frac{1}{2}\left(3F_1 - \frac{1}{2x_T}F_2\right) \tag{2.8}$$

which requires

$$F_1 = \frac{1}{2x_T}F_2 \tag{2.9}$$

or

$$\lim_{Bj} \sigma_L/\sigma_T = 0 \ . \tag{2.10}$$

which is the famous Callan-Gross relation and follows from the quark spin being 1/2. It is well-verified experimentally: Even though R is difficult to measure and still a subject of debate, all experiments agree that for $Q^2 > 1\ \mathrm{GeV}^2$, $R < 0.2$.

Eq. (2.7) can be reduced to a one-dimensional integral. First, we use the δ-function to perform the $\xi_\perp^2$ integral leaving

$$\begin{aligned} F_2 = 2x_T \lim_{q^-\to\infty} \frac{iq^-}{8\pi} \int d\xi^+ d\xi^-\ e^{i(q^+\xi^- + q^-\xi^+)} \times \\ \times \left[\theta(\xi^+)\theta(\xi^-) - \theta(-\xi^+)\theta(-\xi^-)\right] \times \\ \times \left. \langle p|\overline{\psi}(\xi)\gamma^+ \mathcal{Q}^2\psi(0) - \overline{\psi}(0)\gamma^+\mathcal{Q}^2\psi(\xi)|p\rangle_c \right|_{\xi_\perp^2 = 2\xi^+\xi^-} , \end{aligned} \tag{2.11}$$

then integrate by parts on ξ^+ keeping only the leading term at large q^-

$$F_2 = \frac{x_T}{4\pi}\int d\xi^-\ e^{iq^+\xi^-} \left. \langle p|\overline{\psi}(\xi^-)\gamma^+\mathcal{Q}^2\psi(0) - \overline{\psi}(0)\gamma^+\mathcal{Q}^2\psi(\xi^-)|p\rangle_c \right|_{\xi^+ = \vec{\xi}_\perp = 0} . \tag{2.12}$$

So, F_2 is a dimensionless function of $q^+ = -M_T x_T/\sqrt{2}$, *i.e.*, $F_2 \to F_2(x_T)$, which is Bjorken scaling.

Before converting Eq. (2.12) into the most familiar parton model form, we should note that $F_2(x_T)$ measures a particular quark correlation function in the target ground state. The first term in Eq. (2.12), for example, measures the amplitude to remove a quark from the target at some point, ξ_1, and replace it at ξ_2 with $\xi_2 - \xi_1 \equiv \xi$, $\xi^+ = \vec{\xi}_\perp = 0$ and $|\xi^-| < 1/q^+$, leaving the target in the ground state. The shape of the structure function teaches us about a correlation function in the target ground state. One must be careful, however, not to use one's intuition from non-relativistic quantum mechanics: this is not an equal time correlation function, but instead a light-cone correlation function. More about these later.

To further simplify (2.12) we must study light-cone γ-matrices. The matrices

$$P^\pm = \frac{1}{2}\gamma^\mp\gamma^\pm = \frac{1}{2}(1 \pm \alpha^3) \tag{2.13}$$

are projection matrices: $P^+ + P^- = 1$, $P^{\pm 2} = P^\pm$, $P^\pm P^\mp = 0$, and if we define

$$\psi_\pm \equiv P^\pm \psi \ , \tag{2.14}$$

then (2.12) may be rewritten

$$\begin{aligned} F_2(x_T) = \frac{x_T}{2\sqrt{2}\pi} \int d\xi^- \, e^{iq^+\xi^-} \times \\ \times \langle p|\psi_+^\dagger(\xi^-)\mathcal{Q}^2\psi_+(0) + \psi_+(\xi^-)\mathcal{Q}^2\psi_+^\dagger(0)|p\rangle_c \Big|_{\xi^+=\vec{\xi}_\perp=0} \end{aligned} \tag{2.15}$$

where we used Eq. (2.4) to interchange the quark fields in the second term. If we now insert a complete set of states between quark fields, translate the ξ^- dependence out of ψ_+ or $\psi_+^\dagger$, integrate over ξ^- and sum explicitly over quark flavors ($a = u, d, s \ldots$), we get

$$F_2^T(x_T) = x_T \sum_a \mathcal{Q}_a^2 \sum_n \frac{1}{\sqrt{2}}\delta(p^+ + q^+ - p_n^+)\{|\langle n|\psi_{a+}|p\rangle|^2 + |\langle n|\psi_{a+}^\dagger|p\rangle|^2\} \ . \tag{2.16}$$

I have added a superscript T to F_2 to remind us that F_2 depends on the target. For a target of mass M_T, $q^+ = -x_T M_T/\sqrt{2}$ where $x_T = Q^2/2M_T\nu$, and $p^+ = M_T/\sqrt{2}$ (we are working in the target rest frame), so

$$F_2^T(x_T) = x_T \sum_a \mathcal{Q}_a^2 (f_{a/T}(x_T) + f_{\bar{a}/T}(x_T)) \ , \tag{2.17}$$

where

$$\begin{aligned} f_{a/T}(x_T) &= \frac{1}{\sqrt{2}}\sum_n \delta(p^+ - x_T p^+ - p_n^+)|\langle n|\psi_{a+}|p\rangle|^2 \\ f_{\bar{a}/T}(x_T) &= \frac{1}{\sqrt{2}}\sum_n \delta(p^+ - x_T p^+ - p_n^+)|\langle n|\psi_{a+}^\dagger|p\rangle|^2 \end{aligned} \ . \tag{2.18}$$

This is the familiar parton model, except it is written in the target rest frame rather than the "infinite momentum" frame. $f_{a/T}(x_T)$ is the probability (per unit x_T) to remove from the target a quark of flavor a with "momentum" (*i.e.*, p^+) fraction x_T, leaving behind a physical state ($|n\rangle$) with $p_n^+ = (1-x_T)p^+$. Similarly, $f_{\bar{a}/T}(x_T)$ is the probability to remove an antiquark with p^+-fraction x_T leaving behind a physical state with $p_n^+ = (1-x_T)p^+$. $f_{a/T}(x_T)$ and $f_{\bar{a}/T}(x_T)$ are shown graphically in Figs. 8b and 8c. [Notice that p^+ appears in the target rest frame formulation where the "infinite momentum" P_∞ appears in the more familiar formulation.]

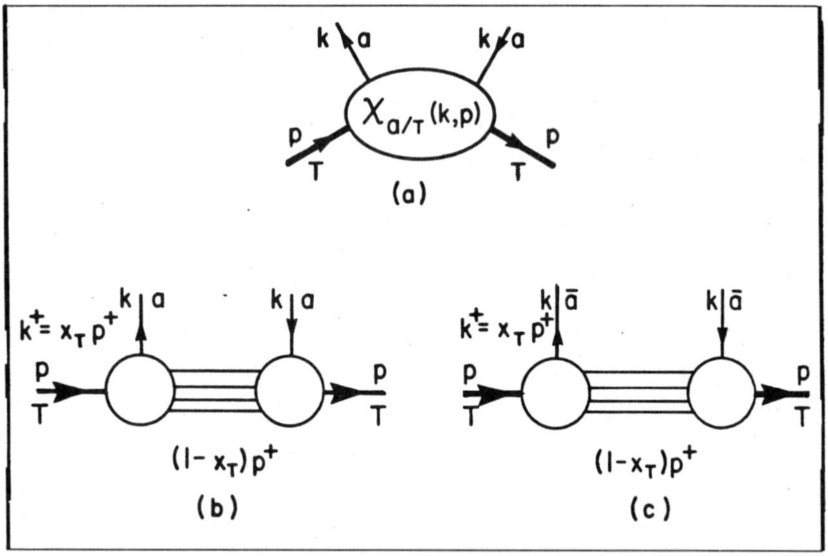

Fig. 8: a) The virtual-quark hadron scattering amplitude; b) The quark distribution function $f_{a/T}(x_T)$. Note that k^+ is fixed but k^- and $\vec{k}_\perp$ are integrated out; c) The antiquark distribution function $f_{\bar{a}/T}(x_T)$.

2.2 Properties of the Distribution Functions

$f_{a/T}(x_T)$ and $f_{\bar{a}/T}(x_T)$ obey important positivity, spectral and normalization constraints. The state n in Eq. (2.18) is physical and must have $p_n^+ > 0$, *i.e.*, $E_n > |\vec{p}_n|$, thus $f_{a/T}(x_T) = f_{\bar{a}/T}(x_T) = 0$, for $x_T \geq 1$ or

$x \geq \frac{M_T}{M}$ ($\approx A$ for a nuclear target). For $0 < x_T < 1$, $f_{a/T}(x_T)$ defined by Eq. (2.18) is manifestly positive. Next, let us consider $f_{a/T}(x_T)$ for $x_T < 0$. Returning to Eq. (2.15)

$$\begin{aligned} f_{a/T}(x_T) &= \frac{1}{2\sqrt{2}\pi}\int d\xi^- \; e^{iq^+\xi^-}\langle p|\psi^\dagger_{a+}(\xi^-)\psi_{a+}(0)|p\rangle_c\Big|_{\xi^+=\vec{\xi}_\perp=0} \\ &= -\frac{1}{2\sqrt{2}\pi}\int d\xi^- \; e^{iq^+\xi^-}\langle p|\psi_{a+}(0)\psi^\dagger_{a+}(\xi^-)|p\rangle_c\Big|_{\xi^+=\vec{\xi}_\perp=0} \end{aligned} \tag{2.19}$$

where we've used Eq. (2.4). Replacing ξ^- by $-\xi^-$ and translating the matrix element we find

$$f_{a/T}(-x_T) = -f_{\bar{a}/T}(x_T) \ . \tag{2.20}$$

Although $f_{a/T}(x_T)$ does not vanish for $x < 0$ (it does vanish for $x_T < -1$) it is determined by $f_{\bar{a}/T}(x_T)$ for $x > 0$. The literature is very confused on this point: there is a lot of talk about the need to "prove" that "no partons can go backwards in the infinite momentum frame", *i.e.*, $f_{a/T}(x_T) = 0$ for $x_T < 0$. We see here that $f_{a/T}(x_T)$ is <u>defined</u> in such a way that measurements in the physical region $0 < x_T < 1$ determine it everywhere. For a more extensive discussion of this issue, see Ref. 3.

Now let us integrate $f_{a/T}(x_T)$ over all x_T. Using $x_T = -\sqrt{2}q^+/M_T$ and Eq. (2.19) and (2.20) we find

$$\begin{aligned} \int_{-\infty}^{\infty} dx_T \; f_{a/T}(x_T) &= \int_0^1 dx_T \; \left(f_{a/T}(x_T) - f_{\bar{a}/T}(x_T)\right) \\ &= \frac{1}{M_T}\langle p|\psi^\dagger_{a+}(0)\psi_{a+}(0)|p\rangle_c \\ &= N_{a/T} - N_{\bar{a}/T} \end{aligned} \tag{2.21}$$

where the last step follows from the fact that $\psi^\dagger_{a+}\psi_{a+} = \frac{1}{\sqrt{2}}j^+_a$ and j^μ_a is a conserved current whose expectation value measures the number of quarks (minus the number of antiquarks) of flavor a: $\langle p|j^\mu_a|p\rangle = 2p^\mu(N_{a/T} - N_{\bar{a}/T})$. Clearly, we may intepret $f_{a/T}(x_T)\left(f_{\bar{a}/T}(x_T)\right)$ as a probability per unit x_T to find a quark (antiquark) of flavor a with $k^+ = x_T p^+$ in the target T:

$$\frac{dP_{a/T}}{dx_T} = f_{a/T}(x_T) \tag{2.22}$$

From this interpretation follows a host of parton model sum rules for the structure functions. The most important for our purpose is the "momentum sum rule".

$$\int_0^1 x_T dx_T \big(f_{a/T}(x_T) + f_{\bar{a}/T}(x_T)\big) = \epsilon_{a/T} + \epsilon_{\bar{a}/T} \tag{2.23}$$

where $\epsilon_{a/T}(\epsilon_{\bar{a}/T})$ is the fraction of the target's p^+ carried by quarks (antiquarks) of flavor a. The derivation mimics the derivation of Eq. (2.21) except the quark stress tensor $i\overline{\psi}_a\gamma^\mu\partial^\nu\psi_a$ appears instead of j_a^μ. If hadrons contained only quarks $\sum_a \epsilon_{a/T}$ would be 1. Instead, it is typically $\sim \frac{1}{2}$ (at large Q^2) indicating that substantial momentum and energy are carried by other, neutral quanta, namely gluons.

If two targets are related by a symmetry, their quark distributions are similarly related. Isospin relates the neutron and proton and gives

$$\begin{aligned} f_{u/n}(x) &= f_{d/p}(x) \\ f_{d/n}(x) &= f_{u/p}(x) \\ f_{s/n}(x) &= f_{s/p}(x), \text{ etc.} \end{aligned} \tag{2.24}$$

For completeness, I record the structure functions for electron and (charged current) neutrino scattering from nucleon targets assumming isospin symmetry:

$$\begin{aligned} F_2^{ep}(x) &= x\bigg[\frac{4}{9}\big(f_{u/p}(x) + f_{\bar{u}/p}(x)\big) \\ &\qquad + \frac{1}{9}\big(f_{d/p}(x) + f_{\bar{d}/p}(x) + f_{s/p}(x) + f_{\bar{s}/p}(x)\big)\bigg] \\ F_2^{en}(x) &= x\bigg[\frac{4}{9}\big(f_{d/p}(x) + f_{\bar{d}/p}(x)\big) \\ &\qquad + \frac{1}{9}\big(f_{u/p}(x) + f_{\bar{u}/p}(x) + f_{s/p}(x) + f_{\bar{s}/p}(x)\big)\bigg] \\ F_2^{\nu p}(x) &= F_2^{\bar{\nu} n}(x) = 2x\big(f_{d/p}(x) + f_{\bar{u}/p}(x)\big) \\ F_2^{\nu n}(x) &= F_2^{\bar{\nu} p}(x) = 2x\big(f_{u/p}(x) + f_{\bar{d}/p}(x)\big) \ . \end{aligned} \tag{2.25}$$

I have ignored heavy quarks (c, b, t) and the Cabibbo angle ($\sin^2\theta_c \cong 0.05$) and used the shorthand notation $\nu p(\bar{\nu}p)$ for $\nu p \to \mu^- X$ ($\bar{\nu}p \to \mu^+ X$), etc. I have also left out the (important) parity violating structure function, F_3, which arises in neutrino scattering. One important implication of Eqs.

(2.25) is that in the absence of any nuclear effect in deuterium one would expect

$$\begin{aligned} F_2^{ed}(x) &= \frac{5}{9}x\big(f_{u/p}(x) + f_{\overline{u}/p}(x) + f_{d/p}(x) + f_{\overline{d}/p}(x)\big) \\ &\quad + \frac{2}{9}x\big(f_{s/p}(x) + f_{\overline{s}/p}(x)\big) \\ &\cong \frac{5}{18}\left[F_2^{\nu p}(x) + F_2^{\overline{\nu}/p}(x)\right] \ , \end{aligned} \tag{2.26}$$

because strange quark pairs are suppressed in the nucleon and weighted by 2/5 relative to non-strange quarks. This relation allows one to look for a nuclear effect in the deuteron in a model independent manner (see Ref. 3.2).

To further simplify the structure function of nucleons, it is customary to distinguish between quarks which must be present to account for the target's quantum numbers (u and d quarks for nucleons) known as "valence quarks" and those which may be present in pairs due to relativistic, quantum effects, known as "ocean quarks". Thus, for an isospin averaged nucleon

$$\begin{aligned} f_{u/N}(x) &= f_{d/N}(x) = f_V(x) + f_O(x) \\ f_{\overline{u}/N}(x) &= f_{\overline{d}/N}(x) = f_O(x) \ . \end{aligned} \tag{2.27}$$

Strange and heavier quarks may also be present. Typically, one assumes

$$\begin{aligned} f_{s/N}(x) &= f_{\overline{s}/N}(x) \lesssim f_O(x) \\ f_{c/N}(x) &= f_{\overline{c}/N}(x) \approx 0, \text{ etc.} \end{aligned}$$

Before leaving this general discussion of the parton model, it is useful to relate the quark distribution function to the amplitude for quark-target scattering. We define the (connected) virtual quark-target forward scattering amplitude by

$$\chi_{a/T}(k,p) \equiv \int d^4\xi\ e^{-ik\cdot\xi}\langle p|T\big(\overline{\psi}_a(\xi)\psi_a(0)\big)|p\rangle_c \tag{2.28}$$

as illustrated in Fig. 8a. χ is a matrix in color and Dirac spaces, but we have suppressed those indices. The distribution function $f_{a/T}(x_T)$ can be projected out of $\chi_{a/T}$ by integrating over all components of k except k^+ which is held fixed, $k^+ = x_T p^+$, and tracing the Dirac indices with γ^+:

$$f_{a/T}(x_T) = \int \frac{d^4k}{(2\pi)^4}\delta\left(\frac{k^+}{p^+} - x_T\right) Tr\ \left[\gamma^+\chi_{a/T}(k,p)\right] \ . \tag{2.29}$$

It is easy to verify that Eqs. (2.28)–(2.29) lead to Eq. (2.18) provided one used Eq. (2.4) to relate the T-product to the ordinary product.

It is often convenient, when studying electroproduction from nuclei, to define quark distribution functions depending on a universal variable $x = Q^2/2M_N\nu$. To preserve their probabilistic interpretation, it is necessary to rescale them:

$$F_{a/T}(x) \equiv \frac{dP_{a/T}}{dx} = \frac{M}{M_T} f_{a/T}(x_T) \ . \tag{2.30}$$

Then,

$$\int_0^{M_T/M} dx \big(F_{a/T}(x) - F_{\overline{a}/T}(x)\big) = N_{a/T} - N_{\overline{a}/T} \tag{2.31}$$

and

$$\int_0^{M_T/M} x dx \big(F_{a/T}(x) + F_{\overline{a}/T}(x)\big) = \frac{M_T}{M}(\epsilon_{a/T} + \epsilon_{\overline{a}/T}) \ . \tag{2.32}$$

At the same time, it is convenient to introduce a structure function per nucleon, $\overline{F}_2^T(x)$

$$\overline{F}_2^T(x) \equiv x \sum_a \mathcal{Q}_a^2 \big(\overline{F}_{a/T}(x) + \overline{F}_{\overline{a}/T}(x)\big) \tag{2.33}$$

where $\overline{F}_{a/T}(x) = F_{a/T}(x)/A$. In the analysis of nuclear targets, I will try to preserve this notation: lower case for intrinsically defined distribution functions, upper case for functions of x and barred upper case for functions of x "per nucleon". Also, the label "A" will denote a nucleus, "T" a generic target and "a" a quark of flavor a. Note that $\overline{F}_2^A(x, q^2)$ is defined so that it would reduce to the (isospin weighted) nucleon structure function $\frac{Z}{A}F_2^P(x, q^2) + \frac{N}{A}F_2^N(x, q^2)$ if the nucleons in the nucleus were non-interacting.

At this point, it would be appropriate to discuss the phenomenology of the neutron and proton structure functions; however, no time would be left for nuclear targets. So the reader will have to consult the references[4] for more information about nucleons. Here, I will mention only a few properties of importance for future work. Near $x = 0$ both $F_{O/N}(x)$ and $F_{V/N}(x)$ are expected to diverge: $F_{O/N} \sim 1/x$ and $F_{V/N}(x) \sim 1/\sqrt{x}$. As $x \to 1$, $F_{O/N}/F_{V/N} \to 0$. The neutron and proton structure functions differ significantly at large x leading to the observation that $f_{d/p}/f_{u/p} \to 0$ as $x \to 1$. Some quark distributions extracted from electron and neutrino scattering experiments are shown in Fig. 9.

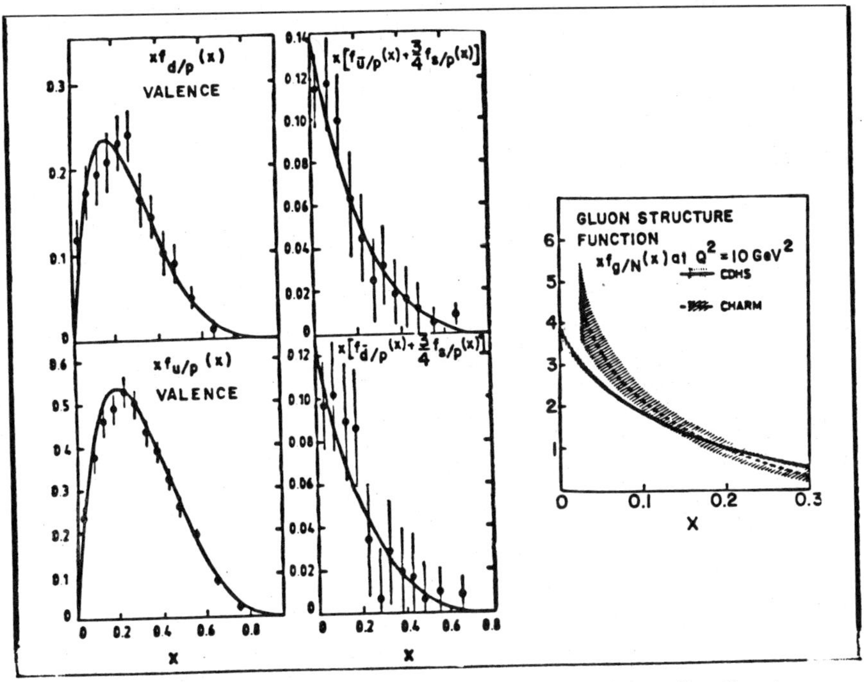

Fig. 9: Examples of quark, antiquark and gluon distributions in the nucleon, from F. Dydak.[2.4]

§3 DATA FROM NUCLEAR TARGETS AND THEIR IMPLICATIONS FOR QUARK DISTRIBUTIONS IN NUCLEI

Before 1982, little attention had been paid to deep inelastic lepton scattering from nuclei. Interest in the subject was awakened by experimental results from CERN. In order to put the rest of these lectures in proper context it is necessary to present those results, discuss the (dis)agreement among experiments and present the most rudimentary parton model analysis of the data, which already has important implications for the quark substructure of nuclei. The basic parton model of Bjorken and Feynman and the material of §2 are the only prerequisites for this analysis. It will be impossible, however, to avoid some reference to QCD, asymptotic freedom and other issues which will be introduced later in these lectures.

The first precise comparison of deep inelastic scattering from a nuclear target with scattering from a nucleon was made by the EMC Collaboration.[1] They compared iron and deuterium targets. They assumed that the weakly bound deuteron approximates the isospin averaged nucleon. Subsequent analysis by Bodek and Simon[2] has confirmed this assumption.

In the absence of any nuclear effect at large Q^2 one expects $F_2^A(x,Q^2) = AF_2^D(x,Q^2)/2$ up to a small correction for the neutron excess in the nucleus. To display deviations from this naive expectation and to minimize systematic errors, it is conventional to plot $S^A(x,Q^2) \equiv 2F_2^A(x,Q^2)/AF_2^D(x,Q^2)$. The deviation of $S^{Fe}(x,Q^2)$ from unity shown in Fig. 10 caught nearly everyone by surprise.[3] The effect at large x was particularly surprising because $\overline{F}_2^A(x,Q^2)/F_2^N(x,Q^2)$ must go to infinity as $x \to 1$. [The denominator vanishes for $x > 1$, the numerator vanishes only for $x > A$.] The "EMC effect", as the deviation of S^{Fe} from unity came to be known, was quickly confirmed by a reanalysis of old SLAC data with iron, aluminum and deuterium targets.[4] Last year a dedicated SLAC experiment (E-139) measured the EMC effect on a sequence of nuclear targets.[5] Their data are shown in Fig. 11.

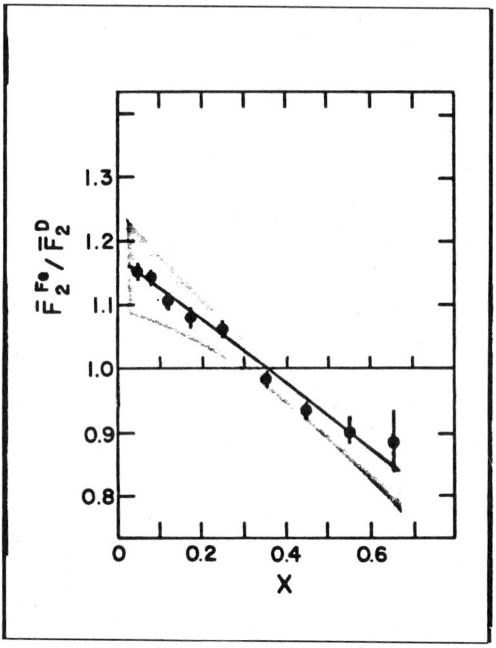

Fig. 10: EMC data on $\overline{F}_2^{Fe}/F_2^D$.[3.1]

The SLAC and EMC data agree reasonably well for $x > 0.3$ but the enhancement seen at low-x by EMC was not observed at SLAC. Recently,

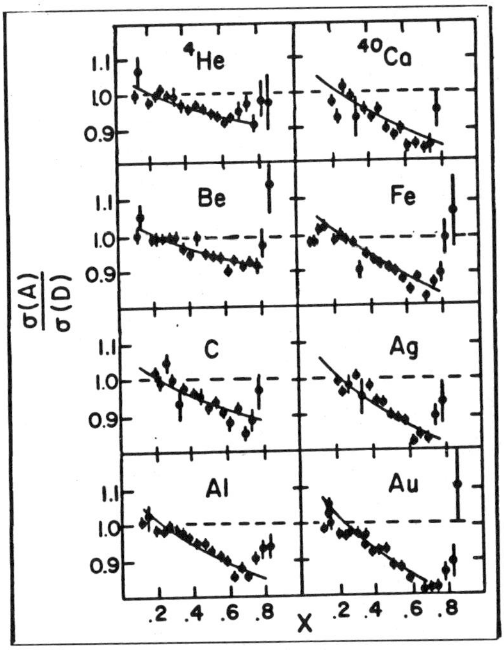

Fig. 11: SLAC data on σ^A/σ^D.[3.5] The fits are from Ref. 6.4.

there has been input from other groups. The BCDMS collaboration at CERN[6] has measured deep inelastic muon scattering from iron, nitrogen and deuterium targets. Their iron data are restricted by detector geometry to $x \geq 0.2$ where they agree with both SLAC and EMC. All available data on iron are shown in Fig. 12a.

The nitrogen data include a point at $x = 0.1$ which is significantly above 1 but below the trend of the EMC iron data. [See Fig. 12b.]

Several neutrino detector collaborations have measured ratios of cross sections from nuclear targets.[7] In general, their results are subject to larger statistical and systematic uncertainties than the electron and muon data. It is fair to say, however, that at low-x, where their statistics are best, the neutrino experiments fail to confirm the enhancement seen by the EMC group. For example, the CDHS group presented some data on S^{Fe} at this year's Moriond meeting.[8] Their data are also shown in Fig. 12c. It is of great importance to sort out the apparent disagreement among experiments at low-x. This can only be done for certain by future experiments. However, we can catalog some of the possibilities while we wait. To help, some of the salient features of the experiments are summarized in Table I. Among the

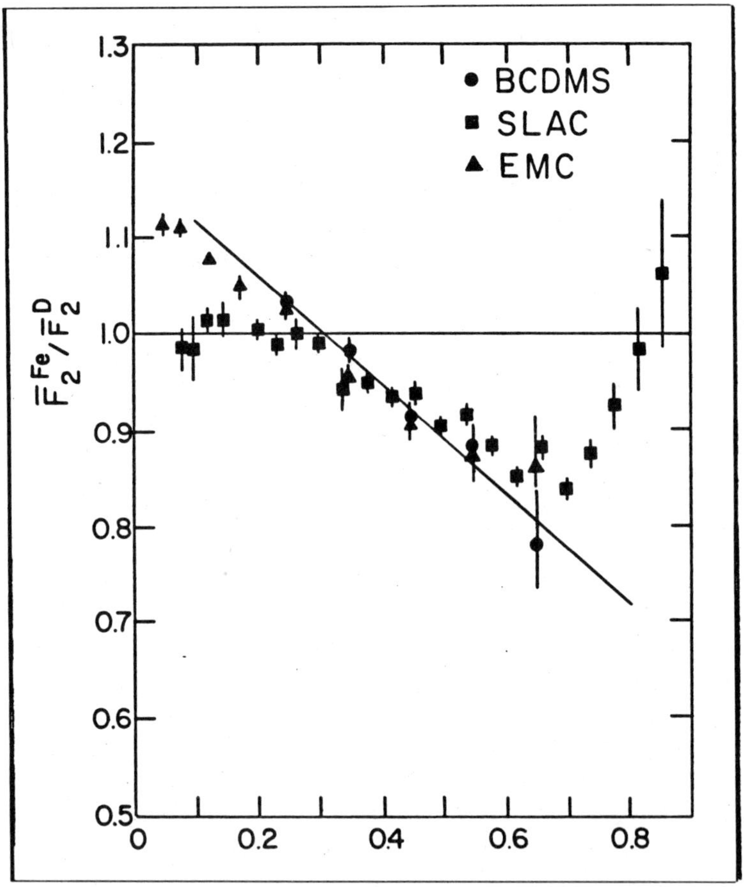

Fig. 12a: Compilation of data on $\overline{F}_2^{Fe}/\overline{F}_2^{D}$.[3.6]

possibilities are:

1. Systematic errors: The EMC collaboration quote a considerable systematic error in the slope of the straight line fit to their data. They also quote a 7% systematic uncertainty in the normalization.[1] Rotating their data to smaller slope and lowering it by 7% improves its agreement with SLAC and BCDMS considerably. Merlo quoted a systematic error on the $x = 0.03$ CDHS point equal to the statistical error. Moving the CDHS point up by that amount largely removes the

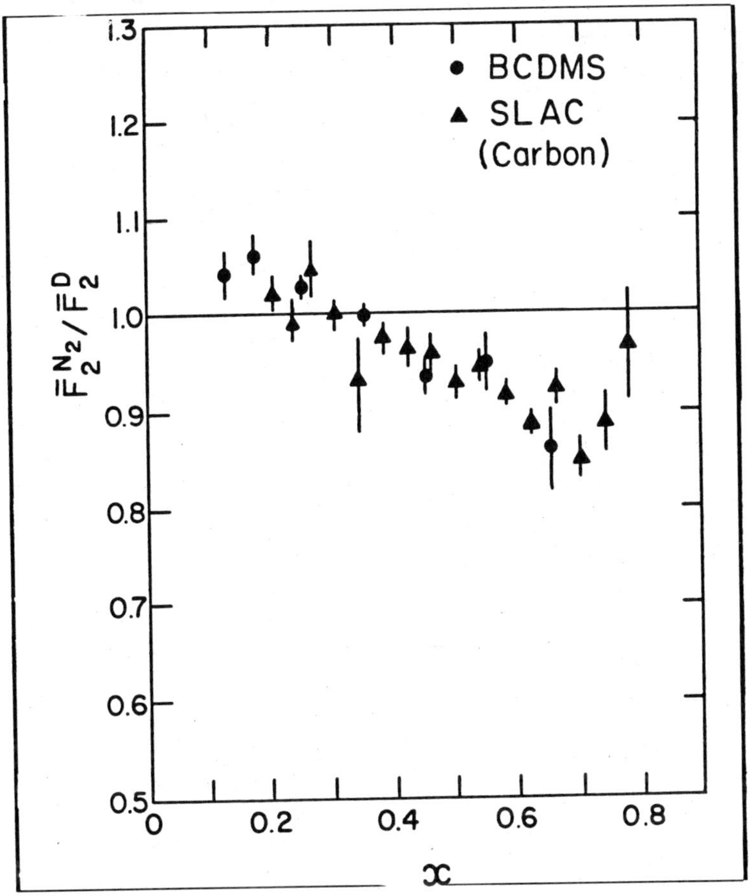

Fig. 12b: Comparison of BCDMS N_2 data with SLAC C data.[3.6]

discrepancy.

2. An A dependence of $R(\sigma_L/\sigma_T)$: The SLAC E-139 experiment measured R on several nuclear targets and found it large ($R \cong 0.2$) and possibly A dependent. If they extract S^A from their data with an A dependent R it agrees better with the EMC measurement. This reconciliation is only superficial, however. It generates a much greater disagreement in the measurement of F_1^{Fe}/F_1^D. [EMC do not, in fact, measure F_1 because the beam energy is high ($\epsilon \approx 1$, see Eq. 1.34)) but

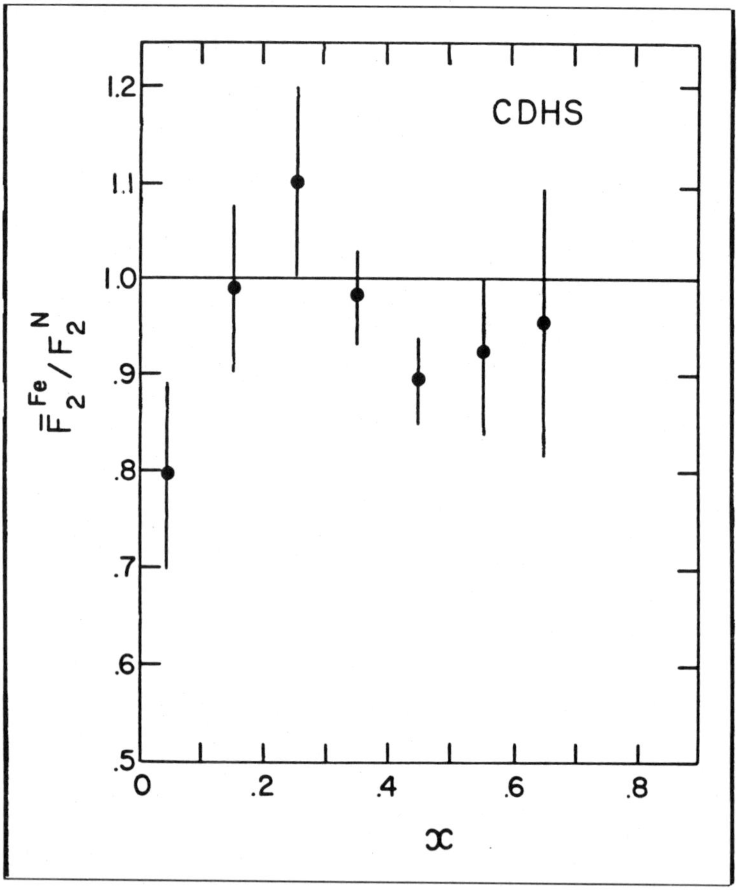

Fig. 12c: CDHS (neutrino scattering) measurement of $\overline{F}_2^{Fe}/F_2^N$.[3.8]

at EMC energies Q^2 is quite large ($Q^2 \geq 9$ GeV^2 for all x-bins) and R is expected to be very small at large Q^2 (even in QCD where the Callan-Gross relation is not exact) so $F_1^A/F_1^N \cong F_2^A/F_2^N$.]

3. Strong Q^2, x and A dependence at low-x: Perhaps the differences among experiments are due to the fact that their bins average in different ways over a rapidly varying function. The likely source of this variation is "shadowing" which is expected to be important at low-x

and will be discussed further in §6.4.

TABLE 1

Group	Trend of $\overline{F}_2^{Fe}/\overline{F}_2^{D}$ at low − x	x−Values	Q^2 Range	Treatment of $R = \sigma_L/\sigma_T$
EMC	High (> 1)	0.05, 0.08...	≥ 9 GeV^2 at x=0.05	$R = 0$
SLAC	Medium (≈1)	≥0.08	≥ 2 GeV^2 at x=0.08	$R = .18$
BSDMC	Medium (>1)	0.1[a]	≥ 30 MeV^2	$R = 0$
CDHS[b]	Low (<1)	0.03...	$< Q^2 >= 3$ GeV^2	− − −[c]

a. Nitrogen data
b. ν scattering.
c. Extraction of F_2 indepdent of R.

For the remainder of these lectures, I will assume that the EMC data for $x > 0.3$ are correct, but for $x < 0.3$, I will assume the truth lies somewhere between the EMC and SLAC results.

The parton model ideas developed in the previous section can be applied directly to the nuclear structure function.[9] In the parton model and the Bjorken limit, $\overline{F}_2^A$ is independent of Q^2. In QCD, the notion of parton distribution functions and other aspects of the parton model are preserved, but $\overline{F}_2^A$ and the parton distribution functions develop a weak but important Q^2-dependence. To allow for this, we occasionally keep the Q^2-label explicit

$$\overline{F}_2^A(x, Q^2) = x \sum_a \mathcal{Q}_a^2 (\overline{F}_{a/A}(x, Q^2) + \overline{F}_{\bar{a}/A}(x, Q^2)) \tag{3.1}$$

and the difference between a nucleus and deuterium is defined by

$$\overline{\Delta}_A(x, Q^2) \equiv \overline{F}_2^A(x, Q^2) - \overline{F}_2^D(x, Q^2) \tag{3.2}$$

$\overline{\Delta}$ depends only on the difference of quark distributions in nucleus A and deuterium. In the valence parton model described in §2,

$$\overline{\Delta}_A(x,Q^2) = x\left(\frac{5}{9}\delta\overline{F}_{V/A}(x,Q^2) + \frac{4}{3}\delta\overline{F}_{O/A}(x,Q^2)\right) \tag{3.3}$$

and $\delta\overline{F}_{V/A} \equiv \overline{F}_{V/A} - \overline{F}_{V/D}$, etc.

The EMC data for $\overline{\Delta}_{Fe}$ are shown in Fig. 13, which was constructed from the published EMC iron data and the data on S^{Fe}. Experimentalists caution that systematic errors are more treacherous in the difference $\overline{\Delta}_{Fe}$ than in the ratio S^{Fe}. Nevertheless, the qualitative features of $\overline{\Delta}_{Fe}$ are probably reliable (modulo our caveats about $x < 0.3$) and they tell us what has happened to the quark distributions in passing from deuterium to iron.

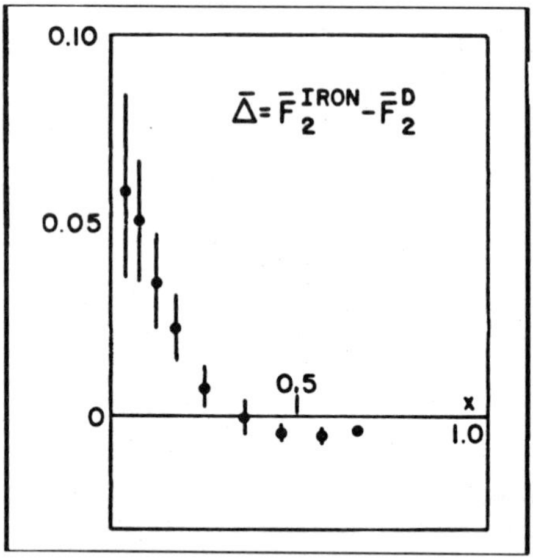

Fig. 13: The difference of the structure function per nucleon of iron and deuterium.[13]

The EMC effect – at least as seen in the EMC data – has several different aspects which can be extracted from a parton model analysis of the data. The same analysis applied to the SLAC data leads to somewhat different conclusions which I will mention along the way:

1. The valence quarks in iron are "degraded" – shifted to lower p^+ – relative to those in deuterium. [All experiments agree on this.]

2. There is an increase in the number of ocean quark pairs in iron compared to deuterium. [The increase is large if we believe the EMC data, smaller or even absent if one accepts the SLAC data at low-x.]

3. The fraction of momentum (p^+) per nucleon on quarks and antiquarks in iron relative to deuterium can be extracted. [It increases slightly if one believes the EMC data but the SLAC data are ambiguous.]

These results hold for other nuclei as well in proportion to the size of the measured "EMC effect". The extraction makes use of parton model sum rules and positivity constraints. For $x > 0.35$, $\overline{F}_{O/N}(x)$ (the ocean quark distribution in the isolated nucleon) is known to be negligible, thus $\delta\overline{F}_{O/N}(x) \geq 0$ for $x > 0.35$. Since $\overline{\Delta}_A(x) < 0$ for $x > 0.35$, we see from Eq. (3.3) that $\delta\overline{F}_{V/A}(x)$ must be negative for $x > 0.35$. The valence quark distribution is conserved, *i.e.*,

$$\int_0^A dx\, \delta\overline{F}_{V/A}(x) = 0 \tag{3.4}$$

because the number of valence quarks (per nucleon) in all nuclei is three. $\delta\overline{F}_{V/A}(x)$ must therefore be positive for some x values where it has not been observed. The SLAC data show $\delta\overline{F}_{V/A}(x)$ turning positive for $x > 0.8$ by which point $\overline{F}_{V/A}(x)$ is very small. I assume that $\int_{0.8}^A dx\delta\overline{F}_{V/A}(x)$ is negligible. This leaves $x < 0.35$ as the region in which $\delta\overline{F}_{V/A}(x)$ is positive.

In §1 we saw that x is conjugate to $1/M\xi^3$ in the laboratory, where ξ^3 is the spatial separation of the quark fields in the correlation function defining $F_{a/A}(x)$. Thus, the shift to lower x is indicative of a shift to longer range quark correlations in the target ground state. This result is independent of whatever "microscopic" explanation of the EMC effect might eventually be forthcoming. Whether it is attributed to "dynamical rescaling", N-quark bags, quark percolation, or more prosaic sources like pion admixtures in the nuclear wavefunction or other binding effects, the EMC effect directly measures an increased quark light-cone correlation length in nuclei. For more discussion of the space-time interpretation of the EMC effect, see Ref. 1.8. In retrospect, it is not surprising that measures of the quark correlation length increase in nuclei.[10] It is believed that quark/nuclear matter, regarded as a function of density at zero temperature, undergoes a deconfining phase transition at some ρ_{critical}. For densities below ρ_{critical}, quarks are confined in nucleons but for densities above ρ_{critical}, they move about more or less freely in a degenerate quark gas. One support for this is that QCD is known to become asymptotically free at large chemical potential (equivalent to high density), so at high enough density a quark gas will become free. We identify the nucleon as the zero density limit of quark matter. As A increases, the mean density of the nucleus increases (as the surface to volume ratio goes to zero), so we may regard the increased quark correlation length in iron as a consequence of its increased mean density and as a precursor of a deconfining phase transition where the correlation length would become very large. This view of the EMC effect is supported by the A dependence observed at SLAC which correlates very closely with nuclear densities. It is discussed at length in §6.3.

Point 2, the measurement of ocean quark pairs, is obtained by examining $\int dx\, \delta\overline{F}_{O/A}(x)$ over the range of the measured data ($x_{\min} \le x \le x_{\max}$):

$$\int_{x_{\min}}^{x_{\max}} dx\, \delta\overline{F}_{O/A}(x) = \frac{3}{4}\int_{x_{\min}}^{x_{\max}} \frac{dx}{x}\overline{\Delta}_A(x) + \frac{5}{12}\int_0^{x_{\min}} dx\, \delta\overline{F}_{V/A}(x) \quad (3.5)$$

where we have used Eq. (3.4) and assumed $\delta\overline{F}_{V/A}(x)$ negligible for $x_{\max} < x < A$. If $\delta\overline{F}_{V/A}(x)$ does not change sign twice, that is, if it remains positive for $x < x_{\min}$, then

$$\int_{x_{\min}}^{x_{\max}} dx\, \delta\overline{F}_{O/A}(x) > \frac{3}{4}\int_{x_{\min}}^{x_{\max}} \frac{dx}{x}\overline{\Delta}_A(x) \quad (3.6)$$

Eq. (3.6) could fail only if $\overline{F}_{V/A}(x)$ behaves as shown in Fig. 14.[11]

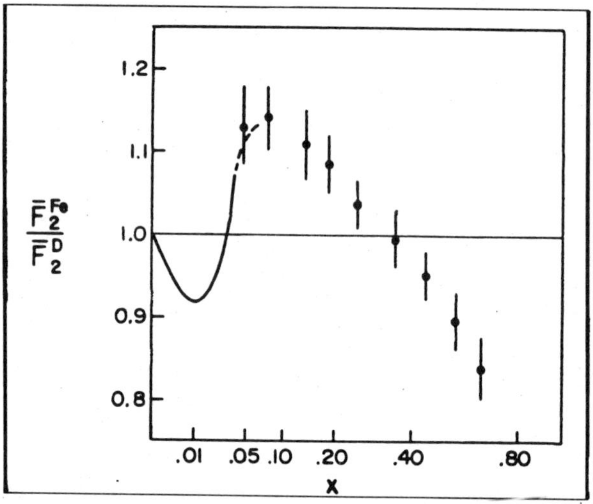

Fig. 14 Behavior of $\overline{F}_2^{Fe}/\overline{F}_2^{D}$ required by the EMC data at low-x if only valence quarks are involved in the effect.

Shadowing is expected to produce a depletion in $\overline{F}_2^A(x, Q^2)$ at low-x and low-Q^2, but it should affect primarily the ocean quark distribution and a shadowing of the valence quarks sufficient to invalidate Eq. (3.6) at large Q^2 would be surprising. According to Eq. (3.6), the ocean quarks are enhanced to the extent the x^{-1} weighted integral of $\overline{\Delta}_A$ is > 0. If one accepts the EMC data, the effect is quite large:

$$\int_{x_{\min}}^{x_{\max}} dx\, \delta\overline{F}_{O/Fe}(x) > 1.5 \int_{x_{\min}}^{x_{\max}} F_{O/N}(x) \tag{3.7}$$

where the right-hand side can be extracted from inelastic neutrino and anti-neutrino scattering from a proton target. If one accepts the SLAC data one finds little or no enhancement; probably the truth lies somewhere in between.

Point 3, the change in momentum carried by all quarks and antiquarks, follows from the momentum sum rule.

$$\begin{aligned}\delta\epsilon_A &= 2\int_0^A x dx[\delta\overline{F}_{V/A}(x) + 3\delta\overline{F}_{O/A}(x)] \\ &= \frac{9}{2}\int_0^A dx\, \overline{\Delta}_A(x) - \frac{1}{2}\int_0^A x dx\, \delta\overline{F}_{V/A}(x)\end{aligned} \tag{3.8}$$

If we ignore the small contributions from $x > x_{\max}$, and $x < x_{\min}$ the EMC data give $5.0 \pm 1.5 \times 10^{-2}$ for the first term on the right. The second term is small and positive (see Eq. (3.4) and subsequent discussion) so we obtain a bound $\delta\epsilon_{F_e} > 5.0 \pm 1.5 \times 10^{-2}$. The SLAC data, on the other hand, do not give a positive $\overline{\Delta}_A(x)$ for $x < 0.3$ so the sign of $\delta\epsilon_{Fe}$ cannot be determined although its magnitude is certainly small.[12]

At the level of the parton model, these features of the data appear logically independent. Some models, notably the rescaling model, are able to correlate a modest increase in the number of ocean quarks with the degradation (*i.e.*, shift to lower x) of the valence quarks. Other models are able to account for only one feature of the data or invoke several effects in concert to account for the different aspects of the EMC effect.

§4 THE CONVOLUTION MODEL AND FERMI MOTION

It is intuitively appealing to regard inclusive electroproduction from nuclei as a two step process. First, the nuclear wave function is decomposed into some basis of constituents, nucleons in the first instance, nucleons and pions in more elaborate schemes, and later perhaps including more exotic objects like Δ's, multiquark conglomerates and so on. Then, the structure functions of the constituents are added incoherently to give the structure function of the whole nucleus. This is the "convolution" model.[1] The simplest version, which includes only nucleons, gives what are known as "Fermi motion" corrections to the free nucleon structure function.[2] These were calculated long before the present excitement about electroproduction from nuclei. More recently, the model has been extended to more exotic constituents[3,4,5] in an attempt to "explain" the EMC effect. There is no adequate derivation of the convolution model. The parton analyses of the § 2 will provide a framework in which the assumptions which lead to the convolution model may be analyzed and criticized.

4.1 Deriving the Convolution Model

The two steps of the convolution model are summarized diagrammatically in Fig. 15. The nucleus, with baryon number A and momentum P, contains a constituent, label T and momentum p, which in turn contains a quark, flavor a and momentum k. The quark absorbs the virtual photon while the fragments of the nucleus and the constituent propagate into the final state without interaction or interference. Other diagrams in which fragments of the nucleus or the constituent T interact or interfere are ignored. Some are shown in Fig. 16.

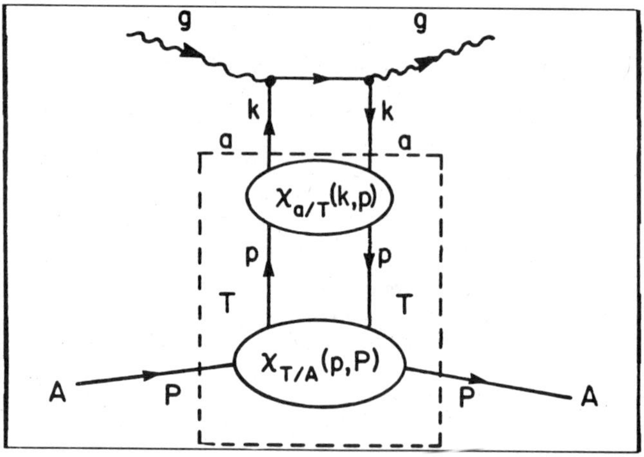

Fig. 15: The convolution model. The amplitude within the dashed box has no large external momentum flowing in it.

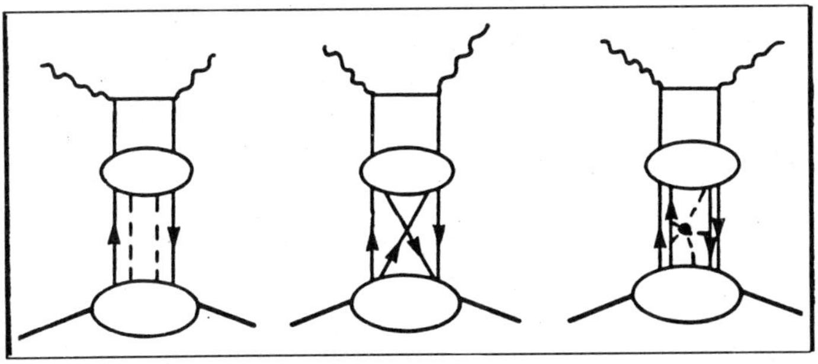

Fig. 16: Contributions which are ignored in convolution models.

Superficially, they resemble the diagrams of Fig. 6b-d which were ignored in the parton model, but there is an important difference: Fig. 6b-d can be dropped because $Q^2 \to \infty$, so $\xi^2 \to 0$ and the operator product expansion in QCD can be used to *prove* that they are $0(1/Q^2)$ compared to Fig. 6a. There is no analogous large mass scale characterizing the process enclosed in the dashed line in Fig. 15, so there is no *a priori* justification for ignoring the additional processes of Fig. 16. Furthermore, the fragments of the nucleus and the constituent have a long time, $\xi^0 \lesssim 1/Mx$, to interact while awaiting the return of the active quark. Nevertheless, for certain constituents under favorable kinematic conditions (*e.g.*, for nucleons in the weak binding limit, or for pions near $x = 0$)[1.8] ignoring final state interactions and interference may be justifiable. For the moment, we will simply ignore the problem and proceed.

The quark distribution for nucleus A is defined in analogy to Eq. (2.29):

$$f_{a/A}(x_A) = \int \frac{d^4k}{(2\pi)^4} \delta\left(\frac{k^+}{P^+} - x_A\right) Tr\left[\gamma^+ \chi_{a/A}(k,P)\right] \tag{4.1}$$

where $x_A = Q^2/2M_A\nu$. The content of Fig. 15 is a convolution form for $\chi_{a/A}(k,P)$:

$$\chi_{a/A}(k,P) = \sum_T \int \frac{d^4p}{(2\pi)^4} \chi_{a/T}(k,p)\chi_{T/A}(p,P) \tag{4.2}$$

where the sum covers all constituents of the nucleus. We have treated the constituent T as a scalar. The generalization to spin 1/2 is straightforward.

Note that the quark and/or constituent legs in Fig. 15 have been properly included in Eq. (4.2): From Eq. (2.28) it is apparent that $\chi_{a/T}$ ($\chi_{T/A}$) is untruncated in the quark (constituent) legs but truncated in the constituent (nucleus) legs. Only the + components of momentum really matter in Eq. (4.1)-(4.2). To make this manifest, we substitute the identity:

$$1 = \int dy\, \delta(y_A - p^+/P^+) \int dz_T \delta(z_T - k^+/p^+) \tag{4.3}$$

where the ranges of the y_A and z_T integrals are yet to be determined. After some algebra we find

$$\begin{aligned} f_{a/A}(x_A) = \int dy_A dz_T \delta(y_A z_T - x_A) \sum_T \\ \int \frac{d^4p}{(2\pi)^4} \delta(y_A - p^+/P^+) \chi_{T/A}(p, P) \\ \int d^4k\, \delta(z_T - k^+/p^+)\, Tr\left[\gamma^+ \chi_{a/T}(k, p)\right] \end{aligned} \tag{4.4}$$

The k-integration defines a Lorentz invariant function of z and p^2 which is the off shell generalization of $f_{a/T}(x_T)$ defined in Eq. (2.29):

$$f_{a/T}(z_T, p^2) \equiv \int \frac{d^4k}{(2\pi)^4}\, \delta(z_T - k^+/p^+)\, Tr\left[\gamma^+ \chi_{a/T}(k, p)\right] \tag{4.5}$$

Let us save the p^2 integral for last by inserting $\int dp_0^2 \delta(p^2 - p_0^2) = 1$. Then the d^4p integration gives

$$f_{a/A}(x_A) = \sum_T \int dy_A dz_T \delta(y_A z_T - x_A) \int dp_0^2 f_{a/T}(z_T, p_0^2) f_{T/A}(p_0^2, y_A) \tag{4.6}$$

where

$$f_{T/A}(p_0^2, y_A) = \int \frac{d^4p}{(2\pi)^4} \delta(p^2 - p_0^2) \delta(y_A - p^+/P^+) \chi_{T/A}(p, P) \tag{4.7}$$

Note that $f_{T/A}(p_0^2, y_A)$ is the probability to find a constituent T in nucleus A with momentum fraction y_A and invariant mass p_0^2, whereas $f_{a/T}(z_T, p_0^2)$ is the probability to find quark a with momentum fraction z_T in an off shell target with invariant mass p_0^2. So $f_{a/T}(z_T) = f_{a/T}(z_T, M_T^2)$

but $f_{T/A}(y_A) = \int dp_0^2 f_{T/A}(p_0^2, y_A)$. In practice, no one uses a p_0^2 dependent quark distribution $f_{a/T}(z_T, p_0^2)$. The rationale for this must be that $f_{T/A}(p_0^2, y_A)$ peaks very strongly at some $\overline{p}^2$ so $f_{a/T}(z, p_0^2)$ can be replaced by $f_{a/T}(z, \overline{p}^2)$. Then the p_0^2 integration in Eq. (4.7) can be performed, but it is an additional assumption that $f_{a/T}(z_T, \overline{p}^2) = f_{a/T}(z_T, M_T^2)$ unless of course $\overline{p}^2 \approx M_T^2$. The result

$$f_{a/A}(x_A) = \sum_T \int dy_A dz_T \delta(y_A z_T - x_A) f_{a/T}(z_T) f_{T/A}(y_A) \tag{4.8}$$

together with the equations which define $f_{a/T}(z_T)$ and $f_{T/A}(y_A)$ comprise the convolution model. The range of the y_A and z_T integrations can be determined by arguments identical to those used in §2.2 to show $0 < x_A < 1$. The result is $0 < y_A,\ z_T < 1$. If the constituents in question are nucleons and the binding is not too strong, then $\chi_{N/A}(p, P)$ does peak strongly and near mass shell (compared to the scale for variation in $f_{a/T}(z_T, p^2)$ which is $\Lambda \sim 200$ - 300 MeV). However, the model is applied to other constituents which are far off shell. Pion contributions are believed to be largest for $p^0 \approx 0$ and $|\vec{p}| \approx 300$ - 400 MeV[6] giving $\overline{p}^2 \approx 0.1$ - 0.2 GeV compared to $m_\pi^2 \cong 0.02$ GeV. Whether the distribution of quarks in a pion so far off shell is the same as the distribution on shell is anybody's guess. In any case, advocates of pion and other convolution based models ignore any p^2 dependence of the constituents' quark distributions.

The assumptions leading to a convolution model are arguable at best and probably can only be supported on a case by case basis. Despite this, I believe they have substantial value as qualtitative guides to nuclear effects and especially as a formalism for treating Fermi motion and other "trivial" sources of nuclear modifications of structure functions.

Here is a convenient summary of the convolution model formulas:

$$\begin{aligned} f_{a/A}(x_A) &= \sum_T f_{a/T/A}(x_A) \\ &= \sum_T \int dy_A dz_T\ \delta(x_A - y_A z_T) f_{a/T}(z_T) f_{T/A}(y_A) \end{aligned} \tag{4.9}$$

$$\begin{aligned} f_{a/T}(z_T) &= \frac{dP_{a/T}}{dz_T}, \quad 0 < z_T < 1, \quad z_T = k^+/p^+, \\ f_{T/A}(y_A) &= \frac{dP_{T/A}}{dy_A}, \quad 0 < y_A < 1, \quad y_A = p^+/P^+, \end{aligned} \tag{4.10}$$

$$\int_0^1 dz_T f_{a/T}(z_T) = N_{a/T} \qquad \int_0^1 dy_A f_{T/A}(y_A) = N_{T/A}$$
$$\int_0^1 dz_T z_T f_{a/T}(z_T) = \epsilon_{a/T} \qquad \int_0^1 dy_A y_A f_{T/A}(y_A) = \epsilon_{T/A} \tag{4.11}$$
$$\int_0^1 dx_A f_{a/A}(x_A) = \sum_T N_{a/T} N_{T/A}$$
$$\int_0^1 dx_A x_A f_{a/A}(x_A) = \sum_T \epsilon_{a/T} \epsilon_{T/A}$$

where $N_{T/A}$ is the number of constituents of type T in nucleus A and $\epsilon_{T/A}$ is the fraction of the nucleus' P^+ carried by constituents T. It is useful to introduce quark and constituent distributions depending on Bjorken's variable $x = M_A x_A/M$ and $y = M_A y_A/M$. We leave z_T as is since the constituent T is in general not at rest and M_T plays no special role. Thus, $x = \sqrt{2}k^+/M$, $y = \sqrt{2}p^+/M$ and $k^+ = z_T p^+$. We define $F_{a/A}(x)$ as in Eq. (2.30) and $F_{T/A}(y) = \frac{M}{M_A} f_{T/A}(y_A)$. Then, Eq. (4.7) becomes

$$F_{a/T}(x) = \sum_T \int_0^{M_A/M} dy \int_0^1 dz_T \delta(x - yz_T) f_{a/T}(z_T) F_{T/A}(y) \tag{4.12}$$

and the sum rules analogous to Eqs. (4.11) are

$$\int_0^{M_A/M} dx \overline{F}_{a/A}(x) = \sum_T N_{a/T} N_{T/A}/A \tag{4.13}$$

$$\int_0^{M_A/M} dx\, x\overline{F}_{a/A}(x) = \frac{M_A}{MA} \sum_T \epsilon_{a/T} \epsilon_{T/A} \tag{4.14}$$

where $\overline{F}_{a/A} = F_{a/A}/A$ is the quark distribution per nucleon. Throughout this discussion, I have been careful not to make the approximation $M_A \cong MA$. Convolution has some unexpected and important kinematic effects on the quark distribution of some constituents. Consider, for example, the contributions of pions in a model like Erikson and Thomas:[6] $p^0 \approx 0$, but $|\vec{p}| \approx 300$ - 400 MeV, so $y_\pi \sim |\vec{p}| \cos\theta/M$ which is typically less than $\sim 1/3$. The valence quark distribution in the pion is expected on theoretical grounds and found experimentally to be quite "hard", $f_{V/\pi}(z_\pi) \sim (1 - z_\pi)$. Convolution converts this into $F_{V/\pi/A}(x) = \int_x^{M_A/M} \frac{dy}{y} F_{\pi/A}(y)\, f_{V/\pi}(x/y)$ which is negligible for x much larger than 1/3. So the hard pion distribution has been mapped to small x. No such fate befalls the nucleon: $p^0 \approx M$, so $y_N \approx 1 + p\cos\theta/M$ which peaks near $y_N = 1$. Nevertheless, motion does affect the quark distribution in nucleons in the nucleus.

4.2 Fermi Motion

The most obvious source of a nuclear effect in deep inelastic electron scattering comes from the fact that the nucleons in the nucleus are in motion. As a "baseline" model for nuclear targets we assume the nucleus consists exclusively of nucleons and that the quark distribition in the nucleons are the same as in isolation. One might naively think if their kinetic energies are small with respect to ν, the motion could be neglected as $\nu \to \infty$. This is not correct, as we shall see.

We begin with

$$f_{N/A}(y_A) = \int \frac{d^4p}{(2\pi)^4}\, \delta(y_A - p^+/P^+)\chi_{N/A}(p,P) \tag{4.15}$$

where $\chi_{N/A}(p,P)$ is the forward, possibly virtual, nucleon nucleus scattering amplitude defined in analogy to Eq. (2.28).

$$\chi_{N/A}(p,P) = \int d^4\varsigma e^{-ip\cdot\varsigma}\langle P|T(\Phi^+(\varsigma)\Phi(0))|P\rangle_c \tag{4.16}$$

where Φ is a nucleon interpolating field. Φ is not uniquely defined, and when $p^2 \neq M_N^2$ different choices yield different results. This reflects an inherent uncertainty when one attempts to use field-theoretic methods to manipulate composite objects. Spin has been suppressed in Eq. (4.16). If we substitute Eq. (4.16) into Eq. (4.15) and perform as many ς and p integrations as possible we obtain a form analogous to Eq. (2.18):

$$f_{N/A}(y_A) = \sum_n \delta(y_A - 1 + P_n^+/P^+)|\langle n|\Phi|P\rangle|^2 \tag{4.17}$$

where the sum is on all states which can be obtained by removing a single nucleon from the nucleus leaving behind a state with $P_n^+ = (1-y_A)P^+$. The problem is how to calculate $|\langle n|\Phi|P\rangle|^2$. This must be done inclusively, *i.e.*, all possible states $\{|n\rangle\}$ must be included and they must be physical states. A variety of approximations can be made but great care must be taken to ensure that the number and p^+ sum rules (Eqs. (4.11)) remain valid. The most naive approach, to replace the nucleus by an independent particle model in which nucleons occupy energy eigenstates in some potential, obeys neither sum rule and must be altered in some *ad hoc* way before it can be used in this context.

When this form for $f_{N/A}(y_A)$ is inserted into Eq. (4.12) to obtain $F_{a/N/A}(x)$ we obtain an independent nucleon model for $F_{a/A}(x)$ which includes what are generally known as "Fermi motion corrections":

$$\overline{F}_{a/N/A}(x) = \int_x^{M_A/M} \frac{dy}{y} f_{a/N}(x/y)\overline{F}_{N/A}(y) \tag{4.18}$$

$[\overline{F}_{a/N/A} = F_{a/N/A}/A$ and $\overline{F}_{N/A} = F_{N/A}/A.]$ $F_{a/N/A}$ has several important features independent of the explicit form of $\overline{F}_{N/A}(y)$. First, as already noted $\overline{F}_{a/N/A}(x)/F_{a/N}(x)$ diverges as $x \to 1$ so Fermi motion correction to the ratio are large and positive near $x = 1$. The divergence of the ratio is deceptive because both $\overline{F}_{a/N/A}(x)$ and $F_{a/N}(x)$ are very small for $x \sim 1$. Second, Fermi motion corrections cannot change the number of quarks of any flavor: $\int_0^{M_A/M}(\overline{F}_{a/N/A}(x) - F_{a/N}(x)) = 0$, which can be obtained from Eq. (4.13) with $T = N$ and $N_{T/A} = A$. Third, the effect of nuclear binding appears to be to decrease slightly the p^+ carried by the quarks (and antiquarks) even though the quark distribution in the nucleon is not altered. This can be seen from Eq. (4.14): $M_A/MA = 1 - \delta$ where δ is the binding energy per nucleon in units of the nucleon mass, and $\epsilon_{N/A} = 1$ (the nucleons carry all the nucleus P^+ in this simple model), so $\int_0^{M_A/M} dx\, x[\overline{F}_{a/N/A}(x) - F_{a/N}(x)] = -\delta\epsilon_{a/N}$. This goes in the right direction toward explaining the EMC effect but explicit calculations with "realistic" nuclear wave functions fail to get a large enough shift in the valence quark distribution and get the wrong shape (*i.e.* x-dependence) of the effect. Also, the model cannot produce an increase in the ocean quark pairs. Recently, it has been claimed that a model of the form we have been discussing can account for the valence quark part of the EMC effect.[7] In that model, $\epsilon_{N/A} < 1$ so P^+ has been lost, presumably to the constituents responsible for nuclear binding, and the quark content of those constituents has not been included in the calculation.

It is interesting to explore the effect of binding in a simple model. Let us assume the nucleons form a relativistic, degenerate free Fermi gas (FG) with Fermi momentum k_F. Then

$$|\langle n|\Phi|P\rangle|^2 \equiv \frac{dN}{d^3p} = \frac{3A}{4\pi k_F^3}\theta(k_F - |\vec{p}|) \ . \tag{4.19}$$

The constant $3A/4\pi k_F^3$ is chosen so $\int^{k_F} dN = A$. The bound nucleons must have an effective mass $M^* < M_N$ in this model otherwise the sum over the energies of the nucleons would exceed M_A. Substituting into Eq. (4.18) and evaluating the d^3p integral,

$$f^{FG}_{N/A}(y_A) = \frac{3}{4}\frac{AM_A}{k_F}\left(1 - \frac{M_A^2}{{k_F}^2}\left(\frac{y_A^2 - M^{*2}/M_A^2}{2y_A}\right)^2\right) \quad , \quad y_- < y_A < y_+ \ , \tag{4.20}$$

where $y_\pm$ are the values for which $f^{FG}_{N/A}(y_\pm) = 0$. One can check that $f^{FG}_{N/A}(y_A)$ satisfies both $\int_0^1 dy_A f^{FG}_{N/A}(y_A) = A$ and $\int_0^1 dy_A y_A f^{FG}_{N/A}(y_A) = 1$ provided M^* is chosen so the energy of the Fermi gas is M_A. For non-relativistic nucleons $\left(M^* = M_A/A + O(k_F^2/M)\right)$ a quadratic approximation suffices

$$\overline{F}^{FG}_{N/A}(y) = \frac{M}{AM_A} f^{FG}_{N/A}(y_A) \approx \frac{3}{4\lambda}\left(1 - \frac{(y-\eta)^2}{\lambda^2}\right) \quad , \quad \eta - \lambda < y < \eta + \lambda \ , \tag{4.21}$$

where $\lambda = k_F/M$ and $\eta = M_A/MA$ $(\lesssim 1)$. Since $\overline{F}^{FG}_{N/A}(y)$ has maximum height $\sim 1/\lambda$ and width $\sim \lambda$ and since it is convoluted with a smooth function, $f_{a/N}(x/y)$, in Eq. (4.18), it is convenient to approximate it as a generalized function:

$$F^{FG}_{N/A}(y) \cong \delta(y - \eta) + \frac{\lambda^2}{10}\delta''(y - \eta) \tag{4.22}$$

for $\lambda << 1$. Note that $F^{FG}_{N/A}(y)$ in this form satisfies the required sum rules trivially. In particular, $\int_0^{M_A/M} dy\, y F^{FG}_{N/A}(y) = \eta = M_A/MA$ which leads to the decrease in the quark's momentum noted above. Substituting into Eq. (4.18), we obtain

$$\overline{F}_{a/N/A}(x) = \frac{1}{\eta} f_{a/N}(x/\eta) + \frac{\lambda^2}{10} d^2/dy^2 \frac{1}{y} f_{a/N}(x/y)\bigg|_{y=\eta} \ . \tag{4.23}$$

In this model the effects of Fermi smearing are small (for $k_F = 300$ MeV, $\lambda^2/10 \approx 0.01$, for typical nuclei $1 - \eta \lesssim 0.01$). A sample calculation of $\overline{F}_{a/N/A}(x)/F_{a/N}(x)$ is shown in Fig. 17. It clearly cannot account for the EMC effect.

Within the context of convolution models, there are only two alternatives: either other constituents must be present (*e.g.*, pions, Δ's, 6 quark bags, etc.)[3, 4] or the structure function of the nucleons must be modified by the nuclear medium.[5] I will not discuss either alternative in these lectures. The interested reader should consult the references for some work in these

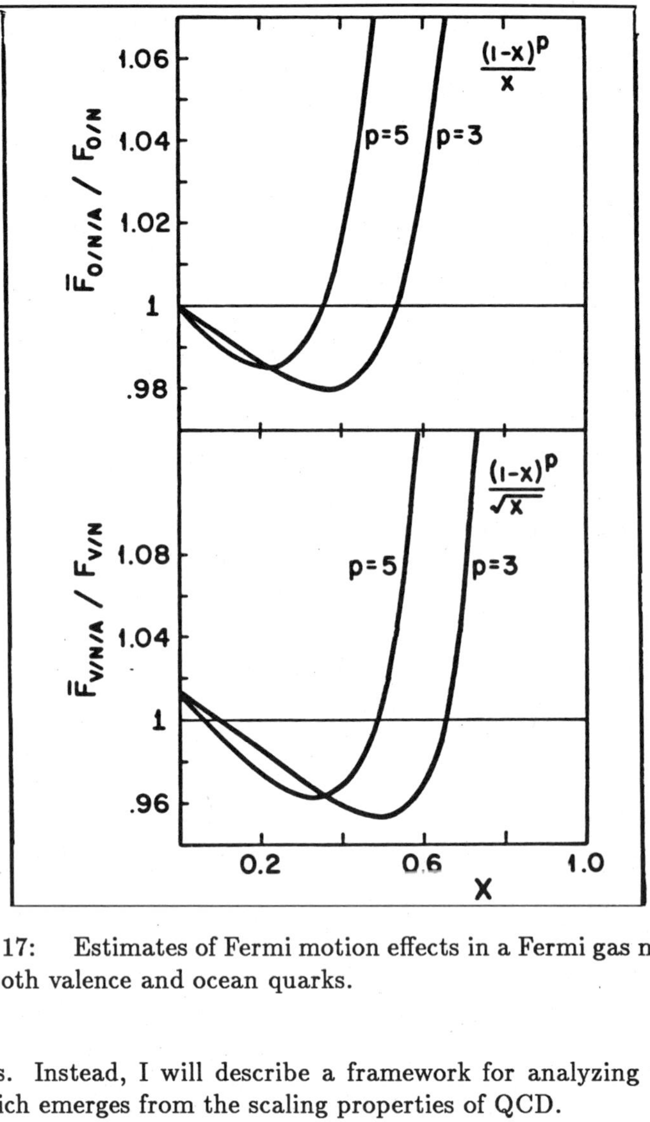

Fig. 17: Estimates of Fermi motion effects in a Fermi gas model for both valence and ocean quarks.

directions. Instead, I will describe a framework for analyzing the EMC effect which emerges from the scaling properties of QCD.

§5. AN INTRODUCTION TO SCALING VIOLATION IN QCD

In interacting field theories the structure functions F_1 and F_2 depend on both x and Q^2 even at very large Q^2. The Q^2-dependence will give us a new handle on the distance scales characterizing the target. To understand the Q^2-dependence we must take another excursion into formalism.

At present, the x-dependence of $F_2(x, Q^2)$ at fixed Q^2 cannot be predicted for any hadronic target in a rigorous fashion. It depends on the non-perturbative dynamics which confines quarks. However, given $F_2(x, Q^2)$ at some large $Q^2 = Q_0^2$, QCD perturbation theory enables one to predict $F_2(x, Q^2)$ at all $Q^2 > Q_0^2$ and at $Q^2 < Q_0^2$ down to some minimum which appears to be of order 1 GeV^2. In QCD, $F_2(x, Q^2)$ depends logarithmically on Q^2 at large Q^2. The logarithmic Q^2-dependence has been verified experimentally and constitutes one of the major quantitive tests of the theory. In addition to $ln\ Q^2$ corrections there are expected to be $0(1/Q^2)$ and higher order corrections which become important at low-Q^2.

In this chapter, I first give a heuristic "derivation" of the logarithmic Q^2-dependence of quark distributions. Next, I will describe some of the formalism behind the Q^2-dependence. Then, I catalogue and discuss the $0(1/Q^2)$ corrections. Finally, I compare the quark description of hadrons in QCD with more naive quark models. This analysis applies equally well to any target, so the target subscripts T and A will generally be suppressed. Until further notice the variable x is x_T with $0 < x_T \leq 1$.

5.1 Logarithmic Scaling Violations

In QCD, quarks are coupled to gluons in much the same fashion that electrons are coupled to photons in QED. The fundamental vertex is shown in Fig. 18a.

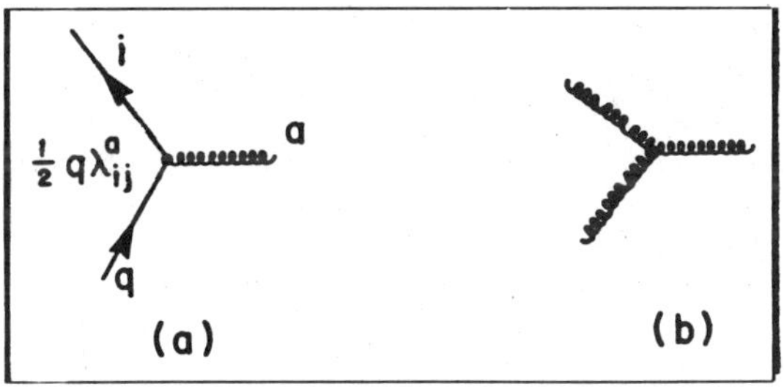

Fig. 18: QCD vertices: a)quark-gluon; b) three gluon.

Instead of the electric charge e, one has $g\lambda^a_{ij}/2$, where g is the QCD coupling ($\alpha_c \equiv g^2/4\pi$). λ^a_{ij} are Gell-Mann's matrices (normalized so $tr[\lambda^a]^2 = 2$) describing the coupling of quark with color i to quark with color j ($i, j = 1, 2, 3$) by emitting a gluon with color a ($a = 1, \ldots 8$). In QED it is well-known that the electric and magnetic fields of a relativistic electron are predominantly transverse and look like the fields of a photon.[1] This is the basis of the Weizsäcker-Williams or "equivalent photon" approximation in QED. A careful study of the analogous process in QCD will lead to the logarithmic scaling violations we seek.[2]

The equivalent photon approximation is usually formulated in a frame in which the electron is extremely relativistic (moving, for definiteness, in the z-direction) – the "infinite momentum" frame. The number of photons associated with the electron depends on the energy of the photon ($E_\gamma = xE_e$) and the impact parameter at which one probes the electron's electromagnetic field:

$$\frac{dN_\gamma}{dx dA} \sim \frac{\alpha}{\pi^2 b^2 x} + \text{terms less singular in b and x} \tag{5.1}$$

where $dA = 2\pi b db$ and the variables are defined in Fig. 19.

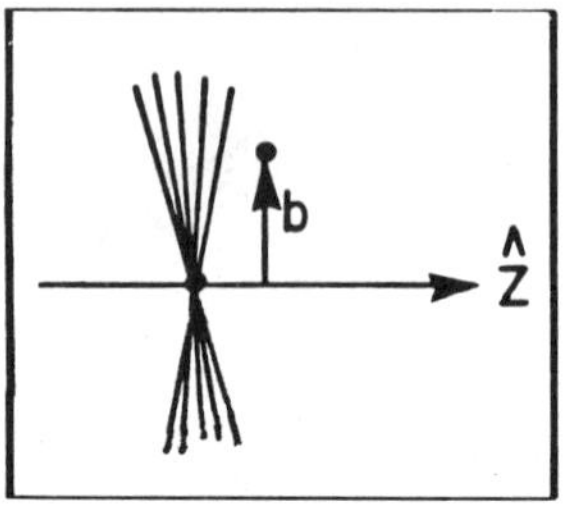

Fig. 19: Weizsäcker-Williams kinematics.

The proper interpretation of Eq. (5.1) is that a measurement sensitive to the intensity of the electron's electromagnetic field performed at impact parameter b and absorbing energy $E_\gamma = xE_e$ could not differentiate between a passing electron and the equivalent number of photons. Eq. (5.1) is derived, for example, by Jackson. The derivation neglects the recoil of the electron, which must lose energy when a photon with $E_\gamma/E_e = x$ is absorbed. This is reasonable for $x \approx 0$. For larger x, one must include recoil and electron spin effects, which requires calculation of Feynman graphs. The lowest order graph corresponding to the measurements I've described is that of Fig. 20a. The calculations can be found in Ref. 5.3. The result is $\frac{1}{x} \to \frac{1}{2x}(1 + (1-x)^2)$. QED and QCD are identical to this order except $e \to g\lambda^a_{ij}$. So the number of gluons (summed over all colors) with momentum fraction x at impact parameter b in a quark (averaged over colors) is

$$\frac{dN_g}{dx db} \sim \frac{2\alpha_c}{\pi b}\frac{1}{3} Tr \sum_a \left(\frac{\lambda^a}{2}\right)^2 \left[\frac{1+(1-x)^2}{2x}\right] \tag{5.2}$$

to leading order in b and α_c. Since

$$\frac{1}{3}\,Tr\,\sum_a\left(\frac{\lambda^a}{2}\right)^2=\frac{4}{3} \tag{5.3}$$

we have

$$\frac{dN_g}{dxdb}=\frac{8\alpha_c}{3\pi b}\left[\frac{1+(1-x)^2}{2x}\right]\ . \tag{5.4}$$

If one probes a quark (or electron) by delivering a momentum transfer $Q_\perp=\sqrt{Q^2}$, then $b_{\min}=1/\sqrt{Q^2}$ plays the role of a resolution: all gluons (or photons in QED) with $b\geq b_{\min}$ will appear distinct from the quark (electron), but the rest ($b<b_{\min}$) will be lumped into what one calls the quark (electron) when probed with that resolution. This is evident, for example, in the Weizsäcker-Williams calculation of bremsstrahlung: if a very high energy electron scatters (*e.g.*, from a heavy nucleus) with momentum transfer $q^2\approx -Q_\perp^2$ then the energy radiated as bremsstrahlung (per unit x) is $\approx x\ E_e\int_{b_{\min}}^{\infty}db\frac{dN_\gamma}{dxdb}$ with $b_{\min}=1/\sqrt{Q_\perp{}^2}$. Thus,

$$\frac{dN_g}{dxd\ ln\ Q^2/Q_0^2}=\frac{2\alpha_c}{3\pi}\frac{1+(1-x)^2}{x} \tag{5.5}$$

is the rate of change with $t\equiv ln\ Q^2/Q_0^2$ of the probability to find a gluon of momentum fraction x in a quark. [Note, Q_0^2 is an arbitrary scale for the logarithm.] In the following, I will use the variables t and Q^2 interchangeably. So we define $P_{g/q}(x,t)=dN_g/dxdt$. [$P_{g/q}$ appears to be independent of t, but in QCD α_c depends logarithmically on t]. When a quark radiates a gluon it leaves behind a quark of lower-momentum. So we can determine the rate of change with t of the probability that a measurement on a quark with energy E_q will detect a quark with $E_q'=xE_q$:

$$P_{q/q}(x,t)=\frac{dN_g(1-x)}{dxdt}=\frac{2\alpha_c}{3\pi}\frac{1+x^2}{1-x} \tag{5.6}$$

corresponding to the Feynman graphs of Fig. 20b.

The three gluon coupling of QCD shown in Fig 18b leads to a probability for a gluon to contain gluons of lower momentum, and the $q\overline{q}g$ vertex of Fig 18a leads to a probability for a gluon to contain quark-antiquark pairs. If we confine our attention to valence quarks we need not study these other processes since they produce only ocean $q\overline{q}$ pairs. Eqs. (5.5) and (5.6) lead to the notion of a quark distribution "evolving" with increased $t\sim ln\ Q^2$. As $ln\ Q^2$ grows, valence quarks emit gluons, gluons

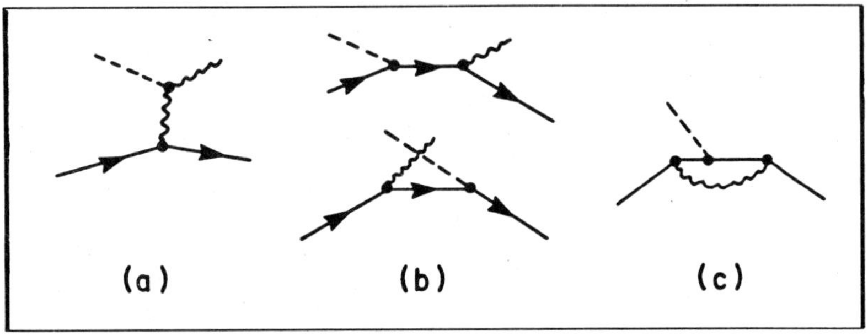

Fig. 20: Feynman diagrams corresponding to Weizsäcker-Williams processes: a) An electron (quark) radiating a photon (gluon); b) an electron (quark) "radiating" an electron (quark); c) Vertex correction required at $x = 1$.

in turn emit quark-antiquark pairs and more gluons. This "evolutionary" view of the Q^2 dependence of quark (and gluon) distributions was first suggested by Kogut and Susskind[4] and was developed for QCD by Altarelli and Parisi.[2]

The idea that the definition of a quark or gluon depends upon the mass or distance scale at which one probes it is a familiar one in quantum field theory. It is related to renormalization. A quark (or electron) propagating in isolation couples to quantum fluctuations in the vacuum. When we perform some measurement on the quark we lump some of the fluctuations (and all of the very short distance ones which lead to divergences) into the definition of the quark and treat the remainder as radiative corrections. To do this we must introduce a mass (or length) scale (μ) known as the "renormalization point" into the theory, which distinguishes, roughly speaking, those fluctuations we treat explicitly ($\Delta x > \mu^{-1}$) from those we incorporate into the definition of the particles ($\Delta x < \mu^{-1}$). It is well-known that μ is arbitrary – nothing physical depends on it – but the scale invariance of QCD allows us to trade μ^2-dependence for Q^2-dependence in a fashion which will be outlined briefly below (§5.2).

Up to now, we've worked in the infinite momentum frame. x measures the z-component of a daughter particle's momentum relative to its parent: $p^3_\infty = xP^3_\infty$. The subscript "∞" denotes quantities measured in the infinite momentum frame. The particles are not far from mass shell and have limited transverse momenta (we can choose $P^3_\infty >> \sqrt{Q^2}$) so $p^0_\infty \cong xP^3_\infty$

and

$$p_{\infty}^{+} = xP_{\infty}^{+} \ . \tag{5.7}$$

The ratio p^{+}/P^{+} is invariant under Lorentz transformations along the z-axis. This follows from the Lorentz transformation in the form $p^{\pm\,'} = e^{\pm y}p^{\pm}, \vec{p}\,'_{\perp} = \vec{p}_{\perp}$, where $\beta = \tanh y$ is the relative velocity of frames along the z-axis. Thus, Eq. (5.7) implies

$$p^{+} = xP^{+} \tag{5.8}$$

in the laboratory, and we can identify x as the variable we've been using all along.

$P_{q/q}(x,t)$ appears to be singular at $x = 1$. This presents a problem because we would like the number of valence quarks in a target to be independent of Q^2:

$$\frac{d}{dt}\int_0^1 dx \frac{dN_q}{dx} = \int_0^1 dx P_{q/q}(x,t) = 0 \tag{5.9}$$

with Eq. (5.6) as it stands, the integral diverges. The resolution of this problem requires methods outside the scope of these lectures.[5] The proper prescription is to interpret $1/1-x$ as a distribution: $1/1-x \rightarrow (1/1-x)_{+}$ defined by

$$\int_0^1 dx \frac{f(x)}{(1-x)_{+}} \equiv \int_0^1 dx (f(x) - f(1))/(1-x) \tag{5.10}$$

and to add a δ-function to $P_{q/q}(x,t)$ so that Eq. (5.9) is satisfied.

Since $\int_0^1 dx(1+x^2)/(1-x)_{+} = -\frac{3}{2}$ we find

$$P_{q/q}(x,t) = \frac{2\alpha_c}{3\pi}\left(\frac{1+x^2}{(1-x)_{+}} + \frac{3}{2}\delta(x-1)\right) \tag{5.11}$$

The rate of change of the quark distributions in a target will be determined by the functions $P_{q/q}(x,t)$ and $P_{q/g}(x,t)$ and by the quark and gluon distributions at t: A quark or gluon with momentum fraction y will be observed to consist of quarks with momentum fraction zy according to $P_{q/q}(z,t)$ and $P_{q/g}(z,t)$ respectively. Clearly, evolution is described mathematically by convolution in precise analogy to §4.2:

$$\frac{df_{N.S.}(x,t)}{dt} = \frac{\alpha_c}{2\pi}\int_x^1 \frac{dy}{y}\overline{P}_{q/q}(x/y) f_{N.S.}(y,t) \tag{5.12}$$

where $P_{q/q} \equiv \frac{\alpha_c}{2\pi}\overline{P}_{q/q}$. The subscript $N.S.$ denotes non-singlet and indicates a quark distribution with non-trivial flavor quantum numbers such as the valence quark distribution of §2. A non-singlet distribution is one which cannot be populated by the quark-antiquark pairs evolved from gluons. Singlet distributions, on the other hand, satisfy coupled integro-differential equations in which the gluon distribution appears as well.

To describe the evolution of $F_2(x, Q^2)$ it is necessary to solve these differential equations involving valence and ocean quark distributions and a gluon distribution $f_g(x,t)$. Starting values of all distributions at $t = 0$, *i.e.* $Q^2 = Q_0^2$, are required as input. The general features of the t-dependence are clear from the Weizäcker-Williams approximation. As t increases, valence quarks lose momentum by emitting gluons – *i.e.*, probed with higher resolution more of a quark momentum appears to be carried by its gluon field. Gluons, in turn, lose momentum to quark-antiquark pairs. The net result, with increasing t, is a transfer of the valence quark momenta to newly created pairs. The gluons are caught in the middle – valence quarks emit gluons but the gluons turn into $q\overline{q}$ pairs. Not surprisingly, the process saturates: at very large $ln\ Q^2$ the fraction of the momentum of any target carried by gluons saturates at ≈ 0.47.[6] The Q^2-dependence of $F_2(x, Q^2)$ is illustrated schematically in Fig. 21. Close, Roberts and Ross were struck by the similarity of Fig. 21 to the shape of the EMC effect at fixed Q^2, and were led to the "rescaling" analysis which is the subject of the next chapter.

Integrodifferential equations like Eq. (5.12) are not easily solved directly. We can gain considerable insight, though, by taking moments in x on both sides and using the special properties of convolutions. Let

$$M^n_{N.S.}(t) \equiv \int_0^1 dx\ x^{n-1} f_{N.S.}(x,t) \quad . \tag{5.13}$$

Then, it is easy to show, from Eq. (5.12), that

$$\frac{dM^n_{N.S.}}{dt} = \frac{\alpha_c}{2\pi} B_n M^n_{N.S.}(t) \tag{5.14}$$

where

$$B_n = \int_0^1 dz\ z^{n-1}\overline{P}_{q/q}(z) \quad . \tag{5.15}$$

Eq. (5.14) is easier to solve. The character of the solution depends on the t-dependence of α_c, which I have not yet specified. In quantum field theories, the effective coupling has a definite and calculable t-dependence.[7]

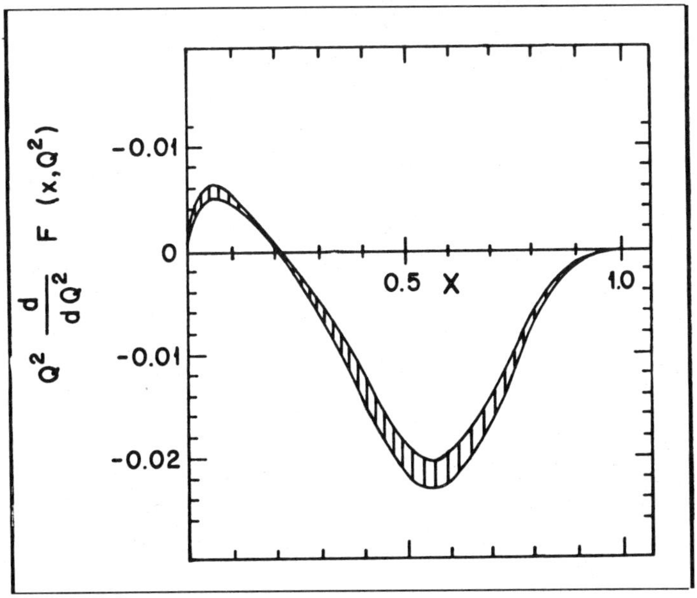

Fig. 21: Q^2-dependence of the structure function $\left(dF_2(x,Q^2)/d\ ln\ Q^2\right)$ at $Q^2 = 3.2\ \mathrm{GeV}^2$ using $F_2(x,Q^2)$ from SLAC data.[5.22]

In free field theory $\alpha_c = 0$ and $M^n_{N.S.}$ is constant – corresponding to exact Bjorken scaling. In some model field theories the effective coupling becomes constant at large t, $\lim_{t\to\infty} \alpha_c(t) \to \alpha_0$.[8] This behavior is known as an "ultraviolet fixed point". In such a theory, the solution of Eq. (5.14) at large Q^2 is

$$M^n_{N.S.}(Q^2) \sim M^n_{N.S.}(Q_0^2)\left(\frac{Q^2}{Q_0^2}\right)^{\frac{\alpha_0 B_n}{2\pi}} \quad , \tag{5.16}$$

i.e., scaling is violated by powers of Q^2. The coefficient B_n is known as an "anomalous dimension". In QCD, the effective coupling vanishes as $t \to \infty$ but it vanishes too slowly to use the approximation $\alpha_c = 0$ at large t: $\alpha_c(t)$ can be shown to have an expansion of the form

$$\frac{\alpha(0)}{\alpha_c(t)} = 1 + b\alpha_c(0)t + 0\left(\alpha_c(0)^2\right) \quad , \tag{5.17}$$

or to leading order,

$$\alpha_c(t) \sim \frac{\alpha_c(0)}{1 + bt\alpha_c(0)} \quad . \tag{5.18}$$

The coefficient b is

$$b = \frac{1}{4\pi}\left(11 - \frac{2}{3}N_f\right) \quad , \tag{5.19}$$

where N_f is the number of quark flavors with masses small compared with $\sqrt{Q^2}$, in practice $N_f \sim 3 - 4$. Eq. (5.18) can be rewritten

$$\alpha_c(Q^2) \sim \frac{1}{b\ ln\ Q^2/\Lambda^2} \tag{5.20}$$

where $\Lambda^2 = Q_0^2 \exp(-1/b\alpha_c(0))$. This well-known behavior of $\alpha_c(Q^2)$ is known as asymptotic freedom. It is very special to non-Abelian gauge field theories and it makes it possible to calculate with QCD at large Q^2.[9] If $\alpha_c(t)$ behaves as in Eq. (5.18), the moments evolve logarithmically:

$$M^n_{N.S.}(t) = M^n_{N.S.}(0)\left(\alpha_c(0)/\alpha_c(t)\right)^{B_n/2\pi b} \tag{5.21}$$

A more familiar form of Eq. (5.21) is obtained by changing variables to Q^2 and taking the logarithm

$$ln\ M^n_{N.S.}(Q^2) \sim ln\ M^n_{N.S.}(Q_0^2) + \frac{B_n\alpha_c(Q_0^2)}{2\pi b}\ ln\ Q^2/Q_0^2 \tag{5.22}$$

where we've replaced $ln\ \left(1 + bt\alpha_c(0)\right) \sim bt\alpha_c(0)$ since we are working to lowest order in α_c. The coefficients $\{B_n\}$, still known as (non-singlet) anomalous dimensions, can be computed from Eqs. (5.11) and (5.15)

$$B_n = \frac{4}{3}\left\{\frac{1}{n(n+1)} - 2\sum_{j=2}^{n}\frac{1}{j} - \frac{1}{2}\right\} \tag{5.23}$$

B_1 is zero, which corresponds to the fact that the number of valence quarks does not change with Q^2. All the others are negative, so all other moments decrease monotonically with Q^2 at large Q^2.

In principle, a particularly simply way to test QCD is to plot the moments of the valence quark distribution versus $ln\ Q^2$: the slope is predicted by QCD up to a single parameter, $\alpha(Q_0^2)$ or Λ. In practice, the situation is quite complicated: $F_2(x, Q^2)$ is not well-measured at all x, so moments are hard to compute; non-singlet quark distributions must be extracted from

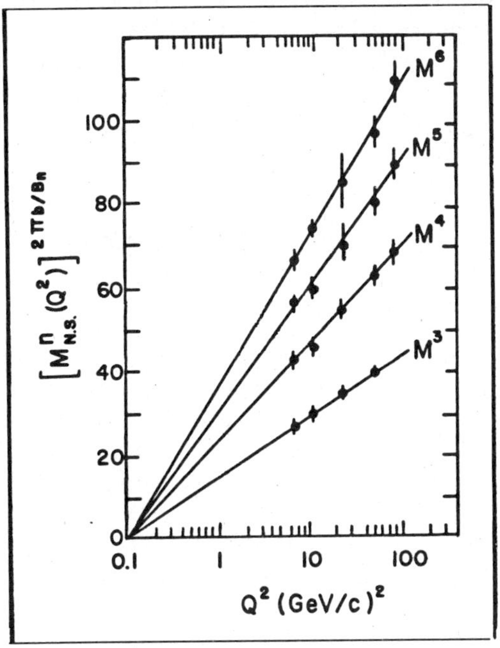

Fig. 22: Testing QCD in inelastic lepton scattering. At large $Q^2[M^n]^{2\pi b/Bn}$ lies on a straight line with a predicted slope when plotted versus $ln\ Q^2$.

$F_2(x, Q^2)$ in order to apply the simple analysis I've described; $\alpha_c(t)$ is not so small where there is good data (remember $d\sigma/dE'd\Omega$ falls like $1/Q^4$) so higher corrections to Eq. (5.17) must be included at least at the low-t end. Nevertheless, the moments have been extracted and compared with theory and the agreement is excellent as shown in Fig. 22.

For our purposes, it is important to remember that the logarithms of the moments depend linearly on $ln\ Q^2$ and that the slope is negative and independent of the target. To get more insight into the Q^2-dependence of deep inelastic structure functions it is necessary to return to coordinate space.

5.2 The Operator Product Expansion and Scaling Violations

In §2 we derived the parton model by studying the spacetime dependence of the product of currents. This gave us considerable insight into the relation between the x-dependence of $F_2(x)$ and the quark correlation function in the target ground state. Now I want to look for similar insight into the Q^2-dependence we've found in QCD and to make sure our earlier insight is not spoiled by the interactions we've added [it isn't].

The starting point is Wilson's operator product expansion[10] (OPE). The idea is that the product of operators simplifies in the limit that their arguments coincide:

$$\lim_{\xi^\mu \to 0} A(\xi)B(0) \sim \sum_{\{\beta\}} C_{\{\beta\}}(\xi)O_{\{\beta\}}(0) \tag{5.24}$$

where $C_{\{\beta\}}$ are C-number functions and $O_{\{\beta\}}$ are finite local operators. $C_{\{\beta\}}$ may depend on any other parameters in the theory such as masses and coupling constants, in addition to ξ; $\{\beta\}$ are all labels which might occur in such an expansion (Lorentz and internal symmetry indices, etc.). The content of Eq. (5.24) is that the singularities are factored out from the operators and that the terms in the expansion can be organized in decreasing order of singularity as $\xi_\mu \to 0$. Eq. (5.24) is useful only for $\xi^\mu \approx 0$, the short distance limit. This will translate into information about the Compton amplitude in the limit $Q^2 \to \infty$ $\underline{\text{and } \omega \to 0}$ ($\omega = 1/x$). Deep inelastic scattering probes the light-cone and requires $\omega \geq 1$, but the dispersion relations discussed in §1.2 will provide the necessary connection between the two regimes.

It is easy to check that Eq. (5.24) works in free field theories. For example,

1. In a free, massless scalar field theory

$$\begin{aligned} \phi(\xi)\phi(0) &= \langle 0|\phi(\xi)\phi(0)|0\rangle + :\phi(\xi)\phi(0): \\ &= \frac{i}{4\pi^2(-\xi^2 + i\epsilon\xi^0)} I \\ &\quad + \sum_{n=0}^{\infty} \frac{\xi^{\mu_1}\ldots\xi^{\mu_n}}{n!} :\phi(0)\overleftarrow{\partial}_{\mu_1}\ldots\overleftarrow{\partial}_{\mu_n}\phi(0): \end{aligned} \tag{5.25}$$

   I, the identity, and $:\phi(0)\ldots\phi(0):$ are the regular operators. Their "coefficient functions" are C-numbers. Note that the first term in Eq. (5.25) diverges as $\xi_\mu \to 0$ but the rest vanish with successively higher powers of ξ_μ.

2. In a free massless Dirac theory let $J(\xi) =: \overline{\psi}(\xi)\psi(\xi):$, then in analogy

to Eq. (2.1) *et seq.*

$$
\begin{aligned}
[J(\xi), J(0)] = \frac{1}{2\pi}\left[\partial^{\rho}\delta(\xi^2)\epsilon(\xi^0)\right] \sum_{n=0}^{\infty} \frac{\xi^{\mu_1}\ldots\xi^{\mu_n}}{n!} : \\
\overline{\psi}(0)\overleftarrow{\partial}_{\mu_1}\ldots\overleftarrow{\partial}_{\mu_n}\gamma_{\rho}\psi(0) \\
-\overline{\psi}(0)\partial_{\mu_1}\ldots\partial_{\mu_n}\gamma_{\rho}\psi(0) :
\end{aligned}
\qquad (5.26)
$$

In a free field theory, the Taylor expansions in Eq's. (5.25) and (5.26) can of course be summed to give a "bilocal operator" like : $\overline{\psi}(\xi)\gamma_{\rho}\psi(0)$: but in interacting theories the coefficient function will differ slightly for each term in the Taylor expansion preventing resummation. This, it will turn out, makes the difference between exact Bjorken scaling as in free field theory, and logarithmic scaling violation as in QCD. It is clear from these examples that the form the operator product expansion takes in a particular field theory depends upon the procedure necessary to remove the singularity which arises when one tries to bring operators to the same space time point. In free field theories the singular piece is a C-number and can be isolated by normal ordering the operator product. In interacting field theories, the divergences are worse and normal ordering does not suffice. Instead, it is necessary to renormalize the operators. There is a certain arbitrariness in renormalization: to define finite operators it is necessary to introduce a mass-scale μ^2 – loosely speaking, it is the scale at which the operator's matrix elements have the values they would have in free field theory – but μ^2 is arbitrary, nothing physical can depend on it. Nevertheless, both the operators $O_{\{\beta\}}$ and the coefficient functions $C_{\{\beta\}}$ will separately depend on the mass scale introduced by the neccessity of renormalization. To allow for this we replace $C_{\{\beta\}}(\xi)$ by $C_{\{\beta\}}(\xi,\mu^2)$ and $O_{\{\beta\}}(0)$ by $O^{(\mu^2)}_{\{\beta\}}(0)$. In anything physically measureable the μ^2 dependence of $C_{\{\beta\}}$ will cancel that of $0_{\{\beta\}}$. μ^2 is known as the "renormalization point".

To make use of the OPE, it is convenient to make explicit the factors of ξ_{μ} which accompany an operator carrying Lorentz indices. To this end we rewrite Eq. (5.24) as

$$
A(\xi)B(0) \sim \sum_{\{\beta\}} C'_{\{\beta\}}(\xi^2,\mu^2)\xi_{\mu_1}\ldots\xi_{\mu_{n_\beta}} O^{(\mu^2)\mu_1\ldots\mu_{n_\beta}}_{\{\beta\}}(0) \quad . \qquad (5.27)
$$

The operators $O_{\{\beta\}}$ can always be defined so they are symmetric traceless in all Lorentz indices: $0^{\mu_1\ldots\mu_i\ldots\mu_j\ldots\mu_{n_\beta}} = 0^{\mu_1\ldots\mu_j\ldots\mu_i\ldots\mu_{n_\beta}}$ and $g_{\mu_i\mu_j}0^{\mu_1\ldots\mu_i\ldots\mu_j\ldots\mu_{n_\beta}} = 0$.[11] Then, n_β is called the "spin" of the operator.[12]

If A and B carry Lorentz indices then the form of Eq. (5.27) becomes more complicated. To avoid writing complicated equations, I will suppress the indices on the currents J_μ and J_ν or equivalently study deep inelastic scattering by a particle coupled to a hypothetical scalar current $J(\xi) = \overline{\psi}(\xi)\psi(\xi)$.

Let us expand the product $T\big(J(\xi)J(0)\big)$ in the fashion of Eq. (5.27) and calculate the contribution to $T(q^2,\omega)$ (see Eq. (1.12)). We must carry out the Fourier transform:

$$\tilde{C}_{\mu_1\ldots\mu_{n_\beta}}(q,\mu^2) \equiv \int d^4\xi e^{iq\cdot\xi} C'_{\{\beta\}}(\xi^2,\mu^2)\xi_{\mu_1}\ldots\xi_{\mu_{n_\beta}} \quad . \tag{5.28}$$

The Lorentz indices on $\tilde{C}$ can only be in the form of q_{μ_k} or $g_{\mu_i\mu_j}$. The latter vanish when contracted with the traceless operator $O_{\{\beta\}}$. In effect, then,

$$\tilde{C}_{\mu_1\ldots\mu_{n_\beta}}(q,\mu^2) = \frac{q_{\mu_1}\cdots q_{\mu_{n_\beta}}}{(q^2)^{n_\beta}}(-1)^{n_\beta}\tilde{C}_{\{\beta\}}(q^2,\mu^2) \quad . \tag{5.29}$$

The phase and the factors of $(q^2)^{n_\beta}$ have been introduced for later convenience. Note that $\tilde{C}_{\{\beta\}}(q^2,\mu^2)$ has the same dimension as $\int d^4\xi\ e^{iq\cdot\xi}C'_{\{\beta\}}(\xi^2,\mu^2)$. Now, $T(q^2,\omega)$ can be written as

$$T(q^2,\omega) \sim \sum_{\{\beta\}} \tilde{C}_{\{\beta\}}(q^2,\mu^2)(-1)^{n_\beta}\frac{q_{\mu_1}\cdots q_{\mu_{n_\beta}}}{(q^2)^{n_\beta}}\langle p|O_{\{\beta\}}^{(\mu^2)\mu_1\ldots\mu_{n_\beta}}|p\rangle_c \tag{5.30}$$

The matrix element in Eq. (5.30) must carry Lorentz indices $\mu_1\ldots\mu_{n_\beta}$. We can write

$$\langle p|O_{\{\beta\}}^{(\mu^2)\mu_1\ldots\mu_{n_\beta}}|p\rangle = \Theta_{\{\beta\}}(\mu^2)(p^{\mu_1}\ldots p^{\mu_{n_\beta}}+\ldots) \quad , \tag{5.31}$$

where the other terms are determined by the fact that $O_{\{\beta\}}$ is symmetric and traceless. They all contain at least one factor of $g^{\mu_i\mu_j}$. Thus, for example, if $n_\beta = 2$, one has $p^{\mu_1}p^{\mu_2} - \frac{M_T^2}{4}g^{\mu_1\mu_2}$. Combining Eqs. (5.30) and (5.31) we find

$$T(q^2,\omega) \sim \sum_{\beta}\tilde{C}_{\{\beta\}}(q^2,\mu^2)\Theta_{\{\beta\}}(\mu^2)\left[\left(\frac{\omega}{2}\right)^{n_\beta} + 0\left(\frac{1}{q^2}\right)\right] \quad , \tag{5.32}$$

($\omega \equiv -2p\cdot q/q^2$) because $p^\alpha p^\beta q_\alpha q_\beta/(q^2)^2 = \omega^2/4$ but $g^{\alpha\beta}q_\alpha q_\beta/(q^2)^2 = 1/q^2$. In fact, only even powers of ω occur in Eq. (5.32) because $T(q^2,\omega)$ is crossing symmetric (see §1.2).

Since ω ($\equiv 1/x$) scales in the Bjorken limit, the importance of any particular operator $O_{\{\beta\}}$ is determined by the large-q^2 behavior of $\tilde{C}_{\{\beta\}}(q^2,\mu^2)$. This can be calculated to all orders in perturbation theory using the methods of the renormalization group. The first step is to determine the dimension of $\tilde{C}_{\{\beta\}}(q^2,\mu^2)$. We define the "naive" or "canonical" dimension (or, for short, simply dimension) of an operator to be the units in which it is measured as a power of mass (with $\hbar = c = 1$). We use the notation $[O_{\{\beta\}}] = d_{\{\beta\}}$. Here are some examples: $[J_\mu] = 3$ because $\int d^3x J_0 =$ charge which is dimensionless; $[\phi] = 1$ because the action, $S = \int d^4x \frac{1}{2}(\partial_\mu \phi)^2$, is dimensionless; likewise $[\psi] = \frac{3}{2}$. Using these dimensions and remembering $|p\rangle$ is covariantly normed, $[|p\rangle] = -1$, we see that $T(q^2,\omega)$ is dimensionless. Let us suppose the operator $O_{\{\beta\}}$ has dimension d_β, then from Eq. (5.27),

$$[C'_{\{\beta\}}] = 6 - d_\beta - n_\beta \tag{5.33}$$

and

$$[\tilde{C}_{\{\beta\}}(q^2,\mu^2)] = 2 - d_\beta + n_\beta \ . \tag{5.34}$$

The dimensions of $\tilde{C}_{\{\beta\}}$ could, in principle, be provided by q^2, μ^2 or quark masses. Weinberg's theorem[13] can be used to show that at large $-q^2$ to each order in perturbation theory quark masses can be ignored provided μ^2 is fixed and not zero. Also, in each order of perturbation theory the μ^2 dependence of a renormalizable field theory is at most logarithmic, reflecting at worst logarithmic divergences in such theories. So we can write

$$\lim_{q^2 \to -\infty} \tilde{C}_{\{\beta\}}(q^2,\mu^2) = \left(\frac{1}{q^2}\right)^{d_\beta - n_\beta - 2} c_{\{\beta\}}(ln\ q^2/\mu^2) \ . \tag{5.35}$$

The next step is to calculate the leading large $-q^2$ dependence of $c_{\{\beta\}}(ln\ q^2/\mu^2)$ to all orders in perturbation theory. In each order, $c_{\{\beta\}}$ grows like a power of $ln\ q^2/\mu^2$. When summed to all orders the logarithms may yield a power, *i.e.*, $c_{\{\beta\}}$ may depend exponentially on its argument. The actual calculation of the asymptotic behavior of $c_{\{\beta\}}$ requires renormalization group methods beyond the scope of these lectures. But as will be seen below, we have performed an equivalent calculation using Weizsäcker-Williams methods in §5.1. If $c_{\{\beta\}}$ does not go like an exponential of its argument, then operators with $d_\beta - n_\beta = 2$ give rise to Bjorken scaling modulo powers of $ln\ q^2$. $d_\beta - n_\beta$ plays such a central role in the analysis of large q^2 effects that it is given a name, "twist",

$$t_\beta = d_\beta - n_\beta \ . \tag{5.36}$$

Operators with $t_\beta < 2$ would give contributions which diverge in the Bjorken limit, but the only operator with $t_\beta < 2$ which can couple to the product of the two currents is the identity, $t_I = 0$, and the identity has no connected matrix elements. Operators with $t_\beta > 2$ give contributions to $T(q^2\omega)$ which vanish in the Bjorken limit (provided $c_{\{\beta\}}$ is not exponential in its argument). Only even twists occur in the expansion of two currents in the limit of zero quark mass[14] so the next important case is $t_\beta = 4$. Twist-4 or $0(1/q^2)$ corrections to scaling are a rich and fascinating, if technically complicated subject, in themselves.[15]

The twist-two operators in QCD come in two classes: quark operators

$$O_{n,a}^{(\mu^2)\mu_1\ldots\mu_n} = S\overline{\psi}_a\gamma^{\mu_1}\partial^{\mu_1}D^{\mu_2}\ldots D^{\mu_n}\psi_a{}^{(\mu^2)} \quad , \tag{5.37}$$

where D_μ is the (color) gauge covariant derivative,[16] "a" labels flavor and S symbolizes the operation of making $O_{n,a}$ traceless and symmetric; and gluon operators which we needn't write out, since they contribute only to singlet distribution functions. It is easy to check that $O_{n,a}$ indeed has $t = 2$ for all n. It should be emphasized that it requires an infinite tower of operators of increasing spin to describe $T(q^2, \omega)$. This is not at all surprising. First of all, if the sum in Eq. (5.30) stopped at some $n_{\max}$, $T(q^2, \omega)$ would be a polynomial in ω with no cut on the the real axis for $|\omega| > 1$. Second, we know that deep inelastic electroproduction probes the light cone, not just short distances. The OPE is a short distance expansion ($\xi_\mu \to 0$) and no finite number of terms in the short distance expansion gives information about light-like separations.

To summarize: the OPE approach leads to a simultaneous expansion of $T(q^2, \omega)$ in q^2 and ω. The expansion in q^2 is an asymptotic expansion and is ordered by the twist quantum number of the operators, the expansion in ω is a Taylor expansion (which converges for $|\omega| < 1$, see §1) and is ordered by the spin of the operators. Ignoring gluon operators – which is adequate if we are interested in non-singlet quark distributions alone – we have

$$T_{N.S.}(q^2, \omega) = \sum_{\text{n even}} c^n_{N.S.}(ln\ q^2/\mu^2)\Theta^n_{N.S.}(\mu^2)\omega^n + \text{higher twist} \quad , \tag{5.38}$$

where we have replaced the generic label β by the spin (n) of the twist-two, non-singlet quark operator. $c^n_{N.S.}(q^2/\mu^2)$ depends only on $ln\ q^2/\mu^2$ and explicit calculation shows that it equals 4 when $q^2 = \mu^2$. Comparing Eq. (5.38) with the Taylor expansion of $T(q^2, \omega)$ developed in §1, (Eq. 1.18), we identify

$$M^n_{N.S.}(q^2) = \frac{1}{4}c^n_{N.S.}(ln\ q^2/\mu^2)\Theta^n_{N.S.}(\mu^2) \text{ for n even} \quad . \tag{5.39}$$

At this point, we can make a connection with the "evolutionary" approach of §5.1 and comment on the derivation of the parton model in QCD. The ordering of contributions at large Q^2 by twist has given a result, Eq. (5.38), which looks like scaling, modulo logarithms. This is deceptive because in general $c^n_{N.S.}$ depends expontentially on its argument, giving rise to power law violations of scaling. Only in asymptotically free theories like QCD does $c^n_{N.S.}$ go like a power of its argument, giving scaling up to logarithms. The derivation of the dependence of $c^n_{N.S.}$ on $ln\ q^2/\mu^2$ is outside the scope of these lectures, but the result,

$$c^n_{N.S.}(ln\ Q^2/Q_0^2) = \left(\alpha_c(0)/\alpha_c(t)\right)^{B_n/2\pi b} \tag{5.40}$$

(where $\mu^2 \equiv -Q_0^2$) should not be surprising since it converts Eq. (5.39) into Eq. (5.21) which we already derived using the more heuristic, Weizäcker-Williams approach. Comparing Eqs. (5.39), (5.40) and (5.21) we see that the moments of the structure functions are directly related to the matrix elements of specific local operators, $M^n_{N.S.}(q^2) = \Theta^n_{N.S.}(q^2)$ where q^2 is the mass-scale at which the operator is $O^{\mu_1 \dots \mu_n}_{n,N.S.}$ is renormalized. The quark operators $O_{n,a}$ which determine the moments are the (gauge invariant) terms in the Taylor expansion of the operator product $\overline{\psi}(\xi)\gamma_\rho\psi(0)$ which determined the quark distribution function in the parton model. So we see that the modification of the parton model required by the interactions in QCD is that each moment of the quark distribution (*i.e.* each term in the Taylor expansion) scales modulo a slightly different power of $ln\ q^2$. Earlier, I remarked that the renormalization point, μ^2, was arbitrary, that nothing physical could depend on it. There is no contradiction here: Eq. (5.39) is independent of μ^2, specifically

$$\begin{aligned}\Theta^n_{N.S.}(\mu^2)d/d\mu^2\ c^n_{N.S.}(ln\ q^2/\mu^2) \\ = -c^n_{N.S.}(ln\ q^2/\mu^2)d/d\mu^2\Theta^n_{N.S.}(\mu^2)\ .\end{aligned} \tag{5.41}$$

But, $c^n_{N.S.}$ depends only on $ln\ q^2/\mu^2$, so

$$d/d\mu^2 c^n_{N.S.}(ln\ q^2/\mu^2) = -d/dq^2 c^n_{N.S.}(ln\ q^2/\mu^2) \tag{5.42}$$

and therefore

$$d/dq^2 M^n_{N.S.}(q^2) = d/d\mu^2\Theta^n_{N.S.}(\mu^2)\Big|_{\mu^2=q^2}\ . \tag{5.43}$$

so the renormalization point dependence of the operator matrix element determines the q^2-dependence of the moment of the structure function.

Combining the OPE analysis with the evolutionary picture of the previous section we recognize that the renormalization point introduced in the OPE analysis is the same as the transverse resolution in the Weizäcker-Williams method. In both cases, it is necessary to define how much of the gluon field is to be lumped into the definition of a quark. Whatever way you look at it, the fact that this definition changes with Q^2 gives rise to the logarithmic scaling violation of QCD.

5.3 Other (Power) Corrections to Scaling

In addition to the $ln\ Q^2$ corrections we've uncovered, there are expected to be $0(1/Q^2)$ and higher order $0(1/Q^{2n})$ corrections which become important at small Q^2. The $0(1/Q^2)$ corrections take several forms. They are easy to distinguish using the language of OPE. First are "target mass" corrections. As will be come clear, this is an unfortunate name. These are apparent $0(M_T^2/Q^2)$ terms which arise from the $g^{\mu_i\mu_j}$ factors in the matrix elements of traceless operators. It has been shown[17] that these corrections may be completely absorbed by replacing the variable x_T by the variable $\xi = -q^+/P^+$ everywhere in the definition of $f_{a/T}(x_T)$. This isn't surprising since q^+ is the variable which emerged automatically in the derivation (see, *e.g.*, Eq. (2.18)). ξ is written in many forms:

$$\xi = (\sqrt{\nu^2 + Q^2} - \nu)/M_T \quad , \tag{5.44a}$$

$$\xi = 2x_T/(1 + \sqrt{1 + Q^2/\nu^2}) \quad , \tag{5.44b}$$

$$\xi = 2x_T/(1 + \sqrt{1 + 4M_T^2 x_T^2/Q^2}) \quad . \tag{5.44c}$$

It is clear from Eq. (5.44b) that the target mass corrections do not, in fact, depend on the target mass![18] They are kinematic corrections (which do not grow like A^2 for nuclear targets), and are well-understood. Second are "quark mass" corrections. These are important for heavy quarks (c, b, t) but negligible for up and down quarks whose masses, $m_{u,d} < 20$ MeV, are tiny.[19] Finally are the dynamical $0(1/Q^2)$ corrections associated with operators of twist-4. Although they are complicated, twist-4 corrections to inelastic electron (and neutrino) scattering have been completely analyzed in QCD.[15] Typically, they are small because the natural mass scale associated with a target is one upon its radius (once "target mass" corrections have been incorporated via ξ-scaling), $1/R \sim 1\ \text{fm}^{-1} \sim 200$ MeV. It is therefore not surprising that scaling, modulo logarithms and using the ξ variable, sets in a very low value of Q^2 (< 1 Gev). For precisely this reason higher twist contributions to $F_2(x, Q^2)$, which measure matrix elements of interesting local operators, are hard to extract from available experimental data.

5.4 QCD and the Quark Model

We have seen that the quark, antiquark and gluon content of a hadron changes with the scale at which it is probed. In more naive quark models the nucleon, for example, is treated as (approximately) three quarks in some confining "bag" with no reference to the scale at which this description might hold. Certainly, if the nucleon were three quarks at some scale μ_0^2 ($\mu_0^2 \sim 1$ GeV^2) then it would become more complicated, containing antiquarks and glue at larger scales $Q^2 > \mu_0^2$ by virtue of QCD radiation. In the early days of QCD it was recognized that if the nucleon's quark, antiquark and gluon distributions measured at large Q^2 were evolved back to lower Q^2, then quark-antiquark pairs and gluons are reabsorbed into the valence quarks, so that at some $\mu_0^2 \sim 1$ GeV^2 all of the $q\bar{q}$ pairs and most of the glue would be gone leaving a nucleon made of three valence quarks alone.

G. Ross and I checked this quantitatively in the M.I.T. version of the bag model.[20] Using QCD evolution to second order, which is necessary because $\alpha(\mu_0^2)$ is not small, we found that measured non-singlet nucleon stucture functions evolved backwards to a μ_0^2 of order 1 GeV^2 indeed gave valence quark x-distributions in agreement with earlier bag calculations.[21] μ_0^2 is then interpreted as a parameter of the quark model: It is the mass scale (or resolution) at which quark fields should be defined in order that the nucleon should be made of three quarks. A recent reevaluation of this program with modern values for structure functions and the QCD Λ parameter (*c.f.* Eq. (5.20)) gave $\mu_0^2 \cong 0.75$ GeV^2.[6.4] Notice that the structure function predicted by simple quark models cannot be compared directly with experimental measurements of F_2 at $Q^2 = \mu_0^2$ because at such a low Q^2 higher twist effects are large but have not been included in the quark model calculations. It seems best to regard quark models as models for the twist-two matrix elements at a renormalization point μ_0^2, which must then be evolved to $Q^2 >> \mu_0^2$ in order to be compared with experiment.

§6 QCD ANALYSIS OF ELECTRON SCATTERING FROM NUCLEI

Close, Roberts and Ross[1] realized that the scale (Q^2) dependence of quark distribution functions in QCD could be used to parameterize and, to some extent, explain the A dependence of structure functions.[2] In §2, we learned that the shift in the valence quarks observed in nuclei could be understood as an increase in the quark correlation length in the nuclear ground state. The increase in ocean quark pairs appeared to be an independent phenomenon. In the QCD inspired analysis, I shall describe both aspects of the EMC effect have a single origin: a dynamical change in scale of the twist-2 matrix elements in nuclei. In the last chapter, we saw that QCD evolution reduced the momentum on valence quarks and increased the number of pairs. Suitably adapted, evolution can explain the EMC effect. This method of analysis has come to be known as "dynamical rescaling" or simply "rescaling". Its virtues are first, it gives a unified description of all aspects of the EMC effect; second, it avoids the dubious assumptions of the constituent convolution models of §4; and third, it gives us insight into the reason other superficially quite different "explanations" of the EMC effect work. Its drawback is that it does not provide a microscopic enough explanation to satisfy most of us: it is not clear exactly what the quarks and gluons are doing differently in a nucleus which gives rise to the effect.

In this chapter I will work from the general toward the specific. First, I will merely use QCD as an aid to present the data in a new way. This presentation will lead to a surprising conclusion and suggest rescaling as a mechanism behind the EMC effect. Then, I will analyze rescaling in some detail. Next, I will describe a calculation of the A dependence motivated by, but perhaps more general than rescaling.[1,8]. Finally, I will close with some remarks about shadowing and future experiments. These have little to do with QCD and less to do with rescaling, but they follow naturally upon the discussion of A dependence.

6.1 A QCD Motivated Presentation of the Data : Rescaling

In §2, we compared the structure functions of different nuclei at fixed Q^2, as functions of x. This is the way the data come from the experimenters. QCD provides an alternative. Consider the moments:

$$M_A^n(Q^2) \equiv \int_0^A dx\, x^{n-2}\overline{F}_2^A(x, Q^2) \tag{6.1}$$

($\overline{F}_2^A = \frac{1}{A}F_2^A$). According to Eq. (5.21), the moments are monotonically falling functions of Q^2 [This analysis like that of §5 is restricted to non-singlet structure functions but a similar conclusion applies as well to singlets]:

$$M_A^n(Q^2) = \left(\frac{\alpha_c(Q^2)}{\alpha_c(Q_0^2)}\right)^{d_n} M_A^n(Q_0^2) \tag{6.2}$$

where $d_n \equiv -B_n/2\pi b > 0$. If QCD is correct, and if Q^2 is large enough so leading order in perturbation theory suffices and $O(1/Q^2)$ corrections are negligible, then $ln\ M_A^n(Q^2)$ must lie on a straight line when plotted versus $ln\ [\alpha_c(Q_0^2)/\alpha_c(Q^2)]$ and the slope must be $-d_n$. Such a plot is shown schematically in Fig. 23 for two different targets with baryon numbers A and A'.

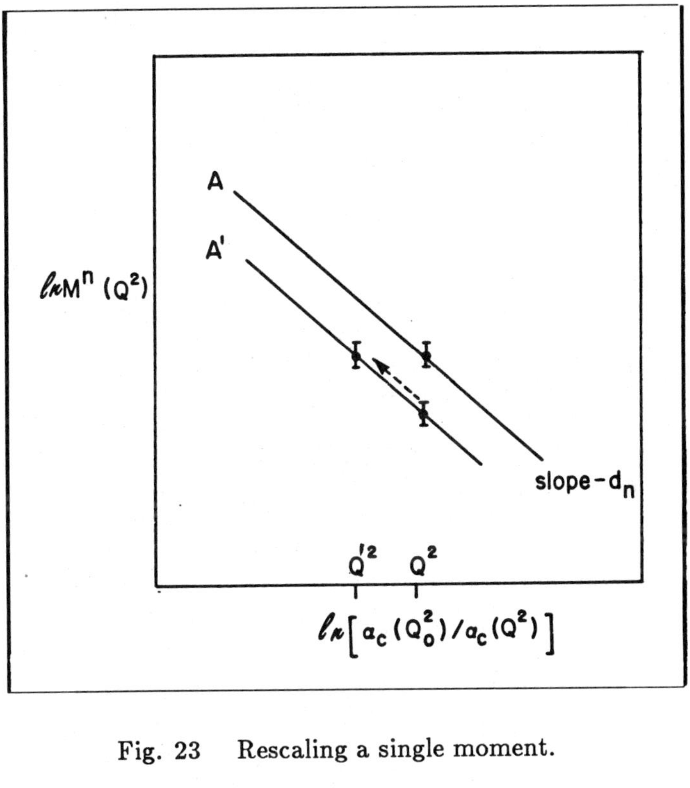

Fig. 23 Rescaling a single moment.

At fixed Q^2, the EMC effect appears as the observation that $M_{A'}^n(Q^2) < M_A^n(Q^2)$ for $A' > A$. On the other hand, it is clear that there is a value of Q^2, call it Q'^2, such that $M_{A'}^n(Q'^2) = M_A^n(Q^2)$. As illustrated in Fig. 23, $Q'^2 < Q^2$. The value of Q'^2 depends, in principle, on

A, A' and Q^2 and on n. Let us define $\xi^n_{AA'}(Q^2)$ (not to be confused with the coordinate ξ_λ or the modified scaling variable ξ of §5) so

$$M^n_{A'}(Q^2) = M^n_A(\xi^n_{AA'}(Q^2)Q^2) \ . \tag{6.3}$$

From its definition

$$\xi^N_{AA'}(Q^2) > 1 \quad \text{for} \quad A' > A \tag{6.4}$$

and

$$\begin{aligned} \xi^n_{AA''}(Q^2) &= \xi^n_{AA'}(Q^2)\xi^n_{A'A''}(Q^2) \\ \xi^n_{AA'}(Q^2) &= 1/\xi^n_{A'A}(Q^2) \end{aligned} \ . \tag{6.5}$$

The Q^2-dependence of $\xi^n_{AA'}(Q^2)$ is determined entirely by QCD, and is independent of A and A'. Consider Eq. (6.2) first for A' at Q_0^2 and Q^2

$$M^n_{A'}(Q^2) = \left(\frac{\alpha_c(Q^2)}{\alpha_c(Q_0^2)}\right)^{d_n} M_{A'}(Q_0^2) \tag{6.6}$$

and then for A at $\xi^n_{AA'}(Q_0^2)Q_0^2$ and $\xi^n_{AA'}(Q^2)Q^2$:

$$M^n_A(\xi^n_{AA'}(Q^2)Q^2) = \left(\frac{\alpha_c(\xi^n_{AA'}(Q^2)Q^2)}{\alpha_c(\xi^n_{AA'}(Q_0^2)Q_0^2)}\right)^{d_n} M^n_A(\xi^n_{AA'}(Q_0^2)Q_0^2) \ . \tag{6.7}$$

Now use Eq. (6.3) to eliminate all reference to moments

$$\frac{\alpha_c(\xi^n_{AA'}(Q^2)Q^2)}{\alpha_c(\xi^n_{AA'}(Q_0^2)Q_0^2)} = \frac{\alpha_c(Q^2)}{\alpha_c(Q_0^2)} \ . \tag{6.8}$$

To lowest order in α_c,

$$\alpha_c(\xi Q^2) = \alpha_c(Q^2)/(1 + \alpha_c(Q^2)b \ln \ \xi) \ , \tag{6.9}$$

which, together with Eq. (6.8), implies

$$\xi^n_{AA'}(Q^2) = [\xi^n_{AA'}(Q_0^2)]^{\alpha_s(Q_0^2)/\alpha_s(Q^2)} \ . \tag{6.10}$$

The extension to next order in α_c is quoted in Ref. 4. Suppose $Q_0^2 < Q^2$, then $\alpha_c(Q_0^2)/\alpha_c(Q^2) > 1$ and

$$1 < \xi^n_{AA'}(Q_0^2) < \xi^n_{AA'}(Q^2) \quad \text{for} \quad Q_0^2 < Q^2 \ . \tag{6.11}$$

The implication of this result is that a small change of Q^2-scale at low Q^2 gets magnified into a large change when observed at large Q^2.

Eqs. (6.4), (6.5) and (6.10) summarize the properties of $\xi^n_{AA'}(Q^2)$ which can be determined from general considerations alone. The surprise came when Close, Roberts and Ross used this method to analyze the EMC data and found that to a good approximation $\xi^n_{AA'}(Q^2)$ appears to be independent of n: $\xi^n_{AA'}(Q^2) \Longrightarrow \xi_{AA'}(Q^2)$, at least for the values of n sensitive to the x-range of the EMC data. Actually, CRR did not construct moments but made an equivalent observation about the structure function itself. Namely, if $\xi^n_{AA'}(Q^2)$ is independent of n then the structure functions themselves as functions of x are related by a universal scale change in Q^2:

$$\overline{F}_2^{A'}(x, Q^2) = \overline{F}_2^{A}(x, \xi_{AA'}(Q^2)Q^2) \tag{6.12}$$

Eq. (6.12) has become known as "rescaling". CRR were led to it by the observation we referred to in §5, that the EMC effect resembles QCD evolution. In fact, the EMC data are not in complete agreement with Eq. (6.12). The excess at low x is more than can be produced by the amount of evolution that is required to fit the depletion at large x. Their analysis, with $\xi_{DFe} \cong 2$ at $Q^2 \approx 20$ GeV^2 is shown in Fig. 24.

The newer SLAC data on iron and deuterium data have a smaller enhancement at low-x and a lower cross over point (where $\overline{F}_2^{Fe}/\overline{F}_2^{D} = 1$), both of which improve the agreement with the rescaling analysis.[3,4] The reader might wish to look back to Fig. 11 to see the present state of the rescaling fits.

Several comments and caveats are in order:

1. QCD is subtle: Changing the scale creates quark-antiquark pairs. If, after all the discussion of §5, this still seems unreasonable, perhaps it would help to remember that a similar thing happens in a Bogoliubov transformation. By redefining the vacuum, annihilation and creation operators get mixed up with one another and a state which originally contained only particles, appears after the transformation, to contain both particles and antiparticles.
2. Rescaling predicts that the EMC effect should vanish at $x \approx 0.2$ where QCD evolution vanishes (see Fig. 21). The EMC data cross unity at $x \approx 0.35$ in disagreement with this prediction. Once again, however, the SLAC data look better: $\overline{F}_2^{A}/\overline{F}_2^{D}$ is definitely below unity for $x >$ 0.3. A careful test must await better data at small-x.
3. Rescaling cannot work for $x \to 1$, or equivalently for $n \to \infty$. As $x \to 1$, the structure function of the nucleon vanishes but that of a

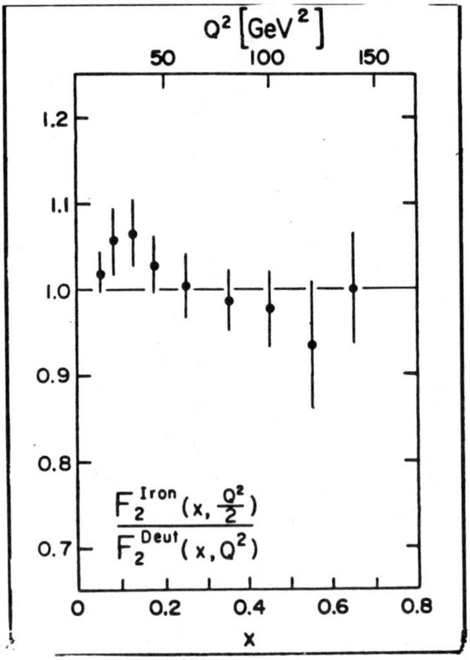

Fig. 24 Rescaling the EMC iron data.[6.1]

nucleus does not. Eq. (6.12) fails to reproduce this behavior. The reason for this failure will become clear soon.

4. Rescaling does not mean that an iron nucleus can be "mapped" to a nucleon by a universal change of scale. The complete description of a nucleus requires matrix elements of all twists. The gross structure of the EMC effect only involves a few operators of twist-two.

6.2 Rescaling and Quark Models

In §5.4 we learned to associate a mass scale μ_0^2 with the quark model description of a hadron. It is the scale at which, approximately, the hadron consists of valence quarks alone. It is an intrinsic characteristic of each hadron. The measured structure function of the nucleon is consistant with this notion with $\mu_0^2 \cong 0.75$ GeV^2. Clearly, any nucleus related to the nucleon by rescaling, as in Eq. (6.12), also admits a valence quark description, but at a shifted mass scale μ_A,

$$\mu_A^2 = \frac{\mu_0^2}{\xi_{NA}(\mu_0^2)} \tag{6.13}$$

Since $\xi_{NA} > 1$, μ_A^2 is less than μ_0^2. This can be checked by the following argument: At any fixed Q^2, the nucleus appears more highly evolved than the nucleon (more pairs, softer valence quarks). So to reabsorb all pairs it is necessary to devolve the nucleus further than the nucleon.

It has become customary to express Eq. (6.13) as a ratio of length scales. Let us define $\lambda_A = c/\mu_A$, $\lambda_N = c/\mu_0$, then

$$\lambda_A/\lambda_N = \mu_0/\mu_A = \sqrt{\xi_{NA}(\mu_0^2)} \tag{6.14}$$

$\xi(\mu_0^2)$ is typically much smaller than $\xi(Q^2)$, because of Eq. (6.10). For example, ξ_{FeN} (20 GeV^2) $\cong$ 2.0, but $\xi_{FeN}(\mu_0^2) \cong 1.33$ making $\lambda_{Fe}/\lambda_N \cong$ 1.15. Thus, an increase of only $\sim$ 15% in the intrinsic length scale of the twist-two matrix elements for iron compared to the nucleon can give rise to the EMC effect.

If there is anything fundamental about rescaling, rather than being merely an accident, the basic relation must be defined at the scale intrisic to the target, μ_0^2, not at some arbitrary Q^2.[1] At μ_0^2, however, the structure function contains large contributions from higher twists, so an equation like Eq. (6.12) with Q^2 replaced by μ_0^2 cannot be written down. Instead, we must write relations between matrix elements of twist-two operators using the formalism of §5.2. We define

$$\langle P|O_{n,a}^{(\mu^2)\mu_1\ldots\mu_n}|P\rangle = P^{\mu_1}\ldots P^{\mu_n}\Theta_{n,a}^A(\mu^2) \tag{6.15}$$

in analogy to Eq. (5.31). Then the operator equivalent of Eq. (6.12) is

$$A^{n-2}\Theta_{n,a}^A(\mu_A^2) = A'^{n-2}\Theta_{n,a}^{A'}(\mu_{A'}^2) \tag{6.16}$$

This is the basic rescaling relation between targets. The powers of A and A' are only kinematic. It leads to Eq. (6.12) at large Q^2 provided $\xi_{AA'}(Q^2)$ is obtained from $\mu_A/\mu_{A'}$ via QCD evolution (Eq. (6.10) plus higher order improvements) and provided the moments can be reliably evolved from μ_A^2 to Q^2 using perturbative QCD. This turns out to be an important proviso. In Ref. 5.20, we found that only the moments with $n < 8$ could be reliably devolved from large Q^2 to a μ_0^2 as small as 1 GeV^2. The problem is that higher order (in α_c) corrections typically contain $ln\ n$ factors making perturbation theory worse for large n. Suppose, then, that Eq. (6.16) were exact for all n. Nevertheless, at large Q^2 only the moments with $n < 8$ would be likely to show uniform rescaling. This explains why rescaling fails near $x = 1$ (as noted earlier) since large n moments probe exclusively large x. In fact, one can estimate the x values for which rescaling should be reliable when observed at large Q^2 by using the Mellin transform relation.[4]

$$n \sim ln\ \alpha_s / ln\ x \tag{6.17}$$

which for $n = 8$ and $\alpha_s = 0.2$ gives $x \sim 0.8$ as the upper limit of reliability.

Turning this argument around one can see that exact uniform rescaling for all x at large Q^2 would be very hard to understand, since it would imply a complicated and non-uniform relation among twist-two matrix elements with $n > 8$ at μ_0^2.

The task for someone trying to understand the EMC effect from the point of rescaling, then, is two-fold. First, one must explain why rescaling should be uniform at the intrinsic scale, *i.e.*, why should $\lambda_A/\lambda_{A'}$ be independent of n. Second, one must predict the A dependence of the rescaling parameter, *i.e.*, of λ_A/λ_N. Close, Roberts, Ross and I have argued that rescaling is a rather natural prediction of quark models with only one length scale. An example is the MIT bag model with only u and d quarks, which may be taken to be massless, so the only dimensionful parameter is the bag constant B. Such models are very close in spirit to QCD itself in which the only dimensionful parameter is Λ. In a model like this, quarks carrying momentum p confined within a radius λ transform into quarks carrying momentum $p' = (\lambda/\lambda')p$ when the confinement scale is changed to λ' – the dimensionless quantity $p\lambda$ is constant. The intrinsic scale μ^2 is then proportional to p^2, there being no other scale in the problem. It is hard to turn this heuristic argument into a proof of uniform rescaling. That would probably require a consistant formulation of QCD perturbation theory (including renormalization) in a bag model, something which exists only in fragments.[5] On the other hand, as Llewellyn Smith has noted, it seems clear that other models such as the non-relativistic quark model have no hope of giving rescaling unless the quark masses are assumed (rather unnaturally) to scale with the inverse confinement radius.[1,8] A satisfactory theoretical understanding of rescaling will have to await a more powerful QCD-based theory of confinement. The second task – determining the A dependence of λ_A/λ_N – is more straightforward. It is the subject of the next Section.

6.3 *A* Dependence

It hardly needs saying that the EMC effect derives from the proximity of nucleons within the nucleus, and that it would vanish if one could arrange that the nucleus was very dilute. Fortunately, nature has given us one very dilute nucleus – the deuteron – and Bodek and Simon[3.2] have shown that $F_2^D(x,Q^2)/F_2^N(x,Q^2)$ is very close to unity. It seems reasonable to assume, therefore, that the magnitude of the EMC effect should be proportional to the probability that nucleons approach each other or overlap within the nucleus.

Close, Roberts, Ross and I[3,4] defined the simplest measure of this effect we could imagine: we defined an "overlapping" volume for a nucleus which is the integral over the nucleus of the two body density $\rho_A(\vec{r}_1,\vec{r}_2)$ multiplied by an overlap factor $V_0(|\vec{r}_1-\vec{r}_2|)$ which we took to be the overlapping volume of two spheres of radius a:

$$\begin{aligned} V_0(d) &= 1 - \frac{3}{4}\left(\frac{d}{a}\right) + \frac{1}{16}\left(\frac{d}{a}\right)^3 \qquad d \leq 2a \\ &= 0 \qquad\qquad\qquad\qquad\qquad\quad d > 2a \ . \end{aligned} \tag{6.18}$$

Then, the overlapping volume per nucleon is

$$V_A = (A-1)\int d^3r_1 d^3r_2 \rho_A(\vec{r}_1,\vec{r}_2)V_0(|\vec{r}_1-\vec{r}_2|) \ . \tag{6.19}$$

$\rho_A(\vec{r}_1,\vec{r}_2)$ is normalized to $\int d^3r_1 d^3r_2 \rho_A(\vec{r}_1,\vec{r}_2) = 1$. Saturation of the nuclear density at large A implies $\rho_A \sim 1/A^2$. With this behavior of ρ_A and the finite range of V_0 it is easy to see that V_A saturates at large A, *i.e.*, $\lim_{A\to\infty} V_A = \text{constant}$. The choice of a geometrical form for $V_0(|\vec{r}_1-\vec{r}_2|)$ was in fact quite arbitrary. Any function which goes to unity as $\vec{r}_1 - \vec{r}_2 \to 0$ and to zero when $|\vec{r}_1-\vec{r}_2| > 2R_{\text{nucleon}}$ and which respects the three dimensional geometry of the problem would do. We calculated the overlapping volume for nucleons with a chosen so $a_{\text{rms}} = 0.9\text{fm}$ ($a_{\text{rms}} = \sqrt{\frac{3}{5}}a$, so $a = 1.16$ fm). $\rho_A(\vec{r}_1,\vec{r}_2)$ was written in terms of the single particle density $\rho_A(\vec{r})$ and a corrrelation function $f(\vec{r}_1-\vec{r}_2)$:

$$\rho_A(\vec{r}_1-\vec{r}_2) = \rho_A(\vec{r}_1)\rho_A(\vec{r}_2)f(\vec{r}_1-\vec{r}_2) \ . \tag{6.20}$$

$\rho_A(\vec{r})$ was taken from experimental measurements of nuclear charge densities. $f(\vec{r}_1-\vec{r}_2)$ should, in principle, depend on A but there is little or no direct information on it from experiment. We took $f(\vec{r}_1-\vec{r}_2)$ from models

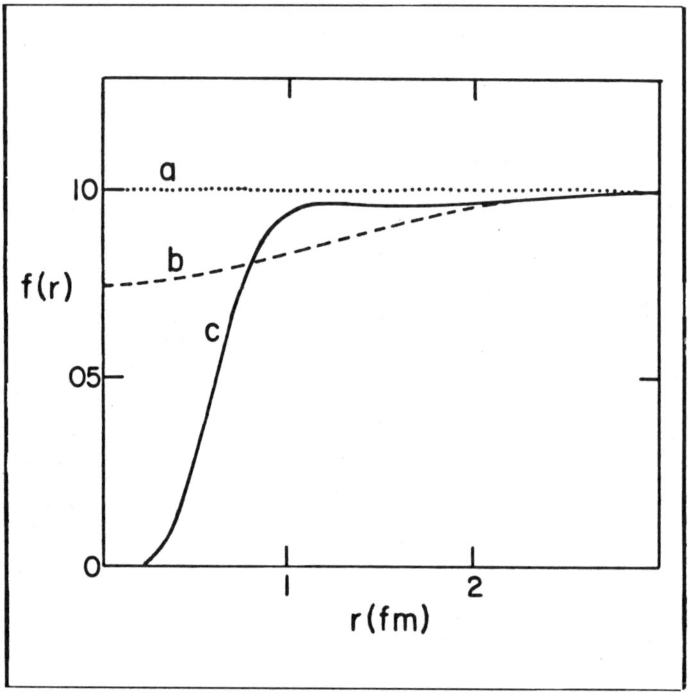

Fig. 25 Choices of the correlation function $f(r)$: a) no correlation; b) Fermi gas correlation; c) Reid soft core correlation.

of nuclear matter, the most realistic probably being one based on a Reid soft core potential[6] shown in Fig. 25.

We then assumed that the effective confinement size in a nucleus A will be intermediate between that for an isolated nucleon and some limiting value λ_{tot} associated with two totally overlapping nucleons. We assumed a linear interpolation in V_A leading to

$$\frac{\lambda_A}{\lambda_N} = 1 + V_A\left(\frac{\lambda_{tot}}{\lambda_N} - 1\right) . \tag{6.21}$$

λ_{tot} might be viewed as a parameter. Instead, we estimated its value from the bag model, where a spherical baryon number two system must have volume at least twice as large as a single nucleon (otherwise it would be stable against decay into two nucleons), so $R_6 \geq 2^{1/3}R_3$. This led us to take $\lambda_{tot}/\lambda_N = 2^{1/3}$, and to the values of λ_A/λ_N given in Table 2.

TABLE 2

Nucleus	λ_A/λ_N (a)	λ_A/λ_N (b)	λ_A/λ_N (c)	$\xi_A(Q^2=20)$
^{2}D	1.018	1.015	1.015	1.07
^{3}He	1.047	1.042	1.040	1.20
^{4}He	1.092	1.082	1.079	1.43
^{6}Li	1.054	1.045	1.045	1.23
^{7}Li	1.075	1.064	1.063	1.33
^{9}Be	1.088	1.074	1.074	1.40
^{12}C	1.124	1.105	1.104	1.60
^{16}O	1.128	1.109	1.108	1.63
^{20}Ne	1.122	1.104	1.104	1.60
^{27}Al	1.165	1.140	1.140	1.89
^{32}S	1.157	1.134	1.134	1.84
^{40}Ca	1.161	1.137	1.137	1.86
^{48}Ca	1.196	1.166	1.166	2.14
^{56}Fe	1.180	1.153	1.154	2.02
^{63}Cu	1.181	1.154	1.154	2.02
^{107}Ag	1.198	1.168	1.169	2.17
^{118}Sn	1.205	1.175	1.176	2.24
^{197}Au	1.229	1.196	1.195	2.46
^{208}Pb	1.220	1.188	1.188	2.37

Values of the confinement size relative to that for the free nucleon for a range of nuclei. The three values (a), (b) and (c) correspond to the three choices of the correlation function $f(r)$ shown in Fig. 25. ξ_{NA} (20 GeV^2) is shown for the Reid correlation function. The others are similar.

These translate into values of ξ_{NA} (20 GeV^2) which are much larger, and these in turn give predictions for the EMC effect in a variety of nuclei. The predictions of the rescaling model are compared with the SLAC data in Fig. 11. The agreement is fine for $x < 0.7$ above which Fermi motion becomes important. One thing which is not obvious from Fig. 11 is that the data correlate well with idosyncracies in the periodic table. Fig. 26 shows predictions for several x values and $Q^2 = 4.98$ GeV^2 compared with SLAC data.

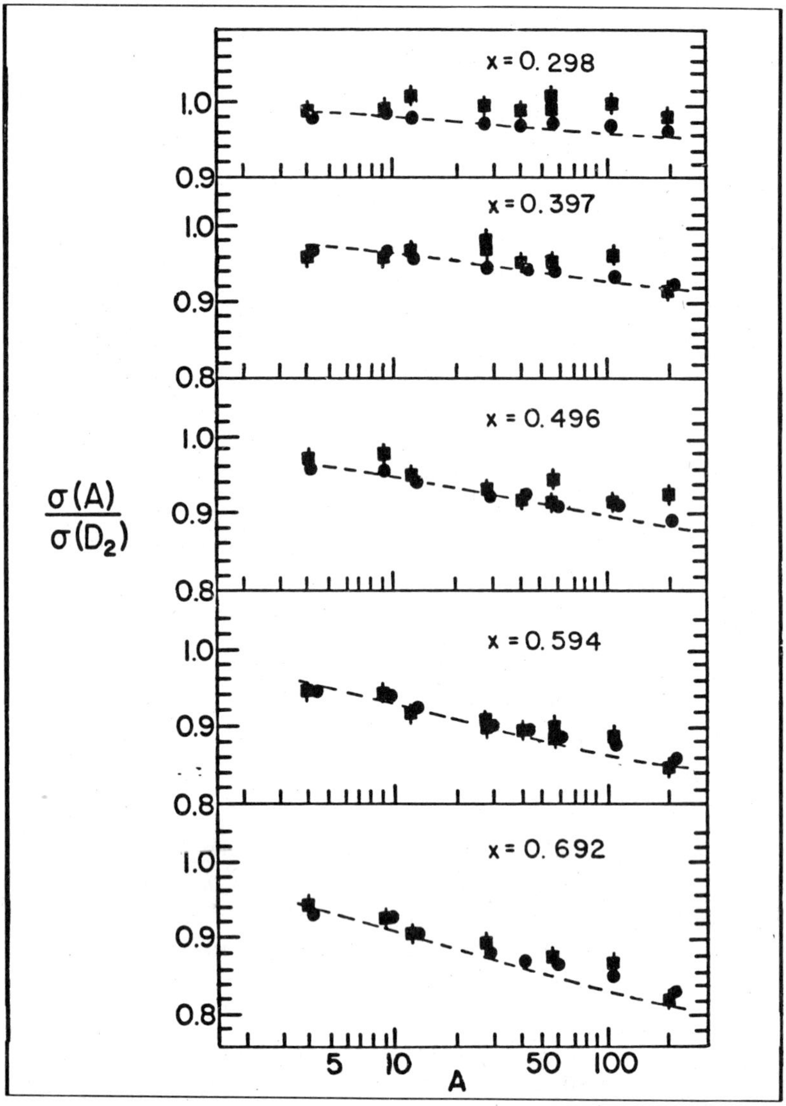

Fig. 26 *A* dependence of the EMC effect at fixed x. The data are squared with error bars. The predictions of rescaling are solid dots.[6.4]

The fluctuations in the rescaling predictions reflect variations in nuclear densities, *e.g.*,^4He is more tightly bound than ^9Be, and generally follow the data. Some predictions for the future are shown in Fig. 27.

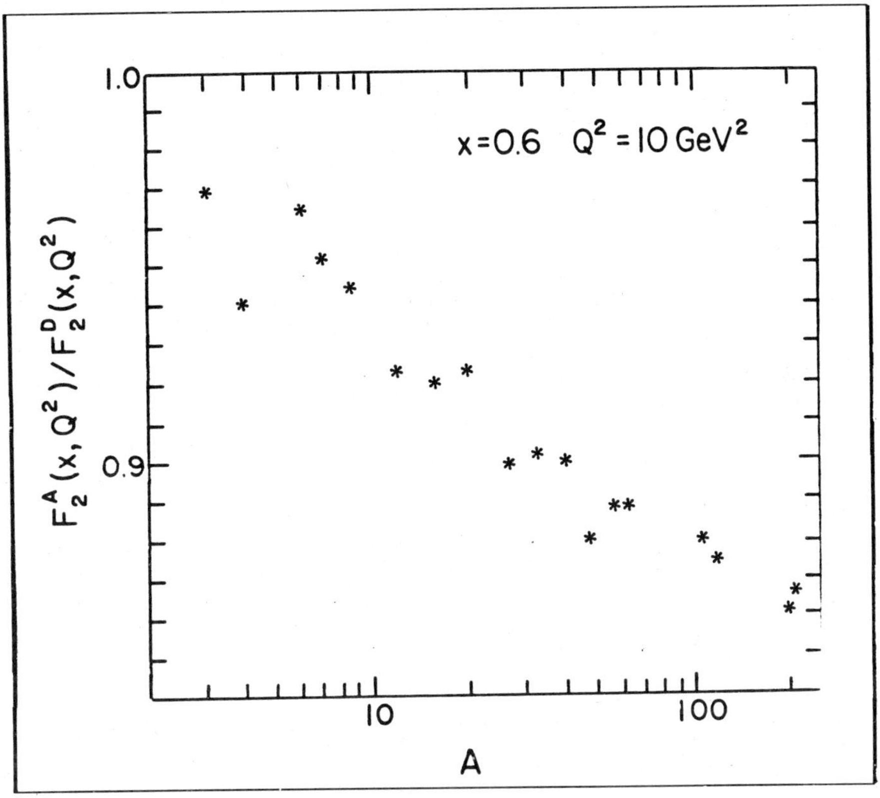

Fig. 27 A dependence at $x = 0.6$, $Q^2 = 10$ GeV predicted by rescaling.

Several comments are appropriate before leaving the discussion of A dependence:

1. Llewellyn Smith[1.8] has argued that the calculation of the A dependence we have given is much more general than the rescaling model. Any scheme which fits the EMC effect in iron and in which the effect is linear in V_A will agree as well with the SLAC data. This is qualitatively true but the assumption of linearity is non-trivial. We argued that λ_A/λ_N should be linear in the overlapping volume. Had we instead

(erroneously) assumed $\xi_{NA}(Q^2 = 20 \text{ GeV}^2)$ to be linear in V_A the A dependence would have come out wrong.

2. The rescaling model is not related to convolution models nor does it ascribe the EMC effect to any particular exotic component in the nuclear wavefunction (*e.g.*, six quark bags). The scale change might originate from a change in the size of individual nucleons, from quark percolation between nucleons, or from multiquark or meson admixtures in the nuclear wavefunction. Personally, I suspect that to the extent that these notions can be well-defined they will turn out to correspond to the same underlying physics – partial deconfinement.

3. The rescaling fit to the A-dependence depends on several parameters: the nucleon radius a, the formula chosen to relate V_A to λ_A/λ_N, the scale μ_0 and the choice of two nucleon correlation function. In fact, none of these were treated as free parameters in Ref. 4. Instead, they were fixed at "reasonable" values from other aspects of hadron dynamics. Of course, the precise choice of parameters is not particularly central to the explanation of the EMC effect.

4. The value of V_A is very large for large A. It is ≈ 0.72 for ^{208}Pb with the Reid correlation function. It has been remarked that such a large value of V_A is "unreasonable" and in some sense contradicted by the many successes of conventional nuclear physics. The overlapping volume of nucleons with $a_{\text{rms}} = 0.9$fm is very large. That is a fact, not a shortcoming of the rescaling model. As for the idea that large V_A is inconsistent with nuclear physics in general – I believe it is an unwarranted concern: It costs ≈ 1 GeV (the string tension) to separate colored sources by 1fm. It even costs 300 MeV to flip a quark spin. These energetic considerations, not naive classical, geometrical considerations determine the quantum mechanics of nuclei.

6.4 Shadowing

It has long been expected that at very low values of x deep inelastic electron scattering from nuclei would behave like hadron scattering from nuclei and exhibit "shadowing".[7] At high energies hadron nucleus total cross sections grow like $A^{2/3}$. The simple explanation of this effect is that the incoming hadron doesn't "see" the nucleons at the back of the nucleus. They are in the shadow of those in front. The cross section then grows like $\pi R^2 \sim A^{2/3}$. Real photon-nucleus interactions are shadowed.[8] This finds a simple explanation in the framework of vector meson dominance: the hadronic interactions of a real photon are well-approximated by supposing the photon converts with probability $\sim \alpha$ into a vector meson (ρ, ω or ϕ) which then interacts hadronically and experiences shadowing. Vector meson dominance fails for large Q^2 inelastic lepton scattering, or at least an infinite tower of vector states with precisely tuned couplings is required, but the prejudice that shadowing occurs there too is strong.

My own thoughts on shadowing are still in flux. Since much of the future experimental work in the field will be carried out in the kinematic region where shadowing might be important, I would like to outline the problem here. Perhaps some reader will solve it! The kinematic relation

$$\xi^3 < 1/Mx \tag{6.22}$$

tells us that for very small values of x the struck quark may propagate over very large distances in the target. For $x \approx 0.5$, $\xi^3 \lesssim 0.4$fm, but for $x \approx 0.05$, $\xi^3 \lesssim 4$fm and for $x \approx 0.005$, $\xi^3 \lesssim 40$fm! The parton model diagram dominant at large Q^2 is shown relative to an iron nucleus in Fig. 28 where the struck quark is associated with a single nucleon.

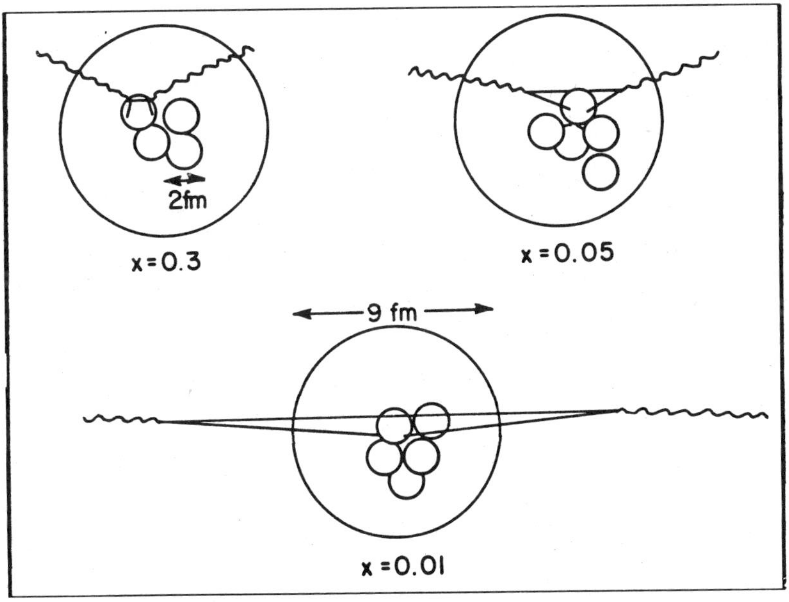

Fig. 28 Cartoons representing the ξ^3 values probed by deep inelastic lepton scattering at low-x from an iron target.

Other possibilities represent EMC-like corrections. The question is whether a quark propogating over such distances is or is not strongly absorbed in the nuclear medium. Hadron interactions at very high energies are predominantly absorptive, but the state which is propagating in Fig. 28 is not an ordinary hadron. In particular, the propagating quark, which is a color $\underline{3}$, is always close to an antiquark[9] which is a color $\underline{\overline{3}}$ and may neutralize some or most of its strong interactions. Furthermore, the invariant mass of the $q\overline{q}$ pair is $-Q^2$.

Suppose for the sake of discussion, we characterize quark propagation in the nuclear medium (accompanied by an antiquark as in Fig. 28) by an absorption length λ which may depend on Q^2. For a given nucleus when $1/Mx < \lambda(Q^2)$ there is little shadowing. Shadowing sets in when $1/Mx \sim \lambda(Q^2)$, but when $1/Mx$ exceeds twice the radius of the nucleus (R_A) shadowing saturates: decreasing x further does not put more matter along the propogating quark's path. So there are two x-values characterizing shadowing, $x_0(Q^2) = 1/M\lambda(Q^2)$ marking its onset and $x(A) = 1/2MR_A$ marking its saturation. For $x < x(A)$, only the front skin of the nucleus down to a depth $\lambda(Q^2)$ participates in the scattering, so shadowing increases with A but goes away as $\lambda(Q^2) \to \infty$. All of this can be summarized by the purely phenomenological formula

$$\overline{F}_2^A(x,Q^2) = \frac{F_2^N(x,Q^2)}{\left[1 + \frac{x_0(Q^2)}{x+x(A)}\right]} \tag{6.23}$$

in which for simplicity I have ignored all other nuclear effects such as rescaling and Fermi motion. Eq. (6.23) does not satisfy the quark number sum rule. It is not intended to be valid at all x, only for $x \sim 0$ where shadowing may be important. According to Eq. (6.23), there is an asymptotic shadowing curve for nuclear matter ($A \to \infty$, $x(A) \to 0$), shown for example in Fig. 29. Finite nuclei track along $\overline{F}_2^{\infty}(x,Q^2)$ until $x \sim x(A)$ below which they depart above $\overline{F}_2^{\infty}$.

A determination of $\lambda(Q^2)$ is crucial.[10] Unfortunately, it is largely unknown. Naively, one might expect $\lambda(Q^2)$ to be a typical meson mean free path, corresponding to a cross section of order 30 mb. This is too naive. The propagating $q\overline{q}$ pair in Fig. 28 must have invariant mass $-Q^2$. This can be generated if both quark and antiquark have transverse momentum of order $\frac{1}{2}\sqrt{Q^2}$ or by a large mismatch in their longitudinal momentum. In the former case, the transverse dimension of the $q\overline{q}$ system is very small, $\sim 1/\sqrt{Q^2}$. The system looks like a small color dipole and has a small absorption cross-section. In the latter case, the transverse dimensions are not

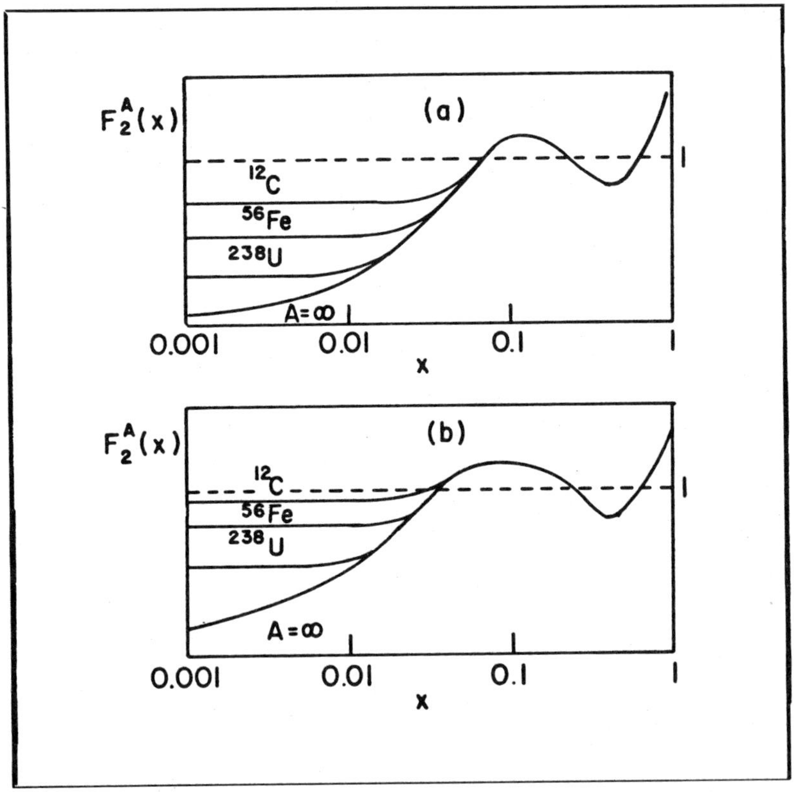

Fig. 29: Scenarios for shadowing. a) and b) correspond to two different values of $\lambda(Q^2)$ with $\lambda_a > \lambda_b$.

small, color screening is not so effective and the absorption cross section may be large. The importance of shadowing depends on which region of phase space dominates. This is not yet known. But at least some of the shadowing seen at $Q^2 = 0$ should disappear at large Q^2. The behavior of $\overline{F}_2^A(x, Q^2)$ at small x expected if shadowing indeed disappears at large $Q^2(x(Q^2) \to 0)$ is shown in Fig. 29.

The discussion in this section has been very qualitative and phenomenological. Eq. (6.23) should be taken with a grain of salt. There is much to do on the subject of shadowing and little of substance to report here. Nevertheless, I though it might be appropriate to end by whetting the reader's appetite for the next round of experiments.

REFERENCES

1.1 Throughout these lectures we use the metric $g_{00} = -g_{ii} = 1$, covariantly normalized states, $\langle p|p'\rangle = (2\pi)^3\, 2E\, \delta^3(\vec{p}\,' - \vec{p})$ and the γ-matrices and other conventions of J. D. Bjorken and S. D. Drell, Relativistic Quantum Mechanics (McGraw-Hill, New York, 1964).

1.2 In fact, the operator product in Eq. (1.8) may be so singular as $\xi_\lambda \to 0$ that the time ordered product is not Lorentz-invariant. In this event the proper definition of $T_{\mu\nu}$ makes use of the T^*-product which differs from the T-product by local terms (distributions at $\xi_\lambda = 0$) which restore Lorentz invariance. It is possible to derive paradoxical (and incorrect) results by ignoring this sublety. For a discussion, see the book by Itzykson and Zuber referenced in the preface.

1.3 J. D. Bjorken, *Phys. Rev.* **148**, 1467 (1966); **179**, 1547 (1969).

1.4 For an introduction to the methods of dispersion theory, see G. Barton, Dispersion Techniques in Field Theory (W. A. Benjamin, New York, 1965).

1.5 L. D. Landau, *Nucl. Phys.* **13**, 181 (1959); J. D. Bjorken, Ph.D. dissertation, Stanford University, 1959.

1.6 For introductory treatments of the P_∞ approach to deep inelastic electron scattering see the books by Close and Feynman referenced in the preface.

1.7 M. J. Lighthill, Introduction to Fourier Analysis and Generalized Functions (Cambridge University Press, Cambridge, 1958).

1.8 C. H. Llewellyn Smith, Oxford University preprint, to be published in the Proceedings of the 1984 PANIC Conference, Heidelberg.

2.1 See, for example, the lectures by Gross and by Ellis and the book by Itzykson and Zuber referenced in the preface.

2.2 The methods employed here are those of light-cone current algebra. More detailed accounts can be found in H. Fritzsch and M. Gell-Mann in Broken Scale Invariance and the Light Cone, 1971 Coral Gables Conference on Fundamental Interactions at High Energy, M. Dal Cin, G. J. Iverson and A. Perlmutter, eds., (Gordon and Breach, New York, 1971); see also the lectures by Ellis referenced in the preface.

2.3 R. L. Jaffe, *Nuclear Physics* **B229**, 205 (1983).

2.4 The books by Close and Feynman and lectures by Ellis have extensive discussions of the parton phenomenology. For recent experimental summaries see F. Dydak, in Proceedings of the 1983 International Symposium on Lepton and Photon Interactions at High Energies (Newman

Laboratories, Ithaca, 1983), D. G. Cassel and D. L. Kreinich, eds., p. 634.

3.1 J. J. Aubert, *et al.*, *Phys. Lett.* **123B**, 275 (1983).

3.2 A. Bodek and A. Simon, University of Rochester preprint UR-906, April 1985, to be published in *Z. Phys.*

3.3 Some aspects of the EMC data, notably an enhancement at low-x were anticipated by A. Krzywicki, *Phys. Rev.* **D14**, 152 (1976) and R. M. Godbole and D. V. L. Sarma, *Phys. Rev.* **D25**, 120 (1982). The majority of papers prior to 1982 treating inelastic electron scattering from nuclei emphasized shadowing ($S^A < 1$) at low-x, an effect which has yet to be observed at large Q^2.

3.4 A. Bodek, *et al. Phys. Rev. Lett.* **50**, 1431 (1983), **51**, 534 (1983).

3.5 R. Arnold, *et al.*, *Phys. Rev. Lett.* **52**, 727 (1984).

3.6 A. Argento, (BCDMS Collaboration) talk presented at the XX Rencontre de Moriond, Les Ares, 1985.

3.7 A. M. Cooper, *et al.*, *Phys. Lett.* **141B**, 194 (1984). For further references see the review by A. Bodek in *Neutrino 84*, University of Rochester Preprint UR-884, June 1984.

3.8 J. P. Merlo, (preliminary) data presented at the XX Rencontre de Moriond, Les Ares, 1985.

3.9 R. L. Jaffe, *Phys. Rev. Lett.* **50**, 228 (1983).

3.10 R. L. Jaffe, F. E. Close, R. G. Roberts and G. G. Ross, *Phys. Lett.* **134B**, 449 (1984).

3.11 L. I. Frankfurt and M. I. Strickman, LNPI preprint 886 (1983).

3.12 Recently, West has argued on rather general grounds that at truly asymptotic Q^2 ($\log Q^2/\Lambda^2 \to \infty$ where Λ is the QCD scale parameter) in QCD, $\delta\epsilon_{Fe}$ should vanish. See G. B. West, *Phys. Lett.* **54**, 2576 (1985). [Such astronomical values of Q^2 are not obtained experimentally, *e.g.*, $F_2(x) = 0$ for all $x \neq 0$ in this limit, but West uses the structure of the corrections to this limit to argue rather convincingly that the EMC data should be renormalized downward at low-x.]

4.1 The origins of the convolution model are obscure. It has been well-known to workers in the field for many years. It has been discussed recently by G. B. West, Los Alamos preprint, LA-UR-84-241, C. H. Llewellyn Smith, *Phys. Lett.* **B128**, 112 (1983), in Refs. 3.9 and 3.10 and in R. L. Jaffe, *Comments Nucl. Part. Phys.* **13**, 39 (1984).

4.2 G. B. West, *Ann. Phys.* (NY) **74**, 646 (1972); W. B. Atwood and G. B. West, *Phys. Rev.* **D7**, 773 (1973); L. I. Frankfurt and M. I.

Strickman, *Phys. Lett.*, **64B**, 433 (1976); **65B**, 151 (1976); **76B**, 333 (1978); *Nucl. Phys.* **B148**, 107 (1979); *Phys, Lett.* **83B**, 407 (1979), **94B**, 216 (1980); A. Bodek and J. L. Ritchie, *Phys. Rev.* **D23**, 1070 (1981), **D24**, 1400 (1981).

4.3 C. H. Llewellyn Smith, *Phys. Lett.* **128B**, 107 (1983); M. Erickson and A. W. Thomas, *Phys. Lett.* **128B**, 112 (1983); E. L. Berger, F. Coester and R. B. Wiringa, *Phys. Rev.* **D29**, 398 (1984).

4.4 C. E. Carlson and T. J. Havens, *Phys. Rev. Lett.* **51**, 261 (1983); for further references and some discussion see N. N. Nikolaev, Oxford University preprint OU-TP-58/84 (1984).

4.5 K. Heller and J. Szwed, *Acta. Phys. Pol.* **B16**, 157 (1985); H. Faissner and B. R. Kim, *Phys. Lett.* **130B**, 321 (1983).

4.6 M. Erickson and A. W. Thomas, reference 3.

4.7 B. L. Birbrair, A. B. Gridnev, M. B. Zhalov, E. M. Levin and V. E. Starodubsky, LNPI preprint 1031 (1985); S. V. Akulinichev, S. A. Kulagin and G. M. Vagradov, NBI preprints NBI-85-08 and NBI-85-20 and references therein.

5.1 For a summary of the equivalent photon approximation see, for example, J. D. Jackson, Classical Electrodynamics, 2nd edition (J. Wiley and Sons, New York, 1975).

5.2 G. Altarelli and G. Parisi, *Nucl. Phys.* **B126**, 298 (1977).

5.3 M. S. Chen and P. Zerwas, *Phys. Rev.* **D12**, 187 (1975).

5.4 J. B. Kogut and L. Susskind, *Phys. Reports* **8C**, 77 (1973).

5.5 Near $x = 1$ the radiative process $q \rightarrow qg$ becomes indistinguishable from the elastic process $q \rightarrow q$. To order α_c one must include the interference between the vertex correction of Fig. 18c and the zeroth order coupling of the probe to the quarks. As usual with such infrared singularities, the divergence in the vertex correction cancels the divergence in the emission amplitude yielding a finite result.

5.6 For a discussion see G. B. West, *Phys. Rev. Lett.* **54**, 2576 (1985).

5.7 M. Gell-Mann and F. E. Low, *Phys. Rev.* **95**, 1300 (1954).

5.8 For a general analysis of the running coupling and its relation to deep inelastic scattering, see N. Christ, B. Hasslacher and A. H. Mueller, *Phys. Rev.* **D6**, 3543 (1973), see also Ref. 5.4 and references quoted in the preface.

5.9 It has been shown that the only renormalizable field theories in 4 dimensional space-time which are asymptotically free are non-Abelian theories (see S. Coleman and D. J. Gross, *Phys. Rev. Lett.* **31**, 851 (1973)).

5.10 K. G. Wilson, *Phys. Rev.* **179**, 1499 (1979). The approach described here is now well-treated in textbooks. See, for example, Cheng and Li and Itzykson and Zuber referenced in the preface.

5.11 See, for example, R. J. Jaffe and M. Soldate, *Phys. Rev.* **D26**, 49 (1982).

5.12 $0^{\mu_1 \cdots \mu_n}$ actually form an irredicible representation of the Lorentz group labelled $(n/2, n/2)$. It contains representations of the rotation subgroup up to a maximum spin-n.

5.13 S. Weinberg, *Phys. Rev.* **118**, 838 (1960).

5.14 Operators of odd twist (*e.g.*, $\overline{\psi}\psi$) are not invariant under the chiral transformation $q \rightarrow \exp(i\alpha\gamma_5)q$ which is a symmetry of QCD with zero mass quarks, so they appear in OPE multiplied by quark masses.

5.15 R. L. Jaffe and M. Soldate in Proceedings of the 1981 Tallahassee Conference on Perturbative QCD, D. Duke and J. Owens, eds. (AIP, New York, 1981), and Ref. 5.11; E.V. Shuryak and A. Vainshteyn, *Phys. Lett.* **105B**, 65 (1981); R. K. Ellis, V. Furmanski and R. Petronzio, *Nucl. Phys.* **B207**, 1 (1982); **B212**, 29 (1983).

5.16 Color indices have been suppressed in Eq. (5.37). ψ_a is a three component complex vector in color: ψ_{ai}. D_μ is a 3×3 martix $(D_\mu)_{ij} = \delta_{ij}\partial_\mu - ig\sum_{a=1}^{8} \frac{\lambda^a_{ij}}{2} A^a_\mu$ where A^a_μ is the four vector potential for the gluon field. It is easy to see that only the + component of $O_{n,a}$ *i.e.*, $0^{(\mu^2)++\cdots+}_{n,a}$ determines $\Theta_{n,a}(\mu^2)$. It is conventional to use "light-cone gauge", $A^+ = 0$, so $D_\mu \rightarrow \partial_\mu$.

5.17 O. Nachtmann, *Nucl. Phys.* **B63**, 237 (1973), **B78**, 455 (1974); R. Barbieri, J. Ellis, M. K. Gaillard and G. G. Ross, *Phys. Lett.* **64B**, 171 (1976), *Nucl. Phys.* **B117**, 50 (1976); H. Georgi and H. D. Politzer, *Phys. Rev. Lett.* **36**, 1281 (1976), **37**, 68E (1976); A. DeRujula, H. Georgi and H. D. Politzer, *Phys. Lett.* **64B**, 428 (1977), *Ann. Phys.* (NY) **103**, 315 (1977); V. Baluni and E. Eichten, *Phys. Rev. Lett.* **37**, 1181 (1976), *Phys. Rev.* **D14**, 3045 (1976).

5.18 Target mass corrections enter the operator product expansion when the operators are made traceless and symmetric. If the complete tensor in Eq. (5.31) is retained through the analysis, the polynomial in M^2/q^2 which results generates the target mass corrections.

5.19 These are "current quark" masses. The constituent masses which are larger and appear in non-relativistic quark models are non-perturbative in origin and appear here among higher twist effects.

5.20 R. L. Jaffe and G. G. Ross, *Phys. Lett.* **93B**, 313 (1980).

5.21 R. L. Jaffe, *Phys. Rev.* **D11**, 1953 (1975).

5.22 L. Baulieu and C. Kounnas, *Nucl. Phys.* **B155**, 429 (1979).

6.1 F. E. Close, R. G. Roberts and G. G. Ross, *Phys. Lett.* **129B**, 346 (1983).

6.2 A superficially similar but fundamentally quite different approach was proposed by O. Nachtmann and H. J. Pirner, *Z. Phys.* **C21**, 277 (1984).

6.3 R. L. Jaffe, F. E. Close, R. G. Roberts and G. G. Ross, *Phys. Lett.*, **134B**, 449 (1984).

6.4 F. E. Close, R. L. Jaffe, R. G. Roberts and G. G. Ross, *Phys. Rev.* **D31**, 1004 (1985).

6.5 T. H. Hanssen and R. L. Jaffe, *Phys. Rev.* **D28**, 882 (1983).

6.6 J. Negele, private communication.

6.7 V. Gribov, in Proceedings 4th Winter Seminar on the Theory of the Nucleus and the Physics of High Energies (Academy of Sciences, U.S.S.R., 1969); J. D. Bjorken, in Particle Physics (Irvine Conference, 1971), M. Bander, G. L. Shaw and D. Y. Wong, eds., (AIP Conference Proceedngs No. 6, AIP, New York, 1972); S. J. Brodsky in Quarks and Nuclear Forces, Springer Tracts in Modern Physics, No. 100 (Springer-Verlag, Berlin, 1977); V. I. Zakharov and N. N. Nickolaev, *Yad. Fiz.* **21**, 434 (1975), *Sov. J. Nucl. Phys.* **21**, 227 (1975); A. H. Mueller, in Proceedings of the XVII Rencontre de Moriond, J. Tran Thanh Van, ed. (Editions Frontières, Dreux, 1982); J. D. Bjorken, (unpublished, private communication from A. H. Mueller).

6.8 M. S. Goodman, *et al.*, *Phys. Rev. Lett.* **47**, 293 (1981).

6.9 This description can be formulated in an infinite momentum frame. There the virtual photon does not create pairs but instead scatters from a sea of quark-antiquark pairs. See, for example, the paper by Mueller under Ref. 7 for a discussion of the relation between the two pictures.

6.10 The following material is derived largely from discussions with A. H. Mueller and unpublished notes by J. D. Bjorken.

NUCLEAR PHYSICS FROM THE QUARK MODEL WITH CHROMODYNAMICS

Nathan Isgur
University of Toronto
Toronto, Canada M5S 1A7

It now seems nearly certain that nuclear physics can be derived from quantum chromodynamics. In these lectures I describe the ideas and methods underlying a possible route to such a derivation through the quark model, and some of the progress that has already been made along this path.

PREFACE

The ideas and work described in these lectures are not presented in chronological order. In fact, if they have any value, it will be because the opportunity to lecture on the topics discussed here in La Troisieme Cycle de la Physique en Suisse Romande in the summer of 1984, at the Schladming Winter School in 1985, and at the Los Alamos Summer School in 1985 forced me to recast these developments in a more logical and pedagogical form. I offer my thanks to the organizers of these schools (Pierre Extermann at Geneva, Heinrich Mitter and Willie Plessas at Graz, and Mikkel Johnson at Los Alamos) for their interest and encouragement in this task.

Since these lectures are mainly a reordering and integration of published work, I would like to explicitly acknowledge at the outset my principal collaborators in the areas described here. In chronological order these are Gabriel Karl (baryons), Roman Koniuk (decays), John Weinstein (the $qq\bar{q}\bar{q}$ system), Kim Maltman (the 6q system), Steve Godfrey and Simon Capstick (the relativised quark model for mesons and baryons), and Jack Paton (the flux tube model).

Finally, I apologise in advance for the fact that these lectures will concentrate on developments as seen from Toronto; they do not do justice to the work of many others who have contributed to this field, although occasional references to very closely related work are given. For more complete references the interested student should refer to the works cited.

Nathan Isgur
Toronto, March 1985

I. INTRODUCTION AND OUTLINE

It now seems nearly certain that nuclear physics can be derived from quantum chromodynamics (QCD). In these lectures I will describe a possible route to this derivation via the quark model, and some preliminary (but very encouraging) results which have already been obtained.

I will begin at the beginning: with QCD. From these foundations I will discuss a flux tube model for hadrons based on the strong coupling limit of Hamiltonian lattice QCD. This model will provide a framework which guides (and may eventually justify) the remainder of our discussion. The first stop along our route is to consider the nature and validity of the quark model in QCD. These considerations naturally lead to a recollection of the great successes of the quark model in helping us to understand mesons and baryons; I will take this opportunity to point out some recent progress in removing the derogatory "non-relativistic" prefix usually attached to the model. The remainder of the lectures will be devoted to the extension of the quark model to multiquark systems. These systems are characterized by the fact that they allow many internal colour couplings, in contrast to $q\bar{q}$ and qqq where this coupling is unique. We will first discuss such systems in general, using the simplest multiquark system $qq\bar{q}\bar{q}$ as an example occasionally. With this preparation we will finally turn to the 6q system, from which we will see that we can begin to understand many of the basic features of nuclear physics from the quark model.

II. QCD AND THE QUARK MODEL

While it now seems likely that quantum chromodynamics (QCD) is the correct dynamics underlying the strong interactions, it has not been possible to put this theory to many very severe tests. The most rigorous tests to date have been those in which, via asymptotic freedom, perturbative techniques can be applied, but even in such "simple" processes as deep inelastic scattering these tests remain imprecise. When we turn to phenomena in which the strong interactions are really strong, then aside from some very general features we have almost no rigorous tests of this theory: the calculation of hadronic masses, intrinsic moments, and couplings is in a very rudimentary stage of development. Even many qualitative features of this strong interaction problem remain unclear: for example, do pure glue states exist, and if so are they narrow or broad?

Some progress in the strong interaction problem has been made, however. On one front, exact analytic relations can be derived in the form of sum rules which, with suitable techniques, can reveal many features of the physics of QCD. However, at the moment the greatest hope for eventually extracting precise predictions from QCD seems to lie with methods for solving the theory numerically via lattice techniques. Lattice simulations have, for example, already provided us with convincing (if not conclusive) evidence that QCD is a confining theory. Also, numerous attempts have been made to numerically calculate on the lattice the low-lying hadronic spectrum, with some qualitatively encouraging results. It is very unclear, however, how much further these ab initio calculations can be taken with forseeable computer capacity.

By contrast, the quark model has had considerable success in systematizing many of the details of meson and baryon spectra, intrinsic moments, and couplings, especially after the naive quark model was supplemented with certain simple dynamical features characteristic of QCD.

In this first lecture we will discuss a model for hadrons based on ideas gleaned from the strong coupling lattice Hamiltonian formulation of QCD. The model contains the ordinary quark model as a special case, but includes as well pure glue states, hybrids (which have both quark and gluonic degrees of freedom in evidence), and other exotics. This model is not QCD but it may be useful as a guide in the present period when rigorous solutions for QCD are unavailable; even later, given the inevitable complexity, it is possible that the model will continue to be useful as a simple qualitative picture of the dynamics of QCD.

A. A Review of QCD

QCD is a non-abelian generalization of QED in which local U(1) phase invariance is replaced by a local SU(3) invariance acting on a new three dimensional internal degree of freedom called colour.[1] If $\psi_\alpha(x)$ (α=1,2,3) is a Dirac field, then under a local (i.e., x-dependent) SU(3) transformation,

$$\psi_\alpha(x) \to \psi'_\alpha(x) = \{\exp(-ig\sum_{i=1}^{8} \Theta^i(x) \frac{\lambda^i}{2})\}_{\alpha\beta}\psi_\beta(x) \quad (1)$$

where the λ^i are the usual eight 3x3 generators of SU(3) and $\Theta^i(x)$ are "phases". If the Dirac Lagrangian is to be invariant under such transformations, then there must be a set of eight minimally coupled compensating vector fields G_i^μ (i =1,...8) (the gluons); by this reasoning one is led to

$$L_{QCD} = \sum_{\text{flavour}} \bar{q}\{\gamma_\mu[i\partial^\mu - g\frac{\lambda^i}{2}G_i^\mu] - m_q\}q - \frac{1}{4}F_{i\mu\nu}F_i^{\mu\nu} \quad (2)$$

where $F_i^{\mu\nu}$ is the QCD field strength tensor

$$F_i^{\mu\nu} = \partial^\mu G_i^\nu - \partial^\nu G_i^\mu - g f_{ijk} G_j^\mu G_k^\nu \quad (3)$$

in which the f_{ijk} are defined via

$$[\frac{\lambda^i}{2}, \frac{\lambda^j}{2}] = i f_{ijk}\frac{\lambda^k}{2} \quad (4)$$

i.e., they are the structure constants of SU(3). Notice that the "extra term" in (3) arises here and not in QED because the gauge group is non-abelian. Thus under the gauge transformation which leads to (1) the gluon fields are both "shifted" and "rotated" so that the QED gauge shift $A^\mu(x) \rightarrow A'^\mu(x) = A^\mu(x) + \partial^\mu\Theta(x)$ becomes here (in the form appropriate to differential phases $\delta\Theta^i(x)$)

$$G_i^\mu(x) \rightarrow G_i'^\mu(x) = G_i^\mu + \partial^\mu\delta\Theta^i + gf_{ijk}\delta\Theta_j G_k^\mu \qquad (5)$$

It is this extra non-linear term which responsible for the remarkable properties of confinement and asymptotic freedom; it is also this innocent-looking term which makes QCD so difficult to solve.

Although it is not the regime in which we are most interested here, we briefly review the phenomenon of asymptotic freedom, which has allowed QCD to be applied with success using perturbative techniques to certain processes involving short-distance physics. At first sight it seems paradoxical that a theory of strong interactions, in which, naively, g would seem required to be large, could ever be treated by perturbation theory. The first step in understanding this is to realize that the coupling constant g appearing in (2) is the "bare" coupling constant and, as such, is not necessarily simply related to the couplings one will observe in various physical processes. Quantum chromodynamics, like any quantum field theory, only makes sense if it is defined (regulated) in terms of some renormalization scheme since the point-like operators appearing in L_{QCD} are highly divergent in character. In the next section we will define QCD in terms of a lattice regularization scheme, but it is more normal to cut the theory off at some very large mass scale M. A renormalizable theory, like QCD, has the property that its predictions for processes with intrinsic scales much less than M are independent of M. Consider, in this light, the quark-gluon coupling at some scale $Q^2 \ll M^2$. We would calculate

$$\alpha_s(Q^2) = f(g,Q^2,M^2) \equiv \alpha_s(g,Q^2,M^2) \qquad (6)$$

where f is, for example, obtained by the sum of all relevant Feynman diagrams with integrations cut off at the scale M. Now for a given bare g and cut off M we see that we can calculate $\alpha_S(Q^2)$ at any Q^2; consequently, any measurement of α_S can be used to fix the bare value of g. This choice of g will, however, depend on M since, crudely speaking, g is acting not as the bare coupling of the point-like L_{QCD} but rather as the bare coupling of the M-regulated theory. Thus, given an observed $\alpha_s(Q^2)$, we must choose $g = g(M^2)$; it is this observation that leads to the renormalization group equations. In QED such an analysis indicates that the bare electric charge $e(M^2)$ must be increased with increasing M^2 to maintain α at its observed value of 1/137 at low Q^2; this fact can be attributed to the screening of the bare charge by vacuum polarization, much as an ordinary charge is screened in a dielectric medium. In QCD one finds that the bare colour charge $g(M^2)$ must be decreased with increasing M^2 in order to maintain the physical coupling $\alpha_s(Q^2)$ at fixed values. This observation then leads to the conclusion that for sufficiently small scale phenomena, QCD will behave like a theory with a small coupling and so may be studied by perturbation theory. Conversely, analysis leads to the conclusion that $\alpha_s(Q^2)$ inevitably grows as Q^2 is decreased, until at some characteristic scale Λ_{QCD} one finds $\alpha_s(\Lambda^2_{QCD}) \sim O(1)$, at which point the perturbative analysis fails and along with its failure the possibility that QCD confines emerges.

In the perturbative regime the colour analogs of electrons and photons, namely quarks and gluons, are the obvious degrees of freedom to employ in calculations, and all of the familiar tools of perturbative quantum field theory rather naturally apply. In the low Q^2, strong coupling regime it turns out to be useful to abandon these familiar elements in favour of others.

B. Strong Coupling Hamiltonian Lattice QCD

Most numerical studies of QCD have been done using a four dimensional lattice on which Feynman path integral techniques can be employed. We are interested here not in numerical but rather in conceptual developments and so will concentrate instead on a Hamiltonian version of QCD in which only space and not time is discretized.[2)] In this formulation the spatial lattice spacing a will play the role of the regulator mass M in standard quantum field theory. Latticizing the theory also has another advantage: it will allow us to set up a strong coupling perturbation expansion in which the expansion parameter for lattice QCD is 1/g instead of g. We may expect to be able to learn more about the strongly coupled regime of the theory in terms of such an expansion, and indeed this seems to be the case: for example, confinement is an automatic property of the $g \to \infty$ limit of lattice QCD. Moreover, the natural degrees of freedom of the strong coupling regime are not quarks and gluons, but rather quarks and flux tubes, the latter being more in accord with various qualitative ideas on the nature confinement in QCD. Of course space is not coarse-grained (at least not on the scale of 10^{-15} meters), so that to relate lattice QCD to real QCD we must consider the limit $a \to 0$. In this limit, as we have discussed, $g \to 0$ so that a strong coupling expansion must fail; this is just the other side of the failure of the weak coupling expansion for small Q^2. If, however, it can be shown that the two regimes "match" around g=1, thereby proving that lattice QCD as $a \to 0$ is QCD, then one would nevertheless expect the strong coupling expansion to be useful in many situations where large scales dominate, just as the weak coupling expansion is useful for short distance physics.

A simple analogy may be useful. Consider approximating a continuous one dimensional harmonic oscillator by a particle hopping along a one dimensional lattice of points $x = na$ $(n = \ldots,-2,-1,0,1,2,\ldots)$ with lattice spacing a. The lattice Hamiltonian could be chosen to be

$$H_{mn} = \left(\frac{1}{ma^2} + \frac{1}{2}ka^2n^2 \right)\delta_{mn} - \frac{1}{2ma^2}\left(\delta_{m,n+1}+\delta_{m,n-1}\right) \quad (7)$$

since then the Schrodinger equation

$$i\frac{\partial\psi_m}{\partial t}(t) = H_{mn}\psi_n(t) \quad (8)$$

becomes

$$i\frac{\partial\psi(x,t)}{\partial t} = \left[- \frac{1}{2m}\frac{\partial^2}{\partial x^2} + \frac{1}{2}kx^2\right] \psi(x,t) \quad (9)$$

as $a \to 0$. Now for $a \to \infty$ with k and m fixed, the potential energy term $\frac{1}{2}ka^2n^2\delta_{mn}$ dominates and the eigenstates correspond to the particle sitting on single lattice sites; corrections to this limit are of relative order $\chi = 1/kma^4$ and one can proceed to systematically do perturbation theory in this hopping strength. Since the characteristic scale of the harmonic oscillator is

$\alpha^{-1} = (km)^{-1/4}$, one will not get realistic wave functions or eigenenergies for the harmonic oscillator for $a >> \alpha^{-1}$ where lowest order perturbation theory applies, but for $\chi \sim 1$ one begins to get good approximations to the solutions of the continuum problem if one works to sufficiently high order in χ. By contrast, starting with free particle solutions to the continuum Hamiltonian and treating $\frac{1}{2}kx^2$ as a perturbation is hopeless. (The difference, of course, is that the hopping parameter expansion for the ground state, for example, will be accurate if a matrix of dimension of order $\frac{1}{a\alpha}$ is diagonalized.)

We now turn to the formulation of QCD on a (cubic) spatial lattice. In this formulation the quark degrees of freedom of the theory "live" on the lattice sites while the gluonic degrees of freedom "live" on the links between these sites (see Figure 1). Let's consider first the theory without quarks: we describe this theory in terms of link variables U_ℓ which (before quantization) are 3x3 SU(3) group elements. The pure gauge field Hamiltonian is then the sum of two parts, one involving only the U's and one which has non-trivial commutation relations with the U's:

$$H_{glue} = \frac{g^2}{2a} \sum_\ell C_\ell^2 + \frac{1}{ag^2} \sum_p Tr[2-(U_{\ell 4}U_{\ell 3}U_{\ell 2}U_{\ell 1}+h.c.)] \qquad (10)$$

with a the lattice spacing and g the corresponding coupling constant. Here C_ℓ^2 is defined in terms of the eight generators $E^a_{\ell\pm}$ of SU(3) transformations of U_ℓ at the beginning (-) or the end (+) of the link ℓ

$$[E^a_{\ell+} , U_\ell] = - \frac{\lambda^a}{2} U_\ell \qquad (11)$$

$$[E^a_{\ell-} , U_\ell] = + U_\ell \frac{\lambda^a}{2} \qquad (12)$$

by $C_\ell^2 = \Sigma_a (E^a_{\ell+})^2 = \Sigma_a (E^a_{\ell-})^2$. In the second term the product of the U's is taken in order around the plaquettes p. To complete lattice QCD one simply adds to (10) a lattice Hamiltonian H_{quark} for the quarks interacting with the glue. With the quark fields as site variables we have

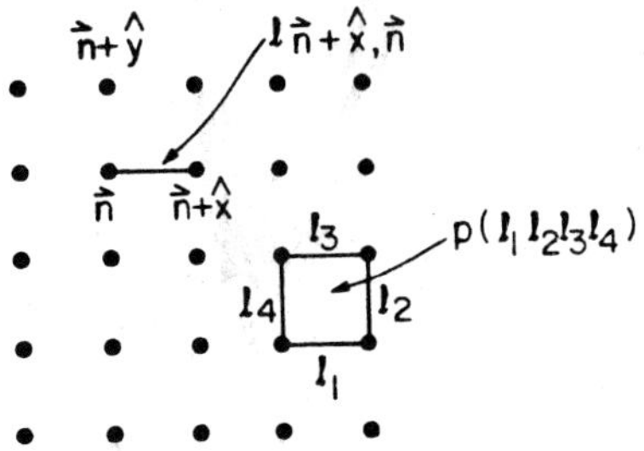

FIG. 1. a two dimensional (x,y) slice of the lattice showing a typical lattice point $\vec{n}=(n_x,n_y,n_z)$, a typical link $\ell_{\vec{n}+\hat{x},\vec{n}}$ from $\vec{n}$ to $\vec{n}+\hat{x}$, and a typical plaquette $p(\ell_1\ell_2\ell_3\ell_4)$

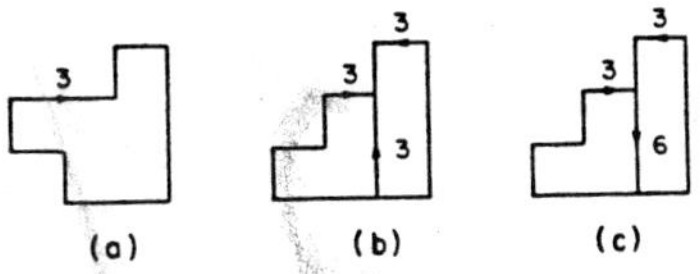

FIG. 2. some primitive pure glue states

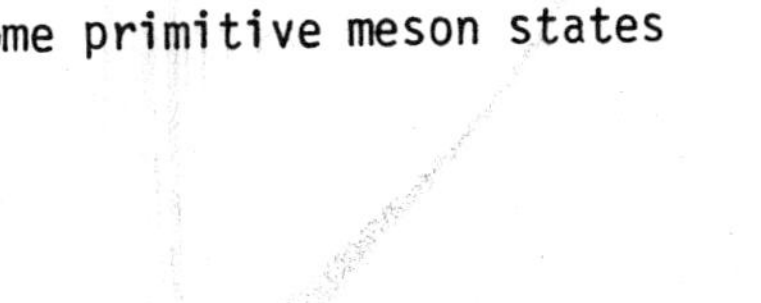

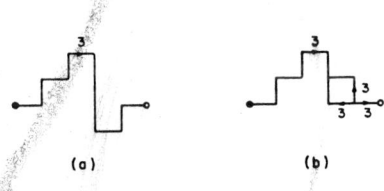

FIG. 3. some primitive meson states

$$H_{quark} = \sum_{flavour} m_q \sum_{\vec{n}} q^\dagger_{\vec{n}} q_{\vec{n}} + \frac{1}{a} \sum_{\substack{flavour \\ \vec{\ell}_{ji}}} q^\dagger_j U_{\ell_{ji}} \alpha_{\ell_{ji}} q_i \qquad (13)$$

where $\alpha_{\ell_{ji}}$ is the Dirac matrix in the direction of the link ℓ_{ji}.

Our complete Hamiltonian $H^{lattice}_{QCD} = H_{glue} + H_{quark}$ has H_{QCD} in $A^0 = 0$ gauge as its naive continuum limit; it is, furthermore, invariant under arbitrary gauge transformation at the lattice sites. Gauss' law takes the form of a constraint in the theory that the only physically relevant states are those which are gauge invariant. Finally, we mention that the quark Hamiltonian as written suffers from what is known as the "species doubling problem", but there are several methods for correcting this problem and this technical point needn't concern us here.

C. The Spectrum of QCD

We are now ready to consider the properties of $H^{lattice}_{QCD}$. We note first that in the strong coupling limit where a (and as we shall see, therefore g) is large

$$H^{lattice}_{QCD} \to H_{sc} = \frac{g^2}{2a} \sum_{\ell} C^2_{\ell} + \sum_{flavour, \vec{n}} m_q q^\dagger_{\vec{n}} q_{\vec{n}} \qquad (14)$$

The eigenvalues of C^2_ℓ are just those of the square Casimir of SU(3): 0 for the singlet, 4/3 for 3 or $\bar{3}$, 10/3 for 6 or $\bar{6}$, 3 for the octet, etc. The quark part of H_{sc} is, on the other hand, diagonalized by an arbitrary number of quarks and antiquarks at arbitrary lattice sites. Since, however, the only physically relevant eigenstates are those which are gauge invariant, the strong coupling eigenstates may be classified as follows:

1) the strong coupled vacuum: In this case all links are unoccupied ($C^2_\ell = 0$) and there are no fermions; the total energy is zero.

2) the pure glue sector: There are still no quarks, but links are excited in such a way that gauge invariant states are produced. The simplest such pure glue states ("glueloops") have a closed path of links in the 3 (or $\bar{3}$) representation. These have energy $\frac{2g^2 L}{3a^2}$ where L is the length of the path; the simplest such state just has the links around the perimeter of an elementary plaquette excited: Tr $[U_{\ell_4} U_{\ell_3} U_{\ell_2} U_{\ell_1}]|0\rangle$, where $|0\rangle$ is the vacuum. Of course more complicated configurations are

allowed, including those with non-triplet flux and those with more complicated topology. For example, three flux links can emerge from a single lattice site since a gauge invariant combination can be formed with the ε_{ijk} invariant tensor. See Figure 2.

3) <u>the meson sector:</u> The simplest quark-containing state consists of a quark and antiquark on the lattice joined by a path of flux links (for gauge invariance). These will have energy $m_q + m_{\bar{q}} + \frac{2}{3}\frac{g^2 L}{a^2}$ so that we automatically have quark confinement in strong coupling. See Figure 3.

4) <u>the baryon sector:</u> The next simplest quark-containing state consists of three quarks connected by an ε_{ijk}-type flux junction. Such quarks will also be confined. See Figure 4.

5) <u>multiquark sectors</u>: When there are more quarks than those required for a meson or baryon, then in general the system will not be completely confined. The simplest such system consists of two quarks and two antiquarks. See Figure 5.

With these examples, the general structure of the eigenstates of the strong coupling limit is clear: it consists of "frozen" gauge invariant configurations of quarks and flux lines. Of course these are not the eigenstates of QCD, but they do form a complete basis (in the limit $a \to 0$) for the expansion of the true strong interaction eigenstates.

The full eigenstates of QCD can be found (in principle!) by considering corrections to the strong coupling limit from the terms we have neglected so far. These terms can induce a variety of effects. Consider first of all the $q^+ U \alpha q$ term. It can, among other things,

1) annihilate a quark at one point and recreate it at a neighbouring point with an appropriate flux link (Figure 6a),

2) break a 3-flux line and create a pair (Figure 6b)

This term thus plays a role analogous to both the usual quark kinetic energy term and the quark-gluon coupling term of the weak coupled theory. Next consider the $\mathrm{Tr}[2-(U_{\ell_4} U_{\ell_3} U_{\ell_2} U_{\ell_1} + h.c)]$ term. It can, among other things,

1) allow flux to hop across plaquettes (Figure 7a)

2) change flux topology (Figure 7b)

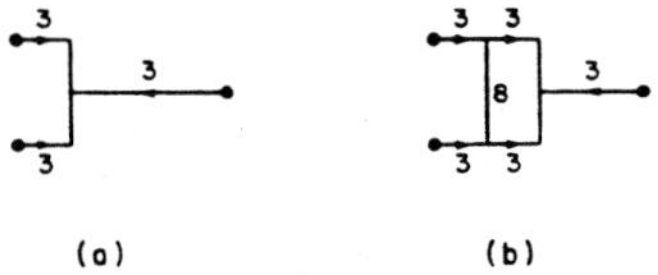

FIG. 4. some primitive baryon states

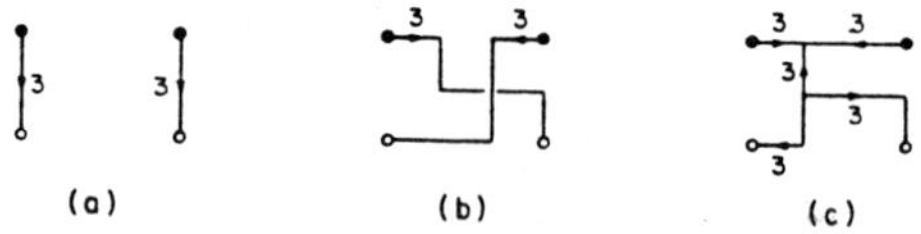

FIG. 5. some primitive $qq\bar{q}\bar{q}$ states

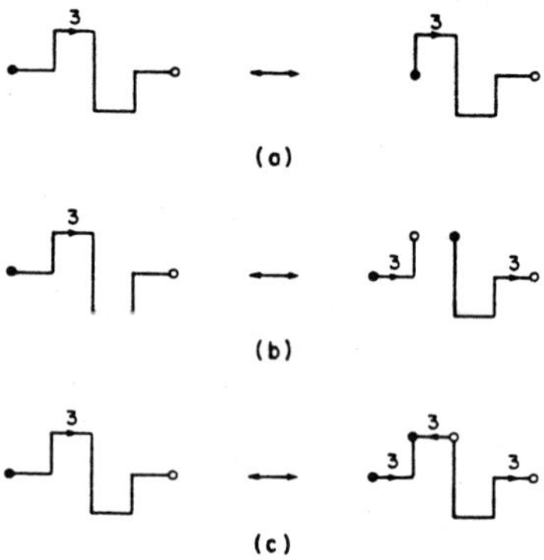

FIG. 6. some effects of $q^{\dagger}U\alpha q$: (a)quark hopping, (b)flux breaking pair creation, (c)$q\bar{q}$ "seeding"

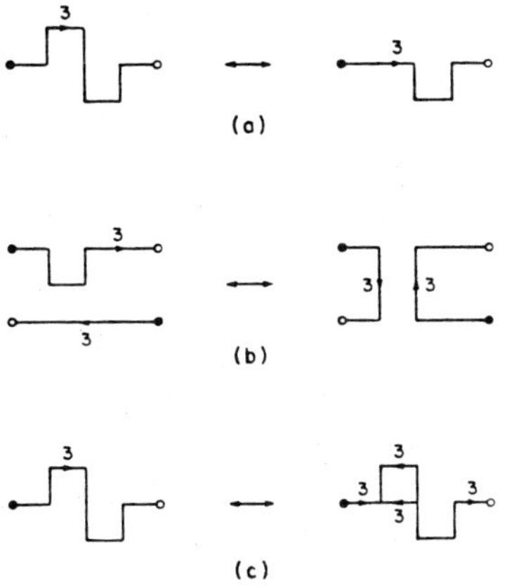

FIG. 7. some effects of Tr[2-(UUUU+h.c.)]: (a)flux tube hopping, (b)flux tube topological mixing by rearrangement, (c)flux tube topological mixing by "bubble formation"

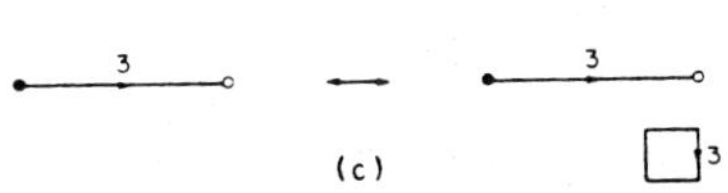

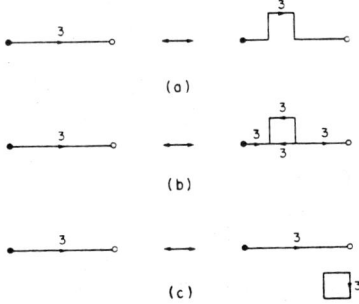

FIG. 8. some lowest order corrections to the string tension

We now consider an instructive calculation on the departures from the strong coupling limit: we will examine the effects of such departures on quark confinement by considering two infinitely massive quarks separated by N lattice sites along an axis of the lattice. In the strong-coupled limit, ignoring the constant $2m_Q$,

$$E_{Q\bar{Q}} - E_{vac} = \frac{2g^2}{3a^2} L = \frac{2g^2}{3a} N \qquad (15)$$

Since we wish to associate this energy with the linear potential bL of quark models (to be discussed below), we would require that $\frac{2g^2}{3a^2}=b$. This in turn shows that the bare coupling g must be chosen to depend on a: $g^2(a) = \frac{3}{2} ba^2$, and we see already that as we let a decrease so that we can recover the continuum limit, we will eventually encounter values of g that will require consideration of corrections to the strong coupling limit. In the case at hand the lowest order corrections will consist of

1) corrections to E_{vac} via mixing between the strong coupled vacuum and states with single excited plaquettes

2) corrections to $E_{Q\bar{Q}}$ via mixing between the straight line of flux and flux configurations with a) kinks, b) bubbles of flux, and c) disconnected excited plaquettes (Figures 8a,b, and c).

The result of such a second order calculation is that

$$E^{(2)}_{Q\bar{Q}} - E^{(2)}_{vac} = \frac{2g^2}{3a^2} L \left[1 - \frac{11x^2}{204}\right] \qquad (16)$$

where $x=\frac{2}{g^4}$ is the strong coupling expansion parameter. We see that we still have a linear confinement potential with

$$b^{(2)} = \frac{2g^2}{3a^2}\left[1 - \frac{11}{51g^8}\right] \quad . \qquad (17)$$

Note that to maintain the same physical string tension we must now choose

$$g^2(a) = \frac{3}{2} ba^2\left[1 + \frac{176}{4131b^4a^8}\right] \quad . \qquad (18)$$

Clearly this calculation can be carried out in principle to arbitrarily high order. When one does this one will find

$$b = \frac{1}{a^2} f(g^2) \tag{19}$$

where $f(g^2)$ is a pure power series expansion in $1/g^2$. If QCD is to confine then we see that as $a \to 0$ we must be able to choose $g(a)$ in such a way that b remains fixed at its physical value. This requirement, along with the requirement that, as $a \to 0, g(a)$ must behave as deduced from weak coupling, shows that confinement, while natural, is not trivial in lattice QCD. Incidentally, from the fact that $\frac{db}{da}=0$ and from knowledge of $g(a)$ from weak coupling, one can deduce that as $a \to 0$

$$f(g^2) = \frac{b}{\Lambda^2} [\beta_0 g^2]^{-\beta_1/\beta_0^2} \exp(-1/\beta_0 g^2) \tag{20}$$

where Λ is a constant, and where $\beta_0 = \frac{11}{16\pi^2}$ and $\beta_1 = \frac{102}{256\pi^4}$ are the first two coefficients in the expansion of the perturbative "beta function" of asymptotic freedom fame. Note that $f(g^2)$ has no power series expansion near $g^2=0$ proving that confinement cannot be seen in QCD via perturbation theory about weak coupling. This clearly makes the strong coupling expansion the more attractive starting point for understanding hadron physics.

D. The Flux Tube Model

Even after successful numerical calculations within QCD are possible, it will still be useful to have models which summarize the very complex structure of this theory. Many such models have already been proposed, including potential models, bag models, string models, etcetera. We will base our discussion instead on a simple model[3)] which emerges from the considerations of the previous sections on Hamiltonian lattice QCD.

To illustrate this model in the simplest possible context, consider first a heavy $Q\bar{Q}$ system and ignore light quarks. Departures from the strong coupling limit will have a number of effects, but among them will be flux hopping terms which will convert the frozen flux lines of the strong coupling limit into dynamical strings which can sustain transverse oscillations. (The conversion from frozen to fluid strings corresponds to the "roughening transition" seen in numerical lattice work). We accordingly first imagine summing all such terms so that we can deal with the $Q\bar{Q}$ pair connected by a dynamical string. Next consider departures from strong coupling that involve "topological mixing" either to other connected string states or to states with disconnected vacuum fluctuations. If L is large we can hope to approximate the state of the system in terms of a lattice of scale a with $L \gg a \sim \lambda_0$, where λ_0 is the scale where g=1 and topological mixing becomes important. Our model for this system is thus that of a discrete string.[4] The ground state of this system for infinitely heavy quarks will consist of the $Q\bar{Q}$ pair with a ground state string stretched between them, the first excited state will be doubly degenerate with either a right-handed or left-handed phonon excited in the lowest string mode, etcetera. As the distance R between the Q and $\bar{Q}$ is varied slowly, the eigenenergy of the string eigenstate S will trace out an adiabatic potential $V^{(S)}(R)$; we associate such potentials with an adiabatic approximation to the physics of mesons. When the $Q\bar{Q}$ pair move in the adiabatic potential bR of the ground state of this QCD string, one recovers the usual spectrum of mesons in the quark model. When the pair moves in the adiabatic potential of an excited string, the resulting hadrons correspond to a new species not contained in the usual quark model: hybrid mesons with both quark and gluonic degrees of freedom in evidence.

The baryon sector, while more complicated, is analogous to the mesons: the ordinary baryons of the quark model correspond to three quarks moving in the adiabatic potential of the ground state of the three junction (Y) string, while excited strings will lead to hybrid baryons. The phenomenological implications of hybrid mesons and baryons will be discussed later.

In multiquark systems we are forced for the first time to go beyond the simple vibrating string picture to consider topological mixing. This is because in such systems adiabatic surfaces will always cross in the absence of mixing, as can be seen by considering Figure 7(b) for the case when the $qq\bar{q}\bar{q}$ system is arranged at the corners of a square. Such systems are consequently considerably more complicated than those we have already considered. They are the main subjects of the last two Chapters.

Of course the simple classification scheme we have presented here is both incomplete and inaccurate. For example, one must expect to find (at high masses) "topological meson hybrids" in which the $Q\bar{Q}$ system moves in the potential of more complicated string topologies. There will also be mixing between topologies via the $\mathrm{Tr}[U_{\ell_1}U_{\ell_2}U_{\ell_3}U_{\ell_4}+h.c.]$ and $q^{\dagger}U\alpha q$ terms and deviations from the adiabatic limit that mix states with different degrees of phonon excitation. While all such effects can be described within the context of this model, its utility will depend on the degree to which they can be ignored, at least for low-lying states. Perhaps the very success of the naive quark model for mesons and baryons provides some evidence in favour of the viability of such a separation.

E. The Interface between Strong and Weak Coupling

While the model just presented may be useful for large distance QCD, it must fail at short distances. Consider the $Q\bar{Q}$ system once again, but now for distances $R = O(\lambda_0)$. To describe the physics of this regime will clearly require a lattice spacing $a < \lambda_0$ where g is small so that a strong coupling expansion becomes, at best, less useful. In particular, at these distance scales, the vacuum will be full of fluctuations and one must expect strong topological mixing as well. It is, therefore, natural to expect that at such scales the string will be mixed by new physics beyond simple flux hopping so that it is plausible that the linear potential bR will be pushed to lower energies in the small R regime. This picture fits with what we know must happen: at short distance the $Q\bar{Q}$ potential must become the Coulombic potential $-\frac{4\alpha_s}{3R}$. It is thus reasonable to supplement our string potentials for mesons by the effects of one gluon exchange (as we do), although it is possible that the transition between these two regimes is significantly more complicated than this simple picture suggests. A similar argument applies to the short distance interquark potential in a baryon, but unfortunately there is no analogous argument to suggest the exact form of the short-distance potential in the pure gluon sector.

Since all presently known hadrons straddle these two regimes (although some high orbital angular momentum light quark mesons are mainly in the confining regime and the lowest-lying $b\bar{b}$ states are mainly in the Coulombic region), we spend a moment here on the effective potential from one gluon exchange.

Consider $q\bar{q}$ scattering in the centre-of-mass frame from momenta $\vec{p}_q = -\vec{p}_{\bar{q}} = \vec{p}$ to $\vec{p}'$. With one gluon exchange we get a certain transition amplitude. If we now require a potential V_{eff} which gives exactly the same on-shell transition amplitude, then we get

$$\chi^{\dagger}_{s'}\chi^{\dagger}_{\bar{s}'}V_{eff}\chi_s\chi_{\bar{s}}=\frac{1}{(2\pi)^3}\int d^3Qe^{i\vec{Q}\cdot\vec{r}}\bar{u}(\vec{p}'s')\bar{v}(-\vec{p}\bar{s})G(Q^2)\gamma^{\mu}_q\gamma^{\bar{q}}_{\mu}u(\vec{p}s)v(-\vec{p}\bar{s}') \tag{21}$$

with $\vec{Q}=\vec{p}'-\vec{p}$ and

$$G(Q^2)=\frac{4\pi\alpha_s(Q^2)}{3Q^2}\vec{F}_q\cdot\vec{F}_{\bar{q}} \tag{22}$$

where $F^i_q=\frac{\lambda^i}{2}$, $F^i_{\bar{q}}=\frac{\lambda^{c\,i}}{2}=\frac{-\lambda^{i*}}{2}$ are the q and $\bar{q}$ colour matrices. Multiplying out the spinors and defining $P\equiv(p+p')/2$ then gives

$$\begin{aligned}V_{eff}(\vec{P},\vec{r})=\frac{1}{(2\pi)^3}\int d^3Qe^{i\vec{Q}\cdot\vec{r}}G(Q^2)\Big(&\Big[1-\frac{Q^2}{4E(E+m)}+\frac{i\vec{Q}\times\vec{P}\cdot\vec{S}_q}{E(E+m)}\Big]\\ &\times\Big[1-\frac{Q^2}{4\bar{E}(\bar{E}+\bar{m})}+\frac{i\vec{Q}\times\vec{P}\cdot\vec{S}_{\bar{q}}}{\bar{E}(\bar{E}+\bar{m})}\Big]\\ &+\Big[\frac{\vec{P}-2i\vec{Q}\times\vec{S}_q}{2E}\Big]\cdot\Big[\frac{\vec{P}-2i\vec{Q}\times\vec{S}_{\bar{q}}}{2\bar{E}}\Big]\Big)\end{aligned} \tag{23}$$

where m and $\bar{m}$ are the quark and antiquark masses, $E = (p^2 + m^2)^{1/2}$ and $\bar{E} = (p^2 + m^2)^{1/2}$ their energies, and $\vec{S}_q$ and $\vec{S}_{\bar{q}}$ their spins.

In the non-relativistic limit this expression gives the familiar Breit-Fermi interaction:

$$V^{ij}_{eff} = V^{ij}_{coul} + V^{ij}_{hyp} + V^{ij}_{so} \tag{24}$$

where

$$V^{ij}_{coul} = \frac{\alpha_s(\vec{r}_{ij})}{r_{ij}} \vec{F}_i \cdot \vec{F}_j \qquad (25)$$

is the colour Coulomb interaction,

$$V^{ij}_{hyp} = - \frac{\alpha_s(r_{ij})}{m_i m_j} \times \left\{ \frac{8\pi}{3} \vec{S}_i \cdot \vec{S}_j \delta^3(\vec{r}_{ij}) + \frac{1}{r^3_{ij}} \left[\frac{3\vec{S}_i \cdot \vec{r}_{ij} \vec{S}_j \cdot \vec{r}_{ij}}{r^2_{ij}} - \vec{S}_i \cdot \vec{S}_j \right] \right\} \vec{F}_i \cdot \vec{F}_j \qquad (26)$$

is the hyperfine interaction, and where

$$V^{ij}_{so} \equiv V^{ij}_{so(cm)} + V^{ij}_{so(tp)} \qquad (27)$$

is the spin-orbit interaction with

$$V^{ij}_{so(cm)} = - \frac{\alpha_s(r_{ij})}{r^3_{ij}} \left(\frac{1}{m_i} + \frac{1}{m_j} \right) \left[\frac{\vec{S}_i}{m_i} + \frac{\vec{S}_j}{m_j} \right] \cdot \vec{L}_{ij} \; \vec{F}_i \cdot \vec{F}_j \qquad (28)$$

its colour magnetic piece and with

$$V^{ij}_{so(tp)} = - \frac{1}{2r_{ij}} \frac{\partial V^{ij}_{coul}}{\partial r_{ij}} \left[\frac{\vec{S}_i}{m_i^2} + \frac{\vec{S}_j}{m_j^2} \right] \cdot \vec{L}_{ij} \qquad (29)$$

being the Thomas precession term.

These interactions, especially the colour hyperfine interaction, play a very important role, particularly in the physics of light quark systems.

References for Chapter II.

1. The idea of colour was introduced by O.W. Greenberg, Phys. Rev. Lett. 13, 598(1964). The idea that the strong interactions might be described by a non-Abelian gauge theory of colour originates in H. Fritzsch and M. Gell-Mann, Proc. XVI Int. Conf. on High Energy Physics (National Accelerator Laboratory, 1972) p. 135; H. Fritzsch, M. Gell-Mann, and H. Leutwyler, Phys. Lett. 74B, 365(1973); S. Weinberg, Phys. Rev. Lett. 31, 494 (1973).

2. For reviews of lattice gauge theories and calculations, see, for example, M. Bander, Physics Reports 75, 206(1981); J. Kogut, Rev. Mod. Phys. 55, 775(1983); L. Susskind, in Weak and Electromagnetic Interactions at High Energies, editors R. Balian and C.H. Llewellyn Smith (North Holland, Amsterdam, 1977). Hamiltonian lattice gauge theories were introduced via the SU(2) theory by J. Kogut and L. Susskind, Phys. Rev. D11, 395 (1975).

3. The flux tube picture described here was introduced in N. Isgur and J. Paton, Phys. Lett. 124B, 247(1983). A more complete discussion and references to related models may be found in N. Isgur and J. Paton, "A Flux Tube Model for Hadrons in QCD", to appear in Phys. Rev. D.

4. String theory has a venerable history: see, e.g., S. Mandelstam, Physics Reports 13C, 261(1974).

III. MESONS AND BARYONS

A. Heavy Quarkonia: the Hydrogen Atoms of QCD

Although, as we have seen, progress is being made in solving strongly coupled QCD, quantum chromodynamics is only understood with any precision in its application to short distance phenomena. One place where this understanding can be applied to hadron spectroscopy is in the study of the low-lying states of a heavy $Q\bar{Q}$ (heavy quarkonium) system: for sufficiently heavy quarks such states are so small that the effects of confinement become unimportant. In this situation the system is dominated by the one gluon exchange potential of Section II.E. and so, apart from some colour factors, becomes almost exactly analogous to the e^+e^- (positronium) system.

Figure 9 compares the spectrum of positronium (m_e=0.511 MeV, α=1/137) with a hypothetical heavy quarkonium (m_Q = 0.511 TeV, α_s = 1/13.7). Note that the Bohr levels differ by a factor of 10^8, while fine and hyperfine effects differ by a factor of 10^{10}. Since the same forces at work here prove very important throughout hadron spectroscopy, we briefly review the origin of various energy levels:

1) Bohr levels: these levels result from the Coulombic potential (25) and correspond to a Bohr radius of $(\frac{2}{3}m_Q\alpha_s)^{-1} \simeq 10^{-2}$fm.

2) $^3S_1$ - $^1S_0$ splittings: these splittings arise from the $\vec{S}_i \cdot \vec{S}_j$ part (the "Fermi contact term") of the hyperfine interaction (26) and are the analogs of the 21 cm line of atomic hydrogen; the remainder of the hyperfine interaction (the "tensor interaction") averages to zero in S-waves since it is a tensor of rank two in space.

3) the $^3L_{L+1}$, ^3L_L, ^1L_L, $^3L_{L-1}$ splittings for L ≠ 0: these splittings arise from two sources, the now-operative tensor force of the hyperfine interaction (26) and the spin-orbit interaction (27).

In addition to the above real analogies between these two systems there is also one apparent analogy which is spurious: the 2S-1P splitting. In positronium this "Lamb shift" receives two contributions: a small negative contribution from vacuum polarization (in QED the effective coupling seen by 2S is stronger than that seen by the more spread out 1P state, so its energy is lowered slightly) and a much larger positive contribution from infrared effects associated with the

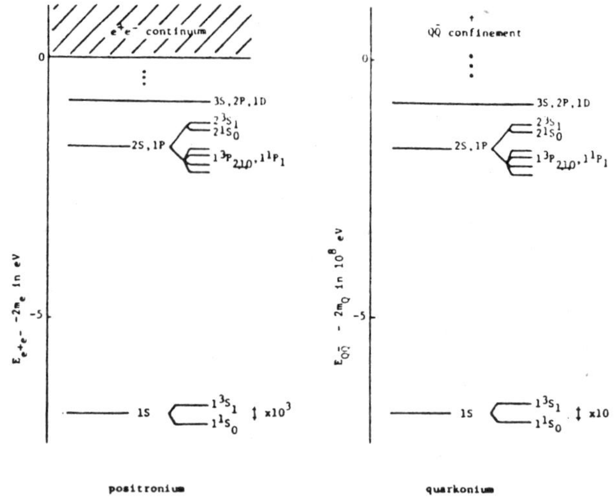

FIG. 9. a comparison of the spectra of positronium and a hypothetical heavy quarkonium with $\frac{4}{3}\alpha_s=10\alpha$ and $m_Q=10^6 m_e$; Bohr levels are shown to scale, but the internal splittings of these levels are only shown very schematically

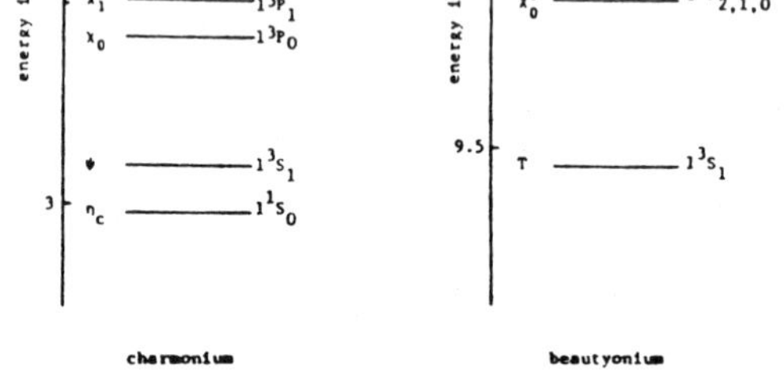

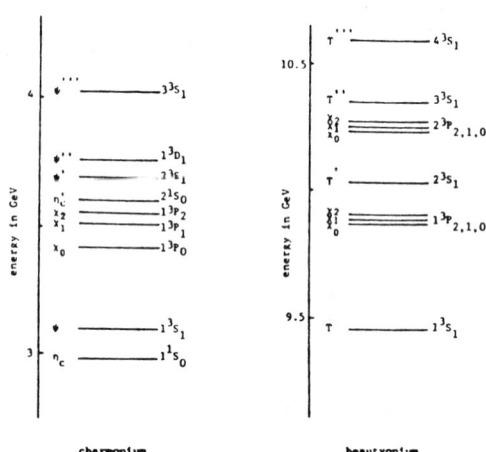

FIG. 10. the experimental spectra of charmonium ($c\bar{c}$) and of beautyonium ($b\bar{b}$); levels are labelled by both their standard names and their quarkonium interpretation

interaction of the system with soft photons. In QCD the same ordering of the states has a quite different origin: now the vacuum polarization makes a positive contribution and is dominant, while since gluons are confined the analog of the strong infrared mixing of e^+e^- with $e^+e^-\gamma$ is suppressed. We mention this minor point since the perturbative 2S-1P splitting of heavy quarkonium provides an important link to our next topic: light quarkonia.

As the quark mass m_q is decreased, the "Bohr radius" of the quarkonium system varies as $(\frac{2}{3}m_q\alpha_s)^{-1}$ (note that both m_q and α_s will vary in this expression since as m_q decreases the running coupling constant will increase). Examining equations (25), (26) and (27) we can see that the ratio of hyperfine and spin-orbit splittings to Bohr splittings will vary like α_s^2 and so as we decrease m_q to regions where $\alpha_s \sim O(1)$ we can expect spectra in which, for example, the $^3S_1$ - $^1S_0$ splitting has become comparable with the 1P-1S splitting. However, more significant than these quantitative changes is the fact that the small 2S-1P splitting, which we think of as characteristic of a Coulombic system (and which is normally in the ratio α_s^3 to the Bohr splittings), will become comparable to the 2S-1S splitting, signalling a large departure of the potential from Coulombic form. This is of course just the perturbative signal for the onset of confinement: in a linear (quadratic) potential the level splittings 1P-1S and 2S-1S are in the ratio of roughly 2 to 3 (exactly 1 to 2).

B. The Extension to Light Quarkonia

As already implied in Section II.E., since all known quarkonium systems straddle the one-gluon-exchange and confinement regimes, their treatment is at the moment dependent on models. We do not attempt to do justice here to the many other alternatives for such models --- foremost among which is the bag model --- but rather stick to the picture we have built up in the previous sections and discuss quarkonium in the confinement regime in terms of a $q\bar{q}$ pair moving in the lowest adiabatic potential of a string. It is then natural to treat quarkonium by considering its full potential to be a two-component potential: the sum of a one gluon exchange potential ($-\frac{4}{3}\frac{\alpha_s}{r}$ which will dominate at small r) and the ground state string potential (br, which will dominate at large r). Indeed, this approach has proved to be immensely successful; the main topic of this chapter will be to briefly review these successful applications.[1)]

Figure 10 shows the observed spectra of the two heaviest quarkonium systems: beautyonium ($b\bar{b}$) and charmonium ($c\bar{c}$). Comparison with Figure 9, in light of our comments on the variation of 2S-1P and of hyperfine and spin-orbit splittings with m_q, reveals immediately the qualitative success of this picture. In fact, the potential model has proved to be much more than a qualitative tool for these systems: it predicts energy levels, total decay rates, decay rates to lepton pairs, and photon transition rates between levels with typically $\pm$ 20% accuracy. There are many problems associated with our understanding of the physics of these systems (some of which will be discussed below) but it seems safe to conclude that the correct basic picture for them is in hand and that future understanding will lead to refinements of this picture.

When we turn our attention to systems which contain light (u,d, or s) quarks, the situation is less clear. One of the main reasons for this obscurity is that the non-relativistic treatment appropriate to $b\bar{b}$ and $c\bar{c}$ becomes itself a serious source of error when light quarks are involved so that it becomes difficult to distinguish inadequacies of the model from inadequacies in the treatment of relativistic effects. Nevertheless, Figure 11 offers what might be considered to be strong evidence that the qualitative physics of quarkonium systems persists even for light quarkonia: the smooth transition of the analogous energy levels from heavy to light quarkonia certainly argues for continuity of the underlying physics. We see first of all from this figure that the $1^3P_2-1^3S_1$ "Bohr level" splitting remains rather constant from the heaviest to the lightest quarkonia. In a Coulombic potential such a splitting would increase in direct proportion to the reduced mass μ while in a linear potential it would decrease like $\mu^{-1/3}$; the approximate constancy in the case at hand emerges automatically as a compromise between these two extremes from the mixture of Coulomb and linear potentials present in the two-component model. Also interesting is the evolution of the $1^3S_1 - 1^1S_0$ splitting which, as expected, increases smoothly from a strength relative to 2S-1S of order $\alpha_s^2 \ll 1$ in $b\bar{b}$ to a strength of order unity in light quark systems where the scales are such that $\alpha_s \sim O(1)$. In addition to these qualitative successes, quarkonium models of mesons containing light quarks have, in studies of both spectra and decays, had considerable quantitative success. Although less accurate than

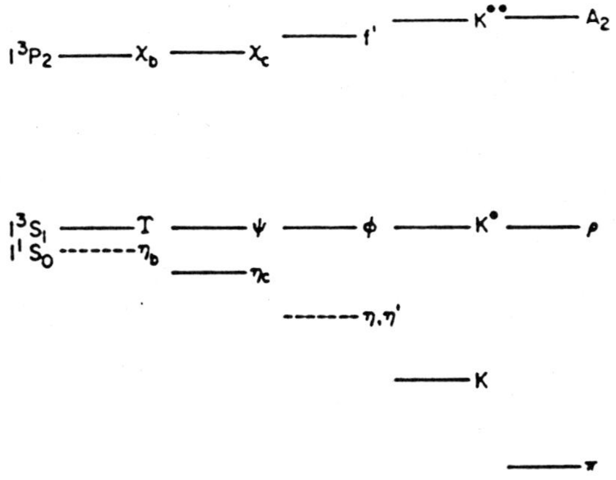

FIG. 11. a graphic illustration of the universality of meson dynamics from the π to the Υ, showing the splittings of 1^3P_2 and 1^1S_0 from 1^3S_1 in the $b\bar{b}$, $c\bar{c}$, $s\bar{s}$, $u\bar{s}$, and $u\bar{d}$ families.

when applied to $b\bar{b}$ and $c\bar{c}$, such models once again seem to give a fairly good account of the masses, decay rates, and static properties of such mesons.

The success of the non-relativistic quark model for mesons (and, as we will soon see, baryons) can probably be understood: while relativistic effects are important in these systems, it seems that they can for the most part be hidden (at least for low-lying states) in the choice of such potential model parameters as the constituent quark mass. When all the dust settles, apparently it is "better to have the right degrees of freedom moving at the wrong speed than the wrong degrees of freedom moving at the right speed".

Of course it would be even better to have the right degrees of freedom moving at the right speed. We have recently shown that such a relativization of the quark model for mesons can be carried out.[2] Some of the features of such a model are:

1) all mesons from the π to the $b\bar{b}$ are treated in the same framework,
2) relativistic kinematics,
3) a semi-relativistic potential including realistic momentum dependences, non-localities, and coordinate smearing.

The resulting picture of mesons is very satisfactory: it provides in a unified framework a good description of the spectroscopy and decays of essentially all known mesons.

Nevertheless, let me conclude this section by mentioning briefly some of the problems of the potential picture which we temporarily "swept under the rug" in our account of its successes. On the theoretical side these problems include:

1) The transition region is unlikely to be simple enough to strictly allow an addition of the two limiting potentials.

2) Stating that the long range potential is linear is insufficient: we must also specify its Lorentz properties or, alternatively, its momentum - dependent properties. These are often taken to be those of a Lorentz scalar interaction in which case a calculation analogous to that leading to (23) leads (with the replacement of $\gamma^\mu \times \gamma_\mu$ by 1×1 and $G(Q^2)$ by an $S(Q^2)$ corresponding to a linear potential) to only one spin-dependent term: a Thomas precession spin-orbit term which tends to cancel the one-gluon-exchange spin-orbit effect. While this

prescription leads to reasonable results, it remains controversial.

3) Two kinds of Fock-space mixing have been neglected: from $q\bar{q}$ to $q\bar{q}q\bar{q}$ via pair creation (including mixing with real and virtual decay channels) and non-adiabatic mixing from the string ground state to excited string states. Preliminary studies of both types of effects indicate that they probably do not change things very drastically, but this remains to be established.

4) The relationship of quarkonium models to the picture of the pseudoscalar mesons as the almost-Goldstone-bosons of broken chiral symmetry remains obscure (although some progress has been made in this direction).

On the phenomenological side there are no major problems with the $b\bar{b}$ and $c\bar{c}$ systems, but for light quarks there may be several. Some examples include:

1) The lightest scalar ($J^P = 0^+$) mesons, the S^* (980) with I = 0 and the $\delta(980)$ with I = 1, do not seem to be 1^3P_0 $q\bar{q}$ states. One possibility, to be mentioned in Chapter IV, is that they are really $q\bar{q}q\bar{q}$ states (for example, $K\bar{K}$ bound states), but the whole scalar meson sector is both experimentally and theoretically confused at the moment.

2) The $J^P = 1^-$ sector, which shows the expected 1^3S_1, 2^3S_1, and 1^3D_1 states in the $c\bar{c}$ sector, is confused for light quarks: two nonets of such states corresponding to 2^3S_1 and 1^3D_1 should have nonstrange members around 1.6 GeV, but at present the signals from various experiments are often apparently inconsistent both theoretically and experimentally.

3) There are at the moment reports of various sorts of unusual states: e.g., the $\iota(1440)$, the $\theta(1690)$, and the $\xi(2200)$. Whether these are $q\bar{q}$ mesons or more exotic objects like meson hybrids or glueballs remains very controversial.

Despite these shortcomings and uncertainties, it is reasonable to consider the potential model for mesons an outstanding success. With it we can make sense of (almost?) all of the hundreds of known mesons and in most cases the model can go well beyond a qualitative description of these states to produce detailed and reasonably accurate predictions for masses, decay rates, and static properties for mesons from the lightest to the heaviest known.

C. Baryons with Light Quarks

It is ironic that the more complex qqq baryon system is better understood experimentally, and for light quarks possibly even theoretically, than the mesons. On the experimental side the reason for this is clear: non-strange and strangeness -1 (S=0 and -1) baryons can be formed as resonances in πN and $\bar{K}N$ scattering, respectively, while no such s-channel resonance production is possible for mesons (except in $J^{PC} = 1^{--}$ via e^+e^-). Why the simple two component potential models work as well as they do for light quark baryons is unclear, but their effectiveness may be related to the fact that the "multibody" charater of these systems makes them somewhat less relativistic than mesons.

In any event, a very simple and analytically solvable model[3)] for qqq has proved quite useful for understanding baryons and so, since an understanding of qqq is obviously essential to our eventual goal of treating 6q, we will devote some time in this section to developing various details of the model. First, however, let us go back to basics.

We expect the lowest-lying baryons to correspond approximately to three quarks moving in the adiabatic potential generated by the ground state of three flux tubes emerging from the quarks and fusing in an ϵ_{ijk} junction (see Figure 4). When two of the quarks are close together, then the flux tube configuration is essentially identical to that of a meson and so we expect the same linear potential br to be in operation. In terms of an effective two-body potential we would say that in this orientation each of the two nearby quarks is attracting the third with a potential 1/2 br. Although the Y junction certainly has three body components, numerical studies have in fact demonstrated that replacing it by a sum of two body potentials with half the strength of mesons provides a good approximation to the spectrum of excited states, and for now we will adopt this approximation in our discussion. This same factor of 1/2, as it turns out, applies to the one-gluon-exchange part of the interquark potential since the expectation values

$$< \frac{\vec{\lambda}_i}{2} \cdot \frac{-\vec{\lambda}_j^{*}}{2} >_{q\bar{q} \text{ in a meson}} = -\frac{4}{3} \qquad (30)$$

and

$$< \frac{\vec{\lambda}_i}{2} \cdot \frac{\vec{\lambda}_j}{2} >_{qq \text{ in a baryon}} = -\frac{2}{3} \tag{31}$$

have the same ratio, so we can actually apply this "$qq = 1/2\ q\bar{q}$" rule uniformly.

Let us now turn to our naive but solvable model: a non-relativistic Hamiltonian with

$$H = H_{si} + H_{sd} \tag{32}$$

where

$$H_{si} = \sum_i \left(m_i + \frac{p_i^2}{2m_i}\right) + \sum_{i<j} \left(\frac{1}{2} br_{ij} + C - \frac{2\alpha_s}{3r_{ij}}\right) \tag{33}$$

is spin-independent and where H_{sd} is a spin-dependent interaction to be introduced later. The quark masses in (33) are the usual non-relativistic constituent quark masses $m_u \simeq m_d \simeq 0.35$ GeV, $m_s \simeq 0.58$ GeV, $m_c \simeq 1.5$ GeV, and $m_b \simeq 5.0$ GeV; the various couplings could in principle be taken from meson spectroscopy.

i) The Spin-Independent Spectrum

We initially concentrate on H_{si} and write

$$H_{si} = \sum_i \left(m_i + \frac{p_i^2}{2m_i}\right) + \sum_{i<j} \left(\frac{1}{2} kr_{ij}^2 + U(r_{ij})\right) \tag{34}$$

$$\equiv H_0 + \sum_{i<j} U(r_{ij}) \tag{35}$$

where $U(r_{ij}) = 1/2\ br_{ij} + C - \frac{2\alpha_s}{3r_{ij}} - 1/2\ kr_{ij}^2$ so that equations (34) and (35) are simply restatements of equation (33). However, written in this form we can solve exactly for the eigenstates of H_o and then, choosing k to minimize the perturbation U in the low-lying states, treat U by perturbation theory. Our first step is therefore to find the eigenstates of H_o. It is convenient in doing so to first distinguish between two cases: the case where all three quarks have equal masses (the S = 0 and -3 sectors uuu, uud, ddu, ddd, and sss), and the S = -1 and -2 sectors (uus, uds, dds, ssu, and ssd)) where one quark has a mass different from the other two.

In the case $m_1 = m_2 = m_3 \equiv m$, the change of variables to

$$\vec{R} = \frac{m_1\vec{r}_1+m_2\vec{r}_2+m_3\vec{r}_3}{m_1+m_2+m_3} \tag{36}$$

$$\vec{\rho} = \frac{1}{\sqrt{2}}(\vec{r}_1-\vec{r}_2) \tag{37}$$

$$\vec{\lambda} = \frac{1}{\sqrt{6}}(\vec{r}_1+\vec{r}_2-2\vec{r}_3) \tag{38}$$

converts H_0 to

$$H_0 = \frac{P^2}{2M} + \left(\frac{p_\rho^2}{2m} + \frac{3}{2}k\rho^2 \right) + \left(\frac{p_\lambda^2}{2m} + \frac{3}{2}k\lambda^2 \right) \tag{39}$$

where M = 3m and $\vec{P}$, $\vec{p}_\rho$, and $\vec{p}_\lambda$ are the momenta conjugate to $\vec{R}, \vec{\rho}$, and $\vec{\lambda}$. Ignoring centre-of-mass motion, the low lying eigenstates of these two degenerate oscillators may be chosen to be of the form $\psi^\sigma_{LM} = \Psi^\sigma_{LM} \frac{\alpha^3}{\pi^{3/2}} \exp(-1/2\alpha^2\{\rho^2+\lambda^2\})$ where

$$N=0 \qquad \Psi^S_{00} = 1 \tag{40}$$

$$N=1 \qquad \Psi^\rho_{11} = \alpha\rho_+ \tag{41}$$

$$N=1 \qquad \Psi^\lambda_{11} = \alpha\lambda_+ \tag{42}$$

$$N=2 \qquad \Psi^{S'}_{00} = \frac{1}{\sqrt{3}}\alpha^2(\rho^2+\lambda^2-3\alpha^{-2}) \tag{43}$$

N=2 $$\Psi^{\rho}_{00} = \frac{2}{\sqrt{3}}\ \alpha^2\ \vec{\rho}\cdot\vec{\lambda} \tag{44}$$

N=2 $$\Psi^{\lambda}_{00} = \frac{1}{\sqrt{3}}\ \alpha^2(\rho^2-\lambda^2) \tag{45}$$

N=2 $$\Psi^{S}_{22} = \frac{1}{2}\ \alpha^2(\rho^2_+ + \lambda^2_+) \tag{46}$$

N=2 $$\Psi^{\rho}_{22} = \alpha^2\ \rho_+\lambda_+ \tag{47}$$

N=2 $$\Psi^{\lambda}_{22} = \frac{1}{2}\ \alpha^2(\rho^2_+ - \lambda^2_+) \tag{48}$$

N=2 $$\Psi^{A}_{11} = \alpha^2(\rho_+\lambda_z - \rho_z\lambda_+) \tag{49}$$

where we have explicitly displayed only the highest state Ψ^{σ}_{LL} of a multiplet of states Ψ^{σ}_{LM} (the others following by normal Clebsch-Gordanry) and where

$$\alpha \equiv (3km)^{1/4} \tag{50}$$

$$\rho_{\pm} \equiv \rho_1 \pm i\rho_2 \tag{51}$$

$$\lambda_{\pm} \equiv \lambda_1 \pm i\lambda_2 \quad . \tag{52}$$

Note that the states have energies $(N+3)\omega$ where

$$\omega \equiv (3k/m)^{1/2} . \tag{53}$$

The notation for σ used here is based on the group S_3 of permutations of three objects: S states transform totally symmetrically, A states transform totally antisymmetrically, while pairs of states (M^{ρ}, M^{λ}) have mixed symmetry under permutations and transform like the two relative coordinates ($\vec{\rho}$,$\vec{\lambda}$), e.g.,

$$(213)\ M^{\rho} = -M^{\rho} \tag{54}$$

$$(213)\ M^{\lambda} = +\ M^{\lambda} \tag{55}$$

$$(132)\ M^{\rho} = \frac{1}{2}\ M^{\rho} + \frac{\sqrt{3}}{2} M^{\lambda} \tag{56}$$

etcetera. This classification is particularly useful since from such states one can readily construct product states of definite symmetry. The relevant multiplication rules are

1) SR transforms like R for R = S, A, M
2) AM^{ρ} and AM^{λ} transform like M^{λ} and M^{ρ} while AA transforms like an S.
3) $\sqrt{\tfrac{1}{2}}(M_1^{\rho}M_2^{\rho}+M_1^{\lambda}M_2^{\lambda})$ transforms like S, the pair $\sqrt{\tfrac{1}{2}}(M_1^{\rho}M_2^{\lambda}+M_1^{\lambda}M_2^{\rho})$, $\sqrt{\tfrac{1}{2}}(M_1^{\rho}M_2^{\rho} - M_1^{\lambda}M_2^{\lambda})$ transform like a mixed symmetry pair M^{ρ}, M^{λ} and $\sqrt{\tfrac{1}{2}}(M_1^{\rho}M_2^{\lambda} - M_1^{\lambda}M_2^{\rho})$ transforms like A.

These rules are easily verified by explicit calculation. To construct totally antisymmetric baryon wavefunctions we require similarly constructed flavour and spin wavefunctions (the colour wavefunction is automatically the antisymmetric colour singlet $C^{A} = \sqrt{\tfrac{1}{6}}$ (RBY - BRY + BYR - YBR + YRB - RYB) so that we need to construct totally <u>symmetric</u> states in flavour x space x spin). As spin wave functions we choose

$$S=\frac{3}{2}\ (\text{with}\ \sigma=S)\ :\qquad \chi^{S}_{3/2} = \uparrow\uparrow\uparrow\ ,\ \text{etc.} \tag{57}$$

$$S=\frac{1}{2}\ (\text{with}\ \sigma=M^{\rho}):\qquad \chi^{\rho}_{+} = \frac{1}{\sqrt{2}}(\uparrow\downarrow\uparrow-\downarrow\uparrow\uparrow)\ ,\ \text{etc.} \tag{58}$$

$$S=\frac{1}{2}\ (\text{with}\ \sigma=M^{\lambda}):\qquad \chi^{\lambda}_{+} = -\frac{1}{\sqrt{6}}(\uparrow\downarrow\uparrow+\downarrow\uparrow\uparrow-2\uparrow\uparrow\downarrow)\ ,\ \text{etc.} \tag{59}$$

where we once again display only the top state of an angular momentum multiplet. As flavour wavefunctions we take

plet

$$\phi^{S}_{\Delta++} = uuu \tag{60}$$

$$\phi^{S}_{\Sigma+} = \frac{1}{\sqrt{3}}(uus+usu+suu) \tag{61}$$

decu-

$$\phi^S_{\Xi^0} = \frac{1}{\sqrt{3}}(ssu+sus+uss) \tag{62}$$

$$\phi^S_{\Omega^-} = sss \tag{63}$$

octet

$$\phi^\rho_p = \frac{1}{\sqrt{2}}(udu-duu) \tag{64}$$

$$\phi^\rho_\Lambda = \frac{1}{\sqrt{12}}(2uds-2dus+usd-dsu-sud+sdu) \tag{65}$$

$$\phi^\rho_{\Sigma+} = \frac{1}{\sqrt{2}}(suu-usu) \tag{66}$$

$$\phi^\rho_{\Xi^0} = \frac{1}{\sqrt{2}}(sus-uss) \tag{67}$$

octet

$$\phi^\lambda_p = \frac{-1}{\sqrt{6}}(udu+duu-2uud) \tag{68}$$

$$\phi^\lambda_\Lambda = \frac{1}{2}(usd-dsu+sud-sdu) \tag{69}$$

$$\phi^\lambda_{\Sigma+} = \frac{1}{\sqrt{6}}(suu+usu-2uus) \tag{70}$$

$$\phi^\lambda_{\Xi^0} = \frac{-1}{\sqrt{6}}(sus+uss-2ssu) \tag{71}$$

singlet

$$\phi^A_\Lambda = \frac{1}{\sqrt{6}}(uds-dus-usd+dsu+sud-sdu) \tag{72}$$

where in this case we have displayed only the top states of isospin multiplets. We have listed all flavour wavefunctions, not just those associated with S = 0 and -3, for future use.

We are finally ready to construct the baryons, beginning with the S = 0 sector. In the notation $|X\,{}^{2S+1}L_\pi J^P\rangle$ where X = flavour, S is the quark spin, L = S, P, D. ... is the orbital angular momentum, π= S, M, or A is the permutational symmetry of the wavefunction, and J^P is the total angular momentum and parity, we have at N = 0, for S = 0, only the two possibilities

$$|N^2S_{S}\tfrac{1+}{2}\rangle = C_A\,\psi^S_{00}\,\frac{1}{\sqrt{2}}\,[\chi^\rho_+\phi^\rho_N+\chi^\lambda_+\phi^\lambda_N] \tag{73}$$

$$|\Delta^4S_{S}\tfrac{3+}{2}\rangle = C_A\,\psi^S_{00}\,\chi^S_{3/2}\,\phi^S_\Delta \tag{74}$$

At N = 1 there are more possibilities like

$$|N^4P_{M}\tfrac{5-}{2}\rangle = C_A\,\chi^S_{3/2}\frac{1}{\sqrt{2}}\,[\phi^\rho_N\psi^\rho_{11}+\phi^\lambda_N\psi^\lambda_{11}] \tag{75}$$

$$|N^2P_{M}\tfrac{3-}{2}\rangle = C_A\,\frac{1}{2}\,[\chi^\rho_+\phi^\rho_N\psi^\lambda_{11}+\chi^\rho_+\phi^\lambda_N\psi^\rho_{11}+\chi^\lambda_+\phi^\rho_N\psi^\rho_{11}-\chi^\lambda_+\phi^\lambda_N\psi^\lambda_{11}] \tag{76}$$

$$|\Delta^2P_{M}\tfrac{3-}{2}\rangle = C_A\,\phi^S_\Delta\,\frac{1}{\sqrt{2}}\,[\chi^\rho_+\psi^\rho_{11}+\chi^\lambda_+\psi^\lambda_{11}] \tag{77}$$

and at N = 2 even more like

$$|\Delta^4D_{S}\tfrac{7+}{2}\rangle \doteq C_A\,\psi^S_{22}\,\chi^S_{3/2}\,\phi^S_\Delta \tag{78}$$

$$|\Delta^2D_{M}\tfrac{5+}{2}\rangle = C_A\,\phi^S_\Delta\,\frac{1}{\sqrt{2}}\,[\chi^\rho_+\psi^\rho_{22}+\chi^\lambda_+\psi^\lambda_{22}] \tag{79}$$

Note that in these examples we have chosen for simplicity states that are the top states of "fully stretched" states of a given L and S; states of lower J from a given L and S may be constructed by the usual Clebsch-Gordan techniques. Of course the S = -3 states are all identical to the S = 0 Δ-like states with the replacement of ϕ^S_Δ by $\phi^S_{\Omega^-}$ and of m_d by m_s.

If we were to neglect the mass difference m_s-m_d, then construction of the S = -1 and -2 states would also proceed in exactly parallel fashion to those for S = 0. In fact in this limit the states $|X^{2S+1}L_\pi J^P\rangle$ we have defined would be in exact correspondence with the usual SU(6) states $|X_\alpha{}^{2S+1}[\mu,L^P]J^P\rangle$ where μ = 56,70, or 20 is the SU(6) multiplicity and corresponds respectively to π = S, M, and A, and where α is an SU(3) label to distinguish between Λ_1 and Λ_8, Σ_8 and Σ_{10}, and Ξ_8 and Ξ_{10}. Thus, the wave functions (40), (43), and (46) are of the types S_S, S'_S, and D_S and give the states of the $[56,0^+]$, $[56',0^+]$, and $[56,2^+]$

SU(6) supermultiplets, the pairs ((41), (42)), ((44), (45)), and ((47), (48)) are of the mixed symmetry types P_M, S_M, and D_M and lead to the states of the $[70,1^-]$, $[70,0^+]$, and $[70,2^+]$ supermultiplets, while the lone state (49) is of the type P_A and leads to the $[20,1^+]$ supermultiplet.

In the harmonic limit associated with H_0 and in the SU(3) limit, all of the N=2 supermultiplets $[56',0^+]$, $[70,0^+]$, $[56,2^+]$ $[70,2^+]$ and $[20,1^+]$ would be degenerate with a spacing ω above the $[70,1^-]$ equal to the spacing of the $[70,1^-]$ above the $[56,0^+]$. We will see below that the anharmonicity potential U has the effect of breaking this high degree of degeneracy of the harmonic limit.

We now complete our construction of baryon states by turning to the S=-1 and -2 sectors. In these cases we have either

$$m_1 = m_2 = m_d < m_3 = m_s$$

or

$$m_1 = m_2 = m_s > m_3 = m_d.$$

With the same change of variables to $\vec{\rho}, \vec{\lambda}$, and $\vec{R}$, H_0 once again separates to the form

$$H_0 = \frac{P^2}{2M} + \left(\frac{p_\rho^2}{2m_\rho} + \frac{3}{2}k\rho^2 \right) + \left(\frac{p_\lambda^2}{2m_\lambda} + \frac{3}{2}k\lambda^2 \right) \tag{80}$$

where now

$$M=m_1+m_2+m_3 \; , \quad m_\rho=m_1=m_2, \quad m_\lambda= \frac{3m_1m_3}{2m_1+m_3} \tag{81}$$

The eigenstates of this Hamiltonian are quite distinct from those of the equal mass case, since the ρ-λ degeneracy has been broken:

$$\omega_\rho = (3k/m_\rho)^{1/2} \; , \qquad \omega_\lambda = (3k/m_\lambda)^{1/2} \tag{82}$$

so that for S =-1 $\omega_\lambda < \omega_\rho$ and for S = -2 $\omega_\rho < \omega_\lambda$. The low-lying eigenfunctions of H_0 are now, in an obvious notation with $\psi \equiv \Psi \left(\frac{\alpha_\rho \alpha_\lambda}{\pi}\right)^{3/2} \exp(-\frac{1}{2}\{\alpha_\rho^2 \rho^2 + \alpha_\lambda^2 \lambda^2\})$, given by

$$N_\rho=0 \quad N_\lambda=0 \qquad \Psi_{00} = 1 \tag{83}$$

$$N_\rho=1 \quad N_\lambda=0 \qquad \Psi_{11}^{\rho} = \alpha_\rho \rho_+ \tag{84}$$

$$N_\rho=0 \quad N_\lambda=1 \qquad \Psi_{11}^{\lambda} = \alpha_\lambda \lambda_+ \tag{85}$$

$$N_\rho=0 \quad N_\lambda=2 \qquad \Psi_{00}^{\lambda\lambda} = \sqrt{2/3}\, \alpha_\lambda^2 (\lambda^2 - \frac{3}{2} \alpha_\lambda^{-2}) \tag{86}$$

$$N_\rho=1 \quad N_\lambda=1 \qquad \Psi_{00}^{\rho\lambda} = \sqrt{4/3}\, \alpha_\rho \alpha_\lambda \vec{\rho} \cdot \vec{\lambda} \tag{87}$$

$$N_\rho=2 \quad N_\lambda=0 \qquad \Psi_{00}^{\rho\rho} = \sqrt{2/3}\, \alpha_\rho^2 (\rho^2 - \frac{3}{2} \alpha_\rho^{-2}) \tag{88}$$

$$N_\rho=0 \quad N_\lambda=2 \qquad \Psi_{22}^{\lambda\lambda} = \sqrt{1/2}\, \alpha_\lambda^2 \lambda_+^2 \tag{89}$$

$$N_\rho=1 \quad N_\lambda=1 \qquad \Psi_{22}^{\rho\lambda} = \alpha_\rho \alpha_\lambda \rho_+ \lambda_+ \tag{90}$$

$$N_\rho=2 \quad N_\lambda=0 \qquad \Psi_{22}^{\rho\rho} = \sqrt{1/2}\, \alpha_\rho^2 \rho_+^2 \tag{91}$$

$$N_\rho=1 \quad N_\lambda=1 \qquad \Psi_{11}^{\rho\lambda} = \alpha_\rho \alpha_\lambda (\rho_+ \lambda_z - \rho_z \lambda_+) \tag{92}$$

with

$$E = (N_\rho + \frac{3}{2})\omega_\rho + (N_\lambda + \frac{3}{2})\omega_\lambda \tag{93}$$

so that much of the degeneracy of the harmonic oscillator levels has disappeared. We will see that this has the consequence that (at least in the absence of spin-dependent interactions) SU(6) is maximally violated in these sectors.

The lack of symmetry of the $m_\rho \neq m_\lambda$ wave functions at first seems to pose a paradox: it is impossible using such functions to construct totally antisymmetric states as required by the generalized Pauli principle. The way out is simple but perhaps a little surprising: the exclusion principle only requires that identical fermions be antisymmetrized, and so obviously does not apply in the unequal mass case. The confusion arises because in the equal mass case the Hamiltonian has S_3 symmetry so that its wave functions can be chosen to belong to the definite representations S, M = (M^ρ,M^λ), A; if one then elevates flavour to the status of a quark variable, states satisfying the generalized Pauli principle will automatically be eigenstates of not only the energy, but also of flavour SU(3).

In the S = -1 and -2 sectors it is therefore more appropriate to use flavour wavefunctions like

$$\Phi_\Sigma = \frac{1}{\sqrt{2}}(ud+du)s \qquad (94)$$

$$\Phi_\Lambda = \frac{1}{\sqrt{2}}(ud-du)s \qquad (95)$$

$$\Phi_\Xi = ssu \qquad (96)$$

which single out the third quark which has a different mass, and to retain only the S_2 subgroup of S_3 as a symmetry group. (We shall see below that even this is unnecessary!). By isospin invariance these three flavour wavefunctions are in fact sufficient for S = -1 and -2 spectroscopy and in terms of this "uds basis" for flavour we can construct from the usual colour and spin states and the $m_\rho \neq m_\lambda$ spatial wavefunctions (all of which still have explicit S_2 symmetry) a complete set of states antisymmetric under 1↔2. We label these states by $|Y\ {}^{2S+1}L_\sigma J^P\rangle$, where Y = Σ, Λ, or Ξ, and S, L, and J are as in the equal mass sectors, and where σ= ,ρ,λ, ρρ, ρλ, λλ etcetera denotes the level of excitation of the ρ and λ oscillators. For example, for $N_\rho = N_\lambda = 0$ we can construct only

$$|\Lambda^2S\,\tfrac{1}{2}^+\rangle = C_A\,\Phi_\Lambda\,\chi^\rho_+\,\psi_{00} \tag{97}$$

$$|\Sigma^2S\,\tfrac{1}{2}^+\rangle = C_A\,\Phi_\Sigma\,\chi^\lambda_+\,\psi_{00} \tag{98}$$

$$|\Xi^2S\,\tfrac{1}{2}^+\rangle = C_A\,\Phi_\Xi\,\chi^\lambda_+\,\psi_{00} \tag{99}$$

$$|\Sigma^4S\,\tfrac{1}{2}^+\rangle = C_A\,\Phi_\Sigma\,\chi^S_{3/2}\psi_{00} \tag{100}$$

$$|\Xi^4S\,\tfrac{1}{2}^+\rangle = C_A\,\Phi_\Xi\,\chi^S_{3/2}\psi_{00} \tag{101}$$

which correspond to the ordinary Λ, Σ, and Ξ of the octet and the Σ^* and Ξ^* of the decuplet. For $N_\rho + N_\lambda = 1$ we can construct the "fully stretched" Λ states

$$|\Lambda^2P_\lambda\,\tfrac{3}{2}^-\rangle = C_A\,\Phi_\Lambda\,\chi^\rho_+\,\psi^\lambda_{11} \tag{102}$$

$$|\Lambda^2P_\rho\,\tfrac{3}{2}^-\rangle = C_A\,\Phi_\Lambda\,\chi^\lambda_+\,\psi^\rho_{11} \tag{103}$$

$$|\Lambda^4P_\rho\,\tfrac{5}{2}^-\rangle = C_A\,\Phi_\Lambda\,\chi^S_{3/2}\psi^\rho_{11} \tag{104}$$

plus four states with J<L+S: $|\Lambda^2P_\lambda{}^1/2^-\rangle$, $|\Lambda^2P_\rho{}^1/2^-\rangle$, $|\Lambda^4P_\rho{}^3/2^-\rangle$, and $|\Lambda^4P_\rho{}^1/2^-\rangle$; the "fully stretched" Σ states

$$|\Sigma^2P_\lambda\,\tfrac{3}{2}^-\rangle = C_A\,\Phi_\Sigma\,\chi^\lambda_+\,\psi^\lambda_{11} \tag{105}$$

$$|\Sigma^2P_\rho\,\tfrac{3}{2}^-\rangle = C_A\,\Phi_\Sigma\,\chi^\rho_+\,\psi^\rho_{11} \tag{106}$$

$$|\Sigma^4P_\lambda\,\tfrac{5}{2}^-\rangle = C_A\,\Phi_\Sigma\,\chi^S_{3/2}\psi^\lambda_{11} \tag{107}$$

plus the four states $|\Sigma\, {}^2P_\lambda\, 1/2^-\rangle$, $|\Sigma\, {}^2P_\rho\, 1/2^-\rangle$, $|\Sigma\, {}^4P_\lambda\, 3/2^-\rangle$, and $|\Sigma\, {}^4P_\lambda\, 1/2^-\rangle$, with $J < L+S$; and finally the Ξ states which are identical in form to the Σ's (since both Φ_Σ and Φ_Ξ are symmetric under 1↔2 interchange) with the replacement $\Phi_\Sigma \to \Phi_\Xi$ and $m_d = m_u \leftrightarrow m_s$. These states correspond exactly in number to the expected $S = -1$ and -2 states of the $[70,1^-]$ SU(6) supermultiplet which has the $[SU(3), J^P]$ decomposition

$$[70,1^-]=[1,\tfrac{3}{2}^-]+[1,\tfrac{1}{2}^-]+[8,\tfrac{5}{2}^-]+2[8,\tfrac{3}{2}^-]+2[8,\tfrac{1}{2}^-]+[10,\tfrac{3}{2}^-]+[10,\tfrac{1}{2}^-] \tag{108}$$

leading one to expect 7 states of each of the types Λ, Σ, Ξ along with 5 N's, 2Δ's and 2 Ω's. However, we can now see that the true eigenstates of H_0 are far from being eigenfunctions of SU(6). Consider, for example, the state $|\Lambda\, {}^2P_\lambda\, 3/2^-\rangle$ of equation (102). In terms of SU(6) states it has the composition (for $\alpha_\rho = \alpha_\lambda$)

$$|\Lambda\, {}^2P_\lambda\, \tfrac{3}{2}^-\rangle = \frac{1}{\sqrt{2}}\, \{|\Lambda_1\, {}^2[70,1^-]\tfrac{3}{2}^-\rangle + |\Lambda_8\, {}^2[70,1^-]\tfrac{3}{2}^-\rangle\} \tag{109}$$

and we see a maximal violation of, in this case, SU(3) symmetry. We will discuss the effects of these violations below. One consequence, which is very transparent in the uds basis, may suffice for now however: the $\Sigma\, 5/2^-$ and $\Lambda\, 5/2^-$ which would be expected to be degenerate in the SU(3) limit will (in this approximation) be split by $\omega_\rho - \omega_\lambda \sim O(50\ \mathrm{MeV})$. Note that this effect is expected to make $\Lambda 5/2^-$ heavier than $\Sigma 5/2^-$ as observed and in reversal to the ground state ordering (which we will see below is a spin-dependent effect). The construction of the $N_\rho + N_\lambda = 2$ states proceeds along similar lines and has similar effects. As examples, we list just the possible fully stretched Λ - like states

$$|\Lambda\, {}^2S_{\lambda\lambda}\, \tfrac{1}{2}^+\rangle = C_A\, \Phi_\Lambda\, \chi^\rho_+\, \psi^{\lambda\lambda}_{00} \tag{110}$$

$$|\Lambda\, {}^2S_{\rho\lambda}\tfrac{1}{2}^+\rangle = C_A\ \Phi_\Lambda\ \chi_+^\lambda\ \ \psi_{00}^{\rho\lambda} \tag{111}$$

$$|\Lambda\, {}^2S_{\rho\rho}\tfrac{1}{2}^+\rangle = C_A\ \Phi_\Lambda\ \chi_+^\rho\ \ \psi_{00}^{\rho\rho} \tag{112}$$

$$|\Lambda\, {}^2D_{\lambda\lambda}\tfrac{5}{2}^+\rangle = C_A\ \Phi_\Lambda\ \chi_+^\rho\ \ \psi_{22}^{\lambda\lambda} \tag{113}$$

$$|\Lambda\, {}^2D_{\rho\lambda}\tfrac{5}{2}^+\rangle = C_A\ \Phi_\Lambda\ \chi_+^\lambda\ \ \psi_{22}^{\rho\lambda} \tag{114}$$

$$|\Lambda\, {}^2D_{\rho\rho}\tfrac{5}{2}^+\rangle = C_A\ \Phi_\Lambda\ \chi_+^\rho\ \ \psi_{22}^{\rho\rho} \tag{115}$$

$$|\Lambda\, {}^2P_{\rho\lambda}\tfrac{3}{2}^+\rangle = C_A\ \Phi_\Lambda\ \chi_+^\lambda\ \ \psi_{11}^{\rho\lambda} \tag{116}$$

$$|\Lambda\, {}^4S_{\rho\lambda}\tfrac{3}{2}^+\rangle = C_A\ \Phi_\Lambda\ \chi_{3/2}^S\psi_{00}^{\rho\lambda} \tag{117}$$

$$|\Lambda\, {}^4D_{\rho\lambda}\tfrac{7}{2}^+\rangle = C_A\ \Phi_\Lambda\ \chi_{3/2}^S\psi_{22}^{\rho\lambda} \tag{118}$$

$$|\Lambda\, {}^4P_{\rho\lambda}\tfrac{5}{2}^+\rangle = C_A\ \Phi_\Lambda\ \chi_{3/2}^S\psi_{11}^{\rho\lambda} \tag{119}$$

there are an additional 9 Λ states with $J < L + S$. (In total there are 13N's, 8Δ's, 19Λ's, 21Σ's, 21 Ξ's, and 8 Ω's at $N_\rho + N_\lambda = 2$). These eigenstates of H_0 not only violate SU(3) but also SU(6); e.g., for $\alpha_\rho = \alpha_\lambda$

$$|\Lambda\, {}^2D_{\lambda\lambda}\tfrac{5}{2}^+\rangle = \tfrac{1}{\sqrt{2}}|\Lambda_8\, {}^2[56,2^+]\tfrac{5}{2}^+\rangle - \tfrac{1}{2}|\Lambda_8\, {}^2[70,2^+]\tfrac{5}{2}^+\rangle + \tfrac{1}{2}|\Lambda_1\, {}^2[70,2^+]\tfrac{5}{2}^+\rangle \tag{120}$$

is a mixture of the SU(6) multiplets $[56,2^+]$ and $[70,2^+]$ and of SU(3) representations within the $[70,2^+]$.

At this point one might wonder why we have introduced SU(6) into this discussion at all. There were two main reasons: 1) in some calculations it is technically useful to use the SU(6) basis, and 2) for historical reasons it is still very common to use SU(6) language in describing many aspects of light baryon physics. Nevertheless, we stress that from the modern point of view SU(6) has very little to do with baryons.

As an amusing and possibly useful aside, we note that it is not usually necessary to use S_3 symmetry at all! Consider, for example, the S = 0 sector. Since by isospin invariance it is sufficient to consider only states of the form uud and since up and down quarks are not identical, we can discuss such states in exact parallelism to the states $\Xi \equiv$ ssu. At $N_\rho + N_\lambda = 0$, for example, the only such states are

$$C_A \text{ (uud) } \chi_+^\lambda \;\; \psi_{00}^S \tag{121}$$

$$C_A \text{ (uud) } \chi_{3/2}^S \psi_{00}^S \tag{122}$$

which we can identify with the proton and the Δ^+. For $N_\rho + N_\lambda > 0$ this method has the disadvantage that it will only automatically diagonalize isospin if the I = 1/2 and 3/2 states are not degenerate, but this ambiguity is always resolved in practice once one considers spin dependent interactions.

With our zeroth order eigenfunctions established, to complete our discussion of H_{si} it remains to do perturbation theory in U of equation (35). We begin by discussing the equal mass case where in the harmonic limit we have only three equally spaced energy levels. Since H_{si} is in this case S_3 symmetric even when $U \neq 0$, the degeneracy of all pairs of wavefunctions $(\psi^{M\rho}, \psi^{M\lambda})$ remains unbroken. This leaves us with still a unique ground state and first excited state energy corresponding to the SU(6) supermultiplets $[56,0^+]$ and $[70,1^-]$ respectively, but with in general five independent N=2 levels associated with $\psi_{00}^{S'}, (\psi_{00}^{M\rho}, \psi_{00}^{M\lambda})$, $\psi_{22}^S, (\psi_{22}^{M\rho}, \psi_{22}^{M\lambda})$, and ψ_{11}^A, corresponding to the five SU(6) supermultiplets $[56',0^+]$, $[70,0^+]$, $[56,2^+]$, $[70,2^+]$, and $[20,1^+]$, respectively. Given these seven independent energies, one would naturally suppose that without specifying U and doing explicit calculations it would be necessary to introduce these seven energy levels via seven parameters. Since it would be necessary to introduce four parameters --- the common quark mass, b, C, and α_s --- even in a first principles calculation, expanding to seven does not seem too bad. However, it turns out that these seven energy levels are constrained by a remarkable theorem: in first order perturbation theory any potential U will split the harmonic oscillator spectrum into exactly the same pattern, and this pattern can be described in terms of only <u>three</u> parameters. Namely

$$E\,[56,0^+] = E\,[S_S] = E_0 \tag{123}$$

$$E\,[70,1^-] = E\,[P_M] = E_0 + \Omega \tag{124}$$

$$E\,[56',0^+] = E\,[S'_S] = E_0 + 2\Omega - \Delta \tag{125}$$

$$E\,[70,0^+] = E\,[S_M] = E_0 + 2\Omega - \frac{1}{2}\Delta \tag{126}$$

$$E\,[56,2^+] = E[D_S] = E_0 + 2\Omega - \frac{2}{5}\Delta \tag{127}$$

$$E\,[70,2^+] = E\,[D_M] = E_0 + 2\Omega - \frac{1}{5}\Delta \tag{128}$$

$$E\,[20,1^+] = E[P_A] = E_0 + 2\Omega \tag{129}$$

as shown in Figure 12. This theorem is easily proved by noting that the expectation value of U in the seven sets of independent wavefunctions can be expressed in terms of the first three (n = 0,1,2) of the moments

$$a_n = \frac{3\alpha^{3+2n}}{\pi^{3/2}} \int d^3\rho\, U(\sqrt{2}\rho)\, e^{-\alpha^2\rho^2} \rho^{2n} \tag{130}$$

in terms of which

$$E_0 = 3m+3\omega+a_0 \tag{131}$$

$$\Omega = \omega - \frac{1}{2}a_0 + \frac{1}{3}a_1 \tag{132}$$

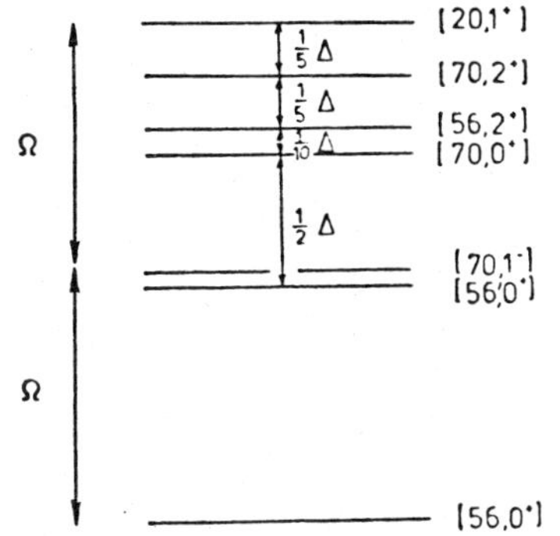

FIG. 12. the pattern of low-lying SU(6) supermultiplets from first order perturbation theory in the anharmonic potential U

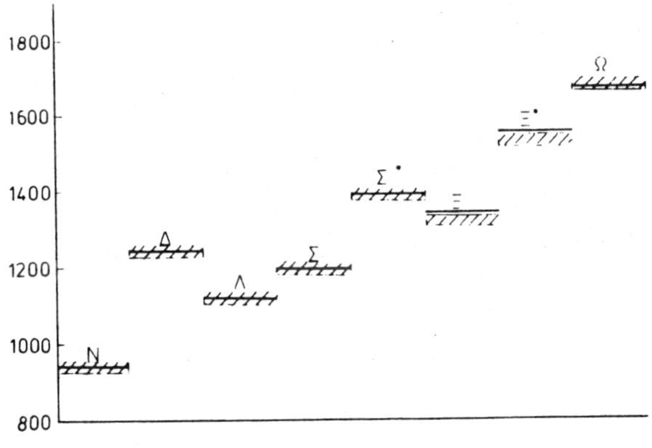

FIG. 13. the experimental spectrum (hatched boxes) of ground state baryons compared to theory (horizontal bars)

$$\Delta = -\frac{5}{4}a_0 + \frac{5}{3}a_1 - \frac{1}{3}a_2 \tag{133}$$

The actual pattern of levels corresponding to the spin-unperturbed S = 0 multiplets seems to be in accord with this theoretical pattern and indicates $E_0 \simeq$ 1150 MeV, $\Omega \simeq$ 440 MeV, and $\Delta \simeq$ 400 MeV. That Δ turns out to be positive is exactly what one expects from the actual form of U: both the attractive Coulomb potential and the fact that the true potential is linear instead of quadratic give positive contributions to Δ. The well-known fact that the $[56',0^+]$ lies far below the $[56,2^+]$ is thus naturally explained. Of course first order perturbation theory can only be expected to be reliable for $\Delta \ll 2\Omega$, but explicit calculations show that 1) this pattern is generally respected for realistic potentials, and 2) the required values for E_0, Ω, and Δ are consistent with the values of the fundamental parameters deduced from meson spectroscopy.

The case U ≠ 0 is somewhat more complicated in the unequal mass case $m_3 \neq m_1 = m_2$, since now the ρ and λ oscillators behave differently. Of course the energies of the states before turning on U are as previously discussed:

$$\frac{\omega_\lambda}{\omega_\rho} = \left(\frac{m_\rho}{m_\lambda}\right)^{1/2} . \tag{134}$$

Although in the presence of a U perturbation this relationship will change, it should remain a reasonable approximation to the actual situation and can be used to correct the relations (123) to (129). The generality of the important qualitative feature here that modes involving heavy quarks will have a lower frequency than those involving light quarks can most easily be seen in operation in an extreme case like $m_3 \gg m_1 = m_2$. In this limit baryons behave like a helium atom with inverted statistics. In the absence of interactions between quarks 1 and 2 this system will have two degenerate negative parity excited spatial wavefunctions above the ground state, namely,

$$|\tilde{\lambda}> \equiv \frac{1}{\sqrt{2}}\{|sp> + |ps>\} \tag{135}$$

$$|\tilde{\rho}> \equiv \frac{1}{\sqrt{2}}\{|sp> - |ps>\} \tag{136}$$

where s and p are the lowest-lying single particle s and p orbitals about quark 3. If one now turns on the residual attractive potential energy of the two light quarks (from any confinement potential), then $|\tilde{\lambda}\rangle$ will end up lying lower than $|\tilde{\rho}\rangle$ since in $|\tilde{\rho}\rangle$ the two quarks are unlikely to be near to each other. Thus though our numerical results will depend on the use of the harmonic oscillator approximations sketched above, our principal conclusions are more general.

This concludes our discussion of the spectrum and eigenstates of H_{si}. There is a considerable body of interesting physics contained just in this spin-independent spectroscopy; we have already mentioned several features of the baryons which are specifically related to the expected form of H_{si}. However, we should also note that the spin-independent eigenstates are sufficient as well to give us a very good zeroth order understanding of several non-spectroscopic features of baryons. The best example is probably the baryon magnetic moments which, assuming Dirac magnetic moments for the usual constituent quark masses, are predicted within $\pm 0.2\mu_N$ in terms of these eigenstates.

In the following section we will see that other physics which is at least as interesting emerges from spin-dependent effects.

ii) Spin-Dependent Effects in Baryons

In a heavy $Q\bar{Q}$ or QQQ system the nonrelativistic reduction of one gluon exchange given in Section II. E. should be applicable. However, in light quark systems m/E will be very different from unity and we must expect that spin dependent effects could be significantly modified from this simple form. After all, as we indicated above, such effects are themselves relativistic corrections to the potentials. We accordingly will begin by taking a more inductive approach to these effects in the light baryons than we would in, say, the $b\bar{b}$ system.

We first consider the ground state baryons in which we can completely isolate the hyperfine interaction (26). (Since $\vec{L}_\rho$ and $\vec{L}_\lambda$ annihilate ψ_{00}, the spin-orbit interaction does not even contribute any off-diagonal effects to the ground states). It is completely straightforward to calculate the diagonal hyperfine contribution to these states which comes entirely from the $\vec{S}_i \cdot \vec{S}_j$ contact interaction. Using (31), the fact that

$$\langle\psi_{00}^S|\delta^3(\vec{r}_{ij})|\psi_{00}^S\rangle = \alpha^3/(2\pi)^{3/2}$$

and

$$\langle\vec{S}_i\cdot\vec{S}_j\rangle = \begin{cases} +1/4 & \text{if i and j have spin 1} \\ -3/4 & \text{if i and j have spin 0} \end{cases} \tag{137}$$

one readily obtains all of the ground state diagonal splittings in terms of a single parameter

$$\delta \equiv \frac{4\alpha_s\alpha^3}{3\sqrt{2\pi}\, m_d^2}\, r_{hyp} \tag{138}$$

where r_{hyp} is a factor we introduce into (26) as a crude measure of relativistic corrections to the naive hyperfine interaction. The relations one finds are (ignoring the fact that $(\alpha_\rho \neq \alpha_\lambda)$

$$\delta M_\Delta = +\frac{1}{2}\delta \qquad \delta M_N = -\frac{1}{2}\delta \tag{139}$$

$$\delta M_{\Sigma^*} = [\frac{x}{3} + \frac{1}{6}]\delta \qquad \delta M_\Sigma = [\frac{1}{6} - \frac{2x}{3}]\delta \qquad \delta M_\Lambda = -\frac{1}{2}\delta \tag{140}$$

$$\delta M_{\Xi^*} = [\frac{x}{3} + \frac{x^2}{6}]\delta \qquad \delta M_\Xi = [\frac{x^2}{6} - \frac{2x}{3}]\delta \tag{141}$$

$$\delta M_\Omega = +\frac{x^2}{2}\delta \tag{142}$$

where we have defined $x \equiv m_d/m_s \simeq 2/3$ for brevity. Equation (139) shows that with (26) the Δ is automatically heavier than the N. Another noteworthy feature is that the mass of the Λ is lowered by as much as the N; this SU(3)-like equality comes about since the Λ has a χ^ρ spin wavefunction (see Equation (97)) which makes the suppressed strange quark chromomagnetic moment irrelevant. Since the same is not true for the χ^λ spin function of the Σ, one gets a $\Sigma-\Lambda$ splitting even though both have the same quark content. In fact one gets

$$\Sigma - \Lambda = \frac{2}{3}(1-x)[\Delta - N] \tag{143}$$

which works very well indeed. The complete results of this programme, including mixing and $\alpha_\rho \neq \alpha_\lambda$ effects, is shown in Figure 13, in which δ has been adjusted to fit the Δ-N splitting. In addition to this spectroscopic evidence, there is other evidence for the role of the hyperfine interaction in the ground states. Most interesting perhaps is that it gives rise to a charge radius for the neutron. Formally this occurs via an off-diagonal mixing with the N=2 $|N\,^2S_M\,^1/2^+\rangle$ state, but its physical origin is simply visualized. In the (ddu) flavour basis, the neutron has the spin wavefunction χ^λ. Since the two down quarks therefore have spin 1 they repel (see Equation (137)) each other, but the two up-down pairs of quarks are in an attractive state (recall that the sum of all three pairwise interactions is attractive). Thus the two down quarks are pushed to the periphery of the neutron and the up quark is pulled to its centre, producing a negative charge radius. Detailed calculations automatically produce the observed magnitude of this effect in terms of the same parameter δ. Note that, unlike the pure ground state spectroscopy, this effect is sensitive to the short range nature of H_{hyp}. This is in accord with the evidence from mesons and also with other effects to be mentioned below. In conclusion, there seems to be excellent evidence from the ground states (in support of similar conclusions from elsewhere) for an interaction in light quark baryons very similar in form to (26). Finally, we mention that these techniques have been applied with apparent success to $m_d \neq m_u$ isospin violating effects and to ground state baryons containing a heavy quark.

When we turn to excited baryons we must expect to encounter not only the well-established hyperfine interaction, but also effects due to spin-orbit interactions. Here we meet a surprise which greatly simplifies the description of excited baryons but which is quite mysterious in this context: spin orbit effects are very small. It requires no great subtlety to see this; the situation can be read right out of the Particle Data Group tables via the masses of the four candidate L = 2 states $\Delta 7/2^+$ (1935±25), $\Delta 5/2^+$ (1905±15), $\Delta 3/2^+$ (2010±150), and $\Delta 1/2^+$(1900±50). Spin-orbit effects would, if present, split these states since, roughly speaking, they can all be associated with S = $^3/2$ coupled with L = 2. A possible explanation of this observation will be discussed below, but for now we note that this situation is not so different from light mesons where the ρ-π hyperfine splitting is 640 MeV, but the P-wave spin-orbit splittings are small: e.g., A_2-A_1 ≃ 45±35 MeV and A_2-B ≃85±15 MeV, with some part of these splittings attributable to hyperfine interactions. In the case of mesons, it seems plausible that these small values are due to three effects: i)the radial dependence of the hyperfine interaction is more singular than that of the spin-orbit interactions so that the former is more effective for a given α_S, ii)the full Thomas precession interaction automatically tends to cancel colour magnetic spin-orbit terms, and iii)the relativistic corrections factors for spin orbit effects are smaller than their hyperfine analog r_{hyp} in (138). The second of these mechanisms can (probably) be seen at work in charmonium. Thus in terms of mesons spin orbit effects in baryons do not seem so unreasonable: if they were present in the same proportion to the Δ-N splitting as they are in mesons to the ρ-π splitting, then their strength would be only about 30 MeV, well within the "noise" of the light quark models. For now we will therefore adopt the usual approach and ignore spin-orbit forces, postponing to subsection iv) a discussion of the mechanism for this suppression.

With this approximation, one need only calculate hyperfine matrix elements and diagonalize. We will avoid most of the resulting gory details, but use the S=0 negative parity baryons as an example. The P-wave baryons are in a sense a perfect testing ground for hyperfine interactions since they simultaneously have one relative coordinate in an S-wave and one in a P-wave. When two quarks are in a relative S-wave, the contact force can operate but the tensor force vanishes, while when two quarks are in a relative P-wave the contact force vanishes (since there is zero amplitude to find the quarks "in contact") while the tensor force operates. Thus in P-wave baryons the two pieces of the hyperfine interaction operate side by side and their relative strength, as well as the short range character of the contact force, is testable.

Another feature of the P-wave baryons is that there are many states with the same IJ^P quantum numbers which are degenerate in the SU(6) limit; such states will in general be very strongly mixed by the SU(6)-violating terms in our Hamiltonian. Apart from the anomalously large matrix element responsible for the interband mixing of the N=2 level S_M state into the nucleon, the resulting intraband mixing angles, which can be "measured" via an analysis of baryon decays, provide the first real test of whether the model is correctly reproducing the internal structure as well as the spectroscopy of baryons.

The calculation is straightforward and gives, in terms of the same parameter δ,

$$<N^2P_M|H_{contact}|N^2P_M> = -\frac{1}{4}\delta \tag{144}$$

$$<N^4P_M|H_{contact}|N^4P_M> = <\Delta^2P_M|H_{contact}|\Delta^2P_M> = +\frac{1}{4}\delta \tag{145}$$

independent, of course, of J^P and

$$<N^4P_M\begin{pmatrix}5/2^-\\3/2^-\\1/2^-\end{pmatrix}|H_{tensor}|N^4P_M\begin{pmatrix}5/2^-\\3/2^-\\1/2^-\end{pmatrix}> = \begin{pmatrix}-1/20\\+1/5\\-1/4\end{pmatrix}\delta \tag{146}$$

$$<N^4P_M\begin{pmatrix}3/2^-\\1/2^-\end{pmatrix}|H_{tensor}|N^2P_M\begin{pmatrix}3/2^-\\1/2^-\end{pmatrix}> = \begin{pmatrix}-1/4\sqrt{10}\\+1/4\end{pmatrix}\delta \tag{147}$$

with all other matrix elements zero. Note that the $^2\Delta$ states have the same contact interaction as the ^4N states because of the short range character of this force. Indeed, at first glance these seven states do seem to split into two bands separated by about $1/2(\Delta-N)$ and with the Δ's in the upper band as predicted (see Figure 14). When tensor forces are considered the SU(6)-like $N^1/2^-$ and $N^3/2^-$ states become mixed, and we see that the resulting mixing angles are even independent of δ! These mixings can have dramatic effects on the decays of these states and in this way they can be determined experimentally, as we will discuss below. In both sectors the sign and magnitude of the measured mixing angles are in good agreement with those expected.

This is not the place to proceed to a blow by blow description of every sector of baryon spectroscopy. Suffice it to say that when applied to the negative parity hyperons and to the positive parity baryons of the N=2 band, the same simple picture continues to work quite well. This is illustrated in Figure 15 which shows the results for the N=2, including not only spectroscopy but also indicating on the basis of an analysis of meson couplings which states ought to have been seen in πN elastic scattering.

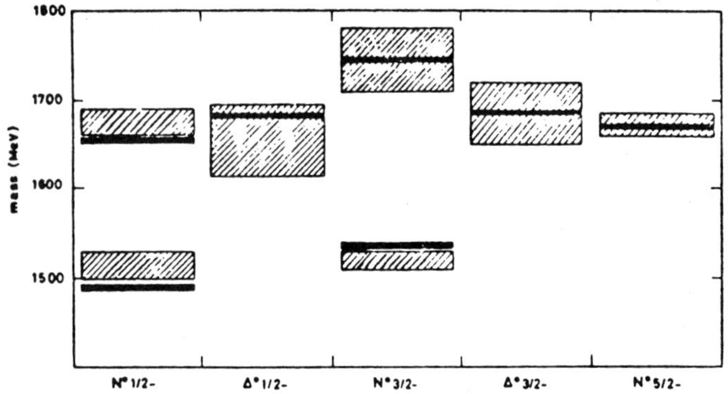

FIG. 14. the predicted S=0 negative parity baryons compared to experiment; the experimentally determined regions in which the masses lie are denoted by shaded boxes, theory is given by the horizontal bars

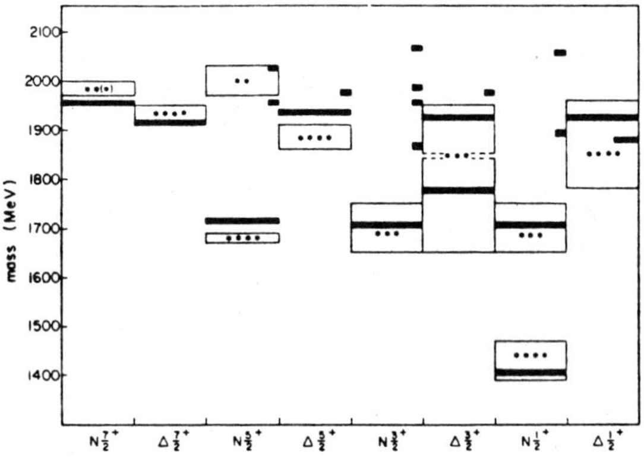

FIG. 15. the predicted S=0 positive parity baryons compared to experiment; the legend is as for Fig. 14, with now in addition states that are predicted to decouple from πN being shown as short horizontal bars

iii) Baryon Couplings

The last comment above leads us to discuss the importance of the study of hadronic couplings as a test of any model of hadron structure. It is clearly insufficient to predict just the masses of hadrons: any complete model should also predict their internal structure. Of course such gross features as the J^P of a state reflect this internal structure but by far the most sensitive indicators of internal composition are decay amplitudes since they can vary dramatically with interferences between different components of a state. Especially when working with a model and with data of limited accuracy, such tests are crucial.

Let me illustrate this point with an example. Notice the two $\Delta {}^5/2^+$ baryons in Figure 15 at about 1950 MeV. In the SU(6) limit, to mention just two points, each of these states is strongly coupled to πN, and the lower one decays to a $\pi\Delta$ P-wave more strongly than to the F-wave. Tensor forces completely change this state of affairs and the model we have been discussing predicts that only the lower state will be observable in πN elastic scattering and that its $\pi\Delta$ F-wave decay will dominate its P-wave. This is in accord with observation. There has, moreover, been a further very recent development: while the upper state is predicted to be practically "invisible" in $\pi N \to \pi N$, it was predicted that it should be visible in $\pi N \to \rho N$. This previously "missing" upper state now seems to have been found in precisely this channel.

As another example, let us now consider S=-1 baryons. To the extent that we can neglect U and hyperfine perturbations of the ρ-λ excitation structure, we can see trivially that ρ excitations will only couple weakly to the usual $\bar{K}$ N resonance channel. This is illustrated in Figure 16 which shows that, in the spectator approximation, $\bar{K}$ emission will leave the ρ oscillator excited and therefore with no overlap with a ground state nucleon. Figure 17 shows the N=2 states with strangeness -1, where the fact that many of the states are missing is attributable to this effect (though a full understanding requires effects like those responsible for decoupling the upper $\Delta {}^5/2^+$ in the previous example).

In concluding this brief sketch of our understanding of the light baryons, we first state that it appears that the simple model we have outlined here is a rather good guide to the physics of baryons with light quarks. It manages to predict

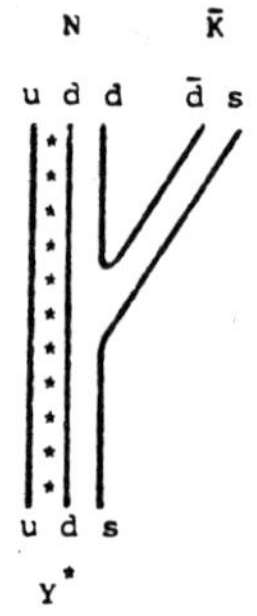

FIG. 16. a graphic demonstration of the decoupling of S=-1 ρ excitations from the K̄N channel

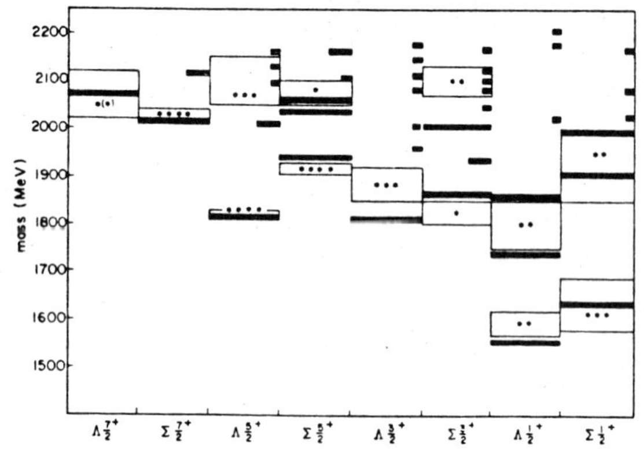

FIG. 17. as in Fig. 15, but for the S=-1 positive parity baryons, and with short bars representing states which are predicted to decouple from K̄N

masses with an accuracy of roughly 50 MeV and in hundreds of cases it rather consistently predicts the signs and magnitudes of the couplings of baryons. Nevertheless, it is important to remember that the model cannot be right. On the theoretical side, ignoring spin-orbit forces must be just an approximation and one cannot be satisfied with the model until this approximation is better understood. There also seem to be a number of places (like the $\Lambda\,3/2^-$ (1520) $-$ $\Lambda\,1/2^-$ (1405) splitting, the photocouplings of the $N\,3/2^+$ (1720), etcetera) where the model seems unacceptably inaccurate. Problems such as these indicate that it is essential to test the model in new regimes (like Ξ^* and Ω^* spectroscopy) in the hope that it will fail and in doing so provide us with clues as to how to proceed to improve it.

v) Relativization and Unification with Mesons

As with mesons (see above), the quark model for baryons not only survives relativization, but is improved by it. Recent work[4)] has shown that when the baryon problem is solved with the potential parameters and relativistic modifications described above for mesons, some remarkable results are obtained, including:

1) baryon spectroscopy emerges automatically: e.g., the Δ-N splitting and the P-wave multiplet centre-of-gravity are reproduced within a few percent, confirming the expected relationship between Δ-N and ρ-π and between, e.g., $N^*(1675)5/2^-$ - Δ and $A_2(1320)$ - ρ.

2) spin orbit forces are small automatically, thereby rationalizing the success of the usual non-relativistic model discussed above.

3) some old puzzles (like the $\Delta\frac{3}{2}^-$ - $\Delta\frac{1}{2}^-$ splitting) are solved.

There is thus very strong evidence now that essentially all of the low-lying states of $q\bar{q}$ and qqq can be understood in terms of the QCD based quark model.

References for Chapter III

1. For a recent discussion of the "QCD-improved quark model" see, e.g., N. Isgur in Testing the Standard Model, AIP Conf. Proc., vol. 81, editors C. Heusch and W.T. Kirk (American Inst. of Physics, New York, 1982), p.1.
2. The semirelativistic treatment of $q\bar{q}$ states can be found in S. Godfrey and N. Isgur, to appear in Phys. Rev. D. This paper has an extensive list of further references on mesons in the quark model.
3. The model of baryons discussed here was developed in collaboration with Gabriel Karl. For a pedagogical review of this model, see N. Isgur in The New Aspect of Subnuclear Physics, edited by A. Zichichi (Plenum, New York, 1980), p. 107 and references therein.
4. The relativized quark model for baryons can be found in S. Capstick and N. Isgur, University of of Toronto preprint.

IV. MULTIQUARK SYSTEMS

A. General Considerations

We should not be misled into believing that we understand hadron dynamics in QCD on the basis of our being able to understand mesons and baryons. The $q\bar{q}$ and qqq systems are very special cases since for these states the lowest adiabatic potential (corresponding to the gluon ground state in the presence of fixed quark sources) is very well isolated from excited glue adiabatic potentials. In this sense the whole of the ordinary quark model spectrum is analogous to the vibrational and rotational bands of a molecule moving in the lowest adiabatic surface of the electronic ground state of the molecule. We have yet to find experimentally the "hybrid" mesons and baryons which are the analogs of the vibrational and rotational molecular bands built on excited electronic (versus gluonic) states.

While the hybrid states depend on the properties of the gluonic fields, their main characteristics should be quite similar to those of ordinary mesons and baryons. In contrast, we should expect multiquark states to behave in entirely new ways[1)]: multiquark states characteristically have many low-lying adiabatic surfaces which cross. Figures 18 and 19 illustrate this characteristic behaviour in the case of the $qq\bar{q}\bar{q}$ system. Figure 18 illustrates very clearly the fact that the state of a $qq\bar{q}\bar{q}$ system depends not only on the positions of the quarks and antiquarks, but also on the state of the flux which flows between them. This is also true for simpler systems like $q\bar{q}$ (where the state of the glue distinguishes the ordinary mesons of the quark model from hybrid mesons), but in multiquark systems we cannot expect to be able to segregate the dynamical states into sectors with an approximately fixed (i.e., adiabatically evolving) glue state. Figure 18(b) shows why this is true: the two flux configurations of Figure 18(a) will be dynamically mixed and this mixing will always be strong in those regions of the lowest adiabatic surface of the $qq\bar{q}\bar{q}$ system where unmixed adiabatic surfaces of a given topology would have crossed. This effect is illustrated more explicitly in Figures 19(a) and 19(b), which show slices through the three lowest adiabatic potentials of the $qq\bar{q}\bar{q}$ system for two different rectangular arrangements of the quarks and antiquarks. They are shown as a function of the length x of one of the sides of the rectangle, with the other side kept at fixed length r. (We will return to Figures 19(a')

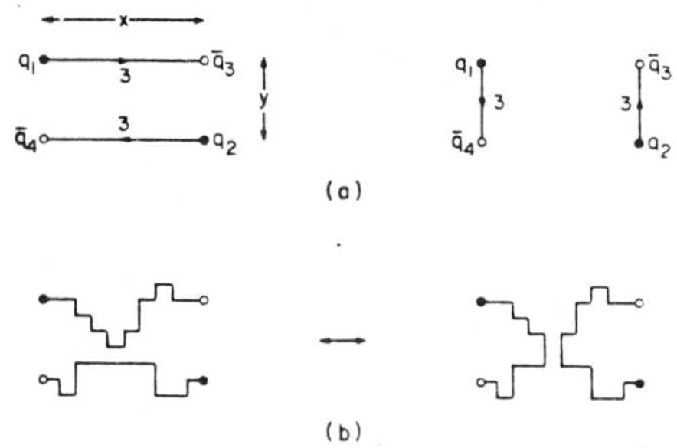

FIG. 18. (a)two low-lying $qq\bar{q}\bar{q}$ configurations, (b)topological mixing between these two configurations

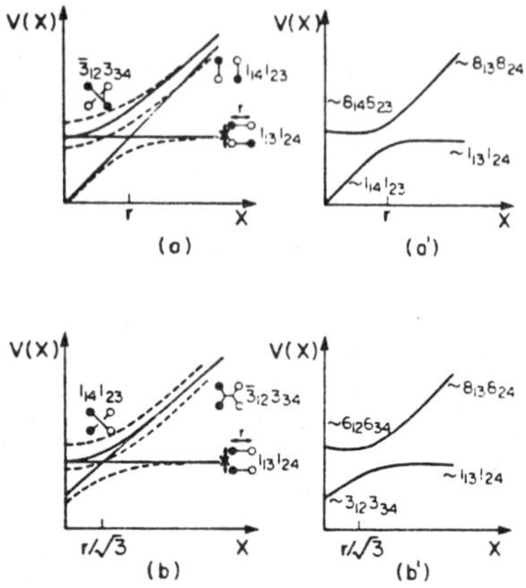

FIG. 19. the adiabatic potentials of the flux tube model and of the $\vec{F}_i\cdot\vec{F}_j$ potential model for two $qq\bar{q}\bar{q}$ geometries: (a) and (b) show the flux tube potentials before (solid curves) and after (shown schematically by dashed curves) topological mixing, while (a') and (b') show the related $\vec{F}_i\cdot\vec{F}_j$ potentials

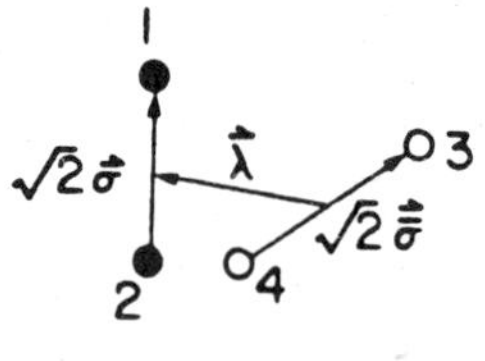

FIG. 20. the relative coordinates of the system $q_1q_2\bar{q}_3\bar{q}_4$

and 19(b′) below when we discuss possible models for this complex behaviour). We note that the lowest adiabatic surface in this prototypical multiquark system has completely different flux tube topologies dominant in different regions of $qq\bar{q}\bar{q}$ configuration space. In a dynamical situation, we cannot expect to be able to describe this system in terms of only a quark wavefunction: for a given $qq\bar{q}\bar{q}$ configuration we will in general have a quantum state that is a strong coherent superposition of the possible states of the glue.

The physics of this complex situation is, as indicated in Figure 18, controlled by flux tube topological mixing. Since this is an effect that becomes important for $g < 1$ where strong coupling perturbation theory is not useful, we cannot provide a very quantitative description of this process. Nevertheless, there are some qualitative features of this mixing which we believe we can safely expect and which are sufficient to form a basis for discussing most of its effects. Consider the example of $qq\bar{q}\bar{q}$ depicted in Figure 18(a) once again. We would like to know the amplitude for conversion between these two topologies as a function of x and y. To estimate this function we first imagine that both x and y are much greater than λ_0 so that a lattice spacing $a \sim \lambda_0$ where $g \sim 1$ can be used to describe the system. In this region topological mixing is not especially weak, but we can expect that a lowest order treatment will give a qualitatively sensible picture of the physics. Such lowest order mixing can occur when two (discrete) strings find themselves, in the tail of their vibrational wavefunctions, in a situation where they each have an element occupying a link on opposite sides of an elementary plaquette (see Figure 18(b)). They can then mix with an amplitude of order $1/ag^2 \sim \lambda_0$. The amplitude that their wavefunctions will allow them to mix is, however, damped like a Gaussian in the required displacements of the flux tubes from equilibrium so that, even after summation over all plaquettes, the mixing will be very small if x and y are both much larger than the transverse size $b^{-1/2}$ of the string wavefunctions. Conversely, if there is good overlap between all four relevant string wavefunctions, then this mixing amplitude will be of order $1/\lambda_0$ and will be strong.

There is an immediate application of these ideas to the question of residual forces between hadrons. If y corresponds to typical hadronic dimensions but $x \gg y$, then flux tube topological mixing between the ground states of the two short (vertical) flux tubes to any states of the two long (horizontal) flux tubes will be will be suppressed. We conclude that the residual forces between colour singlet hadrons will have no long range (power law) tail, in accord with experiment! Conversely, for distances $x \sim y \sim 1$ fermi we should expect strong topological mixing between colour singlets leading to medium range van der Waal's forces which could play a role in nuclear binding. Indeed, we will see below that such effects are important in explicit calculations, and they may indeed be responsible for the bulk of nuclear binding energies!

A full-fledged treatment of the gluon dynamics of multiquark systems along the lines imagined above would obviously be a formidible task. In the meantime much recent work on multiquark dynamics has been based on simpler models. I would say that the minimum requirements for such a model would be:

1) If it is to simplify calculations significantly, it should specify adiabatic potentials that depend only on the quark coordinates (including space, spin, and colour),

2) It should confine colour,

3) It should reduce to the known $q\bar{q}$ and qqq potentials in the appropriate limits,

4) It should produce flux-tube-like potentials between well separated quarks or clusters of quarks with, for example, the correct ratio of the string tension between $3 - \bar{3}$ clusters and $6 - \bar{6}$ clusters, and

5) It should incorporate the known short distance interquark forces from one gluon exchange.

The only model potential I know which meets these minimum requirements is the one which assumes that confinement develops smoothly from Coulomb's law so that

$$V = \sum_{i<j} V_{ij} \tag{148}$$

where

$$V_{ij}(r_{ij}) = -\left[-\frac{\alpha_s(r_{ij})}{r_{ij}} + \frac{3}{4}c + \frac{3}{4}br_{ij}\right] \vec{F}_i \cdot \vec{F}_j \tag{149}$$

where $\vec{F}_i = \vec{\lambda}_i/2$ for a quark and $\vec{\lambda}^c_i/2 = -\vec{\lambda}^*_i/2$ for an antiquark. Aside from meeting the above "minimum requirements", this potential seems to do a remarkable job of imitating the low-lying adiabatic surfaces of the full flux tube picture. This is illustrated in Figures 19(a′) and 19(b′) which show the two adiabatic surfaces (since there are two distinct colour couplings in this system, it has exactly two adiabatic surfaces instead of the infinite number of QCD) of (148) corresponding to the same $qq\bar{q}\bar{q}$ configurations as Figures 19(a) and 19(b). Unfortunately, the resulting imitation is imperfect. The $\vec{F}_i \cdot \vec{F}_j$ potential does not know about the limited transverse extension of the flux lines which form the potential and so does not cut off the analog of topological mixing fast enough to avoid a long range power law tail to hadron-hadron potentials. This flaw, however, by no means makes the $\vec{F}_i \cdot \vec{F}_j$ potential useless: it just means that one must proceed with due caution for possible spurious effects that it might introduce. In many cases in practice this is easy to do.

B. The Multiquark Fiasco

Multiquark physics has a somewhat unfortunate history. A confluence of dubious experimental results and dubious theoretical models in the late 1970's and early 1980's created, indeed, a multiquark fiasco. I am not competent to discuss what went wrong experimentally, but let me review the theoretical side of this fiasco in order to place it in perspective and thereby, I hope, point the way toward a better understanding of multiquark systems.

The story is basically one of throwing caution to the winds. Modelers from at least four different camps were, it seems to me, guilty:

1) <u>bag models</u>: The bag model posits the existence of a confining volume. Thus, by assumption, any number of quarks and antiquarks (so long as they are in an overall colour singlet), when placed in the bag, will produce a multiquark state that appears to be as legitimately confined as the $q\bar{q}$ and qqq states. On the other hand, it is obvious that the bag model should not be used in this naive way: the complicated (and nontrivial) dynamics of confinement may be approximated by a bag for states in which this property is guaranteed (like $q\bar{q}$ and qqq), but it was not intended to be a viable approximation to this dynamics in all situations. Nevertheless, some practitioners published detailed spectra of multiquarks confined to a single bag which they interpreted as having essentially the same legitimacy as bag model predictions for, say, meson spectroscopy.

2) <u>naive potential models</u>: The non-relativistic analog of the error made by some bag modelers was to assume that, in a

multiquark system, the interquark forces would be identical to those confining forces operative in mesons and baryons. This totally unjustifiable assumption will of course also automatically produce a rich spectrum of multiquark states.

3) colour chemistry: A more sophisticated mistake was made by the adherents of the "colour chemistry" school. They recognized that a given multiquark configuration could be connected by a variety of flux tube, or string, topologies. They went on to note that in some of the these topologies (see, for an example, Figure 5(c)) all the quarks and antiquarks would be connected together in a single unit. Based on a false analogy with mesons and baryons, a rich spectrum of such states was then predicted. While closer to the truth, the fatal flaw of this school was, of course, its neglect of topological mixing.

4) $\vec{F}_i \cdot \vec{F}_j$ potential models. The most peculiar story of this set belongs to those modelers who tried to study multiquark systems in the context of the $\vec{F}_i \cdot \vec{F}_j$ potential model. Unlike all of the other modelers, these were actually using a picture for multiquark confinement which, as explained above, should be approximately valid. Yet they too predicted rich spectra of multiquark bound states! In this case the problem was not with the model, but rather that approximations used to solve the $\vec{F}_i \cdot \vec{F}_j$ model were illegitimate. The difficulty and its partial resolution are the subject of the next Section.

To summarize: the multiquark fiasco was created by a series of unfortunate circumstances. Theorists, making a variety of errors from a number of different points of view, supported each other (often with very similar spectra) in predicted multiquark states analogous in their character to ordinary hadrons. This bad psychology/sociology was made even worse by the (temporary) experimental sighting of states. I suspect that the mirror image of this bad psychology/sociology operated on experimenters working in this field.

C. The $qq\bar{q}\bar{q}$ System: A Multiquark Primer

I can think of no better introduction to multiquark systems than to discuss the simplest multiquark system of two quarks and two antiquarks in the $\vec{F}_i \cdot \vec{F}_j$ model.[2)] I have already mentioned that, in early applications of the model, illegitimate approximations were made that led to erroneous predictions of a rich discrete spectrum of multiquark states. We can begin to educate ourselves in the complexity of multiquark systems by learning from these old mistakes. Recall as well in what follows that the $\vec{F}_i \cdot \vec{F}_j$ model is a <u>simple</u> approximation to a much more complicated structure.

In fact, let us begin by considering an even simpler theoretical model which exposes most clearly the basic issues: imagine that the interquark potential were just

$$V_{ij}(r_{ij}) = \frac{1}{2}kr_{ij}^2 \; \vec{F}_i \cdot \vec{F}_j \tag{150}$$

a purely "harmonic" force. (At first one might even imagine that this simple problem could be solved analytically, but it cannot be). With the labels and coordinate system shown in Figure 20, and taking as a colour basis the states $|\bar{3}_{12}3_{34}>$ and $|6_{12}\bar{6}_{34}>$ (where C_{ij} means that particles i and j are coupled to the colour representation C and in each case the system is in an overall colour singlet), we have

$$|\psi> = \psi_{\bar{3}3}(\vec{\sigma},\vec{\bar{\sigma}},\vec{\lambda})|\bar{3}_{12}3_{34}> + \psi_{6\bar{6}}(\vec{\sigma},\vec{\bar{\sigma}},\vec{\lambda})|6_{12}\bar{6}_{34}> \tag{151}$$

and (with "1"$\leftrightarrow|\bar{3}_{12}3_{34}>$ and "2"$\leftrightarrow|6_{12}\bar{6}_{34}>$ in matrices)

$$H = \frac{1}{2m}(p_\sigma^2+p_{\bar{\sigma}}^2+p_\lambda^2)\begin{pmatrix}1 & 0\\ 0 & 1\end{pmatrix} + \frac{1}{2}k\begin{pmatrix} 2\sigma^2+2\bar{\sigma}^2+\frac{4}{3}\lambda^2 & -2\sqrt{2}\,\vec{\sigma}\cdot\vec{\bar{\sigma}} \\ -2\sqrt{2}\,\vec{\sigma}\cdot\vec{\bar{\sigma}} & \sigma^2+\bar{\sigma}^2+\frac{10}{3}\lambda^2 \end{pmatrix} \tag{152}$$

If we now assume that mixing between the $|\bar{3}_{12}3_{34}>$ and $|6_{12}\bar{6}_{34}>$ sectors can be neglected (or at least treated perturbatively) then we will <u>erroneously</u> conclude that both the $\bar{3}3$ and $6\bar{6}$ sectors of the system are confining in all three relative coordinates, so these systems will be analogous to ordinary mesons and baryons. On this basis we would expect to find two separate infinite towers of discrete levels (in the absence of mixing), or, if mixing is treated by perturbation theory, a single mixed tower of discrete states. However, this conclusion <u>must be qualitatively wrong</u> since one can easily show that this Hamiltonian contains solutions corresponding to free mesons in situations where the system is split into two $q\bar{q}$ clusters in that mixture of $|\bar{3}_{12}3_{34}>$ and $|6_{12}\bar{6}_{34}>$ corresponding to having each of these mesonic clusters in colour singlets. We can thus prove that in at least one simple case the $-2\sqrt{2}\ \vec{\sigma}\cdot\vec{\sigma}$ terms cannot be treated perturbatively: at least an infinite number of the states of the (erroneously deduced) discrete tower of states must collapse into a continuum of free mesons. Indeed, matters are even worse: there are an infinite number of such continuous spectra, corresponding to the infinite number of possible internal excitations of the separated meson clusters.

From this perspective, one might expect that if non-trivial multiquark states exist in this system (i.e., states that are not essentially two mesons), then they are likely to correspond to either weakly bound states, or to meson-meson resonances. Certainly, it would seem, the picture of a tower of discrete states analogous to the spectrum of ordinary mesons is very inappropriate.

So far as I am aware, we still do not know the full spectrum of this harmonic multiquark model. However, a recent study[2] of its ground state using variational methods strongly suggests that it does not even have any weakly bound states or resonances; rather, it appears to be best viewed as a theory of two mesons with weak residual interactions.

It would take us too far afield to discuss in detail what happens in more realistic models of $qq\bar{q}\bar{q}$ with an $\vec{F}_i\cdot\vec{F}_j$ potential. Suffice it to say that there are strong suggestions that the physics of such meson-meson systems is roughly analogous to that of the nucleon-nucleon system. Indeed, it may even turn out that the $S^*(980)$ and $\delta(980)$ mesons are $K\bar{K}$ bound states[2,3] roughly analogous to the deuteron which we discuss next.

References for Chapter IV

1. The flux tube picture for multiquark states and its relation to the $\vec{F}_i\cdot\vec{F}_j$ model are discussed in Ref. 4 of Chapter II.
2. The $qq\bar{q}\bar{q}$ system in the $\vec{F}_i\cdot\vec{F}_j$ model is discussed in J. Weinstein and N. Isgur, Phys. Rev. Lett., 48, 659 (1982); Phys. Rev. D27, 588 (1983). References to the $qq\bar{q}\bar{q}$ literature can be found therein.
3. The suggestion that the S^* and δ are $qq\bar{q}\bar{q}$ states originates with R.L. Jaffe, Phys. Rev. D15, 267, 281 (1977); R.L. Jaffe and K. Johnson, Phys. Lett. 60B, 201 (1976); R.L. Jaffe, Phys. Rev. D17, 1444 (1978).

V. "DERIVING" NUCLEAR PHYSICS: SIX QUARKS WITH CHROMODYNAMICS

Given the strong analogy of Section IV.A. and Table 1 between atomic physics and hadron physics, it is natural to try to extend the analogy to make a connection between molecular physics and nuclear physics. This analogy is sketched in Table 2 for the hydrogen molecule and the deuterium nucleus. Of course this basic analogy has some "technical" flaws, the most important being that in this case 6>>4. To see what I mean by this, note that in H_2 an adiabatic approximation (in which the electronic energy is calculated for fixed proton positions and then used as an interproton potential) works well since $m_p >> m_e$. Thus one really only needs to solve a Schrodinger problem for the spatial locations of two particles. In the case of ^2H, no adiabatic approximation is viable, and moreover, the two spins and three colours of each quark play a crucial dynamical role.

There is one further analogy of another sort which I would like to stress. At this time it is safe to say that we understand the hydrogen molecule perfectly: enormous variational calculations have been able to reproduce its properties almost exactly. However, in some sense the very first calculation by Heitler and London in 1927 was the most satisfying since on the basis of that calculation (which only reproduced the dissociation energy and equilibrium radius to about 50%) it was already possible to state the main physics conclusion (which had already been conjectured by Lewis in 1918): the chemical bond is a shared electron. The simplest calculation for the six quark system with chromodynamics is technically of comparable complexity to the latest calculations for H_2, but it should be compared in spirit to the situation in 1927. While such a calculation is, relative to the complexity of the system, very crude, if we are lucky we may still be able to answer some of the basic physics questions posed by the properties of nuclei.

The idea of such a calculation is an old one and I cannot possibly do justice to the history of the subject here. I would rather like to describe the particular approach which we took[1)] and our results. Our calculation was very strongly based on the molecular analogy of Table 2 and the belief that some of the fundamental questions we wished to answer could only be addressed if we could calculate the six quark ground state wavefunction. We accordingly set out to do this instead of to study, for example, scattering via resonating group techniques. As already implied, the calculation was extremely arduous; it would have

TABLE 1: Quark Models in QCD versus Atomic Physics in QED

	QED	QCD
basic potential	$\frac{\alpha}{r}(\frac{e_1}{e})(\frac{e_2}{e})$ electric	$(-\frac{\alpha_s}{r}+\frac{3}{4}c+\frac{3}{4}br)\frac{\vec{\lambda}_1}{2}\cdot\frac{\vec{\lambda}_2}{2}$ chromoelectric
spin-dependence	for example, $\frac{8\pi\alpha}{3m_1m_2}\vec{S}_1\cdot\vec{S}_2\delta^3(\vec{r})(\frac{e_1}{e})(\frac{e_2}{e})$ from magnetic moments	for example, $\frac{-8\pi\alpha_s}{3m_1m_2}\vec{S}_1\cdot\vec{S}_2\delta^3(\vec{r})\frac{\vec{\lambda}_1}{2}\cdot\frac{\vec{\lambda}_2}{2}$ from chromagnetic moments
typical "atom"	e^+e^-, positronium	$c\bar{c}$, charmonium; also non-abelian $q_Rq_Bq_Y$ baryonic atoms
orbital excitations	for example, the Lyman series	for example, B(1235)-π(140) in $q\bar{q}$ and N(1520)-N(940) in qqq
spin splittings	for example, the 21 cm line of atomic hydrogen	for example, ρ(770)-π(140) in $q\bar{q}$ and Δ(1235)-N(940) in qqq
v/c	<<1 in hydrogen; ≃1 in uranium	<<1 in $b\bar{b}$, beautyonium; ≃ 1 in the nucleon

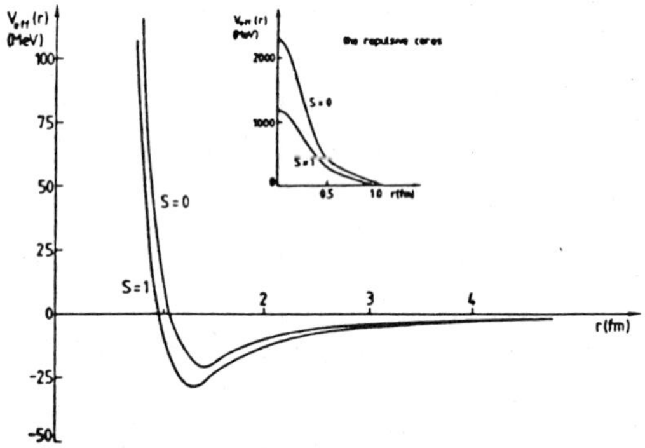

FIG. 21. the effective nucleon-nucleon potentials in the $^3S_1$ and $^1S_0$ channels arising from residual colour forces

TABLE 2: The Hydrogen Molecule in QED versus Deuterium in QCD

	QED	QCD
symbol	H_2	^2H
"atomic" structure	$H_2 \leftrightarrow H + H$	$^2H \leftrightarrow p + n$
basic structure	$H_2 \leftrightarrow 2e^- + 2p$	$^2H \leftrightarrow 3u + 3d$
basic description	four body problem with Coulomb's law	six body problem with chromodynamics
"atomic" dissociation energy	4.48 eV	2.23 MeV
"atomic" separation	0.740 A	3.9 fm
method of solution	variational	variational
"quasiparticle" description	H atoms in effective potential	nucleons in effective potential

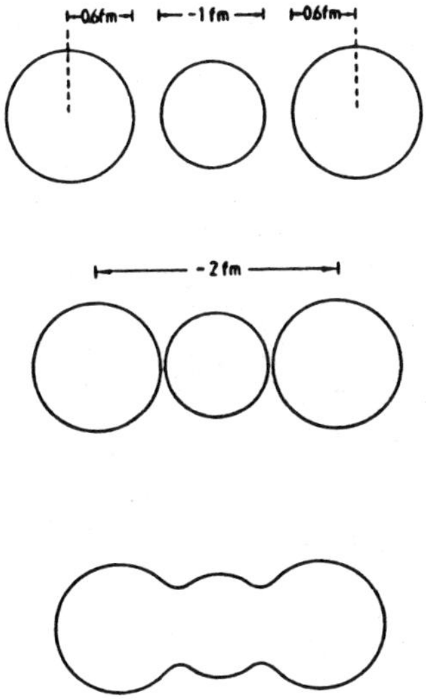

FIG. 22. a cartoon illustrating that for $r \gtrsim 2$ fm meson exchange is unlikely to be appropriate to the description of the inter-nucleon potential

been even more difficult if we hadn't made use of a simple calculational trick. This was to concentrate on the 3u+3d quark sector (I_3=0 to particle physicists, T_3=0 to nuclear physicists) so that we could use $S_3 \times S_3$ symmetry instead of S_6 symmetry (S_n is the permutation group on n objects). This had the disadvantage that the calculation was not automatically diagonalized in isospin. However, the penalty for this was slight: our computer had to diagonalize some larger matrices than it would have otherwise. In exchange, it was possible to do the analytic calculations required in about a year instead of at least double that.

The calculation proceeded via a cluster expansion of the five relative coordinates. Two pairs of coordinates were expanded in a three quark harmonic oscillator basis; the fifth coordinate was described by an intercluster variational wavefunction. The cluster expansion was made in a basis space of 115 dimensions which included all possible colour, spin, flavour, and space combinations possible by allowing up to two units of intracluster spatial excitation; up to 15 parameters were used to expand the intercluster wavefunction which was allowed to be in an S, P, or D-wave. Our goal was to allow the six quark wavefunction to do anything it liked. For example, our wavefunction could certainly reach any reasonable "democratic", "spherical", six quark state clustered around the origin. I emphasize that although we used a cluster expansion, we did not put clustering in by hand.

For our Hamiltonian we used the model of Section III.C. which, as we have seen, provides a reasonably good description of baryon spectroscopy---clearly a minimal requirement if one is to believe the results of a "two nucleon" calculation. I stress that this Hamiltonian is taken over completely from studies of baryon spectroscopy, so that our calculations for the six quark system are purely deductive with no free parameters whatsoever.

At the time we began our very lengthy calculations, this subject was in its infancy,[2)] with early calculations based on an adiabatic approximation seeming to show a repulsive core. In later improved calculations, however, the core disappeared. Eventually it became understood via resonating group techniques that the adiabatic approximation is invalid, and the use of such methods seemed to strongly indicate that a repulsive core emerges naturally from short distance quark dynamics. However, at least within the limited basis spaces possible for such techniques, there seemed to be no trace of nuclear binding! Our approach, while much less elegant than the resonating group methods, is considerably more powerful for studying the properties of the six quark ground states. As a measure of this, one can compare our 115 term cluster expansion with the maximum of 6 terms employed in any resonating group calculation.

Our main results can be summarized as

1) We find that the 3u+3d quark ground state wave functions with I,J = 0,1 show very strong clustering into two three quark systems with neutron and proton quantum numbers and properties. This shows that, within the context of our calculation, nuclear physics is really approximately the physics of nucleon quasiparticles. This is a nontrivial result, as evidenced by the fact that it is quantum number dependent: other six quark systems, when constrained to nuclear densities, do not exhibit such clustering.

2) Given the strong clustering it is natural to project out of the full six quark wavefunction an n-p component. This component, which nearly saturates the full wavefunction, has an amplitude as a function of the n-p separation which we can then interpret as an effective internucleon wavefunction $\psi_{eff}(r)$. This $\psi_{eff}(r)$ can in turn be used to define a bound-state effective potential $V_{eff}(r)$, via the NN Schrodinger equation, which has the following very satisfactory properties:

a) it gives a strong repulsive core for r<1 fm,

b) it provides an intermediate range binding attractive region between 1 and 2 fm, and

c) the $^3S_1$ potential is less repulsive and more binding than the $^1S_0$ potential.

These features are illustrated in Fig. 21. Let me reemphasize that this potential is purely deductive: we have simply solved the six quark problem for a known Hamiltonian. Let me also immediately stress that I do not advocate this potential as a replacement for phenomenological nucleon-nucleon potentials. It is based on an approximate solution to a very rough model for the quark dynamics and so cannot be expected to be accurate. What is relevant is that this calculation may point us in the direction of _eventually_ replacing phenomenological potentials by derived ones, much as the original H_2 calculation of 1927, while quantitatively lacking, captured the basic physics of the situation and eventually led to today's more accurate understanding of the hydrogen molecule. In this spirit, let me attempt, on the basis of our calculation, to provide a picture of the nuclear force analogous to the molecular picture that "the chemical bond is a shared electron."

The attractive region of V_{eff} is the easiest to interpret: it is actually _physically_ analogous in origin to the molecular case. Recall that, at least for separations larger than twice the Bohr radius, there is a van der Waals attraction between two hydrogen atoms which is a weak residue of the electrical forces producing the electrically neutral ep clusters. This weak attractive potential arises because the neutral atoms have a second order perturbation in their energy $-a^2/\Delta E$ from their amplitude a to temporarily excite each other via Coulomb's law into a virtual state with energy ΔE above their ground states (in which the atoms are in P-waves coupled to L=0). Semiclassically we would say that while the atoms are neutral, it is energetically favourable for them to develop small oppositely directed electric dipole moments. Two colour neutral quark clusters can develop a weak (i.e., nuclear strength!) residual attraction by a closely related mechanism: they can temporarily excite each other (via the QCD version of Coulomb's law of Table 1) into a virtual state in which the clusters are in coloured P waves coupled to L=0 and to zero net colour. Semiclassically we would say that while the atoms are colour neutral, it is energetically favourable for them to develop small chromoelectric dipole moments correlated in their orientation in both space and in colour-space.

It is much more difficult to describe the physical origin of the repulsive core. Our calculations, first of all, confirmed the conclusion which had already been reached via resonating group techniques that the repulsive core is a non-adiabatic effect. This is not surprising: unlike the molecular case where the electron motion is fast relative to the motion of the protons, in the six quark problem there are no adiabatic coordinates once the two three quark clusters begin to overlap. As shown in the earliest calculations, a repulsive core does develop if the six quark system is arbitrarily constrained to remain in two purely nucleonic clusters (a consequence of the Pauli principle and the colour hyperfine interactions). The repulsive core apparently emerges from a complete calculation because the NN configuration does not have time to evolve into more energetically favourable configurations available to it: i.e., there is a partial dynamical constraint to the two nucleonic cluster sector. One may describe this constraint in other language by noting that there are kinetic energy terms associated with the mixing angles of non-NN configurations, and these kinetic energy terms create pseudoforces (analogous to centrifugal barrier effects) which produce the NN "constraint". This very dynamical origin for the repulsive core is consistent with its known complex dependence on the spin, orbital, and total angular momenta of the six quark system; note, for example, that the $\vec{S}_i \cdot \vec{S}_j$ dependence of the core has already emerged from our calculation.

There are many weaknesses of the calculation I have just described, and before going on it seems appropriate to mention some of them. This is not the place to discuss these issues fully, however, so I will just make an annotated list of some possible criticisms:

1) non-relativistic and other approximations: Since we were mainly concerned with the properties of the six quark system relative to two separated three quark clusters, many sources of error---including relativistic effects---tend to cancel.

2) the spin-orbit mystery: The baryon data enforce the near absence of spin-orbit effects, and spin-orbit effects of the strength allowed by the data would not significantly effect our results.

3) the $\vec{F}_i \cdot \vec{F}_j$ confinement potential: We have already extensively discussed the fact that such a model for confinement can only approximate the structure expected from QCD. Probably the most serious flaw of this approximation is that it leads to a spurious long-range power-law tail to the effective nucleon-nucleon potential. I will describe its effects on our calculations below.

4) the radial dependence of $H_{s\,ij}$: In the actual calculations the anharmonic potential U was replaced by an attractive δ-function interaction. In baryon spectroscopy this replacement is known to be adequate for low-lying excitations. We have checked that, since it is averaged over clusters, it should also suffice for this problem, but this point deserves further study.

We have checked other possible weaknesses of our calculation and believe that it has no fundamental problems, even though our model for the Hamiltonian is very crude. The reason for this is that, if we are right, the basic features of nuclear physics are not very subtle consequences of QCD: they will emerge from any model which confines colour in a sensible way and which recognizes some of the basic phenomenological constraints from hadron spectroscopy (like the existence of important interquark spin-spin interactions).

Before finally applying our results to an actual nuclear system, we paused to ask ourselves if there were any regions in which our calculation would fail to take into account the dominant physics of the six quark system. We decided that there was one: when the system has segregated into two three quark clusters which are far apart, our infinitely stretchable flux

TABLE 3: Some Properties of the Six Quark Ground States

property	theoretical value	experimental value
E_d(MeV)	$-2.9^{+0.8}_{-0.3}$	-2.23
$(r_d^2)^{1/2}$ (fm)	2.2±0.5	1.95
Q_d (mb)	+2.1±0.5	+2.86
μ_d *(nuclear μ_N)	+0.859±0.003	+0.857
$E(^1S_0)$ (MeV)	$-0.4^{+0.4}_{-0.1}$	unbound

*μ_d is calculated assuming that the departure from $\mu_p+\mu_n$ is due only to the calculated 3.6% D-wave mixing

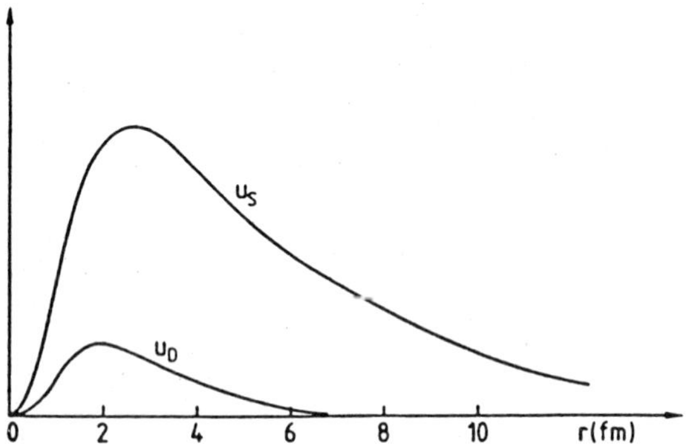

FIG. 23. the deuteron wavefunctions u_S and u_D shown with their correct relative normalization

tubes should be breaking to produce meson exchange. We further argued---on both the naive grounds represented by the cartoon of Figure 22 and on more rigorous grounds---that only the region beyond about 2 fm should be affected. In this region only pion exchange is significant, so to proceed from the results of our calculation to a real nucleus we considered a "hybrid" model in which $V_{NN}=\tilde{V}_{eff}^{NN}+\tilde{V}^{ope}$. Here $\tilde{V}_{eff}^{NN}$ is our derived NN potential cut off at some internuclear separation $r_c \sim O(2fm)$ and $\tilde{V}^{ope}$ is one pion exchange "cut in" at r_c. Our results for the properties of the simplest nucleus, deuterium, were not very sensitive to r_c in the range 2 to 3 fermis: Table 3 compares our "hybrid" calculation with the data. The theory errors shown in the table reflect both sensitivity to r_c and also estimates of the errors arising in the various quantities shown from our basis truncation and our limited variational wave functions. The calculated deuterium wave functions are shown in Figure 23. The results are better than they deserve to be, given the quality of the model.

References for Chapter V

1. K. Maltman and N. Isgur, Phys. Rev. Lett. 50, 1827 (1983); Phys. Rev. D29, 952 (1984).
2. A partial history and a rather complete set of references to related work may be found in the latter of Refs. 1.

VI. CONCLUSIONS

It will be clear from these lectures that the derivation of nuclear physics from the quark model, and ultimately from QCD, is a subject that is still in its infancy. At this time it is even unclear whether the route I have suggested in these lectures will prove to be the best one. Nevertheless, I believe that the picture we have been able to sketch based on this approach is attractive and suggestive of the possibility of further progress.

Among the most important conclusions we have drawn for nuclear physics are these:

1) Nuclear physics is multiquark physics and, as such, involves gluon dynamics in a much less trivial way than mesons and baryons.

2) Nuclear physics is approximately, but far from exactly, multi-nucleon physics. Explicit quark and colour effects exist, but they are subtle effects which will arise from configurations in which the nucleonic quark clusters overlap.

3) It is quite likely that the importance of meson exchange in nuclear physics has been overestimated. For example, it now seems very unlikely that ω exchange is responsible for the repulsive core of the NN interaction.

However, perhaps the most important conclusion that one can draw from this approach would be that it is now appropriate to view the relationship of nuclear physics to hadron physics and QCD as being roughly analogous to that of condensed matter or molecular physics to atomic physics and QED.

HADRONIC MATTER ON A LATTICE

J. Polonyi †
University of Illinois
Urbana, IL 61801

† On leave from the Central Research Institute for Physics, Budapest, Hungary.

1. INTRODUCTION

The description of strongly interactive matter from first principles must give an account of how the quark and gluon degrees of freedom observed in short distance phenomena give place to the conventional degrees of freedom of nuclear physics for long distance processes at low energy. Quantum Chromodynamics, which describes strong interactions, must realize this transmutation of relevant degrees of freedom through its essential nonperturbative features. These lectures give a brief introduction to lattice field theory, which is the most sucsessful systematic way so far to discuss or solve numerically strongly nonperturbative systems [50].

The main concern of these lectures is to show how can we have a qualitative understanding of certain nonperturbative features in terms of classical statistical mechanics and how to obtain quantitative results using simulation methods.

2. PATH INTEGRAL REPRESENTATION

The algebraic realization of the canonical commutation relations of strongly nonlinear quantum field theory is beyond our present capability. But the path integral representation of quantum amplitudes still gives a constructive way to treat such a theory [1]. We begin with the discussion of the path integral representation in the case of one dimensional quantum mechanics.

The system is defined by the Hamiltonian $\hat{H} = 1/2\ \hat{p}^2 + V(\hat{q})$ and the commutation relation is $[\hat{q},\hat{p}] = i\hbar$. Our aim is to give an expression for the matrix element $\langle q|\hat{U}(t)|q'\rangle$ of the time evolution operator $\hat{U}(t) = \exp(-it/\hbar\hat{H})$ between eigenstates of the coordinate operator $\hat{q}$ with eigenvectors $|q\rangle$ such that $\hat{q}|q\rangle = q|q\rangle$. We shall use momentum eigenvectors $|p\rangle$ such that $\hat{p}|p\rangle = p|p\rangle$ as well, which have overlap $\langle q|p\rangle = \exp(iqp/\hbar)$, and the resolution of the identity operator given by $1 = \int dq|q\rangle\langle q| = \int dp/2\pi\ |p\rangle\langle p|$.

Using the one parameter group property of the time evolution $\hat{U}(t+s) = \hat{U}(t)\hat{U}(s)$ or $\hat{U}(Nt) = \hat{U}(t)^N$ we can write this matrix element in the form of an "integral over paths":

$$\langle q|\hat{U}(t)|q'\rangle = \prod_{i=1}^{N-1} \int dq_i \langle q|\hat{U}(t/N)|q_1\rangle\langle q_1|\hat{U}(t/N)\ \ldots\ \hat{U}(t/N)|q'\rangle \tag{1}$$

What is left is to work out an approximation for the short time evolution operator with an error which does not accumulate in (1) when $N \to \infty$. It is easy to see that we can approximate the matrix elements of $\hat{U}(t/N)$ between two overlapping systems of basis vectors

$$\langle q|\exp\{-i\varepsilon\hat{H}\}|p\rangle \sim \langle q|p\rangle \quad [1 - i\varepsilon(\tfrac{1}{2} p^2 + V(q))] \tag{2}$$

Using the completeness of the $|p\rangle$ states we find

$$\langle q|\exp\{-i\varepsilon\hat{H}\}|q'\rangle = \int \frac{dp}{2\pi} \exp[ip(q-q')-i\varepsilon(1/2p^2 + V(q))] \tag{3}$$

The integral can be calculated using the basic formula

$$\int d^n x \exp(-\tfrac{1}{2} xAx+bx) = (2\pi)^{n/2}/\sqrt{\det A} \quad \exp(\tfrac{1}{2} bA^{-1}b) \tag{4}$$

The approximation for the matrix element thus reads

$$\langle q|\hat{U}(\varepsilon)|q'\rangle = 1/\sqrt{(2\pi i\varepsilon)} \quad \exp\{\frac{i\varepsilon}{\hbar} [(\frac{q-q'}{\varepsilon})^2 - V(q)]\} \tag{5}$$

Substituting this result into (1) we obtain

$$\langle q|\hat{U}(t)|q'\rangle = \lim_{N\to\infty}(N/2\pi it)^{N/2} \prod_{i=}^{N-1} \int dq_i \exp\{\frac{i\varepsilon}{\hbar} \sum_i [(\frac{q_{i+1}-q_i}{\varepsilon})^2 - V(q_i)]\}$$

$$= \int D[q(t)] \exp\{\frac{i}{\hbar} S[q(t)]\} \tag{6}$$

The integration over paths with given end points is defined by this limit procedure, called regularization. Unfortunately the proof that the error made in the determination of the short time propagator does not accumulate in the $N \to \infty$ limit can be carry out only for certain class of actions. Matrix elements between states charaterized by wave functions $\psi(q)$ and $\psi'(q)$ are expressed as

$$\langle\psi|\hat{U}(t)|\psi'\rangle = \int dq\, dq' \psi^*(q)\psi'(q') \int D[q(t)]\exp\{\frac{i}{h} S[q(t)]\} \tag{7}$$

For the calçulation of the matrix elements of an operator F(q) we use the expression

$$\langle q|\hat{U}(t_1)F(\hat{q})\hat{U}(t_2)|q'\rangle = \int\frac{dq''}{2\pi}\langle q|\hat{U}(t_1)|q''\rangle F(q'')\langle q''|\hat{U}(t_2)|q'\rangle$$

$$= \int D[q(t)]\exp\{\frac{i}{\hbar}S[q(t)]\}F[q(t)] \tag{8}$$

The domain of integration consists of trajectories covering time evolution $t=t_1+t_2$ and having end points $q(0)=q'$, $q(t_1+t_2)=q$.

It is instructive to study the role of the classical configuration given by $\delta S/\delta q=0$ in the semiclassical limit $\hbar\to 0$. Expanding S[q(t)] around the extremal path we have

$$\langle q|\hat{U}(t)|q'\rangle = \exp\{\frac{i}{\hbar}S[q_{c\ell}]\}\int D[q(t)]$$

$$\exp\{\frac{i}{2}\int dtdt'\big(q(t)-q_{c\ell}(t)\big)\frac{\delta^2 S}{\delta q(t)\delta q(t')}\big(q(t')-q_{c\ell}(t')\big)+O(h)\}$$

$$= \text{const.}\exp\{\frac{i}{\hbar}S[q_{c\ell}]\}\sqrt{\det[\frac{\delta^2 S}{\delta q(t)\delta q(t')}]}\big(1+O(h)\big) \tag{9}$$

The integral variable was rescaled in the second step, $q\to\sqrt{\hbar}\,q$. We see that the path integral is dominated by the classical trajectory and its O(h) neighbours in the semiclassical limit.

In order to get rid of the complex nature of the integrand we make analytical continuation to pure imaginary (Euclidean) time. In this case we have for the trace of the time evolution operator

$$Z=\text{tr}\exp(-\frac{i}{iT}\hat{H}) = \text{tr}\exp(-\beta\hat{H}) \tag{10}$$

Carrying out this analytic continuation in the path integral expressions we obtain

$$Z = \int D[q(t)]\exp\{i\int d(-it)(-\frac{1}{2}\dot{q}^2 - V(q))\}$$

$$= \int D[q(t)]\exp\{-S_{Eucl}[q(t)]\} \tag{11}$$

where the domain of integration consists of periodic paths with period length $\beta=1/T$ ($\hbar=k=1$). We have derived that quantum mechanics in Euclidean space-time or quantum mechanics at finite temperature equilibrium is equivalent with the statistical mechanics of classical one dimensional strings.

Although we lost real time evolution in this analytically continued quantum mechanics, lots of things are still available. Two examples are i. ground state expectation values ii. the spectrum. In fact

$$\text{i. } \lim_{\beta\to\infty} \frac{1}{Z} \text{tr} \exp\{-\beta\hat{H}\}\hat{o} = \lim_{\beta\to\infty} \frac{1}{Z}\sum_n \exp\{-\beta E_n\}\langle n|\hat{o}|n\rangle = \langle 0|\hat{o}|0\rangle$$

Expectation values are dominated by the ground state in the zero temperature limit if the theory has a nonzero gap between the ground state and the first excited state. ii. The function F(E) defined by

$$F(E) = \int d\beta\, Z_\beta \exp\{\beta E\} = \int d\beta \sum_n \exp\{-\beta(E_n - E)\} = \sum_n \frac{1}{(E_n - E)}$$

has poles at the energy levels.

3. QUANTUM MECHANICS

As a simple illustration of the path integral formalism consider the problem of an anharmonic oscillator [2]

$$\hat{H} = \frac{1}{2}\hat{p}^2 + \frac{1}{2}\omega^2\hat{q}^2 + \lambda\hat{q}^4, \qquad [\hat{q},\hat{p}] = i \tag{12}$$

The expectation value of the operator $F(\hat{q})$ in the ground state is given by

$$\langle 0|F(\hat{q})|0\rangle = \lim_{\substack{\Delta\to o \\ N\Delta\to\infty}} \frac{\int \prod_{i=1}^{N} dq_i \exp\{-\Delta \sum_{i=1}^{N} [(\frac{q_{i+1}-q_i}{\Delta})^2 + \frac{1}{2}\omega^2 q_i^2 + \lambda q_i^4]\} F(q_i)}{\int \prod_{i=1}^{N} dq_i \exp\{-\Delta \sum_{i=1}^{N} [\frac{1}{2}(\frac{q_{i+1}-q_i}{\Delta})^2 + \frac{1}{2}\omega^2 q_i^2 + \lambda q_1^4]\}} \tag{13}$$

In practice this average is considered as a canonical ensemble average and configurations distributed according to the Gibbs factor $\exp(-S_{eucl})$ are generated by the Monte Carlo method [3].

Such simulations are carried out for finite Δ and N. An important restriction for these quantities is that a) Δ must be smaller than the inverse energy of the highest excited state which we want to consider as an intermediate state in the calculation, b) ΔN must be larger than the inverse energy of the first excited state

$$\Delta < 1/E_{max}, \quad 1/E_{min} < \Delta N \tag{14}$$

In addition to the usual expectation values we can calculate the absolute value of the ground state wave function

$$|\psi_o(x)|^2 = \lim_{\beta\to\infty} \langle\delta(\hat{q}-x)\rangle \tag{15}$$

which is the distribution of the coordinate q in the simulation. Such histograms are plotted in Fig 1,2.

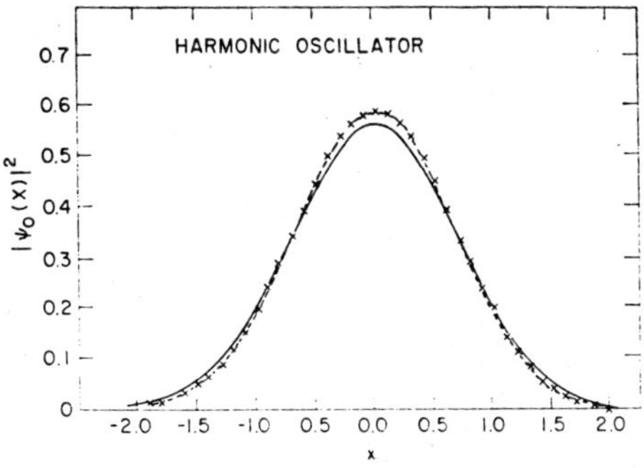

Fig.1. Crosses show the distribution of the coordinate in the simulation when $\lambda = 0$. Dashed and solid lines give the absolute magnitude of the ground state wave function of the transfer matrix (see Section 5) and the continuum harmonic oscillator Hamiltonian respectively.

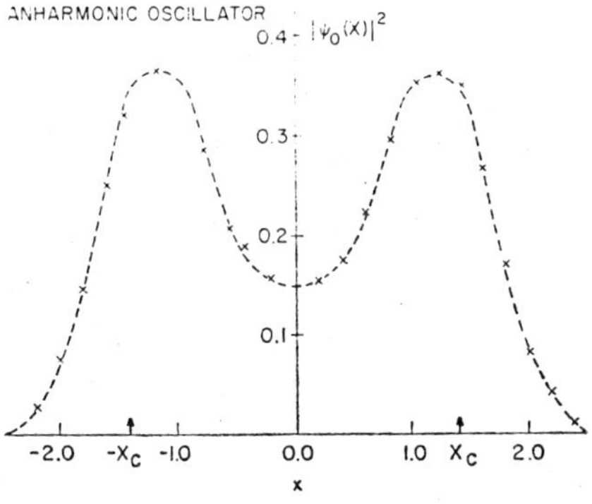

Fig.2. Distribution of the coordinate in the simulation when $V(q) = 2(q^2 - 2)^2$

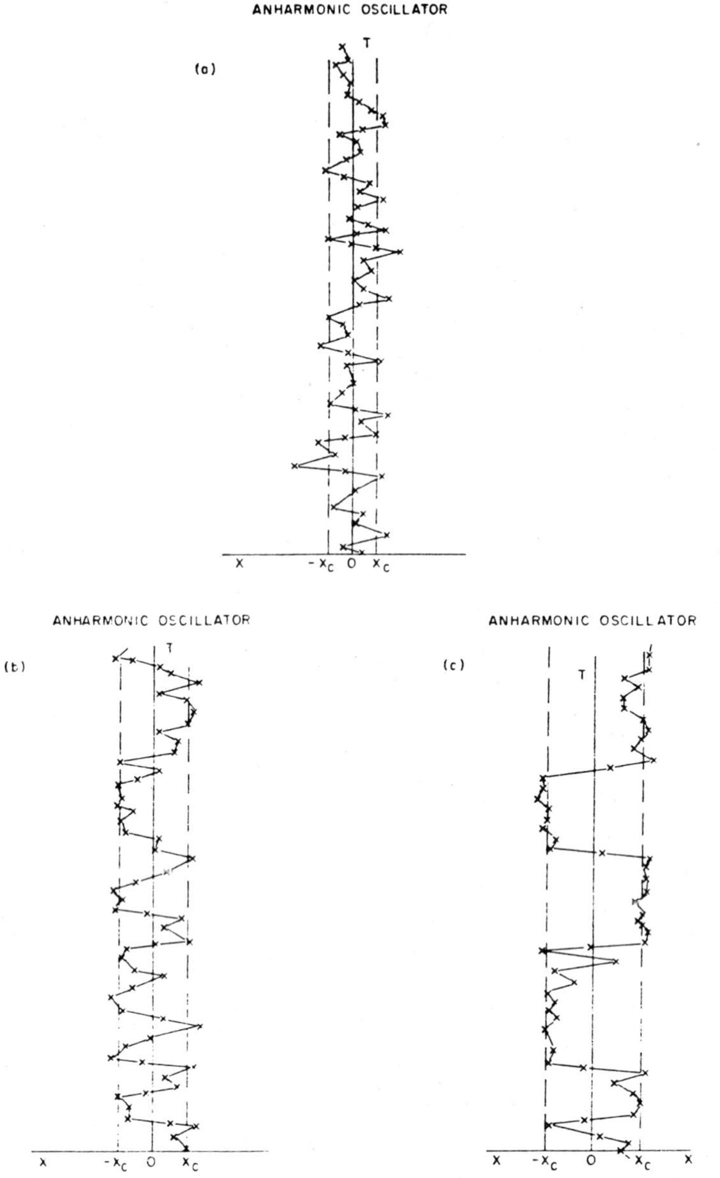

Fig. 3. Typical trajectories in the path integral with potential (a) $V(q)=(q^2-.5)^2$, (b) $V(q)=(q^2-1)^2$, (c) $V(q)=(q^2-2)^2$.

Typical configurations of the simulation are shown in Fig.2 for different values of the parameters in the potential. Observe that as the separation of the two minima of the potential grows the path integral becomes dominated by kinks (trajectories interpolating between different minima) and the lowest lying state becomes a superposition of two naive ground states built around the minima. This is just the behaviour derived in the WKB or steepest descent approximation in the semiclassical limit.

4. SCALAR FIELD THEORY

Consider the generalization of the path integral representation for quantum mechanics in d-dimension. The only change is that a trajectory is now a curve in d-dimensional space $q_i(t), i=1,\ldots,d,\ t_1 < t < t_2$. Expressions (1)-(11) remain valid with obvious modifications like $dq \to d^d q$. We shall apply this formalism for scalar quantum field theory characterized by the Hamiltonian

$$\hat{H}_c = \int d^3x \left\{ \frac{1}{2} \hat{\pi}_c^2 + \frac{1}{2} (\vec{\nabla} \hat{\phi}_c)^2 + V(\hat{\phi}_c) \right\} \tag{16}$$

and commutation relations $[\hat{\phi}_c(x), \hat{\pi}_c(y)] = i\delta(x-y)$. Introducing a three dimensional mesh in space we define lattice field variables $\phi(\vec{n})$ attached to the lattice site $\vec{x}=\vec{n}a$, $\hat{p}(\vec{n}) = a\, \hat{p}_c(\vec{n}a)$ (a is the lattice spacing).

This system is considered as the simple quantum mechanical problem of Section 3 in $d=N^3$ dimension where N is the size of the quantization box in lattice spacing units. Eigenvectors of the field operator are direct products of eigenvectors defined at the sites locally $|\Phi\rangle = \prod_{\vec{n}} |\Phi(\vec{n})\rangle$ and $\phi(\vec{n})|\Phi\rangle = \Phi(\vec{n})|\Phi\rangle$.

Repeating the steps made in Section 2 we can derive the path integral expression for any operator $F[\hat{\phi}]$

$$\langle T(F(\hat{\phi}))\rangle = \frac{\int D[\phi] \exp\{-S_{Eucl}[\phi]\}\, F[\phi]}{\int D[\phi] \exp\{-S_{Eucl}[\phi]\}} \tag{17}$$

where $D[\phi] = \lim_{N\to\infty} \prod_{n^\mu=1}^{N} d\phi(n^\mu)$,

$S_{Eucl} = \sum_n \left\{ \sum_\mu \frac{1}{2} \left(\phi(n+\mu) - \phi(n)\right)^2 + V(\phi(n)) \right\}$. Careful examination of relation (8) with $F[\phi]= \phi(x,t_1)\ \phi\ (y,t_2)\ t_1 < t_2$ reveals the presence of the time ordering operator T: $T((\phi(t_1)\phi(t_2)) = \phi(t)\phi(t_2)$ if $t_2 < t_1$ and $\phi(t_2)\phi(t_1)$ if $t_1 < t_2$ in the expectation values.

There are two complementary expansion methods for these path integrals. The more conventional one is the weak coupling expansion. Consider the theory $V(\phi) = \frac{1}{2} m^2 \phi^2 + \lambda\phi^4$ and expand the exponential function in the integrals in λ

$$\langle T(F[\delta])\rangle =$$

$$\frac{\int D[\phi]\exp^{-\frac{1}{2}\phi G^{-1}\phi}\left[1-\lambda\int\phi^4 d^4x + \frac{\lambda^2}{2!}(\int\phi^4 dx)^2 + \ldots\right]F[\phi]}{\int D[\phi]\exp\{-\frac{1}{2}\phi\ G^{-1}\phi\}\left[1-\lambda\int d^4x\ \phi^4 + \frac{\lambda^2}{2!}(\int d^4x\ \phi^4)^2 + \ldots\right]},$$

$$G^{-1}(n,m) = \sum_\mu (2\ \delta_{n,m} - \delta_{n+\mu,m} - \delta_{n-\mu,m}) + m^2\delta_{n,m} \tag{18}$$

A typical term in this expansion is

$$\langle T(\phi(x_1)\ \ldots\phi(x_n))\rangle = \frac{\int D[\phi]\ \exp\{-\frac{1}{2}\phi\ G^{-1}\phi\}\ \phi\ (x_1)\ \ldots\ \phi(x_n)}{\int D[\phi]\ \exp\ \{-\frac{1}{2}\phi\ G^{-1}\phi\}} \tag{19}$$

To calculate this free Green's function it is useful to introduce the generator functional

$$Z_o\ [J] = \int\ D[\phi]\exp\ \{-\frac{1}{2}\sum_{n,m}\phi(n)\ G^{-1}(n,m)\phi(m) + \sum_n \phi(n)J(n)\} \tag{20}$$

which has the obvious role

$$\langle T(\phi(x_1)\ \ldots\phi(x_n))\rangle = \frac{1}{Z_o[J]}\frac{\delta^n Z_o[J]}{\delta J(x_1)\ldots\delta J(x_n)} \tag{21}$$

(detG drops out by the normalization of $\langle 0|0\rangle$). Application of this expression to a simple case

$$\langle T(\phi(x_1)\phi(x_2)\phi(x_3)\phi(x_4))\rangle =$$

$$2\big(G(x_1 - x_2)\cdot G(x_3 - x_4) + G(x_1 - x_3)\cdot G(x_2 - x_4)\big) \tag{22}$$

shows that (22) is just Wick's theorem. Thus the derivation of the Feynman graph rules (18), Wick's theorem (22) and the causal Green's function (19) is extremely straightforward in the path integral representation.

In the case considered the free action is diagonalized by plane waves $\phi(n)=\sum c(k)\exp(2\pi ik^\mu n^\mu/N)$, $k=1,\ldots,N$, $c(k)=c^*(-k)$, $S_o=N^{-4}\sum_k c^*(k)c(k)G^{-1}(k)/2$ and $G^{-1}(k)=2\sum_\mu(1-\cos(2\pi k_\mu/N) + m^2$. The free propagator G_o (k) has the proper continuum limit in the infrared region k^μ /N<<1. This insures that any perturbatively calculable quantity has the right continuum limit in this field theory.

The other approximation scheme proceeds by expanding in the kinetic energy. First we define the lattice regularized version of the Hamiltonian operator by

$$\hat{H} = \sum_{\vec{n}} \Big\{\frac{1}{2}\,\hat{\pi}^2(\vec{n}) + \frac{1}{2}\sum_{n,i}\big(\hat{\phi}(\vec{n}+i)-\hat{\phi}(\vec{n})\big)^2 + V\big(\hat{\phi}(\vec{n})\big)\Big\}$$

$$= \sum_{\vec{n}} \Big\{\frac{1}{2}\,\hat{\pi}^2(\vec{n}) + \hat{\phi}^2(n) + V\big(\hat{\phi}(\vec{n})\big)\Big\} + \sum_{\vec{n},i}\hat{\phi}(\vec{n})\ \hat{\phi}(\vec{n}+i)$$

$$= \hat{H}_o + \hat{H}' \tag{23}$$

(We shall derive a similar lattice Hamiltonian for lattice QED in a systematic way later). Here we treat the first term which describes noninteracting quantum mechanical systems $\hat{H}_{qm} = \frac{1}{2}\hat{p}^2 + \hat{q}^2 + V(\hat{q})$ at each lattice site as an unperturbed Hamiltonian [4]. Suppose that the eigenvectors of this quantum mechanical Hamiltonian are normalizable and labeled by an integer quantum number m. The unperturbed eigenstates of the lattice Hamiltonian are labeled by the three dimensional configuration of an integer valued field $m(\vec{n})$: $|\{m(\vec{n})\}\rangle = \prod_{\vec{n}} |m(\vec{n})\rangle$.

In the calculation of higher order corrections in H' we need matrix elements like $\langle\{m'\}|H'|\{m\}\rangle$, which are reduced to the product of simple quantum mechanical matrix elements $\langle m'|\hat{q}|m\rangle$.

This strong coupling description of field theory (potential energy is large compared to the kinetic one) suggests the lattice gas picture: Quantum field theory is the set of simple quantum mechanical problems defined at each site, interacting through the kinetic energy. The states of the quantum field theory are characterized by a lattice gas containing $m(\vec{n})$ particles at the site $\vec{n}$. Particle creation or annihilation processes occur when some of these local quantum mechanical systems are excited or deexcited.

If the system considered is nonrelativistic and the total particle number N is fixed, it is more appropriate to use the formalism of 3N dimensional quantum mechanics. In this case the path integral represents integration over 3N dimensional trajectories. The space resolution of this representation is better than in the lattice gas case (the 3N dimensional coordinates are continous variables in contrast to the discrete lattice site coordinates of the lattice model), but particle creation and annihilation are excluded.

5. QUANTUM PHOTODYNAMICS

5.1 Lattice Regularization

Consider the complex scalar field theory characterized by the Lagrangian

$$L = \partial_\mu \phi^* \partial_\mu \phi - V(\phi), \quad V(\phi) = \frac{1}{2} \mu \phi^* \phi + \frac{1}{4} \lambda (\phi^* \phi)^2 \tag{24}$$

which is invariant under the global phase rotation $\phi(x) \to \exp\{i\alpha\} \phi(x)$. One might wonder why is it necessary to make the same chage of phase at different space-time points (especially for points with spacelike separation) in a relativistic local theory. Instead we should require that the change of phase be a local symmetry $\phi(x) \to \exp\{i\alpha(x)\}\phi(x)$. In order to compensate the change of the Lagrangian caused by the gradient term we introduce a compensating vector field $A_\mu(x)$ with the transformation rule: $A_\mu(x) \to A_\mu(x) - \partial_\mu \alpha(x)$ and write

$$L = (D_\mu \phi)^* D_\mu \phi - V(\phi), \qquad D\phi(x) = \left(\partial_\mu + iA_\mu(x)\right) \phi(x) \tag{25}$$

As it is well known, adding the simplest locally invariant action for the compensating field

$$L = -\frac{1}{4 e^2} (\partial_\mu A_\nu - \partial_\nu A_\mu)^2 + (D_\mu \phi)^* D_\mu \phi - V(\phi) \tag{26}$$

we generate Maxwell's equations.

The kinetic role of the vector potential is thus to relate fields at different space-time points which have the same "physical" phase independently of the direction of the local phase conventions: the "physical" phase of the matter field $\phi(x)$ is the same along a curve $x^\mu(s)$ if the field satisfies the parallel transport condition

$$\frac{dx^\mu(s)}{ds} D_\mu \phi(x) = 0 \tag{27}$$

This is a path dependent definition of equivalent phases if the field strength tensor $F_{\mu\nu}$ defined by

$$[D_\mu, D_\nu] = iF_{\mu\nu} \tag{28}$$

is nonvanishing.

These considerations suggest that the path dependent complex unit number

$$U(y,x) = \exp\{-i \int dx^\mu A_\mu(x)\} \tag{29}$$

relating fields with equivalent phases $\phi(y)=U(y,x)\phi(x)$ is more fundamental than the vector field $A_\mu(x)$. We shall replace the variable $A_\mu(x)$ by a collection of $U(y,x)$ in the theory. This step which leads to major changes in the dynamics seems unnecessary at this point. But this will be the only way to maintain gauge invariance in the case of non-Abelian gauge groups [8].

Link variables of the lattice gauge theory are defined by

$$U_\mu(n) = \exp\{iaeA_\mu(n)\} = \exp\{i \int_{na}^{(n+\mu)a} dx^\mu A_\mu(x)\} \tag{30}$$

Local gauge transformations which rotate the phase of the matter field

$$\phi(n) \rightarrow g(n)\phi(n), \quad g(n)=\exp\{i\alpha(n)\} \tag{31}$$

act on the gauge group valued link variables according to the rule

$$U_\mu(n) \rightarrow g(n)U_\mu(n)g^+(n+\mu) \tag{32}$$

Consider the product of link variables $W[C]$ along an oriented curve C starting at the point n and ending at the point m (the convention for links with negative direction is $U_{-\mu}(n) = U_\mu^*(n-\mu)$). The transformation property of $W[C]$ under local gauge transformation is simple:

$$W[C] \rightarrow g(n)\, W[C] g^{+}(m) \tag{33}$$

Important classes of observables are i. Wilson loops W[C] for arbitrary closed loops C ii. matter strings $\phi^{+}(n)W[C]\phi(m)$ for curves C starting at n and ending at m. It is easy to see that these quanitites are gauge invariant. The lattice action contains the shortest Wilson loops and matter strings:

$$S = -\frac{1}{2e^2} \sum_{n,\mu,\nu} \mathrm{Re}U_\mu(n)\, U_\nu(n+\mu)\, U_\mu^*(n+\nu)\, U_\nu^*(n)$$

$$+ \frac{1}{2} \sum_{n,\mu} [\phi^*(n)\big(U_\mu(n)\phi(n+\mu) + U_\mu^*(n-\mu)\,\phi(n-\mu) - 2\phi(n)\big) + V\big(\phi(n)\big)$$

$$\tag{34}$$

This action has at least the proper classical continuum limit for smooth fields.

The change of basic variables from vector fields to phase factors has two new features which seem puzzling at first but will play essential role later: i. Appearance of the coupling strength e in the theory of photons alone. ii.Link variable configurations differing by $2\pi\, m(n)/ea$ ($m(n)$ integer) in phase have the same lattice action.

5.2 Hamiltonian Formalism

The scheme followed in Section 2 was to proceed from the operator formalism of quantum mechanics to the path integral representation. The resulting path integral expressions need regularization in order to define the integrals properly (the ill defined nature of the formal path integral expressions shows up in the form of the usual ultraviolet divergences). The regularisation was to put the theory on space-time lattice and define the continuum theory as the limit when the lattice spacing tends to zero (one naturally has to fine tune the parameters of the theory in order to maintain convergence during this limit).

In order to apply conventional methods of operator quantum mechanics to the well defined regularized version of the theory, the transfer matrix or lattice Hamiltonian formalism was developed [5]. The transfer matrix or the shift operator $\hat{T}$ is the operator which develops the lattice system by time a in Euclidean time and the lattice Hamiltonian H is given by the relation $\hat{T} = \exp\{-a_\tau \hat{H}_L\}$. We shall work out the actual form of $\hat{H}_L$ and the corresponding physical Hilbert space for lattice U(1) gauge theory.

To describe time evolution we have to fix the time dependent gauge degrees of freedom. We choose the condition $A_o(n)=0$ (c.f. Gauss's law below). We shall proceed backward along the line of arguments of Section 2. In this temporal gauge the coordinates of the system are the space-like link variables. We break the action into a sum over time slices and interpret the contribution of one time slice as the matrix element of the time evolution operator T in the field diagonal representation. In the limit $a_\tau \to 0$ we shall obtain H_L.

The lattice action is the sum of space-space and space-time like plaquettes. Space-time like plaquettes connect links with the same space location and direction at two neighbouring time slices. The contribution of time slice n_τ to the action is half of the space-space like plaquette contributions at times n_τ and $n_\tau + 1$ plus the space-time like plaquettes connecting time slice n_τ and $n_\tau + 1$. We can write the partition function (vacuum-vacuum amplitude) in the form

$$Z = \prod_{n_\tau} \int D[U_i(n_\tau,\vec{n})]\exp\{\frac{1}{4e^2} \mathcal{L}_{sp-sp}(n_\tau) + \frac{1}{2e^2} \mathcal{L}_{sp-t.}(n_\tau,n_\tau+1) + \frac{1}{4e^2} \mathcal{L}_{sp-sp}(n_\tau+1)\} \tag{35}$$

The transfer matrix is defined by its matrix elements

$$\langle\{U\}|\hat{T}|\{U'\}\rangle = \exp\{\frac{1}{4e^2} \sum \mathrm{Re}\, UUU^+U^+{}_{sp-sp} + \frac{1}{2e^2} \sum \mathrm{Re} UU'^+ + \frac{1}{4e^2} \sum \mathrm{Re}\, U'U'U'^+U'^+{}_{sp-sp}\} \tag{36}$$

The basis vectors in this formula are the direct products of the local field diagonal vectors corresponding to links

$$\hat{U}_i(\vec{n})|U_j(\vec{n})\rangle = U_i(\vec{n})|U_j(\vec{n})\rangle, |\{U_i(\vec{m})\}\rangle = \prod_{\vec{n}} |U_j(\vec{n})\rangle,$$

$$\hat{U}_i(\vec{n})\ |\{U_j(\vec{m})\}\rangle = U_i(\vec{n})|\{U_j(\vec{m})\}\rangle \tag{37}$$

In order to simplify H_L we take the limit $a_\tau \to 0$. In this case we need the form of the lattice action on an anisotropic lattice $\xi = a_\tau/a_{sp} \to 0$. The action (34) reads on such a lattice as

$$S_L = \frac{a_\tau a_{sp}^3}{2e^2} \sum \{(\frac{1}{a_{sp}} (\nabla_i A_j - \nabla_j A_i))^2 + (\frac{1}{a_{sp}} (\nabla_i A_o - \frac{1}{a_\tau} \nabla_o A_i))^2 + \ldots.$$

$$= \frac{a_\tau a_{sp}^3}{2e^2} \sum \{(\frac{1}{a_{sp}^2} (\nabla_i a_{sp} A_j - \nabla_j a_{sp} A_i))^2$$

$$+ (\frac{1}{a_\tau a_{sp}} (\nabla_i a_\tau A_o - \nabla_o a_{sp} A_i))^2\} + \ldots$$

$$= \frac{1}{2e^2} \sum \{\xi \quad {}_{sp-sp} + \xi^{-1} \quad {}_{sp-t}\} \tag{38}$$

where we have ignored the matter part. With the parametrization $U_i(\vec{n}) = \exp\{iA_i(n)\}$ we have for the matrix elements of the transfer matrix (36)

$$\langle\{U\}|\hat{T}|\{U'\}\rangle = \exp\{\frac{1}{2e^2} \sum_{\vec{n},i} [\frac{\xi}{2} \sum_j \cos\ (\text{curl}^j A_i(\vec{n}))$$

$$+ \frac{1}{\xi} \cos\ (A'_i(\vec{n}) - A_i(\vec{n}))$$

$$+ \frac{\xi}{2} \sum_j \cos\ (\text{curl}^j A'_i(\vec{n}))]\} \tag{39}$$

It is useful to introduce angular momentum operators by the definition $\hat{L}_i(\vec{n}) = -\, i\partial/\partial A_i(\vec{n})$, $[\hat{A}_i(\vec{n}),\hat{L}_j(\vec{m})] = i\delta(i-j)\delta(\vec{n}-\vec{m})$. It is known from elementary quantum mechanics that the momentum operator can be used to shift the eigenvalue of the eigenvectors of the coordinate operator:

$$\hat{A}|A\rangle = A|A\rangle, \quad \exp\{-i\varepsilon\hat{L}\}|A\rangle = |A+\varepsilon\rangle, \quad \exp\{-i\hat{L}\}|U\rangle = |U\exp\{i\varepsilon\}\rangle \tag{40}$$

We can now express the transfer matrix explicitly in terms of the operators $\hat{U}$ and $\hat{L}$:

$$\hat{T} = \exp\{\frac{\xi}{4e^2} \sum_{\vec{n}\; i,j} \mathrm{Re}\hat{U}_i(\vec{n})U_j(\vec{n}+i)\hat{U}_i^+(\vec{n}+j)U_j^+(\vec{n})\}$$
$$\int D[\vec{\varepsilon}(\vec{n})] \exp\{-i \sum_{\vec{n}} [\vec{\varepsilon}(\vec{n})\vec{\hat{L}}(\vec{n}) + \frac{\xi^{-1}}{2e^2}\sum_i \cos \varepsilon^i(\vec{n})]\}$$
$$\exp-\{\frac{\xi}{4e^2} \sum_{\vec{n},i,j} \mathrm{Re}\hat{U}_i(\vec{n})\hat{U}_j(\vec{n}+i)\hat{U}_i^+(\vec{n}+j)\hat{U}_j^+(\vec{n})\} \tag{41}$$

We can carry out the integration in the limit of $\xi \to 0$:

$$\int d\varepsilon \exp\{-i\varepsilon L + \frac{\xi^{-1}}{2e^2}\cos\varepsilon\} \simeq \int d\varepsilon \exp\{-i\varepsilon L - \frac{\xi^{-1}}{4e^2}\varepsilon^2\}(1 + O(\xi))$$
$$= \sqrt{4\pi e^2/\xi}\; \exp\{-2e^2L^2\xi\}\; (1 + O(\xi)) \tag{42}$$

Substituting this result into (41) we obtain the lattice Hamiltonian in leading order in ξ

$$a_\tau H_L = \xi\, a_{sp} H =$$
$$e^2 \sum \hat{\vec{L}}^2(\vec{n}) - \frac{1}{2e^2}\sum \mathrm{Re}\hat{U}\hat{U}\hat{U}^+\hat{U}^+ + O(\xi) = e^2 H_o + \frac{1}{e^2}H' + O(\xi) \tag{43}$$

The first and second terms are identified as the electric ($\vec{E}^2$) and magnetic ($\vec{B}^2$) parts respectively.

5.3 Strong coupling expansion

In the strong coupling limit when the kinetic energy is small compared to the potential, one can make an expansion similar to that used in Section 4. In fact we can treat the electric part as the unperturbed Hamiltonian and the magnetic part as the pertubation for $e \to \infty$ [6]. The unperturbed system is the set of non-interacting rotators at each site so the unperturbed eigenvectors are the direct products of local link eigenvectors

$$\hat{L}|m\rangle = m|m\rangle, \langle A|m\rangle = \exp\{imA\}, |\{\vec{m}(\vec{n})\}\rangle = \prod_{\vec{n}} |\vec{m}(\vec{n})\rangle \tag{44}$$

The unperturbed energy of such a state is $E^0[\{m(n)\}] = e^2 \sum \vec{m}^2(\vec{n})$. The vacuum is the state in which no angular momentum excitation occurs: $|0\rangle = |\{0\}\rangle$. In the calculation of higher order correction in $1/e^2$ one has to use the rule that $\hat{U}_i$ excites the corresponding link $\hat{U}|m\rangle = |m+1\rangle$.

In order to introduce charges in the theory we have to discuss Gauss's law. In fact, in the temporal gauge external charges are introduced through Gauss's law by specifying the physical gauge invariant Hilbert space.

After the gauge fixing $A_0(\vec{n}) = 0$ we still have invariance under space dependent gauge transformations. We shall use these transformations to specify the external charge content of the states. It is easy to see that the operator generating space dependent gauge transformations is of the form $G[\{\alpha(\vec{n})\}] = \exp\{-i\sum \vec{\nabla}\alpha(\vec{n})\vec{L}(\vec{n})\}$. States with no external charges are those which are invariant under such gauge transformations. The operator

$$\hat{P} = \int d[\alpha(\vec{n})]G[\alpha(\vec{n})] = \int D[\alpha(\vec{n})]\exp\{i \sum \vec{\nabla}\hat{\vec{L}}(\vec{n})\alpha(\vec{n})\} \tag{45}$$

projects into this subspace (partial integration was made in the second equation asssuming the absence of excitations at the infinity or periodic boundary condition). In case of nonzero external charges $Q(\vec{n})$ the projection operator is

$$\hat{P} = \int D[\alpha(\vec{n})] \exp\{i\sum(\vec{\nabla}\vec{L}(\vec{n})\alpha(\vec{n}) - Q(\vec{n}))\} \quad (46)$$

Physical states with external charges $Q(\vec{n})$ are the eigenstates of $\hat{P}$ with eigenvalue $1: \hat{P}|\{Q(\vec{n})\}\rangle_{phys} = |\{Q(\vec{n})\}\rangle_{phys}$. We can bring this relation into more familiar form:

$$(\vec{\nabla}\vec{E}-Q)|\{Q(\vec{n})\}\rangle_{phys} = 0 \quad (47)$$

It is easy to see that restoring the integration over $U_o(m)$ in (35) we integrate over all possible space dependent gauge transformations acting on the states at $n_t = m_o$ [35]. In fact $U_o(m)$ can be use to make the gauge transformation $U_i(n) \rightarrow U_o(n)U_i(n)U_o^+(n+i)$ at $n_t = m_o$. Thus we can say that the role of integration over $U_o(m)$ is to insert the projection operator P in the time evolution operator at each time slice. In other words Gauss's law is the equation of motion for $A_o(n)$. In case of external charge we have to modify this projection operator by inserting a source term in $U_o(m)$. In view of this relation it will not be surprising when during the discussion of non-Abelian gauge theory we shall relate non-zero expectation values for $U_o(m)$ to deconfinement (breakdown of Gauss's law for quarks).

Gauss's law (47) shows that the vacuum ($Q(\vec{n}) = 0$) contains electric excitations corresponding to closed electric flux loops. In fact the quantum number $m_i(\vec{n})$ can be interpreted as the electric flux in the direction i and (47) expresses that $\{m_i(\vec{n})\}$ can be written as a sum over closed loops with constant $m_i(\vec{n})$ along them. In the case of an external charge at the point n we need a flux line starting or ending at $\vec{n}$. So the unperturbed state of an electron positron pair in the $1/e^2$ expansion is such that $\vec{m}(\vec{n}) = 1$ along the shortest line connecting the position of the charges and $\vec{m}(\vec{n}) = 0$ otherwise.

The operator which creates this state from the vacuum is $\hat{a}^+(e^+e^-) = \Pi\, \hat{U}_i(\vec{n})$, the product of link operators along the flux line connecting the sources. The unperturbed energy of this state is $E^o = e^2 L$ where L is the length of the flux line. Since the energy is proportional to the separation, these charges will be confined. Naturally this statement refers to the large e^2 region. For smaller e^2 higher order plaquette excitations of the flux line may alter this picture. Observe that it follows from the gauge invariant nature of the two separate terms in the Hamiltonian that any excitation generated by H' will be gauge invariant. As long as the strong coupling expansion converges, Gauss's law is satisfied for the ground state automatically.

We are now in position to give path integral expressions for the static potential for two charges. Consider the matrix element

$$\langle 0|\ \hat{a}(e^+e^-)\ \exp\{-T\hat{H}\}\hat{a}^+(e^+e^-)|0\rangle \tag{48}$$

where $\hat{a}^+(e^+e^-)$ is the creation operator of a state which has nonzero overlap with the exact state of e^+ and e^- separated by distance L. It has the asymptotic behavior for large T

$$\sum_a \langle 0|a\rangle \exp\{-TV_a(L)\}\langle a|0\rangle \sim \text{const.}\ \exp\{-TV(L)\} \tag{49}$$

where a is a label for a complete set of eigenvectors in the e^+ e^- sector of the Hilbert space. (Observe that instead of the vacuum we can use any other state $|b\rangle$ with nonzero overlap $\langle 0|b\rangle$). The matrix element (48) is just the expectation value of the Wilson loop corresponding to the closed loop shown in Fig.4. The t=0 and T edges correspond to the creation and annihilation operators $a^+(e^+e^-)$ and $a(e^+e^-)$ respectively and the x=0 and L edges are the source for A_0 in selecting the proper two charge subspace in the Hilbert space by Gauss's law. We conclude with the important remark that perimeter behavior for the logarithm of large Wilson loop averages indicates a linear confining potential for static charges.

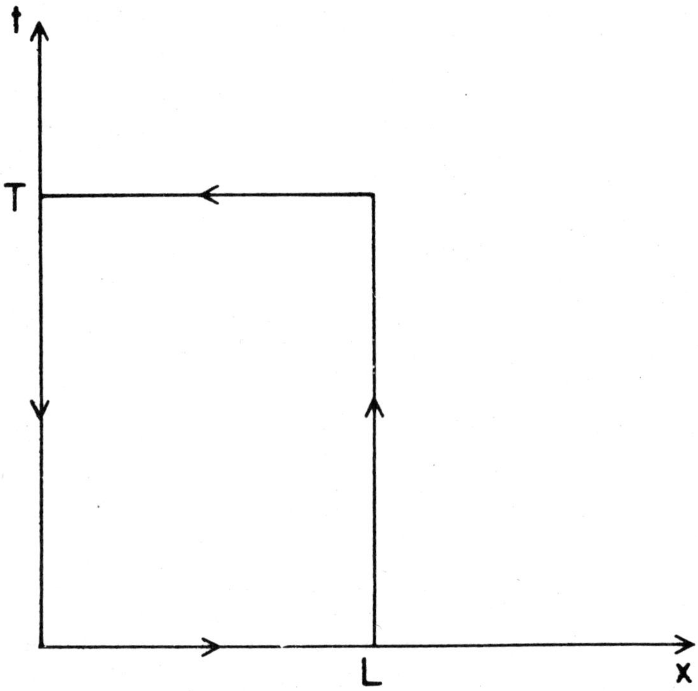

FIG. 4 Contour of the Wilson loop to compute the static potential

5.4 Coninum limit as definition of quantum field theories

Careful Monte Carlo simulations of the U(1) lattice gauge theory in four dimension [7] revealed the phase structure

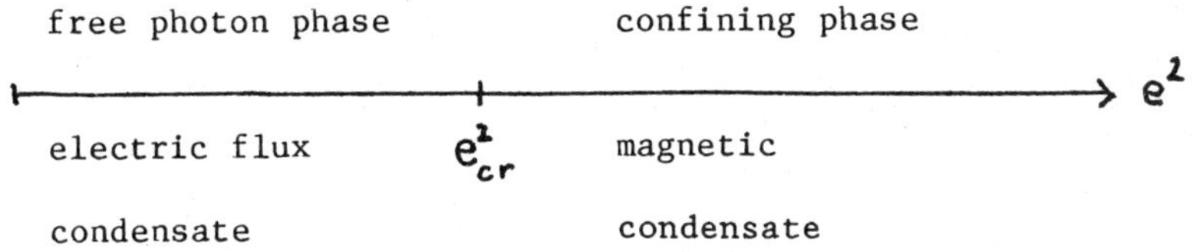

For $e^2 < e^2_{cr} = 1.00$ large Wilson loops show Coulomb's law for charges suggesting that this phase is the usual free photon theory. For $e^2 > e^2_{cr}$ Wilson loops show area behavior and the correlation length ξ diverges as $e^2 \sim e_{cr}^2 + \varepsilon$. This shows the emergence of an unusual strong coupling Quantum Electrodynamics in the continuum limit (the relation between the continuum limit and the diverging correlation length will be spelled out in detail in Section 6). QED, which is a nontrivial interacting theory even for the gauge field alone, confines electric charge.

We can understand this phase structure in a very illuminating way: Consider the partition function of the theory in leading order approximation in $1/e^2$

$$Z = \mathrm{tr}\exp-\left\{-\frac{1}{T}\hat{H}\right\} = \sum \exp\left\{-\frac{e^2}{T}\sum_{\vec{n}} \vec{m}^2(\vec{n})\right\} \tag{50}$$

In this approximation the partition function corresponds to a gas of classical loops. (We are calculating the partition function in the absence of external charge, so Gauss's theorem allows closed electric flux excitations only.) These loops form an interacting gas since the energy is not linear in $\vec{m}(\vec{n})$ (and higher order correction in $1/e^2$ generate interaction as well). A good approximation for the free energy of free loops with length L is $F = -E + TS = -e^2 L + TL\log 5$. The energy is given by the leading order strong coupling formula assuming that the links are excited once only. The additive factor of a link to the entropy of the string is log5 since we have 5 different directions to continue the string. This expression shows that for finite temperature we have a strong and a weak coupling phase: for $e^2 > T\log 5$ the energy dominates and the strings are unlikely, for $e^2 < T\log 5$ entropy wins and electric flux condenses. As the temperature tends to zero the energy contribution in the free energy becomes more important. Since the energy of two static charge (energy of the flux line) increases less rapidly than a linear function of the separation (higher order corrections in $1/e^2$), this simple consideration underestimates the value of the critical coupling $e^2_{cr} = T\log 5$ in the low temperature regime. In addition simple weak coupling expansion indicates the presence of free Coulomb phase at small e.

It was mentioned above that we have two different continuum limits in compact (reminding of the compact nature of the basic variables) QED: a free Coulomb phase for small e and a confining phase as e approaches e_{cr} from above. At this point we change our strategy concerning lattice regularization. We said at the beginning that lattice field theory is a possible regularization procedure for path integral expressions. Since we know no other regularization method which allows nonperturbative calculations, we adopt the view that quantum field theories are defined uniquely only on a lattice.

It is a very important and unfortunately difficult task to relate the information obtained from the operator formalism to regularized path integral results. For example it is well known that systems with an infinite number of degrees of freedom may undergo a phase transition as parameters of the Hamilton operator vary. A popular example is the model (24) when the minimum of the potential $V(\phi)$ moves from zero to nonzero. What happens in the latter case is that the operator representing phase rotation symmetry leads out from the Fock space built around a minimum of the potential $V(\phi)$. Consequently this symmetry will not be realized. This "symmetry breaking" shows up on a lattice as a phase transition in the d+1 dimensional classical statistical mechanical system (17).

There may be several points in the parameter space of this classical system where the correlation length diverges and a continuum theory emerges. Any one of these regions defines a realization of the theory characterized by the same Lagrangian. We shall not discuss the important questions of how one can study and find different continuum limits in a complicated parameter space or how the choice of the integration measure or different "irrelevant"terms in the action may alter the picture. These problems correspond to the realm of the renormalization group method [9,20].

5.5 Magnetic Monopoles on a Lattice

We saw that the formalism of compact QED had two unusual features: appearance of the coupling constant in the case of photons only and the degeneracy of the action for configurations differing by diverging numbers in the continuum limit. The role of the coupling constant is to select the proper realization of QED in the continuum limit. Nature preferred the free photon phase but our world would have been completely different (e.g. confined electrons) if the QED coupling constant had arrived to a large value during the evolution of the universe.

We shall see that the degeneracy of the action leads to the appearance of magnetically charged particles in the continuum limit. These particles play no role in the free photon phase but give the key to the understanding of the confinement mechanism in terms of continuum field theory in the strong coupling phase. Although a linearly rising potential has been derived in the leading order strong coupling expansion, it is not clear at all whether this approximation remains reliable when we take the continuum limit $e^2 \to e^2_{cr}$. The main reason to discuss the confinement mechanism in the case of this unrealistic QED is that there are indications that the same qualitative picture applies to non-Abelian gauge theories as well (see Section 7).

We shall first characterize those lattice configurations which have finite small action due to the compactness of the link variables. A smooth photon field configuration has low action in both the continuum and lattice cases. Increasing the magnetic field in a region will increase the continuum action monotonically. But the lattice variables are compact, so when the magnetic field is close to integer $\cdot\ 2\pi/e$ the action will be small again:

$$S_L = \sum (\cos 2\pi n - 1) = 0 \qquad (51)$$

A plaquette where curl$\vec{A}$ is in the vicinity of $n \cdot 2\pi/e$ is called frustrated.

An instructive modification of such a compact action which should not effect the behavior of the theory at weak coupling is the replacement of the function cosine by a periodic quadratic expression

$$\int D[A_\mu] \exp\left\{\frac{1}{2e^2} \sum \cos(\mathrm{curl} A)\right\} \rightarrow$$

$$\sum_{m(n)} \int D[A_\mu] \exp\left\{-\frac{1}{4e^2}\left(\mathrm{curl}A(n) - 2\pi m(n)\right)^2\right\} \qquad (52)$$

where the variable $A_\mu(n)$ runs from $-\infty$ to $+\infty$ in the integral. Monte Carlo calculations gave the same phase structure for such a modified QED. A(n) corresponds to the photon field and $m_{\mu\nu}(n) \neq 0$ means that the corresponding plaquette is likely to be frustrated if e is small.

A space-space like frustrated plaquette signals the presence of a singular magnetic flux $\phi = 2\pi\, m_{ij}(n)/e$ going through the infinitesimal surface. Calculating $\rho_m = \mathrm{div}\ \mathrm{curl}\ \vec{A}$, we obtain the magnetic charge creating this flux. A straightforward lattice formula for div curl $\vec{A}$ is

$$\exp\{ie\Phi_c\} = \prod_{p\ell} \exp\{ie\Phi_{p\ell}\}, \qquad \exp\{ie\Phi_{p\ell}\} = UUU^+U^+ \qquad (53)$$

where Φ_c is the total flux leaving an elementary cube (=magnetic charge contained by that cube) and the product extends over plaquettes forming the boundary of the cube. The contribution of each plaquette ϕ_{pl} is the product of the link variables along the boundary of that plaquette. Links or plaquettes occuring in the opposite orientation are represented by complex conjugated values. Fig. 5 shows the links participating in the product (53).

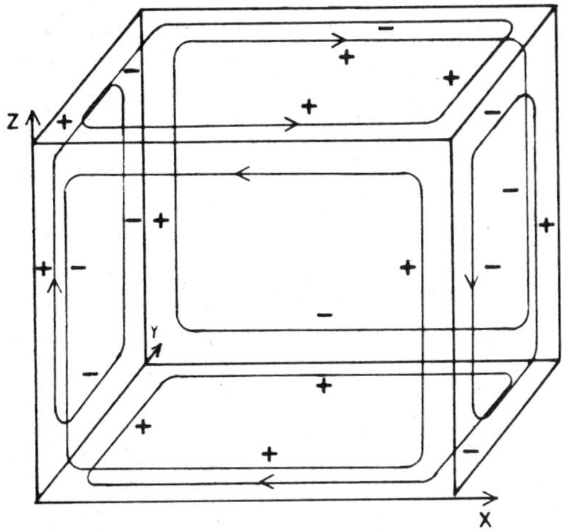

FIG. 5 Construction of div B in an elementary cube. The signs ± represent the sign of the corresponding phase factor in the expression (53).

One can see that each link occurs twice and gives a contribution $U^{\dagger}U$. Thus the product (53) is 1 and the magnetic charges which are allowed in compact QED satisfiy the quantization rule

$$Q_m e = 2\pi n \tag{54}$$

with n integer. This result is not by accident, it is the consequence of a fundamental theorem of homology: the boundary of the boundary is zero [11]. A typical configuration containing a magnetic monopole is depicted in Figs. 6-7. Frustrated plaquettes mark the path of the singular flux line which creates magnetic charge at its end.

We have considered magnetic monopoles at a given time. The world line of a monopole in space-time gives the boundary of the two dimensional sheet swept out by the singular flux line ending at the monopole (see Fig. 8). We define the magnetic current by the expression

$$j_m^\mu = \frac{1}{2} \varepsilon^{\mu\nu\rho\sigma} \nabla^n \Phi_{p\ell}^{\rho\sigma} \tag{55}$$

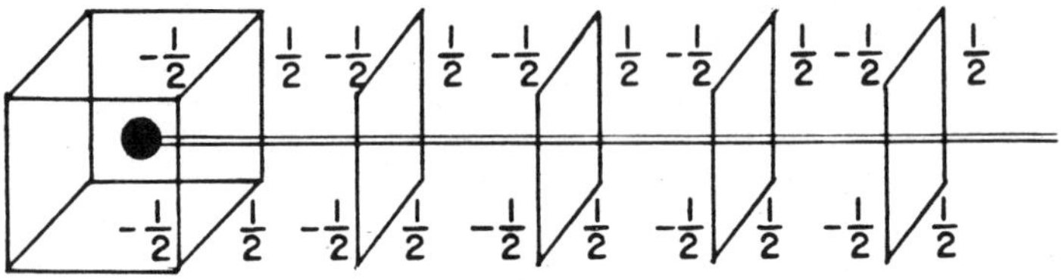

FIG. 6 A string of frustrated spacelike plaquettes representing a singular magnetic flux line. The phase of the links is given in units of 2π

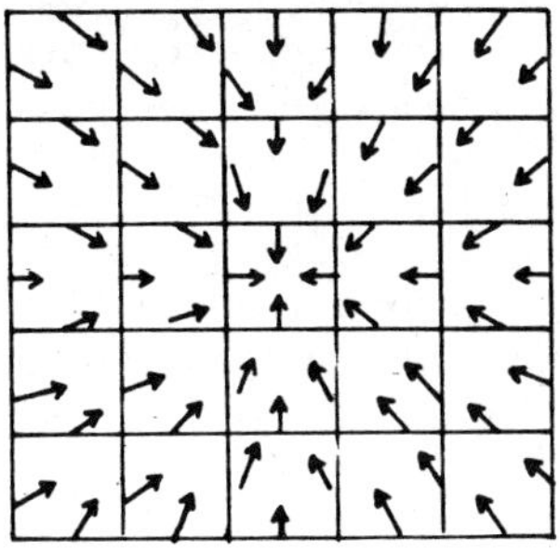

FIG. 7 Cross section of the flux line. The arrows indicate the phase of the link variables.

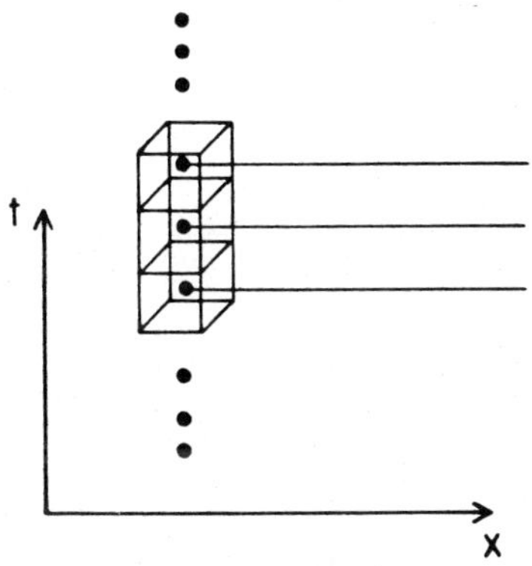

FIG. 8 Space-time picture of a magnetic monopole. The world line of the monopole is the boundary of the world sheet of the singular flux line.

which is conserved by construction, $\nabla^\mu j_m = 0$. In addition the space sum of the zeroth component is an integer multiple of 2π and this integer counts the magnetic charge in units of $2\pi/e$ (for periodic boundary conditions in space).

Conservation of magnetic charge defined by (55) means that the world lines of the monopoles form closed strings. Monte Carlo calculation of the average perimeter of these magnetic charge loops as a function of e shows the condensation of magnetic flux monopoles in the strong coupling continuum limit, $e^2 \to e^2_{cr}$ (see Fig. 9).

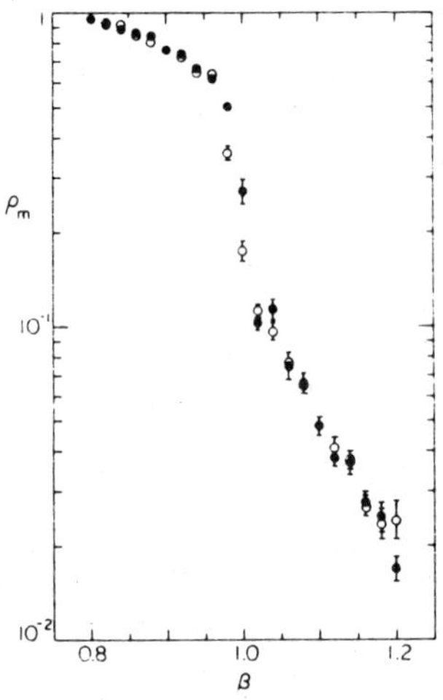

FIG. 9 Average length of the monopoles world lines in a fixed quantization volume as a function of $1/e$ [7].

Summarizing the bulk properties of compact QED: it forms electric flux condensates and gives free photon theory for weak coupling and has magnetic flux condensates and confinement of electric charge in the strong coupling continuum limit. We see now that the two different continuum limits are dual in the sense that they are characterized by the condensation of electric or magnetic flux. The unusual degrees of freedom, magnetic monopoles which are introduced by the compactness, play an essential role in the strong coupling theory.

5.6 Dirac's monopoles in electrodynamics

We shall briefly summarize the construction of magnetic monopoles in continuum theory following Dirac's argument [12]. Magnetic charges are imagined by analogy with electric charges. The magnetic field of the charge Q_m is $\vec{B} = Q_m/4\pi\vec{r}/|\vec{r}|^3$ + a singular flux line. The flux line is necessary since magnetic force lines can not start or end in the vector potential formulation of Maxwell's equations. This flux line is singular because it should be very thin in order to play less role in the dynamics. The simple realization of such a monopole using a very long and thin solenoid which carries the magnetic flux is shown in Fig. 10.

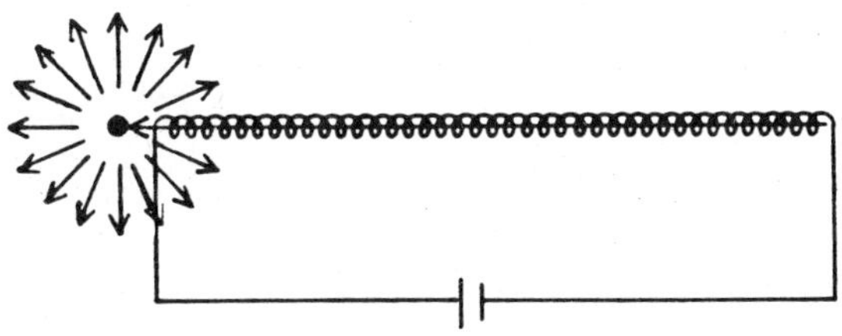

FIG. 10 Theoretical physicists view of a magnetic monopole. The infinitely long and thin solenoid is to provide the magnetic flux.

We would like to make the flux line irrelevant in the dynamics. It will be at least unobservable if its position can be changed by gauge transformations. Imagine a plane perpendicular to the solenoid. This surface is not simply connected from the point of view of the elecromagnetic fields. In fact any closed loop C in the plane which encircles the solenoid has nonvanishing flux $\Phi = \int d\vec{x}\vec{A}(\vec{x})$, indicating that such loops are not contractible to a point. Such a nontrivial topology enlarges the class of allowed gauge transformations on the plane as we can see in the following way: Suppose that we have a charged scalar field ϕ to test electromagnetic forces (or consider the phase factors (29) as fundamental variables). A gauge transformation is allowed if it creates single valued configurations (the notion of single valuedness becomes relevant because the surface is not simply connected) $\phi(x) \to \exp\{i\alpha(x)\}\ \phi(x)$. The nonphysical parameter $\alpha(x)$ may be multivalued; the only restriction is that different sheets must differ by integer multiples of $2\pi: \alpha(\theta = 0) = \alpha(\theta = 2\pi) + 2\pi$. The magnetic flux changes by $\Delta\Phi = \int dx^{\mu} \partial_{\mu} \alpha(x) = 2\pi n$ under such a gauge transformation. We see that Dirac's quantization condition for magnetic charges is equivalent with the requirement that the position of the singular flux line be unobservable. The flux line can be moved by gauge transformation but it can not be transformed away completely.

Now we turn to the discussion of a formal symmetry of electrodynamics containing electric and magnetic charges. It is straightforward to generalize Maxwell's equations to include magnetic charges in a symmetric way:

$$\begin{aligned} \operatorname{div} \vec{E} &= \rho_e & \operatorname{div} \vec{B} &= \rho_m \\ \operatorname{curl} \vec{E} &= -\dot{\vec{B}} - j_m & \operatorname{curl} \vec{B} &= \dot{\vec{E}} + j_e \end{aligned} \tag{56}$$

which can be written in the condensed form

$$\begin{aligned} &\operatorname{div} V = R \\ &\operatorname{curl} V = M(\dot{V} + J), \end{aligned} \qquad V = \begin{pmatrix} E \\ B \end{pmatrix}, \quad R = \begin{pmatrix} \rho_e \\ \rho_m \end{pmatrix}, \quad J = \begin{pmatrix} j_e \\ j_m \end{pmatrix}, \quad M = \begin{pmatrix} 0 & -1 \\ 1 & 0 \end{pmatrix} \tag{57}$$

It is easy to verify in this notation that these equations are invariant under duality transformation which interchanges electric and magnetic components: $V \to MV$, $R \to MR$, $J \to MJ$. It is a question of conventions what we call electric or magnetic charge in electrodynamics. We shall see that this is not true any more for non-Abelian gauge theories.

Although no internal inconsistency was found in the quantum mechanics of electric and magnetic charges, such an extension of QED suffers from the following difficulties: i. Weak coupling expansion is not applicable since one of the charges must be large (c.f. (54)). ii. Despite the quantization law for the charges the unobservable singular flux line generates nonlocal forces (what is happening is that the electric field is expelled from the world sheet of the flux line). This more serious complication is absent in non-Abelian gauge theories.

It is not by accident that magnetic charge quantization is the same in continuum and lattice theories. It was derived by requiring well defined smooth matter field or phase factor configurations in the continuum case. We are working with well defined configurations on a lattice so the condition (54) was obtained trivially. One can not have a lattice field configuration which leads to an ill defined continuum field.

Dirac's monopole has infinte classical magnetic Coulomb energy. It was hoped that quantum fluctations would smear out this singularity as in the case of electric charge. The lesson of Fig. 9 is that this really happens in the strong coupling phase of QED. The more familiar weak coupling phase does not have fluctations enough so the magnetic monopoles are suppressed by their large energy. It is worthwhile to note that the mass of the monopole in the vacuum is in fact zero in continuum strong coupling QED.

We have not discussed the stability of magnetic monopoles. Topological cosiderations [13] show that magnetic monopoles obeying the quantization condition (54) are stable and the magnetic charge is conserved. This might be surprising since there is no symmetry which could generate this conservation law by Noether's theorem. This topological conservation originates from the smoothness of the time evolution.

5.7 Superconductivity, Meissner-effect

We shall discuss condensation phenomena in the context of continuum theory. We may consider the Lagrangian (26) as the original one referring to elementary fields or as an effective one describing the low energy modes of Cooper-pairs bound by phonons in crystals. In the latter case the charge of the scalar field is 2e. Suppose that the coupling λ is weak enough to approximate the vacuum of the theory by a single constant field configuration $\phi_{c\ell}$. Depending on the sign of the quadratic term in the potential this vacuum expectation value $\phi_{c\ell} = \langle\phi\rangle$ is zero or $\sqrt{-\mu/\lambda}$ for $\mu>0$ and $\mu<0$ respectively. For nonzero $\langle\phi\rangle$ the vacuum has less symmetry than the Lagrangian: invariance under phase rotation is broken. We shall discuss this case only.

According to the rules of second quantization $\langle\phi\rangle^2$ is the density of particles in the macroscopically populated lowest lying state. This condensate is different in the relativistic and nonrelativistic cases: Nonrelativistic condensate (e.g.

Cooper-pairs) has nonvanishing charge; $\langle\phi\rangle^2$ gives the density of charge of the condensate as well. In the relativistic case one has the equally populated particle and antiparticle sectors (in the absence of an external potential). $\langle\phi\rangle^2$ is the sum of the particle and antiparticle densities and the net charge is zero. But from the point of view of the Meissner effect both cases are similar.

Shifting the scalar field by its vacuum expectation value (treating the highly populated condensate classically) the Lagrangian is

$$L = -\frac{1}{4}(\partial_\mu A_\nu - \partial_\nu A_\mu)^2 + (D_\mu\phi)^* D_\mu\phi + \phi_{c\ell}^2 e^2 A_\mu^2 + \ldots \tag{58}$$

The appearence of a mass term for photons is called the Higgs-mechanism in the context of relativistic theories. In the case of a nonrelativistic charged vacuum the generation of photon mass can be understood by energy considerations as well: it is favorable for the system to screen the long range Coulomb force and to reduce the diverging electrostatic energy of the charged vacuum.

In order to see the physical degrees of freedoms we have to fix the gauge. We choose the condition $\mathrm{Im}\phi(x)=0$, $\mathrm{Re}\phi(x)>0$ for the original unshifted variables. We may have regions in space time where $\phi(x)$ is far from its vacuum value; moreover it may vanish as well. In the latter case the gauge fixing condition is not well defined and a new degree of freedom, the Nielsen-Olesen vortex, appears [14]. The cross section of such a vortex is shown in Figs. 11–12. It has finite energy if we choose proper asymptotic values

$$\phi(x) \to \phi_{cl}\,(-y + ix)/r,$$

$$\vec{A}(x) \to 1/er^2 \begin{pmatrix} -y \\ x \\ 0 \end{pmatrix}.$$

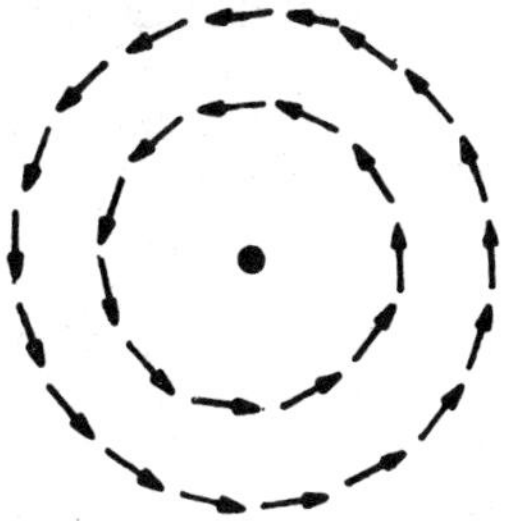

FIG. 11, Cross section of the Nielsen-Olsen vortex. The arrows show the phase of the scalar field.

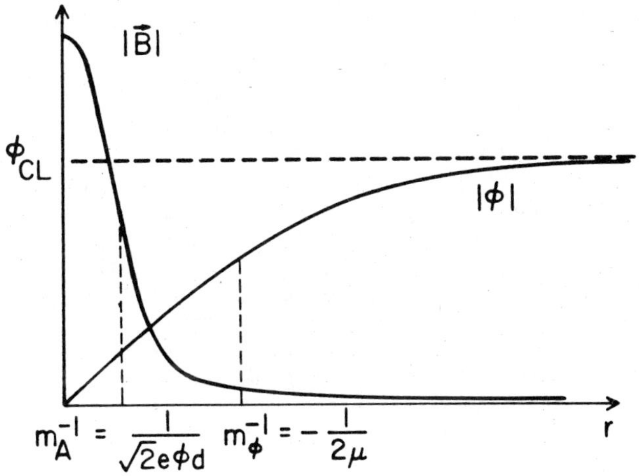

FIG. 12 Schematic dependence of the magnetic field and the magnitude of the scalar field on the distance from the singular line.

The magnetic flux of the vortex is just Dirac's unit $\Phi = \int d\vec{x}\vec{A}(\vec{x}) = 2\pi/e$. Thus this vortex can be imagined as the analogue of the singular flux line in the construction of Dirac's monopole. It is now observable despite the proper flux quantum since the interaction with the matter field generates nonzero energy density in the vicinity of the singular line.

We can now estimate the force law for magnetic monopoles in the symmetry breaking vacuum. Since the monopole antimonopole pair is connected by a Nielsen-Olesen vortex line their interaction energy must be proportional to their separation at least at asymptotically large distances. In fact the photon mass suppresses the vector potential far from the singular line and the energy will be concentrated along a flux tube with diameter $1/m_p \sim -\frac{1}{2\mu}$. This effect makes it seem as if the vacuum would repell magnetic field and is known as the Meissner-effect in the context of superconductivity. Thus magnetic charges are confined in the symmetry breaking vacuum.

After this detour into condensed matter phenomena we return to the problem of strong coupling QED. We have seen that this theory contains magnetically charged particles which are condensed. After separating the photon and monopole degrees of freedom we make a duality transformation on the photon field. The dual electromagnetic field can be derived from a new vector potential. Although the actual Lagrangian can be more complicated than the original one, we know that in the dual picture the scalar field describing monopoles carries electric charge and the lowest lying state is populated macroscopically. So we assume that it is reasonable to approximate the effective Lagrangian which describes the low energy modes of the system by the Lagrangian (26) where $\mu < 0$. Repeating the arguments given before we arrive to a dual Meissner-effect and confinement of electric charge. The condensation of one type of charge generates confining forces in the dual sector [15].

6. NON-ABELIAN GAUGE THEORIES

6.1 Yang-Mills theories in the continuum, Asymptotic Freedom

Deep inelastic scattering of leptons on protons indicates that strong interaction forces become weak at short distances. The only way to generate such relativistic interactions in four dimensional space-time is to describe them by the exchange of massless vector particles with special self interaction. This self interaction which is determined by symmetry considerations is responsible for the strong long range forces necessary to give an account of the absence of quarks in high energy experiments.

Consider the generalization of the model (24) which contains a multicomponent matter field $\phi(x)$ transforming according to a certain representation of the the continous gauge group G (which was U(1) in the case of electrodynamics). The kinetic and potential energy are chosen to be symmetric under the action of G. The extension of this global symmetry to a local one is achieved by the introduction of a Lie algebra valued compensation field $A_\mu(x)=A^a_\mu(x)T^a$. The generators are acting in the matter or gauge field representation. This latter is chosen to be the adjoint one and the normalization of the generators is $\mathrm{tr}T^aT^b=2\delta^{ab}$. The local transformation law and the invariant Lagrangian are given by

$$\phi(x) \to g(x)\,\phi(x), \qquad A_\mu(x) \to g(x)(\partial_\mu + A_\mu(x))g^+(x), \; g(x)\varepsilon G$$

$$L_M = (D_\mu\phi(x))^+ \, D_\mu\phi(x) - V(\phi(x)),$$

$$D_\mu\phi(x) = (\partial_\mu + A_\mu(x))\phi(x) \tag{59}$$

The dynamics of the gauge vector field is generated by adding the locally gauge invariant generalization of Maxwell's action

$$L = L_M + L_{Y-M}, \quad L_{Y-M} = -\frac{1}{2g^2}\,\mathrm{tr}F_{\mu\nu}\,F^{\mu\nu},$$

$$F_{\mu\nu} = [D_\mu,D_\nu] = \partial_\mu A_\nu - \partial_\mu A_\nu + [A_\mu, A_\nu] \tag{60}$$

Physically equivalent directions in the internal, color space are defined by the parallel transport equation (27).

To become familiar with the unique feature of the self interaction in the Yang-Mills action L_{Y-M} we shall study the potential energy of static charges in pure Yang-Mills theory in the perturbation expansion [16]. The derivation of the weak coupling expansion requires special care in this theory. In order to define time evolution and propagators one has to fix the time dependent gauge degrees of freedom. This gauge fixing introduces certain compensating factors described by the Fadeev-Popov field [17] and may allow several gauge equivalent solutions [18]. Assuming that the reader is familiar with the basic concepts of perturbative quantization of Yang-Mills theories [19] we give only the final result up to g^4 for the static potential. It is the logarithm of the expectation value of the Wilson loops of Fig. 4:

$$\exp\{-TV(L)\} = \langle \mathrm{tr} P \exp\{\int dx^\mu A_\mu(x)\}\rangle = \qquad (61)$$

$$1 + g^2 \cdot \boxed{}$$

$$+ g^4 \cdot \left\{ \boxed{} + \boxed{} + \boxed{} + \boxed{} \right\}$$

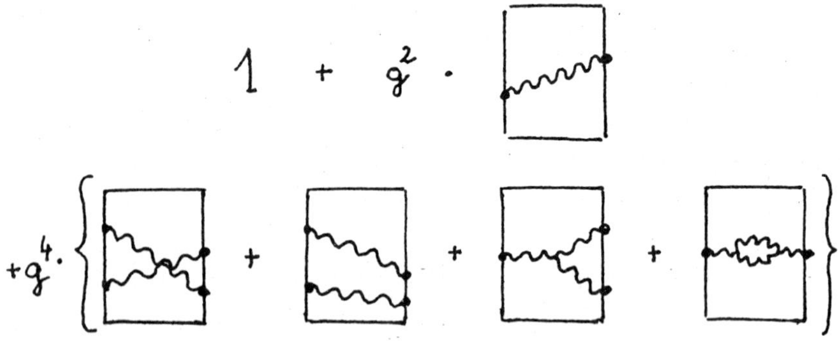

where P stands for path ordering: The exponential function is defined by its Taylor expansion and generators corresponding to smaller values of the parameter along the contour are placed to the right. The solid contours represent the world line of the charges and integration along the contours is included.

Consider first the familiar case of $G=U(1)$. Only the first three graphs are present, leading to the potential

$$V(R) = \frac{g^2}{4\pi} \frac{1}{R} \left(1 - \frac{g^2}{6\pi^2} \log \frac{R}{a}\right) = \frac{g^2_{\mathrm{eff}}(R)}{4\pi} \frac{1}{R} \qquad (62)$$

where the effective coupling strength g_{eff} (R) is defined by analogy to the Coulomb law and an ultraviolet cutoff a (lattice spacing or inverse of the momentum space cutoff) is used. The effective coupling strength is of the form

$$g^2_{eff}(R) = g^2\left(1 - \frac{g^2}{6\pi^2} \log \frac{R}{a}\right) \tag{63}$$

This potential, like any other quantity in the theory, depends on two parameters g and a. We have to remove the cutoff, taking the limit $a \to 0$ while keeping the low energy ($E \ll a^{-1}$) content of the regularized theory unchanged. It is the virtue of renormalizable theories that this process does not generate infinitely many fine tuned terms in the Lagrangian. Symmetry cosiderations show that the introduction of an appropriate function g(a) is enough in the case of Yang-Mills theories. This function is universal in the sense that keeping any low energy quantity fixed in the limit $a \to 0$, we recover the same cutoff-coupling constant relation.

It is easy to compute the actual form of g(a). In fact any low energy observable must satisfy the equation

$$a \frac{d}{da} V(R,a,g) = \left[a \frac{\partial}{\partial a} + a \frac{dg}{da} \frac{\partial}{\partial g}\right] V(R,a,g) = 0 \tag{64}$$

Using the perturbative result for the potential we can read off the beta function of the theory

$$\beta(g) = -a \frac{dg}{da} = \frac{g^2}{6\pi^2} \tag{65}$$

which has the solution

$$a = \Lambda^{-1} \exp\left\{- \int^g \frac{dy}{\beta(y)}\right\} \tag{66}$$

where Λ is a constant of integration.

This way to obtain g(a) from the solution of a differential equation rather than from (62) directly corresponds to the resummation of an infinite subset of diagrams. In fact one can prove that the potential (62) involves powers and multilogarithms of log(R/a). In the approximation where we keep the powers of $\log \frac{R}{a}$, the relation g_{eff} (R=a) = g(a) holds. In the vicinity of R = a higher orders of log(R/a) are negligible and the beta function has no contribution from them. Integration of the differential equation containing perturbative information at R ~ a must reproduce the contribution of these powers of log(R/a) in the effective coupling strength. This results from the existence of a unique flow g(a) in the coupling constant space of the regularized theory which describes the same infrared physics. We see how a formal structure exhibiting renormalizability helps to improve the perturbation expansion by resumming the "leading logarithms".

The constant of integration Λ is the genuine parameter of the theory. In fact any particular value of g refers to the corresponding value of a so only their functional relation in the limit a → 0 characterizes the theory. The cutoff is removed in a way such that the combination

$$\Lambda = a^{-1} \exp \left\{- \int^{g} \frac{dy}{\beta(y)}\right\} \tag{67}$$

remains finite. This way in which a divergent mass scale and the corresponding dimensionless coupling constant give rise to a dimensional constant is called dimensional transmutation.

We turn to the case of G=SU(3) which is the theory of strong interaction, Quantum Chromodynamics. The static potential for SU(3) Yang-Mills theory reads

$$V(R) = \frac{g^2}{4\pi} \frac{4}{R} \left(1 + \frac{22}{16\pi^2} g^2 \log \frac{R}{a}\right) \tag{68}$$

and the leading term in the beta function is $\beta(g) = -\frac{11}{6\pi^2} g^3$. There is no four dimensional renormalizable field theory other than non-Abelian gauge theory with a negative beta function in the perturbative domain. The sign of the beta functions leads to decrease (screening) or increase (antiscreening) in the coupling strength with separation in Abelian or non-Abelian theories respectively. QED (QCD) is an infrared (ultraviolet) stable and perturbative and ultraviolet (infrared) unstable, strongly nonlinear theory. The ultraviolet stable property of QCD is called Asymptotic Freedom to remind one of the vanishing effective strength of the interaction at short distances. (This does not mean that the theory becomes equivalent to a free one in the short distance limit. Other quantities which are related to scale transformations, e.g. dimensions, are modified by the interaction even at short distances).

6.2 Lattice Regularization, Continuum Limit

We shall proceed along the lines followed in 5.1 in defining the lattice regularized version of non-Abelian gauge theories [8]. The Lie algebra valued noncompact field $A_\mu(x)$ is replaced by the group valued field $U_\mu(n)$

$$U_\mu(n) = P\exp\{ \int_{na}^{(n+\mu)a} dxA_\mu(x)\} = \exp\{aA_\mu(n)\} \tag{69}$$

and the lattice action and the action of gauge transformations are given by (34) and (31)-(32). The integral measure is the product of invariant Haar measures for each link $D[U] = \Pi dU_\mu(n)$. This is selected by requiring invariance of the measure under gauge transformations.

Although the transfer matrix formalism and strong coupling expansion methods have not been discussed here, they can be worked out in a way similar to that used in the Abelian case [6]. The only complication is the appearance of Clebsch-Gordan and other coefficients in the expressions for the unperturbed states and energy levels. A formal but more effective strong coupling approximation results when the exponential function of the path integral is expanded in $1/g^2$ and orthogonality relations are used to evaluate the resulting group integrals [20].

One can verify the continuum limit for perturbative and nonperturbative quantities separately. It has been noted in Section 4 that weak coupling graphs have the proper continuum limit for scalar lattice field theories. This leads immediately to the right continuum limit for any perturbative quantity. This does not hold for gauge theories. In fact the lattice action (34) contains infinitely many higher order terms in lattice spacing compared to the continuum action (60). Although the vertices corresponding to these higher order terms contain the lattice spacing with positive power they may lead to new divergent subgraphs in the weak coupling expansion. The detailed calculation of the first few terms in the weak coupling expansion shows that these lattice divergences cancel [21].

The discussion of the continuum limit of nonperturbative quantities requires other methods. Consider a correlation function $\langle A(0)B(n)\rangle$ on the lattice. For simplicity we assume that there is a mass scale related to these operators and the dependence on the separation is $\exp\{-m(A,B)|n|\}$ for large $|n|$. This correlation function defines a mass $m(A,B)/a$ in the continuum limit. We see that the correlation length in lattice spacing units $1/m(A,B)$ must diverge as $1/a(g)$ in order to have a finite continuum limit. The continuum limit is the critical behavior of the classical statistical mechanical system (17).

A second order phase transition in the limit $a \to 0$ is the necessary condition to define a continuum theory. In order to show that the limiting theory has a continuum field theoretical interpretation one has to make sure that correlation lengths diverge appropriately. In fact, forming the ratio of two correlation lengths $m(A,B)/m(C,D)$ or two observables with the same canonical dimensions $\langle\hat{O}\rangle/\langle\hat{O}'\rangle$, one has to see convergence to finite ratios as $a \to 0$. The calculation of these ratios is a commonly applied way to verify the onset of the continuum limit numerically.

We can say more about the coupling constant dependence in the continuum limit in the absence of other dimensional parameters in the theory. The result of a computation for an observable is of the form $\langle O\rangle = a^n f_O(g)$ where the power n is the canonical dimension of the operator $\hat{O}$. (We consider only operators which are independent of the wave function renormalization.) The condition that this expectation value has a finite continuum limit in the case of SU(3) gauge theory is that $f_o(g) \sim \exp\{n8\pi^2/11g^2\}$ (cf. (67)). This behavior signals the onset of the continuum limit in numerical computations. Naturally this result holds in the limit $g \to 0$ only, we have higher order corrections to this formula at intermediate coupling.

Numerical simulations are carried out on finite systems. Thus we have ultraviolet (lattice spacing a) and infrared (lattice size La) cutoffs in the actual computations. If we are interested in the dynamics related to the correlation length $\xi = ca \exp\{8\pi^2/11g^2\}$ we have to ensure that

$$1 < c \exp\{8\pi^2/11g^2\} < L \tag{70}$$

If the coupling constant is too large or small the correlation is too close to the ultraviolet cutoff or the bound state characterized by the correlation length does not fit into the lattice, respectively. The range of g^2 allowed by the inequalities (70) is called the scaling window. The only way to extend this region in order to verify the onset of the continuum limit is to choose larger lattice volume. Fig. 13 shows the correlation length (whose precise definition will be given later) for SU(2) lattice gauge theory in lattice spacing units as a function of the coupling constant. We see that $4/g^2$ must be larger than 2.3 to resolve the correlation length. The other end of the scaling window is given by the size of the lattice, e.g. $4/g^2$ should be smaller than 3.2 for a 6^4 lattice.

The fundamental question for any finite volume computation is the estimate of the smallest lattice size L which allows to see the effects in question. In the case of several characteristc length scales $\{\xi_i\}$ we have max $\xi_i/\xi_j < L$. This circumstance makes the lattice gas type relativistic simulation methods unfeasible for problems in atomic or hadron and lepton physics. Several length scales are separated by orders of magnitude in these cases. This separation allows the application of different adiabatic approximations but makes finite volume simulations difficult. On the other hand, strong interaction physics has a single length scale given by the mechanism of dimensional transmutation (e.g. the confinement radius). So adiabatic approximations are not applicable but numerical simulations can be carried out on relatively small lattices.

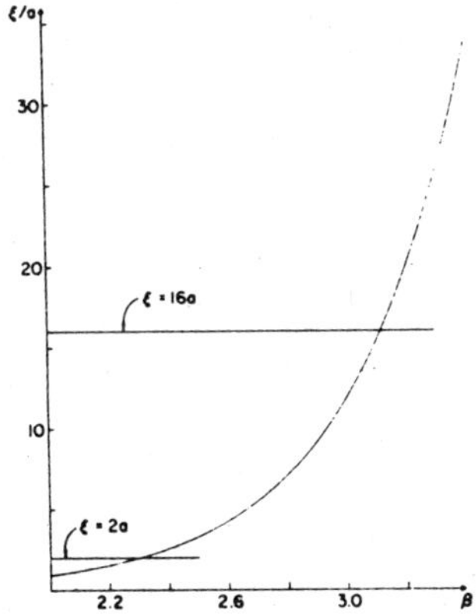

FIG. 13 Correlation length of SU(2) gauge theory in lattice spacing units.

6.3 Monte-Carlo Simulations

We shall briefly review some Monte Carlo results for SU(2) and SU(3) lattice gauge theories. The first important step toward the establishment of the continuum limit was the calculation of the string tension [22]. It was discussed in Section 5.3 how perimeter behaviour of Wilson loops indicates confinement. Denoting the expectation value of an nxm Wilson loop by W(n,m), the static potential is given by the relation $mV(n)=-\log W(m,n)$. We assume that it has the form $mV(n)=\sigma_L mn+p(m+n)$, where σ_L is the string tension in lattice spacing units, $\sigma_L = \sigma/a^2$, and p is the divergent self energy contribution to the potential. σ_L can be determined by fitting this expression to the Monte-Carlo averages or by forming the ratios $W(n,m)/W(n+1,m-1)$. The string tension is plotted for SU(2) gauge theory in Fig. 14. We see that for $4/g^2>2.2$ the numerical points follow an exponential curve predicted by the perturbative beta function. This is explicit evidence for the peaceful coexistence of Asymptotic Freedom (perturbative beta function) and confinement (existence of finite string tension in the continuum limit) on the same lattice.

The knowledge of the numerical value of the string tension in lattice spacing units allows us to compute the proportionality constant $c = \sqrt{\sigma}/\Lambda^{-1}$. In fact using the relation

$$\Lambda = a^{-1}\left(\frac{6\pi^2}{11}\frac{4}{g^2}\right)^{\frac{51}{11}} \exp\left\{-\frac{3\pi^2}{11}\cdot\frac{4}{g^2}\right\},\quad (SU(2)) \tag{71}$$

we obtain c = 79 (the power correction in front of the exponential function in (71) corresponds to the two loop contribution to the beta function. This is the last important term in the ratio Λ/a^{-1}; higher orders modify it by $O(g^2)$). We can compute the value of the lattice spacing in physical units at a given coupling constant using the experimental value of the string tension. Thus we are able to convert any lattice result into physical units.

The ratio of the Λ parameter of SU(2) gauge theory in different regularization schemes can be calculated perturbatively (in fact this ratio is given exactly by calculating the first few terms is the weak coupling expansion) [23]. This ratio can be used to determine the Λ parameter of any commonly used continuum regularization scheme from the experimental value of the string tension (this program has been carried out for SU(3) gauge theory as well). $1/\sqrt{\sigma}$ has

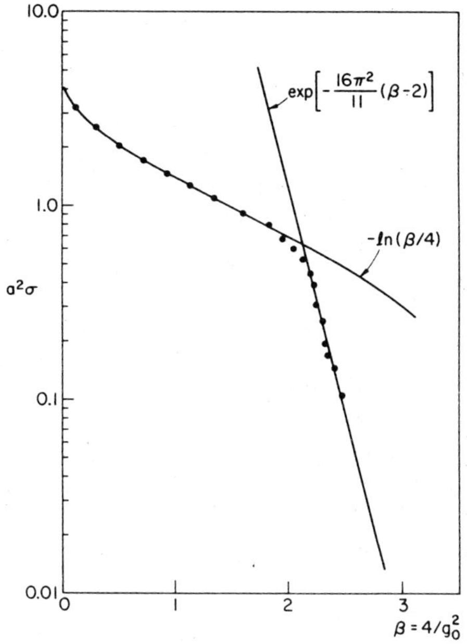

FIG. 14 String tension of the SU(2) gauge theory.

the dimension of length so it can be used as the characteristic length scale of strong interaction physics (confinement radius). It is just this correlation length which is plotted in Fig. 13.

Another dimensional constant which can be used to detect the onset of the continuum limit is the deconfinement temperature T_{dec} (see Section 6.4 for detailed discussion of this phase transition). Fig. 15 shows the critical temperature in lattice spacing units as a function of the coupling constant. We see again the exponential behaviour in the scaling window, indicating that the critical temperature is a finite quantity in the continuum theory.

These and other similar calculations have been carried out for SU(3) gauge theory as well. Very accurate calculations showed that the asymptotic continuum limit (where the beta function agrees with its perturbative form) is reached at a surprisingly small lattice spacing value $a \sim 1/8T_{dec} \sim .12$fm (the corresponding value for SU(2) theory is $1/4T_{dec}$). The use of more sophisticated improved actions may help to see the continuum behaviour at larger lattice spacing values [24].

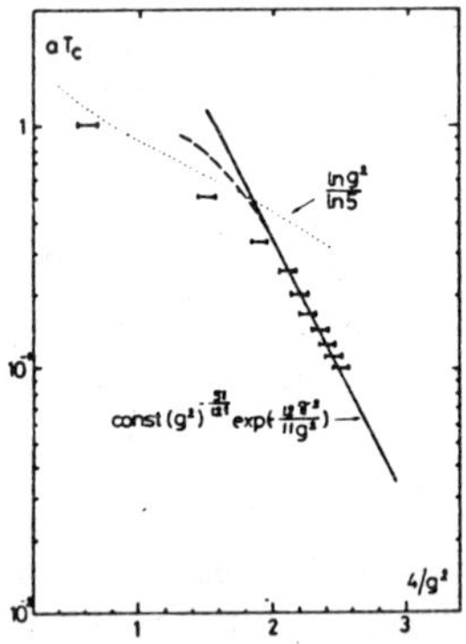

FIG. 15 Deconfining critical temperature of the SU(2) gauge theory [29].

Another important result is the static quark potential. The repetition of the calculation of Wilson loop expectation values at different values of the coupling constant in the scaling window gives the static quark potential at the lattice sites where the lattice spacing is given by (71). Using the parametrization $\xi V(R/\xi)=cR/\xi - \alpha(R/\xi)^{-1}$ for the static potential, the fit to the measured point is shown in Fig. 16 [25]. The correlation length ξ is given by $1/\sqrt{\sigma}$ and the optimal value of c is different from one because of the different definition of the tension. In fact σ_L was determined from the area fit to Wilson loops and the present potential contains the perturbative Coulomb part as well.

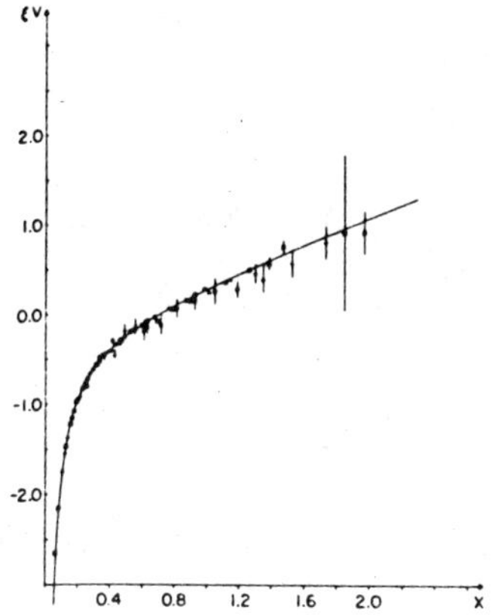

FIG. 16 Static potential of the SU(2) gauge theory. The solid line is the result of the fit with parameters c=0.71, α = 0.164.

6.4 Yang-Mills Theories at Finite Temperature

The basic object of statistical mechanics, the partition function of a quantum system, is given by the Euclidean vacuum-to-vacuum amplitude when the domain of integration in the path integral consists of configurations with period length $\beta = 1/T$ in the imaginary time direction (c.f. Section 2). This circumstance is particularly useful for simulations on finite lattices since the finiteness of the time extent of the lattice has a physical interpretation. We shall now discuss briefly how these physical boundary conditions lead to unusual effects in gauge theories.

Gauge theories contain redundant variables: one has to eliminate time dependent gauge symmetry in order to have unique time evolution. The role of gauge invariance is to select those combinations of the basic variables which are governed by the equations of motion in a well defined way.

Consider a gauge theory in space-time with finite time extent. The gauge symmetry group of this system is the direct product of the local gauge group at each space-time point. Imposing periodic boundary conditions in the time direction the gauge transformations must be periodic as well. So the gauge symmetry group is the usual direct product of the local gauge group but the "last" time slice is skipped in this construction. Since the gauge symmetry group is reduced more gauge invariant physical degrees of freedom appear in the theory.

It is easy to see that the Polyakov lines $\Omega(x) = P\exp\{\int_x^{x+\beta\cdot 0} dt A_0(x,t)\}$ which are Wilson loops corresponding to the world line of a static charge contain these qualitatively new degrees of freedom. In fact the transformation rule (33) for the Polyakov line $\Omega(x) \to g(x)\Omega(x)g^+(x)$ shows that the eigenvalues of this matrix are gauge invariant because the world line is closed by the boundary conditions. The same reduction of the gauge symmetry and appearance of new degrees of freedom results from imposing any gauge invariant boundary conditions for the time evolution in the path integral [25].

The global symmetry group of the finite temperature path integral which remains after a complete gauge fixing is G x C, where G is the gauge group and C is its center (subgroup of G containing elements commuting with any other elements of G). The component G corresponds to global gauge transformations which are still present after fixing the local symmetries. These global gauge transformations are periodic in time automatically. But we can apply different gauge transformations at the initial and final time slice if they differ in a center element only. In fact the transformation law (59) shows that the gauge fields are in the adjoint representation of the gauge group, so they are invariant under the action of the center. The center of any semisimple group is discrete, so smoothness of the fields requires that this symmetry be global.

The Polyakov line is an order parameter for the global symmetry G x C of the system. In fact its gauge transformation is local as in the case of a matter field. Since the world line of a static charge winds around the system in the time direction the Polyakov line changes multiplicatively $\Omega(x) \to \Omega(x)\, c$, $c \in C$, under the action of gauge transformations which are periodic modulo the center of the gauge group. The discrete transformations in C are realized on a lattice by multiplying the time directed links $U_0(n)$ in one time slice by the corresponding center element (this is not a gauge transfomation in the usual sense!).

The trace of the Polyakov line has an important physical interpretation. It is easy to see using the rules of integration for fermionic variables (see Section 7.1) that this expectation value is the ratio of the partition functions with and without a static fermionic charge at the point x:

$$\langle \mathrm{tr}\,\Omega(x) \rangle = e^{-\beta F_q(x)} \qquad (72)$$

Here F_q stands for the free energy of a static quark. Correlation functions $\langle \mathrm{tr}\Omega(x)\ \mathrm{tr}^*\Omega\ (y)\rangle$ give the free energy of a quark-antiquark pair in similar way.

Confinement of quarks means that the free energy of a single charge diverges with the three volume of the system. Since the trace of the Polyakov line transforms multiplicatively under C, confinement is realized in the phase which is symmetric with respect to C. When this symmetry becomes broken its order parameter may aquire a nonzero expectation value and isolated quarks are allowed.

It was noted in Section 5.3 that integration over time directed links corresponds to the projection onto the zero charge density (gauge invariant) subspace. The breakdown of the Z_n symmetry in SU(n) gauge theories (the center of SU(n) is Z_n) means that the projection onto the zero n-ality subspace is prevented by the dynamics. Consequently nonzero n-ality states as quarks are allowed in the system.

There are several hints that high temperature Yang-Mills theories are not confining. Strong coupling calculations on the lattice [26,27] or Asymptotic Freedom (the temperature plays the role of the external mass scale in the effective coupling constant so g_{eff} (T) approaches zero for infinite temperature) suggest that non-Abelian gauge theories have a Debye-screened Coulomb potential for a static quark-antiquark pair at high temperature. (The smallness of the effective coupling constant does not neceserally mean that the system is approaching the naive free gluon-gas limit at high temperature. In fact finite temperature weak coupling expansion around the naive vacuum (A_μ=0) has serious infrared problems [28] preventing the accurate perturbative determination of thermodynamic potentials even at arbitrarily high temperature.)

The appearance of a deconfined phase has been the subject of a large amount of Monte-Carlo simulation [29,30]. The result is that there is a deconfinement phase transition on the lattice and the critical temperature remains finite in the continuum limit (c.f. Fig. 15 for SU(2)). The order of the transition is second and first for SU(n) with n=2 and n>2, respectively according to the predictions of strong coupling results. The critical point is $T_{dec} = 40\Lambda$ and 80Λ for SU(2) and SU(3) theories respectively. Using the value 1Gev/fm for the string tension and the ratio $T_{dec}/\sqrt{\sigma} = 0.45$ found in simulation, we get $T_{dec} = 200$ Mev.

The discrete symmetry with respect to the center is broken in the high temperature phase and realized in the low temperature phase. This is just the opposite of the usual behavior: the low temperature phase used to be the ordered and less symmetrical. The formal reason for this unusual behaviour is Asymptotic Freedom: The coupling constant is decreasing with increase of the temperature. The coupling constant square plays the role of the (unphysical) temperature in the four dimensional classical statistical mechanical system (34) (where we ignore matter fields completely). The physical temperature dependence of this classical system is the opposite of the usual one. We have a similar situation in statistical mechanics when the system is described in terms of the dual variables. This analogy suggests that the real variables of gauge theory are electric rather than magnetic fields. Unfortunately there is no known closed formula for the Lagrangian when expressed in electric variables only.

6.5 Confinement and the Electric Meissner-Effect

It was mentioned in Section 6.4 that the Polykov line is an order parameter for the group G x C and the role of this center C was discussed briefly. Now we turn to the discussion of the role of G in the confinement mechanism.

Monte-Carlo simulation indicates that the symmetry under global gauge transformations becomes broken for $T>T_{dec}$. What happens is that the Polyakov line develops a symmetry breaking expectation value in the deconfined phase [31]. Polyakov lines play a role similar to that of matter fields in the adjoint representation of the gauge group, and quantum fluctations generate a symmetry breaking effective potential for this dynamical Higgs field [32]. One can say that $A_0(x)$ undergoes a Bose-Einstein condensation at $T=T_{dec}$. The unbroken gauge fields commute with the vacuum expectation value $[A_{unbr.}, \Omega_{vac}] = 0$ (cf.(33)).

In the case of SU(3) gauge theory the unbroken symmetry is U(1) x U(1), corresponding to the diagonal generators of the gauge group (λ^3 and λ^8 Gell-Mann matrices) in the gauge where Ω_{vac} is diagonal. The theory describes two photons associated with A^3_μ and A^8_μ and three vector fields

$\vec{u} = \vec{A}^1 + i\vec{A}^2$, $\vec{v} = \vec{A}^4 + i\vec{A}^5$, $\vec{w} = \vec{A}^6 + i\vec{A}^7$ carrying charges with respect to the photon fields. The Polyakov line is the product of the Polyakov lines corresponding to the photon fields $\exp\{a\,\lambda^3 + b\lambda^8\} = \exp\{a\,\lambda^3\}\exp\{b\,\lambda^8\}$. Consequently the confinement properties of static charges, e.g. $\mathrm{tr}\Omega(x)$, can be studied looking into the effective theory of the photon fields alone. Confinement of static charges in SU(3) gauge theory is equivalent with simultaneous confinement in both U(1) gauge theories.

We have discussed in Section 5.7 how U(1) gauge theory shows confining properties if magnetic charges condense. In order to have magnetic charge in electrodynamics we must either introduce singular fields with Dirac's magnetic monopoles or use compact link variables. If the U(1) photon field is embedded in a symmetry broken non-Abelian gauge theory, as in the case of high temperature QCD, stable finite energy magnetically charged particles, 't Hooft-Polyakov monopoles, show up in the theory. The long distance properties of these particles are the same as for Dirac's magnetic monopoles.

We have no self consistent semiclassical approximation to describe the dynamics of the monopole gas in QCD. Instead we assume that the following properties of monopoles in Yang-Mills-matter field systems apply to high temperature QCD:i. The rest mass M of the monopole is proportional to the expectation value of the order parameter, i.e. the strength of the symmetry breaking. ii. The monopole-monopole interaction is attractive [13]. In this case the energy of the monopole pair is $2M-\Delta$ with $\Delta>0$. Thus monopole pairs become bounded and form a condensate when we approach T_{dec} from above. This mechanism is analogous to superconductivity, where Cooper-pairs condense. The repetition of the arguments of Section 5.7 leads to confinement of quarks at $T=T_{dec}$.

This description of confinement as a dual Meissner-effect was carried out for finite temperature QCD and gave an insight into the confinement mechanism at $T=T_{dec}$. The global gauge symmetry is restored in the low temperature phase and we need another mechanism to stabilize magnetic monopoles.

7 FERMIONS

7.1 Grassmann variables

The path integral representation of the matrix elements of the time evolution operator involves commuting c-numbers so Green's functions expressed in this way are symmetric $\langle T(\phi(x)\ \phi(y))\rangle = \langle T(\phi(y)\phi(x))\rangle$. In order to incorporate Pauli's excusion principle for fermions one needs antisymmetric Green's functions $\langle T(\psi(x)\psi(y))\rangle = -\langle T(\psi(y)\psi(x))\rangle$. This antisymmetry is achieved by the inroduction of formal symbols $\psi(x)$, called Grassmann variables with the property $\{\psi(x),\psi(y)\} = \psi(x)\psi(y) + \psi(y)\psi(x) = 0$.

We have to define the usual algebraic and analytical operations on these objects [33]. The algebra of n anticommuting (with respect to the multiplication of the algebra) basis vectors $\psi(k)$, $k=1,\ldots,n$ is defined over the field of complex numbers. Thus the general element of this 2^n dimensional algebra is of the form $\psi = c_o + c(m,n)\psi(m)\psi(n) + c(\ell,m,n)\psi(\ell)\psi(m)\psi(n) + \ldots$, $c(m,n) = -c(n,m)$, $c(\ell,m,n) = -c(m,\ell,n)$, etc. Analytic functions of $\psi(n)$ are defined by their Taylor series, which terminate at most at 2^n-th order. The rule of derivation is $d\psi/d\psi = 1$, $d(\psi\psi')/d\psi = \psi'$, $d(\psi\psi')/d\psi = -\psi'$, etc.

The definition of the integral is obtained from the requirement $\int d\psi f(\psi) = \int d\psi f(\psi+\eta)$ (invariance under the shift of the integration variable is equivalent to the equation of motion $\langle \delta S/\delta q\rangle = 0$ as one can see by shifting the integration variable q(t) in (6)). The only additive definition for integration which satisfies this requirement is $\int d\psi\ c = 0$, $\int d\psi\psi = 1$, $\int d\psi d\psi'\psi'\psi = 1$, etc. The formal application of these ad hoc rules for the case of free fermions leads to the result

$$Z_o[\bar{\eta},\eta] = \int D[\bar{\psi}]D[\psi]\exp\{-\bar{\psi}K\psi + \bar{\eta}\psi + \bar{\psi}\eta\} = \det K \exp\{\bar{\eta}K^{-1}\eta\} \tag{73}$$

The integral can be evaluated either by transforming the quadratic form into a Hopf-form where different $\bar{\psi},\psi$ pairs decouple or by expanding the exponential function. This result justifies the sequence of formal definitions: (73) generates the right antisymmetric Green's functions for free fermions. The generalization for the interactive system

$$Z = \int D[\bar{\psi}]D[\psi]D[A]\exp\{-S_B[A] - \bar{\psi}K[A]\psi\},$$

$$K[A] = i\gamma^{\mu}D_{\mu} - m \tag{74}$$

has the correct weak coupling perturbation series.

7.2 Species doubling

The naive lattice regularization of Dirac's action $S = \int d^dx\bar{\psi}(x)\,(iD_{\mu}\gamma^{\mu} + m)\psi(x)$ is

$$S_L = -\frac{1}{2ia}\sum \bar{\psi}(n)\,\gamma^{\mu}[U_{\mu}(n)\psi(n+\mu) - U^{+}_{\mu}(n-\mu)\psi(n-\mu)] + m\sum \bar{\psi}(n)\psi(n) \tag{75}$$

There are 2^d regions in the Brioullin zone $-\pi < p^{\mu} < \pi$ where the free propagator

$$K^{-1}(p) = \sum_{\mu} \gamma^{\mu}\frac{1}{a}\sin p^{\mu}a + m \tag{76}$$

diverges for $m = 0$. In fact $K^{-1}(p) \sim \varepsilon$ for $p^{\mu} = P^{\mu}_{a} + O(\varepsilon)$, where P^{μ}_{a} is a four vector with components 0 or π and $a=1,\ldots,2^d$.

One can decompose the momenta corresponding to the internal fermion loops in the weak coupling expansion of (75) as $p^{\mu} = P^{\mu}_{a} + \tilde{p}^{\mu}$ where $-\frac{\pi}{2} < \tilde{p}^{\mu} < \frac{\pi}{2}$. Integration over the loop momenta in the weak coupling graph can be written as summation over a and integration over the reduced Brioullin zone $-\frac{\pi}{2} < \tilde{p}^{\mu} < \frac{\pi}{2}$, where there are no infrared lattice modes. Each a gives the same contribution in the continuum limit [34] so every closed fermion loop carries a multiplicative factor 2^d. We have the same factor for $n_f = 2^d$ species of fermions, so we interpret the infrared lattice modes in the perturbative continuum limit as different fermion species.

There are two commonly used modification of the lattice action (75) which reduce the number of species. Wilson's suggestion is to modify the dispersion relation by adding a term to the action which is a second derivative but carries an extra suppression factor a [8]

$$S_W = S_L + \frac{r}{2a}\sum \bar{\psi}(n)[U_{\mu}(n)\psi(n+\mu) + U^{+}_{\mu}(n-\mu)\psi(n-\mu) - 2\psi(n)] \tag{77}$$

The free propagator of this formalism is

$$K^{-1}(p) = \sum_\mu \gamma^\mu \frac{1}{a} \sin p^\mu a + m + \frac{r}{a} \sum_\mu (1 - \cos p^\mu a) \qquad (78)$$

The particle content of the free theory is

$1, \binom{d}{2}, \binom{d}{3}, \binom{d}{4} \ldots$ species with masses m,

$m + \frac{r}{a}, m + 2\frac{r}{a}, m + 3\frac{r}{a}, \ldots$ The massive species decouple in the continuum limit; their only trace is that they break chiral invariance of the regularised theory in a way described by the chiral anomaly structure [36].

Another widely applied modification of (75) is based on the observation that the transformation of the spinor variables

$\psi(n) \to V(n)\psi(n)$, $\gamma^\mu \to V^+(n)\,\gamma^\mu V(n+\mu)$ with

$V(n) = (\gamma^o)^{n_0} (\gamma^1)^{n_1} (\gamma^2)^{n_2} (\gamma^3)^{n_3} \ldots$ diagonalizes Dirac's gamma matrices simultaneously. Keeping only one of the 2^d equivalent decoupled components, we reduce the number of species to $2^{d/2}$. Unfortunately this staggered fermion formalism mixes the spinor and flavor (quantum number labeling the fermion species) space with space-time since different spinor and flavor components of the remaining species are located at neighboring lattice sites. The gauge field which is introduced by the minimal coupling in (75) couples different spinor and flavor components. It is a remarkable result of the weak coupling expansion [37] that these unusual lattice couplings become suppressed by the lattice spacing after integrating the gauge field out perturbatively. Naturally we know nothing about the contributions to these effective vertices which are produced by gauge fields having fluctations at the length scale a. An important role of these higher order vertices is that they generate the proper chiral anomaly equations [38].

The virtue of the staggered fermion formulation is that it allows massless modes. The diagonal part in the second derivative of (78) destroys the formal chiral invariance of Wilson's formulation even for m=0. The staggered formalism is not without difficulties either; the spin and flavor symmetries are extremely complicated. Mostly finite subgroups remain as symmetries of the lattice action and the restoration of the complete continous symmetry group in the continuum limit can be verified numerically only. The chiral U(1) symmetry is reduced to a Z_2 subgroup at finite lattice spacing [39].

There is a general underlying reason for these difficulties. It has been proved using simple and general methods of topology that there is no regularization for a Weyl neutrino field which maintains gauge invariance [60]. The naive "optimal" lattice regularization which preserves chiral invariance and has no species doubling would violate this rule.

7.3 Numerical methods for fermionic systems

Application of the rules of Grassmann integration to the vacuum-to-vacuum amplitude (74) gives the Mathieu-Salam formula (73). It will be assumed in the rest of these lectures that the fermion determinant is real and positive for arbitrary bose field configurations. This condition is satisfied by the usual relativistic systems where chiral invariance causes the eigenvalues of $i\not{D}$ to appear in complex conjugated pairs.

The fermion determinant represents serious difficulties for a Monte-Carlo simulation procedure. The direct exact treatment of the determinant [41] requires unrealistic computing power. In the simplest so called quenched approximation the determiant is ignored completely in the distribution of the bosonic field configurations. Fermionic Green's functions are calculated using (73) before crossing the determinant out from the partition function, e.g. $\langle T(\bar{\psi}(x)\psi(y))\rangle = \langle K^{-1}(y,x)\rangle_B$ where the average $\langle\ldots\rangle_B$ stands for an expectation value computed in the pure bose system (17). These results correspond to the n_f (number of fermion flavors) $\to 0$ limit of the complete theory. In fact the integration over n_f Grassmann fields brings the n_f -th power of the determinant in the path integral.

A better approximation scheme results when the ratio of integrands which is involved in the decision step of a Metropolis-type Monte-Carlo procedure is calculated by a sub-Monte-Carlo process in the small update size limit [42]. The effective action which includes the effect of fermions is $S_{eff}[U] = S_B[U] - \mathrm{Tr}\log K[U]$. We have $\Delta S_{eff} = \Delta S_B - \mathrm{Tr}\{K[U]\Delta K[u]\}$ in leading order in ΔU. The calculation of the nearest neighbor matrix elements of the fermion propagator is carried out by the sub-Monte-Carlo

$$K^{-1}(n,m) = \frac{\int D[\bar{\phi}]D[\phi]\exp\{-\bar{\phi}K^{+}[U]K[U]\phi\}\bar{\phi}(m)\phi(\ell)}{\int D[\bar{\phi}]D[\phi]\exp\{-\bar{\phi}K^{+}[U]K[U]\phi\}} K(n,\ell) \tag{79}$$

Another expansion method results when the effective action is expanded in the fermion kinetic energy, in 1/m [43]. This leads to relatively simple Monte-Carlo procedures but becomes less useful when the continuum limit $m \to 0$ is taken, even in the case when a sub-Monte-Carlo is used to resum the 1/m expansion before each update [44].

There are possible improvements along these lines [45] but the common difficulty is the lack of control over the nonlinear effects of the stochastic or systematic errors made at the decision step.

Another simulation procedure which contains no expansion or sub-Monte-Carlo is based on molecular dynamical methods [46]. The path integral (11) was interpreted as the canonical partition function of a classical system described by the coordinates q and the potential energy S[q]. One can simulate the thermodynamics of a many particle system by solving its equations of motion.

We introduce canonically conjugated momenta p and define the Hamilton function $H = \frac{1}{2} p^2 + S[q]$ and the Gibbs partition function

$$Z = \int D[p]D[q]\exp\{- \frac{1}{T} H\} \tag{80}$$

The time of this hypothetical classical system is nonphysical. It can be viewed as the simulation or computer time. When fermions are present we introduce c-number fields as coordinates Q and their canonically conjugated momenta P and define the Hamilton function [47]

$$H = P(K^+[q]K[q])^{-1}P^+ + \omega^2 Q^+ Q + \frac{1}{2} p^2 + S[q] \tag{81}$$

It is easy to see that the thermodynamical averages of q defined by (80) and (81) are equivalent with the expectation values of the original quantum system.

In general we have the square of the fermion determinant in the Gibbs partition function since the complex generalization of the Gauss-integral (4) contains 1/detA instead of $1/\sqrt{\det A}$. But this extra species doubling can be eliminated in the staggered fermion formalism where the matrix K[U] is of the form K=c E+ih, where h is an hermitian nearest neighbour coupling matrix [48].

We evoke ergodicity to simulate the thermodynamics of the system (81). Ergodicity was postulated in classical statistical mechanics to calculate the measured long time averages as ensemble averages. Now we use this relation in the opposite way: the path integral formulation (11) of quantum mechanics is given in terms of averages over a canonical ensemble and we compute these integrals by solving the corresponding equations of motion

$$\ddot{q} = - \frac{\partial S[q]}{\partial q} + \dot{Q} \frac{\partial}{\partial q} (K^{+}[q]U[q])\dot{Q}, \quad \frac{d}{dt} [\dot{Q}^{+}K^{+}[q]K[q]] = - \omega^{2}Q^{+} \tag{82}$$

and forming long time averages. This simulation samples the microcanonical partition function

$$Z_{M}[E] = \int D[P]D[Q]D[p]D[q] \; \delta(E-H) \tag{83}$$

since the energy is conserved. The coupling strength g^2 (temperature of the canonical distribution) depends on the energy which is an initial condition. The actual coupling constant can be obtained by calculating the average kinetic energy and using the equipartition theorem

$$g^{2}_{eff} = g^{2} \cdot \frac{\langle kin \rangle}{2 \cdot N} \tag{84}$$

The averages computed in the canonical and microcanonical ensembles differ in calculable 1/N (N=number of degrees of freedom) corrections. These corrections are usually negligible in realistic simulations ($N \sim 10^5$) but may become essential in calculating fluctations (e.g. the specific heat). We can eliminate this disadvantage of the simulation by introducing a new degree of freedom which redistributes the energy according to the Boltzmann factor of the canonical ensemble [49]. The canonical distribution can be sampled in this way by means of molecular dynamical methods.

8 QUANTUM CHROMODYNAMICS WITH QUARKS

8.1 Analytic results

The ultimate aim of the simulation of non-Abelian gauge theories is to verify some important nonperturbative features of QCD. These are the hadron mass spectrum and wave functions, the confinement mechanism in the presence of dynamical quarks, spontaneous breakdown of chiral symmetry and the phase structure at finite temperature. These questions have been studied in great detail in the quenched ($n_f \to 0$) approximation. The extension of these works to the full QCD is at the beginning only. The study of zero temperature properties of QCD unfortunately suffers from serious finite size effects. The inclusion of quarks leads to richer spectrum and one has to resolve correlation lengths like m_π^{-1}, m_p^{-1}, etc. in order to describe the low energy region of the theory properly. Bulk properties at finite temperature, like phase structure and symmetry properties are less sensitive to the accuracy of how the higher excited states are treated in the computations. At least the finiteness of the time extent of the system represents physical thermal fluctations in this case.

First we shall review very shortly the importance of the chiral noninvariance in QCD. Consider the theory given by the action (74). Dirac's fermions described by bi-spinors form the reducible (1/2,1/2) representation of the special Lorentz group. The irreducible spinors (1/2,0) and (0,1/2) are coupled by the mass term only. In the absence of such a term we can change the phases of the irreducible spinors independently and the global symmetry is U(1) x U(1). The diagonal subgroup $\psi \to \exp\{i\alpha\}\,\psi$, $\bar\psi \to \bar\psi\,\exp\{-i\alpha\}$ leads to the conservation of fermion number. The antisymmetric combination $\psi \to \exp\{i\beta\gamma^5\}\psi$, $\bar\psi \to \bar\psi\,\exp\{i\beta\gamma^5\}$ is called chiral transformation.

The symmetry of n_f massless fermions is $U(n_f) \times U(n_f)$. Integrating out the gauge field in the corresponding path integral expression for the vacuum-vacuum amplitude we generate the effective theory for the quark fields. There are two significant constraints for this effective theory. Confinement requires that only colorless combinations of the quark fields appear in the effective low energy theory. The other constraint is that the phenomenology of strong interactions and the actual hadron spectrum indicate that the degeneracy of the chiral $U(n_f)$ symmetry is absent from the physical spectrum. The flavored $SU(n_f)$ symmetry must be realized in the Goldstone mode. The absence of the "last" Goldstone mode corresponding to the flavorless U(1) factor was a longstanding puzzle of strong interactions. This mode supposed to decouple from the physical gauge invariant sector of the theory.

We assume that the quarks are massless. The success of the Partially Conserved Axial Current hypothesis suggests the presence of small bare quark mass m in the Lagrangian. This term leads to small perturbations around the spontaneously broken vacuum.

The breakdown of the chiral symmetry can be studied conveniently with the help of the order parameter $\chi_o = \langle\bar{\psi}O\psi\rangle$ where O is a flavor operator. We shall consider the case O=1 for simplicity. A nonzero value of $\langle\bar{\psi}\psi\rangle$ signals the breakdown of chiral U(1) symmetry. Careful study of the other Goldstone modes is then necessary to demonstrate the proper breakdown of the flavored $SU(n_f)$ part.

The picture outlined above may change significantly when the strongly interacting matter is considered at high density or temperature. We shall consider the latter case only here. The confinement may be lost in such extreme conditions since the theory is asymptotically free.

It was mentioned in Section 6 that the infinite free energy of a static quark is related to the presence of the Z_3 symmetry at finite temperature. Gauge transformations which are periodic up to the center elements are not symmetries of the complete theory any more. In fact, fields in the fundamental representation of the gauge group are effected by the center elements so the quark kinetic term breaks this symmetry explicitly. This symmetry breaking is calculable for heavy quarks on the lattice by means of the 1/m expansion [51]. This approximation leads to the upper part of the hypotethical phase diagramm depicted in Fig. 17. The Z symmetry breaking effect of the quarks becomes so strong for intermediate quark mass values that the Polyakov line becomes screened and useless for the study of confinement. The prediction, that the usual signal of the deconfinement phase transition is lost for large but finite quark mass values does not necesseraly mean that this phase transition is absent in the continuum theory. In fact, the transition flattens out when the Compton wavelength of the quarks is comparable with the lattice spacing. This region is irrelevant for the continuum theory where the ultraviolet cutoff should be larger than the physical mass scales.

As the quark mass is decreased from infinity toward zero we leave the region of convergence of the 1/m expansion and arrive to qualitatively new physics controlled by the chiral perturbation expansion (expansion in m).

The double expansion in $1/g^2$ and 1/space-time dimension on the lattice leads to chiral symmetry breaking, nonzero $\langle\bar{\psi}\psi\rangle$ and massless pions at low temperature [52]. Although the strong coupling region is far from the continuum limit, this

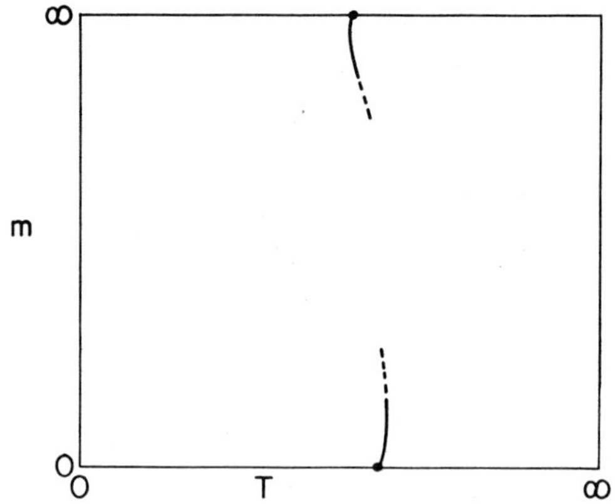

FIG. 17 Schematic phase diagram for lattice QCD.

result is an encouraging signal of the expected phenomenology. This expansion is unfortunately not powerful enough to explore the temperature dependence [53].

One can try different approach in the continuum theory. Assuming that the Goldstone modes are the relevant degrees of freedom at low energy the infrared dynamics can be approximated by a Ginsburg-Landau type phenomenological Lagrangian, called linear σ-model. It gives an account of the phenomenology of chiral symmetry breakdown.

The lower part of the phase diagramm in Fig. 17 is suggested by the σ-model where the chiral invariance is restored at high temperature. It is attempting to use the same σ-model in both phases but this is not necessarily correct. In fact, if confinement is lost in the high temperature phase the σ-model does not include all of the relevant modes of the systems (e.g. colored modes). So the high temperature phase should be described by different effective theory in this case.

The presence of confinement is less easily detectable in the complete QCD than in the pure Yang-Mills theory. Confinement of static charges is equivalent with the area-like behaviour of Wilson loops in the latter case. When dynamical quarks are present they screen Wilson loops and no local order parameter as the Polyakov line can be defined for confinement.

8.2 Numerical results

There are serious difficulties in the simultaion of the complete QCD. The lack of satisfactory fermion formalism and the slowness of the simulation procedure have been mentioned. In addition to these there is a rather practical problem: the approach to the continuum limit becomes complicated when quark mass is present in the Lagrangian. The dimensional argument to establish the simple cutoff-coupling constant relation is not valid in this case.

The rather preliminary calculations carried out so far followed a simple solution of this problem, the limit $m \to 0$ was taken numerically. On has to be careful with this limit when simulating on finite lattice. In fact, the Compton wavelength of the lighter hadrons diverge in this limit and the results become dominated by finite size effects. One has to carry out the calculation with finite bare quark mass and to take the limit $m \to 0$ after the computations. The extrapolations to $m=0$ have to be made in the region where the numerical results show no finite size effects.

The computation of few hadron masses is in a preliminary stage only. What we learned from a more complete calculation of SU(2) gauge theory [54] is that the projection into given flavor states is more difficult in the complete theory than in the quenched case when staggered fermions are used.

The calculations of bulk quantities are in better shape. The chiral condensate $\langle\bar{\psi}\psi\rangle$ of SU(2) theory found to follow the behaviour predicted by the asymtoticaly free beta function of the continuum theory [55]. This result represent a numerical proof of the dynamical breakdown of chiral invariance by the strong interactions.

The question of phase structure at finite temperature has been addressed by several computations [53,56,48,58]. We shall review few of them shortly.

Simulation of the SU(2) gauge theory approved the prediction of the $1/g^2$, $1/m$ expansion, that the deconfinement transition flattens out for large but finite values of the bare quark mass [53]. But it showed too that qualitatively new dynamics show up and a chiral transition emerges when the massless quark limit is taken.

The study of the massless limit of SU(3) gauge theory with $n_f=4$ [48] led to similar results. Naturally the careful verification of the continuum limit and accurate determination of the critical temperature, latent heat require more thorough studies. What is safe to say that the equation of state for hadronic matter undergoes an abrupt transition around the temperature $T_c \sim 240$ MeV. The asymtotic form of the equation of state at very high temperature corresponds to the gas of

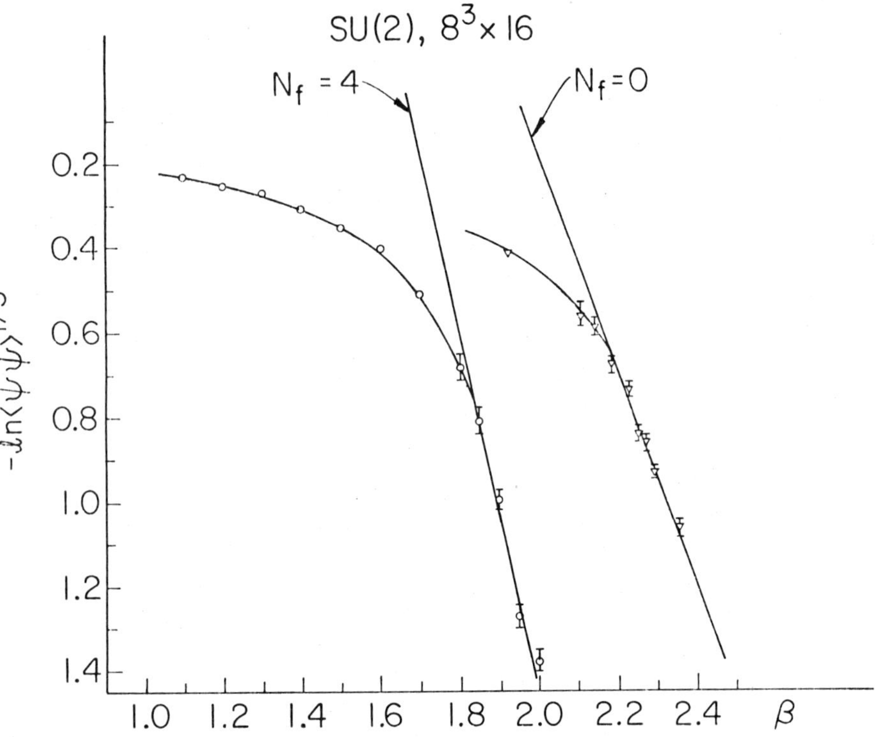

FIG. 18. The chiral condensate for SU(2) gauge theory with $n_f = 4$ and 0.

weakly interacting quarks and gluons. This asymtotics is the only unique signal so far that the chirally symmetric phase should be non-confining. Although it is very difficult to eliminate finite size effects when fermions are present, it seems that light fermionic collective modes give significant contribution to the internal energy in the high energy phase even at much higher temperature than T_c.

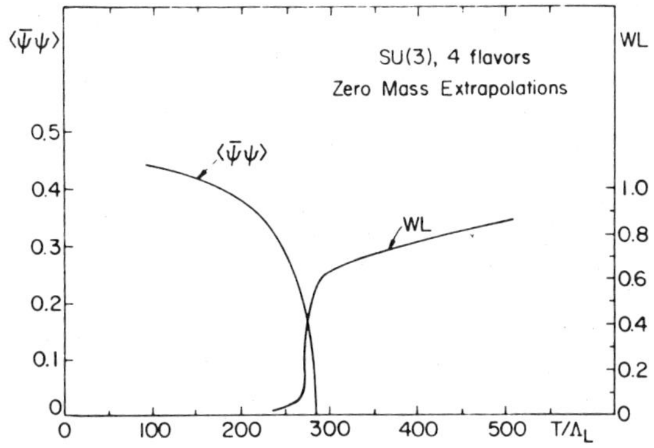

FIG. 19. The chiral condensate $\langle\bar{\psi}\psi\rangle$ and the trace of the Polyakov line $WL=\exp\{-\beta F_q\}$ for QCD with $n_f = 4$. Asymptotic scaling is assumed in converting the temperature into physical units.

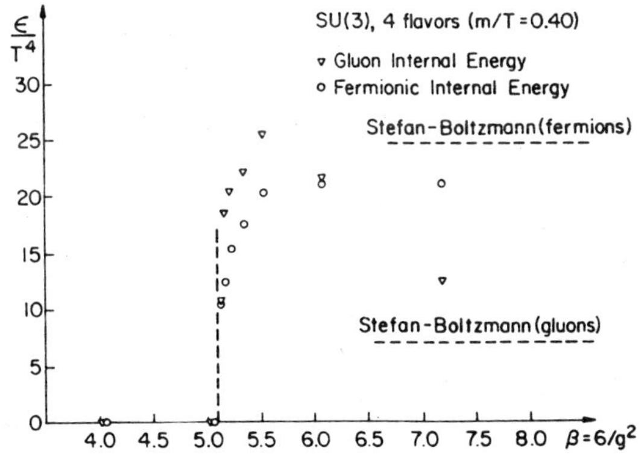

FIG. 20. Internal energy of QCD.

The problem of central importance is to select out the relevant degrees of freedom of the complete theory. Those are the colorless quasiparticles or chromomagnetic monopoles and gluon plane waves in the low or high temperature phase, respectively, of the quarkless QCD. We know much less about the full theory at the present. It is not well established that the chirally symmetric phase is deconfined. The independence of the chiral structure and the confinement can be illustrated by a gauge theory where the fermions are taken in the adjoint representation of the gauge group. Such fermions can not screen Wilson loops corresponding to fundamental test charges. Simulation of SU(2) gauge theory with adjoint quarks [57] revealed the presence of two different critical temperatures $T_{dec} < T_{ch}$.

ACKNOWLEDGEMENT We thank H. W. Wyld for reading the manuscript. The work was founded in part by the grant NSF-PHY82-01948.

REFERENCES

1. R. P.Feynman, A. R. Hibbs, Quantum Mecganics and Path Integrals, McGraw-Hill, New York, 1965

2. M. Creutz, B. Freedman, Ann. of Phys. 132(1981),427.

3. K. Binder in Phase Transitions and Critical Phenomena, C. Domb, M. S. Green ed. Academic, New York, 1976, Vol 5B.

4. S. D. Drell, M. Weinstein, S. Yankielowicz, Phys. Rev. D14(1976), 487, D16 (1977), 1769.

5. M. Creutz, Phys. Rev. D15 (1977), 1128.

6. J. Kogut, L. Susskind, Phys.Rev. D11 (1975), 395.

7. T. A. DeGrand, Doug Toussaint, Phys. Rev. D22 (1980), 2478.

8. K. G. Wilson, Phys. Rev. D10 (1974), 2445.

9. K. G. Wilson, J. Kogut, Phys. Rep. C12,75. K. G. Wilson, in Recent Development in Gauge Theories, Plenum, 1980.

10. K. Symanzik, in the Proceedings of the Trieste Workshop on Nonperturbative Field Theory and QCD, World Scientific, 1982.

11. S. I. Goldberg, Ciurvature and Homology, Academic Press, New York, 1962.

12. P. A. Dirac, Proc. Roy. Soc., Ser. A(1931),133.

13. S. Coleman, Magnetic Monopoles Fifty Years Later, Harvard Preprint, HUTP-82/A032, 1982.

14. H. B. Nielsen, P. Olesen, Nucl. Phys. B61 (1973), 45.

15. S. Mandelstam, Phys. Lett. 118B (1982),125. G. 'tHooft, Phys. Script. 25 (1982), 133.

16. L. Susskind, in Weak and Electromagnetic Interaction at High Energy, North Holland, Amsterdam, 1977.

17. L. D. Fadeev, V. N. Popov, Phys. Lett. 25B (1967), 29.

18. V. N. Gribov, Nucl. Phys. B139 (1978), 1.

19. E. S. Abers, B. W. Lee, Phys. Rep. C9 (1973),1.

20. R. Balian, J. M. Drouffe, C. Itzykson, Phys. Rev. D10 (1974), 3376, D11 (1974), 2098, 2104.

21. A. Hasenfratz, P. Hasenfratz, Phys. Lett. 93B (1980), 165.

22. M. Creutz, Phys. Rev. D21 (1980), 2308.

23. R. Dashen, D. Gross, Phys. Rev. D23 (1981), 2340.

24. K. C. Bowler et al. Nucl. Phys. B257 (1985), 155.

25. J. Polonyi, Symmetry Breaking Thermal Fluctations in Gauge Theories, in the Proceedings of Advances in Gauge Theory, Tallahasse, Florida, 1985.

26. A. M. Polyakov, Phys. Lett. 72B (1978), 477. L. Susskind, Phys. Rev. D20 (1979), 2610.

27. J. Polonyi, K. Szlachanyi, Phys. Lett. 110B (1982), 395.

28. A. Linde, Phys. Lett. 93B (180), 327.

29. J. Kuti, J. Polonyi, K. Szlachanyi, Phys. Lett. 98B (19801), 199.

30. L. McLerran, B. Svetitsky, Phys. Lett. 98B (1981), 195.

31. J. Polonyi, H. W. Wyld, Gauge Symmetry and Compactification, ILL-TH-85-23, submitted to Phys. Rev. Lett.

32. R. Anyshetti, Journ. of Phys. G10 (1984), 439.

33. F. Berezin, The Method of Second Quantization, Academic Press, New York, 1966.

34. W. Kerker, Phys. Rev. D24(1981),1595.

35. D. Gross, R. Pisarski, A. Yaffe, Rev. Mod. Phys. 53 (1981), 43.

36. L. H. Karsten, J. Smit, Nucl. Phys. B183 (1981), 163.

37. H. S. Sharatchandra, H. J. Thun, P. Weisz, Nucl. Phys. B129 (1981), 205.

38. K. Fujikawa, Z. Phys. C25 (1984), 179.

39. H. Kluberg-Stern, A. Morel, O. Napoly, B. Petersson, Nucl. Phys. B220 (1983),447. C. van den Doel, J. Smit, Nucl. Phys. B228 (1983), 122.

40. H. B. Nielsen, M. Ninomiya, Nucl. Phys. B185 (1981), 20, B193 (1981), 173.

41. D. Petcher, D. Weingarten, Phys. lett. 99B (1981), 333. D. Scalapino, R. Sugar, Phys. Rev. Lett. 46 (1981), 519. A. Duncan, M. Furlan, Nucl. Phys. B190 (1981), 767.

42. F. Fucito, E. Marinari, G. Parisi, C. Rebbi, Nucl. Phys. B180 (1981), 369.

43. I. Stamatescu, Phys. Rev. D11 (980), 3594.

44. J. Kuti, Phys. Rev. Lett. 49 (1982), 183.

45. G. Bhanot, U. Heller I. Stamatescu, Phys. Lett. 19B (1982), 440.

46. D. Callaway, A. Rahman, Phys. Rev. Lett. 49 (1982), 613.

47. J. Polonyi, H. W. Wyld, Phys. Rev. Lett. 51 (1983), 2257.

48. J. Polonyi, H. W. Wyld, J. Kogut, J. Shigemitsu, D. Sinclair, Phys. Rev. Lett. 53 (1984), 644.

49. S. Nose, Mol. Phys. 52 (1984), 255. S. Duane, J. Kogut, in preparation.

50. L. Kadanoff, Rev. Mod. Phys. Rev. Mod. Phys. 49 (1977), 267. J. Kogut, Rev. Mod. Phys. 51 (1979), 659, 55 (1983), 775. Yu. M. Makeenko, Usp. Fiz. Nauk. 143 (1984), 161.

51. P. Hasenfratz, F. Karsch, I. Stamatescu, Phys. Lett. 133B (1983), 221.

52. H. Kluberg-Stern, A. Morel, B. Petersson, Nucl. Phys. B125 (1983), 527.

53. J. Kogut, J. Polonyi, H. W. Wyld, J. Shigemitsu, D. Sinclair, ILL-TH-85-20.

54. J. Kogut, J. Polonyi, D. Sinclair, in preparation.

55. J. Kogut, J. Polonyi, H. W. Wyld, D. Sinclair, ILL-TH-85-43.

56. R. Gavai, M. Lev, L. Petersson, Phys. Lett. 140B (1984), 397. U. Heller, F. Karsch, Cern Preprint, TH-4078. F. Fucito, C. Rebbi, S. Solomon, Nucl. Phys. B248 (1984), 615.

57. J. Kogut, J. Polonyi, H. W. Wyld, D. Sinclair, Phys. Rev. Lett. 54 (1985), 1980.

LOW ENERGY PHENOMENOLOGY WITH SKYRMIONS†

I. Zahed and G. E. Brown
State University of New York
Stony Brook, NY 11794

ABSTRACT

We review the recent developments concerning the Skyrme model in the context of QCD, and analyze their relevance to low energy phenomenology. Their implications in nuclear physics are stressed.

† Supported in part by the U.S. Department of Energy under Contract No. DE-AC02-76ER13001 with the State University of New York at Stony Brook.

TABLE OF CONTENTS

INTRODUCTION

Although many believe that QCD is the theory of strong interactions, our quantitative understanding of this theory in the long-wavelength approximation remains still a challenging problem. Aside from tedious but promising lattice gauge calculations, key concepts such as confinement and chiral symmetry breaking are only qualitatively understood. Most of our present low energy predictions stem from models believed to approximate QCD, e.g. bag models, non-relativistic potential models, QCD sum rules, effective chiral models,...

In QCD, the absence of an explicit expansion parameter such as the fine structure constant α in QED, rules out any perturbative scheme upon which most of our present understanding of physics is based. Because of dimensional transmutation[3], the effective quark gluon coupling $g \to g(\Lambda)$ where Λ is the QCD cutoff. Massless QCD is in fact a parameter free theory.

Some years ago 't Hooft[1] and later on Witten[2], proposed to study the $N_c \to \infty$ Yang Mills theory, and use $1/N_c$ as an implicit expansion parameter. They suggested that by comparing the large N_c results with experiment, one would be able to account à posteriori for the relevance of the expansion. Assuming confinement 't Hooft has shown that the large N_c version of QCD is a weakly interacting phase of mesons and glueballs that decouple to leading order. Simple power counting based on the fact that $g^2 N_c$ scales like 1, shows that the correlation function of any bilocal, color singlet operator is dominated by the planar diagrams with exclusive glue insertion as shown in Fig. 1. Quark loops are suppressed by powers of $1/N_c$, and non-planar diagrams by powers of $1/N_c^2$.

$$\langle J(k)J(-K)\rangle = \sum_{\text{glue}} \Big[$$

$$\Big] = \sum_n \frac{\lambda_n}{k^2 - W_n^2}$$

Fig. 1. Dominant planar diagrams in the correlation function of $J(k) = \bar{q}q(k)$.

Insisting on crossing and unitarity, 't Hooft has shown that to leading order in N_c, the scattering amplitudes in QCD are sums of tree diagrams involving exclusively meson resonances. In a remarkable way perhaps, large N_c QCD reduces smoothly to an effective field theory of non-interacting mesons to leading order, prohibits exotics, and turns out to be consistent with Regge phenomenology and Zweig selection rules.

Using the large N_c diagrammatics, Witten[2] has argued that to leading order in N_c, baryons behave as solitons in this weakly coupled phase of mesons. Indeed, a baryon can be thought of as N_c quarks interacting via one-gluon exchange to leading order. If we denote by M_B its ground state energy, then

$$M_B = N_c\,(m+T) + \frac{N_c(N_c-1)}{2}\, g^2 V = N_c\, f\,(g^2 N_c\,, m) \qquad (1)$$

where $f(g^2N_c,m)$ is a smooth function of N_c. Thus, M_B is inversely proportional to $1/N_c$ which is the strength of the 4-meson coupling in the large N_c world. This behavior is typical of a soliton, and can never be obtained in perturbation theory. One can also show that the baryon-baryon interaction is strong and scales like N_c, while the baryon-meson interaction is weak and scales like 1.

Aside from colour, QCD involves different quark flavours u,d,s... For the three flavour case, the relevant dynamics is described by the following Lagrangian*

* Instanton effects are neglected, and consequently the vacuum angle θ is set to zero.

$$\mathcal{L} = \frac{1}{4} \mathrm{TR}[G_{\mu\nu} G^{\mu\nu}] + \bar{q}(i\not{D} - m)q \tag{2}$$

where the quark fields are in the fundamental representation of $SU(3)_f$, and the gluons in the adjoint representation of $SU(3)_c$. The low energy behaviour of QCD, mostly relevant for nuclear physics, is dominated by the u,d quarks whose current masses $m_u \sim 5$ MeV, $m_d \sim 10$ Mev, are small compared to the QCD cutoff $\Lambda \sim 200$ MeV. In the massless limit $m_u = m_d = 0$, the QCD Lagrangian (2) is invariant under $U(2)_L \times U(2)_R$ transformations in flavour space, i.e.,

$$q'_R = e^{iQ_R} q_R$$
$$q'_L = e^{iQ_L} q_L \tag{3.a}$$

where $Q_{R,L} = Q^a_{R,L} T^a$ involve the fundamental generators of $U(2)_{R,L}$ with $T^0 = 1$, and $q_{R,L}$ are the usual right and left quark fields defined through

$$q_{R,L} = \left(\frac{1 \pm \gamma_5}{2}\right) q \tag{3.b}$$

By Noether theorem, the $SU(2)_L$ and $SU(2)_R$ currents are conserved,

$$\partial^\mu \left(\bar{q}\gamma_\mu \left(\frac{1 \pm \gamma_5}{2}\right) T^a q\right) = 0 \qquad a = 1,2,3 \tag{4}$$

while the $U(1)_{L,R}$ currents are plagued with anomalies[4],

$$\partial^\mu \left(\bar{q}\gamma_\mu \left(\frac{1 \pm \gamma_5}{2}\right) q\right) = \pm \frac{N_f g^2}{64\pi^2} \mathrm{TR}[G_{\mu\nu} \tilde{G}^{\mu\nu}] . \tag{5}$$

The absence of parity doublets in the physical spectrum with a pronounced isospin symmetry, suggest that the global $SU(2)_L \times SU(2)_R \times U(1)_V$ is spontaneously broken to $SU(2)_V \times U(1)_V$ via the Nambu-Goldstone mechanism, with the appearance of 3 massless pseudoscalar excitations, the pions: π^0, $\pi^\pm$. In other words, the QCD ground state carries axial charge,

$$Q_V^a |0\rangle = 0$$
$$Q_A^a |0\rangle \neq 0 \qquad (6)$$

and as a result, pions can decay into the vacuum. Since $m_S \sim 150$ Mev, the chiral limit is more tricky when strange quarks are incorporated. So, from now on, we will focus on the two flavour case unless specified otherwise.

2. THE NON-LINEAR σ-MODEL

The relative success of current algebra and PCAC[5] lies in the fact that the QCD ground state breaks chiral symmetry. The central assumption behind the PCAC hypothesis is that at low energy, one can saturate all vacuum expectation values with pionic states. If we denote by $|\pi(p)\rangle$ a pion state of momentum p, and by $A_\mu^a(x)$ the pertinent axial-vector current, then Eq. (6) can be used to define

$$\langle 0| A_\mu^a(x) |\pi^b(p)\rangle = i f_\pi p_\mu e^{ipx} \delta^{ab} \qquad (7)$$

where f_π = 93 Mev is the observed pion decay constant. From (7) follows the PCAC relation

$$f_\pi^2 m_\pi^2 = -(m_u + m_d) \langle 0|\bar{U}U|0\rangle \qquad (8)$$

and the Goldberger-Treiman relation,

$$g_A(0) = g_{\pi NN} \frac{f_\pi}{m_N} \tag{9}$$

where m_π = 139 Mev is the pion mass, g_A = 1.34 the nucleon axial coupling, $g_{\pi NN}$ ~ 13 the πNN-coupling, and m_N = 938 Mev the nucleon mass. Equations (7 - 9) play a fundamental role in low energy phenomenology.

Due to the non-perturbative character of QCD in the long-wavelength approximation, very little is known about f_π and g_A from first principles. The large N_c limit advocated by 't Hooft and witten[2] is suggestive of an effective mesonic description involving the dominant chiral excitations, π^0, $\pi^\pm$, as a substitute to QCD in the low energy realm. Although N_c = 3 and not ∞, one hopes that the $1/N_c$ - expansion, at least in its qualitative form, will provide the starting lines for discussing low energy phenomenology*. In this spirit, the non-linear σ-model stands as the simplest realization of chiral symmetry consistent with phenomenology and in agreement with the large N_c scenario . If we denote by σ the scalar meson field, and by $\vec{\pi}$ the massless pion field, then the resulting dynamics is described by

$$\mathcal{L} = \frac{1}{2}(\partial_\mu\sigma)^2 + \frac{1}{2}(\partial_\mu\vec{\pi})^2$$

$$\sigma^2 + \vec{\pi}^2 = f_\pi^2 \tag{10}$$

Equations (10) are manifestly chiral invariant, since $(\sigma;\vec{\pi})$ corresponds to the (1,0) representation of $SU(2)_V \times SU(2)_A \sim O(4)$. Specifically, Eq. (10) is invariant under the following global set of transformations

---

* The pertinent question at this stage is not how small 1/3 is, but how good are the leading results in the $1/N_c$ - expansion. After all, if $1/N_c$ = 1/3, it might well be that the relevant expansion parameter is $1/N_c^2$ = 1/9 or $1/4\pi N_c^2$ ~ 1/113 etc.

$$\delta_V \sigma = 0$$
$$\delta_V \vec{\pi} = -(\vec{\alpha} \times \vec{\pi}) \tag{11a}$$

$$\delta_A \sigma = -\vec{\beta}\,\vec{\pi}$$
$$\delta_A \vec{\pi} = +\vec{\beta}\sigma \tag{11b}$$

that generate the classically conserved vector and axial-vector currents respectively,

$$V_\mu^a = \epsilon^{abc}\,\pi^b \partial_\mu \pi^c \tag{12a}$$

$$A_\mu^a = \sigma\,\partial_\mu \pi^a - \pi^a \partial_\mu \sigma \tag{12b}$$

To account for the small pion mass, one usually adds an explicit chiral breaking term to (10), in the form

$$\mathcal{L}_B = -m_\pi^2 f_\pi \sigma \tag{13}$$

In the trivial vacuum $\langle\sigma\rangle = f_\pi$ and perturbative pions can be understood as fluctuations around the "tilted Mexican hat". In this framework, one can show that the PCAC relation (7) is fulfilled, and that

$$\partial^\mu A_\mu^a = m_\pi^2 f_\pi \pi^a \tag{14}$$

It should be clear by now, that the non-linear σ-model satisfies the major low energy requirements solely on the basis of chiral symmetry. In fact it does more, as Skyrme[6] noted long ago, it embodies an underlying topological structure that yields non-perturbative field configurations reminiscent of classical baryons. To illustrate this, define the 2×2 unitary field

$$U(x) = \frac{1}{f_\pi}\left[\sigma + i\vec{\tau}\cdot\vec{\pi}\right] , \quad U^{+}U = \mathbb{1} \tag{15}$$

which transforms as the (1/2, 1/2) representation of $SU(2)_L \times SU(2)_R$,

$$U(x) \rightarrow e^{iQ_L}\, U(x)\, e^{iQ_R} \tag{16}$$

with

$$Q_{R,L} = \frac{1}{2}(\vec{\beta} \mp \vec{\alpha})\cdot\vec{\tau} \tag{17}$$

in agreement with the set of transformations (11). In terms of (16), the Lagrangian for the non-linear σ-model (10) reads

$$\mathcal{L} = -\frac{f_\pi^2}{4}\,\mathrm{Tr}\left[L_\mu L^\mu\right] = -\frac{f_\pi^2}{4}\,\mathrm{Tr}\left[R_\mu R^\mu\right] \tag{18}$$

where L_μ and R_μ are the left and right currents on S^3. i.e.

$$\begin{aligned} L_\mu &= U^{+}\partial_\mu U \\ R_\mu &= U\partial_\mu U^{+} \end{aligned} \tag{19}$$

Since det U=1, one concludes that L_μ and R_μ are traceless. Asymptotically, they reduce to simple field gradients,

$$L_\mu \sim -R_\mu \sim \frac{i}{f_\pi} \vec{\tau}\cdot\partial_\mu \vec{\pi} \tag{20}$$

At any fixed time $U(\vec{x})$ defines a map from the 3-dimensional space R^3 onto the group manifold of $SU(2) \sim S^3$, with the natural boundary condition that asymptotically $U(\vec{x})$ goes to the trivial vacuum, i.e. $U(|\vec{x}| \to \infty) = 1$, to ensure that the energy is finite. This implies that R^3 is compactified to S^3. The corresponding set of static maps,

$$U(\vec{x}) : \quad S^3 \to S^3 \tag{21}$$

is known to be non-trivial since $\Pi_3(S^3) = Z$. Consequently $U(\vec{x})$ falls into homotopically distinct classes that are characterized by conserved topological charges or winding numbers. The normalized topological density $B^0(x)$ is just the Jacobian associated to the stereographic projection from S^3 onto R^3. If we denote by $(\Phi^0, \Phi^a) = 1/f_\pi\,(\sigma, \Pi^a)$ an element of S^3, then*

$$B^0(\vec{x}) = -\frac{\epsilon^{abcd}}{2\pi^2}\, \phi^a \partial_1 \phi^b \partial_2 \phi^c \partial_3 \phi^d \tag{22}$$

straightforward manipulations using (15) and (19) give

$$B^0(\vec{x}) = -\frac{\epsilon^{0jkl}}{24\pi^2}\, \mathrm{TR}\,[L_j L_k L_l] \tag{23}$$

which is properly normalized, i.e.,

$$\int_{R^3} d\vec{x}\; B^0(\vec{x}) = \text{winding number} \tag{24}$$

---

*Notice that $2\pi^2$ is just the area of S^3 in R^4.

By construction (24) characterizes the number of times the map $U(\vec{x})$ winds around the group manifold S^3 in the course of describing pion configurations in R^3. The covariant topological current associated to (22), namely

$$B^\mu(x) = -\frac{\epsilon^{\mu\nu\alpha\beta}}{24\pi^2} \, \mathrm{TR}\,[L_\nu L_\alpha L_\beta] \tag{25}$$

is conserved independently of the equations of motion. Unfortunately, the resulting set of static field configurations with non-trivial topology is not stable energetically by Derrick's theorem[7]. Indeed, the energy carried by a static and non-trivial pion configuration is

$$E = \frac{f_\pi^2}{4} \int d^D x \; \mathrm{TR}\,[\partial_i U^\dagger \partial_i U] \tag{26}$$

in D-dimensional space (here D=3). A simple rescaling of $U(\vec{x})$ in space, i.e. $U(\vec{x}) \to U(\lambda\vec{x})$, yields

$$E_\lambda = \lambda^{2-D} E \tag{27}$$

which clearly shows that for D=3, the finite energy configurations of the non-linear σ-model are unstable against scale transformations.

3. THE SKYRME MODEL

To prevent the topological configurations of the non-linear σ-model from collapsing to point-like structures, Skyrme[6] added by hand to (18) a quartic term in the currents L_μ (the Skyrme term)

$$\mathcal{L}_S = -\frac{f_\pi^2}{4} \,\mathrm{TR}\,[L_\mu L^\mu] + \frac{\epsilon^2}{4} \,\mathrm{TR}\,[L_\mu, L_\nu]^2 \tag{28}$$

Here ε^2 is a dimensionless parameter that characterizes the size of the finite field configurations. A simple rescaling in space translates into a rescaling in the ground state energy as follows

$$E_\lambda = \lambda^{2-D} E_2 + \lambda^{4-D} E_4 \tag{29}$$

For $D \geq 3$, (29) exhibits a true minimum in the λ-direction, i.e.

$$\frac{dE_\lambda}{d\lambda} = 0 \quad : \quad \frac{E_2}{E_4} = -\frac{D-4}{D-2} \tag{30a}$$

$$\frac{d^2E_\lambda}{d\lambda^2} = 0 \quad : \quad 2\,(D-2)\,E_2 > 0 \tag{30b}$$

In particular, for D=3 $E_{(2)} = E_{(4)}$, as of course expected from the virial theorem. The energetically stable configurations of (28) with non-trivial topological density, are referred to as skyrmions. The above arguments are of course classical and do not exclude instability at the quantum level. Notice that the skyrmion energy is bounded from below by the topological charge. Indeed, the explicit form of the energy functional associated to (28) reads

$$E = \int d\vec{x} \left(-\frac{f_\pi^2}{4} \,\mathrm{TR}\,[L_i^2] - \frac{\epsilon^2}{4} \,\mathrm{TR}\,[L_i, L_j]^2 \right) \tag{31}$$

Using the Cauchy-Schwarz inequality, one obtains

$$E \geqslant 12\sqrt{2}\,\pi^2 \epsilon f_\pi \,|B| \tag{32}$$

where B is the total topological charge. Equation (32) illustrates in a striking way the mechanism by which geometry induces local stability at the classical level. Although first established by Skyrme[6], it is sometimes referred to as the Bogomol'ny bound[8]. Since there are no self-dual chiral fields, (32) can never be saturated.

The Skyrme term can be understood as a higher order correction to the non-linear σ-model in a gradient-expansion of the Weinberg type[9]. To lowest order in the pion momentum p, this term exhibits the quantum numbers of a massive ρ-meson exchange in the D-wave ππ-channel, i.e.

$$\frac{\epsilon^2}{4} \mathrm{TR}\,[L_i, L_j]^2 \sim 2\frac{\epsilon^2}{f_\pi^2} \left(\partial_i \vec{\pi} \times \partial_j \vec{\pi}\right)^2 \tag{33}$$

suggesting perhaps, that the ϵ^2 term can be obtained from the ππ-data. While the non-linear σ-model constitutes a unique effective action to order $O(p^2)$ consistent with current algebra, the Skyrme term is not to order $O(p^4)$. Indeed, under the general assumptions of locality, Lorentz-invariance, chiral symmetry, parity and G-parity, there are 3-independent candidates to order $O(p^4)$, i.e.

$$L_4 = \alpha\, \mathrm{TR}\,[L_\mu, L_\nu]^2 + \beta\, \mathrm{TR}\,[L_\mu, L_\nu]_+^2 + \gamma\, \mathrm{TR}\,[\partial_\mu L^\mu]^2 \tag{34}$$

Any other combination can be eliminated using the Maurer-Cartan equation

$$\partial_\mu L_\nu - \partial_\nu L_\mu - [L_\mu, L_\nu] = 0 \tag{35}$$

The Skyrme term is however the unique 4th order term that leads to a Hamiltonian second order in time derivatives. While a priori there is no reason to disregard any of these terms, Skyrme's choice corresponds to $\alpha = \frac{\varepsilon^2}{4}$, $\beta=\gamma=0$, and is qualitatively consistent with the ππ-data as noted by Rho[10], and Donoghue et al.[11]

In his pioneering work two decades ago, Skyrme[6] strongly suspected that the non-trivial field configurations of (28) with winding number one were fermions. Moreover, he conjectured that the topological current (25) should be identified with the baryon current, suggesting that skyrmions were classical baryons. These remarkable statements in pre-QCD times, attracted considerable attention recently in the context of QCD, following the work of Balachandran et al[12] and Witten[13].

The classical field configurations associated to (28), are solution to the following Euler-Lagrange equations

$$\partial^\mu L_\mu - 2\frac{\epsilon^2}{f_\pi^2}\partial^\mu [L_\nu,[L_\mu,L_\nu]] = 0 \tag{36}$$

Unfortunately, analytical solutions to (36) are not available. The Bogomol'ny-Prasad-Sommerfield limit[14] well-known in monopole dynamics, cannot be exploited here since (32) cannot be saturated. Equations (36) are highly non-linear, and so far can only be handled under the assumption of maximal symmetry as suggested by Skyrme's hedgehog ansatz

$$U(\vec{x}) = \exp(i\vec{\tau}.\hat{r}F(r)) \tag{37}$$

in which the pion field is radial in both space and isospace. In fact, the hedgehog configuration (37) belongs to a class of degenerate solutions, all reducible to each other through global chiral rotations, i.e.

$$U_A(\vec{x}) = A\exp(i\vec{\tau}.\hat{r}F(r))A^+ = \exp(iA\vec{\tau}A^+.\hat{r}F(r))$$

where A is a global SU(2) rotation.

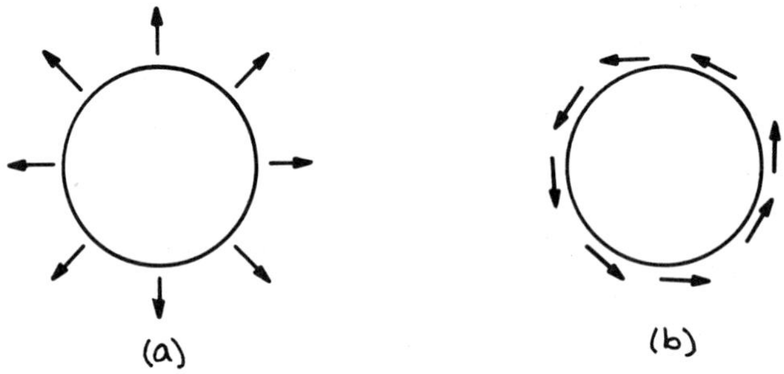

Fig. 2. (a) gives a schematic illustration of the pion field in a hedgehog configuration as described by (37).
(b) describes a combed hedgehog corresponding to a global $\pi/2$ isorotation.

For the hedgehog ansatz (37), isospin (I) and angular momentum (J) are correlated in a way that neither of them is a good quantum number, but their sum $\vec{K} = \vec{J} + \vec{I}$ is,

$$[\vec{K}, U(\vec{x})] = \sin F\,[(\vec{r}\times\vec{\nabla}) + i\frac{\vec{\tau}}{2}, \vec{\tau}\cdot\hat{r}] = 0$$

Since under parity,

$$\hat{\Pi}_{op}\, U(\vec{x})\, \hat{\Pi}_{op}^{-1} = U^{+}(-\vec{x}) \tag{38}$$

one concludes that hedgehog skyrmions carry $K^{\pi} = 0$ assignment, and can be viewed as an admixture of states with I = J and positive parity.

In terms of (37), the variational equations (36) simplify into a radial equation for F(r),

$$F'' + \frac{2}{r}F' - \frac{\sin 2F}{r^2} + \frac{8\epsilon^2}{f_\pi^2}\left(\frac{\sin 2F \sin^2 F}{r^4} - \frac{F'^2 \sin 2F - 2F'' \sin^2 F}{r^2}\right) = 0 \tag{39}$$

and the solution of n winding number satisfies the boundary conditions

$$F(r=0) = n\pi$$

$$F(r\to\infty) = 0$$

By Skyrme's conjecture, this corresponds to an n-baryon system with the following baryon distribution

$$B(r) = -\frac{1}{2\pi^2}\sin^2 F\,\frac{F'}{r^2} \tag{40}$$

Numerical solutions to the above boundary value problem have been obtained by Jackson et al.[15], and Adkins et al.[16]. The former used f_π = 93 Mev and ϵ^2 = 0.00552 to insure the correct value of g^N = 1.34 via the Goldberger-Treiman relation, while the latter used f_π= 64.5 Mev and ϵ^2 = .00421 to reproduce the mass of the nucleon and Δ-isobar. The behavior of F(r) versus r for the parameters of Ref. 15 is shown in Fig. 3, for B=1 and B=2 respectively. Notice that close to the origin $F(r) \sim n\pi - \alpha r$, while at large distance it falls like a power law, $F(r) \to k^2/r^2$. For the specific set of parameters used in Ref. 15

$$F(r) \to \frac{16\epsilon^2}{f_\pi^2}\,\frac{1.078}{r^2} \tag{41}$$

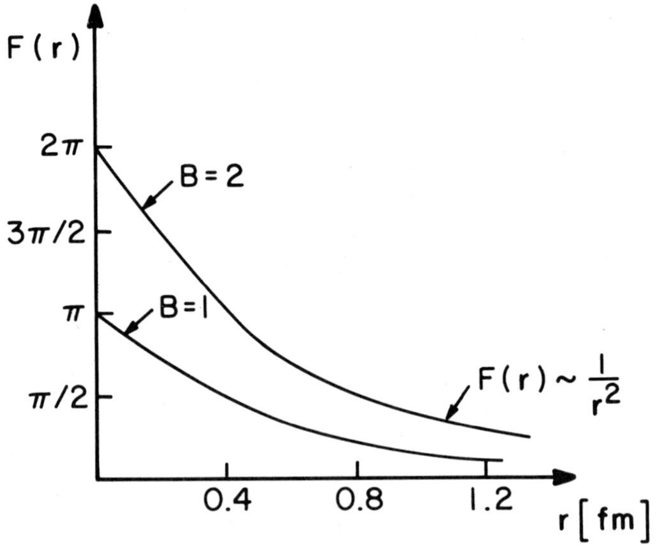

Fig. 3 Behavior of the chiral angle as a function of r for the one- and two-baryon system using the parameters of Ref. 15

To clarify the relationship between g_A^N and ε, it is instructive to notice that while asymptotically F(r) is described by (41), the pion field $\pi^a(r)$ has the following form

$$\pi^a(\vec{r}) = -\frac{f_\pi k^2}{3r^2}\,\hat{r}^i\,\langle\sigma^i\tau^a\rangle \qquad r\to\infty \tag{42}$$

which involves the expectation value in a hedgehog configuration. Old-fashioned soft pion-nucleon effective Lagrangians show that in a nucleon state, the pion field behaves asymptotically like (for more details see Appendix C).

$$\pi^a(\vec{r}) = -\left(\frac{g_{\pi NN}}{8\pi m_N}\right)\frac{\hat{r}^i}{r^2}\,\langle\sigma^i\tau^a\rangle \qquad r\to\infty \tag{43}$$

where $g_{\pi NN}$ is the πNN-coupling and m_N the nucleon mass. Comparison between (42) and (43), and explicit use of the Goldberger-Treiman relation yields

$$g_A^H = \frac{8\pi}{3} K^2 f_\pi^2 \tag{44}$$

which shows that the large distance behavior of the skyrmion is directly tied to the hedgehog g_A^H in the chiral limit. Jackson et al.[15] have used $g_A^H = .80$, while Adkins et al.[16] have predicted $g_A^H = .61$. The former have argued that there are substantial finite N_c corrections. Using simple bookkeeping arguments, they have shown that for a given number of colors N_c, the axial coupling g_A^N is related to the hedgehog g_A^H through,

$$g_A^N = \left(\frac{N_c+2}{N_c}\right) g_A^H \tag{45}$$

so that for $N_c \to \infty$ $g_A^N = g_A^H$, while for $N_c=3$ $g_A^N = 5/3\ g_A^H$.

The mass of the hedgehog configuration can be deduced from (31) using the ansatz (34). Since both quadratic and quartic terms in (31) contribute equally to the energy, thanks to the virial theorem, one obtains

$$M_H = 4\pi\sqrt{2}\ \epsilon f_\pi \int_0^\infty dx\, x^2 \left[F'^2 + 2\frac{\sin^2 F}{x^2} \right] \tag{46}$$

The rms-radius of the hedgehog is just the isoscalar radius constructed using the baryon distribution (40),

$$r_H = \frac{\epsilon}{\pi f_\pi} \left(- \int_0^\infty dx\, x^2 \sin^2 F\, F' \right)^{1/2} \tag{47}$$

Table 1 summarizes the result of Jackson et al. and Adkins et al. for the one-baryon solution. The energy and size of the hedgehod skyrmion does not seem to be unreasonable given the properties of the nucleon and Δ-isobar.

	ε^2	f_π[Mev]	g_A^H	g_A^N	M_H[Mev]	r_H[fm]
J.R.	.00552	93.*	.80*	1.33	1425	.48
A.N.W.	.00421	64.5	.61	1.01	863*	.59

Table 1. Results of Jackson and Rho[15] (J.R.), and Adkins, Nappi and Witten[16] (A.N.W.) for the one-baryon hedgehog. The (*) refers to input parameters.

The addition of a pion mass m_π = 139 Mev, does not yield substantial modifications.

4. ARE SKYRMIONS REALLY BARYONS?

Despite the very appealing character of the soliton scenario and the underlying wisdom of the large N_c-expansion, one is definitely tempted to ask two relevant questions:

i) how non-trivial configurations exclusively made out of mesons can lead to fermions?

ii) why do these fermions if any, have anything to do with the QCD baryons?

The answers to these fundamental questions are tied to the anomaly structure of QCD at low-energy, and any attempt to provide concise and complete arguments will go beyond the course of these lectures. Instead, we will try to provide some qualitative arguments, often at the expense of rigor. (The purists are of course referred to the original literature). For this, we need to introduce and discuss the fundamentals of the Wess-Zumino term.

4.1. The Wess-Zumino term

Consider in this case the $SU(3)_f$ QCD Lagrangian (2) in the chiral limit, i.e. $m_s=m_u=m_d = 0$. We know from earlier arguments in these lectures, that in this case the QCD Lagrangian is globally invariant under $U(3)_L \times U(3)_R$. So, aside from the U(1) axial current which is anomalous, there are 8 conserved Noether currents. In this case also, it is believed that chiral symmetry is spontaneously broken through*

* There are of course substantial m_s-corrections as far as strange particles are concerned.

$$U(3)_L \times U(3)_R / U(1)_A = SU(3) \times SU(3)_R \times U(1)_V \to SU(3)_V \times U(1)_V \quad (48)$$

with the appearance of eight massless pseudoscalar mesons: π, K, η. At low-energy, their dynamics is dictated by a non-linear σ-model of the type described by Eq. (18), where U(x) is now an SU(3) valued field,

$$U(x) = \exp\left(i \frac{\lambda^a \pi^a}{f_\pi}\right) \quad (49)$$

with

$$\pi^a \lambda^a = \sqrt{2} \begin{bmatrix} \frac{\pi^0}{\sqrt{2}} + \frac{\eta}{\sqrt{6}} & \pi^+ & K^+ \\ \pi^- & -\frac{\pi^0}{\sqrt{2}} + \frac{\eta}{\sqrt{6}} & K^0 \\ K^- & \bar{K}^0 & -\sqrt{\frac{2}{3}}\,\eta \end{bmatrix} \quad (50)$$

where $\pi^{\pm} = (\pi_x \pm i\pi_y)/\sqrt{2}$, and the λ's are the ordinary Gell-Mann matrices.

Witten[13] has observed that while the non-linear σ-model (18) is invariant under global $U(3)_L \times U(3)_R$ and the parity operation (38), it exhibits two discrete symmetries which are redundant with QCD, namely $U(+\vec{x},t) \to U(-\vec{x},t)$ and $U(\vec{x},t) \to U^+(\vec{x},t)$. The latter forbids decay processes of the type

$$\begin{aligned} &K^+K^- \to \pi^+\pi^0\pi^- \\ &\pi^+\pi^- \to 2\,\pi^+\pi^- \\ &\cdots \end{aligned} \quad (51)$$

where an even number of mesons decays into an odd number and vice versa. Processes like (51) are mediated by anomalies in QCD, and turn out to be kinematically excluded by the non-linear σ-model. As a remedy, Witten[13] has proposed to modify the equations of motion at low-energy, by adding explicitly a chiral invariant piece that would spoil the redundant symmetries, while preserving the parity operation (38), i.e.

$$\frac{f_\pi^2}{2} \partial^\mu L_\mu + \lambda \epsilon^{\mu\nu\alpha\beta} L_\mu L_\nu L_\alpha L_\beta + \cdots = 0 \quad (52)$$

where the ... stand for higher derivative terms consistent with current algebra. Notice that the added term is linear in the time-derivative, and unique to lowest order in the field gradients. In terms of (52), processes like (51) are no longer forbidden.

Asymptotically, $L_\mu = i\phi_\mu = i\lambda^a \partial_\mu \pi^a / t_\pi + \ldots$, and (52) can be thought of as resulting from variations of an action functional S, through

$$\delta S = -\int d^4x \, \mathrm{TR}\left[\delta\phi \left(\frac{f_\pi^2}{2}\partial^\mu \phi_\mu - i\lambda \epsilon^{\mu\nu\alpha\beta}\phi_\mu \phi_\nu \phi_\alpha \phi_\beta + \cdots\right)\right] \tag{53}$$

Equation (53) can be integrated out order by order in the form of a gradient expansion. Unfortunately, the resulting action has two drawbacks: it does not close, and it does not exhibit explicit chiral invariance. Witten[13] has shown using general topological arguments, that the second term in (53) is related to a closed topological invariant in 5-dimensional space, which is manifestly chiral invariant. Armed with this information, let us specialize in (53) to the set of homotopically equivalent field variations by defining

$$\phi(x,s) = s\,\phi(x) \qquad s \in [0,1] \tag{54}$$

such that

$$\delta S = -\int d^4x \, \mathrm{TR}\left[\phi_s \left(\frac{f_\pi^2}{2}\partial^\mu \phi_\mu - i\lambda \epsilon^{\mu\nu\alpha\beta}\phi_\mu \phi_\nu \phi_\alpha \phi_\beta + \cdots\right)\right] \tag{55}$$

hence,

$$\begin{aligned} S_+ = &\int_0^1 ds \int d^4x \, \frac{f_\pi^2}{4} \partial_s \mathrm{TR}\left[\phi^\mu \phi_\mu\right] \\ &+ i\lambda \int_0^1 ds \int d^4x \, \epsilon^{\mu\nu\alpha\beta} \mathrm{TR}\left[\phi_s \phi_\mu \phi_\nu \phi_\alpha \phi_\beta\right] \\ &+ \cdots \end{aligned} \tag{56}$$

* We will not include the Skyrme term in our arguments, since its presence is mandated by dynamical issues, and does not affect the geometrical character of the arguments that will follow.

Assuming that the result (56) when performed to all orders is chiral invariant yields

$$S_+ = -\frac{f_\pi^2}{4}\int_{R^4} d^4x \, TR[L_\mu L^\mu] + \frac{\lambda}{5}\epsilon^{\mu\nu\alpha\beta\gamma}\int_{R^4\times[0,1]} d^5x \, TR[L_\mu L_\nu L_\alpha L_\beta L_\gamma] \quad (57)$$

with U(x,0) = 1, and U(x,1) = U(x) on the boundary. Modulo the normalization, the term added by Witten involves the topological invariant that characterizes mappings from S^5 onto S^5. It is the closed form expression of the Wess-Zumino terms[17] well-known from current algebra. Notice that (57) vanishes for an $SU(2)_f$ group*.

4.2. Topological quantization of the Wess-Zumino term

The above derivation of (57) contains a subtle ambiguity. Indeed, if instead of extending R^4 to $R^4 \times [0,1]$, we have decided to extend it to $R^4 \times [-1,0]$, the resulting action functional associated to (52) would be

$$S_- = -\frac{f_\pi^2}{4}\int_{R^4} d^4x \, TR[L_\mu L^\mu] - \frac{\lambda}{5}\epsilon^{\mu\nu\alpha\beta\gamma}\int_{R^4\times[-1,0]} d^5x \, TR[L_\mu L_\nu L_\alpha L_\beta L_\gamma] \quad (58)$$

Consistency with quantum mechanics requires that

$$(S_+ - S_-) = \frac{\lambda}{5}\epsilon^{\mu\nu\alpha\beta\gamma}\int_{S^5} d^5x \, TR[L_\mu L_\nu L_\alpha L_\beta L_\gamma] = 2\pi n \quad (59)$$

which shows that for normalized manifolds

* This result reflects the topological fact that $\Pi_5(SU(2)) \sim \Pi_5(S^3) = 0$.

$$\lambda = \frac{-in}{48\pi^2}, \quad n \in Z \tag{60}$$

The Wess-Zumino term is topologically quantized.

The phenomenological implications of (60) on the low energy observables can be investigated by studying anomalous decays such as $\pi^0 \to 2\gamma$, $\gamma \to 3\pi$, $k^+k^- \to 3\pi$, ... In the presence of photons, the Wess-Zumino term yields among others, a π^0-2γ vertex of the form[13]

$$V_{\pi^0 \to 2\gamma} = -\frac{n}{96\pi^2 f_\pi} \epsilon^{\mu\nu\alpha\beta} \pi^0 F_{\mu\nu} F_{\alpha\beta} \tag{61}$$

which coincides precisely with the π^0-2γ vertex obtained from the triangle anomaly in the quark language as shown in Fig. 4b, if we were to identify n with the number of colors N_c.

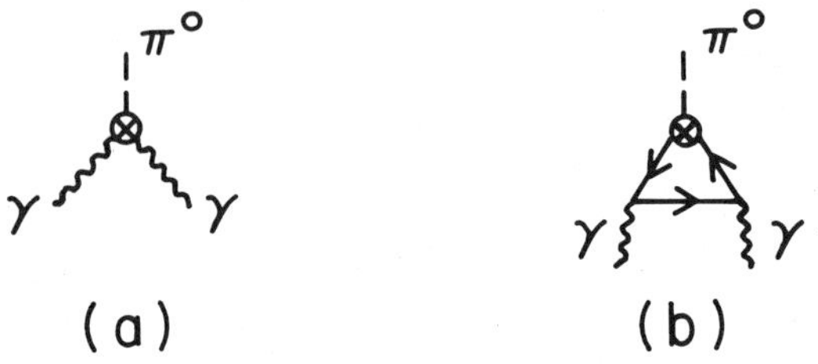

Fig. 4. (a) shows the π^0- 2γ decay in effective Lagrangian language, as compared to the one in the quark language in (b).

To summarize, the pertinent effective chiral Lagrangian at low energy is uniquely given by

$$\mathcal{L} = -\frac{f_\pi^2}{4} \mathrm{TR}[L_\mu L^\mu] - \frac{iN_c}{240\pi^2} \epsilon^{\mu\nu\alpha\beta\gamma} \mathrm{TR}[L_\mu L_\nu L_\alpha L_\beta L_\gamma] + \cdots \tag{62}$$

where ··· involve higher order field gradients of the Skyrme type. For convenience, we will keep calling skyrmions the solitons of (62), although they might not be stable against scale transformations.

4.3. Anomalous Baryon Current

The baryon current described by (62) can be obtained as the Noether current associated to global $U(1)_V$ invariance. At the quark level, the baryon number operator is given by

$$Q = \frac{1}{N_c}\,\mathrm{diag}\,(1,1,1) \tag{63}$$

Straightforward trial and error gauging of (62) using Q as the $U(1)_V$ charge, shows that the baryon current is

$$B^\mu = -\frac{\epsilon^{\mu\nu\alpha\beta}}{24\pi^2}\,\mathrm{TR}\,[L_\nu L_\alpha L_\beta]$$

in agreement with the topological current* (25) proposed by Skyrme[6]. Although we have not derived (62) from QCD, we have gathered consistency arguments based on current algebra and anomalies, that makes the present result plausible.

4.4. Spin-Statistics of Skyrmions

The question of how spin one-half fermions arise from field theories involving exclusively integer spin mesons is rather intriguing. This question was originally addressed by Finkelstein and Rubinstein[19], and also Williams[20], and recently reanalyzed by Witten[13] in the light of the Wess-Zumino term. To illustrate Witten's arguments, consider a skyrmion as described by (62), in 3+1-dimensional space-time with topology $S^3 \times S^1$. To leading order in $\hbar$, the vacuum-to-vacuum expectation value of the skyrmion at rest is

---

* This result was first established by Balanchandran et al.[12] using methods of Goldstone and Wilczek[18] based on vacuum polarization.

$$\langle S(T)|S(0)\rangle = e^{-iM_sT/\hbar}\,(1+O(\hbar)) \tag{64}$$

where M_S is the Skyrmion mass. Now rotate the skyrmion over 2π while moving adiabatically in time (S'). Quantum mechanics tells us that to leading order in $\hbar$, the skyrmion wavefunction acquires a phase factor $e^{-i2\pi J/\hbar}$, where J is the spin of the skyrmion,

$$\frac{\langle S(T)|S(0)\rangle_{2\pi}}{\langle S(T)|S(0)\rangle} = e^{-i2\pi J/\hbar}\,(1+O(\hbar)) \tag{65}$$

which determines J up to an integer. To evaluate J, consider the SU(2) hedgehog configuration (37) embedded in SU(3), i.e.*

$$V(\vec{x}) = \begin{pmatrix} V_H(\vec{x}) & 0 \\ 0 & 1 \end{pmatrix} \tag{66}$$

A 2π-rotation along S' can be described by a rotation in isospace through

$$V(\vec{x},t) = e^{i\lambda_3 t/2}\,V(\vec{x})\,e^{-i\lambda_3 t/2} \tag{67}$$

---

* As pointed out by Balachandran et al.[21] and Jaffe et al.[22], the embedding procedure raises some ambiguities. They can be resolved by requiring the quantum states to have good triality.

In the adiabatic limit, only the Wess-Zumino term feels the rotation, since it is linear in the time-derivatives. Thus, to leading order in the velocity, the action associated to (67) reads

$$S_{cl} = - M_S T - \frac{iN_c}{240\pi^2} \int_{S^3 \times S^1 \times [0,1]} d^5x \, \epsilon^{\mu\nu\alpha\beta\gamma} \, \mathrm{TR} \, [L_\mu L_\nu L_\alpha L_\beta L_\gamma] \tag{68}$$

Simple topological analysis based on the fact that the Wess-Zumino term is none but the topological density associated to $\Pi_5(S^5)$, yields

$$S_{cl} = - M_S T + N_c B \pi \tag{69}$$

where B is the winding number. Injecting (69) into (65) gives

$$J = B \frac{N_c}{2} \qquad (\mathrm{mod}\, 1) \tag{70}$$

For B=1, the skyrmion is a fermion if N_c is odd, and a boson otherwise.

5. PHENOMENOLOGICAL ASPECTS OF SKYRMIONS

5.1. Nucleon and Δ-isobar masses

In order to investigate the properties of the light baryons such as the nucleon and Δ-isobar, it is necessary to extract explicit baryon states from the hedgehog configuration (37). The latter is an admixture of states with equal spin and isospin. In their recent work, Adkins, Nappi and Witten[16] have proposed a systematic scheme that discriminates between states of different spin by quantizing the spinning modes of the skyrmion. Their method is very similar to well-known projective techniques in deformed Hartree-Fock calculations. In short, the idea consists in spinning adiabatically the hedgehog configuration in isospace using the pertinent collective variables, thus generating a classical angular momentum, and then quantizing it using standard canonical rules.

Recall from (37), that the set of isorotated hedgehog configurations is degenerate. At the quantum level, this would translate to a wavefunctional for energy eigenstates that involves a superposition of states with all possible values of A. This superposition can be understood as centrifugal effects in the classical theory if we were to consider a time-dependent isorotation A(t). So, let A(t) be the pertinent collective coordinate, and substitute $U = A(t)U(x)A^{+}(t)$ into (28). In the adiabatic limit, one obtains*

$$\mathcal{L} = -M_s + I\,\mathrm{Tr}[\dot{A}\dot{A}^{+}] \tag{71}$$

where the moment of inertia I is given by

$$I = \frac{8\pi}{3} f_\pi^2 \int_0^\infty r^2 dr\, \sin^2 F \left[1 + \frac{8\epsilon^2}{f_\pi^2}\left(F'^2 + \frac{\sin^2 F}{r^2}\right)\right] \tag{72}$$

Equation (71) describes a spherical top, that can be quantized by standard canonical rules. Since $A \in SU(2)$, it can be locally parametrized by $A = a_0 + i\vec{\tau}\cdot\vec{a}$ with $a_0^2 + \vec{a}^2 = 1$. Thus A in (71) is $A[a(t)]$, while the momentum conjugate to a_k is

$$\Pi_k = \frac{\partial L}{\partial \dot{a}_k} = 4I\dot{a}_k \tag{73}$$

The Hamiltonian associated to (71) reads

$$\begin{aligned} H &= \sum_{k=0}^{3} \Pi_k \dot{a}_k - L \\ &= M_s + \frac{1}{8I}\sum_{k=0}^{3} \Pi_k^2 \end{aligned} \tag{74}$$

Using standard canonical quantization procedure, i.e. $\Pi_k = -i\dfrac{\partial}{\partial a_k}$, one obtains

$$H = M_s + \frac{1}{8I}\sum_{k=0}^{3}\left(-i\frac{\partial}{\partial a_k}\right)^2 \tag{75}$$

Equation (75) involves the Laplacian on S^3 whose eigenstates are the generalized spherical harmonics or Jacobi polynomials, i.e.

* Remember that the Wess-Zumino term vanishes in the SU(2) case.

traceless symmetric polynomials in the a's . The spin ($\vec{J}$) and isospin ($\vec{I}$) operators in the space of collective coordinates can be obtained using the standard Noether construction. From appendix F, we have

$$J_k = \frac{i}{2}\left(a_k \frac{\partial}{\partial a_0} - a_0 \frac{\partial}{\partial a_k} - \epsilon_{klm} a_l \frac{\partial}{\partial a_m}\right) \tag{76a}$$

$$I_k = \frac{i}{2}\left(a_0 \frac{\partial}{\partial a_k} - a_k \frac{\partial}{\partial a_0} - \epsilon_{klm} a_l \frac{\partial}{\partial a_m}\right) \tag{76b}$$

with

$$\hat{J}^2 = \frac{1}{4} \sum_{k=0}^{3} \left(-\frac{\partial^2}{\partial a_k^2}\right) \tag{77}$$

Notice that J_k and I_k are related through parity, i.e. $\Pi_{op} J_k \Pi_{op}^{-1} = I_k$. The explicit energy eigenstates of good spin and isospin are polynomials in the a's of the form $(a_0+ia)^\ell$. The latter carry $I = J = \ell/2$. In general, even polynomials carry integer spin, while odd polynomials carry half-integer spin. To quantize the SU(2) skyrmions as fermions will require odd polynomials. Nucleons with $I = J = 1/2$ correspond to wavefunctions linear in the a's, i.e.

$$|p\uparrow\rangle = \frac{1}{\pi}(a_1+ia_2) \qquad |p\downarrow\rangle = -\frac{i}{\pi}(a_0-ia_3) \tag{78a}$$

$$|n\uparrow\rangle = \frac{i}{\pi}(a_0+ia_3) \qquad |n\downarrow\rangle = -\frac{1}{\pi}(a_1-ia_2) \tag{78b}$$

which are normalized on S^3*. The model generates a tower of states with $I = J = 1/2, 3/2, 5/2...$. Jackson[23] has argued by using a non-relativistic quark model, that for N_c colors in a hedgehog configuration $I_{max} = J_{max} = N_c/2$, suggesting that for $N_c = 3$, only $I = J = 1/2, 3/2$ are relevant, the rest being spurious. In the basis (78), H is diagonal

---

* The normalization on S^3 follows from (95).

$$\langle I=J\, m|H|I=J\, m\rangle = M_S + \frac{J(J+1)}{2I} \tag{79}$$

The nucleon and Δ-isobar masses are given by

$$\begin{aligned} M_N &= M_S + \frac{3}{8I} \\ M_\Delta &= M_S + \frac{15}{8I} \end{aligned} \tag{80a}$$

which shows that the NΔ mass-splitting is rotational in this model, i.e.

$$M_\Delta - M_N = \frac{3}{2I} \tag{80b}$$

with the parameters of Ref. 16 this splitting is set to its experimental value, while with the parameters of Ref. 15 $M_\Delta - M_N$ = 280 Mev. Notice that while M=0 (N_c), $M_\Delta - M_N$ = 0 ($1/N_c$), since $I = 0(N_c)$. The latter follows from the facts that $f_\pi = 0(\sqrt{N_c})$ and $\frac{f_\pi}{\varepsilon} = 0(1)$ through (72). That the splitting goes like $1/N_c$ is consistent with the large N_c limit. The relation of the rotational term (80b) to conventional quark models has been clarified by Brown et al.[24] and Witten[25]. Ground state baryon wavefunctions are composed of N_c quarks in the lowest partial wave. One-gluon exchange between any pair of quarks result in a spin-dependent Fermi-Breit interaction of the form

$$\begin{aligned} V_{1G} &= \frac{v}{8} \sum_{i\neq j} \vec{\sigma}_i \vec{\sigma}_j \\ &= \frac{v}{2} \vec{J}^2 - \frac{3v}{8} N_c \end{aligned} \tag{81}$$

where $\vec{J} = \sum_i \vec{\sigma}_i/2$ is the total angular momentum, and v an overall constant. Comparison between (79) and (81) shows that the moment of inertia $I = (\ v)^{-1}$.

5.2. Axial coupling g_A

The axial coupling g_A measures the spin distribution in the nucleon, and is defined as the expectation value of the axial current A^a_μ in a nucleon state at zero momentum transfer $(q=p_2-p_1)$

$$\langle N(p_2)|A^a_\mu|N(p_1)\rangle = \bar{U}(p_2)\frac{\tau^a}{2}[g_A(q^2)\gamma_\mu\gamma_5 + h_A(q^2)q_\mu\gamma_5]U(p_1) \qquad (82)$$

In the chiral limit $h_A(q^2)$ has a pion pole whose residue is $-2f_\pi g_{\pi NN}$, i.e.

$$h_A(q^2) = \frac{d_A(q^2)}{q^2} \quad \text{with} \quad d_A(0) = -2f_\pi g_{\pi NN} \qquad (83)$$

where $g_{\pi NN}$ is the pion-nucleon coupling. Current conservation implies that

$$2M_N g_A(q^2) + q^2 h_A(q^2) = 0 \qquad (84)$$

Using the non-relativistic limit $(q\to 0)$ in the nucleon rest frame yields

$$\begin{aligned}\lim_{q\to 0}\langle N(p_2)|A^a_i|N(p_1)\rangle &= \lim_{q\to 0}\bar{U}\frac{\tau^a}{2}[g_A(0)\sigma_i + \frac{d_A(0)}{2M_N}\vec{\sigma}\cdot\hat{q}\,\hat{q}_i]U \\ &= \lim_{q\to 0} g_A(0)\,(\delta_{ij} - \hat{q}_i\hat{q}_j)\langle N|\frac{\sigma_j}{2}\tau^a|N\rangle\end{aligned} \qquad (85)$$

where we have used the Goldberger-Treiman relation (9). As pointed out by Adkins et al.[16] the ambiguity in (85) can be circumvented by taking the limit symmetrically, i.e. $\hat{q}_i\hat{q}_j \to \frac{\delta_{ij}}{3}$, so that

$$\lim_{q\to 0}\langle N(p_2)|A^a_i|N(p_1)\rangle = \frac{1}{3}g_A(0)\langle N|\sigma_i\tau^a|N\rangle \qquad (86)$$

For the skyrmion, the axial current A_i can be obtained as the pertinent $SU(2)_A$ Noether current. From appendix E, we have

$$\int d\vec{x}\, A_i^a(\vec{x}) = +\frac{g}{2}\, \mathrm{TR}\,[\tau_i A^+ \tau_a A] \tag{87a}$$

$$g = -\frac{2\pi}{3} f_\pi^2 \int_0^\infty r^2 dr \left[\left(F' + \frac{\sin 2F}{r}\right) + \frac{8\epsilon^2}{f_\pi^2}\left(F'^2 \frac{\sin 2F}{r} + 2F'\frac{\sin^2 F}{r^2} + \frac{\sin^2 F}{r^3}\sin 2F\right)\right] \tag{87b}$$

Between the nucleon state (78a), we obtain

$$\lim_{\vec{q}\to 0} \int d\vec{x}\, e^{i\vec{q}\cdot\vec{x}} \langle N | A_i^a(\vec{x}) | N \rangle = \frac{1}{3} g \langle N | \sigma_i \tau^a | N \rangle \tag{88}$$

Identifying (86) with (88) yields $g_A(0) = g$ as given by (87b). The parameters of Ref. 16 yield* $g_A(0) = .61$, a bit too low in comparison with the experimental value of 1.33. This is expected from the asymptotic behavior of the pion field as discussed above.

5.3. Charge radii and magnetic moments

The distribution of matter in the nucleon can be characterized by the isoscalar charge radius. In terms of the normalized baryon density (40), we have

$$r_0 = \langle r^2 \rangle_{I=0}^{1/2} = \left(-\frac{2}{\pi}\int_0^\infty r^2 dr\, \sin^2 F\, F'\right)^{1/2} \tag{89}$$

Adkins et al.[15] predict $r_0 = .59$ fm, while Jackson et al.[15] have $r_0 = .48$ fm, which are to be compared with the experimental value of .72 fm. It is well known that the isovector electric and magnetic charge radii are infinite for massless pions.[27] The power fall off asymptotically of the pion field in the chiral limit, causes the isovector charge to spread out at infinity, leading to divergent quantities. In the broken phase, the pion field falls off exponentially, giving finite electric and

* See p. 18 for possible N_c corrections, and the recent work by Karl and Paton.[26]

magnetic isovector charge radii as discussed by Adkins and Nappi,[28] and Meissner.[29]

The isoscalar and isovector magnetic moments in the nucleon rest frame are defined respectively as ($e=\hbar=c=1$)

$$\vec{\mu}_{I=0} = \frac{1}{2}\int d\vec{r}\; \vec{r}\times\vec{B} \tag{90}$$

$$\vec{\mu}_{I=1} = \frac{1}{2}\int d\vec{r}\; \vec{r}\times\vec{V}^3 \tag{91}$$

where $\vec{B}$ is the baryon current, and $\vec{V}^3$ the 3rd-component of the isovector current (for more details see appendix E). For an adiabatically rotating skyrmion $\vec{B} \neq 0$,

$$B^i = i\,\frac{\epsilon^{ijk}}{2\pi^2}\,\frac{\sin^2 F}{r}\,F'\,\hat{r}^j\,\mathrm{TR}[\tau^k \dot{A}^+ A] \tag{92}$$

Injecting (92) into (90) and using the spin-up proton state in (78a), yield the isoscalar magnetic moment in the form

$$\begin{aligned}(\mu_{I=0})_3 &= -\frac{2i}{3\pi}\int_0^\infty r^2 dr\; r^2 \sin^2 F\, F' \;\langle p\uparrow|\mathrm{TR}(\tau_3 \dot{A}^+ A)|p\uparrow\rangle \\ &= \frac{i}{3}\langle r^2\rangle_{I=0}\;\langle p\uparrow|\mathrm{TR}(\tau_3 \dot{A}^+ A)|p\uparrow\rangle\end{aligned} \tag{93}$$

The matrix element in (93) can be written as follows

$$\langle p\uparrow|\mathrm{TR}\,(\tau_3 \dot{A}^+ A)\,|p\uparrow\rangle = -\frac{i}{2I}\,\epsilon_{3k\ell}\,\langle p\uparrow|\,a_k \frac{\partial}{\partial a_\ell}\,|p\uparrow\rangle = -\frac{i}{2I} \tag{94}$$

where we have used the canonical prescription, and the fact that

$$\int_{S^3} d\mu(a)\; a_i a_k = \frac{\delta_{ik}}{4}\int_{S^3} d\mu(a) = \frac{\pi^2}{2}\,\delta_{ik} \tag{95}$$

In terms of (94), the isoscalar magnetic moment in a proton state reduces to

$$\mu_p^{I=0} = \frac{\langle r^2\rangle_{I=0}}{6I} = \frac{1}{9}\,(M_\Delta - M_N)\,\langle r^2\rangle_{I=0} \tag{96}$$

The proton and neutron isoscalar magnetic moments being equal, yields

$$\mu_p^{I=0} + \mu_n^{I=0} = \left(\frac{4}{9}(M_\Delta - M_N) M_N \langle r^2 \rangle_{I=0}\right) \mu_B \tag{97}$$

where $\mu_B = \frac{1}{2\mu_N}$ is the Bohr magneton ($e=\hbar=c=1$). In units of Bohr magnetons both Adkins et al.[16] and Jackson et al.[15] predict a value of 1.11, which is to be compared with the experimental value of 1.76.

The isovector magnetic moment can be obtained through similar arguments. Indeed, by using (91) and (E.32) we obtain

$$(\mu_{I=1})_3 = -\frac{\pi}{3} f_\pi^2 \int_0^\infty dr\, r^2 \sin^2 F \left(1 + \frac{8\epsilon^2}{f_\pi^2}\left(F'^2 + \frac{\sin^2 F}{r^2}\right)\right) * \\ * \langle p\uparrow | \mathrm{TR}(\tau_3 A^+ \tau_3 A) | p\uparrow \rangle \tag{98}$$

The matrix element in (98) reduces to

$$\langle p\uparrow | \mathrm{TR}(\tau_3 A^+ \tau_3 A) | p\uparrow \rangle = 2 \langle p\uparrow | (1 - 2(a_1^2 + a_2^2)) | p\uparrow \rangle \\ = 2 - \frac{4}{\pi^2} \int_{S^3} d\mu(a)\, (a_1^2 + a_2^2)^2 \tag{99}$$

Using a polar parametrization of S^3, we can rewrite the remaining integral in (99) in the form

$$\int_{S^3} d\mu(a)\,(a_1^2+a_2^2)^2 = \int_0^\pi d\alpha \sin^2\alpha \int_0^\pi d\beta \sin\beta \int_0^{2\pi} d\gamma\, \sin^4\alpha \sin^4\beta = \frac{2\pi^2}{3} \tag{100}$$

hence

$$\langle p\uparrow | \mathrm{TR}(\tau_3 A^+ \tau_3 A) | p\uparrow \rangle = -\frac{2}{3} \tag{101}$$

In terms of (72) and (101), the isovector magnetic moment (98) in a proton state is

$$\mu_p^{I=1} = \frac{I}{3} = \frac{1}{2}(M_\Delta - M_N)^{-1} \tag{102}$$

Since $\mu_n^{I=1}$ has opposite sign, we deduce

$$\mu_p^{I=1} - \mu_n^{I=1} = \left(\frac{2M_N}{M_\Delta - M_N} \right) \mu_B \tag{103}$$

In units of Bohr magnetons Adkins et al.[16] predict 6.41, while Jackson et al.[15] have 10.61. These results are to be compared with the empirical value of 9.4. The magnetic moments for the proton and neutron measured in Bohr magnetons are

$$\mu_{p,n} = \frac{1}{2\mu_B} \left(\mu_{p,n}^{I=0} + \mu_{p,n}^{I=1} \right) \tag{104}$$

Adkins et al.[16] predict $\mu_p = 1.88$, $\mu_n = -1.32$, $\left| \frac{\mu_p}{\mu_n} \right| = 1.42$

Jackson et al.[15] have $\mu_p = 2.93$, $\mu_n = -2.38$, $\left| \frac{\mu_p}{\mu_n} \right| = 1.23$

These results should be compared with the experimental values $\mu_p = 2.79$, $\mu_n = -1.91$ and $|\mu_p/\mu_n| = 1.46$. In the broken phase ($m_\pi = 139$ Mev), the predicted results are almost unaltered as shown in table 2.

	A.N.W.	A.N.	J.R.	M.	Exp.
M_N[Mev]	939*	939*	1425	1385	939
M_N-M_Δ[Mev]	293*	293*	283	310	293
f_π[Mev]	64.5	54.0	93.0*	94.3*	93
g_A^H	.61	.65	.81*	.74*	1.24
$10^3\varepsilon^2$	4.21	5.34	5.52	5.13	--
$\langle r^2\rangle^{1/2}$[fm]	.59	.68	.47	.43	.72
μ_p[n.m]	1.87	1.97	2.74	2.43	2.79
μ_n[n,m]	-1.31	-1.24	-2.24	-1.98	-1.91
$\lvert\mu_p/\mu_n\rvert$	1.43	1.59	1.22	1.23	1.46

Table 2. Static baryon properties in the Skyrme model. (A.N.W.) and (J.R.) are the predictions of Adkins, Nappi and Witten[16], and Jackson and Rho[15] respectively for m_π=0. (A.N.) and (M.) are the predictions of Adkins and Nappi,[28] and Meissner[29] respectively, for m_π=139 Mev. Quantities with (*) are input parameters. Remember that (J.R.) and (M.) claim g_A^N= 5/3 g_A^H to be the axial coupling.

5.4. The Roper resonance in the Skyrme model

Within 30% accuracy, the Skyrme model seems to give a reasonable description of the static properties of the nucleon and Δ-isobar. It is therefore tempting to use the present framework to investigate dynamical properties of skyrmions, and their possible relevance to the baryon spectrum. These questions have been studied by a number of authors[30,31,32,33,34] using either the semi-classical approximation or variational principles.

In the large N_c-limit, the semi-classical treatment is the relevant method for investigating chiral fluctuations around the skyrmion. Indeed, since $f_\pi \sim \sqrt{N_c}$ and $\varepsilon \sim \sqrt{N_c}$, one can factorize out N_c from the Skyrme Lagrangian (28), so that in the large N_C-limit the vacuum-to-vacuum expectation value is dominated by the saddle point equations. Pion scattering off the skyrmion can be described by standard quadratic fluctuations about the pertinent classical configuration. Since the pion coupling is derivative in nature, higher-order fluctuations (cubic, quartic...) are suppressed by powers of $1/N_C$. At this stage, the nonrenormalizability of the model should not be an issue since we are interested only in energy differences. Moreover, we will ignore the fact that the skyrmion is spinning and neglect possible recoil corrections.

To investigate $K^{\pi} = 0^+$ quantum fluctuations about the Skyrmion, consider the time-dependent hedgehog ansatz

$$U(\vec{r},t) = \exp\left(i\vec{\tau}\cdot\hat{r}\left(F(r) + \xi(r,t)\right)\right) \tag{105}$$

where F(r) refers to the usual skyrmion profile. The Euler-Lagrange equation associated to (105), reduces to a time-dependent equation for the spherical chiral angle $\xi(r,t)$,

$$\left(1 + \frac{16\epsilon^2}{f_\pi^2}\frac{\sin^2 F}{r^2}\right)\ddot{\xi} - \left(1 + \frac{16\epsilon^2}{f_\pi^2}\frac{\sin^2 F}{r^2}\right)\xi'' - \left(1 + \frac{8\epsilon^2}{f_\pi^2}\frac{\sin 2F}{r}F'\right)\frac{2\xi'}{r}$$
$$+\left(2\frac{\cos 2F}{r^2} - \frac{16\epsilon^2}{f_\pi^2}\left(\frac{F''\sin 2F + F'^2\cos 2F}{r^2} - \frac{\sin^2 2F - \sin^2 F}{r^4}\right)\right)\xi = 0 \tag{106}$$

with $\xi(0;t) = \xi(\infty;t) = 0$ for all times, so that the entire baryon number is carried out by the skyrmion. The asymptotic behavior of (106) can easily be worked out using the harmonic ansatz $\xi(r,t) = e^{i\omega t}\xi(r)$. At large distances $F(r) \to \frac{k^2}{r^2}$, and one recovers free pions,

$$\xi(r) \to \alpha(\omega)\, j_1(\omega r) + \beta(\omega)\, n_1(\omega r) \tag{107}$$

where $j_1(x)$ and $n_1(x)$ are the usual spherical Bessel and Neumann functions respectively. The constants α and β determine the phase shift $\delta(\omega)$ of the scattered pions. Indeed, by rewriting (107) in terms of Hankel functions, we have

$$\xi(r) \to \text{Const.} * \left(h_1^-(\omega r) - S_0(\omega) h_1^+(\omega r) \right) \tag{108}$$

where the S-matrix in this channel is given by

$$S_0(\omega) = e^{2i\delta(\omega)} = \frac{\alpha(\omega) - i\beta(\omega)}{\alpha(\omega) + i\beta(\omega)} \tag{109}$$

and lies on the unit circle. At this stage two remarks are in order. First (109) shows a purely elastic scattering in the 0^+-channel. This feature will not hold when the skyrmion is rotated to get states of good spin and isospin.[33] The adiabatic rotation couples K=0 and K=1 channels, leading to inelastic scattering in the P_{11}-channel, as expected. Second, the phase shift $\delta(\omega)$ in (109) sums up both the resonant $\delta_R(\omega)$ and background $\delta_B(\omega)$

$$\delta(\omega) = \delta_R(\omega) + \delta_B(\omega) \tag{110}$$

To extract resonance properties in the 0^+ channel, one has to either solve the scattering equation (106) in the complex energy plane and look for the poles of (109), or equivalently fit a Breit-Wigner profile in the region where the total phase-shift (110) varies rapidly. Although for elastic scattering $\delta_R(\omega)$ has to go from 0 to π, the background phase-shift $\delta_B(w)$ might be strong enough to suppress it. Fig. 5 displays the total phase-shift (110) up to 2 Gev excitation energies, for the parameters of Jackson et al. There is a sharp increase of $\delta(\omega)$ about $\omega \sim$ 300 Mev. Using a Breit-Wigner fit, one obtains a resonance energy $E_R \sim 260$ Mev and a width $\Gamma_R \sim 180$ Mev. Although a bit too low, this resonance has been identified with the $P_{11}(1470)$ Roper resonance observed in the pion-nucleon channel. There are no bound states in this channel, implying that the hedgehog configuration is locally stable against 0^+-deformations.

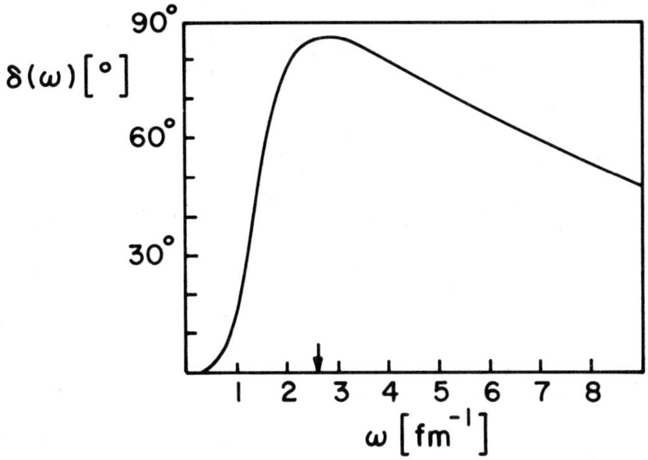

Fig. 5. Pion-skyrmion phase-shift in the $K^{\pi} = 0^{+}$ channel from ref. 30.

A more thorough analysis of the baryon spectrum in the Skyrme model, has been carried out at Siegen University[33] and SLAC[34] and the interested reader is referred to the original work. While the model seems to accommodate rather well the higher baryon resonances, its failure to describe properly the Roper resonance and to account for the low-lying odd parity states is rather troublesome.

5.5 Soft pion theorems and the Skyrme model

In the general context of current algebra, Skyrme's program will remain unsatisfactory in the absence of a consistent quantum description that would account for soft pion-skyrmion dynamics in agreement with soft pion threshold theorems. To achieve this, consider symmetric chiral fluctuations about the hedgehog skyrmion (37),

$$U(\vec{x},t) = e^{i\pi/2f_\pi}\, U(\vec{x})\, e^{i\pi/2f_\pi} \tag{111}$$

where $\pi(\vec{x},t)$ is a perturbative pion field that vanishes for $|\vec{x}| \to 0,\infty$. Injecting (111) into (28) and neglecting hard pion processes, i.e. quadratic forms in Π with large momentum transfer, yield

$$\mathcal{L}_{soft} = \mathcal{L} + \frac{1}{2}(\partial_\mu \vec{\pi})^2 + \frac{1}{2f_\pi}(\partial_\mu \vec{\pi})\cdot\vec{A}^\mu + \frac{1}{4f_\pi^2}(\partial_\mu \vec{\pi}\times\vec{\pi})\cdot\vec{V}^\mu \tag{112}$$

where $\vec{\mathbf{V}}_\mu$ and $\vec{\mathbf{A}}_\mu$ are the $SU(2)_V \times SU(2)_A$ Noether currents derived in appendix F,

$$\vec{V}_\mu = \mathrm{TR}\left[i\vec{\tau}\cdot\left(-\frac{f_\pi^2}{2}L_\mu + \epsilon^2[L_\nu,[L_\mu,L_\nu]] + (L\to R)\right)\right] \tag{113a}$$

$$\vec{A}_\mu = \mathrm{TR}\left[i\vec{\tau}\cdot\left(-\frac{f_\pi^2}{2}L_\mu + \epsilon^2[L_\nu,[L_\mu,L_\nu]] - (L\to R)\right)\right] \tag{113b}$$

As noted by Schnitzer[35], (112) has the appropriate form for a soft pion-skyrmion Lagrangian. Indeed, in the chiral limit (m_π = 0) Weinberg's soft pion-nucleon Lagrangian reads[36]

$$\mathcal{L}_W = \mathcal{L}_N + \frac{1}{2}(\partial_\mu \vec{\pi})^2 + \left(\frac{g_{\pi NN}}{2m_N}\right)(\partial_\mu \vec{\pi})\cdot(\bar{N}\gamma^\mu\gamma_5\vec{\tau}N)$$
$$+ \left(g_A \frac{g_{\pi NN}}{2m_N}\right)^2(\partial_\mu \vec{\pi}\times\vec{\pi})\cdot(\bar{N}\gamma^\mu\vec{\tau}N) \tag{114}$$

where N is the nucleon wavefunction, and L_N the free nucleon Lagrangian. Comparison between (112) and (114) shows that

$$g_A = g_{\pi NN}\frac{f_\pi}{m_N}$$

in agreement with the Goldberger-Treiman relation (9). This shows that the additional Skyrme term does not spoil the soft pion threshold theorems already fulfilled by the non-linear σ-model.

5.6. <u>Soft-pion corrections to the masses.</u>

To estimate the soft pion contribution to the skyrmion, we need to consider the Skyrme Lagrangian beyond tree level .

Unfortunately, the model as it stands is not renormalizable. What this means is that beyond tree graphs, the model requires counter terms of increasing complexity to cure the infinities arising from higher loop calculations. This should not be surprising since the Skyrme Lagrangian (28) is after all a truncated form in a chiral expansion that involves infinitly many higher field gradients. Some years ago, Weinberg[9] has shown that the leading term in (28), i.e. the conventional non-linear σ-model, describes uniquely $\pi\pi$-scattering in the trivial sector to lowest order in the pion momentum $O(p^2)$, and argued that consistency with unitarity requires terms of order $O(p^4)$. Since loop contributions are suppressed by powers of p^2 at low-energy (see Appendix B for details), one needs to consider the 1-loop effect about the quadratic term in (28) along with tree level contributions from the Skyrme-like terms to order $O(p^4)$. In this spirit, it is possible to make the model meaningful order by order in the pion momentum.

To one-loop, the soft pion contributions to Skyrme's effective action are illustrated in Fig. 6. In Euclidean space, they yield[37]

$$S_1^E = S_0^E - \frac{1}{2}\mathrm{Tr}\ln(\partial^2) - \frac{1}{2}\sum_{k=1}^{\infty}\frac{(-1)^{k+1}}{k}\mathrm{Tr}\left[(\partial^2)^{-1}\hat{L}\cdot\partial\right]^k \tag{115}$$

where S_0^E is Skyrme's action, and $\hat{L}_\mu$ is defined through

$$\hat{L}_\mu^{ab} = \mathrm{Tr}\left[\tau^a L_\mu \tau^b\right] \tag{116}$$

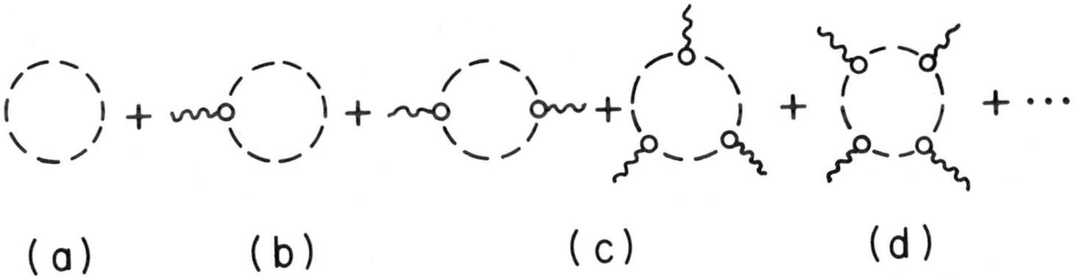

Fig. 6. One-loop soft pion contribution to Skyrme's effective action.

Diagram (a) of Fig. 6 is the free space divergence and can be subtracted away since it does not affect the dynamics. Consistency with Skyrme's action (28) at tree-level shows that only k=1,2,3,4 in (115) contribute. Higher-order pion insertions give rise to higher field gradients that are suppressed at low energy. Using a Pauli-Villars regulator with two cutoff scales $m \sim m_\pi$ and $m_2 \sim m_\eta$ in the infrared and ultraviolet regime respectively,[37] gives

$$S_1 = \int d^4x \left[-\frac{f_\pi^2}{4} \mathrm{TR}\,[L_\mu^2] + \left(\frac{\epsilon^2}{4} + \frac{1}{768\pi^2} \left(1 + \ln\left(\frac{m_\pi^2}{m_2^2}\right)\right) \mathrm{TR}\,[L_\mu, L_\nu]^2 \right. \right.$$
$$\left. \left. + \left(-\frac{1}{256\pi^2} \left(1 + \ln\left(\frac{m_\pi^2}{m_2^2}\right)\right) \mathrm{TR}\,[L_\mu, L_\nu]_+^2 \right] \qquad (117)$$

This summarizes the soft pion corrections to (28). At higher energies, the model must be extended to account for the vector meson resonances such as the ω and ρ.

For hedgehog configurations, the soft pion contribution to the mass term is of the form

$$\Delta M_s = \frac{1}{24\pi} \left(1 + \ln\left(\frac{m_\pi^2}{m_2^2}\right)\right) \int_0^\infty r^2 dr \left(3F'^4 + 4F'^2 \frac{\sin^2 F}{r^2} + 8 \frac{\sin^4 F}{r^4}\right) \qquad (118a)$$

This correction pushes down the skyrmion mass by almost 20% as shown in table 3. The corresponding corrections to the nucleon and Δ-isobar can be obtained by spinning the skyrmion. The decrease in the moment of inertia is given by

$$\Delta I = \frac{1}{9\pi} \left(1 + \ln\left(\frac{m_\pi^2}{m_2^2}\right)\right) \int_0^\infty r^2 dr \, \sin^2 F \left(3F'^2 + 5 \frac{\sin^2 F}{r^2}\right) \qquad (118b)$$

Again, this pushes down the nucleon and Δ-isobar masses as shown in table 3.

ε^2	f_π[Mev]	M_H^π [Mev]	I^π[Mev]	M_N^π[Mev]	M_Δ^π[Mev]	M_Δ^π-M_N^π [Mev]	$g_A^{H,\pi}$
.00421	64.5	605	.82	695	1057	361	.22
.00552	93.0	1101	.90	1183	1511	328	.42

Table 3. Soft pion corrections to the ground state energy M for the hedgehog skyrmion, nucleon and Δ-isobar are quoted along with the moment of inertial I, for the two sets of parameters discussed in the text. See Ref. 37 for more details.

5.7. The skyrmion-skyrmion interaction

Recently, there has been considerable interest in understanding the nucleon-nucleon interaction and dense hadronic matter in the context of the Skyrme model. In fact, these questions were originally addressed and qualitatively answered by Skyrme[6] himself. The complexity of the model at short distances together with the ambiguities to define a potential for strongly interacting skyrmions, urged Skyrme to limit his analysis of the interaction to the long-range part, and to suggest only a lower bound on the binding energy in dense hadronic matter. Most of the recent developments circumvent part of the algebraic complexity using numerical methods. Since many of the present issues are still in the process of being settled, we will be brief in our discussion.

Using the additive character of the winding number, Skyrme[6] has proposed a product ansatz to investigate the asymptotic form of the potential energy of two interacting skyrmions. For two skyrmions centered around $\vec{x}_1$ and $\vec{x}_2$, and far apart from each other, the field configuration can be parametrized as follows

$$U_{ss}(\vec{x};\vec{x}_1,\vec{x}_2) = U_s(\vec{x}-\vec{x}_1)A(\vec{\alpha})U_s(\vec{x}-\vec{x}_2)A^+(\vec{\alpha}) \tag{119}$$

where $A(\vec{\alpha}) = \exp(i\vec{\tau}\cdot\vec{\alpha}/2)$, and $U_s(\vec{x})$ is the undistorted hedgehog configuration (37). For large separations r compared to the skyrmion size, distortions in the skyrmion field can be neglected, since they yield subleading corrections to the energy. As r tends to infinity, U_{SS} approaches the non-interacting configuration, and $E_{SS}(r\to\infty) = 2E_S$ independently of the isospin orientation $\vec{\alpha}$. For a static configuration, the potential energy can be defined as follows

$$V(\vec{x}_1,\vec{x}_2) = \int d\vec{x} \left(\frac{f_\pi^2}{4} \mathrm{TR}\left[L_i(1,2)\, L_i(1,2)\right] + \frac{\epsilon^2}{4} \mathrm{TR}\left[L_i(1,2), L_j(1,2)\right] \right) - E_1 - E_2 \tag{120}$$

Asymptotically, $L_\mu(1,2)$ is

$$L_\mu(1,2) = A\left(L_\mu(2) + U_2^+ A^+ L_\mu(1) A U_2\right) A^+ \tag{121}$$

Simple algebraic manipulations give

$$\begin{aligned} V(\vec{x}_1,\vec{x}_2) = \frac{f_\pi^2}{4} \int d\vec{x}\, \mathrm{TR}\Big(&-A^+L_i(1)AR_i(2) + A^+L_i(1)A\hat{R}_i(2) + A^+\hat{L}_i(1)AR_i(2)\Big) \\ + \frac{\epsilon^2}{2}\int d\vec{x}\, \mathrm{TR}\Big(&\left[R_i(2), A^+L_j(1)A\right]^2 + \left[R_i(2),R_j(2)\right]A\left[L_i(1),L_j(1)\right]A^+ \\ &+ \left[R_i(2), A^+L_j(1)A\right]\left[A^+L_i(1)A, R_j(2)\right]\Big) \end{aligned} \tag{122}$$

where

$$\hat{L}_i = L_i - \frac{2\epsilon^2}{f_\pi^2}\left[L_j,\left[L_i,L_j\right]\right] \tag{123a}$$

$$\hat{R}_i = R_i - \frac{2\epsilon^2}{f_\pi^2}\left[R_j,\left[R_i,R_j\right]\right] \tag{123b}$$

are conserved currents because of (36). Using the asymptotic form (20), and neglecting quartic terms in the field gradients in (122), yield

$$V(\vec{x}_1,\vec{x}_2) \sim -\frac{1}{2} \mathrm{TR}[A^+\tau^a A \tau^b] \int d\vec{x}\, \partial_i^2 \Pi^a(1) \Pi^a(2) \tag{124}$$

Since $\partial_i{}^2 \Pi^a(1)$ vanishes except around the neighborhood of $\vec{x}_1$, we can expand $\Pi^b(2)$ in Taylor's series about $\vec{x}_1$. For the geometry of Fig. 7, we have

$$\Pi^b(2) = f_\pi \,(\widehat{x-r})\, \sin F(|\vec{x}-\vec{r}\,|) \sim f_\pi\, \kappa^2 \partial_r \left(\frac{1}{|\vec{x}-\vec{r}\,|}\right)$$

$$\partial_i^2 \Pi^a(1) = f_\pi\, \partial_i^2 \left(\hat{x}^a \sin F(x)\right) = f_\pi\, \hat{x}^a\, \phi(x) \tag{125}$$

where $\phi(x)$ is a smooth function of x. The asymptotic contribution to (120) is

$$V(r) \sim \frac{\kappa^2 f_\pi^2}{6} \mathrm{TR}[A^+\tau^a A\tau^b]\, \partial^b \partial^a \left(\frac{1}{r}\right) \int_0^\infty dx\, x^3 \phi(x) \tag{126}$$

which is reminiscent of one-pion exchange.

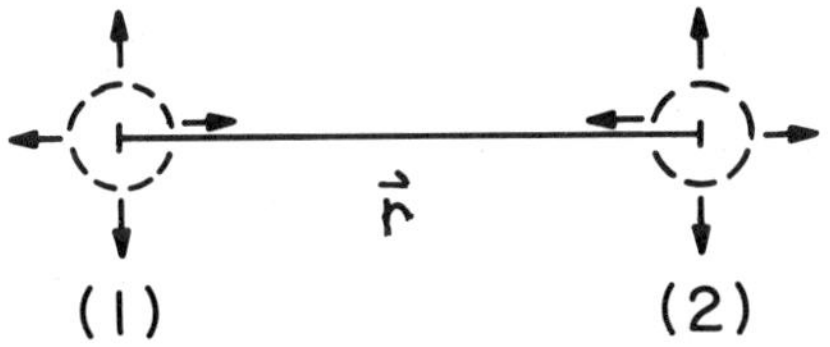

Fig. 7 Two defensive hedghogs at very large distance r.

An extensive analysis of the skyrmion-skyrmion potential has been carried out by Jackson et al.[38] at Stony Brook using the product ansatz (119) for all separations r. Their analysis shows that for a given set of Euler angles $\vec{\alpha} \equiv [\alpha\beta\gamma]$,

$$V(r, [\alpha\beta\gamma]) = V_A(r) + V_B(r) \cos\beta + V_C(r) \cos(\alpha+\gamma) \cos^2\left(\frac{\beta}{2}\right) \tag{127}$$

This simple parametrization is due to the hedgehog character of the solution, which allows for $\vec{\pi}$- and σ-meson exchange, and restricts the ω- and ρ-meson coupling to be vectorial and tensorial respectively. The central part of the skyrmion-skyrmion interaction ($\alpha\equiv0$) coincides with the central potential in the NN-interaction as shown in Fig. 8. Notice the finite range repulsion at zero separation, which is of the order of the skyrmion mass (~ 1 Gev).

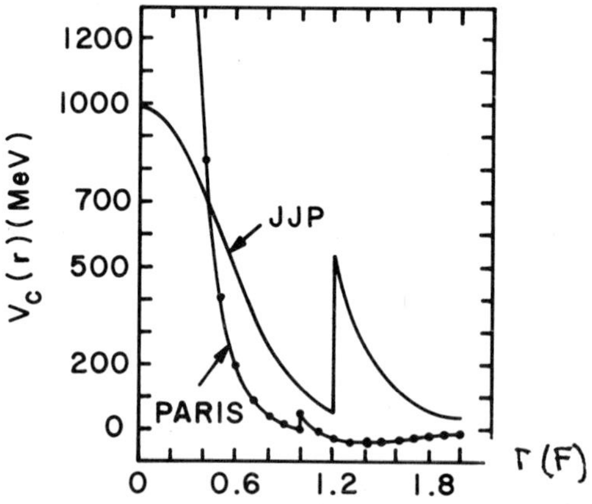

Fig. 8. Comparison of the central potential in the Skyrme model with its corresponding component of the Paris potential from Ref. 38 . The break indicates a scale change by a factor of ten.

The most intriguing feature of the central potential is the absence of the intermediate range attraction, a vital ingredient for nuclear binding. Recently, Jackson[39] and Meissner[40] have shown that by incorporating the soft pion corrections to the skyrmion-skyrmion interaction as discussed in Ref. 37, one recovers the medium range attraction.

Given the simplicity of the calculation, many of the results of Ref. 38 are qualitatively appealing. There is however major concern on the ability of the product ansatz to describe the short distance part of the NN-interaction, where the skyrmions are known to be strongly deformed. The ongoing numerical effort at Urbana[41] to overcome the limitation of the product ansatz is in many respects important, but by no means final. We will comment more on this point next.

5.8. Status of the many-skyrmion problem

Some years ago, Bogomol'ny and Fateev[42] have argued that at short distances the skyrmion-skyrmion interaction is dominated by finite two-body forces, while considering a configuration of N superimposed skyrmions. To see this, consider the variational profile

$$F_N(r) = \begin{cases} N\pi\left(1 - \frac{a_N}{r}\right) & r < a_N \\ 0 & r > a_N \end{cases} \tag{128}$$

where a_N characterizes the size of the multi-skyrmion configuration for which the energy is minimum

$$a_N^2 = 24\frac{\epsilon^2}{f_\pi^2}\left(1 + \frac{1}{N\pi}\int_0^{N\pi} dx \frac{\sin^2 x}{x^2}\right)\left(1 + \frac{3}{\pi^2 N}\right)^{-1} \tag{129}$$

The baryon density associated to (128) has an onion-like structure with a first minimum at $a_{N/N}$. For large N, (129) simplifies into $a_N \sim 2\sqrt{6}\,\epsilon/f_\pi$, independently of N. Using (129) together with the virial theorem yield

$$E_N = 4\pi f_\pi^2 \int_0^{a_N} r^2 dr \left(F'^2 + 2\frac{\sin^2 F}{r^2}\right) \tag{130}$$

For N>>1 this reduces to

$$E_N \sim 8\sqrt{\frac{2}{3}}\,\pi^3 \epsilon f_\pi N^2 \sim 202\,\epsilon f_\pi N^2 \tag{131}$$

which is to be compared with the asymptotic estimate of Ref. 42,

$$E_N \sim \frac{209}{2} N(N + .8726) \tag{132}$$

This large N behavior is reminiscent of two-body interactions obtained by simple pairing of N skyrmions via the combinatoric factor $\frac{1}{2}$ N(N-1).

Recently, Kutshera and Pethick[43] argued that this behavior is specific to configurations in which the characteristic lengths of the constituent skyrmions are different in different directions. Their argument is based on the fact that the energy of a skyrmion depends on the size and shape of the volume in which it is enclosed. For a skyrmion confined to a box of dimensions $\ell_1 \times \ell_2 \times \ell_3$, the energy per unit volume scales like*[43]

$$\mathcal{E} \sim \epsilon^2 \left(\frac{\alpha_1}{\ell_1^2 \ell_2^2} + \frac{\alpha_2}{\ell_2^2 \ell_3^2} + \frac{\alpha_3}{\ell_3^2 \ell_1^2} \right) \tag{133}$$

where the α's are constant. For a fixed density $n = (\ell_1 \ell_2 \ell_3)^{-1}$, (133) reaches a minimum for a cubical configuration, $\ell_1 \sim \ell_2 \sim \ell_3$, thus**

$$\mathcal{E}_{KP} \sim \epsilon^2 n^{4/3} \tag{134}$$

If we were to choose $\ell_1 \sim \ell_2 \sim N\ell_3$, as suggested by the onion-shell behavior discussed above, then[43]

$$\mathcal{E}_{BF} \sim N^{2/3} \mathcal{E}_{KP} \tag{135}$$

which is consistent with (132). Since the energy density (134) is smaller than (135) by a factor of $N^{-2/3}$, Kutshera and Pethick[43] concluded that in dense matter the interaction between skyrmions is limited to nearest neighbours and behaves like r^{-1}***. They claimed that this 'screening' is due to significant many body effects that tend to cancel out the long-range part of the two-body interaction. These results are different from the ones obtained with the product ansatz, where it was found that at short distances the interaction is in fact finite and dominated by two-body forces.

---

* Remember that due to the virial theorem, the quadratic and quartic term contribute equally to the ground state energy. Moreover, the currents L_i's scale like $1/\ell_i$.

** Notice the similarity with the energy of a free quark gas.[44]

*** This follows from the fact that $\mathcal{E}_{KP} \sim n^{1/3} \sim r^{-1}$, where r is a characteristic length.

By now, the alert reader must have noticed that most of the above conclusions are tied to the scaling behavior of the fourth order term. It is clear that by incorporating higher derivative terms in the starting effective Lagrangian, one would end up with a different short distance and high-density behavior, making most of the above results model dependent. Needless to recall that the skyrme model is a low energy approximation to QCD, that can hardly be justified at high densities. So, aside from possibly appealing scenarios, we believe that a reliable statement on central questions such as the short distance behavior of the interaction, the high density limit of hadronic matter and the equation of state, require a quantitative understanding of the non-perturvative character of QCD that goes beyond crude approximations. Of course, this does not concern dilute phases of skyrmions where much can be learned about pion condensation and hadronic clustering at densities that are consistent with the low energy assumptions.

5.9. Exotic skyrmions

So far we have assumed that hadronic matter involves exclusively spherical skyrmions of the hedgehog type. However, it is possible that for reasonable densities, it may become energetically favourable to form new phase configurations involving clusters of skyrmions in the form of tori, cylinders, planes, ... The specific case of periodic baryonic chains involving mappings from non-spherical group manifolds onto S^3 of the form $(S^2 \times S^1; S^3)$ have been discussed in details in Ref. 45.

CONCLUSION

It is rather remarkable that such a simple model embodies much of QCD at low energy. There is no doubt, that further studies of its properties will bring more valuable insights on hadrons in general and nuclear physics in particular, and we hope that the present lectures will provide the necessary stimulus for that.

ACKNOWLEDGEMENTS

We have benefitted from discussion with several of our colleagues in the nuclear theory group at Stony Brook. This work was supported in part by US-DOE under Contract No. DE-AC02-76ER 13001.

APPENDIX A: Baryon number for general textures

A remarkable property of the points where the chiral field $U = \pm 1$, is that the total baryon number can be expressed in terms of the topological singularities of the pion field at these points. Indeed let S_β be a smooth surface surrounding the point, line or surface where $U \pm 1$. (All points on S_β are assumed to be infinitesimally close to the points where $U = \pm 1$). Using current conservation, we can write

$$\frac{dB}{dt} = -\sum_\beta \oint dS_\beta \, \hat{n}_\beta \cdot \vec{B} \tag{A.1}$$

where the sum runs over all the surfaces S_β, including the surface at infinity, and $B^\mu(x)$ is the covariant baryon current defined in (25). Straightforward manipulations give

$$\frac{dB}{dt} = \frac{-i}{8\pi^2} \sum_\beta \oint dS_\beta \, \dot{\theta} \, \epsilon^{ijk} \Big(\sin^4\theta \, \mathrm{TR}\,[\vec{\tau}\cdot\hat{\theta} \; \vec{\tau}\cdot(\hat{\theta}\times\partial_j\hat{\theta}) \; \vec{\tau}\cdot(\hat{\theta}\times\partial_k\hat{\theta})] + \sin^2\theta\cos^2\theta \; \mathrm{TR}\,[\vec{\tau}\cdot\hat{\theta} \; \vec{\tau}\cdot\partial_j\hat{\theta} \; \vec{\tau}\cdot\partial_k\hat{\theta}] \Big) \tag{A.2}$$

where we have used the general parameterization

$$U(\vec{x},t) = \exp\left(i\vec{\tau}\cdot\hat{\theta}(\vec{x},t)\,\theta(\vec{x},t)\right) \tag{A.3}$$

with $\hat{\theta}^2 = 1$. Now let $\hat{e}_k(\beta)$, be an orthogonal, direct, comoving system of coordinates on S_β with $\hat{e}_3 \equiv \hat{n}_\beta$, so that

$$\hat{\theta} = \sum_{k=1}^{3} \hat{\theta}_k \hat{e}_k \tag{A.4}$$

Using the Serret-Frenet relations, and the fact that a fixed surface has a zero torsion, yield

$$\partial_1\hat{\theta} = (\partial_1\hat{\theta}_1 - K_1\hat{\theta}_3)\hat{\ell}_1 + \partial_1\hat{\theta}_2\,\hat{\ell}_2 + (\partial_1\hat{\theta}_3 + K_1\hat{\theta}_1)\hat{\ell}_3$$
$$\partial_2\hat{\theta} = \partial_2\hat{\theta}_1\,\hat{\ell}_1 + (\partial_2\hat{\theta}_2 - K_2\hat{\theta}_3)\hat{\ell}_2 + (\partial_2\hat{\theta}_3 + K_2\hat{\theta}_3)\hat{\ell}_3 \qquad (A.5)$$

where the K_i's are the curvature tensions. In terms of (A.5), Eq. (A.2) becomes

$$\frac{dB}{dt} = -\frac{1}{\pi^2}\sum_\beta \oint dS_\beta\, \dot{\theta}\,\sin^2\theta\;\hat{\theta}\cdot(\partial_1\hat{\theta}\times\partial_2\hat{\theta}) \qquad (A.6)$$

If we assume that the chiral field is turned on adiabatically from its vacuum value, then the amount of baryon number gathered around the surfaces S_β is

$$B = \frac{1}{2\pi^2}\sum_\beta \oint dS_\beta \left[\theta - \frac{\sin 2\theta}{2}\right]_\beta \hat{\theta}\cdot(\partial_1\hat{\theta}\times\partial_2\hat{\theta}) \qquad (A.7)$$

APPENDIX B: Weinberg Scaling Argument

Some years ago Weinberg[9] has argued that loop corrections in effective chiral models were suppressed by powers of p^2 at low-energy, where p is the pion momentum. For that, consider the general form of the S-matrix

$$\langle \pi_{out} | \hat{S} | \pi_{in} \rangle = M \delta^4 (\sum_i p_i) \tag{B.1}$$

In general, M is a function of the external momenta p, the low-energy couplings, and some renormalization scale μ. Since the pion wavefunction scales like -1, we have

$$D_M = 4 - N_e \tag{B.2}$$

where D_M is the naive scale of the transition amplitude M, and N_e the number of external pion wavefunctions. Now, define

$$M \equiv M(p, g, \mu) = p^{D_M - D_g} f\left(\frac{p}{\mu}, g\right) \tag{B.3}$$

where D_M is the dimensions of the various couplings in M. If we denote by N_d the number of vertices formed from effective interactions with d field gradients, then*

$$D_g = \sum_d N_d (4-d) - N_e - 2N_i \tag{B.4}$$

---

* Remember that external wavefunctions scale like -1, and that propagators carry extra factors of f_π^{-2}.

where N_i is the number of internal lines. The number of loops in M is given by

$$N_L = N_i - \sum_d N_d + 1 \tag{B.5}$$

so that

$$D = D_M - D_g = 2 + \sum_d N_d (d-2) + 2N_L \tag{B.6}$$

For a renormalization scale $\mu \sim p$, (B.3) and (B.6) show that the dominant contribution to the S-matrix at low-energy carry the smallest value of D. The leading low-energy contribution is given by d=2 at tree level ($N_L = 0$) and scales like p^2. The next to leading contribution is given by d = 2 at 1-loop ($N_L - 1$) and d= 4 at tree level ($N_L = 0$), and scales like p^4.

APPENDIX C: Asymptotic behavior of the pion field

In the broken phase, Soft-pion-nucleon phenomenology can be described at tree level by a Lagrangian of the form

$$\mathcal{L} = \mathcal{L}_N + \frac{1}{2}(\partial_\mu \vec{\pi})^2 - m_\pi^2 \frac{\vec{\pi}^2}{2} + \left(\frac{g_{\pi NN}}{2m_N}\right)\partial_\mu \vec{\pi}\cdot(\bar{N}\gamma^\mu\gamma_5\vec{\tau}N) \tag{C.1}$$

where L_N summarizes the internal nucleon dynamics. In the static limit, the pion field in a nucleon state is described by an inhomogeneous Klein-Gordon equation of the type

$$(-\vec{\nabla}^2 + m_\pi^2)\,\pi^a(\vec{r}) = \frac{g_{\pi NN}}{2m_N}\,\vec{\nabla}\cdot(\bar{N}\vec{\gamma}\gamma_5\tau^a N) \tag{C.2}$$

whose general solution is given by

$$\pi^a(\vec{r}) = -\frac{g_{\pi NN}}{2m_N}\int d\vec{r}\,'\,\vec{\nabla}_{r'}\,G(\vec{r}-\vec{r}\,')\cdot\bar{N}\vec{\gamma}\gamma_5\tau^a N \tag{C.3}$$

where $G(\vec{r},\vec{r}\,')$ is the Klein-Gordon propagator for massive pions,

$$G(\vec{r}-\vec{r}\,') = \frac{1}{4\pi}\,\frac{e^{-m_\pi|\vec{r}-\vec{r}\,'|}}{|\vec{r}-\vec{r}\,'|} \tag{C.4}$$

Using the convulation property in Fourier space, we can rewrite (C.3) in the following form

$$\pi^a(\vec{r}) = \int \frac{d\vec{q}}{(2\pi)^3}\,e^{i\vec{q}\cdot\vec{r}}\,(i\vec{q}\,G(\vec{q}))\,\vec{S}^a(\vec{q}) \tag{C.5}$$

where $G(\vec{q})$ and $\vec{S}^a(\vec{q})$ are the Fourier transform of $G(\vec{r})$ and the axial current respectively. At large distances, the dominant contribution to (C.5) is due to the long wavelengths ($q\to 0$) emitted by the Source,

$$\pi^a(\vec{r}) = \vec{S}^a(\vec{q}\to 0) \int \frac{d\vec{q}}{(2\pi)^3} e^{i\vec{q}\cdot\vec{r}}\, i\vec{q}\, G(\vec{q})$$
$$= \vec{S}^a(\vec{q}\to 0)\cdot\vec{\nabla}\, G(\vec{r}) \tag{C.6}$$

Straightforward algebra yields

$$\pi^a(\vec{r}) = -\left(\frac{g_{\pi NN}}{8\pi m_N}\right)(1+m_\pi r)\frac{e^{-m_\pi r}}{r^2}\,\hat{r}^i\,\langle N|\sigma^i\tau^a|N\rangle \tag{C.7}$$

with the nucleon matrix element given by

$$\langle N|\sigma^i\tau^a|N\rangle = \int d\vec{x}\, N^\dagger(\sigma^i\otimes 1\!\!1)\tau^a N \tag{C.8}$$

APPENDIX D: Additivity of the Baryon Number

Two well separated skyrmions can be approximately described by the following product ansatze

$$U = U_1 U_2 \tag{D.1}$$

Equation (D.1) is not valid at short distances where strong deformations are expected. Using (D.1), we can define

$$B_\mu^a = \frac{1}{2i} \mathrm{TR}\,[\tau^a (\partial_\mu U) U^+] = B_\mu^a(1) + g^{ab}(1)\, B_\mu^b(2) \tag{D.2}$$

where g^{ab} is an orthogonal matrix,

$$g^{ab}(1) = \frac{1}{2i} \mathrm{TR}\,[\tau^a U_1 \tau^b U_1^+] \tag{D.3}$$

In terms of $U_1 = \exp(i\vec{\tau}.\vec{F}_1)$, we have

$$g^{ab}(1) = \cos 2F_1\, \delta^{ab} + \sin 2F_1\, \epsilon^{abc} \hat{F}_1^c + 2\sin^2 F_1\, \hat{F}_1^a \hat{F}_1^b \tag{D.4}$$

It is easy to show that (D.4) is orthogonal,

$$g^{ab}(1)\, g^{cb}(1) = \delta^{ab} \tag{D.5}$$

and normalized

$$\mathrm{Det}\, g = \frac{1}{3!} \epsilon^{abc} \epsilon^{a'b'c'} g_{aa'} g_{bb'} g_{cc'} = 1 \tag{D.6}$$

The baryon current carried by U as defined in (D.1) reads

$$B^{\mu} = B^{\mu}(1) + B^{\mu}(2) - \frac{1}{4\pi^2} \epsilon^{\mu\nu\alpha\beta} \partial_{\nu} \left[B^{a}_{\alpha}(1)\, g^{ab}(1)\, B^{b}_{\beta}(2) \right] \tag{D.7}$$

where we have used the above properties of g and the Maurer-Cartan equation. So, modulo a total divergence, the baryon current is additive. In open space, the baryon charges add

$$B = \int_{R^3} d\vec{x}\, B^{0}(\vec{x}) = B(1) + B(2) \tag{D.8}$$

a well-known property of product maps.

APPENDIX E: SU(2)×SU(2) Noether Currents

The purpose of this Appendix is to provide the algebraic details that lead to the expressions of the conserved SU(2)×SU(2) Noether currents, using the standard construction of Gell-Mann and Levy. For that consider local and infinitesimal $SU(2)_{L,R}$ transformation as defined through

$$U_L = (1+iQ_L)U$$
$$U_R = U(1+iQ_R) \tag{E.1}$$

in terms of which $L_\mu = U^+\partial_\mu U$ transforms as

$$\delta_L L_\mu = i\,U^+\partial_\mu Q_L U - i\,U^+ Q_L \partial_\mu U$$
$$\delta_R R_\mu = i\,\partial_\mu Q_R + i\,[L_\mu, Q_R] \tag{E.2}$$

Here, the $Q_{L,R}$ are Lie-algebra valued. If we denote by L_Q the locally rotated Skyrme Lagrangian L, given by

$$\mathcal{L} = C_0 \mathrm{TR}[L_\mu L^\mu] + C_1 \mathrm{TR}[L_\mu, L_\nu]^2 \tag{E.3}$$

then the conserved Noether Currents associated to the global $SU(2)_L\times SU(2)_R$ transformations read

$$J^a_\mu = \frac{\partial \mathcal{L}_Q}{\partial(\partial^\mu \epsilon^a)} \tag{E.4}$$

Injecting the left-transformations (E.1) into (E.3) gives

$$\delta_L \mathcal{L} = 2c_0 \, \mathrm{TR}\, [\delta_L L_\mu L^\mu] + 2c_1 \, \mathrm{TR} \left(\delta_L [L_\mu, L_\nu][L_\mu, L_\nu] \right)$$

$$= -2c_0 \, \mathrm{TR}\, [i\partial_\mu Q_L R^\mu] - 4c_1 \, \mathrm{TR} \left([i\partial_\mu Q_L, R_\nu][R_\mu, R_\nu] \right) + \ldots \qquad \text{(E.5)}$$

in terms of which we have the conserved left-handed current $J^a_{\mu,L}$,

$$J^a_{\mu,L} = -2ic_0 \, \mathrm{TR}\, [Q^a_L R_\mu] - 4ic_1 \, \mathrm{TR} \left([Q^a_L, R_\nu][R_\mu, R^\nu] \right) \qquad \text{(E.6a)}$$

Similarly, one obtains the conserved right-handed current $J^a_{\mu,R}$ in the form

$$J^a_{\mu,R} = +2ic_0 \, \mathrm{TR}\, [Q^a_R L_\mu] + 4ic_1 \, \mathrm{TR} \left([Q^a_R, L_\nu][L_\mu, L^\nu] \right) \qquad \text{(E.6b)}$$

The conserved vector and axial-vector currents follow immediately through

$$\mathbb{V}^a_\mu = ic_0 \, \mathrm{TR} \left(Q^a (R_\mu + L_\mu) \right) + 2ic_1 \, \mathrm{TR} \left(Q^a \left([R_\nu, [R_\mu, R^\nu]] + [L_\nu, [L_\mu, L^\nu]] \right) \right) \qquad \text{(E.7a)}$$

$$\mathbb{A}^a_\mu = -ic_0 \, \mathrm{TR} \left(Q^a (R_\mu - L_\mu) \right) - 2ic_1 \, \mathrm{TR} \left(Q^a \left([R_\nu, [R_\mu, R^\nu]] - [L_\nu, [L_\mu, L^\nu]] \right) \right) \qquad \text{(E.7b)}$$

To extract the dependence of (E.7) on the collective coordinate A(t) discussed in the text , it is convenient to define*

$$A(t) = e^{i\vec{\tau}\cdot\vec{\omega}t} \tag{E.8}$$

so that

$$\tilde{Q}^a = A^+ Q^a A = A^+ \frac{\vec{\tau}}{2}\cdot\hat{a}\,A \tag{E.9}$$

$$= \frac{1}{2}\vec{\tau}\cdot\hat{a} - \frac{\sin 2\omega t}{2}\,\vec{\tau}\cdot(\hat{a}\times\hat{\omega}) + \sin^2\omega t\;\vec{\tau}\cdot(\hat{\omega}\times(\hat{\omega}\times\hat{a}))$$

$$\tilde{L}_\mu = (AU^+A^+)\partial_\mu(AUA^+) \tag{E.10}$$

$$= A\,(L_\mu - A_\mu + U^+A_\mu U)\,A^+$$

$$\tilde{R}_\mu = (AUA^+)\partial_\mu(AU^+A^+) \tag{E.11}$$

$$= A\,(R_\mu - A_\mu + UA_\mu U^+)\,A^+$$

$$U^+A_0U - A_0 = i\vec{\tau}\cdot\vec{\alpha}_- \tag{E.12}$$

$$= i\vec{\tau}\cdot\left(\sin 2F\,(\hat{r}\times\vec{\omega}) - 2\sin^2 F\,(\hat{r}\times\vec{\omega})\times\hat{r}\right)$$

$$A_0 - UA_0U^+ = i\vec{\tau}\cdot\vec{\alpha}_+$$

$$= i\vec{\tau}\cdot\left(\sin 2F\,(\hat{r}\times\vec{\omega}) + 2\sin^2 F\,(\hat{r}\times\vec{\omega})\times\hat{r}\right) \tag{E.13}$$

---

*We will not concern ourselves with the ambiguity $\pm\vec{\omega}$ raised by this definition.

$$\begin{aligned}\vec{A}_i &= -\frac{i}{2}\,\mathrm{Tr}\,[\vec{\tau}L_i] \\ &= \hat{r}\,\partial_i F + \frac{\sin 2F}{2}\,\partial_i \hat{r} + (\hat{r}\times\partial_i\hat{r})\sin^2 F\end{aligned} \tag{E.14}$$

$$\begin{aligned}\vec{B}_i &= -\frac{i}{2}\,\mathrm{Tr}\,[\vec{\tau}R_i] \\ &= -\hat{r}\,\partial_i F - \frac{\sin 2F}{2}\,\partial_i \hat{r} + (\hat{r}\times\partial_i\hat{r})\sin^2 F\end{aligned} \tag{E.15}$$

where we have used $A_\mu = A^+\partial_\mu A = A^+\dot{A}\delta_\mu$. Under the transformation $U \to A\,U\,A^\dagger$ the linear term in (E.6) becomes

$$f^L_\mu = \mathrm{Tr}\,[Q_L \breve{R}_\mu] = \mathrm{Tr}\,[\tilde{Q}_L\,(R_\mu - A_\mu + U A_\mu U^+)] \tag{E.16}$$

Since we are ultimately interested in the angle averaged currents, it is convenient to work with angle averaged quantities. With this in mind, we have

$$\begin{aligned}\bar{f}^L_0 = \int d\hat{r}\, f^L_0 &= -i\,\mathrm{Tr}[\tilde{Q}_L\tau^b]\int d\hat{r}\,\alpha^b_+ \\ &= -i\,\frac{16\pi}{3}\sin^2 F\,\mathrm{Tr}\,[\tilde{Q}_L\,\vec{\omega}\cdot\vec{\tau}] \\ &\equiv -\frac{16\pi}{3}\sin^2 F\,\mathrm{Tr}\,[\tilde{Q}_L\,\dot{A}A^+]\end{aligned} \tag{E.17}$$

$$\begin{aligned}\bar{f}^L_i = \int d\hat{r}\, f^L_i &= i\,\mathrm{Tr}\,[\tilde{Q}_L\tau^b]\int d\hat{r}\,B^b_i \\ &= -i\,\frac{4\pi}{3}\,\delta^{ib}\left(F' + \frac{\sin 2F}{r}\right)\mathrm{Tr}\,(\tilde{Q}_L\tau^b) \\ &= -i\,\frac{4\pi}{3}\left(F' + \frac{\sin 2F}{r}\right)\mathrm{Tr}\,(Q_L A\tau_i A^+)\end{aligned} \tag{E.18}$$

Similarly, the trilinear term in (E.6) becomes

$$\begin{aligned} g^L_\mu &= \mathrm{TR}\left([Q_L, \tilde{R}_\nu][\tilde{R}_\mu, \tilde{R}^\nu]\right) = \mathrm{TR}\left(Q_L [\tilde{R}_\nu, [\tilde{R}_\mu, \tilde{R}^\nu]]\right) \\ &= \mathrm{TR}\left(\tilde{Q}_L [R_\nu, [R_\mu, R^\nu]]\right) \\ &\quad - \mathrm{TR}\left(\tilde{Q}_L [R_\nu, [A_\mu - UA_\mu U^+, R^\nu]]\right) + O(\dot{A}^2) \end{aligned} \tag{E.19}$$

where we have dropped higher-order time-derivatives since they involve non-leading N_C-corrections. The time-component of (E.19) gives

$$\begin{aligned} g^L_0 &= \mathrm{TR}\left(\tilde{Q}_L [R_i, [A_0 - UA_0 U^+, R_i]]\right) \\ &= 2\mathrm{TR}\left(\tilde{Q}_L [\vec{\tau}\cdot\vec{B}_i, \vec{\tau}\cdot(\vec{\alpha}_+ \times \vec{B}_i)]\right) \\ &= 2\mathrm{TR}\,[\tilde{Q}_L \tau^b]\,(\vec{B}_i \times (\vec{\alpha}_+ \times \vec{B}_i))^b \end{aligned} \tag{E.20}$$

Using the following results

$$\begin{aligned} \int d\hat{r}\, \alpha_+^b B_i^2 &= \frac{16\pi}{3} \sin^2 F \left(F'^2 + 2\frac{\sin^2 F}{r^2}\right) \omega^b \\ \int d\hat{r}\, B_i^b (\vec{\alpha}_+ \cdot \vec{B}_i) &= \frac{16\pi}{3} \frac{\sin^4 F}{r^2} \omega^b \end{aligned} \tag{E.21}$$

we obtain

$$\bar{g}^L_0 = \int d\hat{r}\, g^L_0 = \frac{64\pi}{3} \sin^2 F \left(F'^2 + \frac{\sin^2 F}{r^2}\right) \mathrm{TR}\left(Q_L \dot{A} A^+\right) \tag{E.22}$$

Similarly, the space-component of (E.19) becomes

$$\begin{aligned} g^L_i &= -\mathrm{TR}\left(\tilde{Q}_L [R_j, [R_i, R_j]]\right) \\ &= -4i\, \mathrm{TR}\,[\tilde{Q}_L \tau^b]\,(\vec{B}_j \times (\vec{B}_i \times \vec{B}_j))^b \end{aligned} \tag{E.23}$$

Using the following results

$$\int d\hat{r}\, B_i^b B_j^2 = -\frac{4\pi}{3}\delta^{ib}\left(F'^2 + 2\frac{\sin^2 F}{r^2}\right)\left(F' + \frac{\sin 2F}{r}\right) \tag{E.24}$$

$$\int d\hat{r}\, B_j^b\,(\vec{B}_i \cdot \vec{B}_j) = -\frac{4\pi}{3}\delta^{ib}\left(F'^3 + \frac{\sin^2 F}{r^2}\cdot\frac{\sin 2F}{r}\right)$$

we obtain

$$\bar{g}_i^L = \int d\hat{r}\, g_i^L = i\frac{16\pi}{3}\left(F'^2\frac{\sin 2F}{r} + 2F'\frac{\sin^2 F}{r^2} + \frac{\sin^2 F}{r^3}\sin 2F\right)\mathrm{TR}\left(Q_L A\tau_i A^+\right) \tag{E.25}$$

Combining (E.17) and (E.22) yields

$$\int d\hat{r}\, J^a_{0,L} = i\frac{32\pi}{3}C_0 \sin^2 F\left(1 - 8\frac{C_1}{C_0}\left(F'^2 + \frac{\sin^2 F}{r^2}\right)\right)\mathrm{TR}\left(Q_L^a \dot{A} A^+\right) \tag{E.26}$$

while combining (E.18) and (E.25) gives

$$\int d\hat{r}\, J^a_{i,L} = +\frac{8\pi}{3}C_0\left(\left(F' + \frac{\sin 2F}{r}\right) - 8\frac{C_1}{C_0}\left(F'^2\frac{\sin 2F}{r} + 2F'\frac{\sin^2 F}{r^2} + \frac{\sin^2 F}{r^3}\sin 2F\right)\right) *$$
$$* \ \mathrm{TR}\left(Q_L^a A\tau_i A^+\right) \tag{E.27}$$

to obtain the angle averaged counterparts of the right-handed currents one can use the substitution F → -F in (E.26) and (E.27). Thus, the angle averaged vector and axial-vector currents become

$$\int d\hat{r}\, V_0^a = -i\left(\frac{32\pi}{3}\right)C_0 \sin^2 F\left(1 - 8\frac{C_1}{C_0}\left(F'^2 + \frac{\sin^2 F}{r^2}\right)\right)\mathrm{TR}\left(\tau_a \dot{A} A^+\right) \tag{E.28}$$

$$\int d\hat{r}\, A_i^a = \left(\frac{8\pi}{3}\right)C_0\left(\left(F' + \frac{\sin 2F}{r}\right) - 8\frac{C_1}{C_0}\left(F'^2\frac{\sin 2F}{r} + 2F'\frac{\sin^2 F}{r^2} + \frac{\sin^2 F}{r^3}\sin 2F\right)\right) \tag{E.29}$$
$$* \ \mathrm{TR}\left[\tau^a A\tau_i A^+\right]$$

For completeness, notice that

$$\int d\hat{r}\,\hat{r}^c B_i^b B_j^i = \frac{4\pi}{3}\,\epsilon^{ibc}\,\frac{\sin^2 F}{r}\left(F'^2 + 2\,\frac{\sin^2 F}{r^2}\right)$$
$$\int d\hat{r}\,\hat{r}^c B_j^b\,(\vec{B}_i\cdot\vec{B}_j) = \frac{4\pi}{3}\,\epsilon^{ibc}\,\frac{\sin^4 F}{r^3} \tag{E.30}$$

so that

$$\int d\hat{r}\,\vec{q}\cdot\hat{r}\,J^a_{i,L} = +\frac{8\pi}{3}\,C_0\,\frac{\sin^2 F}{r}\left(1 - 8\frac{C_1}{C_0}\left(F'^2 + \frac{\sin^2 F}{r^2}\right)\right) *$$
$$* \;\epsilon^{ibc}\,\mathrm{TR}\,(Q^a_L A \tau^b A^+)\,q^c \tag{E.31}$$

hence

$$\int d\hat{r}\,\vec{q}\cdot\hat{r}\,\mathbb{V}^a_i = i\,\frac{16\pi}{3}\,C_0\,\frac{\sin^2 F}{r}\left(1 - 8\frac{C_1}{C_0}\left(F'^2 + \frac{\sin^2 F}{r^2}\right)\right) *$$
$$* \;\mathrm{TR}\,(\vec{\tau}\cdot\vec{q}\;\tau^i A^+ \tau^a A) \tag{E.32}$$

APPENDIX F: Spin and Isospin operators in (SU(2);C).

To construct explicitly the spin/isospin operators in the Hilbert space H of functions α defined on SU(2) and valued in C, (SU(2);C) for short, we will use the conventional Noether construction. In the ansatz

$$U(\vec{x},t) = A(t)\, U(\vec{x})\, A^{+}(t) \tag{F.1}$$

rotations in isospace correspond to A → hA,

$$U_h(\vec{x},t) = h\, U(\vec{x},t)\, h^{+} \tag{F.2}$$

while rotations in space correspond to A → Ah,

$$U_h(\vec{x},t) = A(t)\, h\, U(\vec{x})\, h^{+} A^{+}(t) = A(t)\, U(\mathrm{ad}h\,\vec{x})\, A^{+}(t) \tag{F.3}$$

where adh is the adjoint of h defined by

$$(\mathrm{ad}h)_{kj} = \frac{1}{2}\,\mathrm{TR}\,[\tau_k h \tau_j h^{+}] \tag{F.4}$$

For the hedgehog skyrmion, the Noether currents are

$$J_k = i\, I\, \mathrm{TR}\,[\tau_k \dot{A}^{+} A] \tag{F.5a}$$

$$I_k = i\, I\, \mathrm{TR}\,[\tau_k \dot{A} A^{+}] \tag{F.5b}$$

where I is the moment of inertia (72). J_k and I_k are related by parity, i.e.

$$\hat{\Pi}_{op}\, J_k\, \hat{\Pi}_{op} = I_k \tag{F.6}$$

In other words, rotations of the hedgehog configuration in isospace are equivalent to reverse rotations in space. Since $A^{+}A = 1$, we have

$$\sum_{k=0}^{3} I_k^2 = \sum_{k=0}^{3} J_k^2 = 2I^2 \, \mathrm{TR}\,[\dot{A}^{+}\dot{A}] \tag{F.7}$$

Using the canonical procedure* described in the text one obtains

$$\hat{J}_k = \frac{i}{2}\left(a_k \frac{\partial}{\partial a_0} - a_0 \frac{\partial}{\partial a_k} - \epsilon_{k\ell m}\, a_\ell \frac{\partial}{\partial a_m}\right) \tag{F.8a}$$

$$\hat{I}_k = \frac{i}{2}\left(a_0 \frac{\partial}{\partial a_k} - a_k \frac{\partial}{\partial a_0} - \epsilon_{k\ell m}\, a_\ell \frac{\partial}{\partial a_m}\right) \tag{F.8b}$$

$$\hat{J}^2 = \hat{I}^2 = \frac{1}{4} \sum_{k=0}^{3} \left(-\frac{\partial^2}{\partial a_k^2}\right) \tag{F.8c}$$

It is not hard to show that Eqs. (78) carry the correct spin/isospin asignment and are normalized. Recall that the scalar product in (SU(2);C) is defined by

$$(\alpha|\beta) = \int_{SU(2)} d\mu(A)\, \alpha^{*}(A)\, \beta(A) \tag{F.9}$$

where $d\mu(A)$ is the Haar measure on SU(2).

---

* One might object that since the a's are constrained parameters one cannot use simple canonical procedures. We have convinced ourselves that a correct treatment using a set of independent parameters on S^3, leave all of the naive statements correct if we were to constrain the wavefunctions on S^3 at the end.

REFERENCES

1. G. 't Hooft, Nucl. Phys. B72, 461(1974).
2. E. Witten, Nucl. Phys.B160, 57(1979).
3. S. Coleman and E. Weinberg, Phys. Rev. D7, 1888(1973)
4. S. Adler, Phys. Rev. 117, 2426(1969).
   J. Bell and R. Jackiw, Nuovo Cimento 60A, 47(1969).
5. For a review, see B. W. Lee, "Chiral Dynamics," Gordon and Breach, N.Y. (1972).
6. T. H. R. Skyrme, Nucl. Phys. 31, 556(1962).
   T. H. R. Skyrme, Proc. Roy.Soc. London 260, 127(1961); ibid 262, 237(1961).
7. G. H. Derrick , J. Math. Phys. 5, 1252(1964).
8. E. B. Bogomol'ny, Sov. J. Nucl. Phys. 20, 449(1976)
9. S. Weinberg, Physica 96, 325(1979)
10. M. Rho, Saclay preprint PhT 84-123(1984).
11. J. F. Donoghue, E. Golowich and B. R. Holstein, Phys. Rev. Lett 53, 747(1984).
12. A. P. Balachandran, V.P. Nair, S. G. Rajeev and A. Stern, Phys. Rev. Lett. 49, 1124(1982); Phys. Rev. D27, 1153(1983).
13. E. Witten, Nucl. Phys. B223, 4232(1983); ibid 433.
14. M. K. Prasad and C. M. Sommerfield, Phys. Rev. Lett. 35., 760(1976); and Ref. 8.
15. A. D. Jackson and M. Rho, Phys. Rev. Lett. 51, 751(1983).
16. G. Adkins, C. Nappi and E. Witten, Nucl. Phys. B228, 552(1983)
17. J. Wess and B. Zumino, Phys. Lett. 37B, 95(1971).
18. J. Goldstone and F. Wilczek, Phys. Rev. Lett.47, 986(1981)
19. D. Finkelstein and J. Rubinstein, J. Math. Phys.9, 1762(1968).

20. J. G. Williams, J. Math. Phys. 11, 261(1976).

21. A. R. Balachandran, F. Lizzi and V. G. J. Rodgers, Syracuse Univ. preprint SU-4222(1985).

22. R. L. Jaffe and C. L. Korpa, MIT preprint CTP #1233(1985).

23. A. D. Jackson, private communication.

24. G. E. Brown, A. D. Jackson, M. Rho and V. Vento, unpublished.

25. E. Witten, in "Solitons in Nuclear and Elementary particle physics," edited by A. Chodos, E. Hadjimichael and C. Tze (World Scientific, 1984), and references therein.

26. G. Karl and J. E. Paton, D30 (1984) 238.

27. M. A. B. Beg and A. Zepeda, Phys. Rev. D6, 2912(1972).

28. G. Adkins and C. Nappi, Nucl. Phys. B233, 109(1984).

29. U. G. Meissner, Bochum preprint (1983).

30. I. Zahed, U.-G. Meissner and U. B. Kaulfuss, Nucl. Phys. A426, 525(1984).

31. J. D. Breit and C. R. Nappi, Phys. Rev. Lett. 53, 889(1984).

32. The breathing mode in the Skyrme model is also discussed in

C. Hadjuk and B. Schwesinger, Phys. Lett. 140B (1984);
J. Dey and J. le Tourneux, Montreal preprint (1984);
J. D. Breit, Penn. Univ. preprint UPR-D271T(1984);
K. F. Liu, J. S. Zhang and G. R. E. Black, Phys. Rev. B30 2015(1984); A. Parmentola, Phys. Rev. D30, 685(1984).

33. H. Walliser and G. Eckhart,Nucl. Phys. A429, 514(1984)

34. M. P. Mattis and M. Karliner, Slac preprint (1984).

35. H. J. Schnitzer, Phys. Lett. 139B, 217(1984).

36. S. Weinberg, Phys. Rev. Lett. 28, 188(1967).

37. I. Zahed, A. Wirzba and U.-G. Meissner, SUNY preprint (1985).

38. A. Jackson, A. D. Jackson and V. Pasquier, Nucl. Phys. A432 (1985) 567.

39. A. Jackson, unpublished.

40. U.-G. Meissner, unpublished.

41. M. Sommermann, H. W. Wyld and C. J. Pethick, Urbana preprint (1985)

42. E. B. Bogomol'ny and V. A. Fateev, Sov. J. Nucl. Phys. 37 (1983) 134.

43. M. Kutschera and C. J. Pethick, Nordita preprint 34/49 (1984).

44. M. Kutschera, C. J. Pethick and D. G. Ravenhall, Phys. Rev. Lett. 53 (1984) 1041.

45. I. Zahed, A. Wirzba, U.-G. Meissner, C. J. Pethick and J. Ambjorn, Phys. Rev. D31 (1985) 1114.

NUCLEAR FORCE IN A QUARK MODEL

Makoto Oka
Massachusetts Institute of Technology
Cambridge, MA 02139

ABSTRACT

The short range nucleon-nucleon interaction is discussed in the quark cluster model (QCM) approach. The role of the color-magnetic gluon exchange interaction is stressed. Change of the quark confining scale in the two-nucleon system is investigated in relation to the rescaling model of the EMC effect.

Strong interactions among hadrons are believed to be described by quantum chromodynamics (QCD). Elementary degrees of freedom of QCD are colored quarks and gluons and their interactions are given by color $SU(3)_c$ gauge theory. The ground state baryons, N, Δ, Λ, ... etc., are known to be well represented by three valence quarks which interact with each other by gluon exchanges. Sizes of the baryons are around $0.9 - 1.0$ fm, which is of the same order as typical distance of two nucleons in the nuclear matter. Furthermore, the range of short-distance repulsion between two nucleons is smaller than the radius of the nucleon. Therefore, the substructure of the nucleon is clearly important in the nuclear physics.

Nucleon–nucleon interaction has been investigated in this context for a while. However, gluon exchange among valence quarks would not be able to describe a long-range nuclear force because the color is confined strongly. It is necessary at long distances to introduce color-singlet objects exchanged between the nucleons, which is known as meson-exchange interaction. Therefore, we would like to separate the long-range part from the short-range part, where the meson exchanges are suppressed by the extended form factor. Instead, an exchange interaction due to quark antisymmetrization plays an important role once quark wave functions of the two nucleons overlap with each other (fig. 1). The exchange force is in fact short range because it requires the wave functions overlap. The range is essentially given by the size of the baryon.

The purpose of this report is to discuss the quark exchange interaction between two nucleons. We take the quark cluster model (QCM) approach which satisfies several necessary conditions, *i.e.*, (1) Confinement for color-singlet three-quark clusters (nucleons). (2) Exchange symmetry, which means

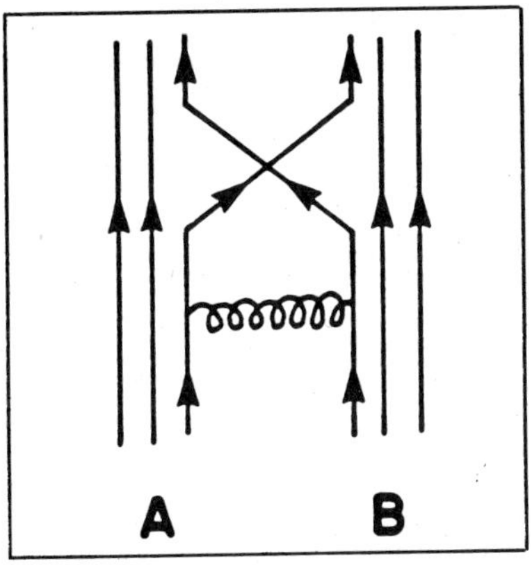

Fig. 1 An example of the quark exchange diagram in the interaction of the baryons A and B.

that the hamiltonian should be symmetric and the wave function antisymmetric under an exchange of quarks. (3) No ambiguity in separating the center of mass coordinate from each baryon. This is important in order to describe a N–N relative motion and to make a scattering problem well defined.

We introduce an effective hamiltonian for six-quark systems,[1]

$$H = \sum_i \left(m_i + \frac{p_i^2}{2m_i} \right) + V_{\rm conf} + V_{\rm res}, \tag{1}$$

where the potential consists of a quark confining part, $V_{\rm conf}$ and a short-range residual interaction $V_{\rm res}$. $V_{\rm res}$ is taken as a static part of the one-gluon exchange potential, which contains spin dependent terms like $\sum_{i<j}(\lambda_i \cdot \lambda_j)(\sigma_i \cdot \sigma_j)\delta(\vec{r}_{ij})$ (color-magnetic interaction). Although we have some ambiguity in choosing the confining potential, the final results indicate that the *short-range* N–N interaction is not sensitive to the choice. In fact, the color-magnetic gluon exchange dominates and gives a strong N–N repulsion at short distances. Here we show some results for a two-body linear potential, $(-a_1)\sum_{i<j}(\lambda_i \cdot \lambda_j)\, r_{ij}$. This potential is well known to give a long-range pathological attraction, or color van der Waals force. However, van der Waals forces do not enter in our calculation because we ignore color-excited baryon states. Recent calculation for a many-body confining potential without pathological long-range forces has shown that it gives qualitatively the same results.[2]

The Schrödinger equation for a six-quark or two-baryon (A and B) system,

$$(H - E)\,\Psi_{AB} = 0 \tag{2}$$

is to be solved with a cluster wave function,

$$\Psi_{AB} = \mathcal{A}\left\{\phi_A(123)\phi_B(456)\chi(R_{123} - R_{456})\right\}, \tag{3}$$

where $\mathcal{A}$ is an antisymmetrizer, ϕ_A (ϕ_B) are the internal wave functions of the baryon A (B) and χ the relative wave function. According to the resonating group method, ϕ_A and ϕ_B are fixed and are integrated over all the internal coordinates, which leaves a nonlocal equation for χ,

$$\int dR' \;\{H(R,R') - E\,N(R,R')\}\;\chi(R') = 0. \tag{4}$$

Solving this equation for appropriate boundary conditions, we may obtain the binding energy and wave function for a bound state, if any, and the S-matrix and phase shifts for scattering problems.

This approach can be applied to any two-baryon systems. It is quite interesting to see whether the strong short-range repulsion observed in N–N interaction is universal for all the two-baryon systems. The quark model provides us with a unified description for all of them. We, in fact, have applied QCM to several two-baryon systems, including $N\Delta$, $\Delta\Delta$, ΛN, ΣN and $\Lambda\Lambda$. We summarize results of the numerical calculations as follows:[1]

(1) The quark exchange force strongly depends on the spin-flavor symmetry structure of the system. For some two-baryon channels like NΔ(S=1, I=1, L=even), ΔΔ(S=3, I=2, L=even), we obtain a strong repulsion, which is almost independent of the detailed dynamics. Those states are characterized by the spin-flavor symmetry [51] and the strong repulsion is entirely due to the quark antisymmetrization. On the contrary, some channels like ΔΔ(S=3, I=0, L=0), NΞ(S=0, I=0, L=0) etc. have no repulsion. This is remarkable because without a short-range repulsion the strong attraction due to the meson exchange may produce a deeply bound (not deuteron-like) dibaryon resonance in these channels. For these channels qualitative behaviors of the short-range interaction are determined only by the symmetry structure of the system.

(2) Unfortunately, for the two-nucleon system, the symmetry structure alone cannot determine the interaction because the system has a mixed spin-flavor symmetry. We found that the exchange force from the color-magnetic interaction (CMI) produces a short range repulsion. Calculated scattering phase shifts (fig. 2) suggest the range of the repulsion is about 0.5 fm for $L = 0$.

The repulsion for ${}^1S_0$ is slightly stronger than for ${}^3S_1$. We would like to stress again that the exchange interaction induced by the CMI is totally responsible for the repulsion. In other words, contribution from the confining potential and the kinetic energy term of the hamiltonian are much less important. This is a favorable result, because the CMI is the most reliable part of

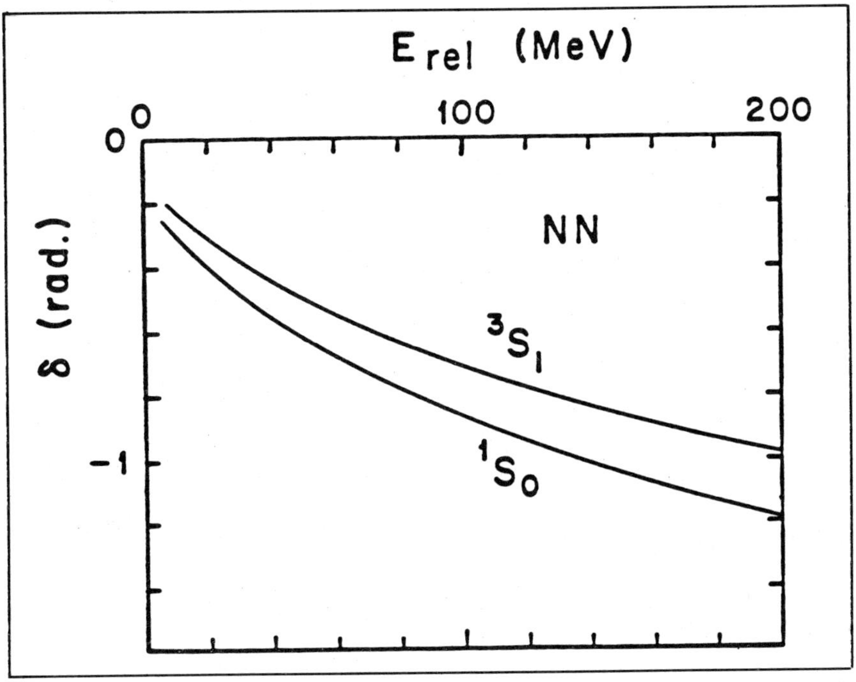

Fig. 2 Scattering phase shifts for the NN $^1S_0$ and $^3S_1$ states.

my hamiltonian. It is common for most of the quark models (for instance, the MIT bag model) and is responsible for the N–Δ mass difference and many other details of the baryon spectrum. The strength of CMI can be determined fairly well by using, for instance, the N–Δ mass difference.

Thus the quark exchange mechanism can describe the short-range baryon-baryon interaction. However, it cannot explain the long-range part, where mesons play the leading role. We would like to supplement it with a meson-exchange interaction in order to demonstrate the role of the short-range part in a realistic model of the nuclear force. Mesons are not explicitly taken into account. Instead, an effective meson exchange potential is introduced[3] by

$$V^{\mathrm{EMEP}}(R) = V_C(R) + V_T(R)S_{12}. \tag{5}$$

Each (central and tensor) part has a one-pion-exchange tail and a phenomenological gaussian with two parameters, $V_0 \exp(-\alpha R^2)$. Choosing the parameters V_0 and α to fit the experimental scattering length and effective range, the nucleon-nucleon phase shifts can be reproduced fairly well (figs. 3-5).

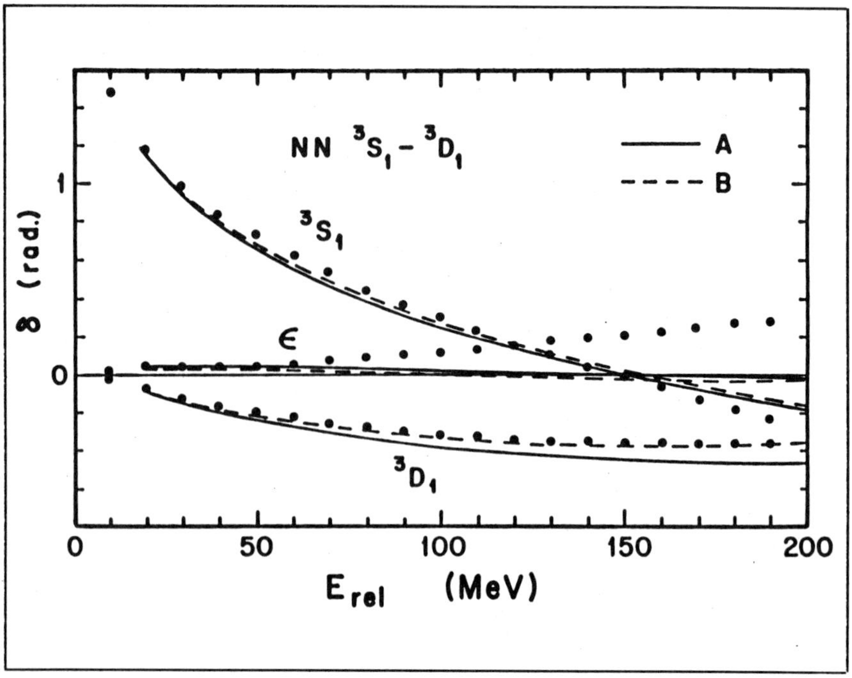

Fig. 3 Scattering phase shifts for the NN $^1S_0$ state in the realistic model. The curves (I)-(III) correspond to different choices of the quark model parameters. The phase shifts calculated without the effective meson exchange potential (EMEP) are also shown for comparison.

In conclusion, the short-range nucleon-nucleon repulsion induced by the color-magnetic gluon exchange with the quark antisymmetrization is consistent with the experimental data.

One of the most exciting phenomena observed recently is the EMC effect or the mass number dependence of the structure functions measured in deep inelastic lepton scattering from nuclei. Although this is a high energy phenomenon, where the description with nucleons and mesons may not be appropriate, an important clue to the low-energy nuclear dynamics was suggested by Close et al.[4] They proposed a rescaling model based on the QCD evolution of the structure function. They pointed out that a valence quark model, for instance, the bag model, is applicable in calculating the leading twist contribution at a low momentum transfer (normalization point) and that the perturbative QCD predicts its evolution to high momentum transfer. The normalization

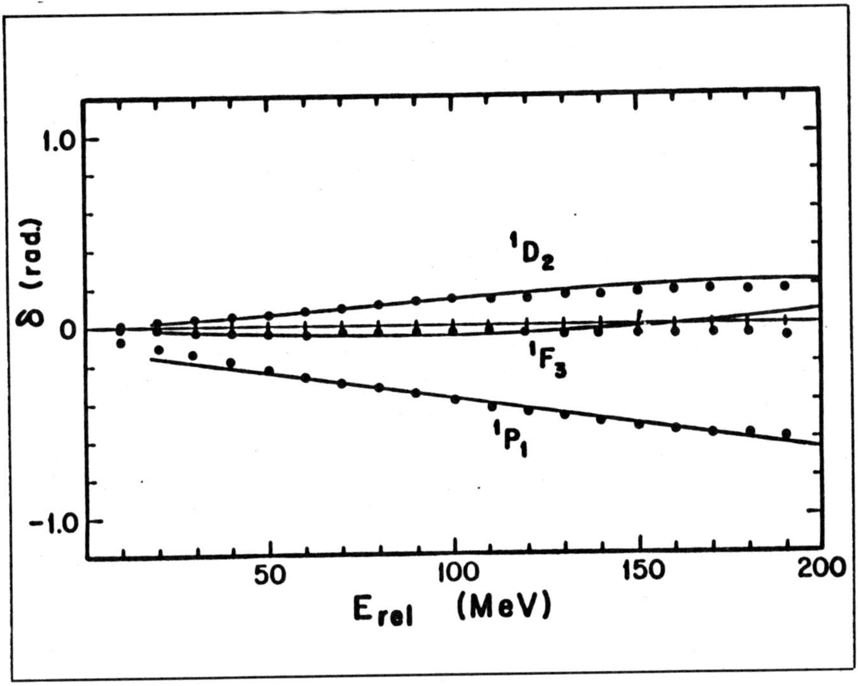

Fig. 4 Scattering phase shifts for the NN $^1P_1$, $^1D_2$, $^1F_3$ states.

point can be determined by the valence quark distribution in the nucleon. A simple assumption was made that the average quark confinement radius or the bag radius of the nucleon determines the normalization point. If a nucleon implanted in a nucleus interacts with the other nucleons, the quark distribution may be modified, which changes the normalization point. A model of a density dependent bag radius was proposed, which shows a partial deconfinement of quarks in nuclei, and explains the observed A dependence very well. In fact, Close et al. expressed the ratio of the quark confinement scale for a nucleus A to that for a free nucleon as an integral

$$\frac{\lambda_A}{\lambda_0} = 1 - \frac{1}{A} \int \rho_A^{(2)}(\vec{r}_1, \vec{r}_2) \left(1 - \frac{\lambda(|\vec{r}_1 - \vec{r}_2|)}{\lambda_0}\right) d\vec{r}_1\, d\vec{r}_2, \qquad (6)$$

where $\rho_A^{(2)}$ is a two-body nucleon density including the two-body correlation. They assumed that $(1 - \lambda(r)/\lambda_0)$ is proportional to the geometrical overlapping volume of the two nucleons. With this crude assumption they could fit

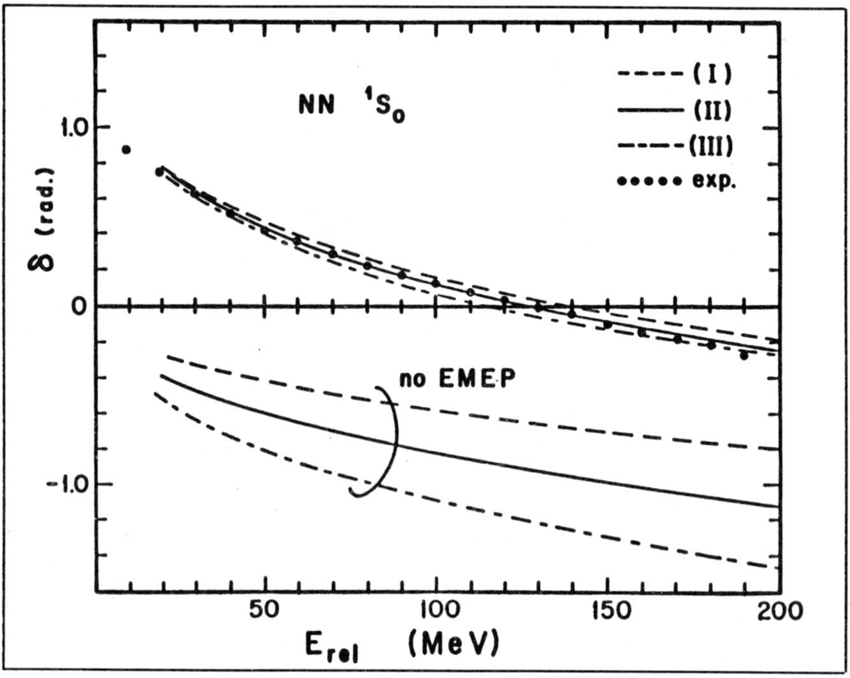

Fig. 5 $^3S_1 - ^3D_1$ scattering parameters in the realistic model. The curves A and B correspond to different choices of the EMEP parameters.

the experimental data remarkably well. It should be noted that in the above equation the modification of the nucleon is written as a function of the relative distance of two nucleons and all complex nucleon dynamics is confined in the two-body density. This separation of the internal modification and the nucleon dynamics, or an adiabatic treatment, is only approximate. In fact, $\lambda(r)/\lambda_0$ might be a complicated function of the density, relative velocity, etc. However, we know from the analysis of the EMC effect that the modification is order of 15% and also that the two-body correlation, or quadratic density dependence, is enough.[4] Therefore we hope that we can make a qualitative argument in the adiabatic approach for the first step of this investigation. This assumption enables us to study the nucleon modification effect in the two-nucleon system.

Here, we wish to discuss the possibility that the one gluon exchange interaction modifies the quark momentum distribution in a two nucleon system.[5] The study is focussed on the following two subjects: (1) role of the quark antisymmetrization in a realistic quark model with the full internal degrees of freedom, and (2) role of the color-magnetic gluon exchanges among the quarks, which is known to be the most important ingredient in the study of the short-range baryon-baryon interaction.

Take a cluster wave function with a fixed relative distance R between the two nucleons. Assuming that the confining scale is proportional to the inverse of the average three momentum, we found that it is almost proportional to b, the size parameter of the gaussian quark wave function in a nucleon. This means that the exchange effect is negligible. In fact, the exchange contribution is suppressed by color, spin and flavor exchange matrix element,

$$C_{\rm ex} \equiv \langle \chi_A \chi_B | \sum_{i \in A, j \in B} P_{ij} | \chi_A \chi_B \rangle, \tag{7}$$

where χ denotes a color-spin-flavor state of a cluster and P_{ij} is an exchange operator. $C_{\rm ex}$ is $-1/9$ for the S-wave N–N system. Thus, if the size parameter b is fixed, we observe negligible rescaling effect.

However, the size b may vary for different R. The calculation of b as a function of R depends on the dynamics for the six-quark system. Let me take the same hamiltonian used above and calculate the energy of the six-quark system as a function of b and R, $H_6(b, R)$. Then $b(R)$ is obtained by taking a variation,

$$\left.\frac{\partial H_6(b, R)}{\partial b}\right|_{\text{fixed } R} = 0. \tag{8}$$

(This is again an adiabatic approximation and the results would be only qualitative.)

At $R = 0$, one clearly sees that the size b or $\lambda(R)/\lambda_0$ is enhanced as much as 30% due to the color-magnetic interaction (CMI). The CMI contribution to the total energy is negative (attractive) for $R \to \infty$ (free nucleons) and is positive and large (strongly repulsive) for $R = 0$, which was the cause of the strong N–N repulsion. Because the CMI matrix element is proportional to $1/b^3$, the repulsion clearly enhances b. At intermediate R, the situation is not clear, but a numerical calculation predicts significant enhancement at $R \leq 1$ fm (fig. 6).

The enhancement, however, is no longer seen at 1 fm $\leq R \leq$ 2 fm, which deviates from the rescaling model analysis of the EMC effect.[4]

In conclusion, we have shown that quark antisymmetrization does not show a significant rescaling effect because of a strong suppression by exchange factors for the color, spin and flavor spaces. However, the size of the individual nucleons is enhanced at short distances by the contribution of color-magnetic gluon exchange for the six quark energy, which has been known as a main cause of the N–N short-range repulsion. We would like to stress here that these two

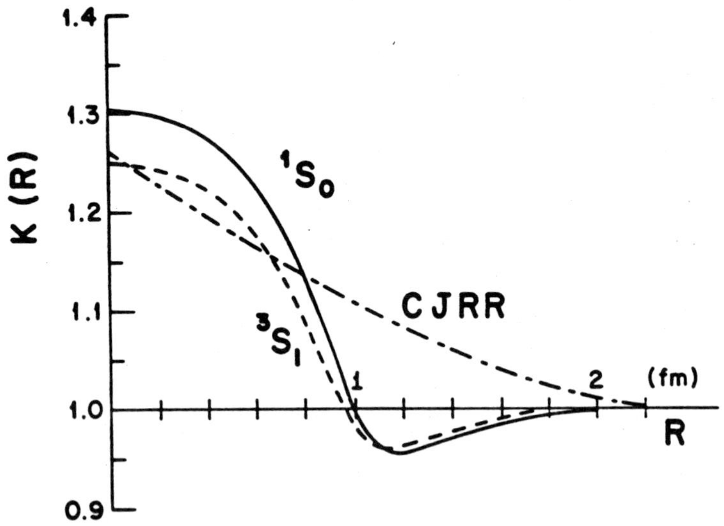

Fig. 6 The ratio $K(R) \equiv \lambda(R)/\lambda_0$ as a function of R for the S-wave $N-N$ systems. The curve CJRR is the parametrization by Close, et al.[4]

conclusions are almost model independent. For the first conclusion, every quark model with the quark antisymmetrization gives the same conclusion. Further, the color-magnetic gluon exchange interaction is common for many quark models including the bag model: It gives N–Δ mass difference, the charge radius of the neutron and many other details of the meson and baryon spectra. Therefore the conclusions are quite important although the model gives only a part of the rescaling effect.

ACKNOWLEDGEMENTS

A part of this work has been done with a collaboration of Koichi Yazaki. This work is supported in part through funds provided by the U.S. Department of Energy (DOE) under contract DE–AC02–76ERO3069.

REFERENCES

1. M. Oka and K. Yazaki, *Phys. Lett.* **90B**, 41 (1980); Prog. Theor. Phys. **66**, 556, 572 (1981); in "Quarks and Nuclei" ed. by W. Weise (1985, World Scientific).

2. M. Oka and C.J. Horowitz, *Phys. Rev.* D **31**, 2274 (1985).

3. M. Oka and K. Yazaki, *Nucl. Phys.*, **A402**, 477 (1983).

4. F.E. Close, R.L. Jaffe, R.G. Roberts and G.G. Ross, *Phys. Lett.* **134B**, 449 (1984).

5. M. Oka, to be published.

TWO BARYON SYSTEMS IN THE NONRELATIVISTIC QUARK MODEL

N. Mankoč-Borštnik
University of Ljubljana, J. Stefan Institute
Ljubljana, Yugoslavia

ABSTRACT

The effective baryon-baryon potential is presented in a model with two three-quark clusters. The high nonlocality and non-adiabaticity of the potential is discussed. The shape of the local adiabatic part strongly depends on the definition of the effective potential and on the subspace used in the calculation. Information about the repulsive core and the attractive part of the potential has to be extracted not only from local, but also from nonlocal terms. The importance of the contributions of different channels including colored-baryon colored-baryon channels is discussed.

INTRODUCTION

When studying a system of atoms (like a fluid) one can usually take the atoms as elementary constituents of the system and forget the electrons and nuclei. The effective interaction between atoms can be either measured or calculated as an effective interaction between two clusters of electrons and a nucleus. One can use a similar procedure in nuclear physics. For nuclei, the constituents of the nucleons, the quarks can largely be neglected. The nucleons are usually successfully treated as point-like and structureless particles. The force between the nucleons is found to contain both local and nonlocal terms.

To calculate the nuclear force as an effective interaction between two clusters of quarks the interaction between quarks has to be known. While in atomic case the force between the two constituents is well known, the quark-quark interaction can only be evaluated for large momentum transfer. For low momentum transfer it can be only roughly estimated. The exchange of one gluon is important at small distances and gives the contribution, $\frac{\alpha(r)}{r}\ \hat{\vec{C}}_1\ \hat{\vec{C}}_2$, the exchange of many gluons causes the confinement and contributes the term, $\beta r^N\ \hat{\vec{C}}_1 \cdot \hat{\vec{C}}_2$. The colour operator term $\hat{\vec{C}}_1 \cdot \hat{\vec{C}}_2$ permits only colour singlet-clusters to escape from colour-singlet system.

When looking for the effective potential between two clusters of particles the same procedure can be used to treat a number of cases. However, one can expect that nonlocal terms are much more important in nuclear then in atomic cases, since the kinetic energy of the relative motion of the two clusters is, due to very massive nucleus in comparison with the electron mass, much smaller than the kinetic energy of the particles in each cluster.

Several attempts have been made to derive the nucleon-nucleon or other hadron-hadron potentials as a residual interaction between quarks.[1-15] The main differences between these approaches are in the choice of the basis wave-functions, in the exactness with which the symmetries of the wave-functions have been treated, in the choice of quark-quark interaction, and in the definition of the effective potential. I discuss in this paper the procedure for obtaining the effective potential when many basis wave-functions are taken to describe the internal motion in the center-of-mass system of each cluster. I discuss some possible choices of local and nonlocal parts of the effective potentials and compare their behaviors.

THE CLUSTER MODEL

In this section I discuss the bound states or the scattering states at low energies of two clusters of quarks in the nonrelativistic approximation, assuming the so-called constituent mass of quarks. The functions $u_\ell(s)$ of the relative motion of the two centers of masses[(12)] have to be determined when solving the Schrodinger equation while the basis wave functions are chosen to discribe the isospin (T) ϕ_{K_T}, spin(S) ϕ_{K_S}, colour (C) ϕ_{K_C} and that spatial part of the wave functions connected with the internal motion in each cluster and the total center of mass motion $\phi_{K_X}(\vec{\xi}_i, \vec{S})$. The coordinates $\vec{s}$, $\vec{S}$, $\vec{\xi}_i$ describe the relative distance between the two centers of masses, the total center-of-mass and the internal coordinates in each cluster, respectively. The wave function of the system is antisymmetric, with spin, isospin, colour and angular momentum as good quantum numbers $I \equiv (S, T, C, L, J)$ and it is a mixture of N-N, Δ-Δ and different octet-octet channels.

The operator $\hat{P}^I_\ell$ is constructed[12,13] which projects the 6-particle wave function into the subspace of the quantum numbers I and chosen permutational symmetry. The index $\ell(\equiv \ell(k, i))$ determines appropriate configurations $K(\equiv k_x, k_T, k_s, k_c)$ in all subspaces and the corresponding permutational symmetries i. Then the trial function:

$$\psi^I = \sum_\ell \hat{P}_\ell^I \{\phi_{k_x}(\vec{\xi}_j^o, \vec{S}^o)\, \phi_{k_T}\phi_{k_S}\phi_{k_c}\, u_\ell(\vec{s})\} \quad , \tag{1}$$

when used in the Schrodinger equation leads to the system of integro-differential equations determining the functions $u_\ell(\vec{s})$ of the relative motion of the two centers of mass of the clusters,

$$\sum_{\ell'} \int H^I_{\ell\ell'}(\vec{s}, \vec{s}')\, u_{\ell'}(\vec{s}')d\vec{s}' = E^I \sum_{\ell'} N^I_{\ell\ell'}(\vec{s}, \vec{s}')\, u_{\ell'}(\vec{s}')d\vec{s}' \tag{2}$$

The kernels $H^I_{\ell\ell'}$ and $N^I_{\ell\ell'}$ have local and nonlocal terms:

$$H^I_{\ell\ell'}(\vec{s}, \vec{s}')=\delta(\vec{s} - \vec{s}')\tau^I_{\ell\ell'}(\vec{s})[-\frac{\hbar^2}{2m_s}\nabla^2_s + U^I_{\ell\ell'}(\vec{s})] + h^I_{\ell\ell'}(\vec{s},\vec{s}'),$$

and

$$N^I_{\ell\ell'}(\vec{s}, \vec{s}') = \delta(\vec{s} - \vec{s}')\, \tau^I_{\ell\ell'}(\vec{s}) + n^I_{\ell\ell'}(\vec{s}, \vec{s}') \quad . \tag{3}$$

One can take a complete basis, $\varphi_p(\vec{s}, \vec{R})$, where $\vec{R}$ is a set of so-called generator-coordinate parameters and p a set of quantum numbers. One can then look for the solution of (2) using the generator-coordinate trial function,

$$u_\ell(\vec{s}) = \sum_p \int f^p_\ell(\vec{R})\, \varphi_p(\vec{s}, \vec{R})\, d\vec{R}. \tag{4}$$

If the basis wave-functions in the spatial space are gaussians with a fixed width b,

$$\phi_{k_x}(\vec{\xi}_j, \vec{S}) = A \cdot e^{-S^2/12b^2} \prod_{j=1}^{4} e^{-\xi_j^2/2b^2} \quad ,$$

then $\varphi_p(\vec{s}, \vec{R})$ is a gaussian,

$$\varphi = (b\sqrt{\tfrac{2\pi}{3}})^{-3/2} e^{-(\vec{s} - \vec{R})^2/2b^2}$$

with only one p (which is omitted).

The trial wave-function of Eq. (4) transforms the integro-differential equations, Eq. (2), into integral Hill-Wheeler equations, which determine the unknown functions $f_{\ell}^{p}(\vec{R})$. In the simple case when the basis wave-functions in the spatial space are gaussians, the Hill-Wheeler equations have a form,

$$\sum_{\ell'} \tilde{H}^{I}_{\ell\ell'}(\vec{R}, \vec{R}')f_{\ell'}(\vec{R}') = E^{I} \sum_{\ell'} \tilde{N}^{I}_{\ell\ell'} (\vec{R}, \vec{R}') f_{\ell'}(\vec{R}') \tag{5}$$

The old and the new kernels are connected by the following transformation,

$$\int \frac{d\vec{s}d\vec{s}'}{(2\pi/3)^{1/2}b} e^{-(\vec{s} - \vec{R})^{2}/\frac{4b^{2}}{3}} \begin{Bmatrix} H^{I}_{\ell\ell'}(\vec{s}, \vec{s}') \\ N^{I}_{\ell\ell'}(\vec{s}, \vec{s}') \end{Bmatrix} e^{-(\vec{s}' - \vec{R}')^{2}/\frac{4b^{2}}{3}} = \begin{Bmatrix} \tilde{H}^{I}_{\ell\ell'}(\vec{R}, \vec{R}') \\ \tilde{N}^{I}_{\ell\ell'}(\vec{R}, \vec{R}') \end{Bmatrix} \tag{6}$$

DEFINITION OF THE EFFECTIVE POTENTIAL

In the simple case that the spatial wave-functions are gaussians Eqs. (2, 3) simplify since then $\tau_{\ell\ell'}(\vec{s}) = \delta_{\ell\ell'}$. The equations can be interpreted as a two-body Schrödinger equations for the relative motion of two clusters with a local potential $U_{\ell\ell'}(\vec{s})$ and a nonlocal potential $V_{\ell\ell'}(\vec{s}, \vec{s}')=h_{\ell\ell'}(\vec{s}, \vec{s}')-E^{I}n_{\ell\ell'}(\vec{s}, \vec{s}')$. The nonlocal part is due to the exchange ofquarks between the two clusters.

Since many channels contribute to the local non-adiabatic part of the potential, one may try to define the local adiabatic part of the potential by using the Born-Oppenheimer approximation. One evaluates first the potential energy part of Eqs. (2, 3) at a fixed distance between the two centers of mass taking: $u_{\ell}(\vec{s}) =\sqrt{\delta}(s - \vec{s}_{o}) a_{\ell}(\vec{s}_{o})$. Then Eq. (2) gives then Born-Oppenheimer equation for the potential energy $E^{jI}(\vec{s}_{o})$ and the coefficients $a_{\ell}^{j}(s_{o})$:

$$\sum_{\ell'} U^{I}_{\ell\ell'}(\vec{s}_{o}) a^{j}_{\ell'}(\vec{s}_{o}) = E^{jI}(\vec{s}_{o}) \sum_{\ell'} \tau^{I}_{\ell\ell'}(\vec{s}_{o}) a^{j}_{\ell'}(\vec{s}_{o}) \quad . \tag{7}$$

The approximate solution of Eq. (2) follows from the Schrödinger equation,

$$-\frac{\hbar^2}{2m_s}\nabla_s^2\tilde{u}_j(\vec{s}) - \frac{\hbar^2}{2m_s}\hat{T}_j^I(\vec{s})\tilde{u}_j(\vec{s}) + E^{jI}(\vec{s})\tilde{u}_j(\vec{s}) = \tilde{E}^{jI}\tilde{u}_j(\vec{s}), \tag{8}$$

with the additional term coming from the $\vec{s}$ dependence of the coefficients

$$a_\ell^j(\vec{s})\ \hat{T}_j^I(\vec{s}) = \sum_{\ell\ell'} a^{*j}_\ell(\vec{s})\tau^I_{\ell\ell'}(s)\ [\nabla^2_{\vec{s}}\ a_\ell^{j'}(\vec{s}) + 2\vec{\nabla}_s\ a^j_{\ell'}(\vec{s})\nabla_{\vec{s}}]\ .$$

One can use the coefficients $a_\ell^j(\vec{s})$ as a basis and look for the solutions of Eq. (2) with the trial function,

$$u_\ell(\vec{s}) = \sum_j a_\ell^j(\vec{s})\ v_j\ (\vec{s}). \tag{12}$$

As a next step one can evaluate the contribution of nonlocal terms[12] since the essential physical features are still hidden in the non-adiabatic and nonlocal terms. In the simple case of the gaussian spatial basis, the parameter $\vec{R}$ represents the distance between the two potentials used to determine the single particle wave functions:

$$X(\vec{x}) = \pi^{-3/4}\ b^{-3/2}\ e^{-(\vec{x}\ +\ 1/2\vec{R})^2/2b^2},$$

and (9)

$$\psi(\vec{x}) = \pi^{-3/4}\ b^{-3/2}\ e^{-(\vec{x}\ +\ 1/2\vec{R})^2/2b^2}.$$

Therefore, one can define a local effective potential by fixing the distance $\vec{R}$ between the two potentials. If the diagonal terms in $\vec{R}$ are then transformed into a diagonal basis and the vectors $g_\ell^j(\vec{R})$ are obtained, the approximate effective local potential has the form:

$$E^{jI}(\vec{R}) = (\sum_{\ell\ell'} g_\ell^{*j}(\vec{R})\tilde{H}^I_{\ell\ell'}(\vec{R},\vec{R})g^j_{\ell'}(\vec{R}))/(\sum_{\ell\ell'} g_\ell^{*j}(\vec{R}),\tilde{N}^I_{\ell\ell'}(\vec{R},\vec{R})\ g_\ell^{j'}(T\vec{R}))$$

$$-(\sum_{\ell\ell'} g_\ell^{*j}\ (\infty)\ \tilde{H}^I_{\ell\ell}(\infty,\ \infty)\ g^j_{\ell'}(\infty)/\sum_{\ell\ell'} g_\ell^{*j}(\infty)\ \tilde{N}^I_{\ell\ell'}(\infty,\ \infty)\ g^j_{\ell'}(\infty)). \tag{10}$$

The potential $E^{jI}(\vec{R})$ includes nonlocal terms concerning the relative motion of the two centers-of-mass, since the relative distance s is smeared about fixed R. This movement is described in the simple case of gaussian spatial functions (Eq. (9)) by the gaussian function

$$(\sqrt{\pi}b)^{-3}\ e^{-(\vec{s}\ -\ \vec{R})^2/4b^2/3}\ .$$

and has, as $R \to \infty$, the value $3/4\ \hbar^2/mb^2$.

The two Eqs. (2) and (5) are equivalent in the sense that, when being solved in the same subspace, both give the same energies E^I or the same phase shifts.

RESULTS

The Hamiltonian of the system is taken to be

$$\hat{H} = \sum_{i=1}^{6} \frac{\hat{p}_i^2}{2m} + \sum_{i>j=1}^{6} (\hat{V}_{ij}^{c} + \hat{V}_{ij}^{ss}) - \hat{T}_t\ , \tag{11}$$

$$\hat{V}_{ij}^{c} = -\hat{\vec{C}}_i\hat{\vec{C}}_j(\tilde{H}\ x_{ij}{}^2 + \delta)\ \text{and}\ \hat{V}_{ij}^{ss} = -\hat{\vec{C}}_i\hat{\vec{C}}_j(1 + \frac{8}{3}\ \hat{\vec{S}}_i\cdot\hat{\vec{S}}_j)\delta(\vec{x}_{ij})\gamma.$$

Here, $\hat{\vec{C}}_i$ and $\hat{\vec{S}}_i$ are the colour and spin operators, respectively. The interaction $\hat{V}_{ij}^{ss}$ provides the splitting between the nucleon and Δ. The parameters δ and γ are chosen so that the observed masses of the nucleon and the Δ are reproduced (See Table 1).

Table 1

Parameters of the quark-quark interaction and quark masses; b is the self consistent gaussian width of the single-particle wave-function.

m (MeV)	$\tilde{H}$ (Mev.fm^{-2})	γ (MeV-fm^3)	δ (MeV)	b (fm)
310	400	295	-711	0.551

Now we consider the three effective local potentials $E^I(\vec{s})$, $E^I(\vec{R})$ and $E(\vec{R})$ from Eqs. (7) and (10). For $E(\vec{R})$ all quantum numbers, except the total angular momentum, are assured, and the relevance of these local potentials are discussed.

A partial test of the nonlocal terms is performed by solving exactly Eq. (2) or (5) for the bound states, and by evaluating the first correction to the nonlocal and non-adiabatic terms. The systems described by the antisymmetric wave-functions with quantum numbers C=0, T=0, S=1, L=0; and C=0, T=0, S=3, L=0 are considered. The adiabatic wave-functions are pure N-N or Δ-Δ channels at large $\vec{s}$ or $\vec{R}$, but at small distances the N-N, Δ-Δ and coloured-baryon-coloured-baryon channels contribute.

The potential $E^I(\vec{R})$ in Figs. 1 and 2 is more repulsive at short distances then the other two, but has a weak attractive part with a minimum around 1.5 fm. The potential $E^I(\vec{s})$ is much less repulsive then the other two in the N-N case and is strongly attractive in the Δ-Δ case. The repulsion is hidden in nonlocal terms. The repulsion in the two case of $E(\vec{R})$ and $E^I(\vec{R})$ is expected to be due to exchange of quarks between clusters.

In Fig. 3 and Fig. 4 the effect of configuration mixing is presented for N-N local effective potentials $E^I(\vec{R})$ and $E^I(\vec{s})$, respectively. The figures show that the effect of configuration mixing is rather weak for the $E^I(\vec{R})$ potential and is extremely strong for the $E^I(\vec{s})$ local potential. The figures illustrate that the local part of the potential can have a very limited meaning.

The effective local Δ-Δ potential $E^I(\vec{s})$ for T=0, C=0, L=0 and S=3 in Fig. 5 is attractive. The approximate Schrödinger equation, Eq. 8, gives for the binding energy of the system as -35 MeV, while the exact solution of Eqs. (2) or (5) gives -40 MeV. The first correction to the Born-Oppenheimer approximation[12] contributes to the repulsion at very small distances and to the attraction for $s \sim 0.5$ fm. The new effective potential does not give any binding.

Figure 6 shows the sensitivity of the effective local potential to the quark-quark interaction and to the mass. For an extremely strong confining potential (κ=1500 MeV fm^{-2}) the repulsive core is narrower and the attractive minimum is deeper and closer to zero. For an extremely weak confining potential the minimum disappears and the repulsive core is wide. The change of quark mass from 310 MeV to 100 MeV has a similar effect. In all the choices the width of the gaussian is self-consistently determined. In general, the qualitative features of the effective potential are not

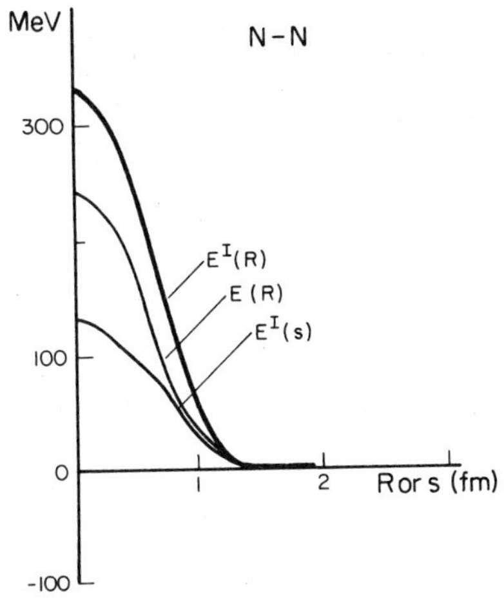

FIG. 1 Effective local N-N potential for T=0, C=0, L=0 and S=1. The effective local potential $E^I(s)$ has no contribution from the exchange of quarks between the two clusters. The two potentials $E^I(R)$ and E(R) contain local and nonlocal terms. The potential $E^I(R)$ has a weak attractive minimum around 1.5 fm. The wave-function contains N-N, Δ-Δ and some coloured-baryon-coloured-baryon channels.

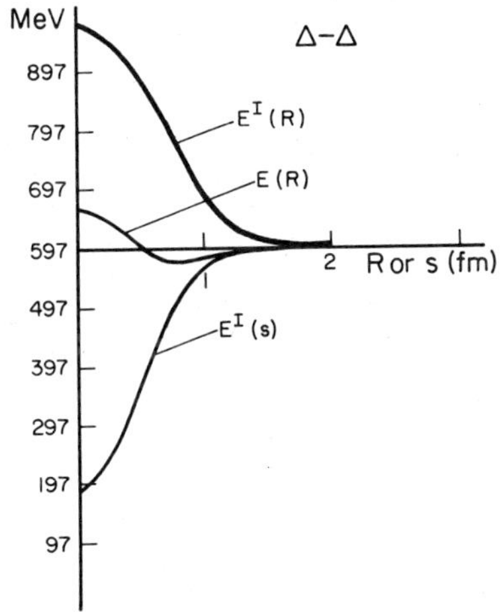

FIG. 2. Effective local Δ-Δ potential for T=0, C=0, L=0 and S=1, as for Fig. 1.

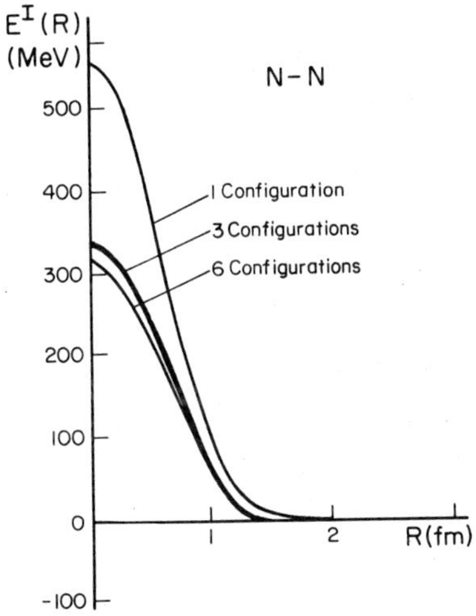

FIG. 3. The effective local N-N potential $E^I(R)$ for T=0, C=0, L=0 and S=1. The curve with one configuration represents the pure N-N channel.

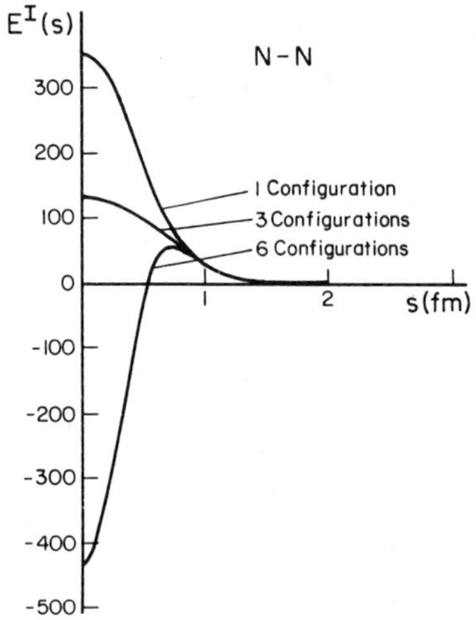

FIG. 4. The effective local NN potential $E^{I}(s)$ for T=0, C=0, L=0 and S=1. The curve with one configuration represents the pure NN channel.

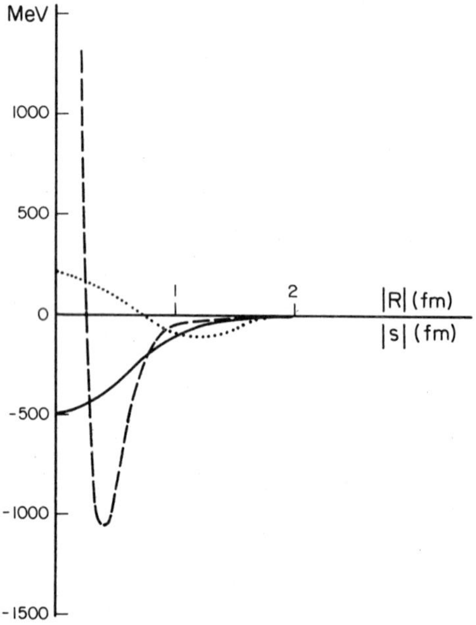

FIG. 5. The effective local Δ-Δ potential $E^{I}(s)$ for T=0, C=0, L=0, and S=3, with (dashed line) and without (solid line) correction from the nonlocal terms. The local $E^{I}(R)$ is also shown by the dotted line.

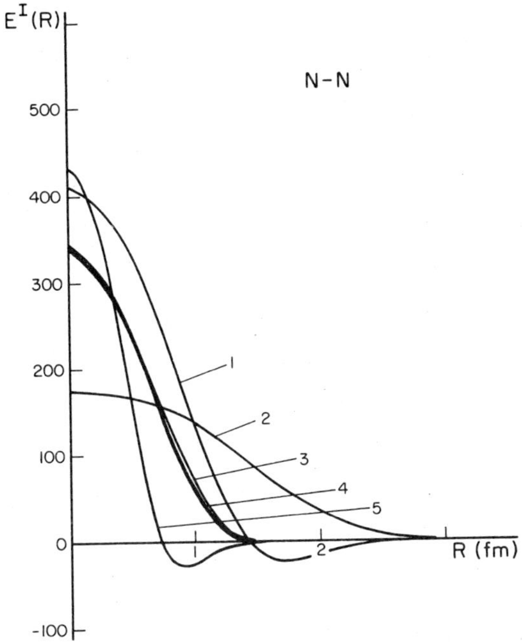

FIG. 6. The effective local N-N potential for T=0, C=0, L=0 and S=1. Curve 1 corresponds to K=400 MeV.fm^{-2} and mc^2=100 MeV; curve 2 corresponds to K=50 MeV.fm^{-2} and mc^2=310 MeV, curve 3 corresponds to different confinement potential of Eq. 11:

$$V_{ij} = -\hat{\vec{C}}_i\,\hat{\vec{C}}_j\left(\frac{-\alpha}{|\vec{x}_{ij}|} + \beta\,|\vec{x}_{ij}| + \delta\right)$$

with α=40 MeV.fm, β=750 MeV.fm^{-1} and mc^2=310 MeV, curve 4 corresponds to K=400 MeV.fm^{-2}, and mc^2=310 MeV, curve 5 corresponds to K=1500 MeV.fm^{-2} and 310 MeV. See Table 1 Ref. 13.

very sensitive to the choice of the parameters. A similar conclusion is also valid for $E^I(\vec{s})$.

CONCLUSIONS

The nonlocal and non-adiabatic terms in the effective potential between two clusters of quarks are strong. The shape of the local adiabatic term depends strongly on its definition and the chosen basis. In the N-N case the local part of the effective potential cannot be directly compared to the corresponding term of the usual N-N potential. Here the N-N potential is determined from a completely different subspace than it is for point-like and structureless nucleons. Different definitions of the effective local potential contain different amounts of nonlocal effects and therefore the results are quite different. The physical reason for the nonlocal terms is the exchange of quarks between clusters. The meaning of different definitions depends on the choice of the basis. Choosing a much wider gaussian would completely change the results unless the parameters of quark-quark interaction are changed so that the N and Δ masses are reproduced.

The local potential $E^I(\vec{s})$ has no contribution from the exchange of the quarks between clusters and the kinetic and potential energies are well separated. The local potential $E^I(\vec{R})$, contains the kinetic energy of the relative motion of the c.m. of the two clusters, as well as the exchange of quarks between clusters. Nonlocal terms can contribute a large amount to the effective potential, and only for specially chosen model spaces, could the local part be dominant. The results for either the local part of the potentials or for the exact solutions are not very sensitive to changes in the parameters of the quark-quark interaction provided we adjust the gaussian width of the single-particle wave functions self-consistently. The model is too rough to reproduce the weakly bound deuteron, even if missing degrees of freedom, like pions were to be added. This is due to the fact that the kinetic and potential energy terms are large while the binding energy of the deuteron is small. In addition, the quark-quark interaction is far from being well defined.

ACKNOWLEDGEMENTS

The author warmly thanks The Ohio State University, Department of Physics for their hospitality and financial support during the time that this paper was prepared.

References

1. D.A. Liberman, Phys. Rev. D16 (1977) 1542.

2. A.N. Mitra, Phys. Rev. D17 (1978) 729.

3. G.W. Barry, Phys. Rev. D16 (1977) 2886.

4. G.S. Warke and R. Shanker, Phys. Lett. 89B (1979) 17; Phys. Rev. C21 (1980) 2643.

5. M. Oka and J. Yazaki, Phys. Lett. 90B (1980) 41.

6. M. Harvey, Nucl. Phys. A352 (1981) 301, 326.

7. C. De Tar, Phys. Rev. D17 (1978) 323.

8. J. Ribeiro, L. Phys. C5 (1980) 27.

9. D. Robson, Nucl. Phys. A308 (1978) 381.

10. T. Bender, H.G. Dosch: Heidelberg reprint 1981.

11. A. Faessler, Springer Tracts in Modern Physics 100 (1982) 214, Springer Verlag, Eol. D. Fries and B. Leutnitz.

12. N. Mankoč-Borštnik, M. Cvetič, J. Phys. G., Nucl. Phys. 7 (1981) 1385.

13. M. Cvetič-Krivec, B. Golli, N. Mankoč-Borštnik, M. Rosina, Nucl. Phys. A395 (1983) 399.

14. M. Cvetic, B. Golli, N. Mankoč-Borštnik, M. Rosina, Phy. Lett. 93B, 489 (1980); 99B, 486 (1981).

15. M. Rosina, M. Cvetič-Krivec, B. Golli, N. Mankoč-Borštnik, Proceed of the Inter. School of Nucl. Phys. Erice, 1981, Progress in Particle and Nucl. Phys., Vol. 8.

16. N. Mankoč-Borštnik, M.V. Mihailovič and M. Rosina, Nucl. Phys. A239 (1975) 321.

COVARIANT CONDENSATE WAVE FUNCTIONS IN THE SKYRME MODEL

John P. Ralston
University of Kansas
Lawrence, KS 66044

ABSTRACT

We construct field theory wavefunctions for Skyrmion states by expanding in a multiparticle Fock space basis. The formalism, which is covariant and free of zero-mode difficulties, allows a calculation of the electromagnetic form factor $F(q^2)$ in terms of a regularization scale μ^2, F_π^2, m_π^2, and the classical Skyrme function $F(r)$. A trial calculation gives $F(q^2) \sim (1/q^4)\exp(q^2/8\lambda^2)$ for $q^2 \to -\infty$.

INTRODUCTION

The conventional motivation for the Skyrme model as an effective action approach to QCD is based[1,2] on the large N_c limit. In this limit F_π^2 scales like $N_c \to \infty$, which leads to a semiclassical expansion[2] dominated by classical physics.

The real world, of course, has F_π = 186 MeV. This is a rather small number as hadronic scales go. Given the asymptotic character of the approximation, which in effect runs $F_\pi^2 \to \infty$ with all other scales fixed[3], it is rather amazing that the phenomenology of the model is so good.

At the same time we know that chiral symmetry is spontaneously broken in the real world, so that if a derivative expansion exists it encompasses the ordinary Skyrme model. So, while the search for the true large N_c theory makes its progress with all deliberate speed, one can independently motivate a study of chiral solitons strictly on the basis of chiral symmetry.[4]

For finite F_π^2, fluctuations will be important. However, one can still[5] employ a variant of the semi-classical expansion by recalling an analogy with statistical mechanics. Given that F_π^2 scales out of the Lagrangian, this constant plays the role of an inverse temperature, β, in the quantum partition function. There is a paradox involved in arguing that $\beta \to \infty$ describes the system at finite β. Nevertheless, there are many systems in nature, e.g. liquid Helium, for which this can be true. The many-body physics of a condensed fraction makes this a good starting point, even in cases where the interactions are too strong and the temperature too high to obviously motivate the picture as an expansion in a small parameter.

From this viewpoint, a condensed fraction may be relevant to the present Skyrme model success. The mathematical use of the large F_π limit would be to put the model into a condensed phase, which persists to some extent as F_π is decreased. The condensate makes its presence felt through an expectation value or classical field, normalized by whatever fraction of the degrees of freedom participate in this collective effect.

From a lattice viewpoint[6] the Skyrmion would represent a topological feature of the continuum limit, quite similar to the monopoles of the Abelian or non-Abelian gauge theory. Exactly as in that case, the theory may be different depending on which side of a critical value of a coupling constant, F_π, one starts with. Moreover, the correlations near the phase transition must be extremely important; since F_π has a dimension, we must compare the size of F_π to some typical fluctuation scale μ at each F_π. It would seem likely that with F_π = 186 MeV the system must be close to a critical value even if we are lucky enough to be above it.

In this paper I will outline a study of the chiral soliton model based on this viewpoint; a more complete discussion is given in Ref. (5). Whether or not a condensed phase is relevant, the same minimal field theory shopping list below needs to be respected. By now, any calculation should:

1. Define matrix elements being calculated.
2. Regulate infinities--you know they are present.
3. Specify states under discussion.
4. Be covariant.

The use of a condensed fraction will inspire the human tendency to linearize, but is not so potent as to eliminate the need for field theory.

Wavefunctions

We will construct states[5] with a position index $|\eta_a\rangle$, so that the expectation of the field operator $\Phi(x)$ satisfies

$$\langle\eta_a|\Phi(\vec{x},0)|\eta_a\rangle / \langle\eta_a|\eta_a\rangle = \phi(\vec{x}-a) \tag{1}$$

where $\phi(x-a)$ represents a classical field given by the model. Rather than explicitly shifting operators (or path integral variables), which ruins translational and other symmetries, one can implement Eq. (1) by expanding $|\eta_a\rangle$ in the basis given by the unshifted operators. Thus in free field theory, using $[a_k, a^\dagger_{k'}] = (2\pi)^3 2\omega_k \delta^3(k-k')$, we consider Bargman or coherent states[7,8]

$$|\eta_a\rangle = \exp\{-\int \widetilde{dk}\ \tfrac{1}{2}|\eta_k|^2 - a_k^\dagger \eta_k e^{ika}\}\ |0\rangle \qquad (2)$$

with $\widetilde{dk} = d^3k/[(2\pi)^3 2\omega_k]$. The parameters η_k determine the shape $\phi(x)$, while the phase ika determines the location in Eq. (2). By choosing η_k one can match (2) to Eq. (1) identically. This does not contradict Goldstone and Jackiw.[9] Such states are essentially identical to those of any other linearization procedure, with the direct substitution of generalized modes for plane waves, but with the advantage of attaching a label (a) to the state rather than an operator. Different modes may also span the space but implement a different time dependence.

The states $|\eta_a\rangle$ are not supposed to be stationary in time; the energy-momentum eigenstates should be smeared over all values of a. To see this we have constructed asymptotic momentum states $|S_p\rangle$ satisfying $P^\mu|S_p\rangle = p^\mu|S_p\rangle$, by taking

$$|S_p\rangle = c_p \int d^4a\ e^{-ipa}|\eta_a\rangle \qquad (3)$$

and constraining $p^2 = m^2$. With $\langle\eta_a|\eta_a\rangle = 1$ in Eq. (2), the conventional normalization $\langle S_{p'}|S_p\rangle = (2\pi)^3 2\omega_p \delta^3(p-p')$ determines c_p. The Fock decomposition is

$$|S_p\rangle = N c_p (2\pi)^4 \sum_J^\infty \frac{1}{J!}\int \prod_{i=1}^{J} \{\widetilde{dk}_i\, \delta^4(p - \sum^J k_\ell)\ \eta_{k_i} a_{k_i}^\dagger\}|0\rangle, \qquad (4)$$

$$N = \exp(-\tfrac{1}{2}\int \widetilde{dk}|\eta_k|^2),$$

which reveals that the superposition is truly stationary.

Now one can calculate legitimate matrix elements by expanding the physical proton in terms of functional sums over the shape parameters η_k:

$$|p\rangle = \int d^2[\eta_\ell]\quad |\eta_\ell\rangle\langle\eta_\ell|p\rangle$$

$$\rightarrow \int d^2[\eta]\quad |S_p(\eta)\rangle\ \ \rho(\eta)$$

This is exact, but truncating $\rho(\eta) \rightarrow Z\delta(\eta-\eta_k)$ will give us $|p\rangle = Z|S_p\rangle$. This replaces the problem by one with the gaussian fluctuations of free field theory.

Finally, the fluctuations will be divergent unless we devise some regularization method to define products of operators. Rather than splitting points, one can smear the operator $\Phi(x) \rightarrow \Phi_f(x)$, defining[8]

$$\Phi_f(x) = \int dx' \, f(x-x') \, \Phi(x') \qquad (5)$$

$$f_k = \int dx \, e^{ikx} f(x); \; f_{k=0} \equiv 1.$$

This introduces a scale $\mu^2 = \int \tilde{dk} |f_k|^2$ and the naive local limit, $f_k = 1$, will be $\mu^2 \to \infty$. Removing any divergences in this limit defines a regularized or normal-ordered operator matrix element.

It should be noted that matrix elements between $|S_p\rangle$ states are determined by off-diagonal matrix elements of $|\eta_a\rangle$. Thus there is generally a continuous sequence of saddle point functions in the gaussian integrals, determined by $\phi(x-a)$ and $\phi(x-a')$.

A Calculation: $F(q^2)$

Now we would like to illustrate the ideas by calculating a matrix element as a function of:

q^2 = momentum transfer,
μ^2 = regulator,
F_π^2 = typical scale of fields,
m_π^2 = if sensitive,
$\phi(x)$= classical parameters for centers of gaussians.

We chose to calculate $\langle S_{p'} | B^\mu(0) | S_p \rangle \propto (p+p')^\mu F(q^2)$, where $F(q^2)$ is the electromagnetic form factor. The choice of B^μ, the topological current defined by

$$B^\mu = \mathrm{Tr}\, \{\varepsilon^{\mu\nu\alpha\beta}(U^\dagger \partial_\nu U)\,(U^\dagger \partial_\alpha U)(U^\dagger \partial_\beta U)\,\} \,/\,(24\pi^2), \qquad (6)$$

$$= -(1/2\pi^2 r^2)\sin^2 F(r) F'(r),$$

using the hedgehog ansatz $U = \exp(iF(r)\hat{r}\cdot\vec{\tau})$, is motivated by the following observations. First, B^μ is the most singular contribution to the electromagnetic current, and second, B^μ is a geometrical quantity that will always be in the theory. Our calculation will depend only on the form of B^μ and $F(r)$ near the origin.

Here we will only report results[5] of an impulse approximation at zeroeth order, using Eqs. (2,3) and (5). The fluctuations coming from the different saddle point functions, associated with $|S_p\rangle$ and $|S_{p'}\rangle$, can be integrated, yielding for <u>large</u> $-q^2$:

$$\langle S_{p'} | B^0 | S_p \rangle = c_p c_{p'} \int da^- e^{-ip^+ a^-} \langle \eta_{a'} | \eta_a \rangle \delta(0) I(q^2, a^-, \mu^2), \quad (7)$$

$$\delta(0)I(q^2,0,\mu^2)| \sim -\int \frac{d^3x\, e^{i\vec{q}\cdot\vec{x}}}{F_\pi^2\, \pi^2}\{\frac{\partial F}{\partial r}f\, \frac{\sin^2 F_f}{r^2}f + \frac{\partial F_f}{\partial r}f\, \frac{\cos 2F_f}{r^2}f\, \Gamma(\mu,F_\pi)\} \tag{8}$$

The fluctuations manufacture $\Gamma(\mu,F_\pi) = 1-\exp(-8\mu^2/F_\pi^2)$. This function is zero as $F_\pi^2 \to \infty$, but re-sums an infinite series of divergences as $\mu^2 \to \infty$. In Eq. (8), the c-number F_f is $F(r)$ regulated in the same way as Eq. (5).

We studied large $-q^2$ for several reasons. Physically the elastic form factor at any q^2 is a quantity crucially dependent on coherence. The soliton model need not be a bad description of such physics. Furthermore, the μ^2 dependence and large q^2 limit might be rather independent of the choice of modes.

By making integral estimates in Eqs. (6,7) we obtain a form factor, $F(q^2)$, which has two distinct parts. A factor of $F_f'^2(0)F_f''(0)/q^4$ comes from the $|\vec{x}|\to 0$ behavior. We interpret this power behavior as scattering from the isospin knot at the core of the Skyrmion. The other part is an exponential coming from the overlap $\langle\eta_{a'}|\eta_a\rangle$. This is a finite size or recoil effect. All terms together give the estimate

$$F(q^2) \sim (\text{const.}/q^4)\exp(q^2/8\lambda^2) \tag{9}$$

where λ^2 is calculable, but dependent on all details of η_k and generally proportional to F_π^2.

Conclusion

Several interesting features have emerged. The dependence on μ^2/F_π^2 is very suggestive and rather general. The procedure is covariant and well-defined. Finally, more detailed calculations must be done to see if $8\lambda^2$ could be as large as a GeV^2. If so, the trial calculation itself might be phenomenologically relevant.[10] A more detailed exposition of the calculations and the relations of the states to those of the canonical procedure is given in Ref. (5).

ACKNOWLEDGEMENTS

This work was supported in part by the U.S. Department of Energy under Grant No. DE-FG02-85ER40214, and by University of Kansas General Research Allocation No. 3603-0038.

REFERENCES

1. A.P. Balachandran, V.P. Nair, S.G. Rajeev and A. Stern, Phys. Rev. D27, 1153(1983). For reviews see I. Zahed and G.E. Brown, Phys. Rep. (to be published) and these proceedings.

2. G.S. Adkins, C.R. Nappi and E. Witten, Nuc. Phys. B228, 552(1983).

3. For consequences of this see E. Braaten and J.P. Ralston, Phys. Rev. D31, 598(1985); R. Ingermanson, "Quantum Fluctuations and the Effective Action Functional in the Skyrme Model," LBL-19122 preprint (1985).

4. N.K. Pak and H.C. Tze, Ann. Phys. (NY) 117, 164(1979); L.F. Fadeev, Lett. Math. Phys. 1, 289(1976).

5. J.P. Ralston, "Two Phase Model of the Skyrmion," Kansas University preprint (1985).

6. See, e.g., J. Kogut, Rev. Mod. Phys. 55, 775(1983); J. Polonyi, these proceedings; R. Saly, "Lattice Skyrme Model," Carlton University preprint (1984).

7. V. Bargman, Proc. Natl. Acad. Sci. 48, 199(1962); S. Schweber, Jour. Math. Phys. 3, 831(1962); R.J. Glauber, Phys. Rev. 131, 2766(1964);Itzykson and Zuber, Ref. (8), p. 118f.

8. Our conventions are that of C. Itzykson and J. Zuber, Quantum Field Theory (McGraw Hill 1980).

9. J. Goldstone and R. Jackiw, Phys. Rev. D11, 1486(1975).

10. See, e.g., G.P. Lepage and S.J. Brodsky, Phys. Rev. D22, 2157(1980); N. Isgur and C. Llewellyn-Smith, Phys. Rev. Lett. 52, 1080(1984).

NUCLEI IN THE SKYRME MODEL

Eric Braaten and Larry Carson
Northwestern University
Evanston, IL 60201

ABSTRACT

The application of the Skyrme model to nuclear physics is discussed. A new approach is presented in which nuclei are identified with static soliton solutions in the appropriate topological sector. When this approach is applied to the deuteron, it yields automatically the correct spin, isospin, and parity quantum numbers.

INTRODUCTION

The Skyrme model has been rather successful in describing nucleons as solitons of the pion field. The correct quantum numbers are obtained automatically, calculations of static properties are accurate to within 30%, and one can even get a reasonable description for certain dynamical properties.[1] It is natural then to ask whether this model can also give a reasonable description of nuclei.

How would one apply the Skyrme model to nuclear physics? One possibility is the "potential approach" pioneered by Jackson, Jackson, and Pasquier.[2] They extracted from the Skyrme model a "Skyrmion-Skyrmion interaction potential." After a suitable projection, they obtained a nucleon-nucleon interaction potential, which reproduced many of the qualitative features of standard nuclear physics potentials.

In this paper, we introduce a new approach which is the exact analog of the standard treatment of the nucleon by Adkins, Nappi, and Witten.[3] In this "soliton approach," a nucleus with atomic number A is identified with a static soliton solution with baryon number A. The quantization of the collective modes associated with this solution yields quantum states of the nucleus.

We begin by discussing the potential approach, pointing out its limitations. We then review the standard treatment of the nucleon, with emphasis on the constraint of Finkelstein and Rubinstein,[4] which implements the fermionic properties of baryons. This "soliton approach" is then generalized, so that it can describe nuclei. Finally we apply it to the deuteron, and show that the correct quantum numbers are obtained automatically.

POTENTIAL APPROACH

In the Skyrme model, the SU(2)-matrix-valued fields $U(\vec{x})$ fall into topological classes labelled by the integer-valued baryon number $\mathscr{B}$. The minimal energy configuration in the $\mathscr{B} = 1$ sector is attained by the static Skyrme solution

$$U_s(\vec{r}) = \exp(iF(r)\hat{r}\cdot\vec{\tau}) \ . \tag{1}$$

Translations and isospin rotations generate other solutions that are degenerate in energy. They can be parameterized by their center of mass $\vec{x}$ and an isospin orientation matrix $A \in SU(2)$:

$$U(\vec{r};\vec{x},A) = AU_s(\vec{r} - \vec{x})A^{\dagger} \ . \tag{2}$$

The energy M_s of these configurations is the classical mass of the Skyrmion.

One can attempt to describe the interaction of two Skyrmions by a potential. Such a "Skyrmion-Skyrmion interaction potential" $V(r,A)$ has been extracted by Jackson, Jackson, and Pasquier.[2] They consider the product ansatz

$$U(\vec{r};\vec{x}_1,A_1)U(\vec{r};\vec{x}_2,A_2) \ , \tag{3}$$

which has baryon number $\mathscr{B} = 2$. It is not a solution to the equations of motion, but its static energy $E(r,A)$ can be computed. It depends only on the separation $r = |\vec{x}_1-\vec{x}_2|$ and on the relative isospin orientation $A = A_1^{\dagger}A_2$. As $r \to \infty$, $E(r,A)$ approaches $2M_s$, the energy of two isolated Skyrmions. The Skyrmion-Skyrmion interaction potential is then

$$V(r,A) = E(r,A) - 2M_s \ . \tag{4}$$

One can extract from this a nucleon-nucleon interaction potential by taking a suitable projection over the isospin orientation A. Jackson, Jackson, and Pasquier found that this potential reproduced many of the qualitative features of the phenomenological Paris potential.

The behavior of $V(r,A)$ at large r has a simple physical interpretation in terms of single pion exchange. It is also fairly unambiguous, since the configuration $U(\vec{r})$ of one Skyrmion will not be severely deformed by the weak asymptotic field of the other. For small r, however, the potential suffers from severe ambiguities. Two overlapping Skyrmions will coalesce into a single configuration with baryon number $\mathscr{B} = 2$, which cannot be easily resolved into two individual Skyrmions with centers of mass $\vec{x}_i$ and isospin orientation A_i. Even if such a separation were possible, there are many deformations which would be energetically more favorable than translations and isospin rotations of the individual Skyrmions. Therefore the dynamics cannot be accurately described by a potential V(r,A). Not only does the validity of the product ansatz break down for small r, but the entire concept of a two-particle potential breaks down as well.

SOLITON APPROACH TO THE NUCLEON

We will present an alternative approach to nuclear physics in the Skyrme model, which yields information about nuclei directly without constructing any potentials. Our "soliton approach" is patterned after the standard treatment of the nucleon given first by Adkins, Nappi, and Witten.[3] We therefore review this treatment, which we can separate into 4 steps.

1. Find a static soliton which minimizes the energy in the $\mathscr{B} = 1$ sector. Such a solution $U_S(\vec{r})$ is given by Eq. (1).
2. Identify the set of solutions which are degenerate in energy. Ignoring the center of mass coordinate, whose quantization leads to momentum eigenstates, these solutions are $AU_S(\vec{r})A^\dagger$, where A is an SU(2) matrix.
3. Quantize the collective modes corresponding to the degenerate solutions. This can be implemented by allowing A to be time-dependent,

$$U(\vec{r},t) = A(t)U_S(\vec{r})A(t)^\dagger \ . \tag{5}$$

Upon substituting this into the Hamiltonian, we obtain

$$\begin{aligned} H &= M[U_S] + \Lambda[U_S]\mathrm{tr}(\dot{A}\dot{A}^\dagger) \\ &= M[U_S] + \frac{1}{2\Lambda[U_S]}\ell(\ell+1) \ , \end{aligned} \tag{6}$$

which is the Hamiltonian for a symmetric top with moment of inertia $\Lambda[U_S]$. It can be diagonalized explicitly, ℓ being the simultaneous eigenvalue of isospin and spin: $\vec{I}^2 = \vec{J}^2 = \ell(\ell + 1)$.

4. Impose the following constraint on the wave-functions:

$$\psi(-A) = -\psi(A) \ . \tag{7}$$

This restricts the eigenvalues ℓ to be half-odd-integer: $\ell = 1/2$, $3/2, \ldots$. This constraint for the SU(2) Skyrme model is the ghostly remnant of the Wess-Zumino term, which was discussed for the SU(3) case by Zahed.[1] It implements the fermionic nature of the nucleon in this theory. It must play a similar role for nuclei, restricting the quantum numbers of the nucleus to conform with the Pauli exclusion principle. It therefore deserves to be discussed in more detail.

This constraint was first understood by Finkelstein and Rubinstein,[4] who observed that if the configuration space for a system is not simply connected, quantum mechanics allows its wavefunction to be multi-valued. A familiar example is the rotation group SO(3). This space is doubly-connected: a 2π rotation corresponds to a closed path in SO(3) which cannot be contracted to a point, while a 4π rotation is contractible. A wavefunction $\psi(\vec{x})$ can therefore be double-valued:

$$\psi(R(2\pi,\hat{n})\cdot\vec{x}) = \pm\psi(\vec{x}) \ , \tag{8}$$

where $R(2\pi,\hat{n})$ is a rotation by 2π about some axis $\hat{n}$.

In the Skyrme model, the configuration space is the set of all SU(2)-valued fields $U(\vec{x})$ which approach 1 as $|\vec{x}| \to \infty$. This space is also doubly-connected. If a configuration $U(\vec{x})$ has odd baryon number $\mathscr{B}$, then $U(R(\theta,\hat{n})\cdot\vec{x})$, where θ runs from 0 to 2π, is a closed path which cannot be contracted to a point. The wavefunctional $\Psi[U(\vec{x})]$ can therefore be double-valued:

$$\Psi[U(R(2\pi,\hat{n})\cdot\vec{x})] = (\pm 1)^{\mathscr{B}}\Psi[U(\vec{x})] \ . \tag{9}$$

Finkelstein and Rubinstein realized that the choices -1 and +1 in Eq. (9) determine completely independent field theories, one in which the Skyrmions are fermions and the other in which they are bosons. Furthermore, if a choice of sign is made for one noncontractible path, then that same sign must be imposed on every other noncontractible path. Thus if one desires the Skyrmions to be fermions, the Lagrangian for the SU(2) Skyrme model must be supplemented by the constraint: $\Psi \to -\Psi$ at the end of every noncontractible closed path.

Eq. (7) can now be understood as the Finkelstein-Rubinstein constraint. A rotation $R(\theta,\hat{n})$ of $\vec{r}$ in Eq. (5) can be factored out of $U_s(\vec{r})$ and absorbed into A, replacing it by $A\exp(i\theta\hat{n}\cdot\vec{\tau}/2)$. Under a 2π rotation, which is a noncontractible closed path in U,

A goes to -A. Therefore the Finkelstein-Rubinstein constraint is simply Eq. (7).

SOLITON APPROACH TO NUCLEI

The soliton approach not only determines the quantum numbers of the nucleon, but it also enables us to compute its static properties as well as certain dynamical properties. If this approach could be extended to a nucleus, its properties would be calculable by the same methods. We have succeeded in making this extension. For a nucleus with atomic number A, this involves the following 4 steps:

1. Find a static solution $U(\vec{r})$ which minimizes the energy in the $\mathcal{B} = A$ sector. Complications arise primarily from the fact that for $\mathcal{B} \neq 1$ $U(\vec{r})$ need not have as much symmetry as the Skyrmion solution in Eq. (1).
2. Identify the set of solutions which are degenerate in energy with $U(\vec{r})$. Such solutions can be generated from $U(\vec{r})$ by rotations of both the isospin and coordinate axes:

$$AU(R\cdot\vec{r})A^{\dagger}, \quad A \in SU(2), \quad R_{ij} = \frac{1}{2}\,Tr(\tau^i B \tau^j B^{\dagger}), \quad B \in SU(2)\ . \tag{10}$$

If the solution $U(\vec{r})$ has special symmetries, the SU(2) matrices A and B may overdescribe the degenerate solutions. This occurs, for instance, in the $\mathcal{B} = 1$ case: every rotation is equivalent to an isorotation, so the matrices B are superfluous.

3. Quantize the collective modes corresponding to A(t) and B(t). This leads to a Hamiltonian of the form

$$H = M + \frac{1}{2}\,(X^{-1})_{ij}K^iK^j - (Z^{-1})_{ij}K^iL^j + \frac{1}{2}\,(Y^{-1})_{ij}L^iL^j\ , \tag{11}$$

where the classical energy M and the inertia tensors X, Y, and Z are functionals of $U(\vec{x})$. The dynamical variables $\vec{K}$ and $\vec{L}$ are the body-centered isospin and spin operators:

$$K^i = -\frac{1}{2}\,Tr(\tau^i A^{\dagger}\tau^j A)\,I^j\ ,$$
$$L^i = -\frac{1}{2}\,Tr(\tau^i B \tau^j B^{\dagger})\,J^j\ . \tag{12}$$

These operators commute with $\vec{I}$ and $\vec{J}$ satisfy $\vec{K}^2 = \vec{I}^2$ and $\vec{L}^2 = \vec{J}^2$, and obey angular momentum algebras.

Thus this Hamiltonian describes two coupled asymmetric tops. Since $\vec{I}$ and $\vec{J}$ commute with H, we can diagonalize $I^2 = i(i + 1)$, $\vec{J}^2 = j(j + 1)$, $I_3 = -i,\ldots,+i$, and $J_3 = -j,\ldots,+j$. We can choose the basis states to also be eigenstates of K_3 and L_3, with eigenvalues $-i,\ldots,+i$ and $-j,\ldots,+j$, so H reduces to a square matrix with $(2i + 1)(2j + 1)$ rows and columns. Its eigenvalues will be energy levels of the nucleus in the (i,j) channel. If the matrices A and B overdescribe the degenerate solutions, the Hamiltonian H must be supplemented by constraints which restrict the allowed states. In the case of the $\mathcal{B} = 1$ Skyrme solution in Eq. (1), the constraint is $\vec{K} + \vec{L} = 0$, which implies i = j and allows the Hamiltonian to be diagonalized explicitly as in Eq. (6).

4. The final step is to impose the Finkelstein-Rubinstein constraint on wave functions. This requires us to identify combinations of the SU(2) matrices A and B which correspond to endpoints of noncontractible closed paths in the space of fields $U(\vec{x})$. For example, a path from A to -A determines a closed path and it is a noncontractible path if the baryon number $\mathcal{B}$ is odd. The same is true for a path from B to -B. We must therefore impose the constraints

$$\psi(-A,B) = (-1)^{\mathcal{B}}\psi(A,B) ,$$
$$\psi(A,-B) = (-1)^{\mathcal{B}}\psi(A,B) . \qquad (13)$$

These imply i,j = 0,1,2,... if $\mathcal{B}$ is even and i,j = 1/2, 3/2,... if it is odd. There may also be other noncontractible closed paths, and the resulting constraints can have dramatic effects on the spectrum. The simplest example is provided by the deuteron.

THE DEUTERON

If the soliton approach discussed in the previous section is to have any relevance to nuclear physics, it should give a reasonable description of the simplest nucleus, the deuteron. In particular, one should get automatically the correct quantum numbers. The deuteron is the lowest energy state with baryon number $\mathcal{B} = 2$. It has isospin i = 0, angular momentum j = 1, and parity P = +1. How will these quantum numbers arise in the Skyrme model?

The isospin and spin i and j must be integers, because of the constraints in Eq. (13). Furthermore, the eigenvalues of the Hamiltonian in Eq. (11) will increase with i and j because the rotational kinetic energy term in Eq. (11) is positive definite. Thus the lowest energy eigenstate of H will have i = j = 0 and the next lowest will either be $i = 0$, $j = 1$ (the quantum numbers of the deuteron) or $i = 1$, $j = 0$. The presence of the $i = j = 0$ state would be disastrous for the soliton approach to the deuteron. The only hope is that this state must be eliminated by a constraint which follows from some special property of the static configuration $U(\vec{x})$. What could this property be?

We can take a hint from the study of the product ansatz (3) by Jackson, Jackson, and Pasquier.[2] If the centers of mass $\vec{x}_1$ and $\vec{x}_2$ are on the z-axis, then the Skyrmion-Skyrmion interaction potential is most attractive if the relative isospin orientation is $A = \exp(i\pi\tau^2/2)$, which flips the τ^3-axis of one Skyrmion relative to the other. The minimum of the potential is found numerically to occur at a separation of about r = 2 fermi and to have a depth of about 40 MeV. The product ansatz for this special value of A has an unexpected symmetry. It is invariant under a simultaneous rotation by π about the z-axis and an isorotation by π about the τ^3-axis:

$$\exp(i\pi\tau^3/2)U(R(\pi,\hat{z})\cdot\vec{r})\exp(-i\pi\tau^3/2) = U(\vec{r}) \ . \tag{14}$$

One can verify that this simultaneous rotation and isorotation defines a closed path which is not contractible. One must therefore impose a Finkelstein-Rubinstein constraint, which can be written in operator form as

$$\exp(i\pi(K_3 + L_3)) = -1 \ . \tag{15}$$

Since the $i = j = 0$ state requires $K_3 = L_3 = 0$, it is eliminated by this constraint. Thus the lowest energy state will be either $i = 0$, $j = 1$ or $i = 1$, $j = 0$. The question of which of these states is lowest in energy can only be answered by a detailed numerical investigation. If we again take the product ansatz as a guide, we find that the state with the deuteron quantum numbers $i = 0$, $j = 1$ is in fact lower in energy.

We next discuss the parity of the deuteron. The parity operator P takes a configuration $U(\vec{x},t)$ into $U^\dagger(-\vec{x},t)$. We will assume that parity has the same effect on the static deuteron configuration $U(\vec{r})$ as it does on the minimal energy product ansatz configuration:

$$PU(\vec{r}) = U^\dagger(-\vec{r}) = \exp(i\pi\tau^2/2)U(\vec{r})\exp(-i\pi\tau^2/2) \ . \tag{16}$$

The parity transform of $AU(R\cdot\vec{x})A^\dagger$ is therefore $A'U(R\cdot\vec{x})A'^\dagger$, where $A' = A\exp(i\pi\tau^2/2)$. The action of the parity operator P on the wavefunction is therefore

$$P\psi(A,B) = \psi(A\exp(i\pi\tau^2/2),B) \ . \tag{17}$$

Since the deuteron has i = 0, its wavefunction is independent of A and it therefore has parity P = +1. Thus the condition on $U(\vec{r})$ in Eq. (16) is sufficient to give the correct parity for the deuteron.

Having understood the quantum numbers of the deuteron, the next step is to compute its static properties. Among the properties that one can calculate are the binding energy, the magnetic moment, the charge radius and the quadrupole moment. These calculations are in progress.

CONCLUSIONS

The soliton approach to nuclei that we have presented is very direct. There is no need to go through an intermediate step of constructing a potential. Once the minimal energy configuration $U(\vec{x})$ is known, the quantum numbers of the nucleus are determined and its static properties can be calculated. Dynamical properties, such as pion-nucleus and nucleus-nucleus scattering, can also be addressed.

In the case of the deuteron, we found that this approach can reproduce its isospin, spin, and parity quantum numbers automatically. The only requirement is that the energy in the $\mathcal{B} = 2$ sector be minimized by a static solution which satisfies the conditions in Eq. (14) and Eq. (16). Numerical investigations using the product ansatz suggest that these conditions are indeed satisfied. More thorough investigations of this question, together with calculations of the static properties of the deuteron, are in progress.

REFERENCES

1. I. Zahed and G. E. Brown, "Low Energy Phenomenology with Skyrmions," to appear in the Proceedings of the Los Alamos Workshop on Relativistic Dynamics and Quarks in Nuclear Physics (1985).

2. A. Jackson, A. D. Jackson, and V. Pasquier, Nucl. Phys. A432, 567 (1985).

3. G. S. Adkins, C. R. Nappi, and E. Witten, Nucl. Phys. B228, 552 (1983).

4. D. Finkelstein and J. Rubinstein, J. Math. Phys. 9, 1762 (1968).